NAUTILUS

The Biology and Paleobiology of a Living Fossil

Reprint with additions

Aims and Scope
Topics in Geobiology Book Series

Topics in Geobiology series treats geobiology – the broad discipline that covers the history of life on Earth. The series aims for high quality, scholarly volumes of original research as well as broad reviews. Recent volumes have showcased a variety of organisms including cephalopods, corals, and rodents. They discuss the biology of these organisms-their ecology, phylogeny, and mode of life – and in addition, their fossil record – their distribution in time and space.

Other volumes are more theme based such as predator-prey relationships, skeletal mineralization, paleobiogeography, and approaches to high resolution stratigraphy, that cover a broad range of organisms. One theme that is at the heart of the series is the interplay between the history of life and the changing environment. This is treated in skeletal mineralization and how such skeletons record environmental signals and animal-sediment relationships in the marine environment.

The series editors also welcome any comments or suggestions for future volumes.

Series Editors

Neil H. Landman, landman@amnh.org
Peter Harries, harries@shell.cas.usf.edu

For other titles published in this series, go to
http://www.springer.com/series/6623

NAUTILUS

The Biology and Paleobiology of a Living Fossil

Reprint with additions

Edited by

W. Bruce Saunders
Bryn Mawr College, Pennsylvania, USA

and

Neil H. Landman
American Museum of Natural History
New York, USA

Springer

Editors
W. Bruce Saunders
Bryn Mawr College
Department of Geology
Bryn Mawr PA 19010
USA
wsaunder@brynmawr.edu

Neil H. Landman
American Museum of Natural History
Central Park West at 79th St.
New York NY 10024
USA
landman@amnh.org

ISBN 978-94-024-0448-7 ISBN 978-90-481-3299-7 (eBook)
DOI 10.1007/978-90-481-3299-7
Springer Dordrecht Heidelberg London New York

Softcover reprint of the hardcover 1st edition 2010
First edition 1987 Plenum Press, New York
Reprint with additions 2009

Cover illustration: *Nautilus belauensis* hatched at the Waikiki Aquarium, October, 1990. Photograph courtesy of Waikiki Aquarium.

Printed on acid-free paper

Springer is part of Springer Science+Business Media (www.springer.com/mycopy)

This volume is dedicated to two individuals,
born nearly a century apart,
whose selfless dedication to *Nautilus*
will provide inspiration for many others to follow

Arthur A. Willey
(1867–1942)

Michael A. Weekley
(1957–1984)

Preface

Few organisms have been as well known to the layman, but as poorly known to science, as the chambered nautilus. Although the shell was known by Aristotle, centuries elapsed before the living animal was first illustrated by Rumpf, in 1705, and its anatomy was not known until Richard Owen's dissection of a specimen captured by sailors in the New Hebrides, published in 1832. Although other accounts followed, virtually nothing was known of the habitat of *Nautilus* until 1895, when Arthur Willey, a young British zoologist, undertook a near-epic three-year quest to decipher the embryology of *Nautilus*, as a clue to the evolutionary history of the cephalopods. Although his goal was not realized, Willey did obtain the first information on the animal's habits, summarized in a major monograph published in 1902, which is still a priceless source of information on *Nautilus*.

With only a few exceptions, there was no further study of this enigmatic animal for almost 60 years. Despite its importance as the only representative of an entire subclass of mollusks, *Nautilus* appears to have been regarded as an inaccessible curiosity by most biologists. On the other hand, paleontologists seemed unwilling to venture into purely biological territory to study *Nautilus* directly. Nevertheless, the considerable paleontological importance accorded the organism is reflected by the detailed treatment of *Nautilus* in the *Treatise on Invertebrate Paleontology* (1964) by H. B. Stenzel.

The hiatus in *Nautilus* research ended abruptly in the 1960s with the outstanding description of the buoyancy mechanism by Eric Denton and J. B. Gilpin-Brown, who had returned to one of Willey's haunts, the Loyalty Islands, armed with modern techniques. Their work seemed to reawaken both zoologists and paleontologists, for scores of articles were published by 1980 and 50 more have appeared in the last five years alone. Greatly revised and, in many cases, entirely new views of *Nautilus* are emerging as a result of new information available, from such diverse sources as telemetric tracking, shell radionuclides, deep-water remote camera sequences, analyses of shell strength, and physiological and aquarium studies.

In 1983, at a Geological Society of America meeting in Indianapolis, Indiana, the idea of assembling a book on *Nautilus* was developed among the "Friends of the Cephalopods," an informal group of paleontologists who had more than passing acquaintance with *Nautilus*. This volume is an outgrowth of that discussion. It constitutes a synthesis of existing information along with a wealth of new material. The mixture is about 50–50. For this, we are appreciative of those contributors who opted to wait patiently for the book to be published, when they justifiably could have resorted to publication in journals.

It is worth noting that although the great majority of living *Nautilus* workers contributed to this effort, a few are not represented. Eric Denton and J. B. Gilpin-Brown (Plymouth Marine Lab), Anna Bidder (Cambridge University), and Norine Haven (Hopkins Marine Station), each provided much-needed stimuli during the "early days" of modern *Nautilus* research, and their contributions stand as important milestones.

In May, 1986, partly to celebrate completion of the book and partly as a means of joining two diverse and seemingly distantly connected guilds—paleontologists and zoologists—a gathering of nautilophiles was held in Philadelphia, followed by a *Nautilus* workshop at Bryn Mawr College. The present volume was the inspiration for these gatherings, not the reverse. Participants included the great majority of zoologists and paleontologists who work on *Nautilus*; that they assembled is a measure of the support and flexibility of the National Science Foundation and the American Association for the Advancement of Science. In recognition of the fact that *Nautilus* research is a multidisciplinary effort, the royalties from this book are being donated to the Paleontological Society, publisher of the journal *Paleobiology*, which has helped foster an interdisciplinary approach to paleontological and biological problems.

The broad span of the contributions presented here, like the international collaboration involved in their preparation, makes credits and acknowledgments difficult. Of greatest importance to the study of *Nautilus* has been the near-limitless number of individuals who have shared our enthusiasm for this animal, most of whom, like us, have had nothing to gain but the satisfaction of curiosity. If *Nautilus* research is to continue, that list must continue to lengthen. A few special mentions are warranted: John Lance, former Director of the Paleontology and Stratigraphy Program, National Science Foundation, encouraged perseverence and assisted in finding means of support for research on a subject that was, according to some, not worthy of support. The Foundation's backing is reflected directly or indirectly in the content of many of the chapters. The National Geographic Society's Committee on Research and Exploration, Edwin W. Snider, Secretary, has been similarly courageous (and liberal) in its risk-taking attitude in funding *Nautilus* research.

We thank a number of people who reviewed chapters in the book: John Baldwin (Monash University), John Chamberlain (Brooklyn College), Kirk Cochran (SUNY, Stony Brook), John Curry (University of York), Roger Hewitt (McMaster University), William Kier (University of North Carolina, Chapel Hill), W. R. A. Muntz (Monash University), J. R. Redmond (Iowa State University), E. A. Shapiro (Georgia Geologic Survey), I. Strachan (St. Andrews University), Andrew Swan (University College of Swansea), Curt Teichert (University of Rochester), Roger D. K. Thomas (Franklin and Marshall College), Peter Ward (University of Washington), and Martin Wells (Cambridge University). We particularly thank Richard Davis (Cincinnati Museum of Natural History) for carefully checking organizational and grammatical details in each chapter.

At Bryn Mawr College, Nancy Weinstein assembled the references and provided a wealth of assistance with manuscript processing; Mitra Fattahipour and Kevin Hefferan aided with drafting.

At the American Museum of Natural History, the following people assisted in proofreading, collating, copying, sorting, drafting, and word-processing: Beverly Heimberg, Susan Klofak, Peter Harries, Stephen Butler, and, especially, Stephanie Crooms. In the final stages of preparation, Douglas Jones (Florida State Museum) smoothed the way to Plenum Press, where Amelia McNamara, Eric Nernberg, and Susan Woolford took over with exemplary team efficiency.

A few personal notes:

Neil Landman especially thanks Niles Eldredge (American Museum of Natural History) for lending support and encouragement toward the completion of the book, the American Museum of Natural History for travel funds to Palau, John Arnold (University of Hawaii) and Bruce Carlson (Waikiki Aquarium) for their kind invitation to join their expedition to collect *Nautilus* in Palau, and Kirk and family.

Bruce Saunders acknowledges Douglas Faulkner's efforts to set the stage in Palau for the long-term research program undertaken there in 1977 with Claude Spinosa, Larry Davis, and the late Michael Weekley. Ron Knight assisted immeasurably in developing the program in Papua New Guinea. The Micronesian Mariculture Demonstration Center in Palau and its directors (J. P. McVey, M. Madranchar, W. M. Hamner, N. Idechong, F. Perron, and G. A. Heslinga) assisted the *Nautilus* research in ways too numerous to mention, and Bruce Carlson and the Waikiki Aquarium have been a bulwark of logistical support. The following individuals offered encouragement at many stages of the project: W. M. Furnish and Brian Glenister (University of Iowa), Ellen Grass (Quincy, Massachusetts), Paul Bond (Bryn Mawr College), R. Tucker Abbott (Melbourne, Florida), and Clyde Roper (Smithsonian Institution). Finally, the patience, support, and endurance of Victoria and Justin Saunders have permitted pursuit of an obsession that now has spanned a decade.

W. Bruce Saunders
Neil H. Landman

Bryn Mawr, Pennsylvania
New York, New York

Contents

VII. Reproduction and Growth

Chapter 31 • Ultrastructure of the *Nautilus* Shell

Charles Grégoire

IX. Swimming and Buoyancy

Chapter 32 • Locomotion of *Nautilus*

John A. Chamberlain, Jr.

Chapter 33 • *Nautilus* Shell Hydrostatics

Earl A. Shapiro and W. Bruce Saunders

Chapter 34 • Buoyancy in *Nautilus*

Lewis Greenwald and Peter D. Ward

X. Aquarium Maintenance

Contributors

John M. Arnold Pacific Biomedical Research Center, University of Hawaii, Honolulu, Hawaii 96822; Marine Biological Laboratory, Woods Hole, Massachusetts 02543, deceased

John Baldwin School of Biological Sciences, Monash University, Clayton, Victoria 3800, Australia; email: john.baldwin@sci.monash.edu.au

Vernon C. Barber Department of Sciences, Roehampton Institute, Whitelands College, West Hill, London SW15 3SN, United Kingdom

George B. Bourne Department of Biological Sciences, The University of Calgary, Calgary, Alberta T2N 1N4, Canada; email: bourne@ucalgary.ca

Bruce A. Carlson Georgia Aquarium, Atlanta, Georgia 96815; email: bcarlson@georgiaaquarium.org

M. Patricia Carpenter Department of Biology, University of California, San Diego, La Jolla, California 92093

John A. Chamberlain, Jr. Department of Geology, Brooklyn College of the City University of New York, Brooklyn, New York 11210; email: jchamberlain@gc.cuny.edu

J. Kirk Cochran School of Marine and Atmospheric Sciences, Stony Brook University, Stony Brook, New York 11794; email: kcochran@notes.cc.sunysb.edu

Desmond Collins Department of Invertebrate Palaeontology, Royal Ontario Museum, Toronto, Ontario M5S 2C6, Canada

Rex E. Crick Department of Geology, University of Texas at Arlington, Arlington, Texas 76019; email: crick@uta.edu

Larry E. Davis Department of Biology, College of St. Benedict, St. John's University, Collegeville, MN 56321; email: ldavis@cbsju.edu

Richard Arnold Davis Department of Chemistry and Physical Sciences, College of Mount St. Joseph, Cincinnati, Ohio 45223-1670; email: R_A_Davis@mail.msj.edu

Yoshio Fukuda Laboratory of Pathology and Animal Physiology, Chiba Prefectural Institute of Public Health, Chiba 280, Japan

Emily Greenfest-Allen Penn Center for Bioinformatics, University of Pennsylvania, Philadelphia, Pennsylvania 19104-6021; email: allenem@pcbi.upenn.edu

Lewis Greenwald Department of Evolution, Ecology, and Organismal Biology, Ohio State University, Columbus, Ohio 43210; email: greenwald.1@osu.edu

Charles Grégoire Centre de Recherches Métallurgiques, Abbaye du Val Benoît, Liege, Belgium, deceased

T. Hamada Department of Earth Science and Astronomy, College of Arts and Sciences, University of Tokyo, Tokyo 153, Japan

Peter J. Harries Department of Geology, University of South Florida, Florida 33620-5201; email: pjharries@gmail.com

Shozo Hayasaka Institute of Earth Sciences, Faculty of Science, Kagoshima University, Kagoshima 890, Japan, deceased

Roger A. Hewitt 12 Fairfield Road, Eastwood, Leigh-on-Sea, Essex SS9 5SB, United Kingdom

P. W. Hochachka Department of Zoology and The Sports Medicine Clinic, University of British Columbia, Vancouver, British Columbia V6T 2A9, Canada

Michael R. House Department of Geology, Southampton Oceanography Centre, European Way, Southampton SO14 3ZH United Kingdom, deceased

William M Kier Department of Biology, University of North Carolina at Chapel Hill, Chapel Hill, North Carolina 27599-3280; email: billkier@bio.unc.edu

Neil H. Landman Division of Paleontology (Invertebrate), American Museum of Natural History, New York, New York 10024; email: landman@amnh.org

Keith O. Mann Department of Geology and Geography, Ohio Wesleyan University, Delaware, Ohio 43015; email: komann@owu.edu

A. W. Martin Department of Zoology, University of Washington, Seattle, Washington 98105, deceased

Tatsuro Matsumoto 1-28-5, Minami-Ohashi, Minami-ku, Fukuoka 815, Japan, deceased

S. Mikami Yomiuri-Land Co. Ltd., Tokyo 206, Japan

W. R. A. Muntz Faculty of Science, Monash University, Clayton, Victoria 3168, Australia

Harry Mutvei Swedish Museum of Natural History, 104 05 Stockholm, Sweden; email: harry.mutvei@nhm.se

Kimihiko Ôki Kagoshima University Museum, Kagoshima 890, Japan; email: oki@sci.kagoshima-u.ac.jp

T. Okutani 714 Takaishi 4-17-1, Asao-ku, Kawasaki City, Japan 251-0003; email: okutani@kir.biglobe.ne.jp

James R. Redmond Department of Zoology, Iowa State University, Ames, Iowa 50011

Toshio Saisho Laboratory of Marine Biology, Faculty of Fisheries, Kagoshima University, Kagoshima 890, Japan

W. Bruce Saunders Department of Geology, Bryn Mawr College, Bryn Mawr, Pennsylvania 19010; email: wsaunder@brynmawr.edu

R. Schipp Institute for General and Special Zoology, Justus Liebig University, D 6300 Giessen, Germany

Earl A. Shapiro Georgia Geologic Survey, 19 Martin Luther King, Jr. Drive, SW Atlanta, GA 30334

Akihiko Shinomiya Laboratory of Marine Biology, Faculty of Fisheries, Kagoshima University, Kagoshima 890, Japan; email: shino@fish.kagoshima-u.ac.jp

Claude Spinosa Department of Geosciences, Boise State University, Boise, Idaho 83725; email: cspinosa@boisestate.edu

Andrew R. H. Swan School of Geography, Geology and the Environment, Kingston University, Kingston upon Thames, Surrey KT1 2EE United Kingdom

Kazushige Tanabe Geological Institute, Faculty of Science, University of Tokyo, Tokyo 113, Japan: email: tanabe@eps.s.u-tokyo.ac.jp

Curt Teichert Department of Geological Sciences, University of Rochester, Rochester, New York 14627, deceased

Junzo Tsukahara Department of Biology, Faculty of Science, Kagoshima University, Kagoshima 890, Japan

Peter D. Ward Department of Biology, University of Washington, Seattle, Washington 98195; email: argo@u.washington.edu

M. J. Wells Department of Zoology, University of Cambridge, Cambridge CB2 3EJ, United Kingdom; email: m.j.wells@zoo.cam.ac.uk, deceased

G. E. G. Westermann Department of Geology, McMaster University, Hamilton, Ontario L8S 4M1, Canada; email: gwestermann@simpatico.ca

Judith E. Winston Virginia Museum of Natural History, Martinsville, Virginia 24112; judy.winston@vmnh.virginia.gov

David S. Woodruff Department of Biology, University of California, San Diego, La Jolla, California 92093; email: dwoodruff@ucsd.edu

J. Z. Young Department of Experimental Psychology, Oxford University, Oxford OX1 3UD, United Kingdom, deceased

Nautilus workshop and symposium participants, Bryn Mawr College, May 29, 1986. Front Row (left to right): Paul Bond*, Bryn Mawr College; Donald Dan, New Friendship, Maryland; Curt Teichert*, University of Rochester; Mrs. R. Schipp, Giessen, Germany; Mrs. W. W. Martin, Seattle, Washington; Arthur W. Martin*, University of Washington; Kazushige Tanabe, Ehime University; William Orr, National Science Foundation, Washington, D. C.; Takashi Hamada, University of Tokyo. Second Row (left to right): Rudolf Schipp, Justus-Liebig-University; Bruce Saunders, Bryn Mawr College; J. Z. Young*, Oxford University; Peter Ward, University of Washington; Eric Denton*, Marine Biological Association (U. K.); David Jacobs, Virginia Polytechnical Institute; Anna Bidder*, Cambridge University; Earl Shapiro, Georgia Department of National Resources; Kirk Cochran, State University of New York at Stony Brook; Roger Hewitt, McMaster University; John Chamberlain, Jr., Brooklyn College, City University of New York. Third Row (left to right): William Kier, University of North Carolina; Martin Wells*, Cambridge University; David Woodruff, University of California, San Diego; Claude Spinosa, Boise State University; George Bourne, University of Calgary; James Redmond, Iowa State University; Bruce Carlson, Waikiki Aquarium; William Muntz, Monash University; John Arnold*, University of Hawaii. Fourth Row (left to right): Desmond Collins, Royal Ontario Museum; Neil Landman, American Museum of Natural History. (* = deceased).

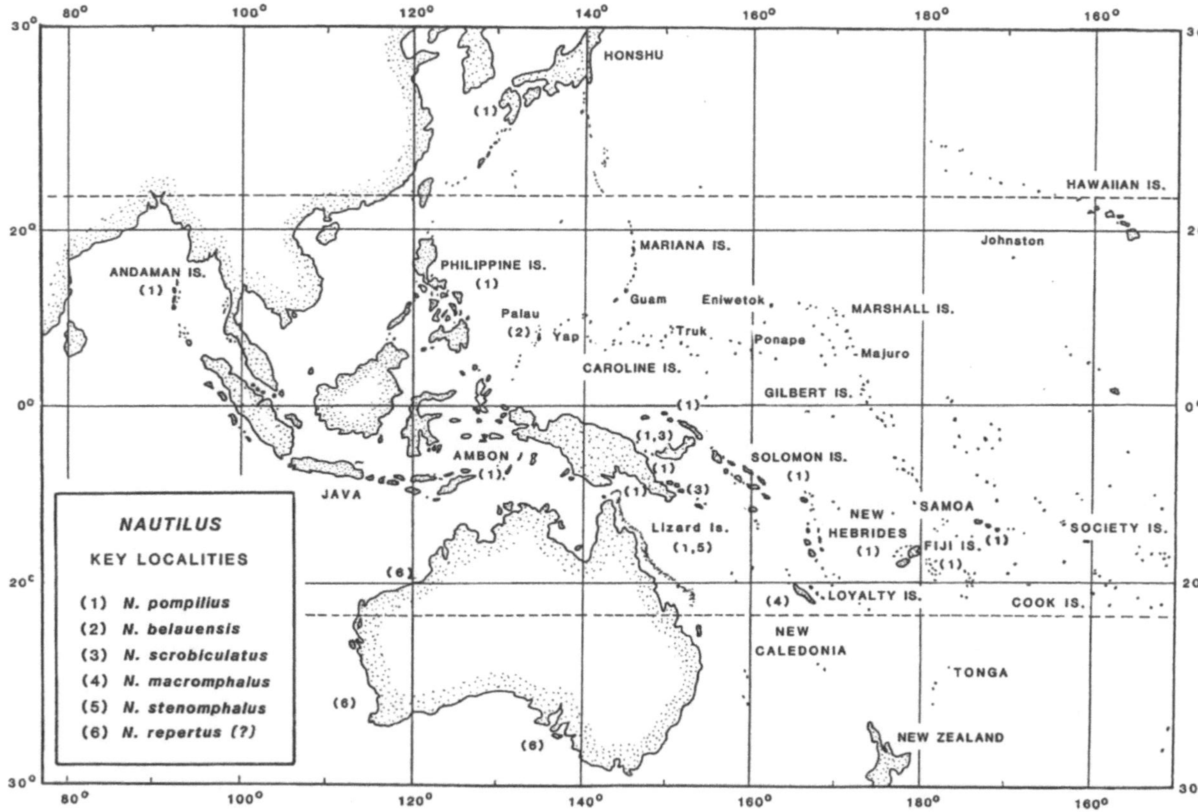

Documented occurrences of living *Nautilus* based on published descriptions of living animals, personal observations, or published reports of stranded specimens with intact soft parts. Only one species, *N. pompilius*, is known from throughout the geographic range, and this species is known to occur together with *N. scrobiculatus* in Manus, Papua New Guinea, and with *N. stenomphalus* off Lizard Island, on the Great Barrier Reef. For sources, see Chapter 3 and Saunders (1981b). (W. B. Saunders.)

Color Plates

Plate I

Plate I. Shells of five species of living *Nautilus* Linnaeus, 1758 and *Allonautilus* Ward and Saunders, 1997 [153]*. *Upper left: Nautilus belauensis* Saunders, 1981 (AMNH 43263), mature male, Mutremdiu Point, Palau, approximately 300 m depth, 1982. *Upper right: N. pompilius* Linnaeus, 1758 (AMNH 43262), specimen Ko 13, mature male, Komuli, Fedarb Islands, Manus, Papua New Guinea, approximately 275 m depth, 1984. *Lower right: Allonautilus scrobiculatus* (Lightfoot, 1786), (AMNH 43261), specimen Nd 102, mature male, Ndrova Island, Manus, Papua New Guinea, approximately 300 m depth, 1984 [114, 109, 153].* *Lower left: N. macromphalus* Sowerby, 1849 (AMNH 133742) Noumea, New Caledonia. *Center: N. stenomphalus* Sowerby, 1849 (AMNH 43260), specimen Lz 24, mature female, Carter Reef, off Lizard Island, Queensland, Australia, approximately 300 m depth, 1985 [112].* All specimens approximately X 1/3. Photograph courtesy of the American Museum of Natural History. * Numbers in brackets [#] refer to annotated references in this edition.

Plate II

Plate II. *Top:* Aerial view looking north over Ngemelis Island, Palau, showing fringe reef. This locale, which is typical of Indo-Pacific *Nautilus* and *Allonautilus* [153]* sites, yielded hundreds of specimens of *N. belauensis* for study, tagging and release, scores of which were recaptured here and elsewhere around Palau. (Note boat, far left center, for scale. See Chapter 9, and Saunders and Spinosa 1978, 1979; Saunders 1983, 1984a.) *Left center: Nautilus belauensis*, a mature specimen, photographed after being tagged and released, off Mutremdiu Point, Palau, in 1982. *Right center:* A trap being retrieved off Ngemelis Island, Palau (see top), June 2, 1982. The trap had been set against the reef face at approximately 180 m for three nights. It contained 44 specimens of *N. belauensis*, including four specimens that had been previously released at this site. Eighty-nine percent of the specimens were male and 82% of the animals were fully mature (see Chapter 9). *Bottom:* Deep-water remote photograph taken off Mutremdiu Point Palau (the type locality for the species), at 217 m depth, using JM 35-2000 still camera (Jay-Mar Engineering, San Pedro CA), showing *N. belauensis* attracted to tuna bait, along with deep-water caridean shrimps, *Heterocarpus ensifer* (see Chapter 9 and Saunders 1984b for details). Photographs top, left center, and bottom: W. B. Saunders; right center: B. C. Carlson. * Numbers in brackets [#] refer to annotated references in this edition.

Plate III

Plate III. *Top: Allonautilus scrobiculatus* (left) [153]* and *Nautilus pompilius* (right), photographed in shallow water, shortly after capture at approximately 270 m. depth, off Ndrova Island, Manus Province, Admiralty Islands, Papua New Guinea. This is the first site where *A. scrobiculatus* was seen alive and represents the first known sympatric occurrence of *Nautilus* and *Allonautilus* [109, 114]*. The shaggy appearance presented by the thick periostracum covering the shell is a unique feature that was unknown until live specimens were obtained in 1984 (see Chapters 3, 9). *Right center:* Apertural view of *A. scrobiculatus,* from the same locale as the previous, showing distinctive, papillose hood texture, and details of the periostracum covering the shell and protruding beyond the aperture (see Chapter 3, and [114, 153]*). *Left center: Nautilus stenomphalus* from Carter Reef, off Lizard Island, Great Barrier Reef, Queensland, Australia. This species, seen alive for the first time in 1985, is distinguished by the absence of umbilical color bands, the lack of an umbilical callus, and by the heavily textured hood (see Chapters 3, 9, and [112]*). *Bottom:* Deep-water remote camera photograph (270m, off Ndrova Island, Manus Province, Papua New Guinea) showing *Nautilus pompilius* and *N. scrobiculatus*, along with deep-water snapper *Etelis carbunculus* attracted to a baited trap [109, 110]*. (Photographs by W. B. Saunders.) * Numbers in brackets [#] refer to annotated references in this edition.

Plate IV

Plate IV. *Top: Nautilus belauensis* hatched at the Waikiki Aquarium, October, 1990. This was one of five hatchlings from brood stock obtained in Palau in 1988. The eggs took ~12 months to develop and hatch, at water temperatures ranging from 22°-24°C. The embryonic shells were 29 mm at hatching, and this photograph was taken several weeks after hatching [6, 31].* (Photograph courtesy of Waikiki Aquarium). *Left center:* Egg capsules of *Nautilus pompilius* from Fiji deposited on aquarium wall at the Waikiki Aquarium, Honolulu. These egg cases, which were infertile, measured approximately 25 mm in diameter (see Chapters 26 and 35). *Right center:* The first live, developing embryo of *Nautilus*, obtained at the Waikiki Aquarium in 1985. This photograph shows *N. belauensis* exposed inside its egg capsule. The cap like embryonic shell had so far formed one chamber; the bright red spot is an eye and the siphon is visible on the side of the shell. Although fertile *Nautilus* embryos had been sought for almost a century, the first ones, including this specimen, were not obtained until 1985 (see Chapters 26, 35, Arnold and Carlson, 1986, and [31]* for details). *Bottom:* Juvenile *N. belauensis,* captured, tagged, and released off Mutremdiu Point, Palau. Specimens as young as this are rare among all populations of *Nautilus* studied to date. The shell diameter of this specimen was approximately 90 mm. or less than half the mean adult size for this species. (Photographs top: Waikiki Aquarium; left and right center: B. C. Carlson; bottom: W. B. Saunders.) *Numbers in brackets [#] refer to annotated references in this edition (see Introduction).

1. Introduction

1.1 *Nautilus* and *Allonautilus*: Two Decades of Progress

W. Bruce Saunders
Department of Geology
Bryn Mawr College
Bryn Mawr PA 19010
wsaunder@brynmawr.edu

Neil H. Landman
Division of Paleontology
American Museum of Natural History
New York, New York 10024
landman@amnh.org

When *Nautilus: Biology and Paleobiology of a Living Fossil* was published in 1987, it marked a milestone in cross-disciplinary collaboration. More than half of the contributing authors (36/65) were paleontologists, many of whom were collaborating with neontological counterparts. Their interest in studying this reclusive, poorly known animal was being driven by a search for clues to the mode of life and natural history of the once dominant shelled cephalopods, through study of the sole surviving genus. At the same time, *Nautilus* offered an opportunity for neontologists to look at a fundamentally different, phylogenetically basal member of the extant Cephalopoda. It was a win-win situation, combining paleontological deep-time perspectives, old fashioned expeditionary zeal, traditional biological approaches and new techniques. The results were cross-fertilized investigations in such disparate fields as ecology, functional morphology, taphonomy, genetics, phylogeny, locomotive dynamics, etc. As one reviewer of the book noted, *Nautilus* had gone from being one of the least known to one of the best understood of living cephalopods.

The 1987 volume quickly went out of print (perhaps more of a commentary on the size of the initial printing than on its popularity), and nothing has replaced it, in spite of much expanded interest in the subject. We have located 180 articles on *Nautilus* and its recently named sister taxon, *Allonautilus,* published since 1987 -- almost half as many articles as were published in the previous two centuries! Why the surge? A number of factors are responsible; one is the recognition that living *Nautilus* is much more accessible and far more common and widely distributed than was once thought; two genera, including perhaps seven species and hundreds, if not thousands, of populations may be scattered across the Indo-Pacific. And the animal is amazingly resilient; it survives capture at depth, retrieval to surface pressures and temperatures, and can live for years in properly maintained surface aquaria. Some wild-caught specimens have been released and recaptured as many as six times. *Nautilus* 'breeding stock' in aquaria have regularly produced living embryos and hatchlings. Perhaps a milestone of sorts is represented by embryos hatched in 2006 in a closed-system aquarium in Nebraska, USA [46]. (Nebraska has been emergent and landlocked since it hosted its last nautiloid during the Cretaceous, more than 65 mya!) "Laboratory animals" can now routinely be ordered from aquarium suppliers—and, eventually perhaps, routinely raised from brood stock. The ready availability of live animals at least partly explains the sharp increase in laboratory studies since 1987.

Arthur Willey remains *the* iconic *Nautilus* researcher; he was a true pioneer among the first generation of *Nautilus* investigators (for a review, see Chapter 1). The summary account of his search to obtain *Nautilus* embryos is a fascinating narrative that mixes ethnology and biology: On the same page with observations of onycophoran development is an account of "..acquiring some personal acquaintance with one or two of the native sorceries.." (namely, experimenting with a local hallucinogen [Willey, 1902, p. 92]). The second generation can be said to include Eric Denton, J. B. Gilpin Brown, Arthur Martin, Anna Bidder and Norine Haven. They were masters of straightforward observation and simple, elegant experiments. Contributors to the 1987 volume fall into a third generation, though some were practicing 20th century approaches in 19th century settings. Here (with some carryovers), we introduce members of a fourth generation. The numbers show that the majority are biologists (more than 80% of the authors of papers reviewed here). Perhaps they are reclaiming "landfall propriety" so rightfully earned by Willey more than a century ago. It would be a mistake, however, to forget the phylogenetic roots of this living fossil, for many things about the living organism remain to be applied to the fossil records of the deep past. The diversity of new contributions is remarkable; some could not even have been imagined just two short decades ago. To begin with, a second genus of living nautiloid was named in 1997; *Allonautilus* (type: *N. scrobiculatus* [Lightfoot, 1786]), which was seen alive for the first time just as the original volume was being published. Realization of an even broader spectrum of new discoveries seems assured, with the explosive rise of new fields such as genomics and "evo-devo," and the development of new technologies such as "critter cams".

Not all news of *Nautilus* is good; with increasing awareness of the two genera and their constituent species, more interest is being expressed by shell collectors and by the general shell trade, which raises the specter of overfishing. Indeed, there have been anecdotal reports of regional depletion of *Nautilus* populations in the Philippines, Indonesia, and New Caledonia. This has led to a proposal to add all species of *Nautilus* and *Allonautilus* to Appendix II of the Convention on International Trade in Endangered Species of Wild Fauna and Flora (CITES). At first glance, this protection might seem desirable. On the other hand, enforcement efforts would be extremely challenging in the remote areas that comprise the animals' range, and any publicity of this new status might indirectly accelerate its overexploitation as a "natural resource," somewhat along the lines of the elephant/ivory dilemma.

Two new color figures are included: One shows the first *in situ* record of living *Allonautilus* along with *Nautilus* in its natural habitat (Plate III); another shows the object of Willey's long but fruitless search, a postembryonic *Nautilus*, hatched at the Waikiki Aquarium in 1990 (Plate IV). We acknowledge the considerable assistance of Steve Thurston, American Museum of Natural History, in scanning the original 1987 plate photographs and the hundreds of illustrations.

1.2 *Nautilus* and *Allonautilus*: Annotated Bibliography of References Published since 1987

Emily Greenfest-Allen
Penn Center for Bioinformatics
University of Pennsylvania
Philadelphia, PA 19104
allenem@pcbi.upenn.edu

Neil H. Landman
Division of Paleontology
American Museum of Natural History
New York, New York 10024
landman@amnh.org

W. Bruce Saunders
Department of Geology
Bryn Mawr College
Bryn Mawr PA 19010
wsaunder@brynmawr.edu

The following review of the recent literature begins with material published in 1987, with some degree of overlap with the book. It largely concludes in 2008, and we apologize for any missed contributions that might have escaped our searches. The organization of reference annotations follows the general layout of the 1987 volume. Numbers in brackets [#] refer to the corresponding citation in the numbered list of references.

2. Occurrence and Distribution

[114] Saunders et al. (1987) reported the first living specimens of *N. scrobiculatus*, which occurs sympartically with *N. pompilius*, at ~200-400 m depth in the Admiralty Islands, off Manus, Papua New Guinea. The shell of live *N. scrobiculatus* is covered with a dense, moss-like periostracum giving a shaggy appearance that is unique among living species of *Nautilus*.

[112] Saunders and Ward (1988) reported *N. pompilius*, *N. stenomphalus*, and possible hybrids that share intermediate features of the two species from 250-440 m depth off Lizard Island, Great Barrier Reef. This is the second known occurrence of sympatric species of *Nautilus*, the first locale for living *N. stenomphalus*, and is the first known example of possible hybridization between two species of *Nautilus*.

[117] Saunders et al. (1989) reported trapping *N. pompilius* at 270-310 m depth in American Samoa; negative trapping results in Western Samoa are inconclusive. Lack of *Nautilus* in traps set at 220-470 m depth in Tonga yielded diverse organisms associated with *Nautilus* elsewhere and suggests that the genus does not occur there. Fijian and American Samoan *Nautilus* exhibit some morphological differences, but show similar population characteristics.

[118] Sawata and Phongsuwan (1994) reported a fresh, necrotic specimen of *N. pompilius* found floating near Raja Island, off Phuket, southern Thailand, Andaman Sea.

[111] Saunders (1998) reviewed occurrences and species characteristics of *Nautilus* in Australian waters, including accounts of living populations of *N. stenomphalus* and *N. pompilius* (and presumed hybrids) off Queensland, Great Barrier Reef, NE Australia, and *N. pompilius* (*aff. N. repertus*) off Rowley Shoals, Western Australia (incl. color illustration of *N. stenomphalus*).

[150] Ward (1998) provided a general account of the discovery of the "king nautilus" in Papua New Guinea and its naming as *Allonautilus* (type species *A. scrobiculatus*) by Ward and Saunders (1997).

[152] Ward (2008) presented a brief, general review of current thought on the distribution, species, and evolutionary status of *Nautilus* and *Allonautilus*, describing current work on *N. pompilius* off Osprey Reef, E. Australia.

3. Phylogeny, Evolution, Systematics, and Genetics

3.1 Evolution and Systematics

[50] Habe and Okutani (1988) described a new subspecies, *N. pompilius suluensis* from the Sulu Sea in the Philippines. It differs from typical

members of *N. pompilius* in having a smaller shell with widely spaced color bands of a purplish hue.

[131] Tanabe et al. (1990) made morphological comparisons of live–caught *Nautilus* from the Philippines, Fiji and Palau, which show that these three populations are similar in shell- and radular morphology, coloration, jaw structures, etc., and are distinguished primarily by mature shell size. These morphological data, combined with known genetic information, suggest that either all three populations belong to a single, widespread and variable species, *N. pompilius*, or that *N. pompilius* and *N. belauensis* from Palau are closely related sibling species.

[58,113,59] Jacobs and Landman (1993) argued that ammonoids are morphologically more similar and thus more closely related to coleoids than to modern *Nautilus*. Therefore, coleoids are a better model for some aspects of ammonoid function and behavior. They suggested that some ammonoids may have been equipped with a coleoid-like mantle, which permitted more efficient jet propulsion than in *Nautilus*. Saunders and Ward (1994) wrote a rebuttal to this paper and argued that *Nautilus* has never been used as a strict analogue for inferring biological function in any group of ammonoids. In reply, Jacobs and Landman (1994) cited several instances in which *Nautilus* served as a proxy for the locomotory properties of ammonoids and urged that the close relationship between coleoids and ammonoids be taken into account in any such interpretation.

[116] Saunders et al. (1996) provided clarification of *N. praepompilius* (Shimansky, 1957) in the Chegan Formation of Kazahstan, confirming that this species is assignable to *Nautilus*, extending the range of the extant genus back to the late Eocene. Morphologically, this species is closest to *N. pompilius*, although there are sutural differences; hatching size was ca. 23 mm diameter, close to that of *N. pompilius*.

[153] Ward and Saunders (1997) erected a new genus, *Allonautilus* for *N. scrobiculatus* and *N. perforatus* (type species *N. scrobiculatus*), based on differences in the morphology of the shell and soft parts. In support of their argument, they also performed a cladistic analysis of all known species of *Nautilus*, plus three additional nautiloid genera.

[53, 153, 151] Harvey et al. (1999) questioned the validity of *Allonautilus* established by Ward and Saunders (1997). They pointed out problems with the cladistic analysis, and used a corrected data set supplied by Ward and Saunders to rerun the analysis. They stated that the results did not support the validity of the new genus. Ward (1999) wrote a rebuttal pointing out that the differences in the morphology of *N. scrobiculatus* and *N. pompilius*, compared to that of other species of *Nautilus*, justified the establishment of the new genus *Allonautilus* (see also [19,21]).

[161] Wells (1999) postulated that the high tolerance of *Nautilus* for low oxygen and hypoxia may have made nautiloids suitable for low oxygen conditions in early oceans and thus contributed to their survival of extinction events like the Permo-Triassic. As oxygen increased in the oceans, nautiloids (and possibly ammonites) could not compete with coleoids and fishes, being hampered by such factors as their large egg yolk, late hatching, and lack of planktonic stage.

[148] Wani et al. (2008) reported a *Nautilus* shell from early Pleistocene deep-water sediments in northwestern Luzon, Philippines. They tentatively referred the specimen to the species *N. pompilius* (Linnaeus, 1758), making it the first and oldest known fossil specimen of that taxon.

3.2 Molecular Phylogeny and Genetics

[140] Vitturi et al. (1990) compared the spermatocyte chromosomes of eight species of cephalopods, including *N. pompilius*. *Nautilus* has the lowest known chromosome value among living cephalopods (n = 26).

[175] Wray et al. (1995) analyzed the phylogenetic relationships of *Nautilus* using mitochondrial and nuclear DNA sequence data plus a suite of morphological characters. Their results indicated that there are three geographically distinct clades consisting of western Pacific, eastern Australia/Papua New Guinea, and western Australia/Indonesia forms. The morphologically and genetically distinct species *N. scrobiculatus* fell outside the three geographically recognized assemblages.

[179] Young and Vecchione (1996) performed a cladistic analysis of living coleoids based on morphological features using *Nautilus* as an outgroup.

[102] Rosenberg et al. (1997) used new sequence data from the D6 region of the 28S rRNA and rDNA of 28 species of molluscs, along with previously published data, to estimate the molecular phylogeny of the major molluscan clades. Their analysis supported the monophyly of the Cephalopoda, Nautiloidea, and Coleoida, but they were unable to resolve relationships among the molluscan classes. The authors used *N. pompilius* and *N. macromphalus* as representatives of the Nautiloidea.

[37] Colgan et al. (2000) used sequence data from two segments of 28S rDNA and histone H3 from several gastropod taxa, a chiton, two bivalves, and *N. scrobiculatus* to test and then review morphology-based hypotheses of gastropod relationships.

[129] Suzuki et al. (2000) isolated and determined the cDNA-derived amino acids of arginine kinases from *N. pompilius*, *Octopus vulgaris*, and *Sepioteuthis lessoni*. They reported three independent mutations in the *Nautilus* arginine kinase, resulting in considerable or complete loss of the enzyme activity. They used these data to estimate the phylogenetic position of the Cephalopoda in the molluscan clade; their analysis supported the monophyly of the bivalves and gastropods, but not of the cephalopods.

[20] Bonnaud et al. (2002) used 18S and 28S ribosomal nuclear genes to investigate cephalopod phylogeny. The authors pointed out that care must be taken in any analysis due to the possible presence of non-homologous forms of 18S rDNA.

[21] Bonnaud et al. (2004) described the complete 18S rDNA gene of *N. macromphalus* and compared its structure to that of *N. pompilius* and *N. scrobiculatus*. The range of interspecific molecular differences supports separation of the present *Nautilus* species into two genera, *Nautilus* and *Allonautilus*.

[72] Lindgren et al. (2004) performed a comprehensive molecular and morphological phylogenetic analysis of the Cephalopoda, using four gene loci and 101 morphological characters. They found that the monophyly of the Cephalopoda, Nautiloidea, Coleoidea, and Decabranchia was well supported; the monophyly of the Octobranchia was not. Vampyromorpha was the most likely sister group to Decabranchia and Teuthoidia was shown to be paraphyletic. The authors used *N. pompilius* and *N. scrobiculatus* as representatives of the Nautiloidea.

[154] Warnke and Keupp (2005) argued that *Spirula* is a better model organism than *Nautilus* to explain the biology, embryonic development, and mode of life of ammonoids. The initial chamber of the embryonic shell of *Spirula* closely resembles that of ammonoids. The authors analyzed the nuclear 18S rDNA to clarify the phylogenetic position of *Spirula* within the Coleoidea. The results of this analysis indicate that *Spirula* is basal to the rest of the Recent Decabranchia.

[136] Uda et al. (2006) determined the sequences of genes coding for arginine kinase (AK) in *N. pompilius* and in combination with similar

data from a variety of invertebrate taxa investigated the evolution and divergence of the AK family. Their analysis showed that the AKs are homologous among the invertebrates; however the AK gene organization is highly divergent, with molluscan AK genes being structurally most similar to that of the Platyhelminthes.

[17] Bergmann et al. (2006) cloned and sequenced the cDNA sequence of the *N. pompilius* hemocyanin. They found that the seven functional units of the *Nautilus* hemocyanin directly correspond to those found in *Octopus*, as does the overall gene structure. They used these data to establish a molecular clock and estimated the divergence of *Nautilus* and *Octopus* at 415 ± 24 Ma.

[22] Boore (2006) reported the complete sequence of the mitochondrial genome of *N. macromphalus*. There is an unusually high number of non-coding regions, whose functions, if any, are unknown. They may contain regulatory signals for transcription and/or replication.

[64] Lafont et al. (2006) investigated the presence and role of calcitonin gene-related peptide (CGRP), a neuropeptide commonly involved in the brain and cardiovascular function of mammals, in two cephalopods, *Sepia officinalis* and *N. macromphalus*. They found CGRP-like, but not calcitonin (CT)-like molecules in the brain, optic-lobes, branchial heart, and kidneys. The location of CRGP binding sites indicates potential autocrine/paracrine and endocrine roles for the peptide. They concluded that the brain-neurotransmitter role of CGRP is a primitive condition, present in the cephalopods and conserved among vertebrates, but the endocrine role, which is present in cephalopods and teleosts, may have been lost in the evolution of the tetrapod lineage.

[65] Lafont et al. (2007) reviewed and performed a meta-analysis of data pertaining to the evolution of the CT//CGRP peptide family, evaluating the role and evolution of the peptides in a variety of clades, including *Nautilus*.

[124] Sinclair et al. (2007) used molecular techniques to examine relationships between two populations of *Nautilus* in the northern sections of the Great Barrier Reef, Australia, and in the Coral Sea. The two populations are distinct. These results reflect the fact that *Nautilus* inhabits the reef-slope up to approximately 500 m but is not an open water swimmer, precluding travel across expanses in excess of 1,000 m depth.

[11, 175] Baratte et al. (2007) reported the first findings of the *engrailed* gene in a cephalopod, *Sepia officinalis,* and discussed the role of *engrailed*

in the evolution of the cephalopod body plan. They made a brief comparison to *Nautilus*, noting that two paralogues of the *engrailed* gene have been found in *N. pompilius* (see Wray et al. 1995).

[19, 175] Bonacum et al. (in press) provided novel insight into the molecular phylogeny and recent phylogeography of *Nautilus*, using two mitochondrial gene regions: 16s rDNA and Cytochrome Oxidase *c* subunit I (COI). They surveyed three populations of *N. pompilius* from the Philippines, Vanuatu (New Hebrides Islands), Ndrova (Admiralty Islands, Papua New Guinea), and the Great Barrier Reef (Australia), as well as populations of *N. macromphalus* from New Caledonia. Samples of *N. repertus*, *N. belauensis*, *N. stenomphalus*, and *Allonautilus scrobiculatus* were also included. They found that *Nautilus* is currently undergoing an evolutionary radiation in the Indo-Pacific region, with divergences driven by geographic isolation. Their results support the earlier findings of Wray et al. (1995), indicating that *A. scrobiculatus* is a well-supported phylogenetic species and sister taxon to all *Nautilus* species. In addition, they showed that *N. macromphalus* is a well-defined phylogenetic species, but *N. pompilius* is paraphyletic, with genetic differences between populations likely correlated with geographic distribution.

4. Ecology, Behavior, and Taphonomy

4.1 Ecology

[101] Reyment (1988) presented a foraging model for ectocochleates that involves balancing the benefits of feeding versus energy expended during foraging, less standard metabolism costs, using data from living *Nautilus* and other organisms. This model suggests that shelled cephalopods were not functionally capable of cruising predation, but could have managed ambush predation, scavenging, and browsing.

[109] Saunders (1990) used remote camera sequences to show *in situ* interactions between *N. pompilius*, *N. scrobiculatus*, caridean shrimps, and deep-water fishes at baited camera sites (147-342 m depth) off Manus Island, Papua New Guinea.

[110] Saunders (1991) provided a narrative account of the discovery of the first living specimens of *N. scrobiculatus* off Manus, Papua New Guinea, in 1984, with *in situ* deep-water camera photographs of *N. pompilius* and *N. scrobiculatus* in their natural habitat.

[115] Saunders et al. (1991) reported that 57% of drifted shells of *N. pompilius* and *N. scrobiculatus* (n = 1,532) from the Admiralty Islands had been bored (and presumably killed) by *Octopus*. These data, plus evidence of sublethal attacks in 2-8% of live-caught animals, indicate that octopod predation is a significant factor in the survival of this ancient lineage.

[103] Roux et al. (1991) explored the deep water environment around New Caledonia. *N. macromphalus* is most abundant at 400±100 m and disappears at 600 m. Accumulations of dead shells were observed with the maximum abundance at 400 m. These shells are commonly encrusted with epizoa.

[93] O'Dor et al. (1993) trapped and released *Nautilus* off southern Papua New Guinea. The animals were fitted with differential pressure transmitters and tracked. The results indicate that *Nautilus* is energetically very conservative. Animals are more active at night and typically migrate up the reef slope at dusk and descend at dawn. *Nautilus* is ideally suited as a vertical scavenger.

[143] Von Boletzky (1999) reviewed cephalopod biogeography and emphasized the role of reproduction and development in determining biogeographic patterns. In comparison with other modern cephalopods, *Nautilus* differs with respect to its long longevity and large egg size. It is restricted to the Indo-Pacific.

[30] Bustamante et al. (2000) analyzed the tissues of *N. macromphalus* to determine the presence of trace elements, presumably reflecting the enrichment of these elements in both coastal and offshore waters due to natural erosion on New Caledonia and enhanced by mining activities. The digestive gland and excretory organs contain the highest percentage of metals.

[180] Zakharov et al. (2006) analyzed the oxygen and carbon isotopic composition of *N. pompilius* from the Philippines and compared the results with those from Cretaceous ammonoids and belemnites. They argued that changes in $\delta^{13}C$ in the *Nautilus* shell are related to annual cycles of phytoplankton production.

[49] Gregory et al. (2006) observed intracytoplasmic inclusion bodies suggestive of iridovirus infection in formalin-fixed tissues from a *Nautilus* that had died without premonitory signs. This represents a novel iridovirus of mollusks.

4.2 Chemosensory and Tactile Behavior

[14] Basil et al. (2000) experimentally examined the behavioral response of *Nautilus* to odor, demonstrating that *Nautilus* efficiently uses chemical information to track odors along a turbulent horizontal plume, and that the behavioral response, including tentacle position, angle of approach, and speed, changed as the animals neared the odor source. They also showed that *Nautilus* uses the paired rhinophores for localization of an odor source, confirming the proposed olfactory sensory function of the organ.

[15] Basil et al. (2002) experimentally determined that female *Nautilus* are capable of discriminating and preferentially recognize males using only odor; males seem unable to discriminate between the sexes using odor.

[16] Basil et al. (2005) experimentally determined the physiological structures linked to different behaviors exhibited by *Nautilus* while odor-tracking. They found that the rhinophores and various tentacles are each associated with specific behaviors that vary as a function of distance from the source. Rhinophores are primarily linked to far-field searching ("cone of search," with digital tentacles extended laterally, revealing down-directed siphon), while near-field (close to source) behaviors (probing bottom of tank/substrate with lateral digital tentacles and contact-chemosensory search by the medial tentacles) are associated most frequently with stimulation of the lateral digital tentacles.

[167] Westermann and Beuerlein (2005) experimentally examined the ability of juvenile male and female and adult male *Nautilus* to detect individuals of the opposite sex. They found that adult males were attracted to the secretions of the rectum of females, and that immature males and females showed no reaction to the homogenates of their conspecifics. The authors also reviewed chemical cues to sexual-selection in coleoids.

4.3 Learning and Memory

[39] Crook and Basil (2008) presented the first evidence of learning and memory in *Nautilus*. Using a Pavlovian conditioning paradigm, they experimentally demonstrated that *N. pompilius* has temporally separated short- and long-term memory stores, producing a biphasic memory curve similar to that found for cuttlefish. Short-term memory generally

persisted for less than 1 hr post-training and long-term memory for 6 to 24 hrs, again similar to what has been reported for other cephalopods. Thus, the authors concluded that the absence of the dedicated neural regions that support learning and memory found in the brains of other extant cephalopods does not limit memory expression in *Nautilus.*

4.4 Stress Response

[61] Jordan et al. (1988) experimentally investigated the behavioral response of juvenile *Nautilus* to increasing hydrostatic pressure. They found that *Nautilus* is able to respond to changes in ambient pressure as small as 1×10^5 N m^{-2} (= 1 atm). They also observed a characteristic behavioral response to pressure increases, which consisted of rapid upward swimming that was continued until the animal became fatigued.

4.5 Taphonomy

[54] Hewitt (1988) reported Eocene nautiloids from the London Clay of England, including a deeper water facies (200-300 m depth) that contains body chambers in life position, one or two complete septa, and probable implosion shell debris. The intermediate facies contains often vertically imbedded shells with broken spar-cemented septa, resulting from septal failure in gas-filled shells sinking to the seafloor. There is some evidence of *Octopus* predation.

[134] Tshudy et al. (1989) performed feeding experiments with live *Nautilus* along with a survey of 767 Cretaceous lobster specimens in museum collections, and suggested that anatomically selective scavenging by cephalopods may explain the differential preservation of lobster cephalothoraxes over abdomens in the fossil record.

[145] Wani (2004) performed experimental postmortem fragmentation of *N. pompilius* shells, indicating that certain breakage patterns may suggest particular postmortem processes.

[147] Based on observations of submerged dead *N. pompilius*, Wani et al. (2005) claimed that the separation of decaying mantle from the shell is required for shells to become negatively buoyant. Once the tissue is lost, phragmocone flooding occurs rapidly due to reduced internal gas pressure (~0.9 atm). They concluded that only shells larger than

~200 mm diameter have postmortem drift potential and that smaller shells would quickly sink near the area of habitat.

[146] Wani and Ikeda (2006) performed flume experiments with flooded *N. pompilius* shells and found that a water flow of 0.20 m/s causes shell reorientation and that shell transport occurs at 0.25-0.37 m/s, resulting in the shell laying on its side with the body chamber located downstream from the phragmocone (dorsal) side of the shell.

5. Physiology

5.1 Nervous System

[176] Young (1988) described the brain of *Nautilus* and other cephalopods. The *Nautilus* brain is simpler than that of any other cephalopod with only 13 lobes. In contrast, the olfactory lobe and organ in *Nautilus* are larger than that in any other cephalopod, reflecting its scavenging habit.

5.2 Circulatory System

[26] Bourne (1987) reviewed the anatomy and histology of the circulatory physiology of *Nautilus* and discussed its relationship to that of other molluscs.

[120, 119] Schipp (1987) reviewed the morphological and functional characteristics of the cephalopod circulatory system, including that of *Nautilus* and compared it to that of other invertebrates. This was followed with a detailed comparative histological and cytological study of the blood vessels of cephalopods, including *Nautilus* (Schipp 1987).

[141] Von Boletzky (1987) compared the ontogeny of the circulatory system in several coleoids to that of *Nautilus*. Initial organogenesis is comparable among the clades, but becomes increasingly divergent through ontogeny, potentially reflecting both clade-specific modifications and adaptations to egg size. The open network of vessels on the surface of the *Nautilus* yolk-sac is unique and seems to be a clade-specific adaptation to the very large yolk mass; cirrate embryos of similar size do not develop a like network.

[122] Schipp et al. (1991) examined the neuroregulatory control of the cephalic aorta in *S. officinalis* and made a brief comparison to that of *Nautilus*.

[160] Wells (1992) contrasted the performance of the *Nautilus* heart with that of coleoids. He found that cephalopod hearts show a five-fold increase in specific power output over the course of their evolution, from 2.7 m Wg^{-1} in *Nautilus* to 5.5 m Wg^{-1} in *Octopus* to 20-30 m Wg^{-1} in *Loligo*, due mainly to a 10-fold increase in heartbeat frequency.

[63] In a detailed histological study, Kleeman and Schipp (1996) showed that the *Nautilus* cephalic aorta has a four-layered wall structure (in contrast to the three-layered structure in coleoids). Nerve endings, innervated in the two peripheral layers, show high, localized activity of acteylchlinesterase, suggesting that nervous control of the vessel wall is partially cholinergic. Several lines of evidence suggest some catcholaminergic control of the muscle regulation.

[125] Springer et al. (2004) used immunohistochemical methods to locate peptides related to various vaso- and cardio-active neuropeptides in the heart of *N. pompilius*. They found that the heart contains the neuropeptide FMRFamide and TKs- and VIP-related peptides. The results also suggested that *Nautilus* possesses a peptidergic innervation that modulates the effect of cholinergic and catcholaminergic neurotransmitters.

[126] Springer et al. (2005) used histological and cytological methods and pharmacological experiments to investigate the distribution and effects of biogenic amines in the *N. pompilius* heart. Their results suggested an important role for amines in the control of the heart; in particular, serotonin likely stimulates excitatory nerve fibers and noradrenaline influences muscle contraction.

5.3 Hemocyanins

[104] Ruth et al. (1988) examined the cellular structure of *N. pompilius* and *N. macromphalus* midgut glands. They described the structure of special basal cells, which they confirm as sites of hemocyanin synthesis, using Zeeman AAS and ASTEM analyses of the content and distribution of copper within the tissue.

[29] Given the relationship between oxygen affinity and cooperativity of oxygen binding, Brix et al. (1994) found that temperature sensitivity and

control of oxygen binding to hemocyanins in *Nautilus* most closely resemble that of the giant squid and octopods.

[105] Ruth et al. (1996) determined that in both *Sepia* and *Nautilus* hemocyanin synthesis is localized in a special type of cell that is cytomorphologically similar in both genera. In *Nautilus* these cells are found in the terminal alveoli of the midgut gland (basal cells).

[34] Chignell et al. (1997) examined the hemocyanin of *Sepioteuthis lessoniana* and compared it to that of other coleoids. They determined that in size, subunit structure, and behavior, *Sepioteuthis* hemocyanin more closely resembles that of other squids than that of *Octopus* or *Nautilus*.

[48] Gatsogiannis et al. (2007) reported the cryo-EM structure of the *Nautilus* hemocyanin at 9.1 Å resolution ($FSC_{1/2\text{-bit}}$ criterion) and described a molecular model of the subunit obtained by rigid-body fitting of its seven individual functional units (FU-a to FU-g). They identified the subunit dimer, the subunit pathway, and the 15 types of inter-FU interface. The authors found that four types correspond to the association mode of the two protomers in the *Octopus* FU-g crystal and drew further conclusions from the *Nautilus* subunit structure about the architecture and function of molluscan hemocyanins.

5.4 Digestive System

[121] Schipp et al. (1990) reported flagellated Gram-negative bacteria residing extracellularly in coelomic cavities and on the pericardial appendages of *N. pompilius* and *N. macromphalus*. These occupy an ecological niche, with high ammonia concentrations, similar to that of the Dicyemida in renal appendages of coleoid cephalopods. This bacterial symbiont may assist in maintaining a steady state in the coelomic fluid and/or urine production.

[165] Westermann and Schipp (1998) performed a detailed histological and cytological study of the digestive track of *N. pompilius*. They found three different cell types in the lamina epithelias mucosae: main cells, goblet cells, and cells with secretory granules. In the caecum, the main cells contain endosymbiotic bacteria and there was some evidence that the organ played a role in absorption. Two types of goblet cells were found in all organs, except the stomach, and likely provide a pathway for food particles. They also discussed the function of secretory cells in the rectum that represent a delimited rectal gland.

[166] Westermann and Schipp (1998) studied the structure and nervous and vascular supply of the digestive tract of *N. pompilius* and *N. macromphalus*. The authors also presented the general morphology of the digestive tract, hypothesized how the digestive cycle may proceed, and made some comparison between the structure of the *Nautilus* digestive organs and those of coleoids.

[106] Ruth et al. (1999) investigated the midgut glands of *N. pompilius* and *N. macromphalus* using light and scanning electron microscopy. The gland shows a significant blood supply, a variety of different enzymatic activities, and tubular organization. Although the primary role of the midgut gland is for digestion, they also found some evidence for excretory processes.

[169] Westermann et al. (2000) used ^{14}C-labelled food to trace absorption of food in *N. pompilius*. The midgut gland is most active in food absorption, followed by the caecum and crop.

[172] Westermann et al. (2002) used X-ray analysis and computational tomography to determine the position and three-dimensional structure of the digestive track in *N. pompilius* and to track the rate and duration of different phases of digestion. Total time between food intake and elimination is ~12 hrs, which is similar to that found in sepioids and octopods, but twice as long as found for squids.

[97] Pernice et al. (2007) identified bacteria producing antimicrobial compounds in the excretory organs of *N. pompilius* from the Philippines and noted the abundant presence of three bacterial groups. These groups are absent from *N. macromphalus* from other geographical areas, suggesting that antimicrobial active Vibrionaceae infect *N. pompilius* by environmental transmission.

[98] Pernice et al. (2007) reported two symbiotic bacterial associations in the highly specialized pericardial appendage of *Nautilus*: high densities of a β-protobacterium and a coccoid spirochaete. They found that these two bacterial phylotypes are phylogenetically distant from any known bacteria and appear to be specific to *Nautilus*.

[99] Pernice et al. (2007) used microbes associated with the excretory organs of *N. macromphalus* as a model system for exploring new methods for the detection of symbiotic bacteria. They compared three methods at the molecular, microbial, and cellular level; their results emphasize the potential of the cellular approach, which facilitated the detection of yet-uncultured bacteria.

[100] Pernice et al. (2008) investigated and compared the concentration of 16 trace elements in the excreting and digestive tissues of *N. pompilius* from Vanuatu and *N. macromphalus* from New Caledonia. They observed that the digestive glands and excreting tissues played key roles in the bioaccumulation and storage of distinct sets of elements. They also observed no significant difference in the concentrations of most elements; however, the concentration of rare earth elements (Ce, La, and Nd) was significantly higher in *N. pompilius* than *N. macromphalus*, potentially because of localized volcanism/upwelling in the Vanuatu region.

5.5 Mouth Parts

[85] Nixon (1988) reviewed what is known of the diet and feeding mechanisms of fossil and living cephalopods; the buccal mass of *Nautilus* is relatively large compared to both living coleoids and to ammonoids.

[178] Young (1993) used serial sectioning to examine the musculature of radulae and then experimentally observed the workings of the radular supports in several cephalopod groups, including *Nautilus*. He revealed that the radular supports work according to the same muscular-hydrostatic principles found for other tissues, such as tentacles, except the action of the supports is purely muscular, the front of the radula supports is rigidly attached to the lips of the radula sac, and the retractors are less developed.

[86] Nixon (1995) presented a standardized nomenclature for identifying the radular elements of living and fossil Nautiloidea, Ammonoidea and Coleoidea.

[74] Messenger and Young (1999) reviewed the radular apparatus of cephalopods. *Nautilus* has an unusual pair of support muscles, or "bolsters" that cause the radular teeth to become erect and splayed, enhancing the ability to rake food into the pharynx. *Nautilus* also has a papillate mass extending over the palate, the "radular appendage," that is partly secretory, but whose function is unknown.

5.6 Statocysts

[177] Young (1988) provided a review and comparison of the structure of cephalopod statocysts and discussed the role and evolution of the statocyst-eye muscle system and its convergence with the vestibule-optic system of fish.

[76] Morris (1989) examined the statoliths of *Nautilus*, revealing that they are novel when compared to those of other cephalopods. *Nautilus* statoliths are small (1.7-17 µm) and numerous, with an ovoid form.

[84] Neumeister and Budelmann (1997) presented an extensive description of the structure of the *N. pompilius* statocyst and used behavioral experiments to investigate the sensitivity of *Nautilus* to horizontal rotatory movements. In addition to being gravity receptor organs, the statocysts are able to detect rotatory movements without the specialized receptor systems found in those of coleoids.

5.7 Visual Sensory Organs

[80] Muntz and Wentworth (1987) made a detailed histological study of the retina of *N. pompilius* and provided some comparisons to the retinal structure of coleoids. The myeloid body differs from that found in other cephalopods.

[77, 79, 78] Muntz (1991) reviewed the differences between the lensless eyes of *Nautilus* and those of *Octopus*, which have a spherical lens. He noted a number of other anatomical differences as well and found that visual acuity and sensitivity are much better in *Octopus*. He later demonstrated (1994) that *Nautilus* is positively phototactic in its natural habitat, using lit- and unlit traps set at ~450 m depth, and suggested *Nautilus* may be adapted to utilize bioluminescence from scavenging shrimps or carrion to locate food. He also showed (1994) that *Nautilus* preferred larger, more intense photic stimuli over smaller less intense ones. This supports studies showing that visual acuity in *Nautilus* is very low and better adapted to sensitivity that to resolution.

[52] Hara et al. (1992) examined the structure of the *Nautilus* retina and experimentally determined the retinal-bearing photopigments. They demonstrated that, like other cephalopods, *Nautilus* visual cells possess rhodopsin in the rhabdomeres and a retinochrome system in the single myeloid body.

[51] Hara et al. (1995) extracted and determined the chemical properties of the photopigments found in the *Nautilus* retina. Properties of *Nautilus* rhodopsin and retinochrome are different from, but share some similarities with, that of squids and conchs. They also argued that some of the photopigment properties are typical of adaptation to a dark, deep-sea environment.

5.8 Tactile Sensory Organs (Tentacles, Rhinophores)

[81] Muntz and Wentworth (1995) performed a histological study of the cirri of digital tentacles. They revealed that the cirri are ridged and covered with columnar epithelium, which is thicker and forms two clear zones on the oral side. The thick epithelium contains granules made up of mucopolysaccharides, which may be responsible for adhesion. The structure of this adhesive surface is different from that reported for other cephalopods.

[107] Ruth et al. (2002) performed a detailed histological and microscopical study of the structure of the tentacles and rhinophores of juvenile *N. pompilius*. They recognized nine different types of ciliated cells, differentially distributed on the epidermis of the sensory organs. The feeding behavior of *Nautilus*, the large-number and specific patterns of ciliation, and specific aspects of tissue histology (e.g., innervation of tentacle trunk, distinct vascular systems) all support the role of the ocular and digital tentacles as sensory organs. Comparison of ciliation and the structure of the oral lamellae of *N. pompilius* to that reported elsewhere for *N. macromphalus* suggests that there are species-specific differences in the functional roles played by these organs.

5.9 Mantle and Siphuncle

[158] Wells (1988) reviewed the structure and function of the mantle in several cephalopod groups, including *Nautilus*.

[170] Westermann et al. (2002) investigated potential neurotransmitters within the nerve endings in the mantle and siphuncle of *N. pompilius* that may be involved in shell formation. Their study provided evidence for a dual cholinerg-aminergic neuroregulation for shell formation, similar to that found in modern coleoids.

[168] Westermann et al. (2005) made a detailed histological and cytological study of the mantle tissue of *N. pompilius* to characterize cells responsible for shell formation.

6. Metabolism

6.1 Ammonia Excretion

[23] Boucher-Rodoni (1989) reported that *N. macromphalus* is an ammonotelic organism, with ammonia representing >70% of excreted organic nitrogen. Although metabolic rates in *Nautilus* may vary in relation to physiological and/or environmental factors, oxygen consumption and ammonia excretion tend to be linear processes. Using the atomic O:N ratio as an indicator of the main metabolic substrates, they argued that *Nautilus* mainly processes proteins under regular feeding conditions. During starvation, lipids contribute more to the energetic needs, and both respiration and ammonia excretion decrease. Comparison to other cephalopods revealed that the metabolic rates of *Nautilus* fall within the lower range of benthic and nekto-benthic species.

[25] Boucher-Rodoni and Mangold (1994) reviewed the physiological and evolutionary implications of ammonia production in cephalopods. *Nautilus* has the lowest metabolic cost among the cephalopods, and, along with *Sepia*, excretes the least amount of ammonia. The rate of ammonia release for *Nautilus* is one-fifth that found in squids, most likely a consequence of the retention of nitrogen gas in the shell chambers.

6.2 Oxygen Uptake

[157] Wells (1987) experimentally investigated oxygen uptake in *N. pompilius*. He found that oxygen uptake is dependent on temperature and that *N. pompilius* can successfully regulate its oxygen uptake to accommodate ambient PO_2, even at nearly hypoxic levels. Feeding has no immediate effect on oxygen uptake, but long-term starvation or regular feeding can affect metabolic rates.

[164, 139] Wells et al. (1992) reviewed the behavior and physiology of *N. pompilius* with respect to oxygen consumption. They concluded that *Nautilus* can survive for long periods in water with very low oxygen, which allows *Nautilus* to exploit habitats unavailable to other animals. The authors hypothesized that this ability may have also characterized extinct shelled cephalopods and that the elimination of hypoxic habitats at the end of the Mesozoic may have contributed to the extinction of most ectocochleates. Vines (1992) gave a popular account of this study.

[24] Boucher-Rodoni and Boucher (1993) experimentally investigated the relationship between respiration and calcification in *N. macromphalus*. $CaCO_3$ exchange rates were linearly related to respiratory quotient. From the highest estimated $CaCO_3$ uptake, maximum growth rate was estimated as 7.1 mg shell wt h^{-1} (= 61 g yr^{-1}).

[45] Eno (1994) estimated the capacity for oxygen uptake in cephalopod gills based on surface areas and thickness of the tissue barrier across which oxygen diffuses in *Nautilus*, *Octopus*, *Sepia*, and *Alloteuthis*. The estimates are close to physiological measurements for these animals, and *Nautilus* is similar to *Octopus* in oxygen diffusing capacity.

[27] Boutilier et al. (1996) investigated the response of *Nautilus* to hypoxia, showing that the animal conserves energy in hypoxic situations by taking prolonged ventilatory and circulatory pauses. During these arrested conditions, they observed some slow transfer of stored oxygen gas from the shell chambers.

[28] Boutilier et al. (2000) further explored the effect of hypoxia on *Nautilus* metabolism by forcing specimens of *N. pompilius* to undergo a period of progressive hypoxia to levels of ~10 mmHg. They found that the animals responded by decreasing their activity, increasing (or ceasing) the periodicity of ventilation, slowing heart rates, and reducing their metabolic rates to 5-10% of those seen in resting, normoxic animals. They showed that *Nautilus* combines metabolic suppression with a down-regulation of systematic oxygen delivery to allow for short bursts of aerobic activity even in a hypoxic situation.

[127] Staples et al. (2000) used closed circuit respirometry to show that rates of oxygen uptake (VO_2) are relatively constant at high ambient PO_2 (oxyregulation) but declines sharply at low PO_2 (oxyconformation). These effects would support extensive daily depth migrations and would also suppress metabolism at low temperature, providing protection from hypoxia encountered in the deepest parts of the habitat of *Nautilus*, or during withrawal of the animal into the shell.

[128] Staples et al. (2003) telemetrically tagged *N. pompilius* and held them in a 4 m × 10.5 m tank with a thermocline and a 3.5 m hypoxic

bottom layer under controlled light-dark conditions. During light phases animals preferred the bottom 2.5 m of the tank despite hypoxic conditions and showed no depth preference during darkness. *Nautilus* apparently is not constrained by low oxygen conditions, at least over the short term. Its hypoxia tolerance may be due to onboard oxygen stores and suppressed metabolism during hypoxia.

7. Reproduction and Growth

7.1 Embryology and Early Growth

[2] Arnold (1988) examined the apical end of embryonic *Nautilus* shells. This part of the shell, known as the cicatrix, is the site of initial calcification. There is no evidence of an organic protoconch attached to the cicatrix, as proposed by Alpheus Hyatt.

[142] Von Boletzky (1988) described cephalopod embryogenesis, emphasizing that the embryonic development of cephalopods is characterized by meroblastic cleavage in contrast to other molluscs, which typically exhibit spiralian cleavage. All cephalopod eggs are very yolky and the largest eggs (2-3 cm in length) occur in *Nautilus*.

[67] Landman et al. (1989) described embryonic shells of *N. belauensis* ranging from a pre-chamber cap-like shell to a three-chambered shell. The cicatrix is the site of initial shell formation. A multilayered calcareous deposit, known as the protoseptum, invests the interior of the shell apex. The microstructure of the embryonic septa and their mode of attachment to the outer shell wall are similar to that of septa in later stages of ontogeny.

[5, 31] Arnold et al. (1990) announced the hatching of the first embryos of *N. belauensis* in the Waikiki Aquarium in 1990. Success was attributed to incubating the embryos in water slightly warmer than 18°C. The development time was approximately 11 months. The newly hatched animals resembled miniature adults and immediately began to feed when bits of shrimp were offered (for details, see Carlson et al., 1992).

[9] Arnold et al. (1990) used the embryonic shells of *N. belauensis* raised in aquaria to investigate the microstructure of the shell utilizing oxygen plasma etching, which removes the organic components without severely damaging the inorganic components. Nacreous tablets of the first

septum are composed of a varying number of crystalline sectors; the central portion is occupied by an organic accumulation.

[4, 6] Arnold and Carlson (1991) observed that the *Nautilus* embryo exhibits four kinds of movements at the time of major organogenesis and formation of the first chamber. These movements may be related to protection, respiration, rearrangement of the yolk and promotion of the extra-embryonic circulation, and respiration of the extra-embryonic yolk sac. Arnold et al. (1991) documented a video presentation of this behavior.

[3] Based on observations of *Nautilus* embryos, Arnold (1992) proposed a new model of shell formation in which the initial components of the prismatic and nacreous layers, called proto-prisms and proto-platelets, respectively, first form intracellularly and are then transported into the periostracal-mantle space where they are assembled into the definitive shell.

[31, 108, 35, 66] Carlson et al. (1992) recounted the history of breeding *N. belauensis* at the Waikiki Aquarium, providing extensive data on shell growth and physical parameters. The first fertile embryos of *Nautilus* were obtained in 1985 (see Plate IV) and were allowed to develop for four months. During 1988-1990, 122 eggs were laid by eight females; 10 eggs segregated in an incubator produced hatchlings ~30 mm shell diameter after ~10-14 months at 22°C (see Plate IV). Hatchling growth rates (~0.4 mm/day) were much higher than reported for larger animals, based on tag-recapture (Saunders, 1983) or radionuclide decay rates (Cochran and Landman, 1984; Landman and Cochran, 1987).

[8] Arnold et al. (1993) documented the predation of *Nautilus* egg capsules by *Nautilus*, using the beak and possibly the radula.

[7] Based on aquarium observations of actual *Nautilus* embryos, Arnold et al. (1993) described the formation of the egg capsule, the manipulation of the egg capsule by the female during its formation, and the hatching of the juvenile after the completion of embryonic development. There is no evidence of a hatching organ. Instead, one seam of the egg capsule gradually opens, and the animal simply pushes its way out.

[83] Mutvei et al. (1993) described the musculature that attaches the body to the shell in embryos and adults of *N. belauensis*. In the early organogenetic stage, the cephalic retractor muscles appear as a dense mass of tissue between the developing eye primordium and the branchial heart. The mantle possesses numerous longitudinal muscles that extend anteriorly and terminate near the mantle margin where they form

a zone of attachment to the shell. In adults, the body is posteriorly attached to the shell along the area of origin of the cephalic retractor muscles, the septal myoadhesive band, and the area of origin of the weakly developed longitudinal mantle muscles. Anteriorly, the mantle is mainly attached to the shell by numerous epithelial extensions, which are housed in pores in the inner layer of the shell at the apertural margin.

[95, 96] Okubo et al. (1995) reported the hatching of *N. belauensis* in the Kagoshima Marineland Aquarium, Japan, in 1988 and 1989 (first reported in the *Chambered Nautilus Newsletter* by Okutani in 1990). The hatchlings survived 32 to 73 days. They analyzed the oxygen and carbon isotopic composition of the shell and observed a large negative shift in $\delta^{13}C$ in the septa that may have been due to partial ingestion of part of the egg capsules prior to hatching.

[130] Tanabe and Uchiyama (1997) analyzed the development of the embryonic shells of *N. macromphalus* and *N. pompilius*. They noted two stages with different shell structure and ornamentation: 1) the cicatrix and 2) the succeeding shell marked by longitudinal growth lines and radial undulations, reflecting shell secretion at the mantle margin.

[135] Uchiyama and Tanabe (1999) reported hatching of *N. macromphalus* in the Toba Aquarium in Japan between 1992 and 1996. The temperature in the tanks was maintained between 22.6 and 24.2°C and the embryonic period ranged between 269 and 362 days. This was shorter than that for *N. belauensis* maintained at lower temperatures in the Waikiki Aquarium, Hawaii.

[56] Isaji et al. (2002) described the ultrastructure of the muscular attachment of the shell in embryos and adults of *N. pompilius*. The inner surface of the shell at the site of muscular attachment is covered with a relatively thin semi-transparent membrane. This membrane may serve to facilitate the periodic peeling off and reattachment of the body during translocation of the organism.

[46] Fields (2006) reported two *N. pompilius* hatched in artificial seawater (23°-25°C) at the Scott Aquarium of the Henry Doorly Zoo, Omaha, Nebraska. Incubation time was ~12 months, and the hatching was recorded on television.

[123] Shigeno et al. (2008) reported the first embryological evidence for the morphological development of the head complex in *N. pompilius*. Their results indicated that the embryonic organs (cephalic compartment, foot, brain cord, mantle, and shell) exhibit bilateral and anterior-posteriorly elongated body plans reminiscent of monoplacophorans and

basal gastropods. They found that the digital tentacles of *Nautilus* later develop from simple serial and spatially patterned bud-like anlagen (clusters of embryonic cells) along the anterior-posterior axis, indicating that the origins of the tentacles develop from the foot; the head complex is formed in middle and late embryos from an anlagen of the foot, cephalic hood, collar, hyponome, and the foot-derived epidermal covers.

7.2 Growth Rate

[70] Landman et al. (1988) analyzed the pattern of bomb-produced radiocarbon in the septa and shell wall of a mature specimen of *N. macromphalus*. According to their calculations, the specimen is 10-12 years old and the timing of septal formation ranges from > 80 to 240 days.

[69] Landman et al. (1989) measured the growth rates of four *N. scrobiculatus* and three *N. pompilius* using a chronology based on the naturally occurring radionuclides ^{210}Pb and ^{210}Po. The age of the last septum in mature specimens of both species was more than one year, whereas the time of septal formation in a submature specimen of *N. pompilius* was ca. 180 days. The $^{210}Po/^{210}Pb$ method was also used to determine the growth rates of four mature *N. pompilius* before and after introduction into an aquarium; the average time of septal formation ranged from 50 to 80 days. These estimates were consistent with those calculated from the $^{210}Po/^{210}Pb$ activity ratios.

[88] Oba et al. (1992) analyzed the oxygen isotopic composition of shells of *N. pompilius* from Fiji and the Philippines. They interpreted the light isotopic values in early ontogeny as indicating that hatching occurs in relatively shallow water (~120 m). Small scale changes in the isotopic composition of a single septum may reflect daily vertical migration.

[36] Cochran and Landman (1993) used naturally occurring radionuclides to determine the growth rate of *Nautilus*. The authors analyzed the ratio of ^{210}Po to ^{210}Pb in *N. belauensis* caught in the wild and in *N. pompilius* both caught in the wild and aquarium maintained. They concluded that the rate of growth decelerates during ontogeny, with the period of chamber formation increasing from approximately 20 days in early ontogeny to approximately 1 year at maturity. The authors also analyzed bomb-produced radiocarbon in a specimen of *N. macromphalus*, demonstrating that the animals lived for about 10 years.

[71] Landman et al. (1994) analyzed the oxygen and carbon isotopic record of *N. belauensis* raised in aquaria under controlled temperature

conditions. They concluded that both embryonic and postembryonic septa are secreted with the same temperature-dependent fractionation of aragonite relative to water as that of other aragonite-secreting molluscs. The $\delta^{16}O$ values of the septa thus provide a reliable means of determining the water temperature, and, by inference, the depth at which the septa form. Using these results, they interpreted the oxygen isotope patterns of *Nautilus* specimens caught in the wild. These analyses indicated that hatching occurs at 22-24°C, corresponding to a depth of 100-200 m, depending on the location. The increase in $\delta^{16}O$ commonly observed in postembryonic septa reflects a migration into colder, deeper water after hatching.

[174] Wood and O'Dor (2000) reported life history data from 18 species of coleoids and *Nautilus*, finding that larger cephalopods take longer to reach maturity ($r^2 = 0.376$, $p = 0.007$). When *Nautilus* is removed from the calculations and time is measured in (water temperature) degree days, the relationship is even stronger ($r^2 = 0.785$, $p \leq 0.001$). Removal of *Nautilus* from the calculations for coleoids can be justified on phylogenetic grounds and because of its very different life history.

[68] Landman et al. (2001) reexamined the occurrence of three newly hatched specimens of *N. pompilius* collected in shallow water (1.25 m deep) in Fiji by Davis and Mohorter (1973). They analyzed the oxygen and carbon isotopic composition of the septa and apertural margin of two of these specimens and concluded that the animals did not hatch in the shallow lagoon in which they were collected. Instead, they must have hatched at a depth of 160-210 m, followed by a descent to a depth of 300-370 m, followed by a final and fatal ascent into shallow water.

[171] Westermann et al. (2004) observed the growth rate of *N. pompilius* in aquaria. They measured the increase in the size of the shell, as well as the number of septa, using X-ray analysis. They calculated that the species grows to maturity in 7.3-8 years.

8. The Shell and Its Architecture

8.1 Shell Chemistry

[38] Crick et al. (1987) investigated the strontium and magnesium chemistry of the shell wall and septa in four species of *Nautilus*: *N. belauensis*, *N. macromphalus*, *N. pompilius*, and *N. scrobiculatus*. In general, the concentration of magnesium is higher and more variable in juveniles

than in adults, suggesting that the physiochemical system becomes more efficient relative to carbonate production with maturation. The four species exhibit sufficiently different concentrations of strontium and magnesium in the shell wall to permit taxonomic discrimination.

[73] Mann (1992) examined the concentration of magnesium and strontium in the shell wall and septa of four species of *Nautilus* including *N. belauensis*, *N. macromphalus*, *N. pompilius*, and *N. scrobiculatus*. The concentration of Sr is higher than that of Mg because all of the shell material is composed of aragonite. There are significant differences in the concentrations of Sr and Mg among the various structural elements, nacreous and prismatic layers in the shell wall, and nacreous material in the septa. In general, the concentrations of Mg and Sr are more variable in shell material deposited early in ontogeny than later in ontogeny. The concentrations of Mg and Sr among the four species of *Nautilus* are not sufficiently different to discriminate species. The concentrations of Mg and Sr differ in the two sympatric species *N. pompilius* and *N. scrobiculatus*.

[132] Tevesz et al. (1992) analyzed the insoluble organic matrix of the shell of *N. pompilius*. Using gas chromatography mass spectrometry, they identified five simple sugars (monosacchirides) in the hydrolyzed matrix. The function of these sugars is unclear but their low concentration does not suggest any structural role.

[12, 13] Barbin (1992) examined shells of *N. pompilius* and *N. macromphalus*, including the septa, under cathodeluminescence. Luminescence is yellow-green in *N. pompilius* and blue to blue-green in *N. macromphalus* and probably reflects fluctuations in the amount of manganese in the aragonite. The intensity increases with ontogeny and is likely due to a deceleration in the rate of growth. Barbin et al. (1995) used similar methods to investigate well-preserved shells of several cephalopods ranging from the Carboniferous to the Recent. The similarity in cathodeluminescence and spectral peaks (again, yellow-green and blue-green) suggest conservation in the chemistry of cephalopod shells since the Carboniferous.

[62] Keith et al. (1993) analyzed the organic macromolecules, including polysacchirides and individual proteins, from the organic matrix in the shells of three molluscs, including *N. pompilius*. The results indicate that N-acetyl glucosamine is present in *N. pompilius* but not in mussels or abalone.

[42] Dauphin and Marin (1995) used infrared analysis to estimate sugar/protein ratios in Recent cephalopod shell carbonates, including *Nautilus*. They found that the insoluble fractions of the shell material

are largely chitin, but the composition of the soluble matrices is more heterogeneous. They also observed an inverse relationship between the concentration of sugars (glucose and galactose), which are more highly concentrated in the most mineralized shells (*Nautilus*), in the soluble matrix and that of proteins (namely glucosamine), which are present in greater amounts in the non-mineralized taxa (*Loligo*). This reflects the evolutionary tendency toward shell reduction and disappearance in cephalopods.

[10] Auclair et al. (2004) investigated the carbon and oxygen isotopic composition of the shells of *N. macromphalus* from New Caledonia. They analyzed a series of samples along the ventral and dorsal edges and a series of samples along two growth lines. The $\delta^{18}O$ record is not affected by either kinetic or metabolic processes. It reflects a nearly constant temperature of precipitation of 18±1°C, corresponding to a mean depth of 250 m.

[87] Nudelman et al. (2006) examined the organic matrix of *N. pompilius* and mapped the distribution of acidic macromolecules and their chemical groups. They documented four different zones underlying a single nacreous tablet: 1) a central zone rich in carboxylates, 2) a surrounding ring-shaped area rich in sulfates, 3) a region between area 2 and the intertabular matrix, containing carboxylates in concentrations lower than area 1, and 4) the intertabular matrix, rich in carboxylates and sulfates. These observations provide insights into the process of crystal nucleation and growth in molluscan nacre.

8.2 Shell Structure

[40] Currey (1988) discussed the mechanical properties of molluscan shell material and referred to the microstructure of the nautilus shell, which consists of a thin outer prismatic layer and a thicker inner nacreous layer. The outer prismatic layer is secreted rapidly and forms a "protective scaffold" for the animal, allowing the slower secretion of the nacre. Nacre is a strong material relative to other shell microstructures, and is effective in stopping cracks. The author discusses the mechanical properties of the *Nautilus* shell and its design as a "buoyancy tank."

[149] Ward (1988) reviewed the functional aspects of the shell of *Nautilus*, including speculation on the role of periostracum, shell sculpture, pellicle layer, and connecting rings in buoyancy control and growth, with extensive inclusion of supporting SEMs.

[18] Blind (1988) compared the shell structure of fossil orthocerid nautiloids and *N. pompilius* and found that the compositional and structural differentiation of the shell and septa, as well as septal necks and connecting rings, are for the most part dissimilar. However, he did find that the development of the outer prismatic layer is comparable, as is the separation of the inner and outer prismatic layers. Both also exhibit an "apical prismatic layer," indicating early ontogenetic differentiation of the genito-intestinal ligament.

[155] Watabe (1988) reviewed shell structure in molluscs, including a section on *Nautilus* describing the shell wall, septa, septal neck, and connecting ring. He also described the nacreous microstructure of the outer shell and septa.

[173] Williams et al. (1989) examined septa from three specimens of *Nautilus* using a proton microprobe to examine whether the Ca/Sr ratio varied over the *Nautilus* lifespan.

[47] Fonseca (1993) demonstrated that *N. pompilius* is an example of a planar logarithmic spiral in nature, whose shape has been inaccurately claimed by artists and designers to be ordained by the Golden Section.

[133] Trego (1993) illustrated an anomalous color band on the body chamber of a mature shell of *N. pompilius* from Indonesia.

[43] Dauphin and Denis (1999) compared well preserved shell material of a Cenozoic nautilid and Jurassic ammonites with that of *N. macromphalus* to determine the extent of diagenesis. The fossil material shows changes in minor element contents and degradation of the organic matrix, implying that it may have suffered some diagenesis even though the original aragonitic mineralogy is preserved.

[32] Castrejón Pita et al. (2003) demonstrated that the shell cross-section of *N. pompilius* possesses a fractal dimension (2.635) that does not depend on the number of chambers.

[41] Dauphin (2006) investigated the microstructure of the nacreous layer in the septa and outer shell wall of *N. macromphalus*. The nacreous tablets in the septa are larger than those in the outer shell wall. The tablets are composed of acicular laths, which are composed in turn of round nanograins surrounded by an organic sheet.

[82] Mutvei and Donovan (2006) investigated the siphuncular wall in the Jurassic phragmoteuthid *Phragmoteuthis huxleyi*, the belemnoid

Megateuthis gigantea, and the aulacoceratid *Mojsisovicsteuthis*?, and compared it with that of Recent *Spirula* and *Nautilus*. The structure of the connecting ring is similar in *Spirula*, *Nautilus*, and fossil nautilid and tarphyceratid nautiloids.

[137] Velázquez-Castillo et al. (2006) studied the ultrastructure of the shell of a species of *Nautilus*. They observed aragonite nanocrystals embedded in the organic matrix, supporting the hypothesis that the proteins and other organic compounds guide the crystal growth.

[138] Velázquez-Castillo et al. (2006) studied the shell structure of *N. pompilius* using scanning (SEM) and transmission electron microscopy (TEM) and X-ray diffraction to better understand how self-assembly works in nature. The results suggest that the organic components of nacre determine crystal growth and keep the shell structure together.

[1] Alvarez-Lloret et al. (2008) reported findings from a preliminary study of the ultrastructure of the shells of *N. belauensis* and the bivalve *Psilunio littoralis* using 2-D X-ray micro-diffraction (μ-XRD) patterns and SEM analysis. They examined the intensity of the aragonite reflections, the calculated crystallinity, and the degree of orientation of aragonite crystals and determined that there exist sharp changes in these parameters at the transitions between microstructurally distinct shell layers. These results indicate that the shell-layer transitions involve a very rapid and precisely controlled microstructural switch.

9. Swimming and Buoyancy

9.1 Locomotion and Energetics

[92] O'Dor et al. (1990) experimentally examined the locomotor energetics of *N. pompilius*. Jet pressures increase linearly with swimming speed, and oxygen consumption increases according to a power function of pressure. They used these relationships to calculate that the cost of transport for *Nautilus* is dramatically lower than that estimated for coleoids, and at very low speeds, even less than that of salmon (an undulatory swimmer). *Nautilus* is a very efficient swimmer in its low-speed, low-energy environment.

[33] Chamberlain (1990) reviewed the mechanism and energetics of jet propulsion in *Nautilus*. He noted that *Nautilus* is an "unspectacular" jetter, moving with thrust and speeds much less than that of squids of

equivalent sizes. He also tracked the diversity of fish and cephalopods locomotory groups thru time, suggesting several reasons for the observed pattern of fluctuating diversity, including the possibility of competitive exchange.

[159] Wells (1990) provided an extensive review and a detailed summary of experimental data from studies of jet pressure, oxygen uptake from the jet stream, and jet pressure in several cephalopod groups, including *Nautilus*. The ventilatory system of cephalopods evolved to minimize the energetic cost of either oxygen extraction or jet propulsion. To efficiently jet propel, cephalopods must be able to alter the rate of oxygen uptake. Squid and *Nautilus* are able to do so, dependent on both environmental conditions and, for *Nautilus* in particular, what "the animal happens to be doing at the time."

[162] Wells and O'Dor (1991) reviewed the history of jet propulsion in cephalopods, reconstructed the likely swimming performance of ammonoids and nautiloids based upon knowledge of modern *Nautilus*, and discussed possible reasons for the replacement of shelled forms with the faster-moving coleoids that couple jet propulsion with the use of undulant fins.

[91] O'Dor and Webber (1991) surveyed locomotor performance along an "evolutionary continuum" of cephalopods (*Nautilus*, *Sepia*, *Loligo*, and *Illex*). They reviewed requirements and origins of specific adaptations for increasing efficiency in lower power density environments (large volumes of unproductive ocean).

[94] O'Dor et al. (1994) reported the results of two expeditions to the Azores to study the behavior and energetics of *Loligo*. *Nautilus* is mentioned in a discussion of the economic consequence of abandoning a shell that provides neutral buoyancy.

[163, 89] Wells and Clarke (1996) reviewed the life history and locomotory energetics of living cephalopods, including *Nautilus*. The slow growth and long life span of *Nautilus* contrast with fast growth and short life span of coleoids. The energetic relationships between transport cost and swimming speed is much lower in *Nautilus* than in coleoids, and more closely approximates the cost of undulatory swimming in salmon, as noted by O'Dor (1989).

[156] Webber et al. (2000) evaluated a variety of cephalopod and fish buoyancy mechanisms and reported new data on the metabolic cost of establishing buoyancy in *Sepia officinalis*. Using field observations, they compared the vertical versus horizontal cost of transport for *N. pompilius*

and *Loligo forbesi*, showing that *Nautilus* expends as much energy climbing or diving as it does swimming horizontally. In contrast, the squid has almost no energetic cost while diving, but consumes considerably more energy holding a vertical position and ascending. The buoyancy mechanisms in *Nautilus* and *Sepia* were compared.

[90] O'Dor (2002) studied coleoid (cuttlefish and squid) swimming energetics in the wild, using telemetry. He also utilized known laboratory-based data on *Nautilus* as a base-line comparison, and found *Nautilus* swims efficiently and close to predicted, but at very slow, speed limits.

[144] To better understand the interplay between swimming and respiration in *N. pompilius*, Vyacheslav (2002) reviewed the morphology, histology, and workings of the mantle cavity complex, funnel, collar, and retractor muscles. The arrangement of muscle fibers in the cephalopodium retractors and funnel is typical of most molluscan groups, but that in the wings and nuchal retractors is simpler. There is a functional conflict between respiratory and locomotory systems; the strong flows produced by the piston-like movements of the head during jetting displace the gills, thus inhibiting jet propulsion and necessitating slow sustained swimming.

9.2 Shell Hydrostatics

[55] Hewitt and Westermann (1988) performed shell implosion experiments on juvenile *Nautilus* shells that indicated tensile membrane strength of ~131 MPA in the septal nacre. Shell strength sets limits to possible vertical migration of actively growing *Nautilus*, estimated at ~300 m depth at hatching to predicted 800 m depth at maturity.

[57] Jacobs (1992) reviewed the history of investigations into the function of the external cephalopod shell, including buoyancy, hydrostatic load, and shell strength in *Nautilus*.

10. Fisheries

[44] Del Norte Campos (2005) reviewed the catch, catch rates, and fishing and marketing practices of the *N. pompilius* fishery of Antique, northwestern Panay island, and the west central Philippines. The author estimated a mean annual catch of 6.6 MT/yr, amounting, on a per capita

basis, to an annual income of US $200 per fisherman. She also found that the industry experiences seasonality due both to weather and availability of baits.

[60] Jereb (2005) provided a richly illustrated, world fisheries-oriented catalogue of taxonomy, morphology, distribution, and general ecologic characteristics of cephalopods, including living nautiloids. Plates I and II include color illustrations of living *Nautilus* and *Allonautilus*.

11. Miscellaneous

[75] Moltschaniwskyj et al. (2007) presented the ethical and welfare considerations when using cephalopods as experimental animals. They reviewed articles in which cephalopods have been used as experimental animals, including *Nautilus*, along with those providing details of methods used in their collection, maintenance, handling, breeding, etc. The well developed nervous system of cephalopods suggests that welfare guidelines for these animals are needed.

1.3 References Published since 1987

[1] Alvarez-Lloret P, Rodriguez-Navarro AB and Checa G (2008) Evolution of the morphology and crystallographic orientation during shell growth in *Psilunio littoralis* and *Nautilus belauensis*. Mineral Mag 72:5-9

[2] Arnold JM (1988) Some observations on the cicatrix of *Nautilus* embryos. In: J Wiedmann and J Kullmann (eds) Cephalopods Present and Past: O.H. Schindewolf Symposium. E. Schweizerbart'sche Verlagsbuchhandlung (Nägele u. Obermiller), Tübingen 1985, pp 181-190

[3] --- (1992) *Nautilus* embryology: a new theory of molluscan shell formation. Biol Bull 183:373-374

[4] Arnold JM and Carlson BA (1991) Living *Nautilus* embryos: in ovo movements. Some comparative aspects. J Ceph Biol 2:1-12

[5] Arnold JM, Awai M and Carlson BA (1990) Hatching of *Nautilus* embryos in the Waikiki Aquarium. J Ceph Biol 1:117

[6] Arnold JM, Carlson BA and Landman NH (1991) In ovo movements of the living *Nautilus* embryo: a video presentation. B Mar Sci 49:660

[7] Arnold JM, Awai M and Carlson BA (1993) *Nautilus* embryology: egg capsule formation, deposition, and hatching. J Ceph Biol 2:51-56

[8] --- (1993) Speculation and comments on predation of *Nautilus* on egg capsules of other *Nautilus*. J Ceph Biol 2:47-50

[9] Arnold JM, Mutvei H, Landman NH et al. (1990) Microstructure of the embryonic shell of *Nautilus belauensis* (Cephalopoda): evidence from oxygen plasma etching. Am Mus Novit:1-9

[10] Auclair A-C, Lecuyer C, Bucher H et al. (2004) Carbon and oxygen isotope composition of *Nautilus macromphalus*: record of thermocline waters off New Caledonia. Chem Geol 207:91-101
[11] Baratte S, Andouche A and Bonnaud L (2007) *Engrailed* in cephalopods: a key gene related to the emergence of morphological novelties. Dev Genes Evol 217:353-362
[12] Barbin V (1992) Fluctuation in shell composition in *Nautilus* (Cephalopoda: Mollusca): evidence from cathodoluminescence. Lethaia 25:391-389
[13] Barbin V, Brand U, Hewitt RA et al. (1995) Similarity in cephalopod shell biogeochemistry since Carboniferous: Evidence from cathodoluminescence. Geobios 28:701-710
[14] Basil JA, Hanlon RT, Sheikh SI et al. (2000) Three-dimensional odor tracing by *Nautilus pompilius.* J Exp Biol 203:1409-1414
[15] Basil JA, Lazeby GB, Nakanuku L et al. (2002) Female *Nautilus* are attracted to male conspecific odor. B Mar Sci 70:217-227
[16] Basil JA, Bahctinova I, Kuroiwa K et al. (2005) The function of the rhinophore and the tentacles of *Nautilus pompilius* L. (Cephalopoda, Nautiloidea) in orientation to odor. Mar Freshw Behav Phy 38:209-221
[17] Bergmann S, Lieb B, Ruth P et al. (2006) The hemocyanin from a living fossil, the cephalopod *Nautilus pompilius*: Protein structure, gene organization, and evolution. J Mol Evol 62:362-374
[18] Blind W (1988) Comparative investigations on the shell morphology and structure of *Nautilus pompilius*, *Orthoceras* sp., *Pseudorthoceras* sp., and *Kionoceras* sp. In: J Wiedmann and J Kullmann (eds) Cephalopods Present and Past: O.H. Schindewolf Symposium. E. Schweizerbart'sche Verlagsbuchhandlung (Nägele u. Obermiller), Tübingen 1985, pp 273-290
[19] Bonacum J, Landman NH, Mapes RH et al. (in press) Evolutionary radiation of *Nautilus.* Am Malaco Bull
[20] Bonnaud L, Saihi A and Boucher-Rodoni R (2002) Are 28S RDNA and 18S RDNA informative for cephalopod phylogeny? B Mar Sci 71:197-208
[21] Bonnaud L, Ozouf-Costaz C and Boucher-Rodoni R (2004) A molecular and karyological approach to the taxonomy of *Nautilus.* C R Biol 327:133-138
[22] Boore JL (2006) The complete sequence of the mitochondrial genome of *Nautilus macromphalus* (Mollusca : Cephalopoda). BMC Genomics 7:182-195
[23] Boucher-Rodoni R (1989) Oxygen uptake and ammonia excretion by *Nautilus macromphalus.* CR Acad Sci III-VIE 309:173-179
[24] Boucher-Rodoni R and Boucher G (1993) Respiratory quotient and calcification of *Nautilus macromphalus* (Cephalopoda: Nautiloidea). Mar Biol 117:629-633
[25] Boucher-Rodoni R and Mangold K (1994) Ammonia production in cephalopods, physiological and evolutionary aspects. Mar Behav Physiol 25:53-60
[26] Bourne GB (1987) Circulatory physiology of *Nautilus.* Experientia 43:484-486
[27] Boutilier RG, West TG, Pogson GH et al. (1996) *Nautilus* and the art of metabolic maintenance. Nature 382:534-536
[28] Boutilier RG, West TG, Webber DM et al. (2000) The protective effects of hypoxia-induced hypometabolism in the *Nautilus.* J Comp Physiol B 170:261-268
[29] Brix O, Colosimo A and Giardina B (1994) Temperature dependence of oxygen binding to cephalopod haemocyanins: ecological implications. Mar Behav Physiol 25:149-162
[30] Bustamante P, Grigoni S, Boucher-Rodoni R et al. (2000) Bioaccumulation of 12 trace elements in the tissues of the Nautilus *Nautilus macromphalus* from New Caledonia. Mar Pollut Bull 40:688-696
[31] Carlson BA, Awai M and Arnold JM (1992) Hatching and early growth of *Nautilus belauensis* and implications on the distribution of the genus *Nautilus.* Proceedings of the Seventh International Coral Reef Symposium, Guam 1992 1:587-592
[32] Castrejón Pita AA, Castrejón Pita JR, Galán AS et al. (2003) The impressive complexity in the *Nautilus pompilius* shell. Fractals 11:163-168

[33] Chamberlain JA, Jr. (1990) Jet propulsion of *Nautilus*: a surviving example of early Paleozoic cephalopod locomotor design. Can J Zoolog 68:806-814
[34] Chignell D, Van Holde KE and Miller KI (1997) The hemocyanin of the squid *Sepioteuthis lessoniana*: structural comparison with other cephalopod hemocyanins. Comp Biochem Phys B 118B:895-902
[35] Cochran JK and Landman NH (1984) Radiometric determination of the growth rate of *Nautilus* in nature. Nature 308:725-727
[36] --- (1993) Using radioisotopes to determine growth-rates of marine organisms. J Chem Educ 70:749-754
[37] Colgan DJ, Ponder WF and Eggler PE (2000) Gastropod evolutionary rates and phylogenetic relationships assessed using partial 28S rDNA and histone H3 sequences. Zool Scr 29:29-63
[38] Crick RE, Mann KO and Chamberlain JA (1987) Skeletal chemistry of *Nautilus* and its taxonomic significance. Biochem Syst Ecol 15:461-474
[39] Crook R and Basil J (2008) A biphasic memory curve in the chambered nautilus, *Nautilus pompilius* L. (Cephalopoda: Nautiloidea). J Exp Biol 211:1992-1998
[40] Currey JD (1988) Shell form and strength. In: ER Trueman and MR Clark (eds) The Mollusca, Volume 11, Form and Function. Academic Press, Inc., San Diego, pp 183-210
[41] Dauphin Y (2006) Structure and composition of the septal nacreous layer of *Nautilus macromphalus* L. (Mollusca, Cephalopoda). Zoology 109:85-95
[42] Dauphin Y and Marin F (1995) The compositional analysis of recent cephalopod shell carbohydrates by Fourier-transform infrared spectrometry and high-performance anion-exchange-pulsed amperometric detection. Experientia 51:278-283
[43] Dauphin Y and Denis A (1999) Diagenèse comparée des phases minérales et organiques solubles dans les tests aragonitiques de nautiles et d'ammonites. B Soc Geol Fr 170:355-365
[44] del Norte-Campos AGC (2005) The chambered *Nautilus* fishery of Panay Island, West Central Phillipines: Fishing practices and yield. In: C Chotiyapputta, EMC Hatfield and C-C Lu (eds) Phuket Marine Biological Center Research Bulletin. Cephalopod Biology, Recruitment and Culture International Symposium and Workshop, 17-21 February 2003.Center Phuket, Thailand, pp 299-305
[45] Eno NC (1994) The morphometrics of cephalopod gills. J Mar Biol Assoc UK 74:687-706
[46] Fields RI (2006) Hatching of *Nautilus pompilius* in a closed system using artificial seawater. Drum and Croaker 37:43-47
[47] Fonseca R (1993) Shape and order in organic nature: The *Nautilus pompilius*. Leonardo 26:201
[48] Gatsogiannis C, Moeller A, Depoix F et al. (2007) *Nautilus pompilius* hemocyanin: 9 Å Cryo-EM structure and molecular model reveal the subunit pathway and the interfaces between the 70 functional units. J Mol Biol 374:465-486
[49] Gregory CR, Latimer KS, Pennick KE et al. (2006) Novel iridovirus in a nautilus (*Nautilus* spp.). J Vet Diagn Invest 18:208-211
[50] Habe T and Okutani T (1988) A new subspecies of living *Nautilus* (Cephalopoda: Nautiloidea) from the Sulu Sea. Venus: Jpn J Malaco 47:91-94
[51] Hara T, Hara R, Kishigami A et al. (1995) Rhodopsin and retinochrome in the retina of a tetrabranchiate cephalopod, *Nautilus pompilius*. Zool Sci 12:195-201
[52] --- (1992) Localization of the retinochrome in the retina of a tetrabranchiate cephalopod, *Nautilus pompilius*. Zool Sci 9:211-217
[53] Harvey AW, Mooi R and Gosliner TM (1999) Phylogenetic taxonomy and the status of *Allonautilus* Ward and Saunders, 1997. J Paleontol 73:1214
[54] Hewitt RA (1988) Nautiloid shell taphonomy: interpretations based on water pressure. Palaeogeogr Palaeocl 63:15-25
[55] Hewitt RA and Westermann GEG (1988) Stress and strain in *Nautilus* shells: some limitations on the buoyancy control and vertical migration of ectocohliates. In: J Wiedmann and J Kullmann (eds) Cephalopods Present and Past: O.H. Schindewolf

Symposium. E. Schweizerbart'sche Verlagsbuchhandlung (Nägele u. Obermiller), Tübingen 1985, pp 705-712

[56] Isaji S, Kase T, Tanabe K et al. (2002) Ultrastructure of muscle-shell attachment in *Nautilus pompilius* Linnaeus (Mollusca : Cephalopoda). Veliger 45:316-330

[57] Jacobs DK (1992) The support of hydrostatic load in cephalopod shells: adaptive and ontogenetic explanations of shell form and evolution from Hooke 1695 to the Present. Evol Biol 26:287-349

[58] Jacobs DK and Landman NH (1993) *Nautilus* -- a poor model for the function and behavior of ammonoids? Lethaia 26:101

[59] --- (1994) *Nautilus* -- model or muddle? Lethaia 27:95

[60] Jereb P (2005) Family Nautilidae. In: P Jereb and CFE Roper (eds) Cephalopods of the world. An annotated and illustrated catalogue of cephalopod species known to date. Volume 1: Chambered nautiluses and sepioids (Nautilidae, Sepiidae, Sepiolidae, Sepiadariidae, Idiosepiidae and Spirulidae). FAO Species Catalogue for Fishery Purposes. No. 4. FAO, Rome, pp 51-55

[61] Jordan M, Chamberlain JA, and Chamberlain RB (1988) Response of *Nautilus* to variation in ambient pressure. J Exp Biol 137:175-189

[62] Keith J, Stockwell S, Ball D et al. (1993) Comparative analysis of macromolecules in mollusc shells. Comp Biochem Phys B 105B:487-496

[63] Kleemann S and Schipp R (1996) The innervation of the cephalic aorta of nautiloids (Tetrabranchiata, Cephalopoda). Zool-Anal Complex Sys 99:258-267

[64] Lafont A-G, Dufour S and Fouchereau-Peron M (2006) Evidence for the presence of molecules related to the neuropeptide CGRP in two cephalopods, *Sepia officinalis* and *Nautilus macromphalus*: comparison with its target organ distribution. Neuroendocrinology 84:138-150

[65] --- (2007) Evolution of the CT/CGRP family: Comparative study with new data from models of teleosts, the eel, and cephalopod molluscs, the cuttlefish and the nautilus. Gen Comp Endocr 153:155-169

[66] Landman NH and Cochran JK (1987) Growth and longevity of *Nautilus*. In: WB Saunders and NH Landman (eds) *Nautilus*: The Biology and Paleobiology of a Living Fossil. Plenum Press, New York, pp 401-420

[67] Landman NH, Arnold JM and Mutvei H (1989) Description of the embryonic shell of *Nautilus belauensis* (Cephalopoda). Am Mus Novit 2960:1-16

[68] Landman NH, Jones DS, and Davis RA (2001) Hatching depth of *Nautilus pompilius* in Fiji. Veliger 44:2001

[69] Landman NH, Cochran JK, Chamberlain JA, Jr. et al. (1989) Timing of septal formation in two species of *Nautilus* based upon radiometric and aquarium data. Mar Biol 102:65-72

[70] Landman NH, Druffel ERM, Cochran JK et al. (1988) Bomb-produced radiocarbon in the shell of the chambered *Nautilus*. Earth Planet Sc Lett 89:28-34

[71] Landman NH, Cochran JK, Rye DM et al. (1994) Early life history of *Nautilus*: evidence from isotopic analyses of aquarium-reared specimens. Paleobiology 20:40-51

[72] Lindgren AR, Giribet G, and Nishiguchi MK (2004) A combined approach to the phylogeny of Cephalopoda (Mollusca). Cladistics 20:454-486

[73] Mann K (1992) Physiological, environmental, and mineralogical controls on Mg and Sr concentrations in *Nautilus*. J Paleontol 66:620-636

[74] Messenger JB and Young JZ (1999) The radular apparatus of cephalopods. Philos T Roy Soc B 354:161-182

[75] Moltschaniwskyj NA, Hall K, Lipinski MR et al. (2007) Ethical and welfare considerations when using cephalopods as experimental animals. Rev Fish Biol Fisheries 17:455-476

[76] Morris CC (1989) Preliminary observations on the ultrastructure of statoliths from *Nautilus*. J Ceph Biol 1:15-20

[77] Muntz WRA (1991) Anatomical and behavioural studies on vision in *Nautilus* and *Octopus*. Am Malaco Bull 9:69-74

[78] --- (1994) Spatial summation in the phototactic behaviour of *Nautilus pompilius*. Mar Behav Physiol 24:183-187

[79] --- (1994) Effects of light on the efficacy of traps for *Nautilus pompilius*. Mar Behav Physiol 24:189-193

[80] Muntz WRA and Wentworth SL (1987) An anatomical study of the retina of *Nautilus pompilius*. Biol Bull 173:387-397

[81] --- (1995) Structure of the adhesive surface of the digital tentacles of *Nautilus pompilius*. J Mar Biol Assoc UK 75:747-750

[82] Mutvei H and Donovan DT (2006) Siphuncular structure in some fossil coleoids and Recent *Spirula*. Palaeontology 49:685-691

[83] Mutvei H, Arnold JM and Landman NH (1993) Muscles and attachment of the body to the shell in embryos and adults of *Nautilus belauensis* (Cephalopoda). Am Mus Novit 3059:1-15

[84] Neumeister H and Budelmann BU (1997) Structure and function of the *Nautilus* statocyst. Philos T Roy Soc B 352:1565-1588

[85] Nixon M (1988) The feeding mechanisms and diets of cephalopods -- living and fossil. In: J Wiedmann and J Kullmann (eds) Cephalopods Present and Past: O.H. Schindewolf Symposium. E. Schweizerbart'sche Verlagsbuchhandlung (Nägele u. Obermiller), Tübingen 1985, pp 641-652

[86] --- (1995) A nomenclature for the radula of the cephalopoda (Mollusca), living and fossil. J Zool 236:73-81

[87] Nudelman F, Gotliv BA, Addadi L et al. (2006) Mollusk shell formation: Mapping the distribution of organic matrix components underlying a single aragonitic tablet in nacre. J Struct Biol 153:176-187

[88] Oba T, Kai M, and Tanabe K (1992) Early life history and habitat of *Nautilus pompilius* inferred from oxygen isotope examinations. Mar Biol 113:211-217

[89] O'Dor RK and Wells MJ (1990) Performance limits of "antique" and "state-of-the-art" cephalopods, *Nautilus* and squid. Am. Malacol. Union Prog. Abstr. 56th Ann. Meeting:52

[90] O'Dor R (2002) Telemetered cephalopod energetics: Swimming, soaring, and blimping. Integr Comp Biol 42:1065-1070

[91] O'Dor RK and Webber DM (1991) Invertebrate athletes: trade-offs between transport efficiency and power-density in cephalopod evolution. J Exp Biol 160:93-112

[92] O'Dor RK, Wells J and Wells MJ (1990) Speed, jet pressure and oxygen-consumption relationships in free-swimming *Nautilus*. J Exp Biol 154:383-396

[93] O'Dor RK, Forsythe J, Webber DM et al. (1993) Activity levels of *Nautilus* in the wild. Nature 362:626

[94] O'Dor RK, Hoar JA, Webber DM et al. (1994) Squid (*Loligo forbesi*) performance and metabolic rates in nature. Mar Freshw Behav Phy 25:163-177

[95] Okubo S, Tsujii T, Watabe N et al. (1995) Hatching of *Nautilus belauensis* Saunders, 1981, in captivity: culture, growth and stable isotope composition of shells, and histology and immunohistochemistry of the mantle epithelium of the juveniles. Veliger 38:192-202

[96] Okutani T (1990) Hatching of *Nautilus* in captivity. Chambered Nautilus Newsletter 59:1-2

[97] Pernice M, Destoumieux-Garzón D, Peduzzi J et al. (2007) Identification of a *Vibrio* strain of producing antimicrobial agents in the excretory organs of *Nautilus pompilius* (Cephalopoda: Nautiloidea). Rev Fish Biol Fisheries 17:197-205

[98] Pernice M, Wetzel S, Gros O et al. (2007) Enigmatic dual symbiosis in the excretory organ of *Nautilus macromphalus* (Cephalopoda: Nautiloidea). Proc. R. Soc. B 274: 1143-1152

[99] Pernice M, Pichon D, Domart-Coulon I et al. (2007) Primary co-culture as a complementary approach to explore the diversity of bacterial associations in marine invertebrates: the example of *Nautilus macromphalus* (Cephalopoda: Nautiliodea). Mar Biol 150:749-757

[100] Pernice M, Boucher J, Boucher-Rodoni R et al. (2008) Comparative bioaccumulation of trace elements between *Nautilus pompilius* and *Nautilus macromphalus* (Cephalopoda: Nautiloidea). Ecotox Environ Safe 72:365-371
[101] Reyment RA (1988) A foraging model for shelled cephalopods. In: J Wiedmann and J Kullmann (eds) Cephalopods Present and Past: O.H. Schindewolf Symposium. E. Schweizerbart'sche Verlagsbuchhandlung (Nägele u. Obermiller), Tübingen 1985, pp 687-703
[102] Rosenberg G, Tillier S, Tillier A et al. (1997) Ribosomal RNA phylogeny of selected major clades in the mollusca. J Mollus Stud 63:301-309
[103] Roux M, Bouchet P, Bourseau JP et al. (1991) Bathyal environment off New Caledonia: Preliminary results of the CALSUB cruise and paleoecological consequences. B Soc Geol Fr 162:675-685
[104] Ruth P, Schipp R and Klussendorf B (1988) Cytomorphology and copper content of the basal cells in the midgut gland of *Nautilus* (Cephalopoda, Tetrabranchiata), a contribution to the localization of hemocyanin synthesis. Zoomorphology 108:1-11
[105] Ruth P, Schimmelpfennig R and Schipp R (1996) Comparative immunohistochemical and immunocytochemical investigations on the location of haemocyanin synthesis in Dibranchiate and Tetrabranchiate cephalopods (*Sepia* and *Nautilus*). In: F Olóriz and FJ Rodríguez-Tovar (eds) Advancing Research on Living and Fossil Cephalopods, Proceedings of the IV International Symposium on Cephalopods: Present and Past. Kluwer Academic, Granada, Spain, pp 189-202
[106] Ruth P, Schipp R and Arnold JM (1999) Organization and function of the midgut glands of *Nautilus pompilius* L. and *Nautilus macromphalus* Sow. (Cephalopoda, Tetrabranchiata). Zoology 102:196-211
[107] Ruth P, Schmidtberg H, Westermann B et al. (2002) The sensory epithelium of the tentacles and the rhinophore of *Nautilus pompilius* L. (Cephalopoda, Nautiloidea). J Morphol 251:239-255
[108] Saunders WB (1983) Natural rates of growth and longevity of *Nautilus belauensis*. Paleobiology 9:280-288
[109] --- (1990) Deep-water camera survey of *Nautilus* in the Admiralty Islands, Papua New Guinea. Natl Geogr Res 6:504-508
[110] --- (1991) Searching for the King *Nautilus* in the Admiralty Islands, PNG. Explorers J 69:21-24
[111] --- (1998) Superfamily Nautiloidea. In: PL Beesley, GJB Ross and A Wells (eds) Mollusca: The Southern Synthesis. Fauna of Austalia, Vol. 5, Part A. CSIRO Publishing, Melbourne, pp 494-498
[112] Saunders WB and Ward PD (1988) Sympatric occurrence of living *Nautilus* (*N. pompilius* and *N. stenomphalaus*) on the Great Barrier Reef, Australia. Nautilus 102:188-193
[113] --- (1994) *Nautilus* is not a model for the function and behavior of ammonoids. Lethaia 27:47
[114] Saunders WB, Davis LE and Knight RL (1987) Sympatric species of *Nautilus* (*N. pompilius* and *N. scrobiculatus*) in the Admiralty Islands, Papua New Guinea. Nautilus 101:93-99
[115] Saunders WB, Knight RL and Bond PN (1991) *Octopus* predation of *Nautilus*: evidence from Papua New Guinea. B Mar Sci 49:280-287
[116] Saunders WB, Shimansky VN and Amitrov OV (1996) Clarification of *Nautilus praepompilius* Shimansky from the late Eocene of Kazakhstan. J Paleontol 70:609-611
[117] Saunders WB, Bond PN, Hastie LC et al. (1989) On the distribution of *Nautilus pompilius* in the Samoas, Fiji and Tonga. Nautilus 103:99-104
[118] Sawata H and Phongsuwan N (1994) Occurrence of *Nautilus pompilius* in the eastern part of the Indian Ocean -- Andaman Sea. Research Bulletin, Phuket Marine Biological Center 59:99-100
[119] Schipp R (1987) The blood vessels of cephalopods. A comparative morphological and functional survey. Experientia 43:525-537

[120] --- (1987) General morphological and functional characteristics of the cephalopod circulatory system: an introduction. Experientia 43

[121] Schipp R, Chung YS and Arnold JM (1990) Symbiotic bacteria in the coelom of *Nautilus* (Cephalopoda, Tetrabranchiata). J Ceph Biol 1:59-74

[122] Schipp R, Jakobs PM and Fiedler A (1991) Monoaminergic peptidergic interactions in neuroregulatory control of the cephalic aorta in *Sepia officinalis* L (Cephalopoda). Comp Biochem Phys C 99:421-429

[123] Shigeno S, Sasaki T, Moritaki T et al. (2008) Evolution of the cephalopod head complex by assembly of multiple molluscan parts: evidence from *Nautilus* embryonic development. J Morphol 269:1-17

[124] Sinclair B, Briskey L, Aspden W et al. (2007) Genetic diversity of isolated populations of *Nautilus pompilius* (Mollusca, Cephalopoda) in the Great Barrier Reef and Coral Sea. Rev Fish Biol Fisheries 17

[125] Springer J, Ruth P, Beuerlein K et al. (2004) Immunohistochemical localization of cardio-active neuropeptides in the heart of a living fossil, *Nautilus pompilius* L. (Cephalopoda, Tetrabranchiata). J Mol Histol 35:21-28

[126] --- (2005) Distribution and function of biogenic amines in the heart of *Nautilus pompilius* L. (Cephalopoda, Tetrabranchiata). J Mol Histol 36:345-353

[127] Staples JF, Hershkowitz JJ and Boutilier RG (2000) Effects of ambient PO_2 and temperature on oxygen uptake in *Nautilus pompilius*. J Comp Physiol B 170:231-236

[128] Staples JF, Webber DM and Boutilier RG (2003) Environmental hypoxia does not constrain the diurnal depth distribution of free-swimming *Nautilus pompilius*. Physiol Biochem Zool 76:644-651

[129] Suzuki T, Fukuta H, Nagato H et al. (2000) Arginine kinase from *Nautilus pompilius*, a living fossil: Site-directed mutagenesis studies on the role of amino acid residues in the guanidino specificity region. J Biol Chem 275:23884-23890

[130] Tanabe K and Uchiyama K (1997) Development of the embryonic shell structure in *Nautilus*. Veliger 40:203-215

[131] Tanabe K, Tsukahara J and Hayasaka S (1990) Comparative morphology of living *Nautilus* (Cephalopoda) from the Philippines, Fiji, and Palau. Malacologia 31:297-312

[132] Tevesz MJS, Schwelgien SF, Smith BA et al. (1992) Identification of monosaccharides in hydrolyzed *Nautilus* shell insoluble matrix by gas chromatography/mass spectrometry. Veliger 35:381-383

[133] Trego KD (1993) Band color pattern on the venter of a mature shell of *Nautilus pompilius* Linnaeus, 1758. Veliger 36:431

[134] Tshudy DM, Feldman RM and Ward PD (1989) Cephalopods: biasing agents in the preservation of lobsters. J Paleontol 63:621-626

[135] Uchiyama K and Tanabe K (1996) Hatching of *Nautilus macromphalus* in Toba Aquarium, Japan. In: F Olóriz and FJ Rodríguez-Tovar (eds) Advancing Research on Living and Fossil Cephalopods, Proceedings of the IV International Symposium on Cephalopods: Present and Past. Kluwer Academic, Granada, Spain, pp 13-16

[136] Uda K, Fujimoto N, Akiyama Y et al. (2006) Evolution of the arginine kinase gene family. Comp Biochem Phys D 1:209-218

[137] Velazquez-Castillo RR, Reyes-Gasga J, Garcia-Gutierrez DI et al. (2006) Crystal structure characterization of *Nautilus* shell at different length scales. Biomaterials 27:4508-4517

[138] Velazquez-Castillo R, Reyes-Gasga J, Garcia-Gutierrez DI et al. (2006) Nanoscale characterization of *Nautilus* shell structure: An example of natural self-assembly. J Mater Res 21:1484-1489

[139] Vines G (1992) *Nautilus* take slow lane to survival. New Sci 135:16-16

[140] Vitturi R, Colombera D, Catalano E et al. (1990) Spermatocyte chromosome study of eight species of the class Cephalopoda (Mollusca). J Ceph Biol 1:101-112

[141] von Boletzky S (1987) Ontogenetic and phylogenetic aspects of the cephalopod circulatory system. Experientia 43:478-483

[142] --- (1988) Characteristics of cephalopod embroygenesis. In: J Wiedmann and J Kullmann (eds) Cephalopods Present and Past: O.H. Schindewolf Symposium.

E. Schweizerbart'sche Verlagsbuchhandlung (Nägele u. Obermiller), Tübingen 1985, pp 167-179
[143] --- (1999) Biology and biogeography of extant cephalopods. Bull Soc Geol Fr 170:205-215
[144] Vyacheslav AB (2002) Reanalysis of functional design of *Nautilus* locomotory and respiratory systems. Am Malaco Bull 17:17-30
[145] Wani R (2004) Experimental fragmentation patterns of modern *Nautilus* shells and the implications for fossil cephalopod taphonomy. Lethaia 37:113-123
[146] Wani R and Ikeda H (2006) Planispiral cephalopod shells as a sensitive indicator of modern and ancient bottom currents: New data from flow experiments with modern *Nautilus pompilius*. Palaios 21:289-297
[147] Wani R, Kase T, Shigeta Y et al. (2005) New look at ammonoid taphonomy, based on field experiments with modern chambered *Nautilus*. Geology 33:849-852
[148] Wani R, De Ocampo R, Aguilar YM et al. (2008) First discovery of fossil *Nautilus pompilius* Linneaus, 1758 (Nautilidae, Cephalopoda) from Pangasinan, northwestern Philippines. Paleontol Res 12:89-95
[149] Ward PD (1988) Form and function of the *Nautilus* shell: some new perspectives. In: ER Trueman and MR Clark (eds) The Mollusca, Volume 11, Form and Function. Academic Press, Inc., San Diego, pp 143-165
[150] --- (1998) Coils of time. Discover 19:100-106
[151] --- (1999) Phylogenetic taxonomy and the status of *Allonautilus* Ward and Saunders, 1997: Reply. J Paleontol 73:1217
[152] --- (2008) Chambers of secrets. New Sci 198:40-43
[153] Ward PD and Saunders WB (1997) *Allonautilus*: A new genus of living nautiloid cephalopod and its bearing on phylogeny of the Nautilida. J Paleontol 71:1054-1064
[154] Warnke K and Keupp H (2005) *Spirula* - a window to the embryonic development of ammonoids? Morphological and molecular indications for a palaeontological hypothesis. Facies 51:65-70
[155] Watabe N (1988) Shell structure. In: ER Trueman and MR Clark (eds) The Mollusca, Volume 11, Form and Function. Academic Press, Inc., San Diego, pp 69-104
[156] Webber DM, Aitken JP and O'Dor RK (2000) Costs of locomotion and vertical dynamics of cephalopods and fish. Physiol Biochem Zool 73:651-662
[157] Wells MJ (1987) Oxygen-uptake and the effect of feeding in *Nautilus*. Veliger 30:69-75
[158] --- (1988) The mantle and muscle cavity of the cephalopods. In: ER Trueman and MR Clark (eds) The Mollusca, Volume 11, Form and Function. Academic Press, San Diego, pp 287-300
[159] --- (1990) Oxygen extraction and jet propulsion in cephalopods. Can J Zoolog 68:815-824
[160] --- (1992) The cephalopod heart: the evolution of a high-performance invertebrate pump. Experientia 48:800-808
[161] Wells M (1999) Why the ammonites snuffed it. Mar Freshw Behav Phy 32:103-111
[162] Wells MJ and Odor RK (1991) Jet propulsion and the evolution of the cephalopods. B Mar Sci 49:419-432
[163] Wells MJ and Clarke A (1996) Energetics: The costs of living and reproducing for an individual cephalopod. Philos T Roy Soc B 351:1083-1104
[164] Wells MJ, Wells J and O'Dor RK (1992) Life at low oxygen tensions: the behaviour and physiology of *Nautilus pompilius* and the biology of extinct forms. J Mar Biol Assoc UK 72:313-328
[165] Westermann B and Schipp R (1998) Cytological and enzyme-histochemical investigations on the digestive organs of *Nautilus pompilius* (Cephalopoda, Tetrabranchiata). Cell Tissue Res 293:327-337
[166] --- (1998) Morphology and histology of the digestive tract of *Nautilus pompilius* and *Nautilus macromphalus* (Cephalopoda, Tetrabranchiata). Zoomorphology 117:237-245

[167] Westermann B and Beuerlein K (2005) Y-maze experiments on the chemotactic behaviour of the tetrabranchiate cephalopod *Nautilus pompilius* (Mollusca). Mar Biol 147:145-151
[168] Westermann B, Schmidtberg H and Beuerlein K (2005) Functional morphology of the mantle of *Nautilus pompilius* (Mollusca, Cephalopoda). J Morphol 264:277-285
[169] Westermann B, Beuerlein K, Ruth P et al. (2000) Tracer studies of food absorption in the digestive tract of *Nautilus pompilius* (Cephalopoda: Tetrabranchiata). Cell Tissue Res 300:173-179
[170] Westermann B, Beuerlein K, Hempelmann G et al. (2002) Localization of putative neurotransmitters in the mantle and siphuncle of the mollusc *Nautilus pompilius* L. (Cephalopoda). Histochem J 34:435-440
[171] Westermann B, Beck-Schildwachter I, Beuerlein K et al. (2004) Shell growth and chamber formation of aquarium-reared *Nautilus pompilius* (Mollusca, Cephalopoda) by X-ray analysis. J Exp Zool Part A 12:930-938
[172] Westermann B, Ruth P, Litzlbauer HD et al. (2002) The digestive tract of *Nautilus pompilius* (Cephalopoda, Tetrabranchiata): an X-ray analytical and computational tomography study on the living animal. J Exp Biol 205:1617-1624
[173] Williams ET, Cadet G, Guardala NA et al. (1989) Pixe analysis of Ca/Sr ratios in *Nautilus* shell. Nucl Instrum Meth B 40-1:624-626
[174] Wood JB and O'Dor RK (2000) Do larger cephalopods live longer? Effects of temperature and phylogeny on interspecific comparisons of age and size at maturity. Mar Biol 136:91-99
[175] Wray CG, Landman NH, Saunders WB et al. (1995) Genetic divergence and geographic diversification in *Nautilus*. Paleobiology 21:220-228
[176] Young J (1988) Evolution of the cephalopod brain. In: ER Trueman and MR Clark (eds) The Mollusca, Volume 12: Paleontology and neontology. Academic Press, New York, pp 215-228
[177] --- (1988) Evolution of the cephalopod statocyst. In: ER Trueman and MR Clark (eds) The Mollusca, Volume 12: Paleontology and neontology. Academic Press, New York, pp 229-239
[178] --- (1993) The muscular-hydrostatic radula supports of *Octopus*, *Loligo*, *Sepia*, and *Nautilus*. J Ceph Biol 2:65-93
[179] Young RE and Vecchione M (1996) Analysis of morphology to determine primary sister-taxon relationships within coleoid cephalopods. Am Malaco Bull 12:91-112
[180] Zakharov YD, Shigeta Y, Smyshlyaeva OP et al. (2006) Relationship between delta C-13 and delta O-19 values of the Recent *Nautilus* and brachiopod shells in the wild and the problem of reconstruction of fossil cephalopod habitat. Geosci J 10:331-345

I

Nautilus Studies—The First Twenty-Two Centuries

Chapter 1

Nautilus Studies—The First Twenty-Two Centuries

RICHARD ARNOLD DAVIS

That lovely brown and white shell, the pearly nautilus, has fascinated humans for millenia. Its precise mathematical spiral and subtle coloration have touched our aesthetic souls. But it is more than that, for the pearly nautilus has titillated our intellects, too. From the time the first shell was brought from its home sea environs, our ancestors speculated as to just what sort of animal makes the shell, what that animal does for a living, and where and how it does it.

The scientific study of *Nautilus* is venerable. It began with Aristotle—maybe.

"The Father of Natural History" was born in 384 B.C. and died in 322 B.C. (Fig. 1). In the last 15 or so years of his life, he produced his *Historia animalium*, a compilation of observations and other information on all manner of animals known to him, including cephalopods. The "paper nautilus," *Argonauta*, was well known to the ancient peoples of the Mediterranean region (Luce, 1969), and Aristotle's description of the animal is unmistakable. The claim made for his being the first Western scholar to know *Nautilus* is based on one cryptic sentence that comes just after his account of *Argonauta* (Peck, 1970):

> There is another that is found in a shell, like a snail; this never comes out, but remains in the shell as a snail does and sometimes puts out its tentacles.

Richard Owen (1832, p. 3) was convinced that this sentence does, in fact, refer to *Nautilus*. He argued that the rest of the cephalopod section of the *Historia animalium* makes it obvious that Aristotle knew cephalopods well enough that he would not have mistaken an animal of another kind for a member of the class. Owen pointed out that *Nautilus* is the only cephalopod that fits Aristotle's description. Dodge (1953, p. 13) also concluded that Aristotle knew *Nautilus*. Others have been less certain. Thompson (1947) suggested that Aristotle was referring to the snail *Ianthina*. Landman (1982) argued that even if Aristotle had been supplied with specimens from lands conquered by his former student, Alexander the Great, he never could have seen a preserved *Nautilus*, let alone a live animal. He then suggested, as Thompson (1910) had done earlier, that Aristotle was endeavoring to describe *Ocythoë*, a small Mediterranean octopus; the male of this genus occupies an empty test of the pelagic tunicate *Salpa* and holds it by fas-

RICHARD ARNOLD DAVIS • Cincinnati Museum of Natural History, Cincinnati, Ohio 45202.

Figure 1. Aristotle (384–322 B.C.). Drawing of a bust of Aristotle. From the American Museum of Natural History.

tening his arms to the inside of the test. This is an intriguing idea, although Aristotle wrote of a "shell, like a snail," which does not sound like the transparent covering of a tunicate.

A difficulty in trying to decipher the history of studies of *Nautilus* is that the word "nautilus" has been used widely for two markedly different cephalopods: the "pearly nautilus" (the subject of this book) and the "paper nautilus" (*Argonauta*). Moreover, other terms have been used as synonyms for the word "nautilus." For example, Aristotle used the term "nautilus" (with the synonym "pontilus"), but the description of the animal so designated is obviously of *Argonauta*. The situation in the *Naturalis historia* of Pliny the Elder is more complicated: In one paragraph, the terms "nautilos" and "pompilos" are used synonymously (and translated as "nautilus" and "pilot-fish," respectively); again, the description is of *Argonauta*. In a later paragraph, Pliny discusses an animal called "nauplius"; its description seems also to be of *Argonauta*. Pliny treated his "nauplius" separately from his "nautilus," but it may be that he was just being cautious in using data not his own—he stated that he got his information about the "nauplius" from someone else. (Linnaeus quoted part of the "nauplius" paragraph when he formally named the genus *Argonauta* in 1758.) The irony, of course, is that when Linnaeus used the name *Nautilus* for the animal we recognize by that name today, he was referring to a creature different from that to which the word had originally been applied and for which the word had been used for well over a millenium (e.g., by Callimachus, Athanaeus, and Oppianus).

In any case, the first obvious description of *Nautilus*, even of the shell, dates from the Renaissance.

Pierre Belon was born near Le Mans in 1517; he died in 1564. Although he studied medicine and became a practicing physician, he is best known as one of the great naturalists of his period and one of the very first workers in the field of comparative anatomy. In a series of works published in the 1550s, Belon described and figured the shell of *Nautilus*. However, he did not seem to have known the

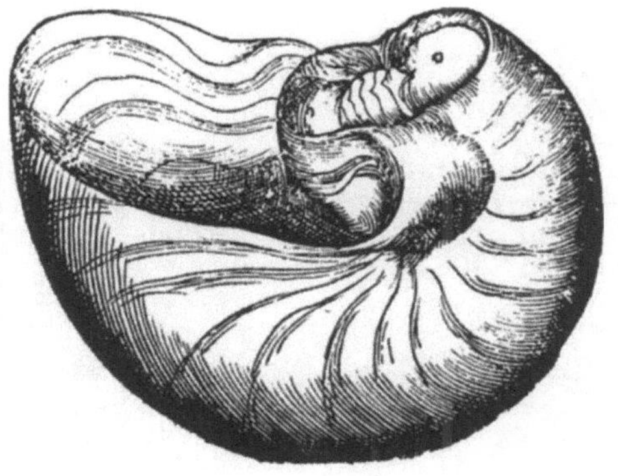

Figure 2. First picture of the *Nautilus* shell—1551. Reproduced from Belon (1555).

animal itself. In 1551, Belon figured *Argonauta*, but called it "nautilus" and pointed out that Pliny had called it "popilus" [a typographical error in Belon's work that should have read "pompilos"] or "nauplius." The shell of what we now call "Nautilus," he said, closely resembled that of his "nautilus" and was commonly called the "grosse Porcellaine" ["thick (or heavy) porcelain"] or "grande coquille de Nacre de Perle" ("great shell of mother of pearl"). By 1555, however, Belon considered *Nautilus* to be a second species of his "nautilus" (Fig. 2). Thus, Belon seems to have been the first person to apply the name "nautilus" to the animal that now bears it as a formal generic name.

Konrad von Gesner was born in Zurich in 1516 and died in 1565. Like Belon, he was a physician who is better known as a naturalist. Gesner (1558) discussed "nautilus" in his great zoological work called *Historiae animalium*. In that work, he described a picture of a shellfish sent him by an English physician, John Fauconerus, and quoted from the letter that accompanied the picture. According to Fauconerus, he saw the animal in England:

> It has a shell on the exterior that is tawny reddish. The interior is so bright and shining that it could vie with the most precious of pearls in beauty of color. The keel of the shell has many storeys that shine with the same color. A sail sticks up out of its hide, and from both sides are let down arms that are meaty and soft like the arms of the octopus, and the remaining part of its body was indistinct and confused as of other types of testaceous creatures.

Of this passage, Owen (1832, p. 5) wrote, "It is, however, so obscure and brief as to render it doubtful whether the animal alluded to really was the inhabitant of the pearly and chambered shell which is described." In any case, Gesner's illustrations of "nautilus" are definitely of *Argonauta*, and there is no indication at all that he personally knew the soft parts of *Nautilus*. (Even his picture of an argonaut in its "shell" is reproduced from Belon!)

Nautilus studies emerged from the age of obscurity and entered a new period in 1705. It was in that year that the first unequivocal description of the *Nautilus* animal was published—complete with illustration. The author was Georg Everard Rumpf (1705) (Fig. 3).

Rumpf was born either late in 1627 or early in 1628 in Hanau, a city near Frankfurt. About 1645, he was recruited by an impoverished nobleman from the

Figure 3. Georg Everard Rumpf (1628–1702). This is a portrait drawn by his son on the occasion of the father's 68th birthday and is from Rumpf (1705).

Hanau region, ostensibly to be a soldier for the Republic of Venice. As it turned out, Rumpf's compatriot was being less than noble; he was, in truth, under contract to the Dutch West-Indies Company to provide men to fight the Portuguese, the then-current enemies of the Netherlands. The ship on which the men embarked was captured, and Rumpf spent nearly three years in Portugal. It may well be that it was there that Rumpf first felt the lure of the "South Seas," having read works of some of the Portuguese explorers (Beekman, 1981). In any case, in 1652, he enlisted in the Dutch East-Indies Company and set sail for Batavia (now called Djakarta), never to see Europe again.

Rumpf had joined up as a fighting man, but soon transferred into the company's civil service. All records indicate that he was a model employee, and he rose rapidly through the hierarchy (Leupe, 1871; Sirks, 1945). Although he was good at languages and had mathematical and engineering skills that served him well in his job, it is for the results of what he called his "lucubrations" that he is best known.

By 1658, he no longer called himself "Jeurian Rumph of Hanau." Instead, he signed "Georgius Everhardus Rumphius," indicating that he was involved in scientific work. By day, he toiled for the company; by night, he worked on what would be his magnum opus, the *Herbarium amboinense*, for it was to botany that Rumpf had decided to devote his life.

Beginning in 1670, Rumpf's life was punctuated by misfortune. He went blind in that year, seemingly from glaucoma. His wife was killed in the terrible earthquake that shook the region in 1674. In 1687, a fire ravaged the entire Netherlands section of Ambon, the city in which Rumpf lived (now called Amboina); his library, collections, and manuscripts were, in large measure, destroyed, as were plates that had been prepared for the *Herbarium*. Then the manuscript of the first half of the work was lost when the ship carrying it to the Netherlands was sunk by the French.

On the other hand, good fortune shone upon Rumpf, too. When he went blind, his immediate superior in the company was prepared to sack him as useless. But Governor-General Joan Maetsuycker gave Rumpf what was, in effect, a lifelong pension at his old rank. In fact, the Dutch East-Indies Company rendered Rumpf tremendous aid in his scientific work—a fact strongly at odds with the company's reputation as being interested only in profits. This help came in the form of clerks, draughtsmen, collectors, equipment, literature, and, most important, the assignment of Rumpf's son, Paulus Augustus, as his paid amanuensis.

By early 1697, a complete version of the *Herbarium amboinense* was in the hands of the company in Amsterdam. Meanwhile, Rumpf had been laboring on the work in which *Nautilus* was to be figured and described. The manuscript of *D'Amboinsche rariteitkamer* was prepared by Paulus Augustus Rumphius and Johan Philip Sipman in 1699 and sent to Dr. Hendrik d'Acquet, Mayor of Delft and one of Rumpf's best friends; it was received in 1701. D'Acquet recognized that the manuscript needed revision to bring it up to the style of Western works and up to date with respect to literature unavailable in the East Indies. He arranged for Simon Schijnvoet to prepare the work for publication.

D'Amboinsche rariteitkamer appeared in print in 1705; unfortunately, Rumpf had died on June 15, 1702, and so did not witness the success that the work enjoyed. The plates were reissued in 1711, along with a Latin introduction, plate explanations, and index, and there was a second edition of the reissue in 1739. A second Dutch edition was published in 1741 and a German version in 1766. In addition, Valentijn edited another Dutch version that was published in 1754. (Ironically, publication of Rumpf's masterwork, the *Herbarium amboinense*, did not begin until 1741, because the company was of the opinion that it contained too much information of interest to the competition.)

Although none of his major works was published until after his death, Rumpf did enjoy a measure of renown during his life. He was in active correspondence with colleagues back in Europe and apparently was well thought of by them, for in 1681 he was elected to the Academia Naturae Curiosorum in Vienna. At that time, the Academia bestowed upon him the title "Plinius Indicus," which appears on the title page of *D'Amboinsche rariteitkamer*.

Rumpf's treatment of *Nautilus* does not indicate that he studied the animal extensively. He gave no specific information as to where the animal could be obtained in the wild; he said merely that it was found in all the seas of the Molucca Islands and near Batavia (now Djakarta) in Java. Nor does it appear that he spent much time, if any, observing live animals, for he recorded that the animal crawls with its hood lowermost. Apparently, Rumpf did regard *Nautilus* as a cephalopod;

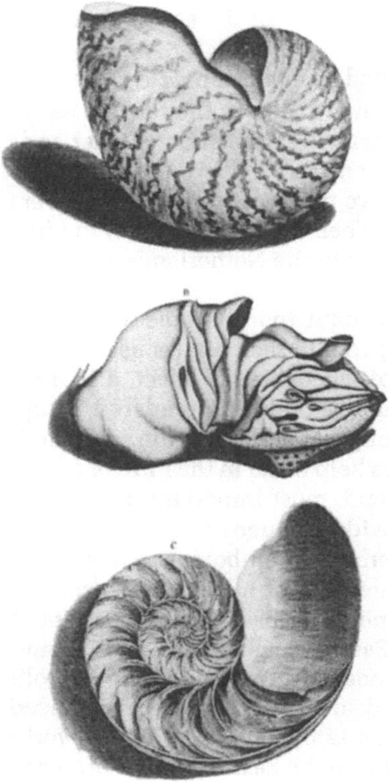

Figure 4. First picture of the *Nautilus* animal. From Rumpf (1705).

he compared it to what he referred to as the "polypus," a Latin word generally used for octopus. He called both *Nautilus* and *Argonauta* by the name "nautilus," but recognized that they are different. The former he called "Nautilus Major sive crassus" ("larger or thick nautilus"), whereas *Argonauta* was designated "Nautilus tenuis" ("thin nautilus"). (Unfortunately, any *Nautilus* specimens that Rumpf may have collected and preserved seem no longer to exist.)

Even though Rumpf produced the earliest recognizable description and illustration of the *Nautilus* animal (Fig. 4), Arthur Willey (1902, p. 737) indicated that Rumpf's work was largely ignored, discounted, and forgotten. This is probably too negative an assessment. True, Rumpf's treatment of *Nautilus* appears to have been unknown to Derham (1726), and Cuvier (1830, p. 18) proclaimed Rumpf's illustration to be "indechiffrable." However, Rumpf's work was known to both prominent biologists and paleontologists in the 1700s and early 1800s. For example, it was cited by Linnaeus (1758, p. 709) at the time the genus *Nautilus* was formally erected and by Parkinson (1811) in his *Organic remains of a former world,* one of the earliest paleontological works in English and a most popular one as well.

Meanwhile, in England, Dr. Robert Hooke was also studying *Nautilus*. Hooke was born in 1635 and died in 1703. For most of his life, he was associated with the Royal Society and was professor of geometry in Gresham College. *Nautilus*

Figure 5. Photograph of George Bennett, who brought back the *Nautilus* specimen studied by Owen. Reproduced by kind permission of the President and Council of the Royal College of Surgeons of England.

was but one of the vast multitude of subjects with which Hooke concerned himself, and, as in the case of Rumpf, his conclusions on the animal were published only after his death (Hooke, 1705, 1726).

Rumpf had dealt only with living *Nautilus;* Hooke, on the other hand, endeavored to discover whether the familiar fossils we now call ammonites were the remains of once-living creatures. To him, it was obvious that the shell of *Nautilus* and that of the ammonite are basically the same—both are planispirally coiled, both have septa, and the septa of both have siphuncular foramina. The fact that *Nautilus* and ammonites are basically the same led Hooke to wonder whether ammonites were extinct or simply waiting to be discovered living in some far-distant corner of the sea or in its depths. (Recall that, in Hooke's day, extinction was a revolutionary idea.) Hooke admitted that he knew nothing of the soft parts of *Nautilus*. This did not deter him from speculating as to its way of life (and, hence, as to the way of life of ammonites). For example, it was Hooke who proposed that *Nautilus* controls its buoyancy by varying the amount of water in the chambers of its shell. The act of filling the camerae he recognized as easy. To account for the emptying—against the water pressure—Hooke (1705, p. 340) suggested that the animal produced "artificial air" by "fermentation of the excrements of the gut, or other juices of the body."

The fourth epoch in the history of *Nautilus* studies began in 1829.

It was the 24th of August. The weather was fine and the wind was calm as the ship lay at anchor off the island of Erromanga, in the New Hebrides Group. The attention of Dr. George Bennett (Fig. 5) was drawn to an object floating on the surface of the water some distance from the ship. To the sailors, it looked like a small, dead tortoiseshell cat. Because the animal they thought it to be would have been a most unusual one in those waters, a boat was sent over to investigate.

Figure 6. Portrait of Richard Owen (1804–1892). From the Peabody Museum of Natural History, Yale University.

The "cat" turned out to be a live pearly nautilus. After observing the animal briefly, Dr. Bennett (1834, p. 407) preserved its soft parts in spirits. Eventually, he took it with him back to England, where the specimen had all the impact of an original discovery (Willey, 1902, p. 737).

In July 1831, Bennett presented his treasure to the Royal College of Surgeons. They, in turn, entrusted the specimen to Richard Owen (Fig. 6) to dissect and describe.

Owen was born in 1804 and, after having studied medicine, was appointed assistant conservator of the Museum of the Royal College of Surgeons. In the course of preparing a series of catalogues of the museum's Hunterian Collection, he acquired an extraordinary knowledge of comparative anatomy. In 1846, he was named Hunterian Professor in the College and, in 1849, Conservator of the Museum. He became superintendent of the natural history department of the British Museum, and it was during his tenure that those collections were removed to become the British Museum (Natural History). He was knighted in 1884 and retired the same year. Death came in December 1892 (Owen, 1894).

Figure 7. Illustration of soft-part anatomy of the *Nautilus* animal removed from the shell. *Left:* dorsal view; *right:* ventral view with mantle turned back. Anatomy: (A) anus; (CR) crescentic ridge on the posterior face of the hood; (DA) dorsal aponeurotic band; (DM) dorsal portion of the mantle; (DT) digital tentacles, composing the cephalic sheath; (DT^2) second digital tentacle; (E) eye; (F) funnel; (Ho. A) auricle of hood; (Ho. C) cirri of tentacles composing the hood; (IP) interbranchial papilla; (O″) postocular tentacle; (O′) preocular tentacle; (P) penis; (PA) preanal papillae; (PP) pericardial pore; (RA) anterior renal pore; (RP) posterior renal pore; (Si) base of siphuncle; (X) backwardly projecting point of dorsal aponeurotic band; (y) constriction of the siphuncle where it passes through a septum. Reproduced from Figs. 2 and 3 of Griffin (1900).

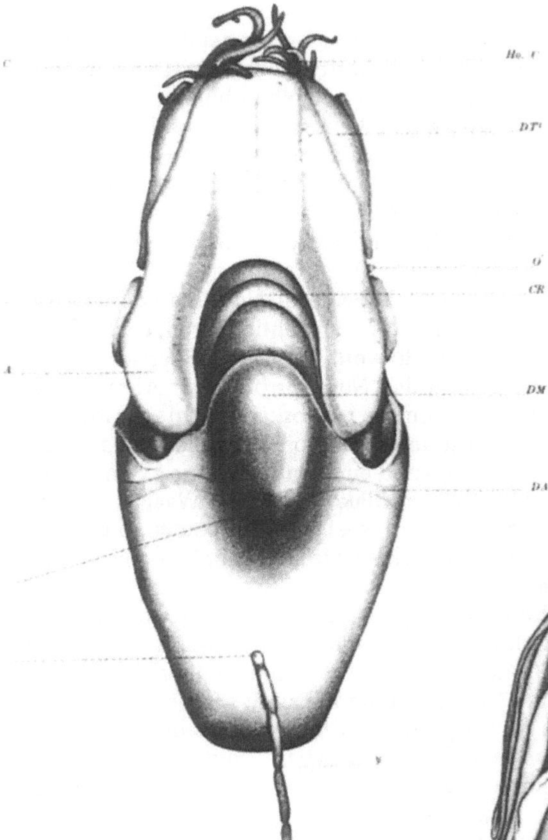

Fig. 2.

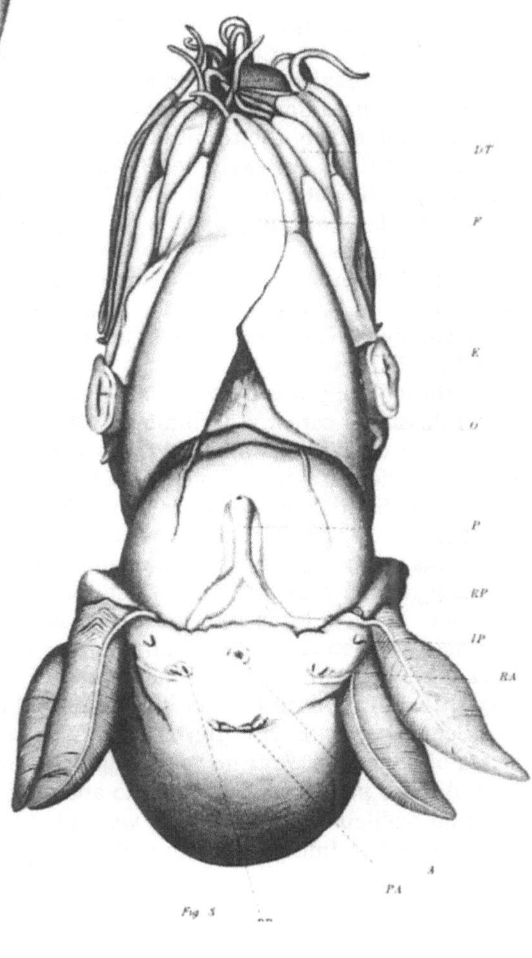

Fig. 3

When Owen received Bennett's specimen of *Nautilus*, he went to Paris to see Cuvier (Willey, 1902, p. 738). Le Baron, however, was on his deathbed and died without seeing the prize. Left as the foremost authority on comparative anatomy in Europe, Owen prepared what was to be his first notable publication, his 1832 *Memoir on the Pearly Nautilus*—basically a detailed description of the anatomy of the individual animal. In describing the soft parts of the specimen obtained from Bennett, Owen scooped the rest of the scientific community, and it was Owen's *Memoir* that signaled the entry of *Nautilus* studies into a new age. (The Bennett–Owen specimen no longer exists; it was destroyed when the Royal College of Surgeons of England was bombed in 1941.)

For the next seven decades, the principal theme of *Nautilus* studies was anatomy. Owen's work was followed by a spate of contributions from others, e.g., de Blainville (1834), Valenciennes, van der Hoeven, Boogaard, Keferstein, Kerr, and Griffin (Fig. 7). (The brief coverage afforded *Nautilus* by Edgar Allen Poe is of interest only because of its author.) A number of workers of this period recognized the importance of *Nautilus* to the interpretation of cephalopods of the ancient past. Hence, a number of them compared and contrasted the extant genus with various fossil forms (e.g., Grant, Bather, Blake, Edwards, Waagen, Foord, Foord and Crick, Appellöf, von Ihering, and Howes). And, of course, there was a series of papers reporting *Nautilus* among mollusks from this or that region (e.g., Angas, 1877; Brazier, 1879; Hedley, 1898).

Such factual information as there was on physiology, ontogeny, ecology, and ethology was mostly the result of chance captures and short observations of the resulting captives. For example, H.M.S. *Challenger* caught a single animal—in a dredge operating at a depth of some 585 m (320 fathoms) off Matuku Island, Fiji (Tizard *et al.*, 1885; Hoyle, 1886; Moseley, 1892). Other sources of information were the humans who inhabited or visited the areas where *Nautilus* occurs (e.g., Bennett, 1859; Bickmore, 1868; Semon, 1899). Data on physiology and other facets were sparse and not always of the greatest accuracy (Pope, 1733–1734; Tenney, 1875; Holmes, 1858) (see Fig. 8). In the waning years of the 19th century, the story began to change.

Arthur Willey was born in 1867 (Fig. 9). Although an Englishman, he began his professional career at Columbia College in New York, where he served as a Tutor in Biology, beginning in 1892. Two years later, he resigned to accept a Balfour Studentship from Cambridge University; his agreed-upon task was to journey to the "South Seas" with the avowed purpose of deciphering the embryology of *Nautilus*. He was there from the autumn of 1894 until the autumn of 1897. His work was not complete when the Balfour Studentship ran out, so he continued working on the project while a lecturer in biology at Guy's Hospital Medical School in London. From 1902 until 1910, he was director of the Colombo Museum in what was then called Ceylon but is now the Democratic Socialist Republic of Sri Lanka. In 1910, he joined the faculty of McGill University in Montreal and stayed there for the rest of his life. He retired in 1932 and died ten years later.

Beginning in 1895, Willey (1895, 1897a–e, 1898a–c, 1899, 1902) produced a series of papers on *Nautilus*, culminating in his monograph of 1902. The greater part of these papers were devoted to anatomy; due to the large quantity of fresh material available to him, Willey was able to improve on the work of his predecessors and go into greater anatomical detail. Unfortunately, in his primary ob-

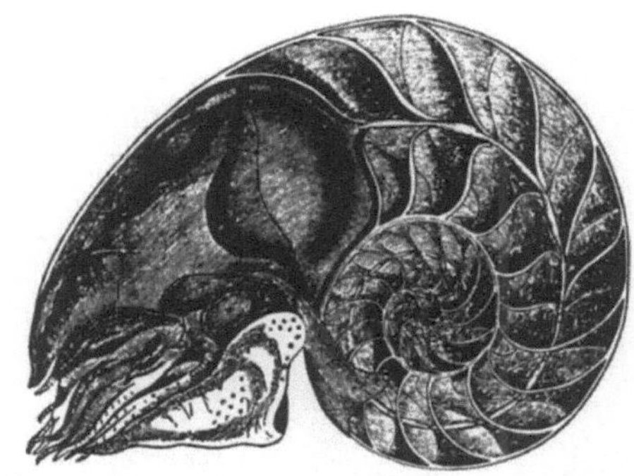

Figure 8. *Nautilus* in life position, at least as presented by Tenney (1875).

jective of revealing the early life history of *Nautilus*, Willey's work was not crowned with success. His captive animals laid eggs, but none hatched, even in those cases in which he had tried artificial insemination. However, Willey did contribute far more information on the way of life and geographic distribution of *Nautilus* than had anyone before him and thereby laid the groundwork for many of the studies of later workers.

(A personal note: The previous two paragraphs make Willey's work with *Nautilus* sound so mundane. In truth, and even ignoring contemporary accounts of cannibalism in the area, Willey's accomplishments in the "South Seas" of

Figure 9. Photograph of Arthur Willey (1867–1942). Photograph taken in the 1902–1910 period. Reproduced from de Silva (1977) with the kind permission of the Director of National Museums, Sri Lanka.

Figure 10. Photograph of Bashford Dean (1867–1928). From the American Museum of Natural History.

nearly a century ago are just short of miraculous, no matter how elegantly understated his own accounts of them are. Only someone who has tried to do field work in the part of the world in which Willey labored can come close to appreciating the monumental nature of Willey's success and the incredible geographic coverage he achieved.)

Even as Willey was wrapping up his project on *Nautilus*, a study on the animal by Bashford Dean (1901) (Fig. 10) was published. During a relatively brief stay in the Philippines, Dean had made some observations on *Nautilus* and had interviewed local residents about the habits and occurrence of the animal. His paper on *Nautilus* is noteworthy as one of the first on live animals. It was really, however, but a brief digression from his life's work—he was a leading expert on fishes. Dean was for some years the curator in charge of fishes at the American Museum of Natural History. (He was also curator of arms and armor at the Metropolitan Museum of Art!) Dean's 1901 paper was dedicated to Willey, and his observations paralleled some of those of Willey. There are other parallels between Dean and Willey, too. Both men were born in 1867, and Dean was teaching vertebrate zoology at Columbia at the very time Willey was a tutor there. Dean died in 1928.

With the coming of the new century, there seemed also to have come a dwindling of interest in *Nautilus*, at least if one judges by the volume of publications. But a glance through the rest of this book and the references at the end will show you that the two-generation dry spell has come to an end. You are living in what may well be the most exciting period of *Nautilus* studies—a veritable renascence.

So read on . . . and enjoy.

Acknowledgments. A historical work such as this chapter depends mightily on the publications of the past. The author of such a work, in turn, depends mightily on those custodians of the printed word, the librarians. Luckily, Cincinnati is blessed with good libraries and better librarians. I commend to you those at the Public Library of Cincinnati and Hamilton County, the Lloyd Library, and the University of Cincinnati Libraries (particularly Biology/Chemistry, Classics, Geology, and Special Collections). Many other individuals deserve credit for numerous kindnesses. Brian F. Glenister, of the University of Iowa, provided me with a copy of Gesner (1558). Richard Thomas, Paul Burrell, and Kees DeJong, of the University of Cincinnati, rendered great aid in my attempts to understand works in Latin and Greek, in French, and in Dutch, respectively. Frederick Veidt, of Cincinnati's Fairview School, likewise helped with German. Several folks provided data and leads about Arthur Willey: Mary McCutcheon, Eleanor MacLean, and R. L. Carroll of McGill University; Eric Holtzman of Columbia University; and Joy Schreiber of Cambridge University. E. Allen of the Royal College of Surgeons of England was of invaluable assistance in obtaining information on Richard Owen and the photograph of Bennett. C. O. van Regteren Altena of Teyler's Stichtung, Haarlem, the Netherlands, provided information regarding Rumpf's collection. Dr. W. Thelma T. P. Gunawardane, Director of National Museums, Sri Lanka, provided the photograph of Willey as a young man; it was conveyed to me through the kind efforts of U.S. Congressman Bill Gradison and U.S. Ambassador to Sri Lanka James W. Spain. Dr. I. Strachan and Dr. R.A. Hewitt reviewed the manuscript and provided valuable suggestions. To all these people, I am grateful.

References and Bibliography

This does not purport to be a complete bibliography of all works on *Nautilus*. Good sources for data on the early works are Gould (1857), Griffin (1897, 1900), and Dodge (1953); for the more recent contributions, consult Stenzel (1964) and the references at the end of this volume.

Angas, George French, 1877, A further List of Additional Species of Marine Mollusca to be included in the Fauna of Port Jackson and the adjacent Coasts of New South Wales, *Proceedings of the Scientific Meetings of the Zoological Society of London for the year 1877*, Part II, pp. 178–194. (*Nautilus* is on page 178.)

Appellöf, A., 1893, Die Schalen von Sepia, Spirula und Nautilus. Studien über den Bau und das Wachstum, *Kongl. Svenska Vetenskaps-akademiens Handlingar* **25**(7):1–106, plates 1–12.

Aristotle, ca. 335–322 B.C., *Historia animalium*, book IV, 525a, 26–29 (translated by A. L. Peck), Harvard University Press, Cambridge, Massachusetts (1970); pp. 14–15 (translated by D'Arcy Wentworth Thompson), Clarendon Press, Oxford, England (1910). (*Argonauta* is discussed in book IV, 525a, 20–26, and book IX, 622b, 5–19.)

Athenaeus, ca. 200, *Deipnosophistae*, book VII, 317–318 (translated by Charles Burton Gulick), William Heinemann (1929), pp. 428–429, London.

Bather, F. A., 1887, The growth of cephalopod shells, *Geol. Mag. N. Ser. Decade 3* **4**:446–449.

Bather, F. A., 1888, Shell-growth in Cephalopoda (Siphonopoda), *Ann. Mag. Nat. Hist. 6th Ser.* **1**(4):298–310.

Bather, F. A., 1888, Professor Blake and shell-growth in Cephalopoda, *Ann. Mag. Nat. Hist. 6th Ser.* **1**(6):421–427.

Beekman, E. M. (ed. transl.), 1981, *The Poison Tree: Selected Writings of Rumphius on the Natural History of the Indies*, University of Massachusetts Press, Amherst.

Belon, Pierre, 1551, *L'Histoire naturelle des estranges poissons marins, avec la vraie peinctvre & description du daulphin, & de plusieurs autres de son espece, obseruee par Pierre Belon du Mans*, R. Chaudiere, Paris.

Belon, Pierre, 1553, *De aquatilibus, libri duo cum eiconibus ad viuam ipsorum effigiem, quoad eius fieri potuit, expressis,* Carolum Stephanum, Paris.

Belon, Pierre, 1555, *La nature & diuersité des poissons, auec leurs pourtraicts, representez au plus pres du naturel,* C. Estienne, Paris.

Bennett, George, 1831, The inhabitant of the pearly nautilus, *London Med. Gaz.* **8**:729. [The name of the author is not given, but the paper is attributed to Bennett by Owen (1832, p. 8).]

Bennett, George, 1834, *Wanderings in New South Wales, Batavia, Pedir Coast, Singapore, and China; being the journal of a naturalist in those countries, during 1832, 1833, and 1834,* Richard Bentley, London.

Bennett, George, 1859, Notes on the range of some species of Nautilus, on the mode of capture, and on the use made of them as an article of food, *Proc. Zool. Soc. London,* Part 27, pp. 226–229.

Bennett, George, 1860. *Gatherings of a naturalist in Australasia; being observations principally on the animal and vegetable productions of New South Wales, New Zealand, and some of the Austral Islands,* J. Van Voorst, London.

Bennett, George, 1877, Notes on the pearly nautilus (*Nautilus pompilius*), *Ann. Mag. Nat. Hist.* 4th Ser. **20**(118):331–334.

Bennett, George, 1878, On the habits of the pearly nautilus, *Report of the forty-seventh meeting of the British Association for the Advancement of Science;* held at Plymouth in August 1877, p. 101.

Bickmore, Albert S., 1868, *Travels in the East Indian Archipelago,* John Murray, London. [*Nautilus* is on pp. 119 and 134–136; Rumpf is on pages 13–14, 134, and 250–252; Beekman (1981) lists an 1869 American edition, published by Appleton, New York.]

Blainville, Henri Marie Ducrotay de, 1825, Nautile, in: *Dictionnaire des Sciences naturelles,* Vol. 34, pp. 285–296, Le Normant, Paris.

Blainville, Henri Marie Ducrotay de, 1834, Anatomie des coquilles polythalames siphonées récentes pour éclaircir la structure des espèces fossiles, *Museum national d'histoire naturelle, Paris, Nouvelles annales 3rd series* **3**:1–25, 2 plates.

Blake, J. F., 1879, On the homologies of the Cephalopoda, *Ann. Mag. Nat. Hist.* 5th Ser. **4**(22):303–312.

Blake, J. F., 1882, *A monograph of the British fossil Cephalopoda.* Part I. Introduction and Silurian species, J. Van Voorst, London.

Blake, J. F., 1888, Remarks on shell-growth in Cephalopoda, *Ann. Mag. Nat. Hist.* 6th Ser. **1**(5):376–380.

Boogaard, J.-A., 1856a, Recherches microscopiques sur les spermophores du nautile, *Ann. Sci. Nat. Paris Ser. Zool.* 4th Ser. **6**:314–318.

Boogaard, J.-A., 1856b, Berigt omtrent mikroskopische onderzoekingen van de spermaphoren van *Nautilus pompilius* [in van der Hoeven (1856)].

Brazier, J., 1879, Mollusca of the Chevert Expedition, *Proc. Linn. Soc. N. S. W.* **2**:143–145.

Callimachus, ca. 250 B.C., Epigram VI, in: *Callimachus Lycophron Aratus* (translated by A. W. Mair), pp. 140–143, William Heinemann, London (1921). [See the note on page 169, *The Greek Anthology: Hellenistic Epigrams,* Vol. II (A. S. F. Gow and D. L. Page, eds.), Cambridge University Press, Cambridge, England (1965).]

Chemnitz, Johann Hieronymus, see Rumpf (1766).

Crampton, Henry E., 1950, A history of the Department of Zoology of Columbia University, *Bios* **21**(4):219–246.

Crouch, Edmund A., 1826, *An illustrated introduction to Lamarck's Conchology; contained in his Histoire Naturelle des Animaux sans Vertèbres: being a literal translation of the descriptions of the Recent and Fossil Genera, accompanied by twenty-two highly finished lithographic plates: in which are given Instructive Views of the various Genera, and their Divisions, drawn from Nature, from characteristic and generally well known* Species, Longman, Rees, Orme, Brown, & Green and J. Mawe, London. (There is also an edition dated 1827; the one copy I have seen appears to be in its original binding, but has no plates.)

Cuvier, Georges, 1830, *Le regne animal distribué d'apres son organisation, pour servir de base a l'histoire naturelle des animaux et d'introduction a l'anatomie comparée,* Tome III, Déterville, Paris; Crochard, Paris.

Dean, Bashford, 1901, Notes on living nautilus, *Am. Nat.* **35**(418):819–837, **35**(419):1029.

Derham, W., 1726, see Hooke (1726).

Dillwyn, Lewis Weston, 1817, *A descriptive catalogue of recent shells, arranged according to the*

Linnaean method; with particular attention to the synonymy, John and Arthur Arch, London. (*Nautilus* is on pp. 338–351, which are in Vol. I.)

Dodge, Henry, 1953, A historical review of the mollusks of Linnaeus. Part 2. The class Cephalopoda and the genera *Conus* and *Cypraea* of the class Gastropoda, *Bull. Am. Mus. Nat. Hist.*, Vol. 103, article 1, 134 pp.

Edwards, Frederic E, 1849, A monograph of the Eocene cephalopoda and univalves of England: First class—Cephalopoda. Cuvier, *Palaeontogr. Soc.* **2**:1–56.

Foord, Arthur H., 1888, *Catalogue of the fossil Cephalopoda in the British Museum (Natural History), Cromwell Road, S.W. Part I. Containing part of the suborder Nautiloidea, consisting of families Orthoceratidae, Endoceratidae, Actinoceratidae, Gomphoceratidae, Ascoceratidae, Poteriocera-tidae, Cyrtoceratidae, and supplement*, British Museum (Natural History), London. (*Nautilus* is on pages xi–xiii. This is only the first volume of the catalogue; subsequent volumes also have much information on *Nautilus*, both modern and fossil.)

Foord, Arthur H., and Crick, G. C., 1889, On the muscular impressions of *Coelonautilus cariniferus*, J. de C. Sowerby, sp., compared with those of the Recent Nautilus, *Geol. Mag. Ser. 3* **6**(11):494–498.

Gesner, Konrad von, 1558, *Historiae animalium. Liber IV. Qui est de Piscium & Aquatilium animantium natura*, Christoph Froschoverum, Tiguri. ("Nautilus" is discussed on pp. 732–736; the picture sent by John Fauconerus is described on p. 734, lines 31–41.)

Gesner, Konrad von, 1603, *Historiae animalium. Liber IV. Qui est de Piscium & Aquatilium animantium natura*, Editio Secunda, Bibliopolio Andreae Cambieri, Francofvrti. [The quote on p. 5 of Owen (1832), the description of the Fauconerus picture, is from p. 623, lines 20–30, of this edition.]

Gesner, Konrad von, 1983, *Beasts & Animals in Decorative Woodcuts of the Renaissance*, Dover Publications, New York. (The illustrations from the "Nautilus" portion of the *Historiae animalium* are reproduced as Fig. 156, 168, 171, and 172.)

Gould, Augustus A., 1857, On the true *Nautilus umbilicatus* of Lister, *Proc. Zool. Soc. London*, Part 25, pp. 20–21.

Grant, Robert E., 1833, Lectures on comparative anatomy and animal physiology. Lecture XI. On the organs of support of the cephalopodous Mollusca, *Lancet 1833–1834* **1**:505–514.

Gray, John Edward, 1833, Some Observations on the Economy of Molluscous Animals, and on the Structure of their Shells, *Philos. Trans. R. Soc. London 1833*, pp. 771–819. (Mention of *Nautilus* is on page 774.)

Gray, John Edward, 1845, On the Animal of Spirula, *Ann. Mag. Nat. Hist.* **15**(98):257–260, plate 15. (Mentions of *Nautilus* are on pp. 259–260.)

Griffin, Lawrence Edmonds, 1897, Notes on the anatomy of Nautilus pompilius, *Zool. Bull.* **1**(3):147–161.

Griffin, Lawrence Edmonds, 1898, Notes on the Tentacles of Nautilus pompilius, *Johns Hopkins Univ. Circ.* **18**(137):11–12.

Griffin, Lawrence Edmonds, 1899, *Ann. Mag. Nat. Hist. 7th Ser.* **3**:170–176 [reprint of Griffin (1898)].

Griffin, Lawrence Edmonds, 1900, The anatomy of Nautilus pompilius, *Mem. Natl. Acad. Sci.* **8**(5th mem.):101–230, 17 plates. (U.S. Senate Document 76, 56th Congress, 2nd session.)

Hadfield, E., 1920, *Among the Natives of the Loyalty Group*, Macmillan, London. (There is a photograph of Arthur Willey on p. 95.)

Hays, H. R., 1972, *Birds, Beasts, and Men: A Humanist History of Zoology*, Penguin, Baltimore (1973 printing).

Hedley, C., 1898, Descriptions of new Mollusca, chiefly from New Caledonia, *Proc. Linn. Soc. N. S. W.* **23**(I):97–105.

Hoeven, J. van der, 1850, Contributions to the knowledge of the animal of Nautilus pompilius, *Proc. Zool. Soc. London*, Part XVIII, pp. 1–9.

Hoeven, J. van der, 1856, Bijdragen tot de ontleedkundige kennis aangaande NAUTILUS POMPILIUS L., vooral met betrekking tot het mannelijke dier, *Verh. K. Akad. Wet.*, Derde Deel, paper 7, 29 pp. 5 plates [includes Boogaard (1856b)].

Holmes, Oliver Wendell, 1858, The Chambered Nautilus, *Atlantic Monthly* **1**(4):468–469. (The poem was first published in the February 1858 installment of Holmes' "The autocrat of the breakfast-table," which was published in book form later in the same year.)

Hooke, Robert, 1705, Lectures and discourses of earthquakes, and subterraneous eruptions: Explicating the causes of the rugged and uneven face of the Earth; and what reasons may be given for the frequent finding of shells and other sea and land petrified substances, scattered over the whole terrestrial superficies, in: *The posthumous works of Robert Hooke, M.D. S.R.S. Geom. Prof. Gresh, & c. Containing his Cutlerian Lectures, and other discourses, read at the meetings of the Illustrious Royal Society* (Richard Waller, ed.), pp. 278–450, Sam. Smith and Benj. Walford (Printers to the Royal Society), London. [In Waller's table of contents, the title is given as: Discourses of earthquakes, their *causes* and *effects*, and histories of several; to which are annext, *physical explications* of several of the fables in *Ovid's Metamorphoses*, very different from other mythologick interpreters. This work is not a coherent whole. It was pieced together by Waller from various notes and drawings that were among Hooke's papers at the time of his death; according to Waller, some bore the date May 29, 1689. A reprint of the book has been produced: *The Posthumous Works of Robert Hooke* (Richard S. Westfall, ed.), Johnson Reprint Corp., New York (1969); another introduced by T.M. Brown, Frank Cass and Co., London, 1971.]

Hooke, Robert, 1726, Dr. Hook's Conjectures about the odd Phaenomena observable in the Shell-Fish called the *Nautilus*, in: *Philosophical experiments and observations of the late eminent Dr. Robert Hooke, S.R.S. and Geom. Prof. Gresh. and other eminent virtuoso's in his time* (W. Derham, ed.), pp. 304–314, W. & J. Innys (Printers to the Royal Society), London. (According to Derham, this work was delivered to the Royal Society as a series of papers in December of 1696. A reprint of the book was produced in 1967 by Frank Cass & Co., London.).

Howes, G. B., 1896, Address of the President, *Proc. Malacol. Soc. London* **2**:57–76.

Hoyle, William Evans, 1886, Report on the cephalopoda collected by H.M.S. Challenger during the years 1873–1876, *Report on the scientific results of the voyage of H.M.S. Challenger during the years 1873–76, Zoology*, Vol. XVI, pp. 199–200.

Humboldt, Alexander von, 1841, Report on work on *Nautilus* by Valenciennes, quoting from a letter from Valenciennes, *Bericht über die zur bekanntmachung geeigneten Verhandlungen der Königl. Preuss. Akademie der Wissenschaften zu Berlin 1841*, pp. 55–58.

Huxley, Thomas Henry, 1853, On the morphology of the cephalous Mollusca, as illustrated by the anatomy of certain Heteropoda and Pteropoda collected during the voyage of H.M.S. "Rattlesnake" in 1846–50, *Philos. Trans. R. Soc. London* **143**(1):29–65. (*Nautilus* is mentioned on pp. 49, 50, 55, and 59.)

Ihering, H. von, 1881, Über die Verwandtschaftsbeziehungen der Cephalopoden, *Z. Wiss. Zool.* **35**(1):1–22.

Keferstein, Wilhelm, 1865, Beiträge zur Anatomie des Nautilus pompilius, *Nachr. K. Ges. Wiss. Georg-Augusts-Univ.* **1865**:356–376.

Kerr, J. Graham, 1895, On some points in the anatomy of *Nautilus pompilius*, *Proceedings of the General Meetings for Scientific Business of the Zoological Society of London* 1895, pp. 664–686.

Landman, Neil H., 1982, Aristotle, Alexander and the Pearly Nautilus, *Discovery* **16**(1):20–23.

Lankester, E. Ray, 1883, Mollusca, in: *Encyclopaedia Britannica*, 9th ed., Vol. 16, pp. 632–695.

Lankester, E. Ray, 1891, Mollusca, in: *Zoological Articles Contributed to the "Encyclopaedia Britannica,"* pp. 95–158, Adam & Charles Black, Edinburgh, Charles Scribner's Sons, New York. [This is a republication of Lankester (1883).]

Leupe, P. A., 1871, Georgius Everardus Rumphius, ambonsch natuurkundige der zeventiende eeuw. *Verh. K. Akad. Wet.*, Twaalfde Deel, paper 3, 63 pp.

Linnaeus, Carolus, 1758, *Systema naturae per regna tria naturae, secundum classes, ordines, genera, species, cum characteribus, differentiis, synonymis, locis, Tomus I, Regnum animale*, Laurentius Salvius, Stockholm [1956 reprint by the British Museum (Natural History), London.]

Luce, J. V., 1969, *The End of Atlantis: New Light on an Old Legend*, Thames and Hudson, London, 224 pp.

Moseley, H. N., 1892, *Notes by a naturalist: An account of observations made during the voyage of H.M.S. "Challenger" round the world in the years 1872–1876*, G. P. Putnam's Sons, New York. (This is the second edition; the first was published in 1879.)

Müller, Johannes, 1841, Bemerkungen des Hrn. Prof. J. Müller, *Bericht über die zur Bekanntmachung geeigneten Verhandlungen der Königl. Preuss. Akademie der Wissenschaften zu Berlin* 1841, pp. 58–59 (translation: *Ann. Mag. Nat. Hist.* **7**:241–245, 1841).

Müeller, Phillip Ludwig Statius, see Rumpf (1766).

Oppianus, ca. 180, Halieutica, Book I, 338–359, in: *Oppian Colluthus Tryphiodorus* (translated by A. W. Mair), pp. 238–241, William Heinemann, London (1928).

Owen, Richard, 1832, *Memoir on the Pearly Nautilus* (Nautilus pompilius, *Linn.*) *with Illustrations of Its External Form and Internal Structure*, Council of the Royal College of Surgeons in London, London.

Owen, Richard, 1837, On the Structure of the Shell of the Water Clam (*Spondylus varius*), *Proc. Zool. Soc. London*, part V, pp. 63–66. (Includes brief mentions of *Nautilus*.)

Owen, Richard, 1842, Observations upon a Specimen of the Pearly Nautilus (*Nautilus pompilius*) obtained by Capt. Belcher, R.N. at Amboina, *Proc. Zool. Soc. London*, Part X, p. 143.

Owen, Richard, 1843, On the structure and homology of the cephalic tentacles in the Pearly Nautilus, *Ann. Mag. Nat. Hist.* **12**(78):305–311.

Owen, Richard, 1855, Lecture XXIII: Cephalopoda, in: *Lectures on the comparative anatomy and physiology of the invertebrate animals, delivered at the Royal College of Surgeons*, pp. 576–603, Longman, Brown, Green, and Longmans, London. (This is the second edition.)

Owen, Richard, 1878, On the relative positions to their constructors of the chambered shells of Cephalopods, *Proceedings of the Scientific Meetings of the Zoological Society of London for the year 1878*, pp. 955–975, plate 40.

Owen, Richard, 1894, *The Life of Richard Owen*, D. Appleton, New York. (The author is the grandson of the subject.)

Parkinson, James, 1811, *Organic remains of a former world: An examination of the mineralized remains of the vegetables and animals of the antediluvian world; generally termed extraneous fossils*, p. 100, Sherwood, Neely, and Jones, London. (The first edition was published in 1804–1808.)

Parkinson, James, 1830, *Outlines of Oryctology. An introduction to the study of fossil organic remains; especially of those found in the British strata: intended to aid the student in his inquiries respecting the nature of fossils, and their connection with the formation of the Earth*, M. A. Nattali, London. (The first edition is dated 1822.)

Peck, A. L., 1970, see Aristotle.

Pelseneer, Paul, 1888, Sur la valeur morphologique des bras et la composition du système nerveux central des Cephalopodes, *Arch. Biol.* **8**:723–756, plates 37 and 38. (*Nautilus* is on pages 730, 732, 734, and 738.)

Pelseneer, Paul, 1899, Recherches morphologiques et phylogénétiques sur les Mollusques archaïques, *Mémoires couronnés et Mémoires savants étrangers, Académie royale des sciences, des lettres et des beaux-arts de Belgique* **57**:1–112, 24 plates. (*Nautilus* is on pages 54–58.)

Plinius Secundus, Gaius (Pliny the Elder), 77, *Natural history*, book IX. xlvii. 88 and xlix. 94 (translated by H. Rackham), pp. 220, 221, 226, and 227, Harvard University Press, Cambridge, Massachusetts; William Heinemann, London (1956).

Poe, Edgar A., 1839, *The conchologist's first book: or, a system of testaceous malacology, arranged expressly for use in schools, in which the animals, according to Cuvier, are given with the shells, a great number of new species added, and the whole brought up, as accurately as possible, to the present condition of the science*, Haswell, Barrington, and Haswell, Philadelphia.

Pope, Alexander, 1733–1734, An essay on Man, in: The Best of Pope (George Sherburn, ed.), lines 177–178 on p. 139, Ronald Press, New York (1940).

Rumpf, Georg Everard, 1705, *d'Amboinsche rariteitkamer, Behelzende eene beschryvinge van allerhande zoo weeke als harde Schaalvisschen, te weeten raare krabben, kreeften, en diergelyke Zeedieren, als mede allerhande Hoorntjes en schulpen, die men in d'Amboinsche Zee vindt: Daar beneven zommige mineraalen, gesteenten, en soorten van Aarde, die in d'Amboinsche, en zommige omleggende Eilanden gevonden worden*, François Halma, Amsterdam.

Rumpf, Georg Everard, 1711, *Thesaurus imaginum piscium testaceorum; quales sunt cancri, echini, echinometra, stellae marinae, & c. ut & cochlearum . . . quibus accedunt conchylia, ut nautilus, cornu Ammonis, &c., conchae univalviae & bivalviae . . . denique mineralia*, P. van der Aa, Lugduni Batavorium.

Rumpf, Georg Everard, 1739, *Thesaurus imaginum piscium testaceorum: quales sunt cancri, echini, echinometra, stellae marinae, &c. ut et cochlearum . . . quibus accedunt conchylia, ut nautilus, cornu Ammonis, & conchae univalviae et bivalviae . . . denique mineralia*, P. de Hondt, Hagae-Comitum.

Rumpf, Georg Everard, 1741, *d'Amboinsche rariteitkamer . . .*, Jan Roman de Jonge, Amsterdam. (This is a second Dutch edition of the 1705 work.)

Rumpf, Georg Everard, 1741–1750, *Herbarium amboinense, plurimas conplectens arbores, frutices, herbas, plantas terrestres & aquaticas, quae in Amboina et adjacentibus reperiuntur insulis . . .*

Omnia . . . belgice conscripsit Georg. Everard. Rumphius . . . Nunc primum in lucem edidit, & in latinum sermonem vertit Johannes Burmannus . . . qui varia adjecit synonyma, suasque observationes, Franciscum Changuion, J. Catuffe, H. Uytwerf, Amstelaedami.

Rumpf, Georg Everard, 1754, *Verhandeling der zee-horenkens en zee-gewassen in en omtrent Amboina en de nabygelegene eilanden, door Francois Valentyn . . . Dienende tot een vervolg van de Amboinsche rariteitkamer, beschreven door Georgius Everhardus Rumphius* (François Valentijn, ed.), J. Van Keulen, Amsterdam.

Rumpf, Georg Everard, 1766, *Amboinsche Raritäten-Cammer, oder Abhandlung von den steinschaalichten Thieren welche man Schnecken und Muscheln nennet, aus dem Höllandischen übersetzt von Philipp Ludwig Statius Müller öffentlichen ordentlichen Lehrer der Weltweissheit zu Erlangen und mit Zusätzen aus den besten Schriftstellern der Conchyliologie vermehret von Johnann Hieronymus Chemnitz* (translated by Philipp Ludwig Statius Müller; Johann Hieronymus Chemnitz, ed.), Kraussischen Buchhandlung, Vienna. (The figures in this edition of Rumpf differ from those in the 1705 and 1741 editions.)

Saville-Kent, W., 1889, Preliminary observations on a natural history collection made in connection with the Surveying Cruise of H.M.S. "Myrmidon," at Port Darwin and Cambridge Gulf—September to November, 1888, *Proc. R. Soc. Queensl.* **6**(5):219–242. (A brief account of a live *Nautilus* on the surface during daylight is on p. 229.)

Seeley, Harry, 1865, On the Significance of the Septa and Siphuncle of Cephalopod Shells, *Report of the Thirty-fourth Meeting of the British Association for the Advancement of Science; held at Bath in September 1864: Notices and Abstracts of Miscellaneous Communications to the Sections*, pp. 100–101.

Semon, Richard, 1899, *In the Australian bush and on the coast of the Coral Sea; being the experiences and observations of a naturalist in Australia, New Guinea and the Moluccas*, Macmillan, London.

Silva, P. H. D. H. de, 1977, *Colombo Museum. 100 years. 1877–1977.* Souvenir, Department of National Museums, Columbo, Sri Lanka.

Sirks, M. J., 1945, Rumphius, the blind seer of Amboina (translated by Lily M. Perry), in: *Science and Scientists in the Netherlands Indies* (Pieter Honig and Frans Verdoorn, eds.), pp. 295–303, Board for the Netherlands Indies, Surinam and Curaçao, New York.

Society for the Diffusion of Useful Knowledge, 1840, Nautilus, in: *The Penny Cyclopaedia of the Society for the Diffusion of Useful Knowledge*, Vol. 16, pp. 108–114, Charles Knight, London.

Stenzel, H. B., 1964, Living Nautilus, in: *Treatise on Invertebrate Paleontology* (R. C. Moore, ed.), pp. K59–K93, Geological Society of America, New York; University of Kansas Press, Lawrence.

Tenney, Sanborn, 1875, *Elements of Zoölogy*, Charles Scribner's Sons, New York.

Thiele, Johannes, 1892, Beiträge zur Kenntnis der Mollusken, *Z. wissenschaftliche Zool.* **53**(4):578–590, pl.23. (*Nautilus* is on pages 583 and 584.)

Thompson, D'Arcy Wentworth, 1910, see Aristotle.

Thompson, D'Arcy Wentworth, 1947, *A Glossary of Greek Fishes*, Oxford University Press, London ("Nautilos" is on pp. 172–175).

Tizard, T. H., Moseley, H. N., Buchanan, J. Y., and Murray, John, 1885, Narrative of the cruise of H.M.S. Challenger with a general account of the scientific results of the expedition, in: *Report on the scientific results of the voyage of H.M.S. Challenger during the years 1873–1876 under the command of Captain George S. Nares, R.N., F.R.S. and the late Captain Frank Tourle Thomson, R.N. Narrative*, Vol. I, *First Part* (*Nautilus* is on pp. 490–491).

Valenciennes, M. A., 1839, Nouvelles recherches sur le Nautile flambé (*Nautilus Pompilius*, Lam.), *Arch. Mus. Hist. Nat.* **2**:257–314, plates 8–11. (The date of this paper is commonly given as 1841; however, van der Hoeven's copy is designated "Exemplaire d'auteur" and is dated 1839.)

Valentijn, François, 1754, see Rumpf (1754).

Vrolik, W., 1843, On the Anatomy of the Pearly Nautilus, *Ann. Mag. Nat. Hist.* **12**(76):173–175.

Waagen, William, 1870, Ueber die Ansatzstelle der Haftmuskeln beim Nautilus und dem Ammonoideen, *Palaeontographica* **17**(5):185–210, plates 39 and 40.

Willey, Arthur, 1895, In the home of the Nautilus, *Nat. Sci.* (*London*) **6**(40):405–414.

Willey, Arthur, 1897a, The embryology of the Nautilus, *Nature* (London) **55**(1426):402–403.

Willey, Arthur, 1897b, Letters from New Guinea on Nautilus and some other Organisms, *Q. J. Microsc. Sci. N. Ser.* **39**:145–180. (The date of the papers by Willey in this volume of the journal is commonly given as 1896; the date on the title page of the volume is 1897, however.)

Willey, Arthur, 1897c, Zoological Observations in the South Pacific. II. On the Nepionic Shell of the Recent Nautilus, *Q. J. Microsc. Sci. N. Ser.* **39**:222–226, 231, plate 13.

Willey, Arthur, 1897d, Zoological Observations in the South Pacific. III. On some Variations in the Shell of Nautilus, with Description of a New Variety (N. pompilius, var. Moretoni, nov. var.), *Q. J. Microsc. Sci. N. Ser.* **39**:227–230, 231, plate 13.

Willey, Arthur, 1897e, The oviposition of *Nautilus macromphalus*, *Proc. R. Soc. London Ser. B* **60**:467–471.

Willey, Arthur, 1898a, The Pre-ocular and Post-ocular Tentacles and Osphradia of Nautilus, *Q. J. Microsc. Sci. N. Ser.* **40**:197–201, plate 10.

Willey, Arthur, 1898b, The Adhesive Tentacles of Nautilus, with some Notes on its Pericardium and Spermatophores, *Q. J. Microsc. Sci. N. Ser.* **40**:207–209, plate 11.

Willey, Arthur, 1898c, Some Zoological Results of a Voyage to Melanesia during the years 1894–1897, *Proc. Cambridge Philos. Soc.* **98**:398–401.

Willey, Arthur, 1899, General Account of a Zoological Expedition to the South Seas during the years 1894–1897, *Proceedings of the General Meetings for Scientific Business of the Zoological Society of London 1899*, pp. 7–9.

Willey, Arthur, 1902, Contribution to the Natural History of the Pearly Nautilus, in: *Zoological results based on material from New Britain, New Guinea, Loyalty Islands and elsewhere, collected during the years 1895, 1896 and 1897, pp. 691–830, plates 75–83, 1 map, Cambridge University Press, Cambridge, England.*

Wit, Hendrik C. D. de (ed.), 1959, *Rumphius Memorial Volume*, Hollandia B. V., Baarn, the Netherlands.

Wit, Hendrik C. D. de, 1977, Orchids in Rumphius' *Herbarium Amboinense*, in: *Orchid Biology: Reviews and Perspectives I* (Joseph Arditti, ed.), pp. 47–94, Cornell University Press, Ithaca, New York. (A brief biography of Rumpf is on pp. 49–50.)

Wynne-Edwards, V. C., 1943, The late Professor Arthur Willey—1867–1942, *McGill News* **24**(3):11–13.

II

The Ancestry of the Genus *Nautilus*

Chapter 2

The Ancestry of the Genus *Nautilus*

CURT TEICHERT and TATSURO MATSUMOTO

1. Introduction

The genus *Nautilus* is known to be represented by at least six extant species that inhabit a well-circumscribed large area in the southwestern Pacific Ocean, the Philippines, Indonesia, and the tropical seas of Australia [see Saunders (1981b) and Part III of this volume]. The most puzzling aspect of these occurrences is that *Nautilus* seems to have no ancestors in the immediately preceding pre-Holocene rocks; i.e., no representatives of *Nautilus*, indeed no nautilids of any kind, have ever been authentically recorded from rocks of Pliocene and Pleistocene age. The only published mention of a Pliocene occurrence is that of a single specimen of a "true *Nautilus*" from the coast of Oregon (Dall, 1909, p. 21), but the specimen and its stratigraphic position were never described, and Miller (1947, p. 9) was inclined to discount this record. At most, it is obscure and undocumented. Also, the definition of a "true *Nautilus*" in 1909 differed from that of today. Since marine Pliocene and Pleistocene rocks are widely exposed in the area of distribution of living *Nautilus*, the fact that no documented records of any nautilids of those ages exist suggests that the immediate ancestors of living *Nautilus* must have been so rare that up until now they have escaped detection and not only because nobody "has taken the trouble to look for them" (Ward, 1984).

Nautilus is the name-giving genus of the family Nautilidae as well as the order Nautilida. In this chapter, we examine the fossil record of the nautilids, but only insofar as it has direct relevance to the problem of the ancestry of *Nautilus* itself (Fig. 1). From the Late Cambrian on, the nautiloids were taxonomically diversified and widely distributed in marine environments. Although greatly diminished in importance after the Paleozoic, they are still ubiquitous in Mesozoic and Early Tertiary strata, where they are represented by many families and numerous genera. However, in this chapter, we will focus as closely as possible on the direct line of ancestry of *Nautilus*.

CURT TEICHERT • Department of Geological Sciences, University of Rochester, Rochester, New York 14627. TATSURO MATSUMOTO • 1-28-5, Minami-Ohashi, Minami-ku, Fukuoka 815, Japan.

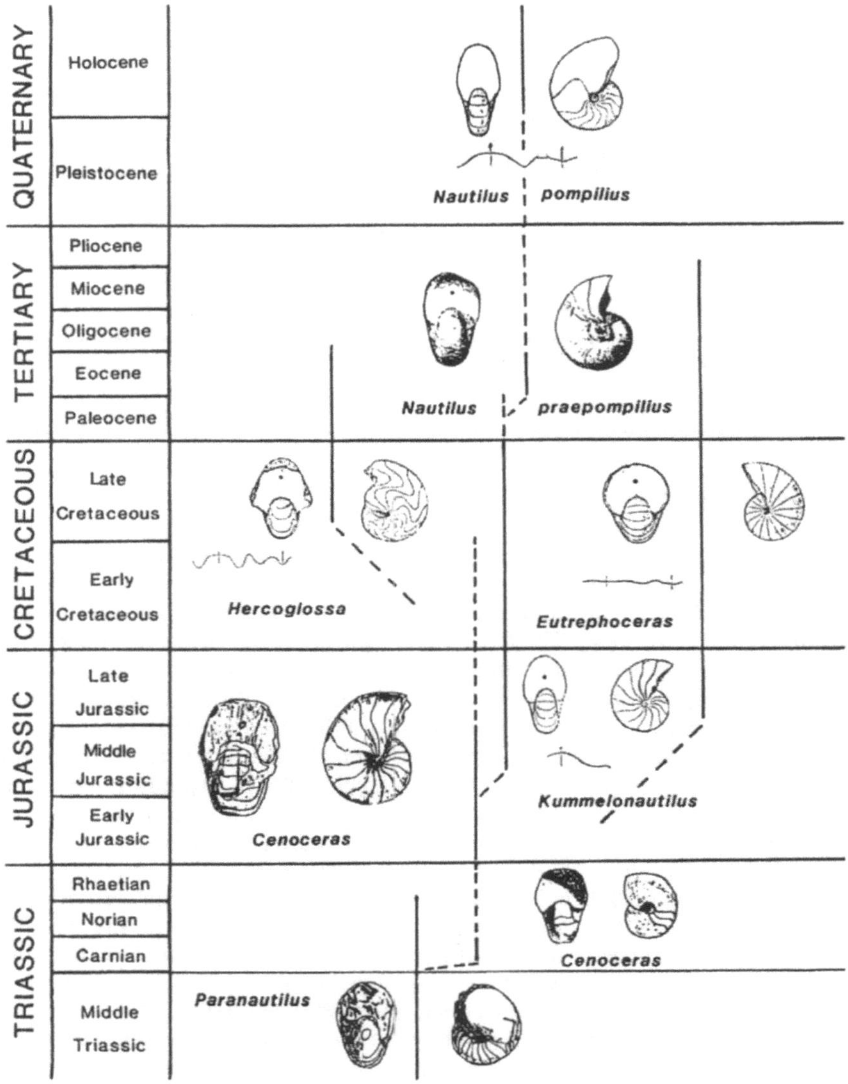

Figure 1. Diagram illustrating the proposed phylogeny of *Nautilus*.

2. The Fossil Record

From the time of its formal establishment by Linnaeus in 1758 until the 1880s, the generic name *Nautilus* was attached by most paleontologists to almost any kind of coiled, septate fossil shell with more or less simple sutures, as opposed to coiled forms with complex sutures, which were referred to a comprehensive genus, *Ammonites*. Although knowledge of the systematics of Mesozoic and Tertiary nautilids was much advanced by the works of Hyatt (1884a) and Spath (1927), who established many new genera and families for these forms, it was customary, until at least the middle of the 20th century, to assign many Tertiary and even Cretaceous species to the genus *Nautilus*. Miller (1947) was the first to publish an accurate description of the suture of *Nautilus pompilius*, the type species of *Nautilus*, and in 1951 he had come to the conclusion that no known fossil species could be assigned to the genus (Miller, 1951, p. 32). Kummel (1956, p. 388) expressed "full agreement" with this interpretation.

About this time, however, two new nautilids of early Oligocene age were assigned to the genus *Nautilus*. These were *N. ucrainicus* Marenko (1956) from a locality near Korostichev, about 100 km west of Kiev in the Ukraine, and *N. praepompilius* Shimansky (1957) from an unspecified locality in the southern part of the Ust-Urt plateau, between the Caspian and Aral Seas. *Nautilus ucrainicus* is an involute shell, as much as 290 mm in diameter, but since Marenko (1956) described its suture as (in translation) "almost perfectly straight with a barely perceptible indentation on the flank of the initial chamber," it belongs quite certainly to the genus *Eutrephoceras* or to one of the subgenera proposed by Schultz (1976).

Nautilus praepompilius Shimansky is more difficult to evaluate. The description by Shimansky (1957) is short and general, but could well suggest a shell of the general appearance of any of the species of living *Nautilus*. The suture is described as having distinct lateral and dorsal lobes, the latter with a small annular lobe. This description corresponds to some extent to the suture of *N. pompilius*, except that the latter also shows an umbilical lobe (Miller, 1947, Fig. 3), which was not mentioned by Shimansky as a feature in *N. praepompilius*. Shimansky (personal communication, 1985) has stated that an umbilical lobe may be present, but if it is, it is very weak. Nevertheless, *N. praepompilius* may well be an ancestral *Nautilus* even though it is separated from the extant species of that genus by some 30 million years (m.y.). According to the latest information (Shimansky, personal communication, 1985), *N. praepompilius* is now also known from rocks of late Eocene age in the same general area as the holotype.

The question then remains: What happened to *Nautilus* in the 30 m.y. between the early Oligocene and the present, and what are the closest relations of this enigmatic animal in the immediate geologic past?

Since there are no nautilids known from Pliocene and Pleistocene rocks, with the exception of an undocumented report from the Pliocene of Oregon, we have to look to the pre-Pliocene record. From the Miocene, only two genera of Nautilida are known, viz., *Eutrephoceras* Hyatt and *Aturia* Bronn. The latter genus, though rarely common, is of worldwide distribution in the Miocene in both hemispheres and is known in places as remote as New Zealand and West Antarctica (this locality from written communication by W. J. Zinsmeister, 1985). *Eutrephoceras*, on the other hand, is found only in the Miocene of southeastern Australia, West Antarctica, and the Mediterranean area. Both genera persisted to the end of Miocene time. Most authors regard *Aturia* as representative of the monogeneric family Aturiidae Chapman 1857, although some (e.g., Wiedmann, 1960; Matsumoto *et al.*, 1984) have included it in the Nautilidae, despite the very specialized configuration of its suture.

The suture of *Eutrephoceras* is almost straight, and the shell is involute and globose. Most authors have included it in the family Nautilidae, but Miller (1951) placed it in a separate monogeneric family, Eutrephoceratidae, which Matsumoto (Matsumoto *et al.*, 1984) regarded as a subfamily of the Nautilidae. Many authors have assumed *Eutrephoceras* to be the direct ancestor of *Nautilus*. However, intermediate forms are not known from the Miocene, but only in older rocks, as, for example, *Hercoglossa*, which did not survive the Eocene.

Mesozoic and Cenozoic nautilids are all believed to have their roots in a single Late Triassic genus, *Cenoceras* Hyatt. During the Triassic, nautilids were highly diversified, represented by seven families with numerous genera, all but one of which disappeared at the end of the Norian Stage, approximately 5 or 6 m.y. before the end of the Triassic Period (Kummel, 1964). The only Triassic nautilid genus known to survive into the Jurassic Period was *Cenoceras*, the first representative of which, *C. trechmanni* (Kummel), is known only from Upper Karnian deposits of the North Island of New Zealand (Kummel, 1953, 1959). Tintant (1984b) doubted the affinities of this species with *Cenoceras*, because he believed that it showed longitudinal striations only in the peripheral area of the whorls. Kummel (1956, p. 427), however, has stated unequivocally that the striations are present "both on the whorl sides and the venter of the inner whorls."

One of us (T.M.) examined the holotype (BM C21947) of *Cenoceras trechmanni* (Kummel) at the British Museum (Natural History) and found that it is a composite internal mold or sculptured steinkern on which the suture lines and the surface ornamentation are both preserved, at least in some parts. Fine longitudinal striations are present on the flank as well as on the venter of the preserved outer whorl, which is still wholly septate. It should be noted that the same kind of reticulate striations are discernible on the shell surface of the juvenile stages in various genera of the Nautilida of later ages, e.g., *Eutrephoceras*, *Cymatoceras*, *Metacenoceras*, and *Nautilus* itself.

Cenoceras is unknown from rocks of Norian and Rhaetian age, which represent an interval of about 12 m.y., but it proliferated and diversified spectacularly in the Early and Middle Jurassic. Kummel (1956) placed five other Jurassic genera in synonymy with *Cenoceras* and recorded 93 species of this comprehensive genus, some 60 of which are restricted to the Early Jurassic. Its geographic distribution was worldwide. Kummel (1956, p. 362) defined *Cenoceras* as including those nautilids of Early and Middle Jurassic age "that are part of the evolving complex which survived the Triassic and which in the great plasticity of the group reflect an extensive adaptive radiation." The significant features of the genus are (1) its surface ornamentation, which consists of longitudinal revolving striae or reticulate striation, and (2) a suture showing a very shallow ventral sinus and a broad lateral lobe. Unfortunately, the internal suture of *C. trechmanni* is not known, but this is also true of most of the Jurassic species of this genus.

Both Kummel (1956, 1964) and Shimansky (1962) suggested an origin of the "*Cenoceras* complex" in the Middle and Late Triassic Syringonautilidae. However, an origin in the Liroceratidae of the superfamily Clydonautilaceae is equally, if not more, convincing, because this family would take the ancestry of the Nautilidae as far back as the Early Carboniferous. According to Kummel (1964), the stratigraphic range of the Liroceratidae is Lower Carboniferous to Upper Triassic.

Kummel regarded *Cenoceras* as the earliest representative of the family Nautilidae. The genera that Kummel (1956) merged with *Cenoceras* were *Digonioceras* Hyatt, *Nautilites* Prinz, *Ophionautilus* Spath, and *Sphaeronautilus* Spath, which differ from *Cenoceras*, according to Kummel, in overlapping minor features of conch form, suture, siphuncle position, and ornamentation. According to Kummel's interpretation, the "*Cenoceras* complex" gave rise to four important family groups: Pseudonautilidae, forms with highly sinuous, "goniatitic" sutures; Paracenoceratidae, characterized by modification of the whorl section; Cymatoceratidae, with ribbed surfaces; and finally Hercoglossidae, having modified, undulatory sutures. There was a proliferation of genera during the Jurassic and Cretaceous, discussion of which would be irrelevant in the context of this chapter. However, one of us (Matsumoto and Muramoto, 1983, p. 90) had previously observed that *N. butonensis* K. Martin (1933) from the Oligocene of Buton Island, Indonesia, could be a Tertiary survivor of *Cenoceras*, unless it is a derived fossil.

In the cephalopod volumes of the two great paleontological encyclopedias of the 1960s, the American *Treatise on Invertebrate Paleontology* and the Russian *Osnovy Paleontologii*, the taxonomy of the terminal period of the Nautilida (Late Cretaceous–Holocene) had been consolidated with significant consensus. The following four families were accepted in each (Shimansky, 1962; Kummel, 1964):

Nautilidae: simple, unornamented shells, more or less involute, simple sutures
Hercoglossidae: involute, unornamented shells, complex sutures
Cymatoceratidae: involute shells, bearing ribs, of highly variable conch shape
Aturiidae: involute, smooth shells, disk-shaped, elaborate suture, siphuncle close to dorsum, including only one genus, *Aturia*

In both works, these four families were placed in a superfamily Nautilaceae that had its origin in the Upper Triassic genus *Cenoceras* Hyatt. However, Schultz (1976) recognized only two families, Nautilidae and Cymatoceratidae, suppressing Hercoglossidae and Aturiidae as synonyms of the former.

A very different taxonomic arrangement was proposed by Kabamba (1983) and Tintant and Kabamba (1983), who recognized the following families in this complex:

Nautilidae: involute nautilicones and spherocones, shell smooth or ornamented, suture straight or slightly sinuous, without external (ventral) lobe
Cenoceratidae: a very plastic group, somewhat evolute, suture slightly sinuous, with ventral lobe, several genera with ornamented shells
Aturiidae: characterized by complex suture, without external lobe

In this arrangement, the genera of the Cymatoceratidae are divided between the Nautilidae and the Cenoceratidae and the family Hercoglossidae is merged with the Aturiidae. Tintant and Kabamba (1983) postulated that costate shells (previously assembled in the family Cymatoceratidae) developed repeatedly from smooth shells (in the other families) in response to such environmental factors as water turbulence (see also Tintant, 1984a).

Of greater importance in our context is a paper by Matsumoto (1983), in which he established a new genus, *Kummeloceras*, for two nautilid species from the lower Upper Cretaceous (Turonian–Coniacian) of Hokkaido, Japan. Since this generic name was found to be preoccupied for a Permian genus, Matsumoto (Matsumoto *et al.*, 1984) proposed the replacement name *Kummelonautilus*, with the type species *Kummeloceras yamashitai* Matsumoto 1983, of Turonian age. As defined by Matsumoto, *Kummelonautilus* includes subdiscoidal to subglobose, nautiliconic, involute shells, with a smooth surface, and sutures characterized by a more or less pronounced ventral saddle, broad lateral lobes, and small umbilical saddles. The internal suture is unknown in the two Japanese species. In a lengthy discussion of the genus, Matsumoto (1983, pp. 12–18) assigned to it about half a dozen species of Jurassic to Eocene age that had been included in either *Eutrephoceras* or *Cimomia* by earlier authors. All these species seem to agree with *K. yamashital* in having a broad ventral saddle and a small umbilical saddle of the suture. The suture of *Cenoceras* differs only in having a shallow ventral sinus rather than a broad saddle.

In the same paper, Matsumoto (1983) stated that the following Cretaceous species might be referable to *Nautilus* (*s. str.*): *Angulithes* (*Cimomia*) *schroederi* Wiedmann from the Cenomanian of Spain and *Nautilus sowerbyanus* D'Orbigny from the Upper Cretaceous of France. However, it now seems to us that the internal sutures of these species are unknown and that, more important, there is no evidence that their sutures possess umbilical and annular lobes. They are therefore perhaps better included in *Kummelonautilus* or regarded as transitional between that genus and *Nautilus*. *Kummelonautilus cookanus* (Whitfield) and *K. bryani* (Gabb) have annular lobes. The taxonomic significance of this feature is, however, under dispute. Foord (1891, p. 180), Spath (1927, p. 24), and Miller (1947, p. 27) suggested that it is of little taxonomic value, and Matsumoto (1983, p. 15) was inclined to agree with them.

The picture that emerges is one of a relatively gradual progression from Upper Triassic to Middle Jurassic *Cenoceras*, through Upper Jurassic to Eocene *Kummelonautilus* to late Eocene to Holocene *Nautilus*. The principal morphological discontinuity occurs between *Cenoceras* and *Kummelonautilus*, in the change of the shape of the suture across the ventral region of the shell. At the same time, the reticulate ornamentation that covers the entire shell in most species of *Cenoceras* is generally restricted, in *Kummelonautilus*, to the young, or at most middle-aged, shell parts (Matsumoto, 1983, p. 15). Interestingly, Tintant (1984b) called attention to the appearance, near the top of the Lower Liassic of France, of *Nautilus inornatus* d'Orbigny, in which reticulate ornamentation is restricted to the first whorl of the shell. He made this the type species of a subgenus of *Cenoceras* that he called *Metacenoceras*. According to Tintant, this subgenus is predominant in the Upper Liassic and in the Middle Jurassic. It may therefore be convenient to accept the family Cenoceratidae Tintant and Kabamba (1983) for these and similar forms, without discussing in detail the scope of this family as proposed by its authors, while *Kummelonautilus* is, without doubt, a member of the Nautilidae proper. In *Nautilus*, the surface reticulation has been further reduced to the juvenile stage and the suture shows further elaboration.

3. Conclusion

Miller (1949) was the first to recognize that the evolution of the Nautilida surged ahead in the Late Cretaceous and Early Tertiary across the Cretaceous–

Tertiary boundary, at which time the Ammonoidea became extinct. Shimansky (1979) listed 11 nautilid genera from the Upper Cretaceous to which *Kummelonautilus* has to be added. Of these 12 genera, 5 continue into the Paleocene (*Eutrephoceras, Deltoidonautilus, Cimomia, Hercoglossa, Kummelonautilus*), where they were joined by 3 new genera, viz., *Teichertia, Aturoidea*, and *Aturia*. All these genera continued into the Eocene, and 3 more genera were added in the late Eocene and Oligocene, viz., *Nautilus, Obinautilus*, and *Neocymatoceras*, each of very restricted distribution. Eocene–Oligocene *Nautilus* is known only from the Ust-Urt plateau. According to Dr. H. Noda (personal communication to the authors, 1985), *Obinautilus* is a member of the Octopoda. After examining the holotype of *Neocymatoceras tsukushiense* Kobayashi (kept at Kyushu University), Matsumoto agrees with Kummel (1956) in regarding *Neocymatoceras* as a synonym of *Cymatoceras*. Thus, *Cymatoceras* did persist until the Oligocene, as evidenced by the aforementioned species and another species from Kyushu (see Matsumoto and Muramoto, 1983, p. 90). The decline of the nautilids in the Miocene has already been described.

Tintant and Kabamba (1983) posed the question: Is *Nautilus* a living fossil or a genus of unknown origin? Ward (1984) speculated that *Nautilus* "may today be engaged in active radiation" from a eutrephoceratid ancestor in the latest Miocene.

The picture that emerges is one of striking parallelism in development of the Nautilida during the Late Triassic on the one hand and the late Cenozoic on the other. In the Norian, there existed about 20 genera of nautilids, about the same number as in the Karnian. *Cenoceras*, the ancestor of all Mesozoic and later nautilids, is represented by one species in the Karnian at one locality only, but it has not yet been found in the Norian. Nautilids are virtually unknown from the Rhaetian [except for one report of rare *Grypoceras* by Kozur (1980)], which represents a time interval of 5–6 m.y. *Cenoceras* is found again, in greater diversity, at the beginning of the Jurassic, about 12 m.y. after its first recorded appearance in the late Karnian.

By comparison, about 10 genera of nautilids are known from the Paleogene, 2 of which survived to the end of Miocene. No nautilids are known from the Pliocene and Pleistocene, representing a time interval of 5–6 m.y. The earliest *Nautilus* is known from the Eocene/early Oligocene, represented by one species in a single area, but it is as yet unknown from the upper Oligocene, Miocene, Pliocene, and Pleistocene (a time interval of about 28 m.y.). *Nautilus* appears in significant numbers, represented by 6, possibly more, species in the Holocene.

From this comparison, we are inclined to conclude that the evolutionary history of the Nautilida is now at a point comparable to that reached in earliest Jurassic time, perhaps the Hettangian. Pia (1914) listed and discussed a total of 69 species of "*Nautilus*" from the Lower Jurassic worldwide, which is about the number of species of *Cenoceras* given by Kummel (1956) for this epoch. Of this total, Pia listed 6 species from the Lias α (= Hettangian + Lower Sinemurian), 1 of which was an "*N.* spec. ind.," while the other 5 were assigned to *Cenoceras* by Kummel. Of these, only *C. striatum* had a reasonably worldwide distribution, whereas the others were from more restricted localities. The Hettangian alone was about 6–7 m.y. long, and it was only after this time interval that the main *Cenoceras* radiation accelerated. Thus, since *Nautilus* is the only surviving genus of the order Nautilida, the members of which were numerous and widely dispersed in the oceans from Jurassic until Miocene times, it may truly be called a "living fossil." On the other hand, it is by no means certain that it is close to extinction; on the contrary, *Nautilus* is indeed a viable genus on the threshold of a new period of radiation that, given favorable climatic and other environmental conditions, may unfold during the next few million years.

III

Nautilus and Its Distribution

Chapter 3

The Species of *Nautilus*

W. BRUCE SAUNDERS

1. Introduction

During the past three centuries, a total of 11 species and 7 variants of living *Nautilus* have been proposed (Table I). It is notable that only one of these, *N. belauensis*, was based on live specimens; the rest were defined solely from drifted shells, the origins of which were either unknown or poorly documented by their authors. Considering this sparsity of data, it is remarkable that the species-level taxonomy of the genus recognized herein remains largely unchanged from that used by such masters as George Sowerby (1849). Nevertheless, there is still considerable confusion regarding the number, the names, and the validity of the variously named species, subspecies, and variants of *Nautilus*.

Malacologists seem to be in general agreement in recognizing three separate species [*N. pompilius* Linnaeus, 1758; *N. macromphalus* Sowerby, 1849; and *N. scrobiculatus* [Lightfoot, 1786] (see, for example, Abbott and Dance, 1983)], whereas some researchers have regarded three to six species as valid [e.g., Miller, 1947; Stenzel, 1964; Toriyama *et al.*, 1965; Japanese Expert Consultation on Living

W. BRUCE SAUNDERS • Department of Geology, Bryn Mawr College, Bryn Mawr, Pennsylvania 19010.

Table I. Status of Described Species and Variants of *Nautilus* and Their Reported Occurrence[a]

Species composition	Status	Reported occurrence
1. *Nautilus pompilius* Linneaus, 1758	Valid	Philippines to Australia
N. pompilius var. moretoni Willey, 1896	Syn.	Papua New Guinea
N. pompilius var. marginalis Willey, 1896	Syn.	Papua New Guinea
N. pompilius var. perforatus Willey, 1896	Syn.	Papua New Guinea
N. pompilius var. pompilia Shimansky, 1948	Syn.	—
N. pompilius var. caudatus Lister, 1685	Syn.	—
N. pompilius var. rumphii Shimansky, 1948	Syn.	—
N. ambiguus Sowerby, 1849	Nom. dub., syn.?	—
N. alumnus Iredale, 1944	Nom. dub., syn.	Queensland, Australia
2. *Nautilus belauensis* Saunders, 1981	Valid	Palau
3. *Nautilus macromphalus* Sowerby, 1849	Valid	New Caledonia, Loyalty Islands
4. *Nautilus scrobiculatus* [Lightfoot, 1786]	Valid	Papua New Guinea, Solomons
N. umbilicatus Lister, 1685	Syn.	—
N. perforatus Conrad, 1849	Syn.	—
N. texturatus Gould, 1857	Syn.	—
5. *Nautilus stenomphalus* Sowerby, 1849	Valid	Queensland
N. stenomphalus var. *stenomphala* Shimansky, 1948	Syn.	—
6. *Nautilus repertus* Iredale, 1944	Valid?	Australia

[a] See also distribution map, p. xxv.

Nautilus (JECOLN), 1980b; Ward, 1984]. In an earlier review of the subject, I concluded that four species were recognizable: *N. pompilius*, *N. macromphalus*, *N. scrobiculatus*, and a newly named species, *N. belauensis* Saunders, 1981. Two other species named from Australia (*N. stenomphalus* Sowerby, 1849, and *N. repertus* Iredale, 1944) were regarded as questionable, and the remainder (Table I) were regarded as invalid or dubious (*N. ambiguus*, *N. moretoni*, *N. alumnus*) or as synonyms [*N. umbilicatus*, *N. perforatus*, *N. texturatus* (all = *N. scrobiculatus*)].

The principal difficulties in evaluating the species of *Nautilus* have been, and continue to be: (1) the lack of knowledge regarding species named and known only from drifted shells, (2) the lack of data on variation within living populations, and (3) the broad geographic range of the drifted shells (see Chapter 4). However, considerable new information derived from the study of living animals is now available, including: (1) morphometric data from populations of *N. pompilius* in the Philippines, Fiji, and Papua New Guinea, and from *N. belauensis* in Palau

(Hayasaka *et al.*, 1982; Hayasaka, 1983, 1985; Saunders and Davis, 1985; Saunders, 1984b) (see also Chapter 6); (2) the first live occurrence of *N. scrobiculatus* has been discovered in the Admiralty Islands, where it is sympatric with *N. pompilius* (Saunders and Davis, 1985; Saunders *et al.*, 1987a); (3) *N. stenomphalus*, together with *N. pompilius*, has recently been obtained from the Great Barrier Reef, Queensland (Saunders and Ward, 1987); and (4) the first electrophoretic enzyme and protein analyses of *Nautilus* have been completed (Masuda and Shinomiya, 1983; Woodruff *et al.*, 1983) (see also Chapter 5). Although studies are still in progress, a number of generalizations are emerging, including the following:

1. The mean size of mature *Nautilus* can vary considerably among isolated populations. Most accounts record size data as shell diameter, and mean mature diameters ranging from 115 to 169 mm have been observed among different populations of the most common species, *N. pompilius*.
2. Patterns of shell color may vary among populations, as measured by degree of coloration, hue, coalescence of stripes across the venter, and the presence or absence of color bands in the umbilical region.
3. The presence of umbilicate forms (those in which an umbilical callus is lacking) has been recognized as a rare occurrence in most populations of callus-bearing species, but this feature is a characteristic of *N. stenomphalus*.
4. Hood texture distinguishes several species.

This chapter is an attempt to synthesize previous knowledge of the species of *Nautilus* with the new data outlined above and to present a synopsis of the characters and distribution of each species as they are understood at present. Judging from the amount of newly emerging information, as well as the introduction of new approaches, it is anticipated that even greater understanding of the species of *Nautilus* and their distribution will be forthcoming in the near future. Moreover, as the range of geographic, morphological, and genetic variation within species becomes better known, and when the amount of overlap among individual species is understood, the recognition of infra- or subspecific taxa of *Nautilus* may prove to be both useful and biologically justifiable. Other interesting and related aspects of this relict organism have yet to be broached, such as the phylogenetic relationships among the various species and the cause—and significance—of their individual distributions within the Indo-Pacific realm.

2. Taxonomy and Description of *Nautilus*

Class:	Cephalopoda Cuvier, 1798
Order:	Ectocochlia Schwartz, 1894
Subclass:	Nautiloidea Hyatt, in Zittel, 1900
Order:	Nautilidae de Blainville, 1825
Genus:	*Nautilus* Linnaeus, 1758

2.1. Diagnosis

Shell compressed and involute to moderately umbilicate, umbilical callus may be present in involute forms; shell exterior smooth to reticulate, with sinuous

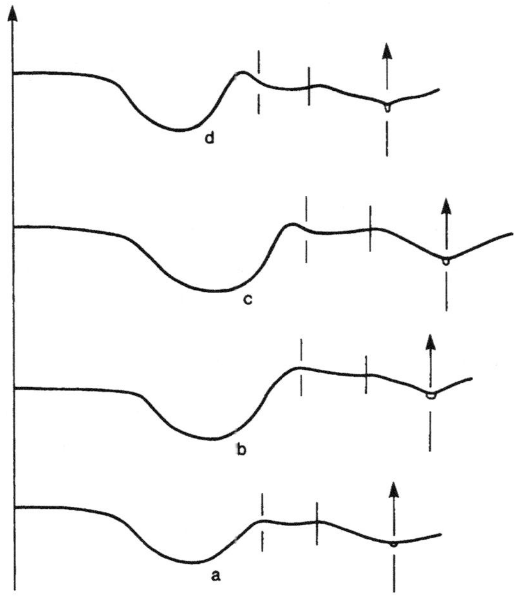

Figure 1. Sutural outlines of *Nautilus*. (a) *N. macromphalus* Sowerby, 1849 (31st septum, at 92 mm diameter, × 0.25); (b) *N. scrobiculatus* [Lightfoot, 1786] (34th septum, at 105 mm diameter, × 0.25); (c) *N. belauensis* Saunders, 1981 (31st septum, at 128 mm diameter, × 0.20); (d) *N. pompilius* Linnaeus, 1758 (30th septum, at 86 mm diameter, × 0.25). Drawn by J. Mussler, Bryn Mawr College, from latex casts made from shells with septa removed. Arrows point toward aperture; large arrow marks venter, smaller arrows mark dorsum.

growth lines in which ocular and hyponomic sinuses are well developed in mature forms. Shell coloration variable, comprising brown to red bifurcating stripes that are restricted to the dorsal portion of the shell at maturity. Internally, approximately 32 septa, with a central, orthochoanitic siphuncle; suture moderately sinuous, with lateral, umbilical, and dorsal lobes (Fig. 1). All living species characterized by a cartilaginous hood enclosing the aperture, 13-element radula, jaws with calcitic denticulation, lack of chromatophores, and lack of tentacular sucker-disks. Species-distinctive features include shell coloration, size, shell form, ornament, radula, and hood texture.

2.2. Composition and Distribution

Five or possibly six species are thought to be recognizable. The entire geographic range is unknown, but the genus appears to be confined to the Indo-Pacific, with living populations ranging from approximately 30°S to 30°N, and approximately 90–185°W (see map on p. xxv).

3. Recognized Species

3.1. *Nautilus pompilius* Linnaeus, 1758 (Fig. 2B,C,D and Fig. 3C,D)

3.1.1. Diagnosis

Mature shell size variable, but typically about 165 mm (Philippines), with a total weight (body plus shell) of about 850 g. Shell has a small umbilicus (≈ 5% of shell diameter), filled with a callus (with rare exceptions); shell coloration variable, but generally comprises irregularly radiating brown stripes that extend from umbilicus to venter.

3.1.2. Discussion

The type species of the genus (and the most common and widely distributed), *N. pompilius* appears to exhibit the greatest range in variation (Fig. 4). The diameters of 234 mature specimens caught during the 1979 ALPHA HELIX Expedition to the Tañon Straits, the Philippines, ranged from 150 mm (mature female) to 188 mm (barely mature male), with a mean of 165 mm. Sexual dimorphism is shown by slightly smaller females (mean diameter 160 mm) compared to males (mean 170 mm). Specimens of this species (provided by D. Dan) purported to have been caught live, off Tubbataha Reef, central Sulu Sea, the Philippines (Fig. 3C and D), are the smallest known representatives of this species, ranging from 103 to 126 mm in mature shell diameter, with a mean of 114 mm ($N = 29$). In Papua New Guinea, the average mature size of geographically isolated populations of *N. pompilius* varies considerably, from 144 mm (Lae, on the north coast of Papua New Guinea) to 169 mm (Kavieng, New Ireland Province), with an overall range in individual (mature) size of 124–199 mm in diameter (Saunders and Davis, 1985). Specimens of this species from Queensland average 153 mm in diameter ($N = 5$). The cause for the wide range in variation in mature size is not known; it may be genetic, or it may be related to ecological factors.

Variation in the pattern of shell coloration is manifest as differences in the amount of coloration, hue, degree of coalescence of banding over the venter, and development of color bands in the vicinity of the umbilicus. There appears to be a trend toward an increased proportion of shells with a white umbilical region, going southward through Papua New Guinea toward Australia (compare Fig. 2B, C, D).

The number of specimens with an open umbilicus (i.e., lacking an umbilical callus) does not appear to vary systematically; typically, fewer than 0.025% of live-caught *N. pompilius* in the Philippines and in Papua New Guinea show this curious abnormality (see also Mapes *et al.*, 1979). In *N. pompilius*, the umbilical callus is secreted during the growth of the second whorl, at approximately 75-mm diameter in Philippine specimens.

3.1.3. Distribution

In his description of *N. pompilius*, Linnaeus stated only that it "inhabits the Indian and African Ocean" (Turton, 1806, p. 305). However, because his reference

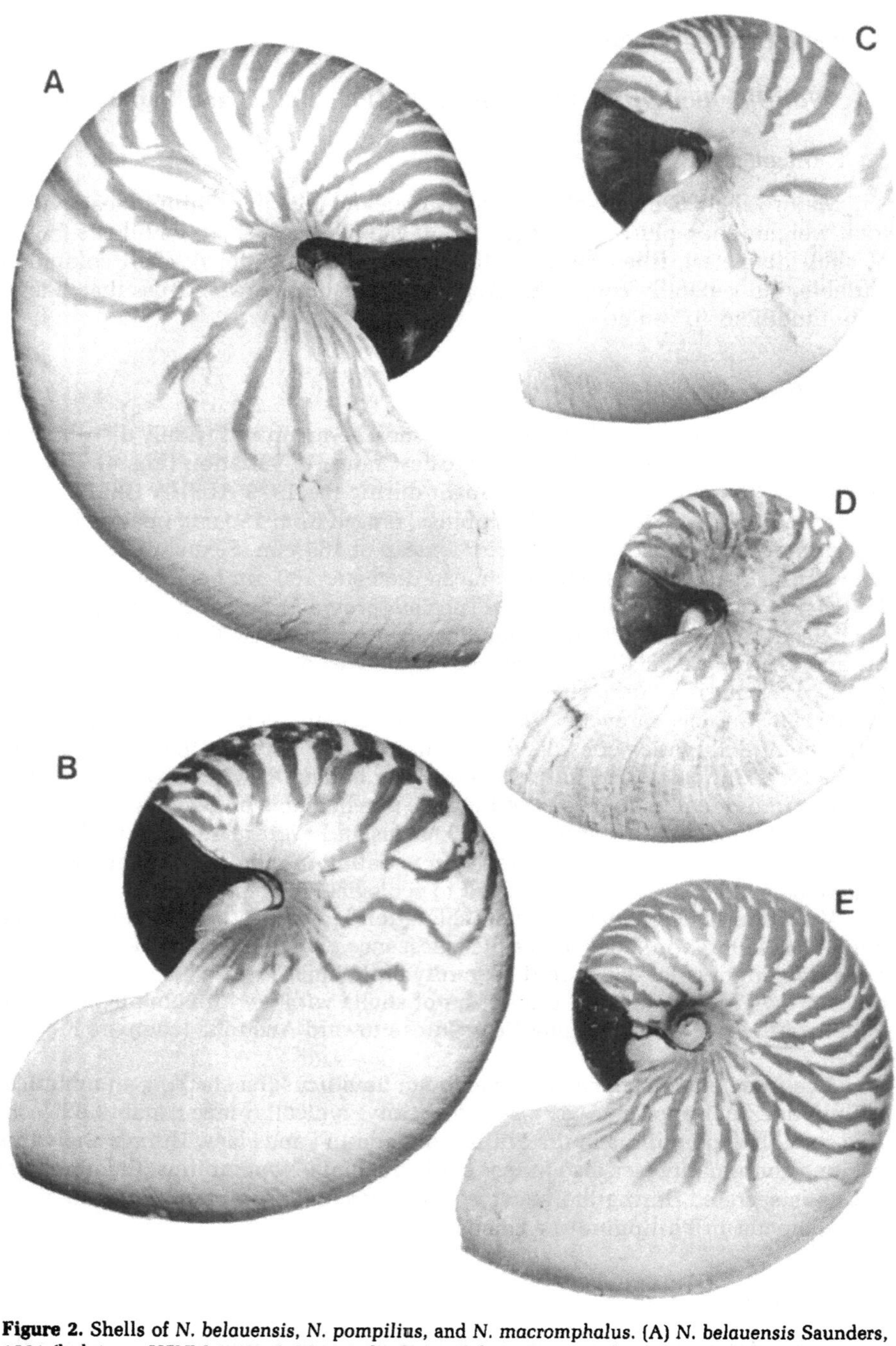

Figure 2. Shells of *N. belauensis*, *N. pompilius*, and *N. macromphalus*. (A) *N. belauensis* Saunders, 1981 (holotype, USNM 730549), Mutremdiu Point, Palau, mature male, approximately 250 m depth, July 1977; (B) *N. pompilius* Linnaeus, 1758 (USNM 730671), Tañon Strait, off Cebu, the Philippines

to an illustration was that of Rumphius (1741, Plate 17, Figs. A–C), Amboina (Ambon), Indonesia (the source of the specimens illustrated by Rumphius), is the type locality for the species. Nevertheless, this species is best known from the Philippines, and it has been the subject of a number of studies (e.g., Griffin, 1900; Dean, 1901; Bidder, 1962; Haven 1972, 1977a,b; Hayasaka *et al.*, 1982; Hayasaka, 1983; Cochran *et al.*, 1981; Ward and Chamberlain, 1983; Arnold, 1985), particularly in the Tañon Straits, between Cebu and Negros, which was the site of research expeditions, based on the R/V ALPHA HELIX, in 1975 and 1979.*

Occurrences of *N. pompilius* in Fiji have been reported by Moseley (1892), Davis and Mohorter (1973), Ward *et al.* (1977), Ward and Martin (1980), Masuda and Shinomiya (1983), Zann (1984), and Hayasaka (1985). A single live animal was captured off Kagoshima Bay, south Japan (Tanabe and Hamada, 1978; JECOLN, 1980b). Other reports substantiate the existence of living populations in the Andaman Islands (Smith, 1887) and the New Hebrides (Owen, 1832; Bennett, 1834), and this species is widely distributed in the Papua New Guinea region, including New Britain, New Ireland, Manus, Lae, and Port Moresby (Willey, 1895, 1896, 1897a,c, 1898a, 1899, 1902; Saunders and Davis, 1985; Saunders *et al.*, 1987a). The existence of living *N. pompilius* from the Great Barrier Reef, Australia, has been established just recently with the capture of six specimens off Lizard Island, Queensland, at depths of 200–400 m (Saunders and Ward, 1987), and in 1986, 39 specimens were trapped outside Pago Pago Harbor, American Samoa (Saunders *et al.*, 1987b). The even wider distribution of drifted shells of *N. pompilius* (see Chapter 4) makes it certain that many more occurrences remain to be discovered.

3.2. *Nautilus belauensis* Saunders, 1981 (Fig. 2A)

3.2.1. Diagnosis

Shell basically the same as that of *N. pompilius*, but distinguished by its larger size (mean mature shell diameter ≈ 200 mm; body plus shell weight ≈ 1300 g); presence of longitudinally crenulated shell sculpture; broadly triangular central rachidian radular tooth [width/height ≈ 65% (Fig. 5)].

3.2.2. Discussion

This species of *Nautilus* was named for the characteristically large forms found in Palau (Belau), in the Western Carolines (Saunders, 1981a). As indicated in Section 3.1.2, mature shell size can vary considerably within isolated populations of *Nautilus*; consequently, shell diameter alone should not be relied on for distinguishing species. However, no other species of *Nautilus* has been doc-

*For results of these expeditions, see the *Journal of Experimental Biology*, Vol. 205 (1978), and *Pacific Science*, Vol. 36 (1982).

(1979 ALPHA HELIX Expedition), mature male; (C) *N. pompilius* (PM 25), off Motupore Island, Central Province, Papua New Guinea, mature male, approximately 250 m depth, October 1984; (D) *N. pompilius* (LA 16), Huon Gulf, off Lae, Morobe Province, Papua New Guinea, mature female, 220 m depth, July 1984 (note dirty shell surface, which is characteristic of specimens from this deltaic setting); (E) *N. macromphalus* Sowerby, 1849 (DMNH 19937), New Caledonia. All figures × 0.6.

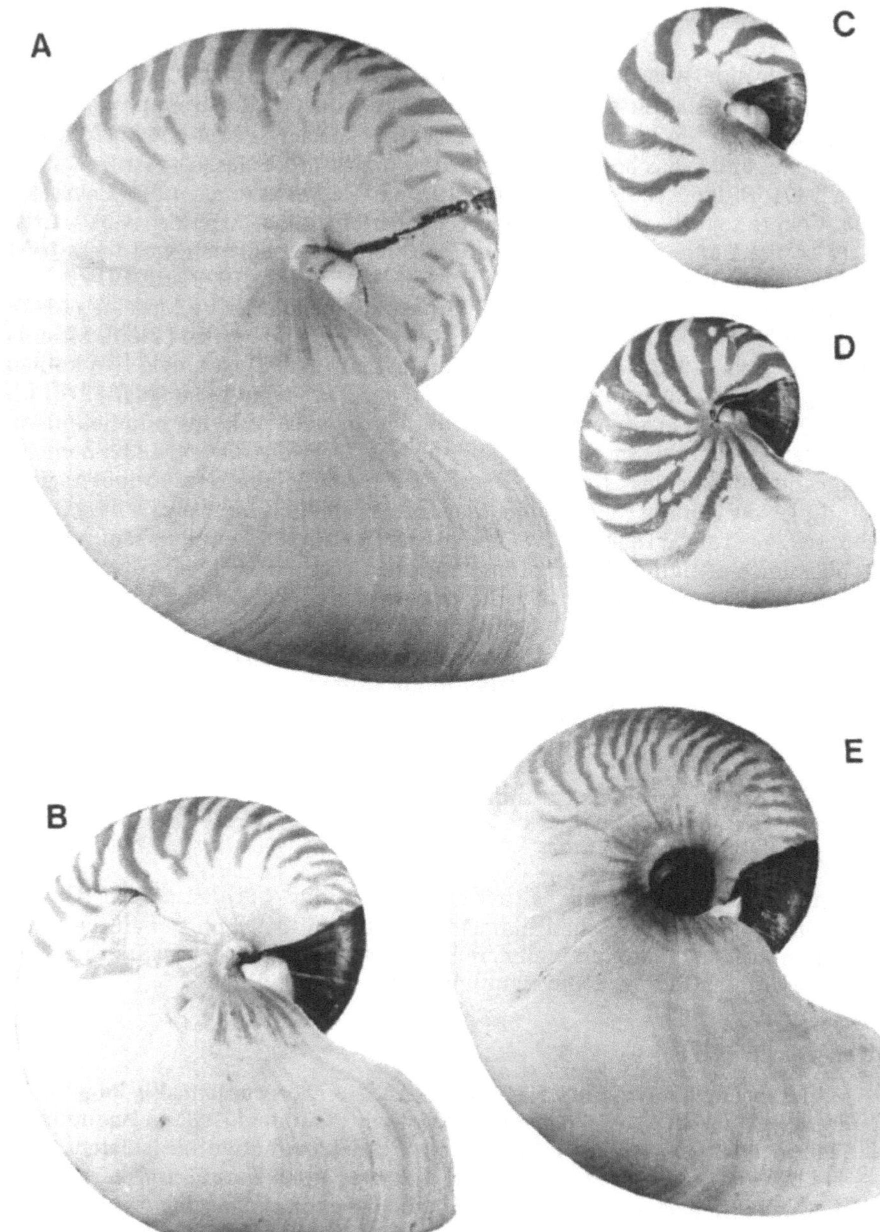

Figure 3. Shells of *N. stenomphalus*, *N. repertus* (?), *N. pompilius*, and *N. scrobiculatus*. (A) *N. repertus* (?) Iredale, 1944, approximately 200 km off Broome, W. Australia, mature male, 380–450 m depth, 1985 (note that black layer on dorsal body chamber surface was removed); (B) *N. stenomphalus* Sowerby, 1849 (LZ 26), off Carter Reef, Lizard Island, Queensland, mature male, approximate 300–450 m depth, July 1985; (C, D) *N. pompilius* (TB 27, TB 1) reportedly from Tubbataha Reef, Sulu Sea, the Philippines, mature males [these are the smallest known mature specimens and show unusual color variations; the specimen in (C) shows filed aperture]; (E) *N. scrobiculatus* [Lightfoot, 1786], drift shell, Rambutsyo Island, Manus Province, Papua New Guinea, mature male. All figures × 0.6.

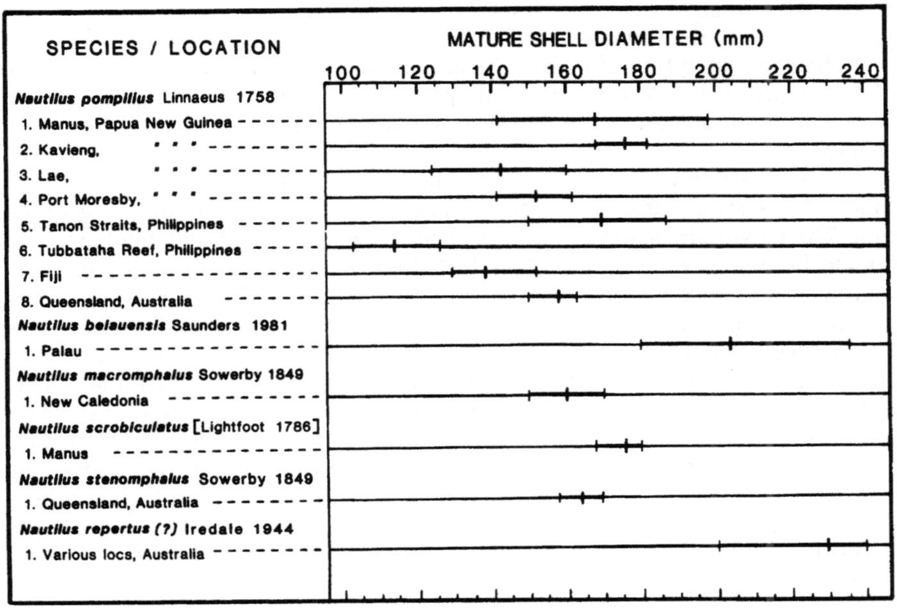

Figure 4. Intraspecific and interspecific variation in mature shell diameter in *Nautilus*. All data are from live-caught animals, except for *N. repertus* (?), which are based on drifted shells.

umented that matches *N. belauensis* in mean overall size, although isolated shells of the questionable species *N. repertus* may exceed the size of *N. belauensis*. The extent to which this is a product of ecological factors, as opposed to allometric or genetic differences, is unknown. Several lines of evidence suggest that the difference between the mature shell size of *N. pompilius* (≈ 170 mm in diameter) and that of *N. belauensis* (≈ 200 mm) is more than just the product of a longer period of growth: (1) The total number of septa appears to be similar in both species (33–37 septa in *N. belauensis*, 32–38 septa in *N. pompilius*) and (2) in *N. pompilius* from the Philippines, the umbilical callus is secreted at approxi-

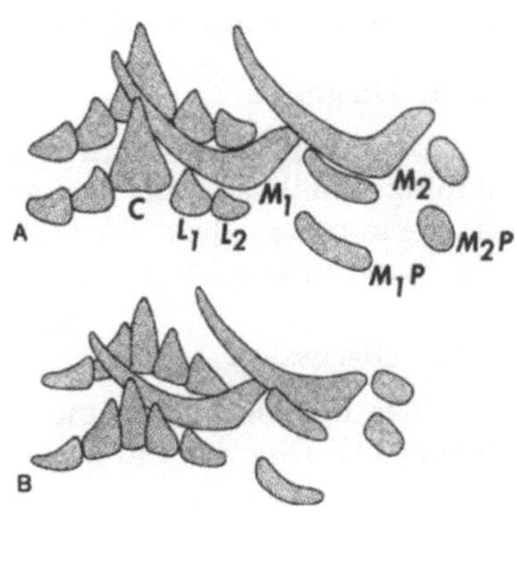

Figure 5. Radulae of *N. belauensis* (A) and *N. pompilius* (B). (C) Central rachidian tooth; (L_1, L_2) first and second lateral teeth; (M_1, M_2) inner and outer marginal teeth; (M_1P, M_2P) inner and outer marginal support plates. Note that each row includes a total of 13 elements, including (1) C, (2) L_1, (2) L_2, (2) M_1, (2) M_2, (2) M_1P, and (2) M_2P. Note the differences between the central rachidian teeth of the two species. Reprinted with permission from Saunders (1981a).

mately 75 mm shell diameter, whereas in *N. belauensis*, closure occurs at approximately 110 mm diameter.

The shell sculpture of *N. belauensis* includes delicate, longitudinally crenulated ridges that produce a reticulate pattern (Saunders, 1981a, Fig. 5) closest to that of *N. scrobiculatus*, although it is more strongly developed in the latter species.

Because study of *N. belauensis* has involved examination of literally thousands of live specimens, a good deal of information is available regarding variation. The maximum observed range in shell size extends from 180 mm (a mature female, body plus shell weight 998 g) to 239 mm diameter (a barely mature male, total weight ≈ 2000 g). Mean mature shell size is 204 mm diameter, with a mean weight (body plus shell) of 1308 g (Saunders and Spinosa, 1978). The characteristic pattern of shell coloration consists of bifurcating brown to red stripes that extend from the umbilicus to the venter, but they do not coalesce across the venter, as they do in many specimens of *N. pompilius* from Papua New Guinea (Saunders and Davis, 1985). A rare color variant seen in the Palauan form is that in which the umbilical region lacks stripes (see Saunders, 1981b, Fig. 11); this particular pattern is characteristic of *N. stenomphalus*, but it also occurs in *N. pompilius*, particularly in the southern part of its geographic range. Specimens of *N. belauensis* in which an umbilical callus is lacking are known (Saunders, 1981b, Figs. 9 and 10).

3.2.3. Distribution

Nautilus belauensis is known only from its type locality, Palau, although there have been no attempts to establish whether its range extends to any of the adjacent Caroline Islands. The fact that a tagged shell of *N. belauensis* was recovered in Mindanao, 1000 km to the west (Saunders and Spinosa, 1979; Saunders, 1981b, Fig. 8), emphasizes that caution should be exercised in extending the geographic range of individual species on the sole basis of isolated shells. Such drift occurrences may account for the unusually large shells that occasionally turn up in the collections offered by shell dealers in the Philippines.

Many aspects of the biology of *N. belauensis* have been described by Saunders *et al.* (1978), Saunders and Spinosa (1978, 1979), Saunders (1981a,b, 1983, 1984a,b, 1985), Woodruff *et al.* (1983), Carlson *et al.* (1984), and Ward *et al.* (1984).

3.3. *Nautilus macromphalus* Sowerby, 1849 (Fig. 2E)

3.3.1. Diagnosis

Distinguished primarily by the prominent, open umbilicus, with inwardly sloping umbilical walls and evenly rounded umbilical shoulder; umbilical diameter approximately 16% of shell diameter. Mature shell size approximately 160 mm; shell coloration pattern similar to that of *N. pompilius*.

3.3.2. Discussion

It is notable that despite the considerable difference in shell morphology between this species and *N. pompilius*, no differences in the soft parts have been reported. In addition, there appears to be little intraspecific variation, although almost no comparative data are available. Size data for this species (provided by Peter Ward) show a mean mature shell diameter of 162.1 mm ($N = 62$), with the males only slightly larger (mean 164.6 mm; $N = 23$) than females (mean 162.1 mm; $N = 15$).

3.3.3. Distribution

Nautilus macromphalus is known only from the New Caledonia–Loyalty Islands region, where the living animal was studied first by Arthur Willey (1897d, 1899, 1902) and later by Bidder (1962), Denton and Gilpin-Brown (1966), Meenakshi *et al.* (1974), A. W. Martin *et al.* (1978), Ward and Martin (1978, 1980), Packard *et al.* (1980), Ward and Wicksten (1980), Ward *et al.* (1980b, 1981), Ward and Greenwald (1982), B. E. Taylor and Ward (1983), and Ward (1986). Aquarium-based studies of *N. macromphalus* include those of Hamada and Mikami (1977), Mikami and Okutani (1977), and JECOLN (1980b).

3.4. *Nautilus scrobiculatus* [Lightfoot, 1786] (Fig. 3E)

3.4.1. Diagnosis

Moderate-size shell (≈180 mm in diameter, total weight ≈ 900 g), distinguished by a large, vertical-walled umbilicus (≈ 20% of shell diameter) that exposes earlier whorls back as far as the shell apex and that may be perforate. Coloration restricted to brown to yellow-brown bands concentrated on upper flanks of shell; bands coalesce across the venter. Shell surface distinctly reticulate or scrobiculate. In life, shell is covered with a thick, shaggy periostracum, unique to this species; brown hood covered by fine to coarse, elevated, conical white papillae.

3.4.2. Discussion

Until recently, knowledge of *N. scrobiculatus* was confined to data derived from drift shells, with but one exception: In 1895, Arthur Willey obtained "a single mutilated specimen . . . accompanied by its shell, which had been picked up from the surface of the sea, not far from Milne Bay in British New Guinea . . ." (Willey, 1902, p. 744). Willey's find was not matched until almost 90 years later, when, in 1984, *N. scrobiculatus*, along with *N. pompilius*, was trapped off Manus, Papua New Guinea, at a depth of approximately 250 m (Saunders and Davis, 1985; Saunders *et al.*, 1987a).

The availability of live specimens of *N. scrobiculatus* makes possible a more complete description of what has been regarded as the rarest and most distinctive species of *Nautilus*. Mature, live-caught specimens from Manus range in size from 168 mm in diameter (a female) to 197 mm (a male); observed total weights (shell plus body) range from 710 g (female) to 1340 g (male), with a mean of 908 g.

One of the most unusual features of this species is the presence of thick leaves of yellow-brown periostracum parallel to the aperture. These cover all but the dorsal portion of the shell and give the live animal a yellowish, shaggy or mossy appearance. Also unique is the texture of the hood; it is covered with a series of conical white bumps, or papillae, up to approximately 10 mm in diameter. No other differences in the soft parts have been noted, except that in life, some of the tissues (mantle, siphon, eyes) are tinted yellow-brown, in contrast to the white tissue in other species.

3.4.3. Distribution

The only two documented occurrences of *N. scrobiculatus* are in Papua New Guinea: in Manus, Admiralty Islands (Saunders and Davis, 1985; Saunders *et al.*, 1987a), and approximately 1000 km to the south, in Milne Bay Province (Willey, 1902). However, drift shells indicate that the species range may extend southward to the Solomons region. It is notable that *N. scrobiculatus* was the first known example of sympatric species of *Nautilus;* it was trapped along with *N. pompilius* in a ratio of approximately 1 : 7 (Saunders *et al.*, 1987a).

3.5. *Nautilus stenomphalus* Sowerby, 1849 (Fig. 3B)

3.5.1. Diagnosis

Moderate size (shell ≈ 165 mm in diameter, weight ≈ 700 g); like *N. pompilius*, but typically lacks an umbilical callus; shell coloration reduced and lacking in umbilical region; hood covered with irregular papillae.

3.5.2. Discussion

The status of this species, one of three proposed by G. B. Sowerby (1849), has been regarded as questionable. I suggested that it might be a variant of *N. pompilius*, because drift shells available from North Queensland show a range of variation that spans both *N. stenomphalus* and *N. pompilius* in terms of coloration and in the presence or absence of an umbilical callus (Saunders, 1981b, Figs. 2–4). However, recent trapping on the Great Barrier Reef, near Lizard Island, has produced specimens of both *N. pompilius* and *N. stenomphalus*, which show that the two species can be distinguished on the basis of hood texture and shell characteristics (Saunders and Ward, 1987).

Although data available at this time are limited, mature shell size ranges from 141.7 to 170.4 mm diameter (mean 156.9 mm; $N = 6$), and total weight ranges from 422 to 730 g (mean 605.3 g). It appears that *N. stenomphalus s.s.* has a white umbilical region, lacks an umbilical callus, and has a hood that is highly textured. By contrast, associated specimens of *N. pompilius* show shell characteristics typical of that species, although some shells exhibit *stenomphalus* coloration (white umbilicus), and some may lack an umbilical callus; however, none seems to combine these features with the distinctive hood texture of *N. stenomphalus*. Of 29 specimens caught off Lizard Island, 7 typified *N. stenomphalus s.s.* and 11 *N. pompilius s.s.;* the other 11 specimens each exhibited some shell characteristic of *N. pompilius* (umbilical callus or shell coloration). These data indicate that it may not be possible to distinguish *N. stenomphalus* from *N. pompilius* on the basis of shell characteristics alone, but that soft-part morphology is required. However, electrophoretic study of this material is under way and may shed light on this situation.

3.5.3. Distribution

Nautilus stenomphalus is known to occur only on the Great Barrier Reef, Queensland, whence most drift shells ascribed to this species are also derived, suggesting that its range may be limited to that region.

4. Questionable Species

4.1. *Nautilus repertus* Iredale, 1944 (Fig. 3A)

4.1.1. Diagnosis

Shell form like that of *N. pompilius*, but distinguished by its large size (≈ 220 mm), reduced, orange-brown coloration, and lack of radial stripes in the umbilical region.

4.1.2. Discussion

This species is somewhat problematical. Its definition was originally based on a single drift shell (and the drawing of another) from Western Australia (Iredale, 1944, pp. 295–296). Its large size (≈ 220 mm), white umbilical region, and orange-brown color bands were cited as distinguishing criteria. Size data are limited, but mean mature shell size appears to be approximately 229 mm diameter (N = 8) (data from Iredale, 1944; P. Ward, C. Teichert, and R. T. Abbott, personal communication, 1985). A number of other large specimens matching Iredale's description have turned up from other regions in Australia (e.g., Cotton, 1957a,b).

The wide distribution of shells referred to this species, sharing common size and coloration characteristics, suggests that it may be a distinctive form. It should be noted, however, that the color pattern of *N. repertus* matches the description and illustration of *N. ambiguus* Sowerby, 1849 (see Section 5.1). If the size of the specimen illustrated by Sowerby becomes known and is compatible with the size range of *N. repertus* (yet to be determined), the latter may be regarded as a synonym of *N. ambiguus*.

4.1.3. Distribution

The specimens described by Iredale (1944) were found at Pelsart Island, Houtmans Abrolhos, and Rottnest Island, Western Australia. A specimen with soft parts was found in Foul Bay, Yorketown, southern Yorke Peninsula, South Australia (Riddle, 1920; figured by Cotton, 1957a,b). Additional drift-shell occurrences have been reported from a number of sites around Australia (see Saunders, 1981b). Most recently, 14 shells of live-caught specimens, obtained by deep-water shrimp bottom-trawls, operating approximately 200 km off Broome, Western Australia, at a depth of 380–450 m, were shown to me by Donald Dan. The coloration of the specimens includes typical *pompilius* patterns, as well as the reduced, yellowish color banding said to be typical of *N. repertus*, and the specimens ranged from 184 to 243 mm in diameter (mean 220 mm). This locale was apparently the source of an immature specimen ($\approx$ 110 mm in diameter), possibly representing *N. repertus*, that was trawled from a depth of approximately 300 m, about 225 km NNW of Port Hedland, Western Australia, and is in the Western Australia Museum collections (S. M. Slack-Smith, personal communication, 1982).

5. Dubious Species

5.1. *Nautilus ambiguus* Sowerby, 1849

5.1.1. Discussion

Sowerby distinguished this species by its wider aperture, compared to that of *N. pompilius*, and by the reduced coloration, which does not extend to the umbilicus. The species was illustrated, but neither size nor locality was given. Sowerby (1849, p. 464) astutely noted that this species might be only a sexual variation of *N. pompilius*, which would explain the difference in apertural width [mature males are wider than mature females (Saunders and Spinosa, 1978)]. Because the size of the specimen illustrated was not noted, the pattern of coloration remains its only distinguishing feature. However, shell coloration is quite variable, and the white umbilical variant is known to occur in *N. pompilius*, *N. belauensis*, *N. stenomphalus*, and *N. repertus*. Thus, there are no characters that permit distinguishing this as a separate species, and *N. ambiguus* should be regarded as a *nomen dubium*.

5.1.2. Distribution

Sowerby did not cite the origin of the specimen he illustrated, but specimens with similar color patterns are found in populations of *Nautilus pompilius* in the Philippines, Papua New Guinea, Australia, and elsewhere. This color pattern is rarely observed in *N. belauensis* from Palau (Saunders, 1981b, Fig. 11), and as noted in Sections 3.5 and 4, it is thought to characterize both *N. repertus* and *N. stenomphalus*.

5.2. *Nautilus alumnus* Iredale, 1944

5.2.1. Discussion

Iredale (1944) regarded drift shells from North Queensland as distinctive enough to warrant specific designation. He described them as ". . . not exceeding 6 inches in diameter, but with the painting very different, only some twelve to fourteen bands clearly separable at the periphery, not continuing to edge . . ." (Iredale, 1944, p. 295). However, he neither designated nor illustrated a type. In addition, the broad range of variation in coloration of *N. pompilius* makes the minor distinctions cited by Iredale insufficient to distinguish the species. *Nautilus alumnus* is regarded as a *nomen dubium*.

5.2.2. Distribution

As noted by Iredale, this form was regarded as characteristic of Queensland, Australia, but its dubious distinction makes its distribution uncertain.

5.3. *Nautilus moretoni* Willey, 1896

5.3.1. Discussion

A number of variants of *N. pompilius* have been named (see Section 6), of which one, *N. pompilius* var. *moretoni*, was subsequently elevated to species rank by Shimansky (1948) (see also Shimansky and Zhuraleva, 1961; Mapes *et al.*, 1979). This form had originally been distinguished by Willey (1896) because it possessed an open umbilicus. However, as noted elsewhere, this character occurs to varying degrees in several other species of *Nautilus* and should not be accorded taxonomic significance unless it can be shown to be a population characteristic, as it is in the case of *N. stenomphalus*.

5.3.2. Distribution

The original description of Willey (1897c) was based on a single drift shell obtained from Papua New Guinea, presumably from the vicinity of Samarai, Milne Bay Province. Other occurrences of specimens showing an open umbilicus are noted above.

6. Variants and Subspecies

Given the considerable range of variation within individual species of *Nautilus*, and the amount of overlap among species, infra- or subspecific taxa of *Nautilus* are not recognized at this time. However, as more data on geographically isolated populations become available, such taxonomic units may become biologically justifiable. It may turn out that different frequencies of occurrence of such features as umbilical perforation, or particular patterns of shell coloration, will be shown to characterize isolated populations of *Nautilus*. If so, it might be logical to recognize these differences taxonomically at the subspecies level. However, their identity should be based on documented population characteristics, rather than solely on aberrant or unusual individual specimens.

The following taxa were established on the basis of isolated shells and appear to represent examples of variation within species of *Nautilus* that do not warrant taxonomic recognition:

1. *Nautilus pompilius* var. *moretoni*, *N. pompilius* var. *perforatus*, and *N. pompilius* var. *marginalis* are three variants named by Willey (1897c), to distinguish differing degrees of umbilical closure in specimens from Papua New Guinea. It is notable that two of the variants (*perforatus* and *marginalis*) were noted to be asymmetrical; i.e., they are unequally developed on opposite sides of the same shell. In the third variant (*moretoni*), the umbilicus was open, and an umbilical shoulder was developed. But, as Willey noted, intermediate varieties exist. These features have been seen as rare occurrences within populations of *N. pompilius*, *N. belauensis*, and *N. stenomphalus* (see above) and are not accorded taxonomic significance.

2. A substantially modified scheme of nomenclature for *Nautilus* was proposed in 1948 by V. N. Shimansky (1948), based on a study of shells available from various Soviet museum and institute collections. He utilized the following characters to differentiate species and variants:

a. Umbilical closure, depression, and shape
b. Number of septa in the first whorl and total number of septa
c. A spiral coiling index
d. Septal shape
e. Shell color and sculpture
f. Shape of whorl cross section

Using these criteria, Shimansky recognized five species of *Nautilus* (*pompilius*, *macromphalus*, *stenomphalus*, *moretoni*, and *umbilicatus*). He included five variants of *N. pompilius* (var. *pompilia* Shimansky, *rumphii* Shimansky, *caudatus* Lister, *perforatus* Willey, and *marginalis* Willey) and two variants of *N. stenomphalus* (var. *stenomphala* Shimansky and *scrobiculatus* Solander). Some of the characters used by Shimansky may well prove to be useful in analysis of variation (e.g., septal shape), but many others are known to vary nonsystematically within populations of *Nautilus*. In addition, the various criteria listed above were applied unevenly in distinguishing the taxa, and only one of the new variants (var. *pompilia*) was illustrated. As a whole, the scheme is inapplicable. Nevertheless, Shimansky's whole-morphology approach was innovative and may be regarded as a precursor to morphometric analysis of variation in *Nautilus*.

7. Isolating Factors, Geographic Differentiation, and Speciation

In reviewing a few years ago what was known of the species and distribution of *Nautilus*, I closed by speculating that although the poor (and apparently contracting) post-Miocene record of *Nautilus* did not seem to portend well for its future, its broad range of intraspecific and geographic variation suggested sufficient genetic plasticity to permit survival, and perhaps even renewed speciation (Saunders, 1981b, p. 15). A wealth of new data on genetic and morphologic variation that seem to support this view is now available, particularly for the widespread species *N. pompilius*. They show that: (1) Morphological differences are manifested by more than just variation in size; coloration, in particular, seems to vary among populations. (2) *Nautilus* exhibits normal levels of genetic variation in electrophoretic markers. (3) Geographically isolated populations exhibit both morphological and genetic differences (Hayasaka *et al.*, 1982; Tanabe *et al.*, 1983, 1985; Ward, 1984; Saunders and Davis, 1985; Masuda and Shinomiya, 1983; Woodruff *et al.*, 1983) (see also Chapters 5 and 6).

There is also now available a considerable body of new information on the animal's role in its natural habitat and its geographic distribution. *Nautilus* is perhaps uniquely adapted to a vertically mobile, deep-forereef scavenging role (Saunders, 1984b; Carlson *et al.*, 1984; Ward *et al.*, 1984) (see also Chapter 9). In addition, it appears that *Nautilus*, particularly *N. pompilius*, is a near-ubiquitous component of oceanic reef settings in the Indo-Pacific, although much confirmation of actual populations versus drifted occurrences is needed.

Given the documented mobility and longevity of *Nautilus* (Saunders and Spinosa, 1979; Saunders, 1983), it is perhaps not surprising that populations are homogeneous internally and apparently randomly mating (Woodruff *et al.*, 1983). However, it also seems clear that water depths greater than 800 m, combined with other ecological and physiological factors, pose significant barriers to migration and gene flow between what may be even geographically adjacent populations. Details of these limiting factors, and how they affect *Nautilus*, are provided elsewhere (Saunders and Wehman, 1977; Saunders and Spinosa, 1979; Chamberlain and Moore, 1982; Saunders, 1984b; Zann, 1984; Ward and Westermann, 1985) (see also Chapters 9 and 12) and may be summarized as follows:

1. The lower depth limits are determined by shell implosion and siphuncle rupture depth ($\approx$800 m); cameral flooding may limit long-term habitat depth to 300–500 m.
2. The upper depth limits are variable, both seasonally and geographically, inasmuch as they are determined by water temperature (>25°C). Thus, in most areas of the Indo-Pacific, migration across shallow shelf areas (<100 m) would be limited to short-term movement (several days at the most). Biological interactions, particularly predation by teleost fishes, are probably also an important limiting factor in shallow water.
3. There is little evidence to support midwater movement by *Nautilus*; it seems rarely to leave the bottom. In addition, the animal is virtually helpless against predatory attacks from fishes, particularly in open water.

In combination, these factors would limit the most active migration and dispersal routes to depth-defined corridors approximately 300 m deep (typically between 100 and 300 m). Movement at depths between 300 and 500 m would be of relatively short-term duration; at depths approaching 800 m, it would be precarious. Migration across extensive stretches of shallow, warm water (<100 m, >25°C) or deep water (>800 m) would be limited to midwater or surface drifting or rafting. An apparent result of such an event is the surface capture of a specimen of *N. pompilius* off southern Japan, which was interpreted as having drifted from the Philippines, approximately 2000 km to the south, on the Kuroshio Current (Tanabe and Hamada, 1978; JECOLN, 1980b) (see also Chapter 4). Although such events would seem to be rare, they must have occurred with sufficient frequency to allow colonization of the majority of the Indo-Pacific island complexes (there are exceptions; for example, no documented record of *Nautilus* has been obtained from the Marianas, although there is no known reason that it should not be there).

Each of these populations would seem to be isolated to its individual island complexes, essentially on arrival, by built-in physical and ecological constraints. The rate, nature, and products of subsequent geographic differentiation, and whether they would lead to speciation, would depend on many variables, both ecological and genetic. Their examination and documentation, which are only now being undertaken, would seem to be a logical next step in *Nautilus* research.

ACKNOWLEDGMENTS. This chapter synthesizes contributions of the many investigators whose works are cited. In addition, a number of new observations are included, based on data and specimens kindly provided by Donald Dan, New Friendship, Maryland, and on recent efforts in Papua New Guinea and Australia, which were supported by NSF Grants EAR 83-18932, BSR 86-08065, and by the National Geographic Society.

Chapter 4

Geographic Distribution of *Nautilus* Shells

MICHAEL R. HOUSE

1. Introduction

The distribution of living *Nautilus* has been a source of interest since earliest times. Their occurrence as rare postmortem drift specimens on African and Asian shores may have led to their reaching ancient civilizations by traders, but whether *Nautilus* was known to Aristotle has been questioned by Landman (1982b) and others (see Chapter 1). The record of illustrations of *Nautilus* on Minoan vases (Luce, 1969) is erroneous; the illustrations are of *Argonauta*. In recent years, records of living *Nautilus* have been assembled and much commented on (Stenzel, 1957; Toriyama *et al.*, 1965; Saunders, 1981b). Paleontologists have been particularly interested in the postmortem distribution, mainly as a guide to how the distribution of chambered cephalopods in the past may have been extended after death, and there have been attempts to assemble these data (House, 1973; Reyment, 1973). This chapter presents a general review and a synthesis of available distribution data.

In 1965, Toriyama *et al.* published a list of *Nautilus* localities, many culled from the literature, using the location numbers 1–73. Their localities and the material reported are marked on an accompanying map (Fig. 1), but the detailed sources are not repeated and reference should be made to their paper for details. Several records noted by these authors were not given numbers because of am-

MICHAEL R. HOUSE • Department of Geology, The University of Hull, Hull HU6 7RX, England.

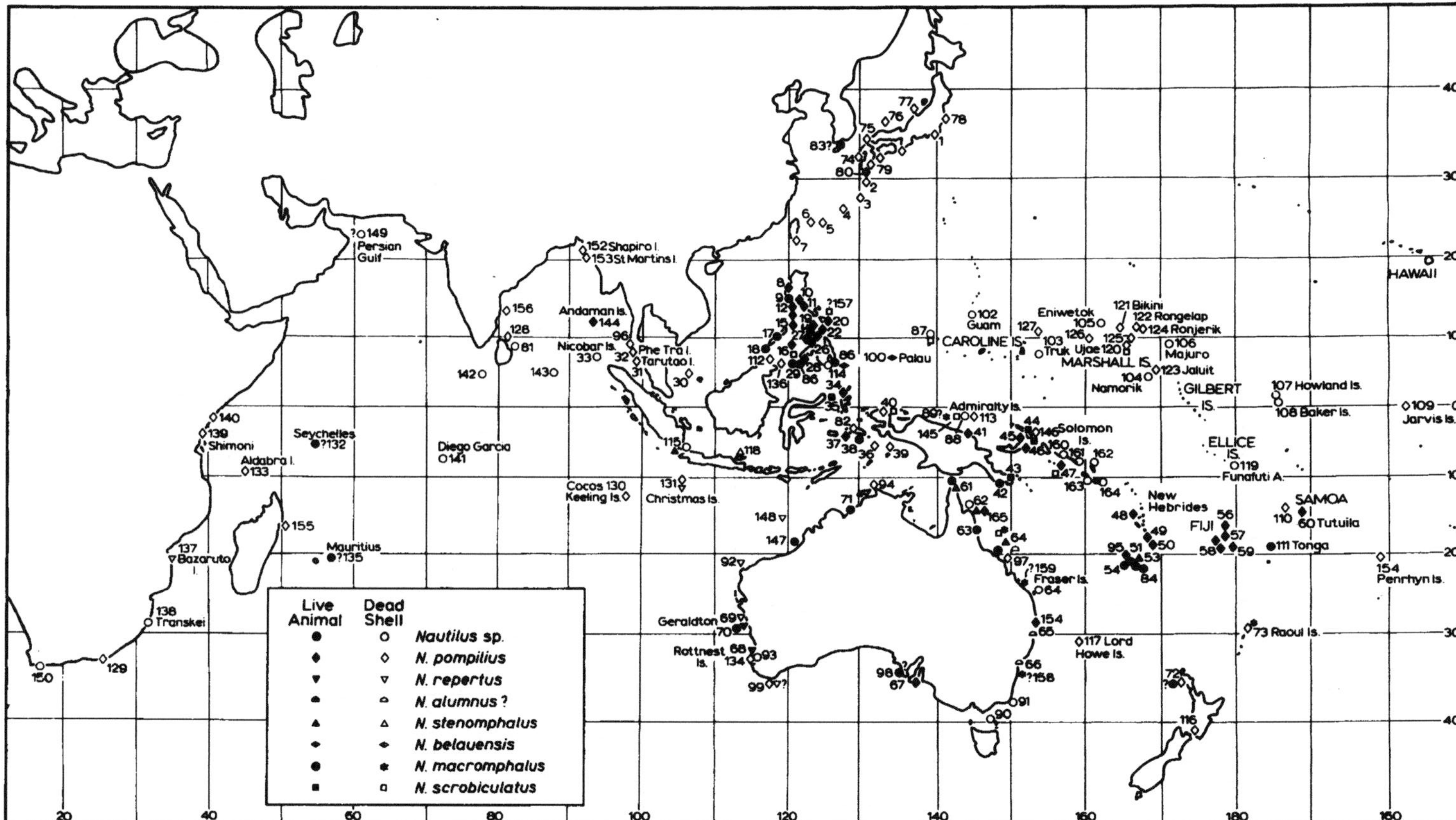

Figure 1. Map of the Indian and western Pacific Oceans showing the distributions of records of living *Nautilus* and drifted shells of *Nautilus*.

biguities, mostly relating to precise localization. A number of additional localities are also presented in Fig. 1 (locality numbers 74–165). These comprise new data, as well as records reported in *The Chambered Nautilus Newsletter* (CNN), an occasional publication that has been edited and distributed since 1974 by H. K. Dugdale, Delaware Museum of Natural History; the CNN numbers cited throughout this chapter are the issue numbers. A large number of records are reported here for the first time; these have resulted from correspondence by the author in the 1970s, mainly conducted through consulates in scattered islands in the Indian and Pacific Oceans, and from some data of the National Museum of Natural History (NMNH), Washington, D.C.

2. Nomenclature

An assessment of the taxonomic status of the many named species and varieties of *Nautilus* awaits a detailed analysis (but see Chapter 3). It would be an excellent subject for computer-based, image-analytical, and statistical techniques. Stenzel (Teichert *et al.*, 1964b) recognized five species: *N. pompilius* Linnaeus, 1758, *N. repertus* Iredale, 1944, *N. stenomphalus* Sowerby, 1849, *N. macromphalus* Sowerby, 1849, and *N. scrobiculatus* [Lightfoot, 1786]. To these five, Saunders has added a sixth, *N. belauensis* Saunders, 1981. The review of living species by Saunders (1981b) (see also Chapter 3) limits valid species to these six.

Because this account of distribution is primarily a compilation of determinations by others, it would be misleading to give the impression that any fundamental taxonomic restudy was undertaken. Hence, Saunders's revision is followed, except that reported occurrences of some varieties of species that he considers to be synonyms are listed here as they were cited, to keep distinct the references and records.

3. Distribution of *Nautilus* Shells

Concerning many records of living *Nautilus*, either they have not been assigned to a species or there is some uncertainty as to whether the material was alive. These include possibly specimens taken off Korea (Loc. 83, CNN 13); Dunk Island (Loc. 63) and Fraser Island (Loc. 64), Queensland; Iluka (Loc. 101, CNN 3), New South Wales; Port Lincoln (Loc. 98, CNN 4), South Australia; Pelsart Island (Loc. 70) and Port Hedland [Loc. 147, (specimen in Western Australia Museum information from T. Whitehead), CNN 25], Western Australia; the Seychelles [Loc. 132 (Reyment, 1973); in correspondence from A. G. Jarrett (1974), CNN 9]; and Mauritius [Loc. 135 (Bidder in Reyment, 1973), but local shops are known to import from the Philippines].

Northward from the Philippines is a string of records of drifted shells: Taiwan (Loc. 7), Okinawa (Loc. 4), Yaeyama Island (Loc. 6), and Japan, where several recent records have been made by Hamada (1966) (Locs. 74–79). Westward, there are records in the South China Sea (Loc. 30) (Reyment, 1973) and along the eastern shore of the Andaman Sea in southwest Thailand [Locs. 31, 32, 96; CNN 4, 5

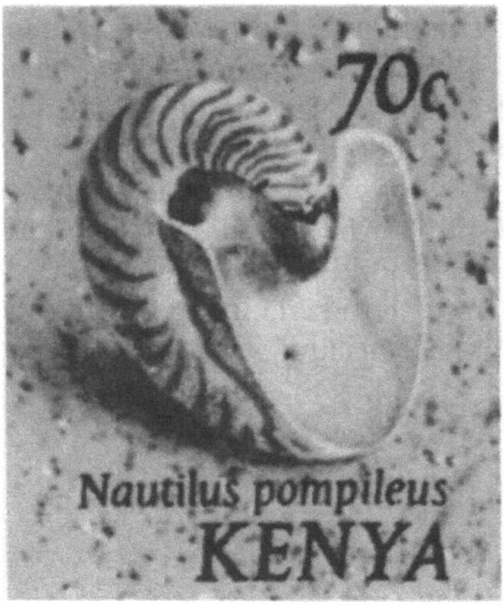

Figure 2. A Kenyan postage stamp from the first shell-definitive issue showing a drifted *Nautilus* from the Kenyan coast and the incorrect specific spelling *pompileus*, which was corrected on the second issue. (21 mm × 25 mm)

(Hamada, 1965; Yancey, 1975)], with an unidentified species of *Nautilus* (presumably *N. pompilius*) in the Nicobar Islands (Loc. 33). Near the head of the Bay of Bengal are several records at Shapiro and St. Martin Islands (Locs. 152, 153) (Teichert, 1970). I am indebted to R. A. Hewitt for drawing my attention to a record from Ennur on the coast of Madras (Loc. 156) (Srinivasan, 1941, p. 196), and there is a note of a damaged dead specimen north of Trincomalee, Sri Lanka (Loc. 128, CNN 3), and a battered shell from elsewhere on the shore of the island (Loc. 81, CNN 11). *Nautilus* was reported southwest of Sri Lanka on the Valdivia Expedition (Loc. 142) and by Brassey (1879, p. 393, information from J. W. Wells) in *Around the World in the Yacht Sunbeam*, where it is reported that at a given location southeast of Sri Lanka (Loc. 143), they encountered "a great deal of driftwood, many large trees, branches, plants, leaves, *Nautilus* shells, backbones of cuttlefish. . . ." *Nautilus* shells have been reported to me by J. D. Taylor from the northern end of Diego Garcia Atoll (Loc. 141), and there is an unlocalized record of *Nautilus* in the Persian Gulf (Loc. 149) (Woodward, 1880, p. 188).

Recently, the record of drifted specimens down the east African coast has been considerably extended. J. A. Whyte has informed me that imperfect shells collected locally in the Malindi and Lamu region (Loc. 140) are for sale in Mombasa (with Philippine specimens also in stock), and at Shelly Bay, Mombasa, a piece of *N. pompilius* was reported (CNN 12). Two former Peace Corps volunteers, Stephen Quinn and David Rucitto, reported that shells were common along a 500-m section of beach in the Kilifi area, north of Mombasa, and were collected by local fishermen for sale in Mombasa. At Shimoni, imperfect shells are reported on the reefs (Loc. 139) (House, 1973). *Nautilus pompilius* is figured on a Kenyan postage stamp (the specific name was incorrectly spelled on the first issue; Fig. 2). At Aldabra Island (Loc. 133), *Nautilus* shells arrive each year (information from R. J. Hnatink and J. D. Taylor), and one shell was reported at the eastern end of the island (House, 1973). Old and broken shells of *Nautilus* from the Transkei region of South Africa (Loc. 138) are in the East London Museum (information from R. N. Kilburn), and a specimen is recorded from False Bay. Six shells of *N.*

pompilius from Algoa Bay, between Port Elizabeth and East London (Loc. 129), are also reported to be in the East London Museum (information from D. H. Kennelly, CNN 3). From Western Cape, South Africa (Loc. 150), there is one badly preserved shell in the Natal Museum, Pietermaritzburg (in correspondence from R. N. Kilburn).

3.1. *Nautilus pompilius* Linnaeus, 1758

Linnaeus (1758, p. 709), when erecting *N. pompilius*, did not give new figures, but quoted the illustrations of Rumphius (1741, Plate 17, Figs. A–C), and these figures constitute the holographs. The illustrated specimen was from Amboina (Ambon), in the Moluccas (Loc. 37), and this has been taken to be the type locality (Iredale, 1944). From that vicinity, shells have also been recorded more recently (Loc. 82; CNN 12,12). This species has been reported widely, but perhaps uncritically, elsewhere, throughout the Philippines (Locs. 8–29), at Antique (Loc. 85, CNN 18), off Mindanao (Loc. 86, CNN 18), and to the south near Halmahera (Loc. 34). A living specimen was taken at Kawajiri, southern Kyushu, Japan (Loc. 80) (Tanabe and Hamada, 1978; Hamada *et al.*, 1980) (CNN 15). Westward from the main area of occurrence, there are odd records of living *N. pompilius* from the Andaman Islands (Loc. 144) (Smith, 1887; Foord, 1888) (R. N. Kilburn writes that there is a fresh specimen of a small *Nautilus* with a small umbilicus in the Natal Museum from there). Eastward from the Philippines, there are records of *N. pompilius* near Mios Woendi and the Schouten Islands (Loc. 40). In Papua New Guinea, Bruce Saunders (in correspondence and Saunders and Davis, 1985) reported trapping *N. pompilius* at Manus, Lae, Port Moresby (Loc. 42), and in New Ireland at Kavieng. There are also reports of this species in the Bismarck Archipelago (Loc. 45), at Ninigo and Alim Islands (Loc. 88, CNN 7).

There are additional records of *N. pompilius* in the Solomon Islands (Loc. 47); in the New Hebrides (Locs. 48–50) at Erromanga Island (Bennett, 1834), Aneityum Island, and others (Bennett, 1859); and at Balade (Loc. 95, CNN 5), New Caledonia. The species is widely recognized in Fiji (Locs. 56–59, CNN 40) and at Amouli, Tutuila Island, in Samoa (Loc. 60).

R. T. Abbott, Melbourne, Florida, has brought to light the following note on *N. pompilius* in Australia, by J. C. Cox to H. A. Pilsbury, editor of *The Nautilus* (1897, Vol. 11, p. 43):

> What surprised me most was to find large numbers of rather broken specimens of *Nautilus pompilius* thrown up in Eden Bay. . . . Can it be possible they are eaten by whales and that the shells (are) extruded as excrement?. . . . great schools of whales come in there, it is said, to rub themselves on the coarse gravel bottom of the bay.

Most recently, live specimens of this species have been captured off Lizard Island, Queensland (Loc. 165) (Saunders and Ward, 1987).

Drift shells, usually referred to *N. pompilius*, are very widely distributed. They are frequently noted around the main area of live occurrences, of course, but here they are recorded only outside the area of known living occurrences. Thus, there are shells in the Western Australia Museum from Balabac Island (Loc. 112) and Doc Can Island (Loc. 136) determined by T. Whitehead as *N. pompilius* (information from S. M. Slack-Smith) and Mindanao (Loc. 114, CNN 3) and Indonesia (Loc. 115, CNN 3).

Other records in the Indian Ocean are on Cocos-Keeling Island (Loc. 130), determined by H. A. Rehder as *N. pompilius* (in correspondence), and at Greta Beach, Christmas Island [Loc. 131 (S. M. Slack, in correspondence informed me that specimens, with apertures extensively damaged, are in the Western Australia Museum, Nos. 215–273)], determined by T. Whitehead to be *N. pompilius*.

Eastward from the Philippines, there are records of drifted *Nautilus*, generally specified as *N. pompilius*, throughout the Caroline, Marshall, Gilbert, and Ellice Islands, as far east as the Line Islands; these records extend considerably the drift range given earlier (House, 1973).

There are many reports from Palau (Loc. 100). H. A. Rehder (in correspondence) noted *N. pompilius* (NMNH 616893) from Kayangel Island; this and other live records of *Nautilus* (CNN 7, 10, 11, 13, 15, 17) from Palau are actually *N. belauensis* (see Section 3.5). Elsewhere in the Caroline Islands, *N. pompilius* is reported on Ulithi Atoll (Loc. 87, found on a cruise of the *R. V. Vityaz*, CNN 7), and L. G. Eldredge (in correspondence) informed me that there is a single beach-drift *Nautilus* in the collections of the University of Guam from that island in the Marianas Group (Loc. 102). There is a specimen from Kapingamaringi Atoll (Loc. 127), determined by H. A. Rehder as *N. pompilius* (NMNH 622352), and R. P. Owen has written that on Truk Island (Loc. 103), *Nautilus* is found not infrequently.

H. A. Rehder (in correspondence) identified *N. pompilius* from the Marshall Islands; from Ujelang Atoll (NMNH 614698, 614719, Loc. 126); from Wotho Atoll (NMNH 614337, Loc. 125); from Bokon Island, Bikini Atoll (NMNH 586114, Loc. 121); from Burok and Labaredj Islands, Rongelap Atoll (NMNH 584166, 583353, Loc. 122); from Bock Island, Ronjerik Atoll (NMNH 583363, 586279, Loc. 124); from Jaluit Island, Jaluit Atoll (NMNH 659691, Loc. 123); and from Ujae Atoll (NMNH 607349, 614579, Loc. 120). J. E. Maragos (in correspondence) informed me that he has noted *Nautilus* shells above the high-water mark on Namorik Atoll (Loc. 104), and L. G. Eldredge (in correspondence) reported *Nautilus* shells seen on a beach on Majuro Atoll (Loc. 106). J. E. Maragos also informed me that he has seen *Nautilus* shells above the high-water mark on Eniwetok Atoll (Loc. 105), and drift shells have been noted there inside the barrier reef (CNN 18).

Farther east, Yoshio Kondo (in correspondence) reported that there are three worn *Nautilus* shells in the Bernice P. Bishop Museum, Honolulu, from Howland Island (Loc. 107), and three similar shells in the same museum from Baker Island, collected in 1924 (Loc. 108). The most easterly occurrence known to me is from Jarvis Island in the Line Islands (Loc. 109); R. A. Rehder (in correspondence) noted an *N. pompilius* shell from that locality in the H. Jewell Collection of the Academy of Natural Sciences, Philadelphia.

From the Ellice Islands, shells have been noted above high-tide level on Funafuti Island (Loc. 119) by J. E. Maragos. They also occur to the southeast from Samoa (Loc. 110), and R. K. Dell has written that a specimen of *N. pompilius* from that location is in the collections of the National Museum of New Zealand. From Tutuila (Loc. 60), live specimens are known. H. A. Arthington-Davy wrote from Tonga (Loc. 111) that the Fisheries Officer there has informed him that "the Pearly Nautilus is found on Tonga. Both live and dead specimens are found, though not very frequently, and they are more likely to be found in the Northern groups of the Kingdom (Vav'u and Hu'apai) than in the south." From Penrhyn Island (Loc. 154), there is a record of *N. pompilius*, and there are records on Raoul Island (Loc. 73) and North Island, New Zealand (Loc. 116), and Lord Howe Island (Loc. 117), which R. K. Dell (in correspondence) has determined to be *N. pompilius* from shells in the National Museum of New Zealand.

Between the Celebes, Indonesia, and New Guinea, there are a number of records [Locs. 36, 39, 82 (CNN 13)], and also in the Admiralty Islands (Loc. 113, three shells in the Western Australia Museum identified as *N. pompilius* by T. Whitehead, information from S. M. Slack-Smith). From farther east, there are seven specimens in the British Museum (Natural History) presented by Anna Bidder from Rabaul (Loc. 146) in the Bismarck Archipelago (CNN 20).

For the Solomon Islands, Reyment (1973) has published a map giving localities for drifted shells of *N. pompilius* on Shortland Island, Choiseus (Loc. 160), Santa Isabel (Loc. 161), New Georgia Islands (Loc. 163), Russel Islands, Florida Islands, Malaita (Loc. 162), Guadalcanal (Loc. 163), and San Cristobal (Loc. 164).

Around Australia, there are a number of records of *N. pompilius* shells: at Brampton, Bushy, and Tryon Islands off Queensland (Loc. 97, CNN 4); Eden, New South Wales (Loc. 91, CNN 5); and Flinders Island, Tasmania (Loc. 90, CNN 5). *Nautilus* shells are noted at Ningaloo (Loc. 92, CNN 5) and Leschenault Inlet (Loc. 93, CNN 5), Western Australia, and *N. pompilius* shells are reported from Denmark and Albany (Loc. 99, CNN 3, 4) and Rottnest Island (Loc. 134, Western Australia Museum, No. 12976, S. M. Slack-Smith, in correspondence), Western Australia, and at Port Essington (Loc. 94, CNN 5), Northern Territory. There is a record from Yorke Bay, South Australia (Loc. 67) (Riddle, 1920), that was assigned to *N. repertus* by Cotton (1957a). This was disputed by Iredale (1944), but has subsequently been accepted, according to Teichert (Stenzel, 1957). However, it should be stressed that no attempt has been made to examine this material, and the taxonomic assignments may be in need of revision.

Willey (1897c) and Shimansky (1948) erected a number of varieties of *N. pompilius*, but Saunders [(1981b) and Chapter 3] considers these varieties synonyms of *N. pompilius*, and they are not separately shown on the accompanying map.

3.2. *Nautilus repertus* Iredale, 1944

Nautilus repertus (Iredale 1944, p. 295) was erected as a replacement name for an unlocalized specimen invalidly named *N. ambiguus* by Sowerby (1849, Plate 97, Fig. 2), using a name that was preoccupied and not available. Its identity has been interpreted following descriptions and an illustration given by Iredale (1944, p. 296) of material from Rottnest Island, Western Australia. Live specimens were not confirmed. There is a possible record from near Denmark (Loc. 99, CNN 20), Western Australia; from Rowley Shoals (Loc. 148, CNN 25), Western Australia, there is a shell noted as *N. repertus* in the Western Australia Museum. The record of *N. repertus* taken live on Palau (Loc. 100, CNN 6, 7) actually referred to specimens later named *N. belauensis* by Saunders (1981a). Damaged shells determined to be *N. repertus* by T. Whitehead are in the Western Australia Museum from Christmas Island [Loc. 131 (S. A. Slack-Smith, in correspondence)]. A drifted, old, and more or less broken shell determined by R. N. Kilburn (in correspondence) as *N. repertus* is reported from Bazaruto Island, Mozambique (Loc. 137). A large shell (230 mm in diameter) in the geology collections at the University of Iowa (SUI 40095) that has *repertus*-like coloration was collected from a beach near Kupang, Timor (B. F. Glenister, in correspondence).

3.3. *Nautilus alumnus* Iredale, 1944

Iredale (1944, p. 295) erected *N. alumnus* for

> *Nautilus pompilius*, of Australian writers, Brazier, Hedley, etc., recording shells from Queensland, drifted specimens from New South Wales. (Locs. 64–66).

He did not illustrate any specimen, but he compared his Queensland material with the *N. pompilius* figure of Rumphius, stating that they

> agree in their somewhat small size, not exceeding 6 inches in diameter, but with the painting very different, only some twelve to fourteen bands clearly separable at periphery, not continuing to edge, and not as wrinkled as in the previous figure, apparently also of different colour, lake not brown.

Since Iredale did not figure the species, critical interpretation is not at present possible, and this species has been regarded as invalid by Saunders [(1981b) and Chapter 3]. Iredale specifically excluded a living record from Yorke Peninsula, South Australia (Loc. 67), which, following Stenzel, is here referred to *N. pompilius*.

3.4. *Nautilus stenomphalus* Sowerby, 1849

Nautilus stenomphalus was named by Sowerby (1849, p. 465, Plate 97, Fig. 3) for a specimen for which no locality was given in the original description, but that Iredale thought to have come from North Queensland (approximately indicated as Locs. 61 and 64 in Fig. 1). Iredale (1944) noted that it had an umbilical perforation and that in color and size it was like *N. macromphalus*, but with an

> umbilicus less than half an inch in diameter, sides steep, umbilical area sloping, coloration lake, peripheral bands separate but tending to merge, not running into the umbilicus, leaving a white circumbilical area.

Saunders (1981b) considered this a possible subspecies of *N. pompilius*, until live specimens of *N. stenomphalus* were obtained off Lizard Island, Queensland (Loc. 165), in 1985 (see Chapter 3). Shells referred to this species are reported from Madura Island, Java (Loc. 118, CNN 3, specimens in the Museum Zoologicum Bogoriense, Bogor), and Panaitan Island (Loc. 115, CNN 3).

3.5. *Nautilus belauensis* Saunders, 1981

This most recently named species of *Nautilus* was described by Saunders (1981a), with the type locality being Palau in the Western Caroline Islands (Loc.

100). It differs from *N. pompilius* in being larger, in having later closure of the umbilicus, and in having crenulated growth lines that give a lirate surface texture, rather like that of *N. scrobiculatus*. It has not been reported outside Palau, with the exception of a drifted shell, tagged off Palau, that was found in Mindanao (Loc. 86) in the Philippines (figured in Saunders 1981b, Fig. 8). Other records previously referred to *Nautilus* in this area refer to this species (CNN 7, 10, 11, 13, 15, 17), according to information provided by Bruce Saunders.

3.6. *Nautilus macromphalus* Sowerby, 1849

Sowerby (1849, p. 464, Plate 98, Figs. 4 and 5) gave no indication of the locality of his material when naming this species, but Bennett (1859) suggested a New Caledonian source. Documented occurrences of live *N. macromphalus* are known only from a confined area around New Caledonia (at Kunie and Mare, Locs. 54, 55) (Sowerby, 1847–1887; Bennett, 1859) (CNN 26, 27) and the Loyalty Islands (Locs. 51–55, 84, CNN 9, 18) (Saunders, 1981b). There is a mistaken report from the south coast of Manus Island (which should be queried in Fig. 1, Loc. 89, CNN 5). Other records are North Island, New Zealand (Loc. 72, CNN 19), and questionably from Wallongong (Loc. 158, CNN 31), New South Wales. There is also a questionable report from north of Brisbane, Queensland (Loc. 158, CNN 31), and recently, several abraded, beach-collected specimens were observed in the collection of Mrs. G. Pini, at Lizard Island, Queensland (W. B. Saunders, personal communication, 1985).

3.7. *Nautilus scrobiculatus* [Lightfoot, 1786]

This specific name was first published in the auction catalogue of the famous shell collection of the Duchess of Portland (Lightfoot, 1786; Item 3906). No one has doubted that this is a manuscript name applied by Daniel Charles Solander, Linnaeus's fair-haired boy, who is known to have done much work on the collection and who was largely responsible for introducing the Linnaean method into England during a long stay; his success in doing so was perhaps influenced by his reportedly urbane manners and his being known to be the favorite pupil of the great master. Rather than record this source in the species paternity, Rehder (1968) has followed the International Code of Zoological Nomenclature and decreed that Lightfoot (who meticulously prepared the auction catalogue) should be the author of the species; the catalogue was published anonymously, hence the brackets around Lightfoot's name. Saunders (1981b) included in synonomy with this species *N. umbilicatus* Lamarck (1822, p. 633), for the specimen figured by Lister (1685, Plate 552, Fig. 4), *N. perforatus* Conrad (1849, p. 213), and *N. texturatus* Gould [1857, p. 20 (named without illustration but with an expression that it was identical with Lister's figure on Plate 522)]. This species has a very large, deep umbilicus and a reticulate or scrobiculate surface ornament (Saunders, 1981b). Willey (1902) recorded a damaged specimen with soft parts from near Milne Bay in Papua New Guinea (Loc. 43), and *N. scrobiculatus* is recorded as drift shells in New Ireland (Loc. 44) and New Britain (Loc. 46). Shells are recorded from the Sulu Sea (Loc. 16, CNN 28) and, questionably, from Eastern Samar (CNN 28). There are two shells in the British Museum (Natural History) presented by Anna Bidder from Manus Island (Locs. 145, 89, CNN 5); living specimens from Manus Island have been reported recently (CNN 34) (Saunders and Davis, 1985). There is an odd record off Ulithi Atoll (Loc. 87) in the Caroline Islands of a shell of *N. scrobiculatus*, heavily overgrown with oysters, reported on a cruise of the *R. V. Vityaz* (CNN 7), and H. A. Rehder (in correspondence) noted a specimen in the NMNH from Ujae Atoll (Loc. 120) in the Marshall Islands.

4. Pattern of Postmortem Drifting

It is clear that adult shells of *Nautilus* may float for considerable periods after death. An empty shell was kept floating in an aquarium tank for over 2 years (House, 1975, p. 483), but the record is held by a floating specimen that drifted ashore after 11 years (Ishii, 1981). Saunders and Spinosa (1979) recorded the postmortem drift of a shell over a distance of 1000 km (between Palau and Mindanao) in a maximum of 138 days, an average of approximately 7 km per day.

In an interesting study of drifting bottles, Satyanarayana Rao (1963) gives westward traveling rates in the Indian Ocean, using the Equatorial Current, as averaging 17 km per day; one bottle averaged 20.9 km per day between Australia and the coast of East Africa. Another, released in the Arabian Sea, traveled down the east coast of Africa and up the west coast to 6°N, traveling 9650 km in 400 days, at a speed of about 24 km per day.

House (1973) related the pattern of drifted shells and the life areas of *Nautilus* to the ocean currents of the Indian and Pacific Oceans, and this map has been republished with modifications in other works (Kennedy and Cobban, 1976; Landman, 1982b). A revised map, incorporating the new data, is presented in Fig. 3. Ocean currents change with the season, so the currents given are rather generalized. Even so, it is clear that there is a relationship between the shell drift pattern and the known main ocean currents.

The drifted shells found northward toward Japan are regarded as products of drift from the Philippines by the Kuroshio Current. It is to be expected that shells will be carried farther to the east-northeast by the North Pacific Current, but there are no records so far of shells reaching the western shores of North America. The action of the North Equatorial Current may explain the absence of records in the Marianas Islands and Ogasawara Gunto, Bonin Islands. The wide swath of records through the Caroline and Marshall Islands seems clearly related to action of the Equatorial Current, and this current might be expected to carry shells to the shores of South America, but there are no records yet known there. The wide spread of living records east-southeast from the Philippines to the New Hebrides, Fiji, and Samoa is against the south Equatorial Current and is presumably evidence either of relics of a former distribution or of active eastward colonization. In such a circumstance, novel types are to be expected, and this may be illustrated by the distribution of *N. macromphalus*. The gyre between Australia and New Zealand may explain distributions in that area. The evidence for live colonization around the southern and western coast of Australia, given the West

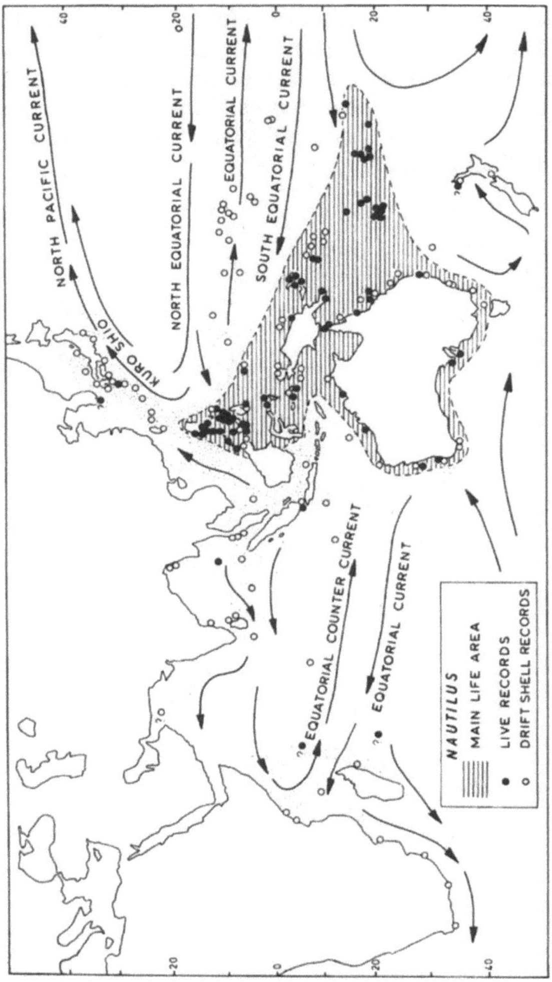

Figure 3. Map showing the relationship of living and postmortem distributions of *Nautilus* to approximate ocean currents. Updated from House (1973).

Wind Drift, would appear to be related to active, if occasional, colonization perhaps down the coasts of both Queensland and Western Australia.

The drift records, and occasional live records in the South China Sea and in the Bay of Bengal, seem clearly related to the main current between Borneo and Indonesia. Whether the records along the east coast of Africa result from continued drifting in the same direction or from westward drift from Western Australia by the Equatorial Current is at present a matter for speculation, but if the record of *N. repertus* from Mozambique (Loc. 137) is confirmed, then an Australian source may be probable. This may also be the source for the live records (Locs. 132, 135) in the western Indian Ocean, or, as suggested earlier (House, 1973), there may be real colonization by *Nautilus*. The new records of shells along the coast of South Africa as far as the Western Cape (Loc. 150) suggest that drifted shells are to be expected even in the South Atlantic.

Chapter 5

Genetic Variation and Phylogeny in *Nautilus*

DAVID S. WOODRUFF, M. PATRICIA CARPENTER, W. BRUCE SAUNDERS, and PETER D. WARD

1. Introduction

The origin and evolution of the genus *Nautilus* have long been obscured by the poor fossil record, by apparent morphological conservatism, and by confusion regarding the taxonomy of the living species. Studies of protein polymorphisms promise a solution to this impasse, since such variation reflects the structure of the genes themselves and should record the molecular phylogeny of the animals more clearly than do phenotypic characters. The preliminary study of the molecular evolution of *Nautilus* reported in this chapter elucidates some of these long-standing problems in a usefully predictive manner and suggests that they are not beyond full resolution with existing techniques.

Teichert and Matsumoto (Chapter 2) trace the ancestry of *Nautilus* to the late-Eocene [≈37 million years ago (Ma)] *N. praepompilius* from the Plato Ust-Urt east of the Caspian Sea. In contrast, Ward (1984) suggested that the genus arose from a eutrephoceratid ancestor in the latest Miocene (≈7–10 Ma) of Australia. In the

DAVID S. WOODRUFF and M. PATRICIA CARPENTER • Department of Biology, University of California, San Diego, La Jolla, California 92093. W. BRUCE SAUNDERS • Department of Geology, Bryn Mawr College, Bryn Mawr, Pennsylvania 19010. PETER D. WARD • Department of Geological Sciences, University of Washington, Seattle, Washington 98195.

absence of *Nautilus* fossils of Miocene, Pliocene, or Pleistocene age, and in the absence of a clear-cut generic boundary, these disparate views are hard to judge. One hypothesis predicts that the genus, and possibly some of its living species, might be as much as 40 million years old. Accordingly, *Nautilus* would be regarded as truly a living fossil—a long-surviving descendant of a once abundant group that arose in the Cambrian (≈570 Ma). On the other hand, if the genus arose in the late Miocene, it is only 7–10 million years old and constitutes a recently successful group of chambered cephalopods that are at present undergoing a period of speciation.

As with the question of the origin and age of *Nautilus*, there is no clear consensus as to the actual number of living species or their relative ages. Saunders (Chapter 3) recognizes five or possibly six species: *N. belauensis* Saunders, 1981, *N. macromphalus* Sowerby, 1849, *N. pompilius* Linnaeus, 1758, *N. scrobiculatus* [Lightfoot, 1786] *N. stenomphalus* Sowerby, 1849, and possibly *N. repertus* Iredale, 1944. One of these, *N. scrobiculatus*, is so different from the others that Stenzel (1964) suggested it might represent a separate subgenus. Another, *N. macromphalus* from New Caledonia, has soft parts that seem to be identical to those of the geographically widespread species *N. pompilius*, despite considerable differences in shell morphology. Two other species have even stronger morphological affinities with *N. pompilius: N. belauensis* from Palau and *N. stenomphalus* from the Great Barrier Reef. The latter is so similar to *N. pompilius* that specimens from Lizard Island display a mosaic of characters typical of each species, suggesting interspecific hybridization (Chapter 3). Traditional morphological studies thus suggest various degrees of phylogenetic relationship among the extant species; these hypothetical affinities can now be tested against the molecular data.

This attempt to prepare a genetic phylogeny of living *Nautilus* presupposes that they are genetically variable, i.e., that their apparent evolutionary inertia is not due to an inherent lack of genetic variation. Our pilot study of *N. belauensis* showed that this species, at least, is moderately variable (Woodruff *et al.*, 1983). The survey of 21 presumptive gene loci revealed that the proportion of loci that are polymorphic (P) is 0.20 and that mean heterozygosity per individual ($\overline{H}$) is 0.09. The species was thus found to be amphimictic, exhibiting levels of genetic variation typical of those seen in hundreds of other horotelic animals (Nevo *et al.*, 1984). That pilot study, like the one presented herein, involved an assessment of genetic variation based on the detection of polymorphic proteins by single gel electrophoresis. Such proteins, primarily enzymes the alternate allelic forms of which are termed *allozymes*, are produced by an individual's structural genes and are thus representative of a significant part of the genome. The electrophoretic technique has been used widely to establish the breeding system, population structure, and genetic phylogeny of various groups of animals (for reviews of the procedure and its theoretical basis and utility, see Avise, 1974; Ferguson, 1980; Oxford and Rollinson, 1983). In our studies, it has proved invaluable in elucidating the evolution of such taxonomically difficult molluscan groups as *Cerion* (Woodruff and Gould, 1980; S. J. Gould and Woodruff, 1986), *Biomphalaria* (Mulvey and Woodruff, 1985; Woodruff *et al.*, 1987), *Oncomelania* (Woodruff *et al.*, 1986a, 1987), and *Crepidula* (Woodruff *et al.*, 1986b). Its application to *Nautilus* should be equally rewarding.

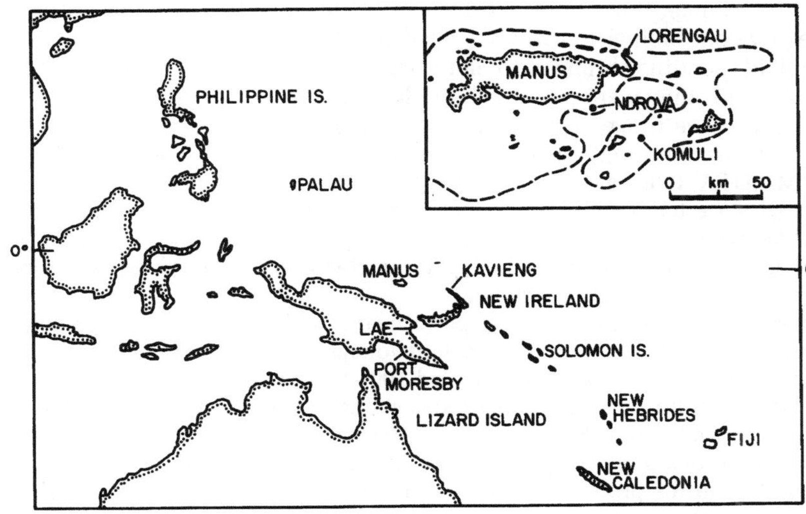

Figure 1. Sample sites and other localities mentioned in the text.

2. Materials and Methods

Specimens of the five species of *Nautilus* studied were collected from ten localities in the Indo-Pacific region (Fig. 1). The species, localities, and numbers of individuals examined electrophoretically are listed in Table I. Animals were collected in baited traps set at depths of 80–300 m, generally along forereef slopes. Whenever possible, samples of both muscle and tentacle tissue were collected

Table I. Populations of *Nautilus* Studied

Population name	Locality (collection year)	N^a
Nautilus belauensis	Ngemelis Island and Mutremdiu Point, Palau, West Caroline Islands (1982)	89
Nautilus macromphalus	Noumea, New Caledonia (1984)	25
Nautilus pompilius		
Lorengau (LO)	Northeast Manus (1984)	5
Ndrova (ND)	Southeast Manus (1984, 1985)	56
Komuli (KO)	Fedarb Is., southeast Manus (1984, 1985)	29
Kavieng (KV)	New Ireland (1984, 1985)	53
Lae (LA)	Huon Gulf, Papua New Guinea (1984, 1985)	32
Port Moresby (PM)	Papua New Guinea (1984, 1985)	49
Lizard Island (LZ)	Carter Reef, Queensland (1985)	6
Fiji (F)	Suva, Fiji (1985)	20
Nautilus scrobiculatus		
Ndrova (ND)	Southeast Manus (1984, 1985)	28
Komuli (KO)	Fedarb Is., southeast Manus (1984)	1
Nautilus stenomphalus	Carter Reef, Lizard Island, Queensland (1985)	7

[a] Number of individual specimens electrophoresed.

for each specimen. Muscle and three or more tentacle tips from an individual gave equivalent biological activity (Woodruff *et al.*, 1983). Tissues were frozen in the field, pending shipment to San Diego on dry ice; samples were then held at −70°C until they were processed for electrophoresis.

Tissues were prepared for electrophoresis by homogenizing 0.5–1.0 g (wet weight) of thawed sample in an equal volume of grinding solution (0.01 M Tris, 0.001 M EDTA, 0.05 mM NADP, pH 7.0). A glass rod was used to homogenize the samples mechanically. The homogenate was centrifuged for 20 min at 15,000g at 4°C. Wicks of Whatman No. 3 chromatography paper (9 mm × 5 mm) were dipped into the supernatant, blotted, and inserted into the gel. Thus, each sample was derived from a single specimen.

Horizontal electrophoresis was performed at 4°C using 12.5% starch gels (50 g Sigma starch in 400 ml buffer). Constant voltage was applied for 15–18 hr, until a bromophenol blue marker had traveled 100–120 mm anodal to the origin. Gels were then cut into horizontal slices and stained at 37°C, using the methods of Shaw and Prasad (1970), Selander *et al.* (1971), and Harris and Hopkinson (1976). The combinations of electrophoretic conditions, stains, and buffers used for resolution of *Nautilus* proteins are shown in Table II. Agar overlays (25 ml of a 2% Bactoagar solution) were used to apply stains for glyceraldehyde-3-phosphate dehydrogenase, α-glycerophosphate dehydrogenase, and phosphoglucose isomerase. The esterase substrate was α-naphthyl acetate.

Electromorph patterns resolved on each gel were interpreted genetically and were photographed at the time of optimal staining. Isozymes were numbered in order of decreasing anodal mobility in multilocus systems, and allozymes were identified alphabetically in the same order. Common abbreviations for each protein are capitalized; abbreviations for loci are italicized. Multilocus genotype data were recorded for each tissue and individual specimen and were analyzed with the BIOSYS-1 computer programs (Swofford and Selander, 1981). Because no tissue-specific differences in genotype were detected, muscle and tentacle data were pooled for each specimen. Allele frequencies at variable loci were examined for evidence of temporal or sexual differences in those populations in which such tests were appropriate. Intrapopulation, year-to-year, and male–female allele frequencies were tested for heterogeneity using the Pearson contingency χ^2 statistic. No significant differences attributable to sex or sampling year were found between subpopulations of sufficient sample size (Woodruff *et al.*, 1983; Woodruff and Carpenter, 1987). Consequently, males and females from the same locality were pooled for subsequent analyses discussed in this chapter, as were samples from the same locality collected in different years.

The proportion of polymorphic loci [(P) a locus was considered polymorphic if more than one allele was detected], the mean individual heterozygosity [($\overline{H}$) by direct count], and the mean number of alleles per locus were calculated. Genotype frequencies for each polymorphic locus were tested for their agreement with panmictic equilibrium expectations using both a χ^2 test, calculated using Levene's (1949) correction for small sample size, and a significance test using exact probabilities (analagous to Fisher's exact test). Multilocus unbiased genetic distance coefficients (D) of Nei (1978) and genetic similarities (S) of Rogers (1972) were calculated for all populations, and the latter were clustered according to Nei's D using the UPGMA algorithm (see Section 4.7). Standard errors on the estimates of D were calculated using the method of Nei *et al.* (1985).

Table II. Enzyme Loci and Electrophoretic Conditions Employed in the Study of Genetic Polymorphism in *Nautilus*

Protein (EC number)	Locus (number of alleles)		Buffer system[a]
Asparate aminotransferase (2.6.1.1)	*Aat-1*	(1)	C
	Aat-2	(4)	C
Coomassie blue protein	*Cbp-1*	(3)	E
	Cbp-2	(3)	E
	Cbp-3	(2)	B
Esterase (3.1.1.1)	*Est*	(3)	D
Glyceraldehyde-3-phosphate dehydrogenase (1.2.1.12)	*Gap*	(2)	A
α-Glycerophosphate dehydrogenase (1.1.1.8)	*Gpd*	(4)	A
Isocitrate dehydrogenase (1.1.1.42)	*Idh-1*	(4)	B
	Idh-2	(2)	B
Lactate dehydrogenase (1.1.1.27)	*Ldh*	(2)	D
Leucine aminopeptidase (3.4.11)	*Lap-1*	(2)	C
	Lap-2	(3)	C
Malate dehydrogenase (1.1.1.37)	*Mdh-1*	(4)	B
	Mdh-2	(1)	B
Malic enzyme (1.1.1.40)	*Me*	(4)	A
Phosphoglucomutase (2.7.5.1)	*Pgm*	(6)	C
6-Phosphogluconate dehydrogenase (1.1.1.44)	*Pgd*	(2)	B
Phosphoglucose isomerase (5.3.1.9)	*Pgi*	(3)	A
Superoxide dismutase (1.15.1.1)	*Sod-1*	(2)	B
	Sod-2	(4)	D

[a] (A) 0.378 M Tris, 0.165 M citrate, pH 6.0; diluted 1:29 for gels, undiluted for electrodes (70 V/40 mA) (Shaw and Prasad, 1970). (B) 0.1 M Tris, 0.1 M maleic acid, 0.01 M EDTA, 0.01 M $MgCl_2$, 0.13 M NaOH, pH 7.4; diluted 1:9 for gels, undiluted for electrodes (50 V/50 mA) (Shaw and Prasad, 1970). (C) 0.5 M Tris, 0.65 M borate, 0.02 M EDTA, pH 8.0; diluted 1:9 for gels, undiluted for electrodes (100 V/35 mA) (Selander *et al.*, 1971). (D) 0.087 M Tris, 0.009 M borate, 0.001 M EDTA, pH 9.0; diluted 1:3 for gels, undiluted for electrodes (200 V/25 mA) (Shaw and Prasad, 1970). (E) TBE pH 9.0 (Buffer D), diluted 1:3 for gels; TBE pH 8.0 (Buffer C), undiluted for electrodes (100 V/20 mA) (Hoagland, 1984).

3. Results

Consistent and genetically interpretable results were obtained for 21 presumptive gene loci from the examination of 14 enzyme and general protein systems (Table III). The results will be described in detail elsewhere (Woodruff and Carpenter, 1987). Of these 21 loci, 2 (*Aat-1*, *Mdh-2*) were monomorphic in all populations examined; 5 loci were fixed for a single allele *within* each population, but exhibited definite interpopulation variability (*Cbp-2*, *Gap*, *Me*, *Sod-1*, *Sod-2*); the remaining 14 loci were polymorphic in six to ten populations (variability may have been either intra- or interspecific). Of the 19 variable loci, 6 had two alleles, 6 had three alleles, 6 had four alleles, and 1 (*Pgm*) had six alleles. Allele frequencies, numbers of individuals examined, and values for P and $\overline{H}$ are given in Table III for all polymorphic loci and are discussed below.

Presumptive genotype frequencies at each variable locus in each sample were tested for departure from predictions based on a model of random mating (*panmixia*). Simple χ^2 tests revealed significant departures from expectations in 18 of 109 cases at the $p = 0.05$ level. However, because the χ^2 test is inappropriate in

Table III. Allele Frequencies for 19 Polymorphic Loci in Five Species of *Nautilus* with Summary Statistics of Genetic Variability

	N. pompilius											
Locus Allele	N.E. Manus (Lo)[a]	S.E. Manus (Nd)	S.E. Manus (Ko)	New Ireland (Kv)	N. New Guinea (La)	S. New Guinea (PM)	Fiji (F)	Lizard Island (Lz)	*N. stenomphalus* Lizard Island (Lz)	*N. macromphalus* New Caledonia	*N. belauensis* Palau	*N. scrobiculatus* S.E. Manus
Aat-2												
(N)	5	30	23	42	32	29	18	6	7	4	24	27
a	—	—	—	0.03	—	—	0.64	0.67	0.64	—	0.79	—
b	0.60	0.50	0.54	0.46	0.44	0.47	—	—	—	—	0.21	0.98
c	0.40	0.48	0.46	0.50	0.56	0.53	0.36	0.33	0.36	1.00	—	0.02
d	—	0.02	—	0.01	—	—	—	—	—	—	—	—
Cbp-1												
(N)	5	32	18	45	32	33	20	6	7	7	25	28
a	—	—	—	—	—	0.02	—	—	—	—	—	1.00
b	0.70	0.59	0.58	0.50	0.55	0.51	0.95	0.33	0.79	0.21	0.60	—
c	0.30	0.41	0.42	0.50	0.45	0.47	0.05	0.67	0.21	0.79	0.40	—
Cbp-2												
(N)	5	45	25	49	32	45	20	6	7	23	63	27
a	1.00	1.00	1.00	1.00	1.00	1.00	1.00	1.00	1.00	—	1.00	—
b	—	—	—	—	—	—	—	—	—	—	—	1.00
c	—	—	—	—	—	—	—	—	—	1.00	—	—
Cbp-3												
(N)	5	34	28	45	32	45	20	6	7	23	17	27
a	1.00	0.88	0.86	0.73	1.00	1.00	0.60	0.67	0.71	1.00	0.71	0.48
b	—	0.12	0.14	0.27	—	—	0.40	0.33	0.29	—	0.29	0.52
Est												
(N)	5	48	29	49	32	46	20	6	7	25	65	29
a	—	—	—	—	—	—	—	—	—	—	0.43	—
b	0.90	0.85	0.64	0.90	0.83	0.84	1.00	1.00	1.00	0.62	0.57	0.71
c	0.10	0.15	0.36	0.10	0.17	0.16	—	—	—	0.38	—	0.29
Gap												
(N)	2	39	28	49	21	40	20	6	7	24	43	29
a	1.00	1.00	1.00	1.00	1.00	1.00	1.00	1.00	1.00	1.00	1.00	—
b	—	—	—	—	—	—	—	—	—	—	—	1.00
Gpd												
(N)	5	28	19	45	32	16	20	6	7	16	28	29
a	—	—	—	—	—	—	—	—	—	—	—	1.00
b	1.00	0.89	1.00	0.89	0.83	0.81	0.58	0.75	0.21	—	1.00	—

c	—	0.11	—	0.11	—	—	—	—	—	1.00	—	—
d	—	—	—	—	0.17	0.19	0.42	0.25	0.79	—	—	—
Idh-1												
(N)	5	26	15	37	32	28	20	6	7	3	35	28
a	—	—	—	—	—	—	—	0.50	—	—	—	—
b	—	0.46	0.47	0.07	0.42	0.14	—	0.08	0.07	1.00	—	—
c	1.00	0.54	0.53	0.93	0.58	0.86	0.55	0.42	0.93	—	1.00	1.00
d	—	—	—	—	—	—	0.45	—	—	—	—	—
Idh-2												
(N)	5	32	20	45	32	31	20	6	7	16	60	29
a	1.00	0.98	0.88	0.98	1.00	1.00	1.00	1.00	0.93	0.53	1.00	0.95
b	—	0.02	0.12	0.02	—	—	—	—	0.07	0.47	—	0.05
Ldh												
(N)	5	20	20	30	21	14	1	3	1	1	12	23
a	0.30	0.20	0.52	0.38	0.12	0.11	—	0.50	0.50	—	0.54	0.09
b	0.70	0.80	0.48	0.62	0.88	0.89	—	0.50	0.50	—	0.46	0.91
c	—	—	—	—	—	—	1.00	—	—	1.00	—	—
Lap-1												
(N)	5	54	28	49	32	48	20	6	7	25	39	29
a	—	0.07	0.09	0.12	0.02	0.01	0.70	0.42	0.36	0.48	0.67	0.14
b	1.00	0.93	0.91	0.88	0.98	0.99	0.30	0.58	0.64	0.52	0.33	0.86
Lap-2												
(N)	5	30	24	49	32	31	20	6	7	19	24	28
a	0.80	0.88	0.71	0.82	0.73	0.79	0.87	0.83	0.50	0.92	0.31	0.98
b	0.20	0.12	0.29	0.18	0.27	0.21	0.10	—	—	0.08	0.69	—
c	—	—	—	—	—	—	0.03	0.17	0.50	—	—	0.02
Mdh-1												
(N)	5	28	11	45	22	29	20	6	7	9	29	25
a	—	—	—	—	—	—	—	—	—	—	0.66	0.86
b	0.80	0.68	0.64	0.98	0.86	0.78	1.00	0.92	1.00	0.56	—	0.14
c	0.20	0.32	0.36	0.02	0.14	0.22	—	0.08	—	0.44	—	—
d	—	—	—	—	—	—	—	—	—	—	0.34	—
Me												
(N)	5	32	27	46	32	44	20	6	7	22	28	27
a	—	—	—	—	—	—	—	—	—	—	—	1.00
b	1.00	1.00	1.00	1.00	1.00	1.00	—	1.00	1.00	—	1.00	—
c	—	—	—	—	—	—	—	—	—	1.00	—	—
d	—	—	—	—	—	—	1.00	—	—	—	—	—

(continued)

Table III. (*Continued*)

Locus Allele	*N. pompilius* N.E. Manus (Lo)[a]	S.E. Manus (Nd)	S.E. Manus (Ko)	New Ireland (Kv)	N. New Guinea (La)	S. New Guinea (PM)	Fiji (F)	Lizard Island (Lz)	*N. stenomphalus* Lizard Island (Lz)	*N. macromphalus* New Caledonia	*N. belauensis* Palau	*N. scrobiculatus* S.E. Manus
Pgm												
(N)	5	32	23	49	32	37	20	6	7	17	30	28
a	—	—	—	—	—	—	—	—	—	—	—	0.68
b	—	—	—	—	—	—	—	—	—	—	0.47	—
c	—	—	—	—	—	—	—	—	—	1.00	—	—
d	—	—	—	—	—	—	—	—	0.79	—	—	0.32
e	—	0.31	0.37	0.05	0.27	0.13	—	—	—	—	—	—
f	1.00	0.69	0.63	0.95	0.73	0.87	1.00	1.00	0.21	—	0.53	—
Pgd												
(N)	2	24	20	44	21	14	20	6	7	13	24	24
a	—	0.02	—	—	—	—	0.03	—	—	—	—	0.04
b	1.00	0.98	1.00	1.00	1.00	1.00	0.97	1.00	1.00	1.00	1.00	0.96
Pgi												
(N)	5	35	22	45	32	37	20	6	7	14	82	28
a	0.70	0.47	0.46	0.37	0.41	0.55	0.25	0.25	0.29	0.61	—	—
b	0.30	0.53	0.54	0.63	0.59	0.45	0.75	0.75	0.71	0.39	1.00	—
c	—	—	—	—	—	—	—	—	—	—	—	1.00
Sod-1												
(N)	5	33	27	49	27	36	20	6	7	20	89	29
a	1.00	1.00	1.00	1.00	1.00	1.00	1.00	1.00	1.00	1.00	1.00	—
b	—	—	—	—	—	—	—	—	—	—	—	1.00
Sod-2												
(N)	5	35	19	37	24	25	16	3	2	24	25	28
a	—	—	—	—	—	—	—	—	—	—	—	1.00
b	—	—	—	—	—	—	—	—	—	—	1.00	—
c	1.00	1.00	1.00	1.00	1.00	1.00	1.00	1.00	1.00	1.00	—	—
d	—	—	—	—	—	—	—	—	—	—	—	—
$\bar{N}$	4.7	34.3	23.0	45.0	29.3	33.8	18.8	5.7	6.5	16.0	40.4	27.5
+S.E.	0.2	1.9	1.1	1.1	1.0	2.3	0.9	0.2	0.4	1.8	5.3	0.4
$\bar{A}$	1.4	1.7	1.6	1.7	1.5	1.6	1.5	1.5	1.6	1.3	1.4	1.5
P	38.1	66.7	57.1	61.9	52.4	52.4	42.9	47.6	52.4	33.3	42.9	47.6
$\bar{H}$	0.11	0.14	0.15	0.12	0.16	0.12	0.10	0.18	0.17	0.12	0.09	0.06

($\bar{N}$) mean sample size (S.E. = standard error)/locus; ($\bar{A}$) mean number of alleles/locus; (P) percentage of loci polymorphic; ($\bar{H}$) mean heterozygosity/locus.
[a] See Table I for population names.

Table IV. Matrix of Genetic Similarity and Distance Coefficients[a]

Population[b]	1	2	3	4	5	6	7	8	9	10	11	12
Nautilus pompilius												
(1) N. Manus (Lo)	—	0.921	0.904	0.911	0.917	0.936	0.714	0.809	0.795	0.533	0.748	0.465
(2) S.E. Manus (Nd)	0.009	—	0.945	0.919	0.953	0.938	0.734	0.834	0.785	0.572	0.734	0.474
(3) S.E. Manus (Ko)	0.014	0.007	—	0.896	0.919	0.897	0.704	0.824	0.787	0.578	0.761	0.450
(4) New Ireland (Kv)	0.014	0.020	0.026	—	0.919	0.931	0.774	0.882	0.838	0.528	0.766	0.479
(5) N. New Guinea (La)	0.012	0.002	0.015	0.018	—	0.962	0.733	0.835	0.785	0.558	0.724	0.465
(6) S. New Guinea (Pm)	0.002	0.007	0.023	0.012	0.004	—	0.732	0.831	0.784	0.557	0.728	0.480
(7) Fiji (F)	0.219	0.210	0.227	0.178	0.213	0.213	—	0.812	0.765	0.553	0.639	0.396
(8) Lizard Is. (Lz)	0.076	0.058	0.066	0.035	0.060	0.063	0.147	—	0.860	0.539	0.727	0.428
(9) *N. stenomphalus*	0.113	0.123	0.131	0.101	0.121	0.122	0.203	0.072	—	0.501	0.710	0.448
(10) *N. macromphalus*	0.590	0.513	0.524	0.584	0.522	0.550	0.514	0.568	0.628	—	0.460	0.341
(11) *N. belauensis*	0.199	0.207	0.186	0.183	0.212	0.210	0.363	0.223	0.272	0.746	—	0.426
(12) *N. scrobiculatus*	0.717	0.728	0.786	0.721	0.727	0.700	0.911	0.805	0.755	1.112	0.834	—

[a] Entries above the diagonal; genetic similarity (*s*) of Rogers (1972); entries below the diagonal: unbiased genetic distance (*D*) of Nei (1978).
[b] See Table I for population names.

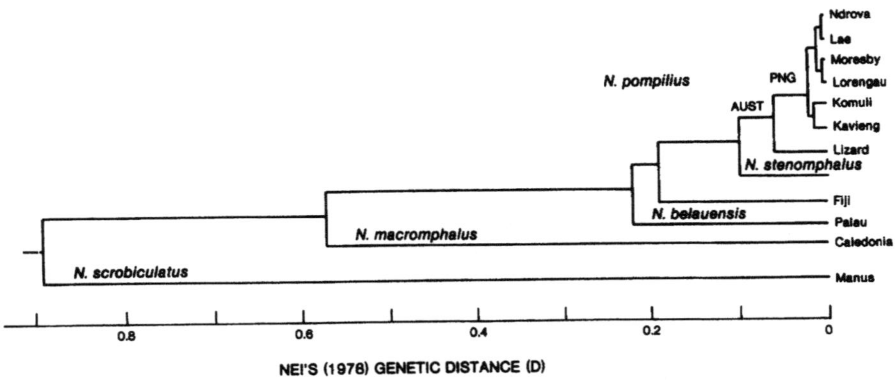

Figure 2. Phenetic tree based on 21 genetic loci. Note the very close similarity of Papua New Guinea (PNG) samples of *N. pompilius*. The Australian (AUST) sample of *N. pompilius* is intermediate between the PNG samples and the sample of *N. stenomphalus* (Lizard Island). The Fiji sample is moderately differentiated and taxonomically problematic (see the text). The cophenetic correlation is 0.98.

cases in which cell-size expectations are less than 1, the more robust Fisher-exact test was used; it revealed only seven cases in which allele frequencies were apparently not in Hardy–Weinberg equilibrium (Woodruff and Carpenter, 1987). These cases were scattered among loci in six samples and, accordingly, would be expected by chance alone in multilocus surveys of this type. We conclude, therefore, that no significant deviation from panmixia was detected in any of the samples or species of *Nautilus* studied.

This finding of panmixia allowed us to employ several standard measures of population structure and interpopulation differentiation. Nei's genetic distances and Rogers's genetic similarities are reported in Table IV and Fig. 2 and are discussed in Section 4.

4. Discussion

4.1. Adequacy of the Results

Since the detailed results of our electrophoretic analyses will be presented elsewhere (Woodruff and Carpenter, 1987), we feel obliged to begin this discussion by commenting on the soundness (or inadequacies, or both) of the data base. As with most studies of this type, the data in Tables I and III reveal considerable variation in sample size per locus and per locality. For those animals represented by tentacle tissue alone, interlocus variation in sample size arose from technical difficulties in obtaining large enough tissue samples (specimens were handled in small boats on the open sea); for other animals, the variation was caused by differential rates of enzyme inactivation associated with occasional failure of refrigeration systems in the field. Intersample variation, on the other hand, reflects local trapping success or animal abundance; it is appropriate to consider how this variation affects the interpretation of our results.

Perhaps the worst case—*N. scrobiculatus* from Komuli, southeast Manus ($N = 1$, 21 loci)—is quite inadequate for estimating P and $\overline{H}$. Empirical and theoretical studies have shown how samples of more than about ten individuals will generally give estimates of these parameters within 1% of their "real" values as determinable by this technique (Gorman and Renzi, 1979; Nei and Roychoudhury, 1974). On the other hand, estimates of genetic distance (*D*) are relatively independent of sample size; a single specimen may be used to represent a population or a species (Gorman and Renzi, 1979) (see also the discussion in Avise, 1983). Accordingly, the small Komuli *N. scrobiculatus* sample is adequate for some purposes, but not for others. Similar arguments can be made for the other small samples (e.g., *N. stenomphalus* and *N. pompilius* from Lizard Island, Great Barrier Reef, and *N. pompilius* from Lorengau, north Manus). With these exceptions, the remaining sample sizes are statistically adequate for all the analyses presented.

Estimates of genetic variability are, in fact, more strongly influenced by the specific loci chosen for electrophoretic analysis than by sample size. Some loci are typically monomorphic in all specimens studied; others (variable substrate esterases and phosphatases) are notoriously polymorphic. The characterization of *Nautilus* discussed herein is based on 20–21 loci (mostly allozymes) representing a range of proteins, including variable substrate and regulatory and nonregulatory enzymes. For theoretical reasons, it has become conventional to include the results of such balanced multilocus surveys in review articles when more than 14 loci are considered (Avise and Aquadro, 1982; Thorpe, 1982, 1983; Nevo *et al.*, 1984). Our study exceeds this level of acceptability.

These two points notwithstanding, it must be noted that our electrophoretic data underestimate true levels of genetic variability and interpopulation genetic differentiation. Conventional single gel electrophoresis, as used in our study, fails to resolve about 20% of the allozymes actually present in a sample (Ayala, 1982; Selander and Whittam, 1983). The effects of this bias will be considered below.

4.2. Genetic Variation in *Nautilus pompilius*

At present, this species is known to range from Japan or the Philippines in the north to Australia in the south, and from the Andaman Islands in the west to Fiji in the east. Our sampling (Table I) focuses on the central part of this wide range, and our genetic characterization of this species (Table III) is, accordingly, preliminary.

Excluding Fiji and the two localities in which the mean sample size per locus was less than ten, we find each of the five Papua New Guinea samples highly variable: $\bar{A} = 1.5$–1.7, $P = 0.52$–0.67, and $\overline{H} = 0.12$–0.16. For comparison, a survey of variability in 361 other species of invertebrates (which are characteristically more variable than vertebrates and plants) revealed mean values of $P = 0.375$ and $\overline{H} = 0.10$ (Nevo *et al.*, 1984).

Such high levels of genetic variation within each population could be sustained only if *N. pompilius* is outbreeding panmictically. Our analysis of genotype frequencies confirms this as the most likely breeding system in this dioecious species (Woodruff and Carpenter, 1987). Such variable, outbreeding species are termed *amphimictic*.

The six Papua New Guinea samples of *N. pompilius* are genetically indistinguishable. The intersample *D* values (Table IV and Fig. 2) are all less than 0.02; these values are exceeded by the standard errors on such estimates. Such very low values of *D* are typical of conspecific populations. A survey of over 7000 comparisons of conspecific populations of plants and animals revealed that only 2% of the intraspecific *D* estimates exceeded 0.10 (Thorpe, 1983). The vast majority of the more than 100 mollusk species studied to date have intraspecific genetic distances of less than 0.10 (Woodruff *et al.*, 1987; Woodruff, 1987).

From New Ireland and Manus Island in the north to Port Moresby in the south (over 1500 km as measured along potential routes of migration), there is only minor genetic differentiation. Given the high vagility and longevity of *Nautilus* (Saunders and Spinosa, 1979; Saunders, 1983; Ward, 1983a) and the retarding effects of gene flow on local differentiation, this result is not surprising. Similarly, the discovery of a slightly greater degree of genetic differentiation ($D = 0.06$) between the Papua New Guinea *N. pompilius* and an Australian sample, from 800 km farther south (Lizard Island), is to be expected for such a widespread and variable species. In fact, within this 2300-km transect from the northern Bismarck Sea to Lizard Island, genetic distances are positively correlated with geographic distances. This result is predicted by Sewall Wright's stepping-stone model of evolutionary differentiation of traits that are relatively neutral with respect to prevailing selection.

In contrast to the genetic homogeneity of the Australia–Papua New Guinea samples of *N. pompilius*, the putative *N. pompilius* sample from Fiji was relatively well differentiated. Alleles unique to Fiji, and occurring at frequencies greater than 0.40, were found at three loci: *Idh-1*, *Ldh*, and *Me*. The multilocus genetic distance between Fiji and the other eight samples was $D = 0.20$. This value corresponds to a degree of differentiation higher than that seen in most comparisons of conspecific populations. For example, Woodruff (1987) found that published interspecific distances among congeneric, sexually reproducing mollusks ranged from 0.05 to more than 1.0 and generally fell in the range 0.20–0.60. Thorpe (1983) surveyed 900 comparisons of interspecific *D* values between congeners representing various groups of plants and animals and reported the mean genetic distance to be about 0.40 (range 0.03 to more than 1.0). Similarly, the study by Avise and Aquadro (1982) of genetic differentiation in vertebrates would suggest that the Fiji sample of *N. pompilius* is distinct enough from the other populations to be afforded the rank of a separate species.

Such a conclusion would be premature. There is, of course, no simple basis for translating a genetic distance value into a taxonomic decision. The processes of speciation are not closely coupled to changes in structural genes, and speciation may occur with minimal genomic reorganization. Although there are few cases in which intraspecific distances exceed 0.10, there are three exceptional cases involving mollusks in which intraspecific distances exceed 0.20 (reported by Chambers, 1980; Nevo *et al.*, 1983; M. S. Johnson *et al.*, 1984). Clearly, the Fiji situation falls into a gray area, in which clear taxonomic interpretation of the genetic distance is at present impossible.

Resolution of the taxonomic status of the Fiji population requires two things. First, we need comparable measures of genetic differentiation for geographically intermediate populations to fill in the 3000-km gap in our record, i.e., samples from such locations as the Solomon Islands and the New Hebrides. Such data may show that the Fiji populations are in full genetic contact with those of the Papua New Guinea region and, as such, represent only one extreme geographic form of the widespread species *N. pompilius*. The slightly reduced level of polymorphism seen in the Fiji sample ($P = 0.43$) may reflect simply its outlying, peripheral location.

Second, other aspects of the Fiji *Nautilus* must be examined, including anatomy, morphology, developmental biology, behavior, and ecology. Sound taxonomic decisions should be based on full consideration of the various types of evidence available. This requirement is especially true of decisions involving allopatric populations in morphologically conservative groups like *Nautilus*, in which sibling species might be expected.

Finally, it should be noted that we have sampled only the central part of the geographic range of *N. pompilius*. Our characterization of this amphimictic species will be incomplete until we can include representatives of the populations known to occur to the northwest, in Indonesia, the Andaman Islands, the Philippines, and possibly as far away as southern Japan and East Africa.

4.3. Genetic Variation in *Nautilus macromphalus*

Nautilus macromphalus is known only from the region of New Caledonia and the Loyalty Islands. It is isolated by a distance of more than 1200 km from the *Nautilus* populations of the Australian Great Barrier Reef and by about 500 km from those of the New Hebrides. Unfortunately, we were unable to resolve genotypes for all specimens collected, since 21 of the 25 samples were tentacle rather than muscle tissue. Nevertheless, it is clear that *N. macromphalus*, like *N. pompilius*, is an amphimictic species. The proportion of polymorphic loci (P) is 0.33, and the mean heterozygosity per individual ($\overline{H}$) is 0.12 (Table III). The former value may not be significantly lower than that found in *N. pompilius* ($P = 0.52$–0.67), because larger sample sizes for *Aat-2*, *Idh-1*, and *Ldh* may show that these loci are also polymorphic in *N. macromphalus*. The various genotypes were segregating in accordance with Hardy–Weinberg expectations for a panmictic population in the five cases in which sample sizes were adequate for testing (Woodruff and Carpenter, 1987).

4.4. Genetic Variation in *Nautilus belauensis*

Woodruff *et al.* (1983) reported on variation in 89 specimens of this species, known only from Palau, located about 800 km north of Papua New Guinea. On the basis of subsequent experience with over 300 additional *Nautilus*, we have reinterpreted the original electromorphic patterns and now conclude that *N. belauensis* is more variable than initially reported (Table III) (Woodruff and Carpenter, 1987). Genetically interpretable variation is now discerned at *Est*, *Ldh*, *Lap-2*, and *Mdh-1*, and at a 21st locus, *Cbp-3*. *Nautilus belauensis* has an average of 1.4 alleles per locus; the P is 0.43; the $\overline{H}$ is 0.09. Thus, the species is only slightly less variable than N. *pompilius*. Our original conclusion, that the animals studied were drawn from a single, large panmictic population, is confirmed.

4.5. Genetic Variation in *Nautilus scrobiculatus*

Living specimens of this species were obtained for the first time in 1984, from Manus, in the Admiralty Islands. We obtained 28 specimens from near Ndrova Island and 1 specimen from near the Fedarb Islands (Komuli), about 25 km to the southeast. Because the latter specimens possessed no unique alleles, we combined the two samples for the purposes of this discussion.

Nautilus scrobiculatus, like the preceding species, is amphimictic: The mean number of alleles per locus is 1.5, the P is 0.48, and the $\overline{H}$ is 0.06. Only the last-mentioned value is significantly different from that seen in other *Nautilus*, in which $\overline{H}$ was generally 0.10–0.16. The reason for the slightly reduced individual heterozygosity in N. *scrobiculatus* is unclear; it does not appear attributable to sampling error, however, because genotypic frequencies are in close agreement with expectations for a panmictic population.

4.6. Genetic Variation in *Nautilus stenomphalus*

We discuss this species last because it is the least satisfactorily sampled to date. The seven animals obtained from near Lizard Island, Great Barrier Reef, show clear signs of hybridization with N. *pompilius* and therefore constitute a poor basis for the genetic characterization of the species. Nevertheless, there is nothing in our preliminary observations to suggest that N. *stenomphalus* is not similar to the other species in its genetic variability (Table III).

4.7. Interspecific Genetic Differentiation

We estimated the relative degree of interpopulation genetic differentiation using the measure of unbiased genetic distance (*D*) of Nei (1978). This is a measure of the number of codon substitutions per locus; it can vary from zero (when two samples are identical) to infinity. Within a given group of related organisms, there is a correlation between taxonomic status and genetic distance. Surveys of thousands of reported cases show that conspecific populations typically have *D* values of less than 0.10 (Thorpe, 1983). It is with these values in mind that we discuss the preliminary results for *Nautilus*.

The intersample *D* values (Table IV) were used to generate a phenetic tree (Fig. 2) by the unweighted pair-group clustering method (UPGMA) (Nei, 1975). In addition, we calculated Rogers's similarity coefficients for the same data set (Table IV); Rogers's metric distance statistic can be used to generate distance-Wagner trees. In this chapter, we will describe only the UPGMA analysis, because Nei's *D*, unlike other electrophoretic distance statistics, is clearly related to the biologically significant mutation rates, and at low to intermediate values (<1.0), it may be linearly related to time (Nei, 1981; Avise, 1983).

The overall genetic similarity of the seven samples of *N. pompilius* from Papua New Guinea and from Queensland has been discussed above. The data indicate that these populations are in full genetic contact with one another. Also noted above was the unexpectedly large *D* value (0.20) that characterizes *N. pompilius* from Fiji. Although these animals are possessed of a sufficient degree of difference, in many groups, to be regarded as a separate species, their taxonomic status is unclear at present. The Fiji population is either a well-marked geographic variety of the widespread *N. pompilius* or a closely related sibling species.

The phenetic diagram (Fig. 2) shows *N. belauensis* to be related to, but well differentiated from, *N. pompilius*. Like the Fiji population (*N. pompilius*) described above, *N. belauensis* has a limited distribution at the edge of the range of *N. pompilius*. It differs from the New Guinea populations of the latter species at over 20% of the loci examined and possesses unique alleles at *Est*, *Mdh-1*, *Pgm*, and *Sod-2*. This degree of genetic differentiation is almost unknown among conspecific populations of sexually reproducing species of animals (Thorpe, 1983), and even allowing for the standard error around our distance estimate (0.23 ± 0.11), these genetic data support our earlier decision (Woodruff *et al.*, 1983) to recognize the Palauan *Nautilus* as a separate species.

Nautilus macromphalus and *N. scrobiculatus* are well differentiated from the other species (Table IV and Fig. 2). Although still inadequately characterized, *N. macromphalus* apparently has fixed differences at three loci, and *N. scrobiculatus* has similar unique and fixed differences at seven loci. The two species are not closely related to one another or to the taxa of the *pompilius* group discussed above. The average distance values shown in Fig. 2 have the following standard errors: $D = 0.57 \pm 0.19$ for *N. macromphalus* and $D = 0.90 \pm 0.26$ for *N. scrobiculatus*. Such *D* values are typical for congeneric animal species.

Due to the limited sample size, *N. stenomphalus* can be only poorly characterized. Nevertheless, it appears to be genetically allied to *N. pompilius*, and it appears that the two are similar enough to be producing viable hybrids. Its taxonomic and evolutionary status cannot be assessed until more specimens are examined from elsewhere around the coast of Australia. The fact that the two taxa are hybridizing in one area does not, of course, preclude their both having reached the status of full taxonomic species. It is now known that species can hybridize without losing their integrity (Woodruff, 1979, 1981), since there are several well-documented cases of natural hybridization occurring between mollusk species that have differentiated only to the level of $D = 0.05$ (S. J. Gould and Woodruff, 1978, 1986).

This interpretation of the taxonomic status and affinities of the *Nautilus* populations examined genetically is remarkably concordant with the morphology-based taxonomic relationships. Nevertheless, it must be pointed out that Fig. 2 does not constitute a definitive phylogenetic tree. Although additional sampling from throughout the range of *Nautilus* may confirm the overall pattern revealed here, a few surprises are perhaps to be expected. We must now establish: (1) how the Australian taxa referred to at present as *N. stenomphalus* and *N. pompilius* are related to one another; (2) whether the questionable form, *N. repertus* Iredale (named only from drifted shells), is a separate species (see Chapter 3); and (3) how Philippine *N. pompilius* differ from those collected 3000 km to the south, around the Bismarck Sea. The basic morphological conservatism of *Nautilus* makes it difficult to make predictions about the affinities of these populations based, for example, on the direction of present ocean currents. The existence of a few additional well-differentiated local populations, or even species, should be expected.

It is also likely that we will be able to gauge the full impact of gene flow on the evolution of *Nautilus* species as more genetic data become available. Long-range dispersal by drifting has undoubtedly led to opportunities for hybridization between otherwise allopatric populations. In this regard, Table III contains some tantalizing observations; e.g., a single copy of Fiji-type *Aat-2* appears in the New Ireland sample, and New Caledonian *Gpd* occurs at low frequency in two samples of *N. pompilius* from 2500 km to the northwest. Such long-distance gene flow would have a retarding effect on local genetic differentiation. Furthermore, if significant gene flow can be demonstrated, the resulting patterns of reticulate evolution will be difficult to analyze using cladistic methods now popular with some workers.

4.8. Phylogenetic Relationships of Living Species

Phenograms based on Nei's genetic distances can be interpreted phylogenetically, if one makes assumptions about the selective neutrality of enzyme variants and the regularity of a molecular clock based on rates of evolutionary divergence in protein structure. The sage opinion of Cain (1983) notwithstanding, numerous theoretical and empirical studies point to the relative near-neutrality of allozymes (Wright, 1978; Milkman, 1982; Hedrick, 1983). Similarly (Thorpe, 1982, p. 140):

> . . . the case for there being a molecular clock—at least to the extent where following the genetic isolation of populations the molecular structure of homologous protein molecules diverges, linearly or otherwise—would appear to be overwhelming.

The notion of an approximately regular but stochastic rate of amino acid substitutions within a clade is now widely accepted. Accordingly, if Nei's genetic distance is linearly related to time over the range of distance values observed in *Nautilus*, we can make a first-order inquiry into the group's evolutionary history.

Nei (1975) estimated that a *D* value of 1.0 indicated a divergence time of about 5 million years. The data in Table IV and Fig. 2 therefore suggest an age of about 4.5 million years for *N. scrobiculatus* and about 3 million years for *N. macromphalus*. *Nautilus belauensis* and the Fiji population (*N. pompilius*) have apparently been diverging from *N. pompilius* for about 1 million years, and *N. stenomphalus* is perhaps only half that age.

These estimates are speculative, however, because calibration of the multilocus genetic distance clock is controversial. Some workers have found Nei's calibration too low and have followed Sarich (1977) in estimating that a *D* value of 1.0 is equivalent to 20 million years. Such a calibration has given geologically satisfactory results when applied to some groups of fish and amphibians, but it would necessitate a fourfold increase in the estimated ages of the *Nautilus* species. Still other workers have used a $D = 1.0$ value as representing only 1 million years divergence. This 20-fold variation in clock calibration (reviewed by Avise and

Aquadro, 1982) raises serious doubts as to the value of these molecular phylogenies. However, recent reports on the rates of silent (neutral) substitutions in pseudogenes do not support the slower calibrations. Britten (1986) found only a fivefold variation in these "free-floating" mutation rates at so-called silent sites. Because such rates are the highest for base substitutions of nuclear DNA, it now seems likely that some workers have seriously overestimated divergence times (for an excellent discussion of evolutionary rates, see Schopf, 1984).

Until these issues are resolved, we will use Nei's original, conservative calibration and recognize that our age estimates are speculative. Nevertheless, it would appear that the living species of *Nautilus* are relatively recent in their origins. Regardless of clock calibration, all taxa are apparently of post-Miocene origin. The technique tells us nothing about the age of the genus itself or about the number of species that may have existed in the past.

Unfortunately, the allozyme data cannot be used directly in qualitative analyses of molecular evolution, due to difficulty in distinguishing ancestral from derived electromorphs. Nevertheless, additional samples may elucidate the history of some of the more isolated populations of *Nautilus*. For example, it is tempting to speculate that several species arose by peripatric speciation, following colonization of Palau, New Caledonia, and Fiji by *N. pompilius*-like ancestors. In relative isolation, speciation by genetic transilience or founder–flush modes could occur rapidly and with minimal changes in the allozymes (Carson and Templeton, 1984). Evidence for such phylogenetic reconstructions may come from a search for patterns of covariation of suites of genetic and morphological characters, as it has in other taxonomically difficult groups of mollusks such as *Cerion* (S. J. Gould and Woodruff, 1986; Woodruff and Gould, 1987).

5. Question: Is *Nautilus* a Living Fossil?

It is not surprising that *Nautilus* is sometimes referred to as a living fossil in both scientific and popular literature. The genus is derived from a lineage extending back several hundred million years and has conchologically similar, 75-million-year-old relatives. Living fossils are generally perceived as species that have persisted unchanged for eons and have somehow ceased to evolve at appreciable or "normal" rates. *Limulus, Lingula, Latimeria, Neopilina*, and *Sphenodon* are among the extant genera commonly referred to as living fossils. Eldredge and Stanley (1984) included case profiles of more than 30 other examples of this phenomenon. Ward (1984), in asking whether *Nautilus* should be considered a living fossil, concluded that the available data were insufficient to frame a definitive answer. Now, with the genetic data at hand, the problem is worth reassessing. It is now possible to test the suggestion of Schopf (1982) that in *Nautilus* and other living fossils, the putative duration and extent of "arrested evolution" may have been exaggerated.

Schopf (1984, pp. 272, 276) reviewed the literature on living fossils and found that the term had been used to describe seven different situations:

1. A living species that has persisted over a very long interval of geologic time.

2. A living species that is morphologically and physiologically quite similar to a fossil species, as seen over long intervals of geologic time.
3. A living species that has a preponderance of primitive morphologic traits.
4. A living species that has one of the above, *and* a relict distribution.
5. A living species that was once thought to be extinct.
6. An extant clade of low taxonomic diversity whose species have one or more of the properties of 1, 2, and 3.
7. A relatively little morphologically modified representative of a relatively archaic lineage with little modern representation.

The last situation differs from the others in that it pertains only to morphology and not to species or speciation.

The genetic evidence reported in this chapter indicates that the species of living *Nautilus* are not ancient; the oldest may be about 5 million years old, and the youngest are less than 1 million years old. Whether this interval (1–5 million years) constitutes a "very long interval of geologic time" [*sensu* (1) above] is obviously debatable. In addition, *Nautilus* does not qualify as a living fossil in the sense of definition (2), (3), (4), (5), or (6). It does satisfy, however, the terms of definition (7); living *Nautilus* species are morphologically conservative, retaining strong resemblances to their ancient forebears. Only in this morphological sense can *Nautilus* unequivocally be called a living fossil.

There is, in fact, little evidence to support the idea that "living fossil" species have persisted for eons and have somehow ceased to evolve at normal rates (Schopf, 1982, 1984). The notion of bradytelic species [extremely slowly evolving species (Simpson, 1953)] is without scientific merit. As Schopf (1984) so clearly demonstrated, "living fossils" do not constitute a problem for speciation or macroevolutionary theory; rather, they pose a challenge to developmental biology. We share the view of Schopf (1984, p. 285):

> Whatever value there may be in the concept of living fossils resides in considering why specific traits may not change over long intervals of geologic time.

Their apparent morphological stasis is suggestive of underlying developmental genetic constraints. Such constraints are not due, however, to a lack of inherent genetic variability. As our study has shown, *Nautilus* species have high levels of polymorphism and individual heterozygosity and are actually more variable than *Homo sapiens*. Thus, *Nautilus* and other living fossils that have been studied [e.g., *Limulus* (Selander *et al.*, 1970) and *Lingula* (Hammond and Poiner 1984)] do not support the hypothesis that phylogenetic relics exhibit very low levels of genetic variation. On the contrary, the genetic data suggest that *Nautilus* may be actively speciating and the genus may be undergoing a minor radiation. If the apparent evolutionary stasis is simply an artifact of low morphological complexity and conservatism, it is perhaps best not to apply the term living fossil at all. In this regard, it is relevant that new analyses of large numbers of morphological characters, using cluster analysis and principal components analysis (Chapter 6), reveal considerable interpopulation variation in shell morphology in *N. pompilius*, suggesting that past interpretations of low morphological diversity were in error.

The actively evolving species of living *Nautilus* have much more to tell about the evolutionary processes that made their ancestors so successful. We should not continue to regard them as terminal species in a final decline. There is life in this old line yet.

ACKNOWLEDGMENTS. The assistance of Larry Davis, Washington State University, and the late Michael Weekley, Waikiki Aquarium, was instrumental in developing the techniques and in securing the tissue samples on which this study was based. We are indebted to Bruce Carlson for the sample from Fiji and for logistic help from Honolulu. Others who assisted in obtaining and transporting material for study include the R. L. Knight family, Lorengau, Manus; fisheries personnel at Port Moresby, Lae, Kavieng, and Manus, Papua New Guinea; P. and L. Colin and N. Polunin, Motupore Island Research Station, University of Papua New Guinea; and B. L. Goldman and P. and G. Pini, Lizard Island Research Station of the Australian Museum, Queensland. Lisa Strong and Peg Mulvey provided invaluable help in the electrophoresis laboratory, and Adina Merenlender assisted with statistical analysis. This study was supported by NSF Grant EAR 83-18932.

Chapter 6

Morphological Variation in *Nautilus* from Papua New Guinea

ANDREW R. H. SWAN and W. BRUCE SAUNDERS

1. Introduction

The generally held consensus has been that *Nautilus* exhibits little morphological variability (e.g., Ward, 1984). However, it is relevant that most studies of morphological variation have been restricted largely to comparisons of mature shell size and weight and to documentation of sexual dimorphism within individual populations (e.g., Hirano, 1977; Saunders and Spinosa, 1978; Hayasaka *et al.*, 1982, 1983; Tanabe *et al.*, 1983; Ward, 1984; Saunders and Davis, 1985) (see also Chapter 3). The first effort to evaluate "whole animal" variation (Tanabe *et al.*, 1983) involved analysis of nine shell and body characters taken from samples of *N. pompilius* from the Philippines and from Fiji. Partly because of sample size limitations, the results of that study only tended to confirm earlier observations of sexual dimorphism and of mature size differences between the two populations. Two expanded studies, by Tanabe *et al.* (1985) and by Tanabe and Tsukahara (Chapter 7), have shown that these two widely separated populations do exhibit a fairly large degree of variation.

The study presented in this chapter represents an effort to examine the degree

ANDREW R. H. SWAN • Department of Geology, University College of Swansea, Swansea SA2 8PP, United Kingdom. W. BRUCE SAUNDERS • Department of Geology, Bryn Mawr College, Bryn Mawr, Pennsylania 19010. *Present address of A.R.H.S.*: School of Geological Sciences, Kingston Polytechnic, Kingston-upon-Thames, Surrey KT1 2EE, United Kingdom.

and type of morphological variation among six populations of *Nautilus* that are geographically isolated, to varying degrees, within the region of Papua New Guinea (Fig. 1) (see also Saunders and Davis, 1985; Saunders *et al.*, 1987a). The study is based on 276 specimens that were trapped in 1984 at depths of approximately 100–400 m at the following six sites: Port Moresby, Central Province (49 specimens); Lae, Morobe Province (38); Kavieng, New Ireland Province (48); and Komuli (41), Ndrova (96), and Lorengau (4), all Manus Province, in the Admiralty Islands. Included in the analysis are 266 specimens of the widely distributed species *N. pompilius* (which was obtained at all sites), and 10 specimens of the rare form *N. scrobiculatus* (from Ndrova and Komuli, Manus Province); given these disproportionate numbers, the study is largely concerned with the former species.

2. Methods

After the specimens were trapped, each specimen was weighed, sexed, measured, judged as to maturity, and photographed (apertural and lateral views). The reliance on photographs for the majority of the data has imposed certain limitations on the analysis—the morphological characters measured were necessarily those that are external and visible. Consequently, the resulting data are inevitably dominated by features of the shell rather than of the soft parts. Although an attempt was made to incorporate as much of the external morphology as was feasible, the results clearly do not provide a complete account of the phenome. However, the subset of the total morphology dealt with here does incorporate characters that are related to soft-part morphology (e.g., body shape and size), and characters of the shell are likely to be significant in relation to the animal's environment.

2.1. Characters Assessed

Initially, there are no grounds for supposing that any morphological feature is more significant than any other, so an attempt was made to record all visible features, with the exception that features of pathological origin (e.g., shell injury/repair) and shell epizoans were not assessed, because they are not a property of the animal itself. The initial data recorded from each specimen included a measurement or numerical assessment of each of 51 characters (Table I).

The coiling of *Nautilus* exposes one whorl to external view, which records a three- to four-fold increase in shell diameter. There is potential significance in different ontogenetic stages within this range, so characters were assessed at various diameters in each specimen. The characters measured have been defined to allow numerical coding (Fig. 2 and Table I).

Assessment of visible characters was foregone only if assessing them would have required an excessive amount of operational effort or difficulty. It is clearly impossible to record every inflection or detail of the color pattern; the coloration characters assessed here reflect its general properties. Similarly, subtle aspects of

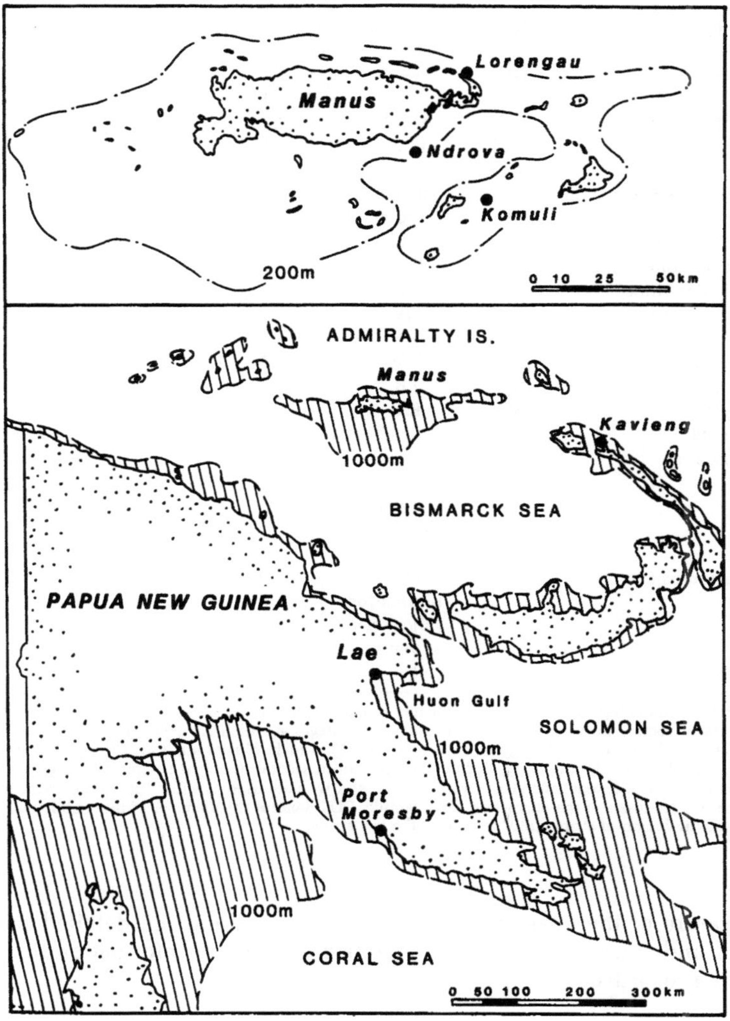

Figure 1. Map showing the locations of six *Nautilus* populations sampled in Papua New Guinea. Trap sites (●) were located along forereef slopes at depths of approximately 100–400 m. Included are three sites in Manus Province (Komuli, Ndrova, and Lorengau) and one each off Kavieng, New Ireland Province; Lae, Morobe Province; and Port Moresby, Central Province. *Nautilus pompilius* was obtained at all sites, but *N. scrobiculatus* was trapped only in Manus Province. The hatching indicates areas of water depths less than 1000 m; these areas should approximate those within which active migration by *Nautilus* can occur.

Table I. *Nautilus* Character Definitions and Usage[a]

Character	Code	Calculation[a]	Ontogenetic stage m	a	b	c	d	w	x	y	z
Diameter	DM	—	3, 4	1	—	—	—	—	—	—	—
Radius	R	—	—	1	1	1	1	—	2	—	—
Diameter of umbilicus	DU	A/DM	3	1, 2	—	—	—	—	—	—	—
Maximum width	WM	B/DM	3, 4	1	—	—	—	—	—	—	—
Width at salient	WS	C/B	3, 4	1	—	—	—	—	—	—	—
Extent of thick periostracum	PE	Angular	3	1	—	—	—	—	—	—	—
Extent of black layer	BL	Angular	3, 4	1	—	—	—	—	—	—	—
Extent of unstriped shell	US	Angular	3, 4	1	—	—	—	—	—	—	—
Depth of hyponomic sinus	HS	D/R	—	1	1	1	1	3, 4	—	2, 3, 4	—
Depth of ocular sinus	OS	E/R	—	1	1	1	1	3, 4	—	2, 3, 4	—
Umbilical projection	UP	F/R	—	1	1	1	1	3, 4	—	2, 3, 4	—
Spacing of stripes at venter	SV	G/R	—	1	1	1	1	—	—	2, 3, 4	—
Stripe separation	SS	H/R	—	1	1	1	1	—	—	2, 3, 4	—
Stripe projection	SP	J/R	—	1	1	1	1	—	—	2, 3, 4	—
Stripe length	SU	K/R	—	1	1	1	1	—	—	2, 3, 4	3, 4
Color at venter	CV	% red	—	1	1	1	1	—	—	2, 3, 4	3, 4
Stripe branching	SB	Stripes at venter (*N*)/stripe terminations (*N*)	—	1	1	1	1	—	—	2, 3, 4	3, 4
Stripe anastomosis	SA	Dorsal bifurcations (*N*)/stripes at venter (*N*)	—	1	1	1	1	—	—	2, 3, 4	3, 4

[a] Refer to Fig. 2. The ontogenetic stages are indicated in Fig. 2A, except for m (mature). The numbers in the "Ontogenetic stage" section indicate the particular ontogenetic stages at which the characters were measured or used in the analyses of characters: (1) original measurement; (2) used in cluster analysis; (3) used in first principal components analysis (including *N. scrobiculatus*); (4) used in second principal components analysis (excluding *N. scrobiculatus*). The ontogenetic stage codes a–d, m, and w–z are used as subscripts to the character codes to specify the precise usage of the characters. For example: DU_m denotes the diameter of the umbilicus at maturity; SU_z denotes the length of stripes 180° back from the last stripe.

the curvature of the aperture escape consideration, and the detailed pattern of tubercles and ridges on the hood contains a quantity of information that cannot feasibly be recorded.

2.2. Cluster Analysis of Individual Specimens within Populations

Initially, the data from each population were subjected to cluster analysis, to resolve morphologically similar subgroups. Subsequent analyses could then treat each cluster, rather than each specimen, as a unit. The advantages of this procedure are: (1) The members of each cluster commonly show features from various ontogenetic stages (mature, submature, immature) that can all be incorporated into the character set of the cluster. (2) The character set that represents

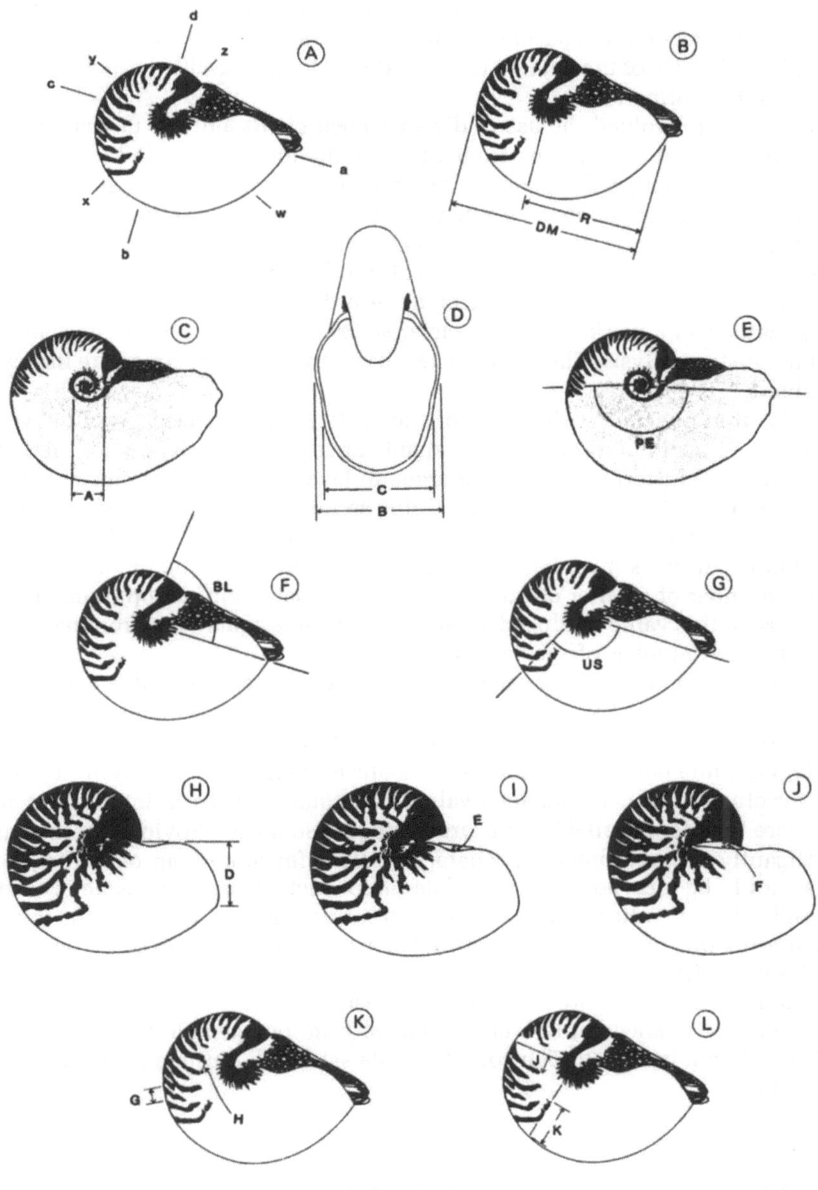

Figure 2. *Nautilus* shell characters, illustrated on lateral views of *N. scrobiculatus* (C, E) and of *N. pompilius* (A, B, F–L) and a generalized apertural view (D). Refer to Table I for further details of character value calculation and usage. (C, E) The stippling symbolizes thick periostracal covering. (A) The letters indicate the ontogenetic stages at which the characters were assessed: (a, b, c, d) positions are at 90° intervals relative to the aperture; (w, y, z) positions are at 90° intervals relative to the position of the last stripe at the venter (x).

a cluster corresponds to an average of the character states of the constituent individuals, so occasional spurious character values do not assume undue significance. (3) The size of the data set can be reduced to manageable proportions with little loss of information.

Clustering involved the use of distance coefficients and the unweighted average method. All individuals from each population were included, and each population was analyzed separately. The characters used in cluster analysis must have a clear correspondence between individual specimens—in this case, it is important that they be of the same ontogenetic stage in each specimen. It was found that the ontogenetic stage 90° back from the final color stripe at the venter (see Fig. 2A) is observable in specimens regardless of maturity and provides a useful ontogenetic marker. The character states at this point were consequently used in cluster analysis. The 12 characters used from each specimen are indicated in Table I.

It is often not clear which level of similarity should be used to split clusters. However, no attribution of objective significance to the clusters is implied. The advantages of clustering the data all argue in favor of decreasing cluster number (and thereby increasing cluster size). Doing so must be compromised against the necessity of retaining good documentation of morphological variation; all major morphological types must be resolved. This resolution is better achieved with a greater number of smaller clusters. A similarity coefficient of approximately 4.0 was chosen; this value produced a number of clusters equal to approximately one third the total number of specimens.

Once the composition of each cluster has been determined, it is necessary to represent the cluster by a suite of characters with values averaged from cluster members. Mature morphology is not represented in all specimens, but nearly all clusters include mature individuals, so character values at maturity were obtainable for clusters. Cluster character values that could be used in intercluster analysis were therefore available for a large ontogenetic range. Provided these change significantly through ontogeny, character values for more than one ontogenetic stage should be incorporated in the data set for each cluster. Ontogenetic scatters for each cluster indicate typical character value changes with growth (Fig. 3). Considering this information, 24 characters were chosen to represent each cluster (see Tables I, II).

Values for these characters for each cluster were derived from simple averaging (mature characters) or were estimated from ontogenetic trajectories (such as those shown in Fig. 3). The complete data set for all clusters is available from the authors.

2.3. Principal Components Analysis of Populations

The critical part of this analysis is the determination and portrayal of variation within and among populations. Previous studies of *Nautilus* (e.g., Saunders and Spinosa, 1978; Tanabe *et al.*, 1985) have illustrated uni- and bivariate characteristics of populations. However, fundamental to our approach is the attempt to

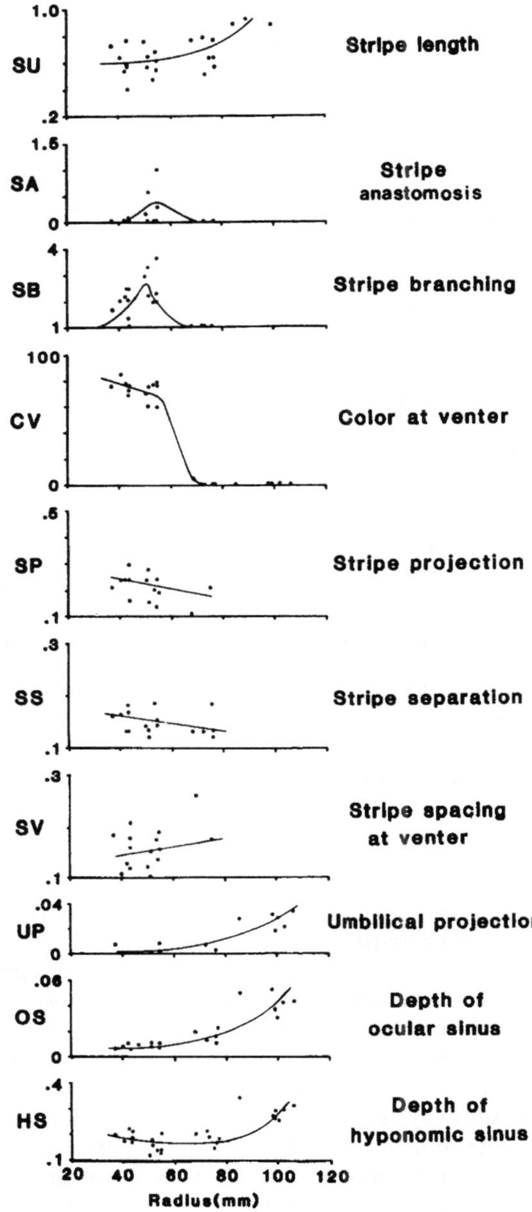

Figure 3. Ontogenetic scatters and estimated trajectories for a representative *Nautilus* cluster. The data were derived from cluster PMH, *N. pompilius*, from Port Moresby. These graphs form the basis for the estimation of cluster character values. Note that these ontogenetic character value plots demonstrate a number of growth-related patterns: (1) A number of characters tend to change only slightly and gradually relative to increased size (= radius), e.g., SU, SV, and UP. (2) Two characters, OS and HS, increase sharply at the largest radius; these are characteristics that indicate maturity. (3) The peaked trajectories shown by SA, SB, and CV reflect a strong prominence of these characters at a particular ontogenetic stage and their subsequent loss; in these instances, it reflects the elimination of shell coloration as the formation of the final (mature) body chamber is initiated.

Table II. Composition of the First Five Principal Components of Variation for the Two Analyses of *Nautilus* Clusters

	Analysis 1: *N. pompilius* and *N. scrobiculatus*					Analysis 2: *N. pompilius*				
Char.	P.C. 1	P.C. 2	P.C. 3	P.C. 4	P.C. 5	P.C. 1	P.C. 2	P.C. 3	P.C. 4	P.C. 5
DU_m	0.32	0.32	−0.04	0.15	0.02	—	—	—	—	—
WM_m	0.01	0.16	0.29	0.40	0.01	−0.08	0.24	−0.42	0.14	0.01
WS_m	−0.20	−0.11	0.31	−0.19	−0.20	−0.08	0.34	−0.17	−0.09	−0.33
BL_m	−0.16	0.01	0.19	0.05	0.10	−0.05	0.14	−0.20	−0.29	−0.22
PE_m	0.32	0.34	−0.02	0.14	0.02	—	—	—	—	—
DM_m	0.27	−0.13	0.00	0.03	−0.32	0.30	0.00	−0.04	−0.30	0.01
US_m	0.22	0.37	0.05	0.12	−0.12	−0.17	0.18	0.03	0.34	0.04
HS_w	−0.04	−0.08	0.08	0.03	0.59	0.05	−0.01	0.08	0.44	0.03
OS_w	0.00	−0.21	0.06	0.47	−0.17	0.14	0.11	0.44	−0.20	0.05
UP_w	0.06	−0.32	−0.13	0.35	0.11	0.25	−0.13	0.40	0.02	0.02
HS_y	0.14	0.11	0.33	0.03	−0.42	0.04	0.40	−0.05	−0.28	0.07
OS_y	−0.06	−0.11	0.41	−0.07	0.02	0.05	0.38	−0.05	0.09	0.40
UP_y	−0.09	−0.06	0.42	0.11	0.02	−0.01	0.41	0.11	0.12	0.41
SV_y	−0.24	−0.02	−0.16	−0.28	−0.18	−0.18	−0.17	−0.21	−0.09	0.20
SS_y	−0.25	−0.05	−0.21	−0.13	−0.30	−0.19	−0.20	−0.10	−0.30	0.41
SP_y	−0.12	0.04	−0.26	0.06	−0.31	−0.19	−0.15	0.03	−0.34	0.24
CV_y	0.23	−0.12	−0.09	−0.33	0.15	0.27	0.15	−0.29	0.15	0.17
SB_y	0.19	−0.33	0.07	0.05	−0.06	0.33	0.10	0.07	−0.11	0.12
SA_y	−0.01	−0.29	0.01	0.11	−0.12	0.18	0.01	0.16	−0.23	−0.29
SU_y	0.31	−0.18	0.14	−0.15	−0.03	0.35	0.11	−0.17	−0.02	−0.04
CV_z	0.31	−0.06	−0.19	−0.32	0.02	0.28	−0.21	−0.37	0.05	0.01
SB_z	0.26	−0.31	0.03	−0.04	−0.01	0.37	0.04	−0.02	−0.03	0.26
SA_z	0.05	−0.21	−0.27	−0.16	−0.04	0.15	−0.27	−0.12	−0.22	0.10
SU_z	0.29	−0.16	0.12	0.12	0.05	0.31	0.10	−0.14	−0.02	−0.18
%	18.17	14.52	10.30	8.81	6.51	19.21	11.88	10.40	9.16	6.59

encompass as many morphological characters as possible. The most satisfactory and objective technique for resolving patterns of variation in multivariate data sets is principal components analysis. This identifies suites of intercorrelated characters that are liable to interpretation as factors that affect variation; this also allows the graphic ordination of data in a way that minimizes data loss.

A result of this approach is the removal of "randomizing" effects to the subordinate components of variation, where they can be ignored. This occurs because "randomizing" effects will normally affect characters in isolation and will not result in systematic character correlations. Effects in this category should include errors in measurement and pathological distortions due to shell repair.

Two principal components analyses were undertaken: one of all clusters for the two species, *N. pompilius* and *N. scrobiculatus* (using the full set of cluster characters), and a second that excluded clusters composed of *N. scrobiculatus* [thereby rendering redundant the characters DU (umbilical diameter) and UP (umbilical projection)].

3. Results

3.1. Analysis of All Clusters (*Nautilus pompilius* and *Nautilus scrobiculatus*)

3.1.1. Description of Principal Components

As shown by the character values recorded in Table II, the first principal component (P.C.1) results primarily from an association of a wide umbilicus (DU) with a thick periostracum [(PE) Fig. 2E]. These are positively correlated with size at maturity and a tendency toward possession of long, closely spaced stripes. P.C.2 represents a tendency for widely umbilicate forms with a thick periostracum to exhibit a large extent of unstriped shell (US) at maturity, with simple, weakly branching stripes over the remainder. Both P.C.1 and P.C.2 reflect, to a large degree, the distinctive umbilicate shell form of *N. scrobiculatus*, with its unique periostracum and reduced coloration (see Saunders *et al.*, 1987a). P.C.3 is close to P.C.2 in the second analysis (see Section 3.2). Further components are increasingly more poorly defined and likely to be of subsidiary importance.

3.1.2. Variation

Plots of the ordination of the *Nautilus* clusters against the first three principal components show a strong polarization of the two species (Fig. 4). There is evidently considerable variation within each species, but this variation is nonoverlapping, and the species are morphologically quite discrete. The morphological discontinuity in principal component morphospace has a clear manifestation in the frequency distribution of the umbilical size and the degree of periostracal covering. The value of both these characters for *N. pompilius* is 0; for *N. scrobiculatus*, the minimum values are 0.14 and 125°, respectively. There is no *a priori* reason that intermediates should not occur; that they do not is the main contributory factor to the discrete morphologies of the two species. These two characters are not, however, the only ones that differ between the species; the descriptions of P.C.1 and 2 (Section 3.1.1) suggest additional differences in stripe patterns and size.

The sample size of *N. scrobiculatus* available is not sufficient to justify further discussion of variability within the species. Variation in *N. pompilius* is better discussed with reference to the second analysis (see Section 3.2).

3.1.3. Interpretation of Principal Components

As noted in Sections 3.2.1 and 3.2.2, the first two components of variation are dominated by the character correlations that result from the contrast between the two species. No explanation for these specific differences emerges from this analysis; i.e., there is no explanation available for the periostracum of *N. scrobiculatus*, nor is the functional significance of its open umbilicus known. At present, the association of the two sheds light on neither.

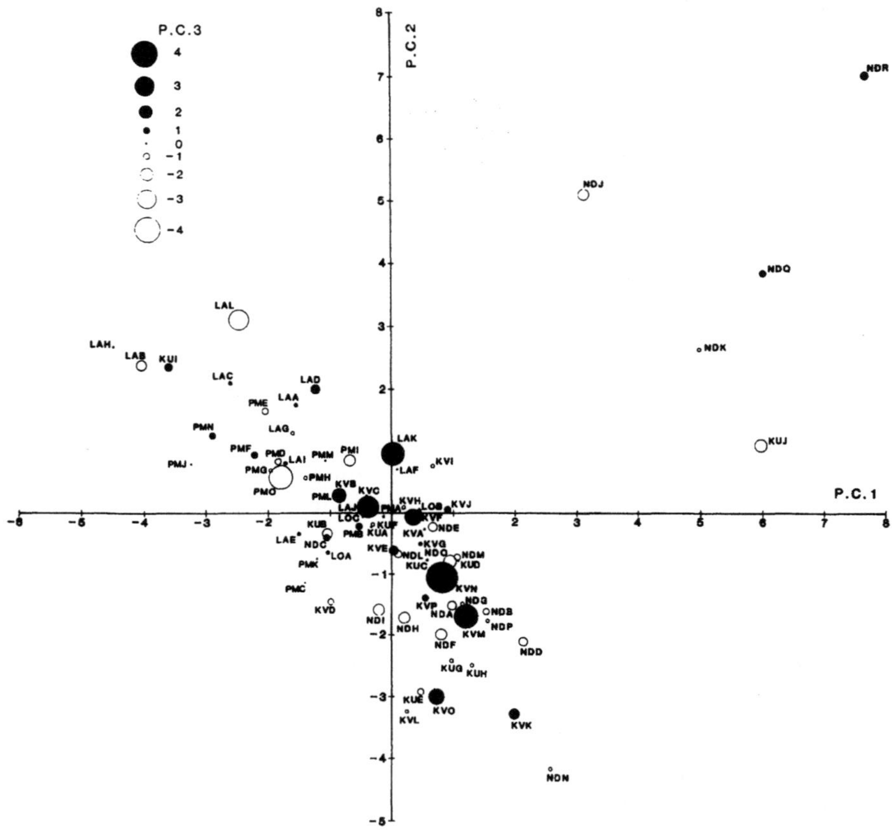

Figure 4. Ordination against the first three principal components of all clusters of *Nautilus* from Papua New Guinea. The composition of the principal components is given in Table II. Clusters KUJ, NDJ, NDK, NDQ, and NDR represent specimens of *N. scrobiculatus* exclusively; all other clusters contain only *N. pompilius*. The cluster-code prefixes indicate the trap sites: (PM) Port Moresby; (LA) Lae; (KV) Kavieng; (KU) Komuli; (ND) Ndrova; (LO) Lorengau.

3.2. Analysis of *Nautilus pompilius* Clusters

3.2.1. Description of Principal Components

The first principal component (P.C.1) results from an intercorrelation of characters that pertain primarily to the color stripes, particularly stripe length (SU) and stripe branching (SB) (Table II). Long stripes tend to branch more and to merge at the venter. Specimens with more strongly developed striping tend also to be larger and to have a stronger umbilical projection at maturity.

The dominant correlation in P.C.2 is that of shell width (especially W_s) with degree of immature (but not mature) apertural sinuosity. There are also correlations with other characters, particularly an association with a lack of anastomozing stripes.

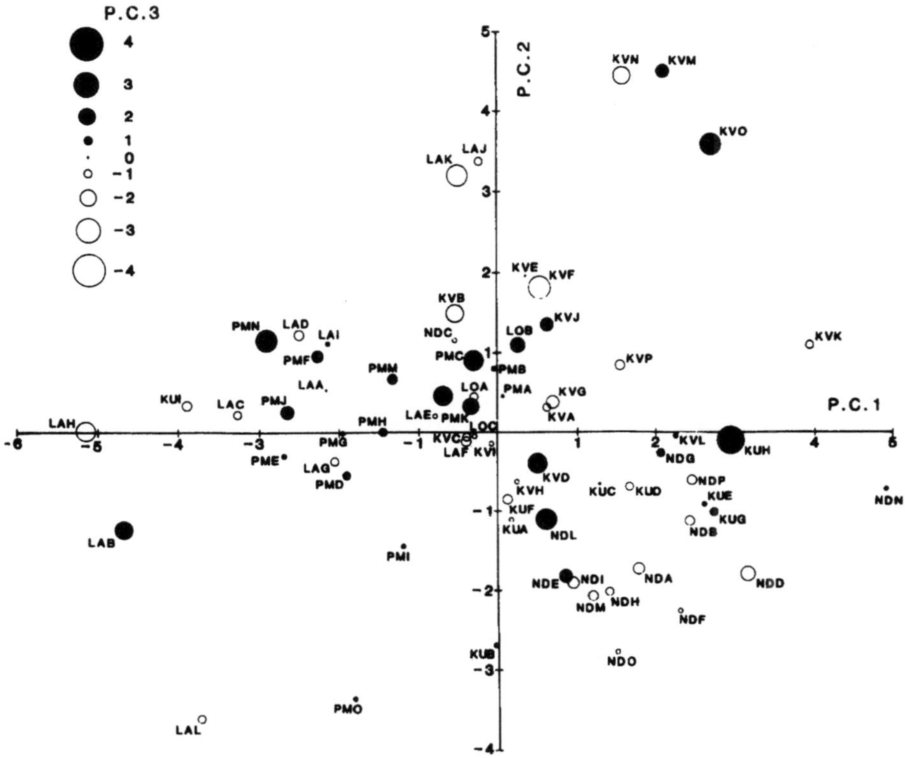

Figure 5. Ordination against the first three principal components of all clusters of *N. pompilius* from Papua New Guinea. The analysis is the same as in Fig. 4, except for the exclusion of all *N. scrobiculatus* clusters, and of characters DU (umbilical diameter) and UP (umbilical projection), which are inapplicable to *N. pompilius*. The composition of the principal components is given in Table II.

The third component of variation (P.C.3) is due to the correlation at maturity of a strong ocular sinus (OS) and umbilical projection (UP) with narrowness of the shell (W_m rather than W_s). Shells with these characters also tend to show relatively wide gaps between the stripes on the venter.

An ordination of clusters of *N. pompilius* against the first three principal components (Fig. 5) shows no clear subdivision within the total sample, although some subgroupings become apparent in density contour plots of P.C.1 and P.C.2 (Fig. 6). On this evidence, there is no morphological justification for distinction of species or subspecies among the different populations analyzed.

3.2.2. Interpretation of Principal Components

3.2.2a. P.C.1. The main characters that contribute to this component pertain to the degree of development of color stripes. Thus, the higher the P.C.1 score, the greater the coverage of red pigment on the shell. This component of variation

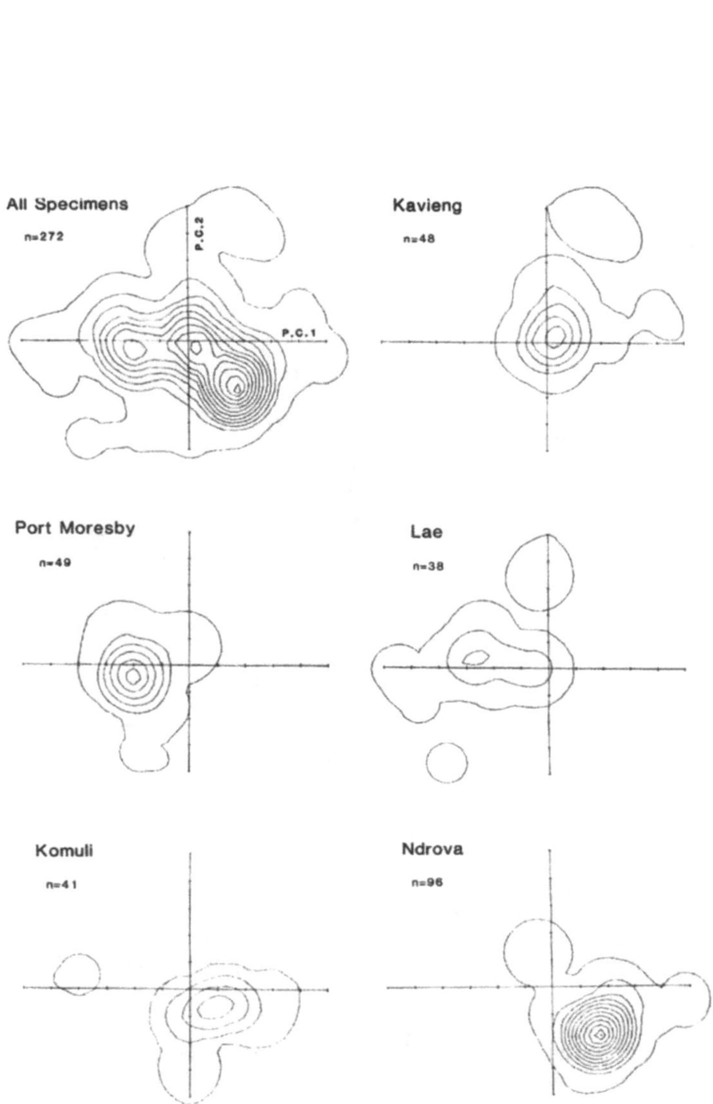

Figure 6. Density contours of the distribution of clusters of *N. pompilius* with respect to the first two principal components, based on the same analysis as Fig. 5. Clusters were weighted according to size. The contours are at equal intervals in all plots; the outer contour encloses all clusters. The top left plot includes all clusters; each of the other plots includes clusters from only the population indicated. The distribution of each population is to some extent unique, and a number of generalizations can be made: (1) Some morphologies are unique to one population. (2) All populations contain specimens with morphologies that are equal to the average of the total sample. (3) Variation is approximately equal in each sample. (4) Some populations (e.g., Ndrova and Komuli) contain specimens that are atypical of their clusters; their presence may be a product of migration from adjacent populations.

is apparently that which led Ward *et al.* (1977) to distinguish two color "polymorphs" in *N. pompilius* from Fiji. Our analysis does not indicate that these "polymorphs" are worthy of any taxonomic status.

It is striking that the Papua New Guinea populations with the most subdued development of red coloration are from the mainland sites. One of these populations (Lae) was trapped within the Markham River Delta, in the Huon Gulf. Bottom sediments at the site comprise fine clay, which stands in considerable contrast to the typically rocky/sandy forereef conditions at other locations. It is possible that a paler shell is an adaptation to conditions of poor visibility or, conversely, that the more strongly striped shells are specifically adapted to conditions of higher visibility.

3.2.2b. P.C.2. One of the characters that contribute strongly to this direction of variation (W_s) has been found to be strongly significant in sexual dimorphism (Saunders and Spinosa, 1978). Its frequency distribution is bimodal; the higher values correspond to the males, which are characteristically larger and broader at maturity. In the analysis discussed herein, high W_s in *N. pompilius* is correlated with strong apertural sinuosity before maturity. Apertural complexity is a characteristic of male ammonite dimorphs; however, the occurrence of strong apertural sinuosity before and not at maturity in the P.C.2 of variation renders a dimorphic explanation less likely for *N. pompilius*. In addition, the clustering process has not systematically segregated the sexes, so sexually distinctive character values (D_m, W_s) in most cases have been "averaged out" in arriving at the cluster character values. Nevertheless, it is conceivable that a residual amount of variation attributable to dimorphism remains and may have been resolved as P.C.2. This could not, of course, account for the segregation of populations (especially KV) along the P.C.2 axis, and no explanation for this is apparent at present.

3.2.2c. P.C.3. Further principal components of variation are increasingly more poorly defined and less readily interpretable. The correlation with P.C.3 of maximum aperture width (W_m) with the development of the ocular sinus (OS) and the umbilical projection (UP) is potentially of sexual significance (W_m is greater in male *Nautilus*), though this interpretation suffers from some of the same problems as P.C.2.

3.3. Variation within and among Populations

Within the total range of variation shown by *N. pompilius*, different populations are seen to exhibit differences in morphology—or, as shown by separate density contour plots, they are seen to occupy different morphospace, with respect to the first two principal components (Fig. 6). The following generalizations can be made:

1. Some morphologies are unique to one population. These may be rare, morphologically isolated individuals [e.g., clusters PMO and PMI (Fig. 5)], or they may constitute a large percentage of the population [e.g., Ndrova clusters (Fig. 6)].
2. All populations contain specimens with morphologies that are average for

Table III. Phenetic Dissimilarity (d) between Each Pair of Populations Based on the Euclidean Distance between Population Centroids[a]

Population	PM	LA	KV	KU	ND
PM	—	5.36	6.56	7.23	8.19
LA	5.36	—	6.58	7.40	8.90
KV	6.56	6.58	—	6.14	5.44
KU	7.23	7.40	6.14	—	5.71
ND	8.19	8.90	5.44	5.71	—

[a] The clusters were weighted according to size, and the data were standardized. All *N. pompilius* cluster data were used except those of the Lorengau population (only 4 specimens). For the definitions of the population codes, see the Fig. 4 caption.

the total sample (corresponding to around P.C.1 = 0, P.C.2 = 0 on the ordination).

3. The extent of variation is approximately equal within each population.
4. The two "mainland" Papua New Guinea populations (Lae and Port Moresby) dominate negative P.C.1 morphospace. The two main Manus Island populations (Komuli and Ndrova) are similar to each other and are strongly localized in positive P.C.1 and negative P.C.2 morphospace.

Table IV. Minimum Distances between Populations[a]

Population	PM	LA	KV	KU	ND
	A. Minimum "direct" distance between sampling sites (km)				
PM	—	1017	1000	1000	1050
	—	0	496	417	417
LA		—	426	235	285
		—	200	313	313
KV			—	139	179
			—	217	217
KU				—	50
				—	0
ND					—
					—
	B. Minimum "coastal" distance between sampling sites (km)				
PM	—	1017	2135	1434	1454
	—	0	75	226	226
LA		—	1143	341	376
		—	75	226	226
KV			—	139	179
			—	217	217
KU				—	50
				—	0
ND					—
					—

[a] The population codes are defined in the Fig. 4 caption. Section A uses the direct, midwater route across the Bismarck Sea; section B uses the coastal route, around the archipelago.

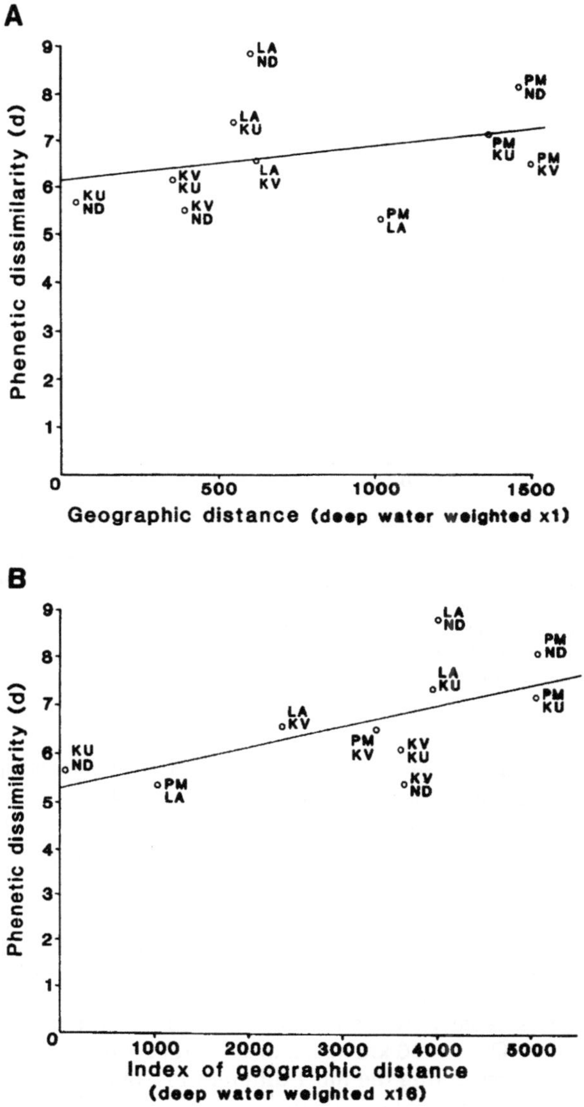

Figure 7. Graphs plotting phenetic dissimilarity against indices of geographic distance. Each point pertains to a pair of localities [indicated by codes (see the Fig. 4 caption)]; the data are from Tables III and IV. A linear regression is given for each graph. (A) Distance (km) across deep water (>500 m) is unweighted. The correlation coefficient is 0.337. (B) Distance (km) across deep water is multiplied by 16, giving the maximum correlation coefficient of 0.657 (see Fig. 8). Similarity between faunas is apparently inversely proportional to this distance index.

It should be noted that intersection of the distance regression with the morphological difference axis (distance = 0) at around d = 6.0 implies that two samples from the same population at the same site will be morphologically different. If the regression genuinely reflects the true situation, then this difference can result only from large random effects due to small sample size. If such effects obtain in this analysis, a greater dispersion of values of d would be expected. The sample size (minimum 38) should be adequate to prevent this type of error. It seems more likely that the regression line extrapolated to the y axis cannot be used to predict differences between populations at short distances. The true differences between populations closer than 50 km may be approximated better by a steep gradient on the graph descending to around 1 = 0, d = 0, which would imply a different process affecting gene flow between closer populations, so that the probability of gene flow declines drastically at a certain distance between populations.

5. Two populations (Ndrova and Komuli) contain specimens (clusters NDC and KUI) that are atypical within their populations, but that would not be out of place in other populations [Kavieng and Lae, respectively (Fig. 6)]. Such occurrences could be explained as products of migration between populations, although there is no direct evidence for this.

To compare the differences and similarities among the Papua New Guinea

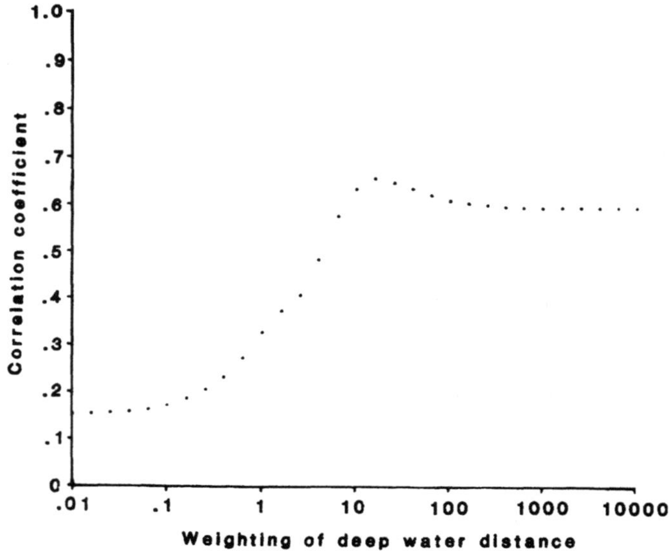

Figure 8. Change of correlation coefficient with deep water distance weighting for the relationship between phenetic dissimilarity and geographic distance index (as in Fig. 7). The distance index consists of distance across deep sea (>500 m) multiplied by the deep water weighting, plus shallow (<500 m) water distance. This suggests that gene flow across a unit distance of deep water is at least an order of magnitude less probable than across shallow water. For these data, correlation coefficients greater than 0.55 are significant at the 5% level.

populations, an index of overall similarity between populations is desirable. The use of principal components for this index is not preferred, due to lost information. The index used here is the simple Euclidean distance between the centroids of each population, based on standardized values of all 22 characters used in the principal components analysis. In calculating centroids, values were weighted according to cluster size. The Lorengau population was eliminated from this analysis due to small sample size (N = 4). The results of these calculations indicate that differences among populations vary by a factor of nearly 1.7 (Table III). It is notable that the two geographically closest populations (ND and KU; ≈ 50 km apart) are among the most similar; the two most distant populations (PM and ND; ≈ 1400 km apart, measured as the shortest distance parallel to the coastline) are among the most different morphologically.

The possible influence of ecological parameters as isolating factors is discussed in Chapter 3. It was suggested that because of such limiting factors as implosion and water temperature, *Nautilus* movement should largely be restricted to a depth-defined corridor less than 800 m deep; most movement should occur between approximately 100 and 300 m, and deeper and shallower migrations would be short-term events. To evaluate whether geographic distance and water depth could be correlated to population differences in the Papua New Guinea region, an index of geographic separation was utilized in the study presented herein. This involved compilation of the geographic distances between sample sites, with distances across seas deeper and shallower than 500 m recorded separately (Table IV). The index of geographic separation used is the distance across "deep" water, multiplied by a variable weighting, plus distance across "shallow" water. For any two sites, there are commonly alternate routes: a "coastal" route and a "direct" route, which spans open water (see Fig. 1). The former results in a lower index, if deep water distance is strongly weighted.

A graph of phenetic distance (dissimilarity) between populations plotted against minimum geographic distance shows a loose positive correlation (Fig. 7A). The direct distance used in this case is probably a poor measure of isolation, for it does not take into account the effects of deep water barriers; i.e., deep water distance is weighted equally to shallow water distance. It is not possible *a priori* to arrive at accurate weightings for depth as a barrier or isolating mechanism, since doing so requires knowledge of the relative probabilities of migration and genetic communication across a unit distance of deep vs. shallow water. Iterative increases in weightings of deep water distance, however, improve the correlation up to a weighting of ×16 (Figs. 7B and 8).

4. Discussion: Variation within and among Populations

Our objective in this analysis was to analyze the degree and type of morphological variation in isolated populations of *Nautilus*. These differences have been summarized by distribution in principal components morphospace (see Figs. 5 and 6), and the degree of their difference has been quantified using phenetic distance between population centroids (see Table III). However, the extent to which causes for the morphological differences can be identified is severely limited. To begin with, it is not known whether the characters used in the analysis are random, genetic differences among populations or whether they reflect adaptations to specific local settings; indeed, almost nothing is known of the deep forereef settings themselves. In addition, genetic information from *Nautilus* is only now becoming available (Woodruff *et al.*, 1983) (see also Chapter 5). Despite these limitations, some amount of speculation regarding the implications—and the possible causes—of the patterns recorded here seems warranted.

Two aspects of morphological variation of *N. pompilius* in the Papua New Guinea region stand in need of explanation: (1) the variation within populations and (2) the differences among populations.

4.1. Variation within Populations

Intrapopulation variation can be derived from various sources, including:

1. Genetic variation
2. Sexual dimorphism
3. Differences resulting from different life histories of individuals (including nutrition and pathology)
4. Migration of forms from other populations

There is evidence from this analysis and from previous analyses that sexual dimorphism contributes to variation. This study has also shown examples of variation that could be explained in terms of migration—e.g., the presence in the Komuli and Ndrova populations of atypical specimens that resemble members of adjacent populations (Kavieng and Lae). However, these two factors probably contribute only a small part of the total variation observed. The other two sources of variation (genetic and individual life histories) together are likely to account for the bulk of the variation, and each is potentially important, but they cannot be distinguished on the evidence considered here.

4.2. Variation among Populations

Processes that tend to increase interpopulation morphological differences can be distinguished from those that encourage homogeneity; among the former are:

1. Different selection pressures or ecophenotypic effects in different environments
2. Genetic isolation due to geographic factors
3. Genetic isolation due to behavioral factors
4. Length of time of genetic isolation
5. Random, genetic changes

Processes that might inhibit differentiation of populations include:

1. Constancy of adaptive response to similar environments
2. Gene flow through overlapping populations
3. Gene flow by migration
4. Lack of genetic variability (shown not to be relevant in *Nautilus*) (see Chapter 5)

Environmental effects are difficult to assess without detailed information about benthic conditions at the trap sites. The only known potentially important ecological factor is the deltaic setting of the Lae population, in the Huon Gulf. This might be a cause for the distinctively low P.C.1 scores for this population (see Fig. 5), but it does not explain a similar morphology in the Port Moresby population, which is more typical steep forereef. Other environmental differences among individual sites must occur inevitably, but the degree of their relevance to *Nautilus* morphology may not necessarily be great. If the differences prove not to be significant, then the residual differences between populations [largely in the P.C.2 and P.C.3 directions (see Figs. 5 and 6)] would seem to be random, adaptively neutral, genetic differences.

The correlation of dissimilarity with an index of geographic distance (see Fig. 7) cannot be explained entirely by purely environmental effects or by random, adaptively neutral changes. If each population were ideally adapted to the peculiarities of its local environment, then this correlation would imply that the differences between environments are proportional to the geographic differences between them. This is unlikely to occur over a number of sites that vary in longitude as well as in latitude and are distributed in widely different geographic settings. If the morphologies of populations differed due to purely random changes, then a correlation would be statistically unlikely. Whatever the "driving force" of morphological evolution of populations, it seems clear that the result has been affected by the degree of geographic isolation.

Complete genetic isolation would allow each population to evolve toward adaptive equilibrium with its local environment. However, the correlation between morphological difference and distance suggests that this equilibrium has not been attained. If isolation were complete, the correlation could be due to insufficient time having passed for the populations to have reached equilibrium, and the geographic distance index could represent length of isolation. For most of the populations studied here, complete isolation is unlikely. It has been shown that large-scale and long-term movement occurs within *Nautilus* populations—at a minimum, 150 km per year in Palau, a setting similar to that studied here (Saunders and Spinosa, 1979). Even a small potential for rafting or midwater movement would permit contact among all the populations sampled in Papua New Guinea. Movement from Manus to Kavieng, for example, would require traversing deep water (>1000 m) for a distance of less than 100 km (see Fig. 1). In turn, migration from Kavieng to Lae (via coastal routes along New Ireland and New Britain) would require crossing deep water for only 50 km. Because *Nautilus* populations could, in principle (and they appear to), inhabit all points along the Papua New Guinea coast, all populations between Lae and Port Moresby (≈ 800 km) could be in at least occasional or indirect contact, and varying degrees of gene flow could occur throughout.

It would seem that the observed differences among *Nautilus* populations probably reflect the amount of gene flow among them. This relationship can be regarded as operating in opposition to selection in local environments; it is possible that populations are prevented from attaining local "adaptive equilibrium" by influx of genetic material from other populations, which, in turn, are isolated to varying degrees and represent different local conditions. The amount of this influx is inversely related to the index of geographic distance.

The analysis presented in this chapter sought to derive as much information as possible about morphological variation in a series of populations of *Nautilus* in Papua New Guinea. Given the restrictions imposed by the amount of available data (sample size and number, habitat profiles), little can be proven. However, the results have revealed trends that invite speculation. There are many gaps in our knowledge, and the tentative explanations presented herein may be regarded as hypotheses, which can be tested when more data are acquired. In particular, the selective significance of various morphological variants may be discerned when more environmental data are available, and the correlation of morphological differences with distance can be assessed further by sampling additional new sites.

ACKNOWLEDGMENTS. The samples on which this study was based were obtained during the summers of 1984 and 1985, with the assistance and support of the many people listed in Chapter 9 of this volume. Renewed acknowledgment, however, is due Larry Davis, Washington State University, and the R. L. Knight family, Manus, Papua New Guinea. Isobel Swan assisted in preparing the figures, and Eric Payne computed the regressions. This study was supported by NSF Grant EAR 83-18932.

Chapter 7

Biometric Analysis of *Nautilus pompilius* from the Philippines and the Fiji Islands

KAZUSHIGE TANABE and JUNZO TSUKAHARA

1. Introduction

Since Moseley (1838) first analyzed the geometry of the *Nautilus* shell, its morphological features and allometric growth have been studied by many authors (Thompson, 1917; Huxley, 1932; Hirano, 1977; Hirano and Obata, 1979; Hirano *et al.*, 1980). Most previous work was based on small collections of individual specimens, the sexes and detailed locality record of which were unknown. In 1981 and 1983, a research project on the ecology of *N. pompilius* was carried out in the Philippines and the Fiji Islands as a joint venture of Japanese, Philippine, and Fijian scientists (under the leadership of Prof. S. Hayasaka of Kagoshima University). During this project, we examined the biometric properties of soft and hard tissues in captured animals (Hayasaka *et al.*, 1982; Tanabe *et al.*, 1983, 1985; Tsukahara, 1985). This chapter summarizes our work on *N. pompilius*, with special reference to sexual dimorphism expressed in allometric relationships and the variation in the shell and soft parts in the two populations from widely separated areas.

KAZUSHIGE TANABE • Geological Institute, Faculty of Science, University of Tokyo, Tokyo 113, Japan. JUNZO TSUKAHARA • Department of Biology, Faculty of Science, Kagoshima University, Kagoshima 890, Japan.

2. Materials and Methods

This chapter is based mainly on three collections of *N. pompilius* from the Philippines and the Fiji Islands. One collection consists of 52 specimens that were captured in the Tañon Strait (depth of 120–310 m) about 2–3 km off Bindoy Village, Negros Oriental, the Philippines, in 1981 by Hayasaka and others (for the actual catch records, see Fig. 3 and Table 10 in Hayasaka *et al.*, 1982). The other two collections consist of 101 and 62 specimens. They were captured from 11 points (depth of 180–550 m) about 10 km off Suva Harbor (Kandabu Passage) and from 6 points (depth of 250–470 m) in Mbengga Passage off Pacific Harbor, Viti Levu Island, Fiji Islands, in 1983 by a team of Japanese and Fijian scientists (for details, see Tanabe *et al.*,1985). Measurements by Tanabe *et al.* (1983) on the shells of 31 specimens of *Nautilus* from the Suva area (collected by Hayasaka and Shinomiya in 1981 from almost the same locations as those of the 101 specimens mentioned above) were also used in this study. The Philippines and the Fiji Islands represent the northwestern and southeastern marginal areas of the range of *N. pompilius* (see Fig. 1 in Chapter 3); therefore, the specimens from these two sites are especially suitable for analysis of geographic variation.

2.1. Field Methods

All the specimens studied were captured using baited traps, which were specially designed for the task by Philippine and Fijian fishermen. After capture, each animal was labeled, weighed, measured, and sexed. Sexual identification was based on the presence of primary reproductive organs (ovaries and testes). However, the sex of some specimens from the Suva area that were tagged and released in their habitat after laboratory observation was determined by the presence or absence of several accessory organs, such as the spadix (found in mature males only) and the nidamental gland (found in females only) (see Haven, 1977a,b; Saunders and Spinosa, 1978). Maximum shell diameter and apertural whorl breadth and height were measured on every specimen with a slide caliper (accuracy ± 0.05 mm). Except for one specimen from the Pacific Harbor area, all live animals were weighed using a dial scale (accuracy ± 1 g). These measurements were used to analyze average relative growth and sexual dimorphism. In addition to these basic characters, fresh gonads from some specimens collected in both the Philippines and the Fiji Islands were weighed to determine the relationship between sexual maturity and shell size.

2.2. Laboratory Methods

Measurements from all the specimens captured were analyzed to investigate the allometric relationships among characters in each sample and the variation in total weight and shell size. Relative growth and ontogenetic changes in shell form were examined in the shells of 28 animals from the Philippines and the Fiji Islands. These shells were cut and polished along the median plane, and the

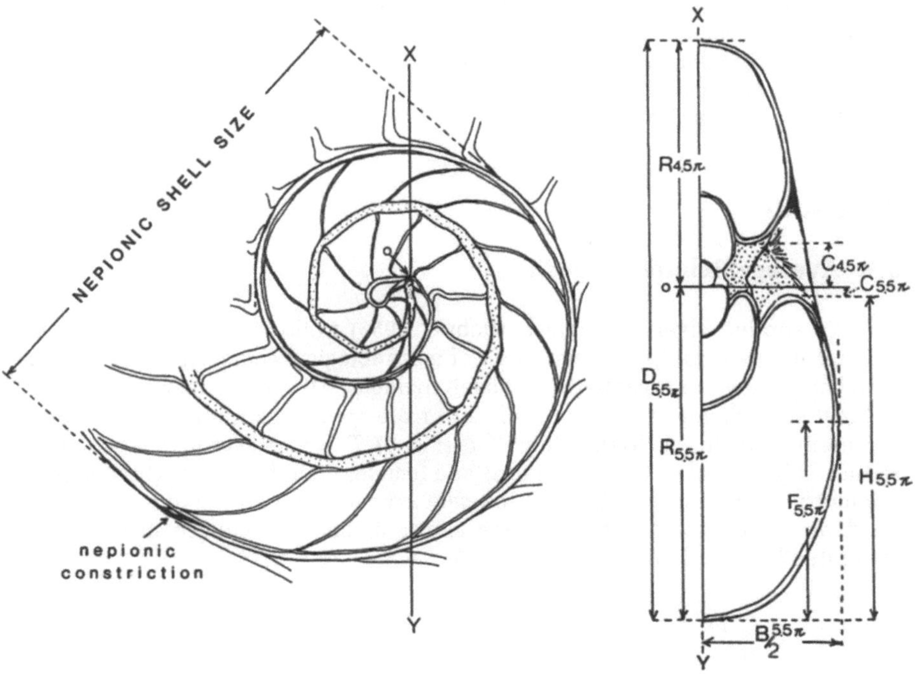

Figure 1. Basic morphology, orientation, and measurements of the shell of *N. pompilius* in median (*left*) and dorsal–ventral (*right*) sections. See the text for discussion. Reprinted with permission from Tanabe *et al.* (1985).

nepionic shell size was measured. Each "half" shell was also cut along the longitudinal axis of the cecum on the median section to produce a dorsal–ventral section (Fig. 1). Shell diameter (D), half-length of the whorl breadth (B/2), whorl height (H), distance from coiling axis to umbilical shoulder (C), and inner whorl height (F) were measured on the halved dorsal–ventral sections at intervals of 180°. Measurements of median and dorsal–ventral sections were made with a profile projector (Nikon V-12), attached to a digital micrometer (accuracy ±1 μm). On the basis of these measurements, the shell-form ratios proposed by Raup (1966) and Chamberlain (1976), such as whorl expansion rate [$(R_n/R_{n-1})^2$; n:whorl number, since $n > 2\pi$], distance of the whorls to the coiling axis (C/R), relative whorl thickness (B/H), and flank position (F/D), were calculated for each specimen at several growth stages. The following abbreviations were used in the biometric descriptions:

- N = Sample size
- $\bar{X}$ = Arithmetic mean
- V = Coefficient of variation
- s = Standard deviation
- O.R. = Observed range

$\sigma_{\bar{X}}$ = Standard error of the sample mean
α = Slope of allometry (growth ratio)
σ_{α} = Standard error of the growth ratio
r = Correlation coefficient
P = Probability level

3. Results

3.1. Average Relative Growth

We applied bivariate analysis (Imbrie, 1956) to this species to express the allometric relationship between a pair of morphological characters. The relationship between whorl breadth (B) and shell diameter (D) and between whorl height (H) and shell diameter (D) for female and male specimens from the Suva area is represented on double logarithmic plots (Fig. 2). The slopes (growth ratios) of the reduced major axes of B vs. D in males and females both indicate slightly negative allometry, while those of H vs. D indicate isometry. According to Kermack and Haldane (1950), the difference in the slopes of the two reduced major axes is evaluated by calculating the statistic z:

$$z = (\alpha_1 - \alpha_2)/\sqrt{\sigma_{\alpha_1}^2 + \sigma_{\alpha_2}^2}$$

where α_1 and α_2 ($\alpha_1 > \alpha_2$) are the slopes of the axes and σ_{α_1} and σ_{α_2} are the standard errors of the two slopes. Using this method, we calculated z values between the two reduced major axes of the male and female samples from the Suva area. The z values between females and males in the Suva sample (0.27 for log B vs. log D; 0.56 for log H vs. log D) are much smaller than the critical value 1.96 at $P = 0.05$; hence, the slopes in both sexes from the same area do not show

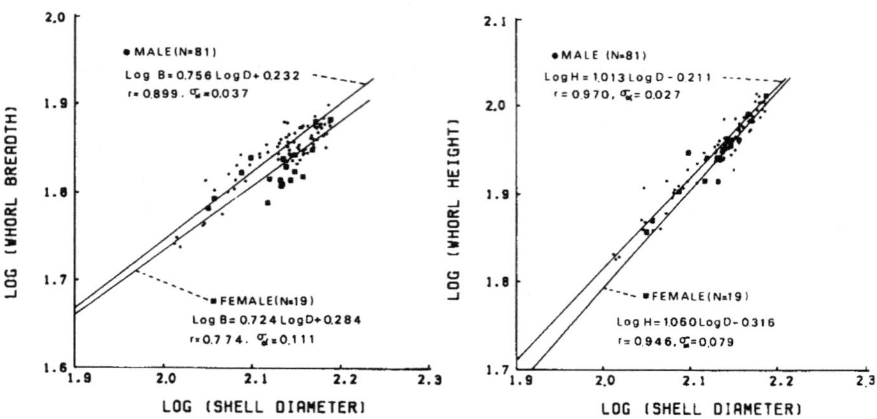

Figure 2. Double logarithmic scatter plots with reduced major axes showing the allometric relationship between whorl breadth and shell diameter (*left*) and between whorl height and shell diameter (*right*) in male and female specimens of *N. pompilius* from the Suva area of the Fiji Islands.

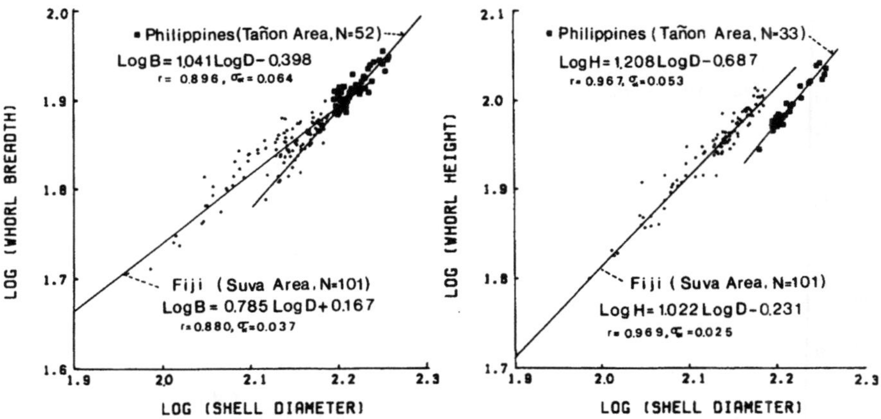

Figure 3. Double logarithmic scatter plots with reduced major axes showing the allometric relationship between whorl breadth and shell diameter (*left*) and between whorl height and shell diameter (*right*) in the samples (males and females) of *N. pompilius* from the Fiji Islands and the Philippines.

any statistically significant difference. Similarly, males and females from the Suva area show no significant difference in the slopes of their reduced major axes between pairs of log B, log H, and log D. Therefore, the linear trends of the reduced major axes between pairs of morphological characters in the samples from the Suva and Pacific Harbor areas of Fiji are essentially equivalent.

On the other hand, comparisons between the Fijian and Philippine collections reveal wide differences (Figs. 3 and 4). A significant difference occurs in the slopes of the reduced major axes between the two samples for whorl breadth

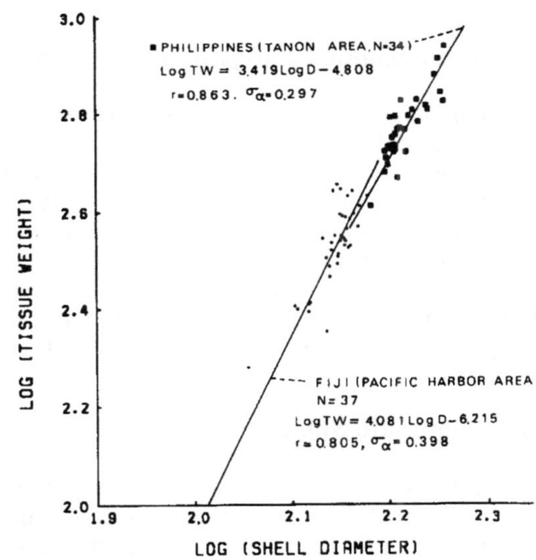

Figure 4. Double logarithmic scatter plots with reduced major axes showing the allometric relationship between tissue weight and shell diameter in the samples (males and females) of *N. pompilius* from the Fiji Islands and the Philippines.

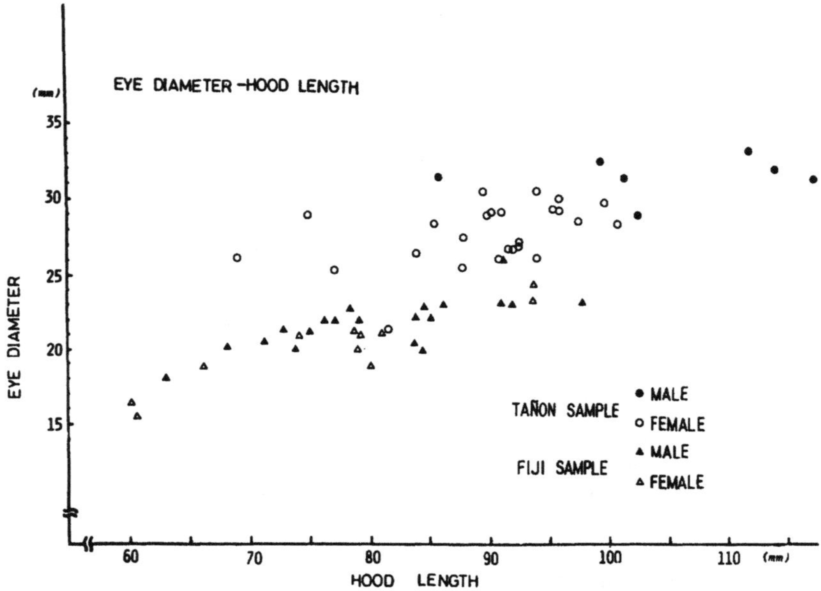

Figure 5. Double scatter plots showing the allometric relationship between eye diameter and hood length in the specimens of *N. pompilius* from the Tañon, Philippines and Suva, Fiji areas.

and whorl height vs. shell diameter (z = 3.46 for log B vs. log D; z = 16.23 for log H vs. log D; $z > 2.58 = Z_{p = 0.01}$; where $P < 0.01$ is significant). In addition, differences occur in the relative growth pattern of eye diameter vs. hood length (Fig. 5).

3.2. Ontogenetic Change of Shell Form

Ontogenetic changes in shell-form parameters were examined in 21 specimens (15 males and 6 females) from the Suva area (Fig. 6). Flank position and the distance of the whorls to the coiling axis remain nearly constant, whereas relative whorl thickness and whorl expansion rate decrease with growth. In the early stages of development, males and females show no significant difference in these shell-form parameters, but in the mature or almost mature stages (5.5π), males have larger values of relative whorl thickness and whorl expansion rate than do females. These larger values indicate that mature males usually have larger whorl radii and whorl breadths than mature females.

3.3. Nepionic Shells

The mean nepionic shell size and its 95% confidence interval in 8 Fijian specimens (27.50 ± 2.04 mm; *s* = 2.44 mm, *V* = 8.86, and *O.R.* = 24.4–32.6 mm) are slightly larger than those in 6 specimens from the Philippines (25.72 ±

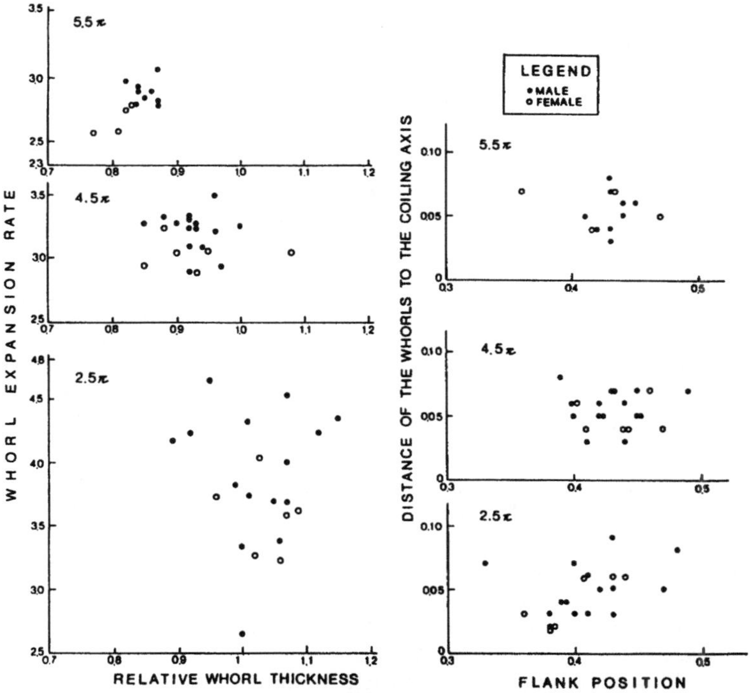

Figure 6. Two series of double scatter plots of the shell-form parameters of Raup (1966) in selected specimens of *N. pompilius* from the Suva area of Fiji showing the ontogenetic changes in shell form between males and females. Reprinted with permission from Tanabe *et al.* (1985).

1.04 mm; s = 0.99 mm, V = 3.84, *O.R.* = 24.7–27.1 mm). A Student's *t*-test, however, does not show a statistically significant difference ($P < 0.05$) in the mean nepionic shell size between the two samples. From the oxygen isotopic composition of *Nautilus* shells, Cochran *et al.* (1981) and B. E. Taylor and Ward (1983) concluded that the nepionic constriction forms at hatching. If this opinion is correct, the foregoing data probably reflect the similarity in the size of the postembryonic young.

3.4. Gonad Development and Sexual Dimorphism

The hypothesis that variation in total live weight and shell size in mature specimens can be attributed to sexual dimorphism has already been confirmed for *N. pompilius* from the Philippines and the Fiji Islands (Haven, 1977a; Ward *et al.*, 1977; Ward and Martin, 1980; Hayasaka *et al.*, 1982; Tanabe *et al.*, 1983), for *N. macromphalus* from New Caledonia (Ward and Martin, 1980), and for *N. belauensis* from Palau, West Caroline Islands (Saunders and Spinosa, 1978). In these species, mature males generally have larger shells and heavier soft parts

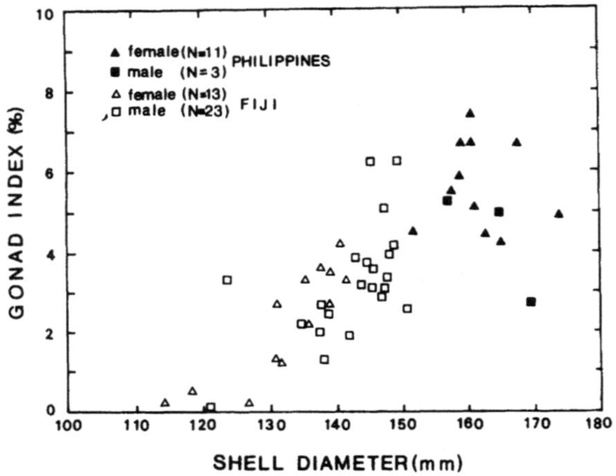

Figure 7. Double scatter plots of gonad index [testis or ovary weight/soft tissue weight (%)] vs. shell diameter in the selected specimens of *N. pompilius* from the Philippines and the Fiji Islands.

than mature females. In addition, Stenzel (1964), Haven (1977a), Saunders and Spinosa (1978), and Arnold (1984) have pointed out that the shells of mature males have a broader and rounder aperture than those of females of similar age. These differences are supported by the specimens collected from the Fiji Islands (Fig. 6), although principal component analysis of mature specimens from the Philippines revealed no statistically significant differences in the basic shell-form ratios between sexes (Tanabe *et al.*, 1983).

In describing shell maturity, the earlier literature has relied mainly on characteristic shell features such as a blackened and constricted aperture, fusing together of the final two or three septa, and a thickened last septum. In this study, we examined the relative gonad weight in selected specimens from the Fiji Islands and the Philippines. Scatter plots of the gonad index [ovary or testis weight/soft tissue weight (%)] vs. shell diameter are shown for two samples in Fig. 7. In the females from the Fiji Islands, ovary weight increases abruptly after a shell diameter of 120 mm and attains a maximum (≈ 15 g) at a shell diameter of approximately 140 mm. In contrast, gonad development is more marked in males, and full development of testes does not occur until a shell diameter of 140–150 mm. If we consider sexual maturity as 3.0 on the gonad index, males have larger shells than females at this point. The large gonad indices calculated for the Philippine specimens indicate that most of them (especially the females) have approached or attained sexual maturity (Tsukahara, 1985). The average shell diameter at maturity for both sexes in the Philippine sample is about 25 mm larger than in the Fijian sample [e.g., 171.64 mm and 147.21 mm for males from the Philippines (N = 15) and Fiji Islands (N = 14), respectively; see also Fig. 7, and Fig. 12 in Chapter 11]. Ward *et al.* (1977) and Hayasaka *et al.* (1982) have also noted that mature or almost mature specimens of *N. pompilius* from the Philippines are much heavier and larger than those from the Fiji Islands.

4. Conclusions

In the Fiji Islands and the Philippines, *N. pompilius* exhibits a distinct sexual dimorphism in weight, shell dimensions, and soft tissues. Mature males generally possess a broader whorl aperture than mature females. The two samples from the Philippines and the Fiji Islands are clearly different in overall weight, size at maturity, and the slopes of the allometric relationships of several morphological characters. The morphological differences between the samples from the two areas are much greater than those between sexes within the same area. In addition, Masuda and Shinomiya (1983) discovered a statistically significant difference in allele frequencies between samples from these two areas. From these lines of evidence, we conclude that populations of *N. pompilius* from the Fiji Islands and the Philippines are differentiated both genetically and morphologically.

ACKNOWLEDGMENTS. We are particularly indebted to Prof. Shozo Hayasaka of Kagoshima University for the use of his facilities and for his encouragement both in the field and in the laboratory. Our thanks to Dr. Alfredo E. Evangelista, Assistant Director of the Philippines National Museum, Prof. Angel C. Alcala, Vice President of Silliman University in the Philippines, and Dr. Uday Raj, Director of the Institute of Marine Resources (IMR) at the University of the South Pacific (Fiji), for their assistance and the use of their facilities for planning and operating our field studies. Thanks also to the other members of the project in the Philippines and Fiji for their assistance in the field and for their helpful discussions. We express our sincere gratitude to the staff of the IMR and Mr. Wilson Vailoces in Bindoy, Philippines, who helped us trap *Nautilus*. Dr. Neil H. Landman of the American Museum of Natural History and Dr. Andrew Swan of the University College of Swansea critically read the manuscript. This work was supported by Grants-in-Aid for Overseas Scientific Research (No. 57043059 for 1982; No. 58041055 for 1983; No. 59043050 for 1984).

Chapter 8

Biomineralization and Systematic Implications

REX E. CRICK and KEITH O. MANN

1. Introduction

Nautilus is the only cephalopod with an external, camerated shell composed of calcium carbonate ($CaCO_3$). Interest in the chemistry of the calcified structures of *Nautilus* has developed slowly during the past three decades, but sufficiently to provide insight into the chemical relationship of calcified structures, cephalopod biomineralization, and the physiology of *Nautilus*. Understanding the skeletal chemistry of *Nautilus* provides a means of studying the cephalopod physiochemical system, which facilitates the exploration of four factors: (1) the differences in the genetic makeup of the physiochemical system responsible for $CaCO_3$ production among species and populations, which forms the basis for biochemical taxonomy and its phylogenetic implications; (2) the ontogenetic variation in the chemistry of shell aragonite and its possible relationship to the life cycle of *Nautilus;* (3) the concentrations of trace elements in skeletal carbonate unrelated to $CaCO_3$ production, which serve as indicators of nutrient concentrations in seawater; and (4) the implications of all aspects of *Nautilus* chemistry for the study of similar problems in extinct cephalopods, the skeletons of which have survived in a chemically unaltered state.

Research prior to 1980. The following summary of investigations prior to 1980 is necessary because advances in technology, trace element chemistry, and knowledge of invertebrate physiological systems, having modified many of these early results, make comparisons with post-1980 work difficult.

REX E. CRICK • Department of Geology, University of Texas at Arlington, Arlington, Texas 76019. KEITH O. MANN • Department of Geology, University of Iowa, Iowa City, Iowa 52242.

In the context of general studies of the geochemistry and biochemistry of strontium (Sr) and magnesium (Mg), the skeleton of *Nautilus* was commonly used to describe the Sr or Mg chemistry of ectocochliate cephalopods. Several studies used a variety of analytical techniques to obtain Sr and Mg concentrations in shell carbonate of *Nautilus*, and without exception, each investigation addressed either Sr or Mg (Noll, 1934; Vinogradov, 1945; Chave, 1954; Odum, 1957; T. G. Thompson and Chow, 1955; Lowenstam, 1963; Price and Hallam, 1967; Hallam and Price, 1968). The importance of the elemental covariance of these basic skeletal components and its relationship to the animal's physiochemical system were not considered.

Chave (1954) and Odum (1957) conducted pioneering works in the Sr and Mg contents of cephalopods. Chave (1954) reported the Mg concentration in one shell of a single specimen of *Nautilus*, and Odum (1957) reported the Sr concentration in the shell of one specimen of *N. pompilius*. Lowenstam (1963) expanded on this work, although he reported the presence of Sr in only a few shells of *N. pompilius*. Price and Hallam (1967) and Hallam and Price (1968) published a series of short papers concerned with the Sr content of a few shells of *N. pompilius* relative to the Sr content of nautiloids and ammonoids. Their work strongly suggested that the Sr concentration varied with ontogeny and that possibly a functional relationship existed between the elemental variations and natural phenomena. The fact that the Sr and Mg concentrations reported by these authors did not agree is not surprising, because each value was based on only one or two specimens, and the consequent differences attest to the variability of the shell chemistry of *Nautilus*.

As renewed interest in the nonchemical aspects of *Nautilus* grew in recent years, investigators also began to study the organic and inorganic chemical components of the skeleton, calcified structures, and body fluids (A. W. Martin, 1975; McConnell and Ward, 1978; R. E. Crick and Ottensman, 1983; Masuda and Shinomiya, 1983; Woodruff *et al.*, 1983; Cochran and Landman, 1984; Crick *et al.*, 1984, 1985; Lowenstam *et al.*, 1984). Post-1980 studies differ from earlier studies in three ways, in that they: (1) consider the covariance of major and minor elements; (2) extend the studies to include the four common species of *Nautilus* from as many populations as possible; and (3) integrate skeletal chemistry and physiology.

2. Biomineralization in *Nautilus*

Like those of other mollusks, the physiological system of *Nautilus* governs the precipitation of calcium carbonate (Wilbur, 1976; Crenshaw, 1982). The physiochemical system of *Nautilus* is possibly the most complex among marine invertebrates, because it involves the production of several minerals: (1) precipitation of aragonite from an extracellular fluid to form the shell; (2) precipitation of aragonite from an extracellular fluid to form the septa; (3) precipitation of high-Mg calcite (4.4% $MgCO_3$) by the buccal papillae to form the beaks of the animal's masticatory apparatus; and (4) precipitation of statoconia, comprised of aragonite and amorphous calcium phosphate, in the statocyst. Although the shell and septum are aragonite, they are deposited by different portions of the mantle epithelium under different conditions and at different rates. The ultrastructure and elemental content of the shell and septum also differ.

The shell consists of three distinct aragonitic layers: the outer prismatic layer, the nacreous layer, and the inner prismatic layer. The outer prismatic layer is thinner than the inner prismatic layer. The nacreous layer forms the largest volume of shell aragonite, with the nacreous/prismatic layer thickness ratio being 3:1 along the venter of the shell.

A specialized area of the mantle secretes a proteinaceous sheet (the periostracum) that becomes tanned and serves as the substrate on which deposition of the outer prismatic layer occurs. The periostracum and mantle edge, or a combination of the periostracum, outer prismatic layer, and mantle edge, serve to isolate a region immediately adapical of the shell margin (the extrapallial space) from the marine environment. The extrapallial space contains an extracellular fluid that serves as a medium for the precipitation of shell carbonate that is out of chemical equilibrium with seawater. The animal maintains the composition of the extrapallial fluid by passing Ca, Sr, Mg, and other ions, as well as bicarbonate and CO_2, through the mantle epithelium to the extrapallial space. The actual production of aragonite occurs under conditions similar to inorganic production of $CaCO_3$ and is controlled by the disequilibrium of the fluid as CO_2 content is varied. The mantle also secretes organic material comprised mostly of protein (the organic matrix of the shell) into the extrapallial fluid, or directly onto the inner shell surface. Nucleation of aragonite takes place on this matrix, or on the existing crystal surfaces where these crystals, or the organic matrix, controls crystal orientation. As crystal growth proceeds, lateral growth continues slowly until the crystals coalesce and displace the surrounding organic matrix, which then becomes sandwiched between individual crystals. It is the organic framework of proteins that gives the shell of *Nautilus* its structural rigidity.

A *Nautilus* septum consists predominantly of nacreous aragonite situated between an adapical spherulitic–prismatic layer of aragonite and an adoral aragonitic, prismatic layer; both prismatic layers are very thin and commonly discontinuous over the surface of the septum (Mutvei, 1972).

An extrapallial space, occasionally present at the rear of the body, contains an extracellular fluid from which septal aragonite precipitates. Unlike shell aragonite, which is deposited in thin bands or layers at the leading edge of the aperture, septal aragonite is deposited simultaneously over the entire surface of the septum. This deposition requires the production of a large volume of extracellular fluid in a short period of time and the maintenance of this fluid for the 2–3 weeks required to grow a septum (Ward *et al.*, 1981).

Uroliths consist of a nucleus of hydroxyapatite [$Ca_5(PO_4)_3OH$] (always present in the renal fluids) surrounded by concentric layers of Mg-oxalate dihydrate ($MgC_2O_4 \cdot 2H_2O$) (Crick *et al.*, 1985). A. W. Martin (1975) and Schipp and Martin (1981) described uroliths as products of secretion rather than excretion. They suggested that uroliths serve as reservoirs for Ca, which is remobilized and used during the precipitation of septal aragonite. Observational and chemical data argue against the function of uroliths as Ca reservoirs and indicate that their occurrence is indeed a by-product of the episodic production of large quantities of extracellular fluids in conjunction with septal formation (Crick *et al.*, 1985).

Two possible mechanisms may provide the processes necessary to produce uroliths in the renal appendages of *Nautilus*. First, following the forward repositioning of the mantle in the living chamber in preparation to form a new septum, the extracellular fluid required for aragonite production must be produced in less than 8 days (Ward *et al.*, 1981). This time constraint results from the 6- to 8-day lapse between the time the animal moves forward (no extracellular fluid is present prior to this move) and the time that the new septum is sufficiently dense to be observed on X-ray radiographs. In view of the discriminative powers of *Nautilus* with respect to Mg [99.99% of the Mg present in seawater is excluded from shell carbonate (Crick *et al.*, 1985)], rapid generation of the volume of Ca-rich extracellular fluid to fill the extrapallial space at the rear of the mantle would result in Mg enrichment of the blood. The blood of *Nautilus* already contains three times more Mg than Ca (Greenwald and Ward, 1982). At the same time, the animal secretes additional quantities of proteins into this extracellular fluid to provide an organic matrix for septal aragonite. Under normal circumstances, renal fluid of *Nautilus* contains high concentrations of organics (e.g., glycoproteins, glycopeptides, polypeptides) of the type that form the organic matrix of the shell and septa. The presence of such proteins and polypeptides in the renal fluid of mammalian kidneys has been shown to suppress the formation of kidney stones either by promoting ion complexing, which lowers the saturation level of the fluid, or by adsorbing onto crystal faces and disrupting crystal growth (Robertson *et al.*, 1982). Kidney stone formation then takes place with these organics either absent or in diminished concentrations. The increased concentration of Mg ions and the possible reduction of organic content in the renal fluid of *Nautilus* may well control urolith production.

Second, the production of cameral liquid [the fluid that fills the chamber vacated by the animal as it moves forward to form a new septum (Greenwald and Ward, 1982)] may provide an additional source of chemical imbalance in the blood and renal fluids. Greenwald and Ward (1982) demonstrated that the Mg content of cameral liquid is significantly less than that of the blood or seawater and could be only an ultrafiltrate of the blood or a secreted body fluid. The organic and inorganic composition of cameral fluid is unknown.

In addition to the variety of mineral precipitates produced by *Nautilus*, equally impressive is the balance in timing maintained among the various physiochemical activities. Continuous precipitation of the shell occurs at a rate that may vary during ontogeny (Saunders, 1983). Septal aragonite is produced episodically in response to growth with the subsequent movement of the animal forward in the living chamber. The production of uroliths in the renal appendages appears to correspond to changes in the physiochemical system as it prepares to produce fluid from which septal aragonite will be precipitated.

The only other mineralized structures of *Nautilus* are the statoconia contained in the statocyst. Their aragonitic composition includes a high concentration of amorphous calcium phosphate (Lowenstam *et al.*, 1984). The statoconia possess considerably greater Sr and Mg contents than shell and septal aragonite, but it is unclear what portion of the Sr and Mg is contained in the calcium phosphate.

3. Trace Elements and Biomineralization

Factors that affect elemental concentrations in skeletal carbonate can be grouped into three major categories: (1) crystallographic controls, (2) environmental controls, and (3) physiological controls.

Variations of Sr and Mg concentrations in shell carbonate, within and among species and populations, cannot be attributed to differences in the $CaCO_3$ polymorph, because the septa and shell of *Nautilus* are comprised of aragonite. In addition, biochemically precipitated aragonite has been shown to accommodate Sr and Mg in higher concentrations than are present in *Nautilus*. Therefore, the levels of these elements in *Nautilus* are not at the limits imposed by crystal chemistry.

Environmental factors, especially temperature, can affect the composition of shell carbonate. While it has been demonstrated that temperature affects trace and minor element concentrations in shell carbonate of many mollusks from shelf environments (Pilkey and Goodell, 1964; Harris and Pilkey, 1966; Dodd, 1967; Wilbur, 1972; Smith *et al.*, 1979), two factors act to diminish the influence of the environment on trace and minor element concentrations within the shell of *Nautilus:* (1) The study of Nansen Cast bottle data for all seasons over a period of four decades illustrates that the salinity and temperature of subsurface waters between the depths of 200 and 500 m (the common depth range of *Nautilus*) are isohaline and isothermal and that differences in salinity and temperature between population sampling areas are slight (see Fig. 3). (2) Brewer (1975) and T. R. S. Wilson (1975) reported that the concentrations of Ca, Sr, and Mg vary linearly with salinity within the range of 25–40ppt. Therefore, if variations in subsurface salinities and temperatures occur, the physiochemical system of *Nautilus* would continue to experience the same relative concentrations of the three elements crucial to $CaCO_3$ production regardless of its geographic location.

There is, of course, some influence of the external environment on physiological functions as they affect shell formation, and potentially some contamination of extracellular fluids by seawater, but there is no evidence that these factors have a significant influence on shell chemistry (Wada and Fujinuki, 1976). Furthermore, it is reasonable to expect that these types of influences would affect all species or populations equally, causing similarities in shell chemistries among species and populations. Such similarities are not consistent with the reported data. It is perhaps because of these isothermal and isohaline conditions that species of *Nautilus* show considerably less variability in elemental concentrations of skeletal carbonate than do other mollusks.

The physiological system of mollusks is controlled genetically (Wilbur, 1972); differences in the ionic composition of extracellular fluid and the elemental concentration of shell carbonate reflect genetic differences among physiological systems. Physiological aspects of *Nautilus* involve both the extrapallial fluid, in which shell aragonite is produced, and the transfer of materials from this fluid to the site of crystal formation (Wilbur, 1972, 1976). Ions that form the aragonitic crystals of the shell and septum of *Nautilus* (and of mollusks in general) pass from the mantle cavity, containing the external medium and food, to the tissue and blood sinus, and finally into one or more extrapallial spaces, located between the growing surface of the shell wall and the mantle epithelium. Skeletal carbonate contains concentrations of elements in proportion to their concentration in the extracellular fluid (Wilbur, 1972, 1976; Wada and Fujinuki, 1976; Crenshaw, 1982). The compositional regulation of the extracellular fluids by the mantle cells is genetically controlled and is expected to vary with species (Wilbur and Owen, 1964; Wilbur, 1972). Such chemical differences have been observed among extant bivalve genera (Wilbur, 1972) and among extinct congeneric species of nautiloid cephalopods (R. E. Crick and Ottensman, 1983). In light of these observations, different concentrations of trace elements in shell aragonite of four extant congeneric species of *Nautilus* would not be unexpected.

4. Materials and Methods

The following review of recent research that we have conducted, both published and unpublished, is based on many laboratory analyses. Sampling, sample preparation, and chemical analyses were uniformly carried out; however, when these procedures were amended or altered, the changes are specified in the text.

Specimens were prepared for sampling by sectioning the shells in the plane of symmetry to expose the septa. Potential contaminants (e.g., surficial films, organic membranes, epibionts) were removed by an air abrasive from the internal and external surfaces of the shell and septa. For the purpose of comparing the chemistry of contemporaneously deposited shell and septal aragonite, a sampling scheme developed by Mann (1983) was used; this scheme allowed the chemical correlation of samples separated in space, but not in time. For each septum sampled, a strip of shell was removed parallel to, and between, curvilinear growth lines corresponding to angles 85° and 105° adoral from each septum. This was done because X-ray radiography of living *Nautilus* revealed that the shell aperture lies at a point approximately 81–85° adoral from that septum at the initiation of septal formation, and septal formation ceases at a point approximately 106–110° adoral from that septum (Ward *et al.*, 1981). The 85–105° interval was chosen as a conservative estimate of the shell aragonite formed during the formation of a coeval septum. These septum–shell pairs represent time–growth increments, with the first formed increment labeled as septum–shell pair 1. Sample preparation for chemical analyses and the methods of chemical analysis were the same as outlined by Crick *et al.* (1984).

5. Ontogenetic Concentrations of Strontium and Magnesium

Mean concentrations of Sr and Mg in parts per million (ppm) of the shell and septa from one specimen from each of four species of *Nautilus* are reported in Table I. Concentrations are categorized as representing septum–shell pairs 7th to last. Coefficients of variation (in parentheses following each standard error) are used as a measure of variation within and among species.

Three of the specimens (*N. belauensis*, *N. macromphalus*, and *N. scrobiculatus*) show considerable differences in the Sr and Mg content of the early and late portions of the shell (Fig. 1). This variation is most pronounced in the Mg

Table I. Means and Standard Errors of Strontium and Magnesium for Septum to Shell Pairs 7th to Last and 20th to Last for Four Species of *Nautilus*[a]

Species[b]	Samples (N)	Strontium Shell	Strontium Septum	Magnesium Shell	Magnesium Septum
N.b.	29	2143 ± 61.3 (15.4)	2414 ± 149.7 (33.4)	182 ± 7.5 (22.3)	189 ± 19.4 (55.2)
	16	2326 ± 60.4 (10.4)	2396 ± 64.8 (10.8)	180 ± 5.6 (12.4)	151 ± 6.8 (18.1)
N.m.	24	1697 ± 22.8 (6.6)	1767 ± 24.5 (6.8)	505 ± 74.8 (72.5)	981 ± 155.2 (77.5)
	11	1767 ± 31.5 (5.9)	1767 ± 35.9 (6.7)	202 ± 8.7 (14.4)	206 ± 12.1 (19.5)
N.p.	25	1618 ± 17.5 (3.6)	1627 ± 64.8 (13.2)	342 ± 30.8 (45.0)	430 ± 37.9 (44.1)
	12	1662 ± 12.3 (2.5)	1653 ± 12.3 (2.6)	304 ± 31.1 (35.5)	449 ± 31.1 (30.0)
N.s.	25	1705 ± 33.3 (9.7)	1670 ± 18.4 (5.5)	833 ± 114.4 (67.7)	1256 ± 100.3 (39.9)
	12	1740 ± 24.5 (4.8)	1653 ± 23.6 (4.9)	639 ± 100.8 (54.7)	1064 ± 70.7 (23.0)

[a] Values are in parts per million. The values in the first row for each species are for septum–shell pairs 7th to last; those in the second row are for pairs 20th to last. The values in parentheses following the standard errors are coefficients of variation (standard deviation as a percentage of the mean).

[b] Species: (N.b.) *N. belauensis*; (N.m.) *N. macromphalus*; (N.p.) *N. pompilius* (Philippine Islands); (N.s.) *N. scrobiculatus*.

content of both the shell and septa of *N. macromphalus*, most erratic in the Mg content of the shell and septa of *N. scrobiculatus*, and only moderately evident in the Sr and Mg content of septa of *N. belauensis*. The shell and septal chemistry of *N. pompilius* differs from that of the other three species in having an Mg concentration that varies, apparently randomly, throughout observed ontogeny, whereas the Sr concentration remains essentially invariant.

The patterns produced by Mg and Sr concentrations for the specimens (Fig. 1) can be explained in only one way. The pronounced variability of Mg in the shell and septa of *N. macromphalus* and *N. scrobiculatus* (and to a much lesser degree in the early septa of *N. belauensis*) must reflect a physiochemical system that "matures" with ontogeny relative to its ability to reduce the level of Mg in skeletal aragonite. The *N. scrobiculatus* specimen never achieves the control over levels of Mg achieved by *N. macromphalus* by the 20th septum–shell pair, whereas the only slightly variable nature of Mg and Sr in early septa of *N. belauensis* ceased after the 10th septum–shell pair.

These observations lead to the consideration of why a secular decrease in the amount of Mg in skeletal aragonite exists. The most probable explanation lies with the kinetics of crystal chemistry and structure. The major inorganic constituents of the *Nautilus* shell, in order of relative abundance, are Ca, Sr, and Mg, with ionic radii of 1.0, 1.13, and 0.72, respectively. Although aragonite is polymorphous with calcite, the aragonitic structure is stable for ions larger than Ca (>1.0 radius), whereas the calcitic structure is stable for ions smaller than Ca (<1.0 radius). This stability factor for aragonite is conferred by the aragonitic structure, which consists of planar trigonal CO_3 groups with Ca ions in positions of hexagonal close packing; this situation gives aragonite its pseudo-hexagonal character. It is important to note that each CO_3 group lies between six Ca atoms and is arranged such that each O is linked to three Ca ions. This produces a radius ratio (radius cation:radius anion) of Ca:O of approximately 0.71, which is intermediate between the coordination numbers 6 and 8. Similar radius ratios of Sr and Mg to O are 0.81 and 0.51, with corresponding coordination numbers of 8

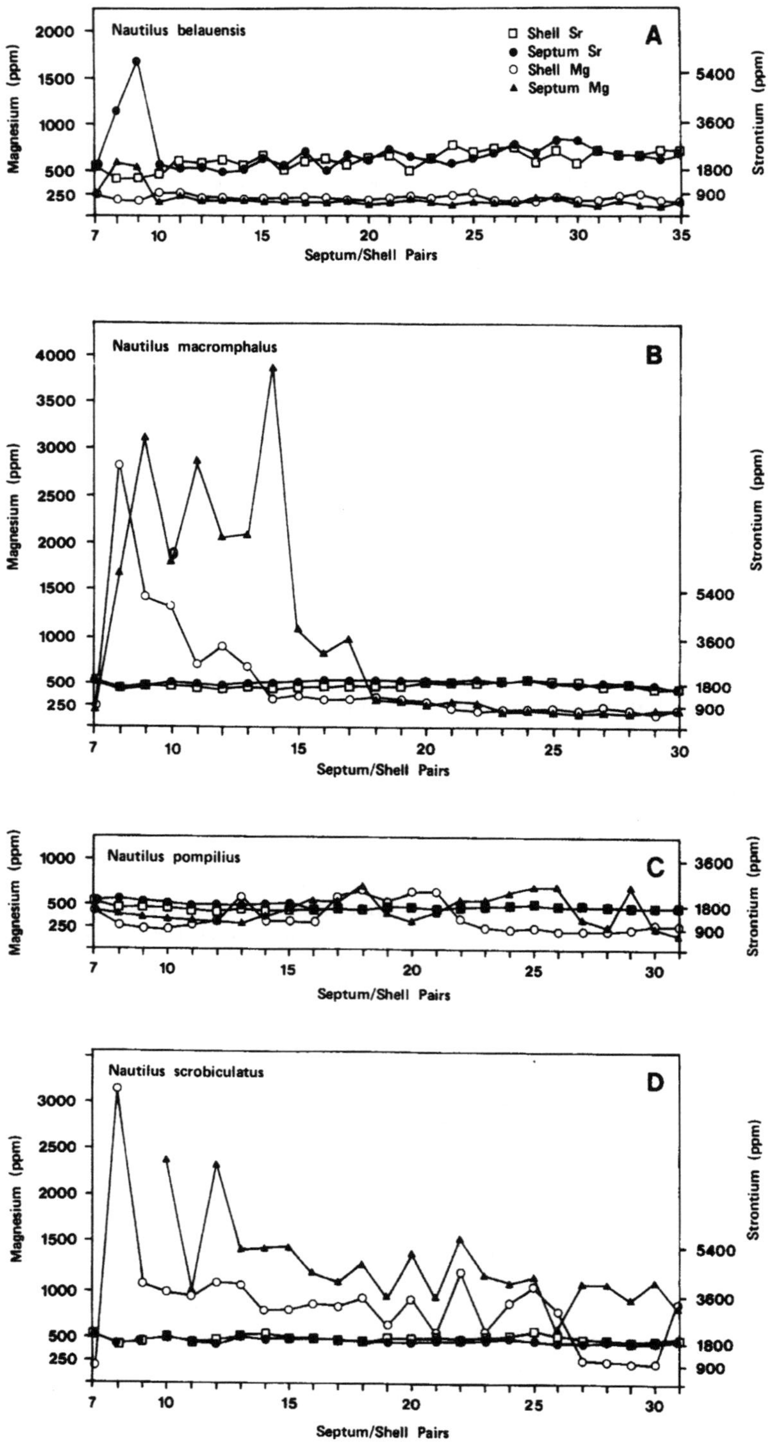

Nautilus belauensis
A
□ Shell Sr
● Septum Sr
○ Shell Mg
▲ Septum Mg
Magnesium (ppm)
Strontium (ppm)
Septum/Shell Pairs
Nautilus macromphalus
B
Nautilus pompilius
C
Nautilus scrobiculatus
D

and 6 (Mason and Moore, 1982). Thus, the Sr ion substitutes for the Ca ion in the aragonitic structure more easily than does the Mg ion. Further, the inclusion of Mg in the crystal lattice at Ca sites distorts the orthorhombic structure as the lattice "adjusts" to the smaller Mg ion; consequently, high-Mg aragonite will be more unstable than low-Mg aragonite (Mason and Moore, 1982).

An important consideration in this reasoning is a comparison between the relative ratios of Ca, Mg, and Sr in world oceans (1.0:3.2:0.02) and the ratios of average Ca, Mg, and Sr concentrations for *Nautilus* (1.99:0.0003:0.002). The body fluids of mollusks are essentially isosmotic with sea water, and for *Nautilus* to produce aragonite with these ionic ratios, quantities of Mg on the order of 100 times the amount of Ca required for aragonite production must be excluded from the makeup of extrapallial fluids. Why *Nautilus* discriminates against Sr is not clear, although the larger ionic diameter of Sr would tend to slightly distort the aragonitic lattice and make Sr less desirable as a skeletal component. Excluding as much Mg as possible, and using available Sr when necessary, creates the most efficient means of producing a shell and septum composed predominantly of nacre (pseudo-hexagonal crystals). Organisms of lower evolutionary grade with skeletons of aragonite commonly do not discriminate against Sr relative to Ca; e.g., taxa as diverse as planktonic foraminifera (Graham *et al.*, 1982) and coelenterates (Amiel *et al.*, 1973) commonly have an Sr concentration an order of magnitude greater in shells and tests than does *Nautilus*. Most important, however, is the magnitude of the exclusion of Mg from the skeleton of *Nautilus*. It seems reasonable to assume that genetic differences in physiochemical systems might be more pronounced with respect to Mg than Sr. This inference is indirectly supported by our data.

6. Strontium and Magnesium Concentrations among Species

Statistical testing of the data in Table I showed that by restricting the sampling interval to septum–shell pairs 20th to last, analysis of as few as five random samples would produce the chemical signature with only a slight increase in the standard error (Table I).

To compare the Sr and Mg concentrations among *Nautilus* species, a total of 16 shells were sampled: 5 shells each of *N. belauensis* and *N. macromphalus*, 4 shells of *N. pompilius*, and 2 shells of *N. scrobiculatus*. The small sample size for *N. scrobiculatus* reflects its rarity. Five randomly selected septum–shell pairs, obtained from septum–shell pairs 20th to last, were analyzed from each of the 16 specimens.

The means listed in Table II and the data plotted in Fig. 2 show that the Sr concentrations of the shell and septa vary within reasonably narrow limits for each species. Concentrations of Sr in the shell and septa are least variable for *N.*

←

Figure 1. Distribution of Mg and Sr in septum and shell-wall pairs throughout the ontogeny of four species of *Nautilus* beginning with the 7th septum–shell pair. Mg and Sr concentrations for the septum of *N. scrobiculatus* (D) begin at septum–shell pair 10 because earlier septa were destroyed during preparation. Key: (□) shell Sr; (■) septum Sr; (○) shell Mg; (●) septum Mg. Reprinted with permission from Crick *et al.* (1984).

Table II. Concentrations and Atomic Ratios of Strontium and Magnesium for Four Species of *Nautilus*[a]

Species	Number of samples	Shell				Septa			
		Sr		Mg		Sr		Mg	
		ppm	Sr/Ca × 10^{-2}	ppm	Mg/Ca × 10^{-2}	ppm	Sr/Ca × 10^{-2}	ppm	Mg/Ca × 10^{-2}
N. belauensis	25	2002	0.229	160	0.066	2118	0.242	164	0.068
		± 41.7	± 0.0048	± 5.8	± 0.0024	± 36.0	± 0.0041	± 9.1	± 0.0037
N. macromphalus	25	1762	0.202	176	0.073	1668	0.193	174	0.072
		± 17.1	± 0.0020	± 9.0	± 0.0037	± 16.2	± 0.0019	± 10.0	± 0.0041
N. pompilius	20	1818	0.208	196	0.081	1829	0.209	273	0.112
		± 41.6	± 0.0048	± 9.1	± 0.0037	± 49.3	± 0.0056	± 38.3	± 0.0158
N. scrobiculatus	10	1917	0.219	378	0.156	1836	0.210	615	0.254
		± 56.6	± 0.0065	± 88.0	± 0.0362	± 66.3	± 0.0076	± 125.6	± 0.0517

[a] The values are means and standard errors of the means.

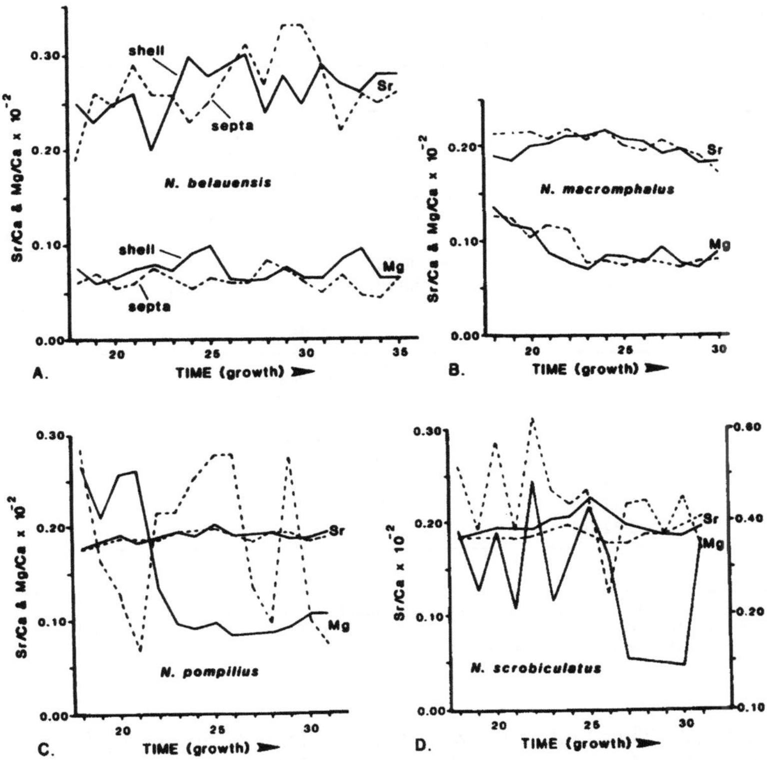

Figure 2. Graphical representation of atomic ratios Sr/Ca and Mg/Ca relative to time–growth increment for four species of *Nautilus* (see the text for explanation of the sampling procedure). Key: (—) shell data; (– – –) septal data. (D) Mg/Ca ratios are plotted on a scale that is twice that of Sr/Ca ratios. Reprinted with permission from Crick *et al.* (1984).

macromphalus and most variable for *N. scrobiculatus*. Concentrations of Mg in shell and septal aragonite are least variable for *N. belauensis* and most variable for *N. scrobiculatus*. The Sr content of shell and septal aragonite is highest in *N. belauensis* and lowest in *N. macromphalus*. The Mg content of the shell and septa is highest in *N. scrobiculatus* and lowest in *N. belauensis*.

The results of Student's *t*-tests (Table III) reveal that neither Sr nor Mg used alone can discriminate among species. The power of discrimination of combined Sr and Mg concentrations in shell and septal aragonite was tested by performing a discriminant function analysis on the chemical data. The data meet all criteria for discriminant analysis (Davis, 1973), and data for each species were treated as a group. Two variables, atomic ratios Sr/Ca and Mg/Ca, were transformed into one discriminant function each for the shell data and for the septal data. The Mahalanobis distance function (D^2) was used to measure the separation of multivariate means of the discriminant functions. D^2, expressed in units of pooled variance, was transformed into an F ratio and used for tests of significance (Davis,

Table III. Matrices of Computed Student's *t*-Test Values for Strontium and Magnesium Contents of Shell and Septa among Species[a]

Species	N.b.	N.m.	N.p.	N.s.
	Values for strontium			
	Shell			
N.b.	—	5.31 (4)	3.08 (2)	1.14 (0)
N.m.	10.89 (4)	—	−1.34 (0)	−3.48 (2)
N.p.	4.85 (4)	−2.96 (2)	—	−1.38 (0)
N.s.	3.99 (3)	−3.06 (2)	−0.09 (0)	—
	Septa			
	Values for magnesium			
	Shell			
N.b.	—	−1.51 (0)	−3.46 (2)	−3.94 (3)
N.m.	−0.74 (0)	—	−1.53 (0)	−3.58 (2)
N.p.	−3.04 (2)	−2.74 (2)	—	−2.91 (2)
N.s.	−5.72 (4)	−5.57 (4)	−3.33 (2)	—
	Septa			

[a] The numbers in parentheses indicate the probability of committing Type 1 error when rejecting null hypothesis: (0) not significant; (1) <0.05; (2) <0.01; (3) <0.001; (4) <0.001. Degrees of freedom = $(n_1 + n_2) - 2$. See Table I for sample sizes and footnote b to Table I for the names of species.

1973) (Table IV). These ratios show that the contents of Sr and Mg in the shell are significantly different among all species, with the exception of *N. macromphalus* and *N. pompilius*, and that the contents of Sr and Mg in the septa are significantly different among all species (Table IV).

The Sr, Mg, and Ca chemistry of the shell and septa of the four species of *Nautilus* records basic differences or similarities in the physiochemical systems of these species with respect to the process of biomineralization. Because previous investigations have shown that the ionic composition of the extracellular fluid

Table IV. Matrix of Computed *F* Ratios for Testing the Significance of Separation of Groups of Species on Discriminant Function[a]

Species[b]	N.b.	N.m.	N.p.	N.s.
	Shell			
N.b.	—	12.49 (4)	6.70 (2)	16.03 (4)
N.m.	42.31 (4)	—	0.94 (0)	19.02 (4)
N.p.	15.54 (4)	8.18 (3)	—	12.84 (4)
N.s.	28.38 (4)	36.75 (4)	15.98 (4)	—
	Septa			

[a] The numbers in parentheses indicate the probability of committing Type 1 error when rejecting null hypothesis; see the footnote to Table III for probabilities. Degrees of freedom = 2.75.

[b] See footnote b to Table I for the names of species.

controls the ionic composition of the shell carbonate, the statistical significance of these chemical differences is interpreted as evidence that the Sr and Mg concentrations in extracellular fluids used in the formation of shell and septal aragonite are different in certain species. These differences can be ascribed only to differences in the ionic concentration of the blood of these animals or to differences in the discriminatory properties of the outer mantle epithelium that produces the extracellular fluid. The latter is believed to be the case, because the blood of marine mollusks is typically isosmotic with seawater (Wilbur, 1972; Wada and Fujinuki, 1976) and because Greenwald and Ward (1982) reported that the concentrations of Ca and Mg in the blood of *N. macromphalus* are similar to those of seawater.

Work of a similar nature, using skeletons of Upper Carboniferous nautiloid cephalopods preserved in bitumen (Crick and Ottensman, 1983), demonstrated that three previously described species of the genus *Mitorthoceras* could be distinguished on the basis of the Mg content of the shell wall, although the Sr content of the shell wall was not significantly different. Unpublished data of Crick on Jurassic ammonites and belemnites and Cretaceous belemnites show similar patterns among species and genera with respect to both Sr and Mg in shell walls, septa, and (in the case of belemnites) rostra.

7. Strontium and Magnesium Differences among Populations

The broad geographic distribution of *N. pompilius* is unique among the known species of *Nautilus;* although it is best known from the Philippines and from the Fiji Islands, there appear to be many populations between these localities (Fig. 3). A major difference among populations of *N. pompilius* is shell diameter; the Fiji population has a mean diameter of approximately 140 mm (Ward et *al.*, 1977; Zann, 1984) and the Philippine population a mean diameter of 165 mm (Chapter 3). Because size by itself is a poor characteristic for distinguishing species, and because of the observed normal variation of characters within *N. pompilius*, Saunders (1981a) suggested that members of these populations might rep-

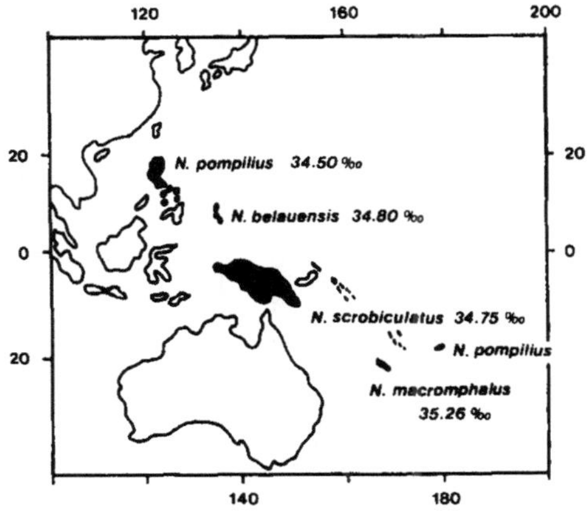

Figure 3. Location map showing populations of *Nautilus* studied. Salinities (ppt) are average values for the depth range of 200–500 m. The salinity for the Fiji populations of *N. pompilius* (20°S latitude) is the same as that given for *N. scrobiculatus*.

Table V. Means and Standard Errors of Strontium and Magnesium (ppm) for Shell and Septum Aragonite of Populations of *Nautilus pompilius* from the Philippines and the Fiji Islands[a]

		Philippines			
		Shell		Septa	
		Sr/Ca	Mg/Ca	Sr/Ca	Mg/Ca
	Fiji	1663 ± 12.3 (3.3)	304 ± 31.1 (45.8)	1654 ± 12.2 (3.3)	449 ± 37.9 (37.7)
Shell					
Sr/Ca	1917 ± 42.0 (10.9)	2.31	—	—	—
Mg/Ca	211 ± 8.9 (21.2)	—	4.99	—	—
Septa					
Sr/Ca	1820 ± 24.5 (6.7)	—	—	2.25	—
Mg/Ca	223 ± 4.6 (10.3)	—	—	—	6.63

[a] The values in parentheses are coefficients of variation (standard deviation as a percentage of the mean). The values along the diagonal are approximate Student's t-test values of corresponding chemistries. All tests are significant at $p < 0.05$ and 43 *df*.

resent subspecies of *N. pompilius*. Subsequently, Masuda and Shinomiya (1983) investigated intraspecific genetic variation in *N. pompilius* by studying allele frequencies for nine gene loci of specimens from the Philippines and the Fiji Islands. Their results showed significant differences in allele frequencies at two loci, and they concluded that gene flow between the two areas is limited to the point that genetic differentiation has taken place.

To further test the hypothesis that the Philippine and Fiji populations are genetically different, Sr and Mg concentrations of five shells from Fiji and four shells from the Philippines were compared. The mean concentrations of 45 shell–septum samples (5 septum–shell pairs were obtained from each specimen) are listed in Table V, together with results of Student's *t*-tests of their differences among populations. The data reveal that the Mg content of shell and septal carbonate of the Fiji population is less variable than those of the Philippine population (lower coefficients of variation), but that the Sr content of the Fiji population is more variable than that of the Philippine population. The shell and septal carbonate of the Fiji population contains approximately 10–15% more Sr and 30–50% less Mg than the Philippine population. The Sr and Mg chemistries among shell walls and septa are significantly different between these two populations (Table V). These differences are comparable in magnitude to those reported above for four species of *Nautilus*. Thus, with regard to biomineralization, the physiochemical systems of Fiji and Philippine populations are as distinctive as those of recognized species of *Nautilus*. We know of no precedent for defining species on the basis of significant differences in the trace and minor elements of skeletal $CaCO_3$; however, these differences, coupled with the geographic isolation of the two populations, the observed morphological differences in shell size, and the reported differences in allele frequencies at two gene loci (Masuda and Shinomiya, 1983), suggest that these two populations of *N. pompilius* represent at least separate subspecies of *N. pompilius*, if not separate species. It is unlikely that two populations of the same species could have evolved physiochemical systems that produce $CaCO_3$ chemistries as statistically different as recognized species without the influence of differing genetic makeups.

8. Chemical Differences in Nacreous and Prismatic Aragonite among Species

Price and Hallam (1967) documented that the Sr content of the nacreous layer of the shell wall for two specimens of *N. pompilius* was greater than that of the prismatic layer. At present, nothing is known of the Mg content or of how Sr and Mg concentrations of nacreous and prismatic aragonite differ among species of *Nautilus*. Toward the end of remedying this situation, the Sr and Mg concentrations of these layers were analyzed separately. To determine the degree to which the chemistry of the shell and septa within and among species is influenced by the chemistry of nacreous and prismatic aragonite, contiguous outer prismatic and nacreous layers of the shell wall at five arbitrary septum–shell sample sites were mechanically separated and analyzed for Sr and Mg. Because of the extreme thinness of the inner prismatic layer of the shell wall, this layer was not analyzed.

Table VI. Means and Standard Errors of Strontium and Magnesium (ppm) for Spherulitic–Prismatic and Nacreous Layers of the Shell and the Nacreous Layer of Septa of Four Species of *Nautilus* and Results of Approximate Student's *t*-Tests to Determine Differences among Chemistries of Layers and Various Structures[a]

	Shell				Septum	
	Prismatic		Nacre		Nacre	
Species[b]	Sr	Mg	Sr	Mg	Sr	Mg
N.b.	2118 ± 58.6 (6.8)	197 ± 12.4 (15.3)	2013 ± 55.1 (6.7)	180 ± 5.1 (6.9)	2634 ± 96.3 (8.9)	155 ± 3.6 (5.7)
N.m.	1750 ± 25.5 (3.5)	199 ± 10.0 (12.1)	1689 ± 26.3 (3.8)	209 ± 10.2 (12.0)	1784 ± 26.3 (3.6)	180 ± 3.6 (5.0)
N.p.	1880 ± 39.4 (5.1)	216 ± 5.8 (6.5)	1539 ± 19.3 (3.1)	321 ± 27.0 (20.6)	1662 ± 12.3 (1.8)	240 ± 16.3 (16.8)
N.s.	1584 ± 15.8 (2.5)	228 ± 13.6 (14.6)	1785 ± 8.8 (1.1)	870 ± 64.4 (18.1)	1663 ± 24.5 (3.6)	1100 ± 15.8 (3.5)

	Strontium							
	Nacre				Prismatic			
Species[b]	*N.b.*	*N.m.*	*N.p.*	*N.s.*	*N.b.*	*N.m.*	*N.p.*	*N.s.*
	Matrix A				Matrix B			
Nacre								
N.b.	—	4.88	7.46	3.72	−0.86	—	—	—
N.m.	—	—	4.26	−3.30	—	−1.44	—	—
N.p.	—	—	—	10.86	—	—	−7.25	—
N.s.	—	—	—	—	—	—	—	10.12
					Matrix C			
Prism.								
N.b.	—	—	—	—	5.28	2.98	7.90	
N.m.	—	—	—	—	—	−2.73	4.92	
N.p.	—	—	—	—	—	—	6.41	
N.s.	—	—	—	—	—	—	—	
	Magnesium							
	Matrix D				Matrix E			
Nacre								
N.b.	—	−2.42	−4.68	−9.77	−1.28	—		—
N.m.	—	—	−3.48	−9.25	—	0.65	—	—
N.p.	—	—	—	−7.20	—	—	—	—
N.s.	—	—	—	—	—	—	3.42—	8.93
					Matrix F			
Prism.								
N.b.	—	—	—	—	—	−0.09	−1.25	−1.45
N.m.	—	—	—	—	—	—	−1.36	−1.50
N.p.	—	—	—	—	—	—	—	−0.65
N.s.	—	—	—	—	—	—	—	—

[a] The value in parentheses beneath each mean is the coefficient of variation (standard deviation as a percentage of the mean) for that mean. Matrices A, C, D, and F compare Sr or Mg of nacreous or prismatic layers *among* species; matrices B and E compare Sr or Mg of nacreous and prismatic layers *within* species. Underlined test values document relationships that are significantly different at $P < 0.05$ and 10 *df*.

[b] See footnote b to Table I for the names of species.

Table VII. Strontium/Magnesium Ratios Illustrating Differences in the Relative Concentrations of Strontium and Magnesium in Spherulitic–Prismatic and Nacreous Layers of the Shell Wall and Nacreous Layer of the Septum

	Shell		Septum
	Prismatic	Nacre	Nacre
Species[a]	Sr/Mg	Sr/Mg	Sr/Mg
N.b.	10.75	11.18	16.99
N.m.	8.79	8.08	9.91
N.p.	8.71	4.79	6.92
N.s.	6.94	2.05	1.51

[a] See footnote *b* to Table I for the names of species.

For the same reason, it was possible to sample and analyze only the nacreous layer of the septum. These results are presented in Table VI with the same conventions as in Table I.

The data of Table VI illustrate several trends in the carbonate chemistry of the nacreous and prismatic layers of the shell and the nacreous layer of the septum:

1. The most general trend documented by these data is that the Sr and Mg concentrations of the prismatic and nacreous layers of the shell wall and nacreous layer of the septum of *N. scrobiculatus* behave inversely to those of the remaining three species.

2. The concentrations of Sr in layers of the shell are much less variable than concentrations of Mg in the same samples. This trend is reversed in the septal nacre of *N. belauensis* and *N. scrobiculatus.*

3. With the exception of *N. scrobiculatus,* the Sr content of prismatic aragonite of the shell wall is greater than that of the nacreous layer, whereas the Mg concentration in the nacreous layer is greater than that of the prismatic layer for *N. macromphalus, N. pompilius,* and *N. scrobiculatus.*

4. With regard to the chemistries of nacreous layers of the shell wall and the septum, the Sr content of septal nacre is greater than the Sr concentration in the nacre of the shell walls of *N. belauensis, N. macromphalus,* and *N. pompilius.* The Mg content is less in septal nacre than in shell nacre only for *N. belauensis* and *N. macromphalus.* The septal nacre of *N. scrobiculatus* contains a very high concentration of Mg relative to that of the other species, whereas the Sr content of the septal nacre is less than that of its shell nacre.

These relationships are summarized in Table VII as Sr/Mg ratios. Only *N. macromphalus* has a reasonably consistent balance between Sr and Mg concentrations in the prismatic and nacreous layers in the shell and the nacreous layer of the septum. The remaining species show considerable variation in the Sr/Mg ratios of one or more of these layers. These trends show that *N. belauensis* and *N. scrobiculatus* have considerably different concentrations of Sr and Mg in prismatic and nacreous layers of the shell and septum (Fig. 4). The pattern of these concentrations in the shell wall and septum of *N. macromphalus* and *N. pompilius* is similar but significantly different. These data were tested for differences

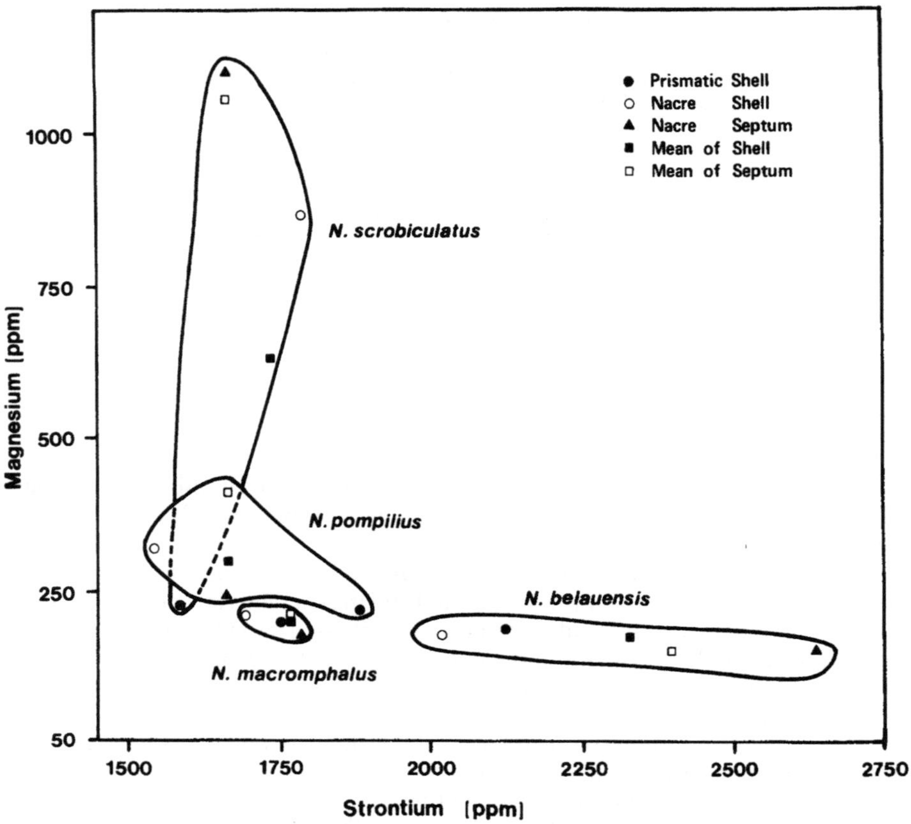

Figure 4. Summary of the bivariate relationship among species of *Nautilus* with respect to the Mg and Sr concentrations of shell spherulitic–prismatic and nacreous layers, the nacreous layer of the septum, and the means of these elements in the shell and septum. Continuous lines enclose these bivariate relationships for each species.

in the mean concentrations of Sr and Mg among species using the approximate Student's *t*-test (see Table VI). The results show that the Sr content of the nacreous and prismatic shell layers (Table VI, matrices A and C) and the Mg content of shell nacre (Table VI, matrix D) are significantly different among all species pairs. The implications of these test results are threefold:

1. The Sr content of nacreous and prismatic shell layers is significantly different among all species, but the additive combination of the two layers may not be significant (see Table II) owing to differences in the volumes of these layers.
2. The Mg content of prismatic aragonite in the shell is not significantly different among species because of the high variability of Mg concentrations in this layer.
3. The nacreous layer exerts a greater influence over the chemical signature

Table VIII. Test Values of Approximate Student's t-Test for Differences among Strontium and Magnesium Concentrations in the Nacreous Layers of the Shell and Septa within Four Species of *Nautilus*[a]

		A. Strontium				B. Magnesium			
		Shell				Shell			
		N.b.	*N.m.*	*N.p.*	*N.s.*	*N.b.*	*N.m.*	*N.p.*	*N.s.*
S	*N.b.*	5.09	—	—	—	−3.66	—	—	—
e	*N.m.*	—	−2.42	—	—	—	−2.46	—	—
p	*N.p.*	—	—	4.93	—	—	—	−2.33	—
t	*N.s.*	—	—	—	−4.50	—	—	—	3.18
a									

[a] Matrix A: comparison of Sr content of nacre of shell and septa; matrix B: comparison of Mg content of nacre of shell and septa. All test values are significant at $P < 0.05$ and 10 *df*. See footnote *b* to Table I for the names of species.

of the shell of species of *Nautilus* than does the prismatic layer, because a higher or lower concentration of Sr and Mg in shell nacre, and the larger volume of this layer, tend to mask these concentrations in prismatic layers.

The results of tests on the data in Table VI, to determine differences among Sr or Mg concentrations in nacreous layers of the shell and septa within four species of *Nautilus*, show that the Sr and Mg concentrations of the shell and septal nacre of each species are significantly different (Table VIII). A comparison of the Mg or Sr content of the nacreous and prismatic layers of the shell within species shows that only *N. pompilius* and *N. scrobiculatus* have concentrations that are significantly different between these layers (Table VI, matrices B and E). These statistical differences in the Sr and Mg content of prismatic and nacreous layers of the shell indicate that these layers were precipitated from extracellular fluids of different compositions. It is most probable that, like that of some bivalves with more than one extrapallial space, the mantle epithelium of *Nautilus* supports three separate extrapallial spaces, one for each shell-wall layer.

9. Effects of Stress on the Physiological System

The effects of physiological stress on skeletal chemistry were investigated by analyzing shell material deposited prior to, and after, the capture of several specimens. Three specimens of *N. pompilius* reared, after capture, at the New York City Aquarium and one specimen of *N. macromphalus* maintained, after capture, at the University of California at Davis were used for this investigation. In each case, the environmental conditions were similar and did not vary significantly in temperature and salinity from those of the natural habitat, with the exception of depth. The leading edge of the shell of each *N. pompilius* specimen was notched on arrival and thereafter at 14-day intervals until death. Only the time of arrival and death were recorded for the *N. macromphalus* specimen.

The concentrations of Mg and Sr in precapture shell material varied only

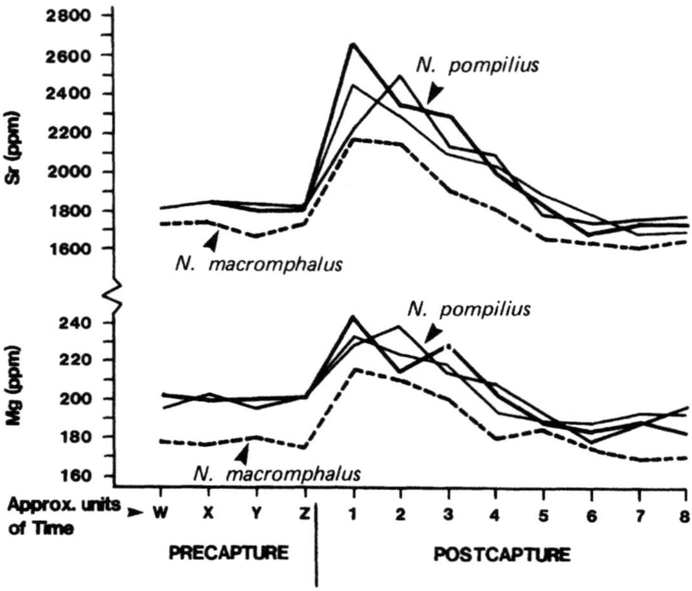

Figure 5. Distribution of concentrations of Mg and Sr in three shells of *N. pompilius* and one shell of *N. macromphalus* with respect to precapture and postcapture shell growth. Postcapture growth is represented by increments of 2 weeks; precapture growth increments are of approximately the same length based on observed growth rates of free animals at the same stage of maturity.

slightly for these specimens (Fig. 5). However, during the first 2 weeks after capture (interval 1 in Fig. 5), Sr concentrations increased by an average of 45% for *N. pompilius* and 29% for *N. macromphalus*. Concentrations of Mg increased an average of 23% for *N. pompilius* and 20% for *N. macromphalus* during the same period. During the following 10 weeks, the Sr and Mg concentrations steadily decreased to precapture concentrations. Because the salinity and temperature of tank waters of the two aquarium facilities remained reasonably constant during periods of captivity, the correlation among the changes in shell chemistry within *N. pompilius* and between *N. pompilius* and *N. macromphalus* indicates that the observed changes in shell chemistries illustrated in Fig. 5 are not random. This correlation suggests that such changes reflect the influence of one or more causal factors that cut across taxonomic lines. The most logical cause of the perturbation in shell chemistry is the physiological stress associated with capture, containment, and transportation. From the observed changes in skeletal chemistry, this stress affected the physiochemical system of aragonitic production, such that both *N. pompilius* and *N. macromphalus* became less efficient at maintaining concentrations of Sr and Mg in the extrapallial fluid used in the precipitation of shell aragonite.

IV

Ecology

Chapter 9

Ecology, Distribution, and Population Characteristics of *Nautilus*

W. BRUCE SAUNDERS and PETER D. WARD

1. Introduction

Until recently, knowledge of the ecology of *Nautilus* was largely based on trapping results, observations of captured animals held in shallow water, and speculation and hearsay. Despite the limitations imposed by such sources, a considerable amount of information regarding the depth range, diet, and geographic distribution of *Nautilus* had been assembled by the turn of the century. This was ably summarized by Henryk Stenzel (1957), in a short contribution that includes an excellent annotated bibliography, in the *Treatise on Marine Ecology and Paleoecology*. This was followed a few years later by the excellent chapter by Stenzel (1964) in the *Treatise on Invertebrate Paleontology*. The latter is a concise synthesis of almost everything known about *Nautilus* at that time.

W. BRUCE SAUNDERS • Department of Geology, Bryn Mawr College, Bryn Mawr, Pennsylvania 19010. PETER D. WARD • Department of Geological Sciences, University of Washington, Seattle, Washington 98195.

During the more than two decades since Stenzel's 1964 chapter was published, considerable effort has been expended on *Nautilus*, as this volume attests. A wealth of new data has been assembled; this chapter comprises a synthesis of the published knowledge available at present as well as new information on various aspects of the ecology of *Nautilus*.

2. Habitat, Depth Range, and Distribution

In the past, knowledge of the habitat of *Nautilus* has been derived largely from such conventional approaches as trapping results, scuba-assisted observations, and aquarium studies. In recent years, a number of new techniques have been used to study the animal and its deep-water habitat, e.g., deep-water remote cameras (Saunders, 1984b, 1985; Hattori *et al.*, 1985), sonic tracking (Carlson *et al.*, 1984; Ward *et al.*, 1984), and stable radioisotope analysis of shell material (Cochran *et al.*, 1981; Oba and Tanabe, 1983; Taylor and Ward, 1983). There follows a summary of these results, plus new information, on the range and distribution of *Nautilus*. For convenience, it is arranged by topic and by locale, including the major sites at which studies of *Nautilus* have been undertaken.

2.1. Trapping and Photosequence Data

The results of deep-water trapping efforts have provided most knowledge of the depth range of the various species of *Nautilus*. However, the accuracy of such data varies, depending on technique, slope, currents, and the means of depth measurement; Bashford Dean (1901), for example, noted that *Nautilus* trapping depths of 450–750 m claimed by fishermen in Tañon Strait, the Philippines, differed considerably from maximum depths (200 m) recorded on hydrographic charts for this region. Overall, however, the general agreement of trapping data from many sites suggests that they are a fairly reliable index of depth distribution.

2.1.1. The Philippines

The first information available on *N. pompilius* in the Philippines was provided in the excellent account by L. E. Griffin (1900) of the anatomy of *N. pompilius*, which was based on 66 specimens stated to have been caught at a depth of 550 m off the southeastern coast of Negros Island by the Menage Expedition, from the Minnesota Academy of Sciences. Griffin (1900, p. 104) quoted a communication from D. C. Worcester, which stated that the natives obtained as many as four or five animals per night at depths of 185–250 m. Bashford Dean (1901) reported that native fishermen regularly obtained specimens in Tañon Strait, located between the islands of Cebu and Negros, at depths that he estimated to be approximately 100–200 m. Talavera and Faustino (1931) provided a brief and general account of *Nautilus* trapping in the Philippines, but included little specific information on depth.

The accounts by Norine Haven (1972, 1977a,b) of *N. pompilius* in Tañon

Strait provided the first substantive ecological data for *Nautilus*. Her information was based on extensive field reports obtained in collaboration with local fishermen, whose trap yields (overnight sets) at depths of approximately 60–250 m ranged as high as 19 animals per trap (average 5 animals). It is notable that during the 1979 ALPHA HELIX Expedition to the same area, local fishermen reported that yields from overnight traps were averaging only 1 animal.

A series of detailed reports have been produced by teams of Japanese biologists working in this area (Hayasaka, 1983; Hayasaka *et al.*, 1982) (see also Chapter 11), providing one of the best overall pictures of the physical habitat of *Nautilus*. Summarily, these reports show that *Nautilus* in this area ranges from approximately 100 to 525 m; most animals are caught at depths of 120–170 m.

Deep-water camera photosequences obtained by Saunders during the 1979 ALPHA HELIX Expedition recorded low numbers of *Nautilus* in Tañon Strait. A maximum of two specimens were recorded in single frames from four photosequences spanning periods of up to 17 hr at depths of 100–220 m. This area has yielded the lowest trapping and camera-recorded frequencies encountered at any of the sites in which *Nautilus* has been studied. It is not known whether this is a product of overfishing or the ecological havoc wrought by reef siltation (caused by forestry and agriculture practices) or a combination of both.

2.1.2. Palau

The presence of *Nautilus* shells in Palau (Belau), Western Caroline Islands, was regarded as a product of postmortem drifting (e.g., Stenzel, 1964) until live animals were trapped there in 1975 by Douglas Faulkner (Dugdale and Faulkner, 1976). Between 1977 and 1982, an extensive trapping–tagging–recapture program was carried out in Palau by W. B. Saunders, C. Spinosa, M. Weekley, and L. E. Davis (Saunders and Spinosa, 1978, 1979; Saunders, 1981a, 1983, 1984a,b, 1985). During the course of these studies, some 2387 specimens of *N. belauensis* were trapped at depths of approximately 100–400 m, processed, tagged, and released by diver at depths of 15–50 m. Traps were generally tied off the reef foreslope and left for 3-day periods. Yields varied, depending on design, circumstance, and luck; in 1982, this technique averaged 34 animals per trap, and as many as 67 animals were obtained in a single trap. Recapture rates varied, but in 1982, 15% of almost 1000 animals released during a 3-month field season were recaptured (some as many as five times), and many traps contained more than 30% recatches. The shallowest depth at which an animal was caught in Palau was 66 m, in a trap set at that depth overnight, and the deepest documented occurrence was recorded by remote camera at a depth of 504 m (see below).

The most complete picture to date of *Nautilus* in its natural habitat was obtained by the use of a battery-powered remote deep-water camera, set at baited bottom sites along the forereef slope inside Mutremdiu Point, Palau (Figs. 1 and 2) (Saunders, 1984b, 1985). Both still and movie cameras were used. They were attached to a carrion-baited trap–frame and covered a field of view measuring approximately 2 m × 1 m. The records provided by the photographic transparencies and movie film permit identification of the numbers and types of organisms attracted to bait sites with *Nautilus*, as well as bait discovery times and nocturnal vs. diurnal activity (Fig. 3) (for details, see Saunders, 1984b).

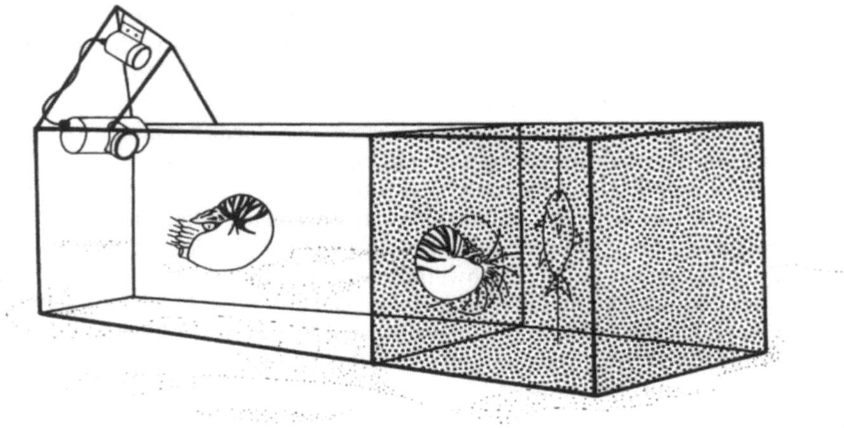

Figure 1. Diagram of the camera–trap combination used to obtain remote deep-water photosequences. The unit is attached to a rope harness and tethered to a surface buoy for the duration of the photosequence. The trap is 1 m wide. Reprinted with permission from Saunders (1984b).

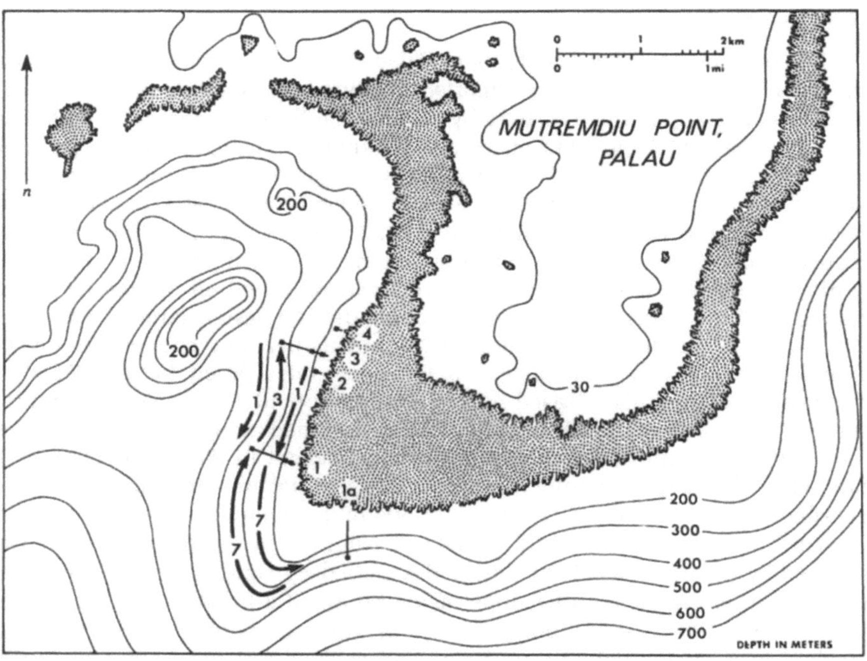

Figure 2. Bottom-contour map of vicinity of Mutremdiu Point, Palau, showing reef reentrant (left side of map) at which bottom-site photosequences were obtained. Also shown are trap sites (Nos. 1,1a–4) and numbers and directions of movement of tagged and recaptured *Nautilus* between trap sites during May–August 1977. Reprinted with permission from Saunders (1985).

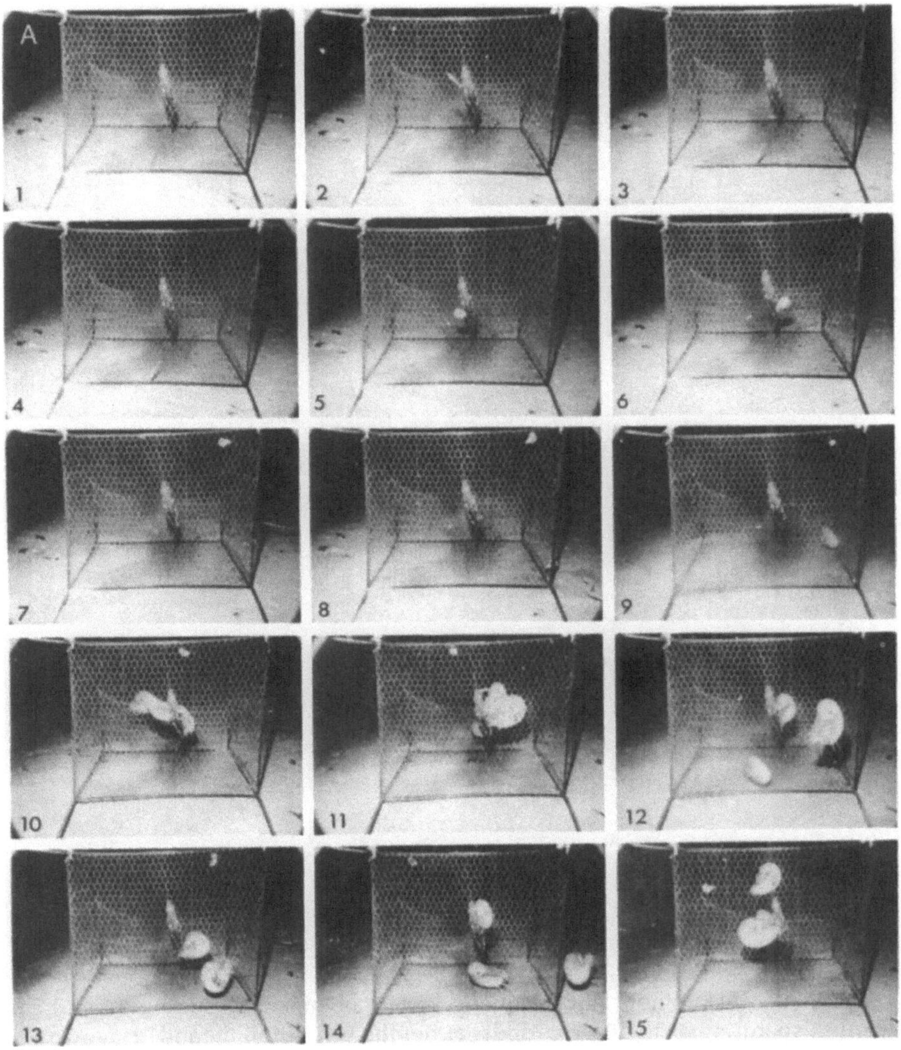

Figure 3. Remote daytime photosequence (No. 4, 6-24-1981) at a depth of 274 m, 15-min intervals (first photograph at 10:00, last photograph 16:30). Frames of note: (4) Juvenile *N. belauensis* approaching right rear of trap within 1 hr of trap's placement on bottom. (5–8) Young animals are also visible as second juvenile is approaching right front of trap. (9) First mature *Nautilus* is in the trap and is joined by two others within 15 min (10). (10, 11, 13) Animal 2207, tagged and released 6 days earlier. (12) An animal approaching front of trap in typical search posture, swimming backward with tentacles extended. (13, 15, 19, 21), Animals swimming anteriorly, in characteristic fashion after scent source has been localized. (21–23) Two *Nautilus* mating. (27) Trap being winched to surface. For additional details, see Appendix 1 and Table 2 in Saunders (1984b). Reprinted with permission from Saunders (1984b).

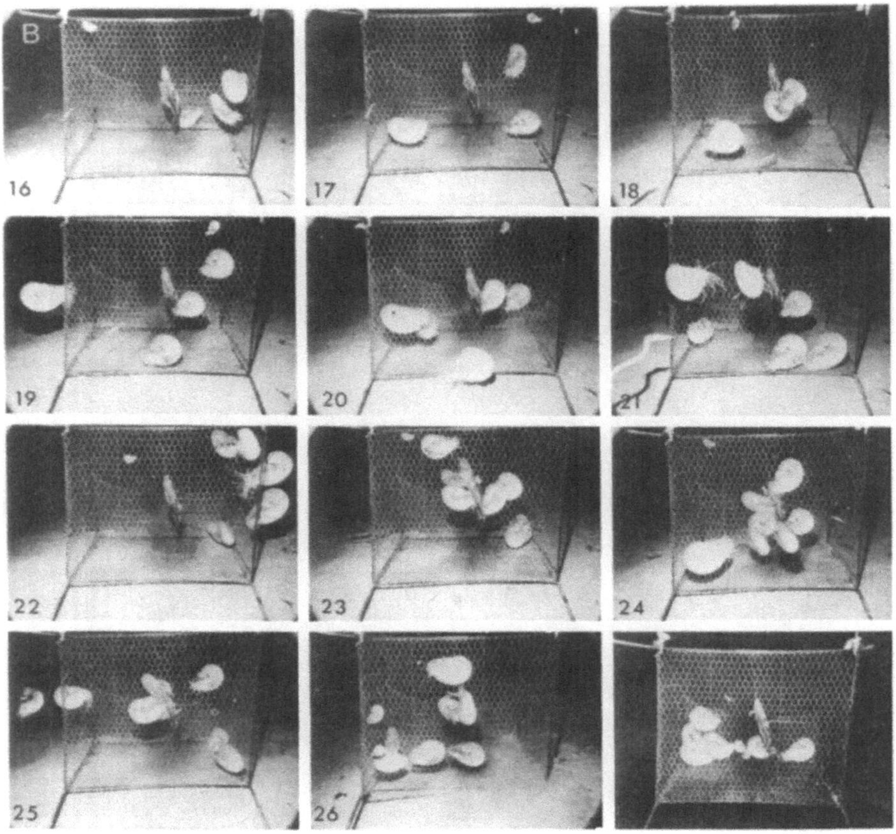

Figure 3. (*Continued*)

A total of 41 remote photosequences were obtained, including 22 still sequences spanning 4- to 16-hr periods, at depths of 73–538 m, and 19 movie sequences taken at 30-min to 2-hr intervals, at depths of 170–396 m. The results, combined with trapping data, indicate a total depth range of approximately 65–500 m for *N. belauensis*. However, a preferred range of 150–300 m is indicated by the greater number of animals (as many as 16 in a single frame) and the shorter bait-discovery times (averaging 1.7 hr) (Fig. 4) (see also Table 1 in Saunders, 1984b). Below 500 m, *Nautilus* is rarely recorded in Palauan photosequences, and then generally only as solitary animals, seen only in occasional frames, commonly late in the sequence. The greatest depth at which a specimen was photographed was 504 m, and none was recorded in the two deepest sequences (528 and 538 m).

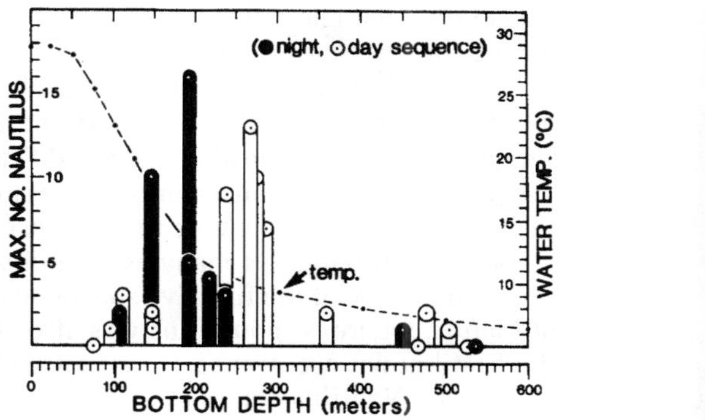

Figure 4. Summary of 22 photosequences taken off Mutremdiu Point, Palau, showing maximum number of animals in a single frame, bottom depth and temperature, and night vs. day sequences. Note the (1) greater number of animals within the depth range of 150–300 m, (2) higher frequency of animals in shallower nocturnal photosequences, and (3) overall depth range of approximately 90–500 m. Reprinted with permission from Saunders (1984b).

2.1.3. Fiji

The first account of *Nautilus* from Fiji is that of a single specimen dredged off Matuku Island at a depth of approximately 570 m, during the Challenger Expedition (Moseley, 1892). There have been a number of subsequent reports of trapping and depth data for *N. pompilius* from the reef outside Suva Harbor, Viti Levu; they are consistently the deepest *Nautilus* to be reported. Ward *et al.* (1977) and Ward and Martin (1980) obtained *Nautilus* at 100 to 600–650 m, with the highest yields from traps set deeper than 300 m. Hayasaka (1985) reported catching 46 animals at 366–549 m in the same area, and Zann (1984) described catching 90 specimens from depths of 400–500 m (averaging 3.1 *Nautilus* per day), with the highest yields from traps set below 450 m. It is worth noting that Zann also employed a bottom-triggering camera, which recorded no *Nautilus* in 150 photographs, covering approximately 3000 m^2 of bottom area. Saunders *et al.* (1987b) reported trapping 40 specimens at 220–470 m in the same area.

2.1.4. Papua New Guinea

The earliest record of *Nautilus* from the region of Papua New Guinea is that of Valenciennes (1841), whose excellent anatomical account was based on a specimen obtained from New Guinea by a Dutch conchologist in Batavia; however, no other details of its origin were given. Arthur Willey's long and futile search for fertile *Nautilus* embryos was based primarily in Papua New Guinea, at Blanche Bay, near Rabaul, New Britain Province, although he also visited New Hanover, New Ireland, and Milne Bay. However, relatively few details regarding the depths at which he obtained *Nautilus* were provided: "It is found in Blanche Bay at a depth of more than 70 fathoms, and is caught in large barrel-shaped baskets made of bamboo . . ." (Willey, 1895, p. 409); "—50 to 70 fathoms—" in the Bismarck Archipelago (Willey, 1899, p. 7); and ". . . from thirty to seventy fathoms . . ." in New Britain (Willey, 1902, p. 698). Anna Bidder (1962) reported trapping a single specimen of *N. pompilius* near Rabaul at a depth of 100 fathoms.

Recent work at widely separated sites in Papua New Guinea has shown that *N. pompilius* is relatively common in the region (Saunders and Davis, 1985; Saunders *et al.*, 1987) (see also Chapter 6). Information on depth and setting obtained during the course of these studies is summarized in Sections 2.1.4a–d.

2.1.4a. Manus. An extensive trapping program was undertaken in the Admiralty Islands in 1984 by Saunders and L. E. Davis, in collaboration with R. L. Knight Pty. Ltd. and the Manus Coastal Fisheries Station. *Nautilus pompilius* was trapped in this area consistently along steep forereef slopes at depths estimated to be 200–300 m. Trap yields varied considerably, but averaged 20 specimens per 3-night set, although as many as 37 animals were obtained in 1-night sets. This program produced the first live specimens of *N. scrobiculatus*, off the south coast of Manus, where *N. pompilius* was also obtained in the same traps, establishing the first known occurrence of sympatry of two species of *Nautilus* (Saunders *et al.*, 1987a). *Nautilus pompilius* far outnumbered *N. scrobiculatus;* the overall ratio was 7:1, but the majority of traps had none of the rarer species (although one overnight trap yielded 7 *N. scrobiculatus* and 11 *N. pompilius*).

During the course of the Manus fieldwork in 1984 and 1985, trapping was conducted in what was interpreted as a submerged volcanic caldera, located within a small island complex in the Fedarb Islands known as Komuli Plantation (see Fig. 1 in Chapter 6 for a map). The flat caldera floor (at a depth of 330 m) appears to be ringed by a shallow rim, approximately 20 m deep. Within the crater, 12 specimens of *N. pompilius* were caught, including 2 specimens trapped along the inside lip at a depth of 80 m in an overnight trap; the remainder were obtained at depths of 200–330 m. Deep-water camera sequences showed the bottom of the crater to be flat and covered with fine silt. Traps set along the outside slopes of the crater yielded *N. pompilius* as well as 1 specimen of *N. scrobiculatus*, at a depth of approximately 250 m. It is intriguing to consider how access to this crater has been obtained by *Nautilus* and whether any isolation has been achieved.

2.1.4b. Kavieng. This site in New Ireland Province is located some 50 km east of New Hanover. When Willey visited in 1896, he reported that *Nautilus* shells were being used as canoe ornaments and bailers (Willey, 1902, p. 701). In 1984, trapping was conducted along the moderately sloping north coast off Kavieng by Saunders, with the assistance of A. Wright, A. Young, and P. Dalzell, of the Kavieng Fisheries Research Station. Three traps set overnight from 145 to 165 m deep yielded only one specimen of *N. pompilius*, but six overnight sets from 240 to 290 m averaged eight animals per trap. Drift shells of *N. scrobiculatus* are found in this area, but no live animals were obtained during the course of trapping.

2.1.4c. Lae. The Huon Gulf, located on the north coast of Papua New Guinea, is the focus of the Markham River delta, which descends steeply from the interior highlands. This setting for *Nautilus* is unique in that it includes both steep slopes and a mud-covered bottom, and it lacks the typical reef development commonly associated with *Nautilus*. A fisheries deep-water trapping program has been under way in this area for several years, led by Norman Quinn, Papua New Guinea University of Technology. During July and November 1984, Saunders and Quinn set six traps overnight at depths ranging from 220 to 300 m. The average yield was five *N. pompilius* per trap, with higher yields (eight to nine animals) in the deeper traps (280 and 300 m).

Mature Huon Gulf *Nautilus* are considerably smaller than those of any other Papua New Guinea population sampled (Saunders and Davis, 1985) (see also Chapter 6). It is tempting to consider that this small size might be an ecological product; that it might be is suggested by the deltaic setting and the associated fauna, which is different from that observed at any other site. The fauna included numerous large isopods (≈ 50 mm), echinoids, starfish, slipper lobsters, and deep-water sharks (as many as seven in a single trap).

2.1.4d. Port Moresby. Located on the south coast of Papua New Guinea, this forereef locale is typical of other *Nautilus* occurrences, with a steep reef face dropping to considerable depths. From June to October 1984 and 1985, traps were set off Motupore Island, approximately 15 km south of Port Moresby, and in 1985, by Saunders, Ward, Davis, and P. and L. Colin, at estimated depths of 200–300 m; yields averaged seven animals per overnight trap.

2.1.5. Australia

The first live *Nautilus* to be documented from the Great Barrier Reef were obtained in 1985, off Lizard Island, Queensland (Saunders and Ward, 1987). Six specimens of *N. pompilius* and seven *N. stenomphalus* were caught in four traps set for one and two nights at depths of 150–400 m. It is notable that this was the first record of living *N. stenomphalus* and only the second record of sympatry of two species of *Nautilus*.

2.1.6. New Caledonia

The shallowest known occurrences of *Nautilus* are those of *N. macromphalus* in New Caledonia and the adjacent Loyalty Islands. This was documented first by Willey, who noted that in ". . . Sandal Bay, Lifu, in the Loyalty Group . . . *Nautilus* migrates at night from deep water into as little as three fathoms" (Willey, 1899, p. 8) and that ". . . the fish-traps . . . are sunk to depths varying from three to eighteen fathoms . . ." (Willey, 1902, p. 732). New Caledonia is the only reported site in which *Nautilus* has been seen regularly in shallow water by divers. Sightings at night along the reef face at depths of approximately 5 m (e.g., Cousteau and Diolé, 1973; Ward and Martin, 1980) are common. It is important to note that these sightings have been made only during the winter months, when surface water temperatures are relatively low (≈ 22°C). Ward and Martin (1980) and Ward (1983a) reported that *N. macromphalus* is most common at depths of 300–400 m, but that it is also caught at much shallower depths, 50–100 m. The specimens of *N. macromphalus* on which Denton and Gilpin-Brown (1966) based their experiments were obtained by natives fishing with hand line, off Lifou, and R. A. Davis (personal communication) caught *N. macromphalus* by hand line at night in Sandal Bay, at a depth of less than 5 m, in June 1975, when the surface water temperature was 24°C. Bidder (1962) reported trapping depths of 60–100 m off Noumea.

2.1.7. The Samoas and Tonga

Trapping in American and Western Samoa and Tonga was undertaken by Saunders, P. N. Bond, and Fisheries personnel, in collaboration with Operation Raleigh (United Kingdom), in July 1986. Traps set outside Pago Pago Harbor, American Samoa, at depths of 270–310 m yielded 39 specimens of *N. pompilius s.l.* (Saunders *et al.*, 1987b). This extended the known eastward range of living *Nautilus* (previously thought to be Fiji) by more than 1600 km. Subsequent trapping efforts in Western Samoa (130 km westward) yielded negative but inconclusive results, due to difficulties in setting traps. Traps set six times at depths of 280–500 m off Tongatapu Island, Tonga (≈ 800 km south of Samoa), yielded a diverse assemblage of deep forereef organisms, but lacked *Nautilus;* this experience seems to delimit the southeastern range of the genus.

2.2. Stable Isotopic Evidence for Habitat Depth and Water Temperature

Information on habitat depth, as well as water temperature, can be derived from the ratios of oxygen isotopes present in *Nautilus* shell material. By measuring the O^{18}/O^{16} ratio in calcium carbonate, the temperature at which the mineral crystallized can be determined. This temperature can be used as an index to the depth at which the shell was secreted, because water temperature changes rapidly with depth at locales where *Nautilus* is found. The first application of this approach to *Nautilus* was conducted by Eichler and Ristedt (1966a,b), who measured the O^{18}/O^{16} ratios in the septa and shell walls of two specimens of *N. pompilius* from the Philippines. Their results showed major changes in the oxygen isotope ratios between the seventh and eighth septa and between the apertural shell secreted at the time of formation of the first seven septa and that formed after the seventh septum. The computed water temperatures before the eight septum were between 25 and 30°C, indicating that the young shells had formed at warm, probably near-surface temperatures. Eichler and Ristedt's isotope data from all subsequent shell walls and septa indicated much cooler temperatures (13–17°C), suggesting that after hatching, the small *Nautilus* migrated into deeper, cooler water. Since the authors provided no locality data for their specimens, accurate determinations about the actual temperature and depth profiles at the capture site could not be made.

In 1981, Cochran *et al.* (1981) analyzed the shell isotopes of the same species, which had been live-caught in the Visayan Sea, the Philippines. They refined the techniques of assigning temperatures to the measured isotope ratio values. The equation for converting O^{18}/O^{16} ratios to temperature values originally described by Epstein *et al.* (1953) had been based on calcite, one of the two mineral morphs of calcium carbonate. The shell of *Nautilus*, however, is composed of aragonite, rather than calcite; a different set of equations must be used to convert the isotope ratios to temperature values.

The two specimens from the Visayan Sea showed the same abrupt isotopic "step" between the seventh and eighth septa reported by Eichler and Ristedt (1966a,b). In this case, the change in isotope ratios corresponded to approximately 23–25°C (<100-m water depth) for the first seven septa and to 18–15°C (150–180 m) for subsequent septa. However, Cochran and colleagues preferred to relate the isotopic and morphological changes to pre- and posthatching isotope fractionation, caused by diffusive transfer across the egg case membrane before hatching. It should be noted that their study showed that the isotope-based water temperatures obtained for posthatching animals were compatible with data available for the region from which the shells were derived and that warm (and shallow) water depths—and, indeed, any isotope-based temperature data—for prehatched *Nautilus* are suspect.

This interpretation was followed by Crocker *et al.* (1985), who also suggested that the high temperatures observed for the first seven septa may be related to fractionation of isotopes that occurs within the egg. They showed that the liquid within *Nautilus* eggs is depleted in O^{18} relative to seawater. Thus, carbonate produced within the egg contains O^{18}/O^{16} values suggestive of temperatures of 30°C or more, when in reality the eggs were laid and developed in much cooler waters. After the eggs are hatched, the shell carbonate is produced in a normal seawater environment, creating the isotopic step observed.

The oxygen isotope method has also been applied to a number of other *Nautilus* occurrences, including: (1) *N. pompilius* from Fiji (Taylor and Ward, 1983), indicating posthatching shell growth at temperatures equivalent to depths of 200–300 m; (2) *N. macromphalus* from New Caledonia, indicating posthatching depths of 200–350 m (Taylor and Ward, 1983); (3) *N. belauensis*, indicating depths of 125–200 m (Ward, 1987); and (4) *N. pompilius* from Tañon Strait, the Philippines, indicating depths between 150 and 300 m (Oba and Tanabe, 1983).

The oxygen isotope method of computing depth of carbonate formation provides only a rough estimate of habitat depth during shell formation. Seasonal temperature changes, salinity fluctuations, and undetected fractionation of the isotopes by *Nautilus* during carbonate formation are all potential sources of error. However, the relative uniformity of results obtained for the various species of *Nautilus*, and from very different oceanographic settings, verifies the utility of this approach. While carbonate deposition in *Nautilus* takes place at a range of temperatures, most of the depths computed from these water temperatures are between 150 and 300 m, which is compatible with trapping and photosequence data.

3. Depth-Limiting Factors

A number of physical and physiological factors limit the vertical range of *Nautilus*. Some limits are absolute (e.g., shell implosion), whereas others are relative or may vary (e.g., water temperature and chamber flooding depth). These factors are summarized briefly in Sections 3.1–3.4.

3.1. Water Temperature

Aquarium maintenance and holding cage experiments have shown that temperatures above approximately 25°C are lethal to *Nautilus*, generally within several days (Saunders and Spinosa, 1979) (see also Chapter 35). This limitation does not prevent the animals from moving into warm, shallow water for brief periods [*N. belauensis* has been trapped in water approaching 28°C (Saunders, 1984b)], but such incursions are probably of short duration (<10 h). In Palau, water temperatures in the preferred depth range of 150–300 m are approximately 9–17°C

(Fig. 4); optimal long-term aquarium maintenance temperatures appear to range as high as 21°C. Similar seawater temperature ranges have been reported from the Philippines, the Fiji Islands, and New Caledonia (Haven, 1972; Hayasaka *et al.*, 1982; Ward and Martin, 1980) (see also Chapter 11).

The implications of temperature as a limiting factor are fairly significant; it restricts the upper limit of habitat and long-term migration to predictable depths, which vary both geographically and seasonally. Thus, extensive movement across shallow-water platforms, such as the Palau Archipelago, are not possible (Saunders and Spinosa, 1979), but shallow-water occurrences of *Nautilus*, as reported in New Caledonia and the Loyalty Islands (see above), may be common during the winter months, when surface temperatures approach habitable levels (Ward and Martin, 1980). What is not clear is why *Nautilus* occurrences are not more common in regions where optimal temperatures are consistently encountered at shallower depths; this may be related to biological factors (see below).

3.2. Biological Factors

Little is known of the nature of interactions between *Nautilus* and associated organisms in the deep forereef. It is clear that in many respects, *Nautilus* compares and fares poorly against its modern counterparts, the dibranchiates and the fishes (see Chapters 12 and 32). It seems worth considering that the extreme vertical mobility of *Nautilus* (approaching 500 m) permits exploitation of the shallower reef face during darkness, followed by retreat to a deep-water refuge during daylight. The physiological constraints of the internal air bladder of most fishes prohibit their following this unique elevator style of scavenging.

3.3. Shell Implosion and Siphuncle Rupture Depth

Inasmuch as gas pressure within the phragmocone of *Nautilus* is about 1 atm (Denton and Gilpin-Brown, 1966) (for a discussion, see Chapter 34), the shell must be strong enough to withstand high hydrostatic pressures without internal equalization of chamber pressure. This aspect of *Nautilus* has been the subject of considerable study, originating with the experimental implosion of sealed shells in hyperbaric chambers by Denton and Gilpin-Brown (1966) and Raup and Takahashi (1966). Saunders and Wehman (1977) expanded the approach to include a large number of specimens, spanning a wide range of sizes. Their experiments gave a variety of implosion pressures, from which maximum depth limits were computed ($\approx$ 700 m), and their data indicated that the shells of juveniles in the early phases of posthatching growth are stronger than mature shells.

Implosion experiments using live animals were described by Ward and Martin (1980). Specimens of *N. pompilius* in the Fiji Islands imploded at a depth of about 750 m, and live *N. macromphalus* imploded between 750 and 800 m. Implosion of living *N. pompilius* in pressure tanks (Kanie *et al.*, 1980, 1981; Kanie and Hattori, 1983) produced implosion depths between 785 and 830 m. It is important to note that dry shells (e.g., Raup and Takahashi, 1966; Saunders and Wehman, 1977) generally implode at lower pressures (by as much as 200–300 m depth equivalent) than fresh or wet shells and live animals (see Table 2 in Saunders, 1984a). It appears that unless kept wet, shells may dry soon after death, leading to serious weakening and implosion at lower pressures than normal.

Implosion, when it occurs, is a sudden and violent event. The shell disintegrates, leaving a few large pieces of body chamber and remnants of the umbilical region, which, in *N. pompilius*, contains the solid umbilical callus. The septa appear to shear away from the shell wall, although each suture zone may retain a small flange of septum. The connecting rings are not preserved. The way in which the shell fails was studied by Saunders and Wehman (1977) and Kanie and Hattori (1983), using strain gauges attached to the shell during implosion experiments. Results showed that the concave septum was subject to tensile stress, whereas the external flanks of the shell were subjected to compressional pressures. Both studies concluded that the last septum is the site of failure. More recently, Chamberlain and Chamberlain (1985) have concluded that it is the septal suture (where the septum joins the shell wall) that is the primary zone of weakness. In their study, fresh shells were subjected to elevated gas and water pressure, and the mode of failure was documented.

A related problem for *Nautilus* is the potential for failure of the siphuncular tube, which is subjected to ambient hydrostatic pressure. This problem has been studied experimentally by Collins and Minton (1967) and by Chamberlain and Moore (1982). The first study indicated that the siphuncle would withstand pressures equivalent to depths of approximately 470 m; the second showed rather consistent rupture at depths equivalent to 800–850 m, which is in close accordance with shell implosion limits.

3.4. Cameral Liquid Emptying

A recently developed line of evidence regarding habitat depth relates to the influence of depth on the removal of cameral liquid. It has been suggested that habitat depths greater than 300–400 m would result in eventual flooding of the chambers; the resultant excessive negative buoyancy would, in turn, immobilize the animal (for a discussion, see Chapter 34). Thus, it may be that osmotic capabilities of the siphuncle, rather than the physical strength of the shell, dictate maximum habitat depth. The relationships involved are reviewed below.

Nautilus obtains neutral buoyancy from the gas- and liquid-filled chambers (the phragmocone) of the shell. Denton and Gilpin-Brown (1966) showed that a new chamber is originally filled with a saline liquid very similar in composition to seawater. Coincident with calcifying a new septum, the siphuncle removes salt ions from this liquid, making it increasingly hypoosmotic compared to liquid (blood) within the siphuncle. The difference in osmolarity sets up an osmotic gradient that causes liquid to move from the chamber into the siphuncle.

When *Nautilus* empties a chamber at surface pressure (1 atm), the osmotic pressure gradient is the only force acting on the liquid. With increasing depth, however, hydrostatic pressure will oppose liquid movement into the siphuncle, because the pressure within the siphuncle is ambient hydrostatic pressure. At 300 m, the difference in pressure between the chamber and the interior of the siphuncle approaches 30 atm; chamber emptying will occur only if osmotic pressure produced by salt concentration is greater than hydrostatic pressure. The rate of emptying will depend on the difference of these two values. If hydrostatic pressure exceeds osmotic pressure, the chamber will fill slowly with liquid forced from the siphuncle into the chamber.

The maximum osmotic pressure differential produced between a liquid of blood salinity and fresh seawater equals the hydrostatic pressure found at 240 m. If *Nautilus* inhabits depths greater than this, a mechanism other than simple osmosis must be used to maintain empty chambers. Denton and Gilpin-Brown (1973) proposed that a modified form of osmosis, called local osmosis, allows local salt concentration increases within the cells of the siphuncle, in such a way that very high osmotic pressures could be developed. Greenwald *et al.* (1980), using simulated hydrostatic pressures, demonstrated that emptying of cameral liquid at depths greater than 240 m could take place, but at very slow rates, and only when the experimental liquid was decoupled from the siphuncle.

The actual rates at which cameral liquid is removed from chambers have been measured and shown to decrease with depth (Ward and Martin, 1978; Ward, 1982, 1987; Ward and Westermann, 1985). At depths greater than 300 m, limited experiments to date have produced refilling, rather than emptying (Ward and Westermann, 1985). This finding suggests that the depth at which refilling outstrips emptying may be far less than the implosion depth of 750–800 m and may be as shallow as 300–400 m. If so, flooding depth would represent maximum long-term habitat depth; it would not prevent *Nautilus* from going below 300 m, but such occurrences would be of short duration. The 300-m depth range seems to be supported by remote camera observations in Palau; below 300 m, *Nautilus* is seen only in occasional frames, often late in the sequence (Saunders, 1984b).

More study of the rate of chamber flooding would be of particular interest, for it might indicate how long an animal could remain at depths greater than 300 m. It might even be worth considering whether short-term shallow-water incursions, such as have been monitored by telemetry (Carlson *et al.*, 1984; Ward *et al.*, 1984), might serve to assist in the removal of excess quantities of cameral fluid gained during deep-water visits.

4. Diet and Feeding Behavior

Most knowledge of the dietary habits of *Nautilus* has been derived from observations of crop–stomach contents or from aquarium-based observations, and a considerable body of data has been assembled. Owen (1832), Willey (1895), and Griffin (1900) all remarked on the presence of fragments of decapod crustaceans within the alimentary canals of dissected specimens, concluding rightly that these creatures represent a major source of food for *Nautilus*. As Willey and Griffin also noted, the crops and stomachs of trapped animals were generally gorged with bait—fish and chicken—suggesting a carnivorous opportunistic diet (Willey, 1895, p. 410; Griffin, 1900, p. 104). Haven (1972) reported that caged *Nautilus* consumed both crabs and fish, but had no evidence that they were capable of capturing their food alive. She also confirmed the conclusion of Willey (1902, p. 773) that the tentacles of *Nautilus* serve a chemosensory function in searching for food.

To these observations can be added a few more. The dissected crops of specimens of *N. pompilius* from Lae, Papua New Guinea, contained many fresh fragments of deep-water regular echinoids—both test fragments and viscera. Other specimens of *N. pompilius*, dissected in Manus, contained coleoid beaks, and *Nautilus* tentacle fragments are occasionally found as well. At every locale where crop contents or fecal residues have been examined, crustacean test fragments were recorded.

There are several problems in interpreting natural diet from crop contents. First, the crops of animals caught in traps are generally gorged with bait—as much as 20% of the body weight can be accommodated in the distensible crop (Ward *et al.*, 1977; Saunders, 1983). Second, crop specimens cannot be taken as evidence of captured prey—molts, for example, are as popular with aquarium *Nautilus* as freshly killed animals (see Chapter 35). In addition, a scavenger's crop is as likely to contain windfall carrion as animals it has captured. Third, potential food sources are attracted into baited traps with *Nautilus*. In a number of cases, fresh or partly consumed crabs, shrimps, sharks, conger eels, and teleosts (some still alive), bearing distinctive *Nautilus* bite marks, were recovered from traps in Palau (Saunders, 1984b). Although this finding indicates that *Nautilus* is able to capture live prey, if the prey is confined and unable to escape, there is no evidence that such capture is otherwise possible (and this conclusion is supported by aquarium observations).

In New Caledonia, the Loyalty Islands, and the New Hebrides (Vanuatu), *Nautilus* can be captured along the reef face by hand at night. Study of the crop contents of specimens captured in this manner provides a more natural indication of the diet of *Nautilus*. Ward and Wicksten (1980) identified 23 different crustaceans from the crop contents of nine such specimens of *N. macromphalus* from New Caledonia. Over two thirds of the animals identified belonged to a single species of hermit crab, *Aniculus aniculus*. In five additional specimens hand-captured since the original study, this species also made up the majority of identifiable remains (Ward, 1987), but no shrimp remains were identified. Other crustacean material included the remains of brachyuran and reninid crabs, pieces of lobster exoskeleton, and the remains of a galatheid crab. Fish bones were recovered from one specimen.

A second source of food for *N. macromphalus* in New Caledonia is the fresh molts of larger crustaceans, such as lobsters. Numerous observations have now been made by divers, who have seen as many as three *Nautilus* eating the same lobster molt on the shallow reef platform at night (Magnier and Laboute, 1978; Ward and Wicksten, 1980). *Nautilus* generally eats the legs, tail portions, and abdomen first, leaving the more skeletonized cephalothorax. In aquariums, *Nautilus* also readily eats molts such as those of the common lobster *Panulirus longipes*, a species the remains of which are found within the stomachs of wild-caught *Nautilus*. In a 2-hr period, the telson, uropods, abdomen, and several of the walking legs were observed to have been ingested by an aquarium-held specimen (Ward and Wicksten, 1980). Saisho and Tanabe (1985), reporting on the crop and stomach contents of 34 specimens of *N. pompilius* trapped off Viti Levu, Fiji, recorded remains of crustaceans, fishes, nematodes, and *Nautilus* tentacle fragments.

Nautilus appears to be a windfall feeder, and the highly distensible crop would be an asset to this mode of life. Equally advantageous are the large, strong jaws, which contain heavy calcareous occlusal surfaces that are unique among living cephalopods (Saunders *et al.*, 1978). And it is relevant that recent experiments with *N. macromphalus* show that food can be stored for up to 2 weeks in the crop before digestion (K. Mangold, personal communication). All these characteristics would be of great aid to an organism foraging in an environment where food supply is neither plentiful nor predictable.

Nautilus searches and locates food by using a well-developed chemosensory system located on the tentacles (see Fukuda, 1980) (see also Chapter 17). In addition, the recent demonstration that *Nautilus* exhibits a positive phototactic response suggests that vision may be a factor in locating bioluminescent activity, such as that displayed by shrimps (Chapter 15). Judging by the speed with which baited traps are found by *Nautilus*, it appears that the animal can detect even small amounts of scent in the water. Twelve photosequences (146–284 m in depth) in Palau showed that bait discovery time for *N. belauensis* averaged 1.7 hr, and in nine sequences, it was 1.2 hr (Saunders, 1984b). By comparison, shrimps commonly located the bait in less than an hour. The photosequences also showed that *Nautilus* characteristically arrived at the bait site swimming backward, with tentacles extended [see Fig. 3 (frame 12)] in a "cone of search," as described by Bidder (1962). When the scent source is localized, the animals commonly swim anteriorly, groping for the bait with widely extended tentacles, until it is located (frames 13, 15, 19, and 21). They may feed for hours at a time, leave the bait (and exit the trap), and return periodically, but they largely abandon the site once bait has been consumed. Observations by divers in New Caledonia show that as soon as a food item has been taken into the tentacles, the animal commonly begins vigorous swimming activity, while simultaneously ingesting the prey. This may carry the animal well away from the bottom.

5. Movement and Activity

5.1. Lateral Migration and Long-Term Movement

The discovery, in 1977, that *Nautilus* could survive trapping to be released and recaught opened new avenues for investigating the animal in its natural habitat. This led to a mark–recapture program that ultimately involved release of more than 2000 specimens of *N. belauensis* in Palau during five field seasons between 1977 and 1982 and provided the first records of the animals' movements over an extended period (Saunders and Spinosa, 1979).

During May–August 1977, 375 tagged specimens were released at two sites in Palau. Of these, 30 were recaught, and 3 were recaught twice within 78 days. Although 11 animals were recaught at their release site, 19 animals had moved 1–2 km [the maximum distance between trap sites (see Fig. 2)]. One year later, trapping was resumed at the same sites, as well as at three other widely separated sites (Fig. 5). Of 460 animals trapped during 1978, 9 animals tagged in 1977 were recaught. Of these, 4 had moved from 40 to 150 km along the fringe reef in the 10- to 12-month period after their release. In addition, in 1978, 1 animal was

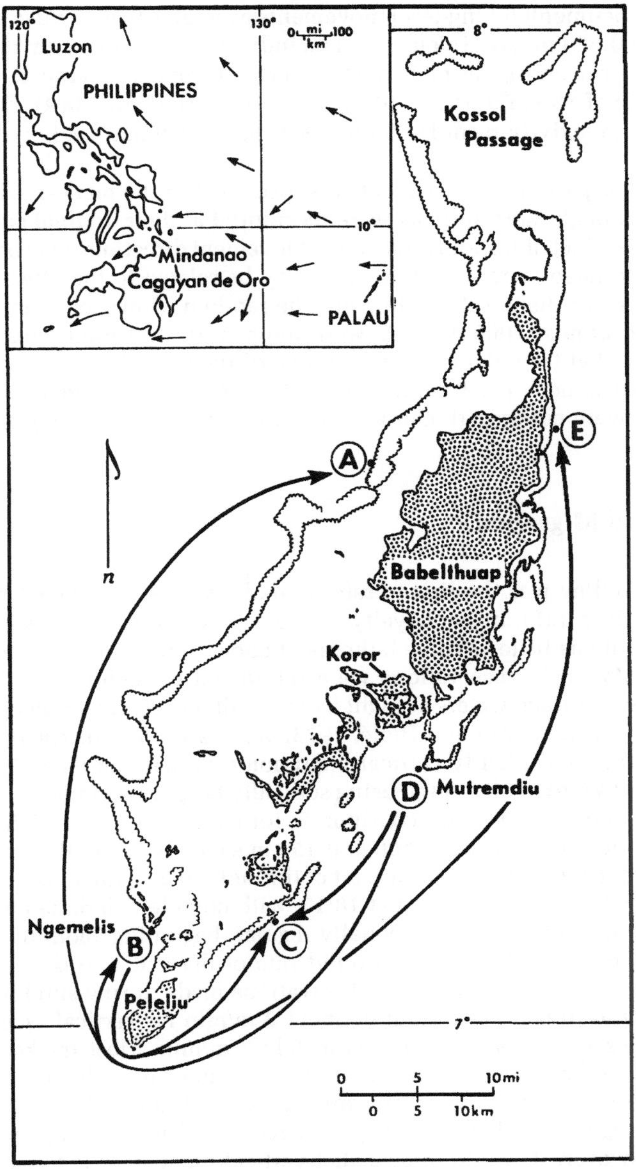

Figure 5. Long-term movement of tagged N. *belauensis* around Palau during 1977–1978. Four animals released at sites B and D were recaptured 10–12 months later, 40–150 km away (sites A, B, C, and E). Warm water temperatures (≈ 29°C) limit traverses across the shallow platform. *Inset:* current directions between Palau and Mindanao, where a tagged shell of N. *belauensis* was recovered 138 days after being released in Palau, documenting postmortem drift of at least 7 km/day. Reprinted with permission from Saunders and Spinosa (1979).

recaptured 51 days after release, 31 km to the south of its original capture site. These data documented long-term movements averaging 0.45 km/day and shorter-term rates of 0.60 km/day. Telemetric data indicate similar (0.43 km/day) (Carlson *et al.*, 1984) or even higher rates of short-term movement; 1 animal was tracked for a distance of 16 km during a 10-day period (Ward *et al.*, 1984). Extrapolating from aquarium activity records, Zann (1984) has calculated 1.6 km of swimming per day.

No regular pattern of movement was discerned in the mark–recapture study in Palau. Although most animals were recaptured at their original sites, this was probably an artifact of trapping locations. Movement directions may be influenced by large-scale deep-water currents, they may be related to the effects of smaller currents on scent dispersal, or they may be random. However, it is notable that during 1982, almost 1000 animals were trapped and released, most at a single site. The fact that trap yields increased toward the end of the 1982 program (averaging 34 specimens per trap, including 15–30% tagged animals) suggests that the animals were congregating around what had become a feeding station.

5.2. Vertical Migration

Arthur Willey was the first to refer to daily vertical movement by *Nautilus*: ". . . Sandal Bay, Lifu, in the Loyalty Group. . . . Nautilus migrates at night from deep water into as little as three fathoms. It comes quite close to shore" (Willey, 1899, p. 8). Willey's observations were reinforced by sightings of *N. macromphalus* in near-surface waters at night by scuba divers, in New Caledonia and the Loyalty Islands (Cousteau and Diolé, 1973; Ward and Martin, 1980).

Confirmation of nightly vertical migration was made in Palau by Carlson *et al.* (1984) and Ward *et al.* (1984), using sonically tagged specimens of *N. belauensis*. One specimen, monitored continually for over 6 days and nights, regularly ascended to depths of between 100 and 150 m each night and then descended to 250–350 m during daylight. The ascent of this animal generally occurred between 17:00 and 18:30; sunset occurred at 18:15, with complete darkness by 19:00. Descent in the morning hours was equally regular, beginning about 04:30 to 05:00. The specimen generally showed a rapid rate of descent into deeper water; the calculated rates, based on frequent observations made during sundown and sunset, were found to be a maximum of about 2 m/min for vertical ascent, whereas maximum descent rates were approximately 3 m/min. Most tracked specimens show the same basic pattern of nighttime movement into shallower water, followed by descent into deeper water during daylight hours. The patterns are not completely regular, in that several specimens ascended in daylight or descended at night. However, it would appear that vertical movement is triggered by fluctuation in light level, because both upward and downward movements occurred during crepuscular periods.

This pattern of movement in Palau was confirmed by trapping and photosequence records (Saunders, 1984b). The shallowest depth at which an animal was trapped was 66 m, in an overnight set, and photosequences recorded more animals during darkness at shallower depths (150–200 m) (see Fig. 4). Overall, it appears that *N. belauensis* in Palau shows a fairly regular pattern of vertical migration, similar to that first proposed by Willey.

In Palau, tracked vertical movements seemed to follow bottom contours, although on several occasions it appeared (from comparison of the transmitter signal depth with bottom depth indicated by depth finder) that specimens were swimming above the bottom. This record may be anomalous, for it appears that *Nautilus* rarely strays far from the bottom, on the basis of evidence from direct observation, photosequences, and suspended trapping experiments (Zann, 1984). Thus, vertical migration is an inshore–offshore movement. In Palau, it appears to have been superimposed on a long-term southward movement of the specimens along the contours of the forereef system, which resulted in daily movement of as much as several kilometers.

A similar pattern of movement was shown by tracking *N. pompilius* in Papua New Guinea, where a single specimen of *N. pompilius* was followed over a 2-day period. As in Palau, this specimen moved in shallower nighttime depths than daytime depths, with major vertical movement occurring at dusk and dawn.

In all cases examined, *Nautilus* tagged with ultrasonic transmitters showed depth ranges between about 100 and 500 m; most time was spent between 200 and 300 m. In this respect, the transmitter data are in agreement with the isotopic data (which indicate that most *Nautilus* secrete septa in water temperatures found at depths of approximately 200–300 m) and with the remote camera observations (indicating an overall range of 100–500 m, but a preferred range of 150–300 m). Telemetry indicates that daily vertical movement of approximately 200 m appears to be common. Confirmation that this vertical range is characteristic of all species and populations of *Nautilus* requires extensive study, for local ecological differences, such as slope and water temperature, may negate any generalization based on the behavior of a single species. Also, the frequency and motivating factors behind vertical movement need to be identified. Vertical movement could be related to reproductive activity, it may involve feeding patterns or avoidance of predation, or it may even be involved with buoyancy regulation or growth (see Chapter 34).

5.3. Nocturnal versus Diurnal Activity

Willey (1902) regarded *Nautilus* as a nocturnal animal, presumably because of the nightly shallow-water migrations that he recorded. Haven (1972) concurred with this view, noting that animals in shallow-water holding cages were largely inactive during the day. The records of Zann (1984) of aquarium activity for *N. pompilius* over extended periods showed greater activity during darkness. Contrary to these conclusions, however, the photosequence records from Palau show that *Nautilus* activity is not restrained during daylight (Saunders, 1984b). Comparison of a series of photosequences, obtained during both daylight and darkness, shows that at depths of 150–300 m, mean bait discovery time during daylight was 1.2 hr (eight sequences), whereas mean nocturnal discovery time was 2.75 hr (four sequences). Additionally, the maximum number of *Nautilus* per frame recorded during daylight vs. darkness is similar (Figs. 4 and 6). It is notable also that the

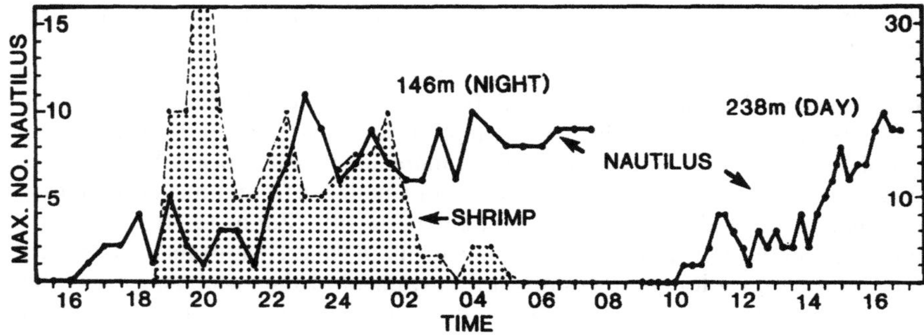

Figure 6. Record of nocturnal and diurnal photosequences, showing time of each photograph, maximum number of *Nautilus* (solid line) and maximum number of shrimps [*Heterocarpus ensifer* (dashed line and stippled pattern)] in each frame. Note the arrival and departure of shrimps during the crepuscular periods; by contrast, *Nautilus* is active during both daylight and darkness within the deeper portion of its range. Reprinted with permission from Saunders (1984b).

activity of associated shrimps (*Heterocarpus ensifer*) is strongly correlated with crepuscular periods (Fig. 6).

The contrasting conclusions regarding nocturnal activity are not in conflict. It would appear that, beginning at dusk, *Nautilus* regularly moves into shallower water to forage for food. This move may be triggered by changes in light level, or it may be in response to the movement of other organisms, such as shrimps (see Chapter 15). It is also known that the number of diurnal reef fishes is replaced by only a fraction of that number of nocturnal species (Goldman and Talbot, 1976), thereby possibly reducing competition and threat of predation. In its normal, deep-water habitat, by contrast, *Nautilus* is equally opportunistic both by night and by day; it may even be the dominant diurnal scavenger within its optimal depth range (150–300 m) (Saunders, 1984b).

6. Population Characteristics

6.1. Sex Ratios

One of the most consistent observations regarding *Nautilus* populations has been the predominance of males. This was noted by Willey (1895, 1902) in *N. pompilius* from New Britain, where 150 of 216 specimens (69%) were male. Haven (1972) reported 95% males among 534 specimens of *N. pompilius* trapped during August–September 1971 in Tañon Strait and noted later (Haven, 1977a) that 92% of almost 3000 animals trapped in 1971–1972 were male (monthly variation 85–97%). Similar proportions were reported for this species in Fiji [74% males, N = 57 (Ward and Martin, 1980); 81% males, N = 162 (Tanabe, 1985)]. Of 167 specimens of *N. pompilius* trapped at widely separated sites in Papua New Guinea (Manus, Kavieng, Lae, and Port Moresby), 75% were males (Saunders and Davis, 1985).

Sex-ratio data for the other species of *Nautilus* are similar. Ward and Martin (1980) reported that 79% of 114 specimens of *N. macromphalus* were male, and Saunders and Spinosa (1978) recorded 72% males among 375 specimens of *N. belauensis* trapped in 1977. Of 30 specimens of *N. scrobiculatus* trapped in Manus, Papua New Guinea, 80% were males (Saunders et *al.*, 1987a).

As these records show, trap yields indicate that *Nautilus* populations typically comprise about 75% males. Causes for these uneven proportions remain unknown. There is no evidence of depth segregation by sexes, nor is there any indication that males are more susceptible to trapping than females.

6.2. Sexual Dimorphism

Dimorphic differences in size and in aperture shape have been well documented in *Nautilus*. Willey remarked on this and figured examples in his first report from New Britain (Willey 1895, p. 411, Fig. 1), but noted (p. 412), "In most adults the shell can be readily identified as belonging to male or female, but often the identification is very difficult, and in young shells impossible." Willey's statement provided impetus for the analysis of dimorphism in *N. belauensis* in Palau (Saunders and Spinosa, 1978). The results of that study have been confirmed and added to by the observations of others (Ward and Martin, 1980; Hayasaka *et al.*, 1982; Tanabe *et al.*, 1983, 1985; Tanabe, 1985; Zann, 1984; Saunders and Davis, 1985). In summary, males are consistently larger, and the male shell aperture is wider (Fig. 7). As Willey noted, the differences are not apparent until relatively late in ontogeny. In *N. belauensis*, differentiation begins at approximately 180 mm diameter, as shown by plots of aperture width against diameter (Fig. 8). The differences in size and width are due, in part, to both the development of the spadix as males approach maturity and the lateral constriction of the aperture in females (Fig. 9).

6.3. Mature/Immature Ratios

As striking as the uneven sex ratios in *Nautilus* is the relative rarity of immature animals. The greatest amount of data is available for *N. belauensis* in Palau. Of a total of 2387 animals, 80% were fully mature (as indicated by a thickened body chamber wall, blackened aperture, accentuation of the hyponomic and ocular sinuses, constriction of the aperture, and elongation of the body chamber) (Saunders, 1983). Only 5% of the specimens measured less than 157 mm in diameter (mean mature male diameter is 209 mm); i.e., the majority of immature animals were relatively large (submature), and very young animals (juveniles) were exceedingly rare (the smallest animal captured measured 77 mm in diameter).

As shown in virtually all published frequency distributions of *Nautilus* shell diameters, the population is markedly skewed to the right, and no indication of size classes (excepting male–female difference) is apparent (see Fig. 7). It has been suggested that this anomaly is due to depth segregation of immature animals. As evidence, Ward and Martin (1980) reported that 25 Fijian *N. pompilius* from

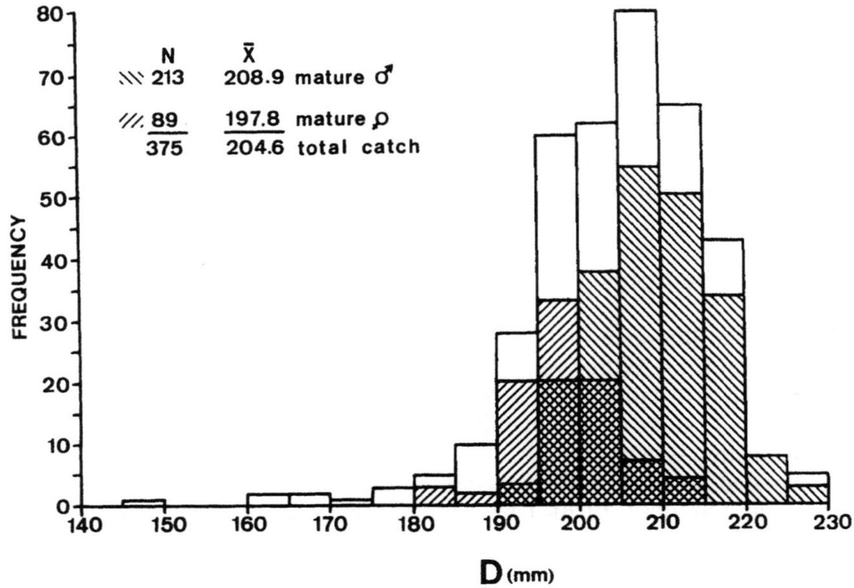

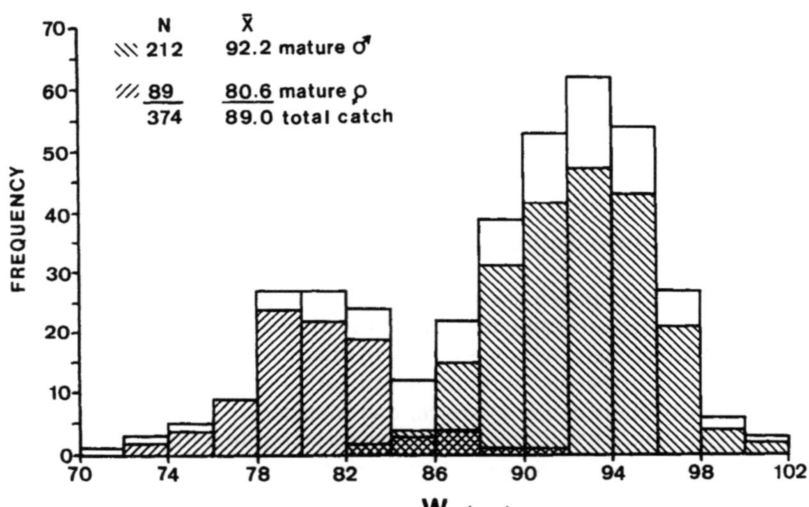

Figure 7. Size frequency distribution of maximum shell diameter (D) and shell width measured beneath the ocular sinus (W_s) in 375 specimens of *N. belauensis* trapped in Palau (May–August 1977). The shaded patterns indicate mature animals and sex; the unshaded tops of the histogram bars indicate immature animals and total catch (includes both mature and immature animals). Although mature males are generally larger (mean 209 mm) than females (mean 198 mm), there is considerable overlap. By contrast, broader apertural width distinguishes mature males from females. Reprinted with permission from Saunders and Spinosa (1978).

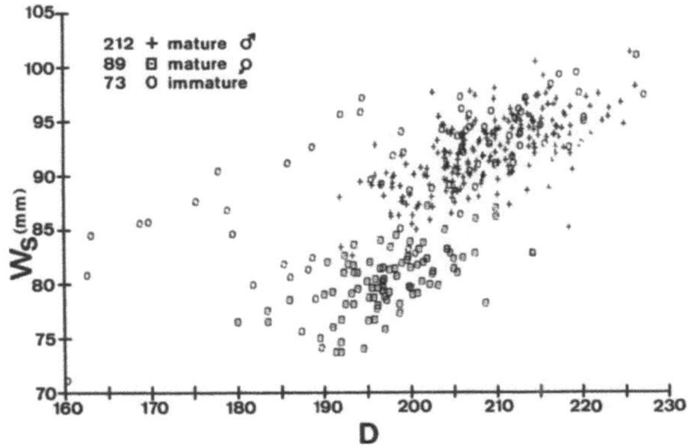

Figure 8. Scatter diagram of aperture width measured beneath ocular sinus (W_s) vs. maximum shell diameter in *N. belauensis* (D). Note that differentiation between males and females begins at about 180 mm diameter, with constriction of the female aperture. The development of the spadix in males requires retention of a broader aperture (see Fig. 9). Reprinted with permission from Saunders and Spinosa (1978).

depths of 100–300 m averaged 500 g body weight in air, compared to 413 g for 32 specimens from depths of 300–600 m. However, we now know that there may be considerable variation in body weight, depending on crop content, and that there is considerable variation in the weight of mature animals.

No evidence to support depth/size segregation was obtained from the Palau trapping data; more significantly, photosequence accounts showed that juvenile *Nautilus*, although rare, do in fact occur with the adults (see Fig. 3). In a number of photosequences, juveniles (≈ 80 mm in diameter) located the bait sites more quickly than did the adults.

As stressed by Saunders (1984b), the lack of size classes and the rarity of very young *Nautilus* should not be regarded as anomalous, if viewed in the context of what is known of deep-sea reproductive strategy (e.g., Dayton and Hessler, 1972; Grassle and Sanders, 1973). Most deep-sea species for which data are available reproduce throughout the year (Rokop, 1974) and produce few but relatively large ova (Grassle and Sanders, 1973; Sanders and Allen, 1973). This is also characteristic of *Nautilus;* aquarium observations indicate that reproduction occurs

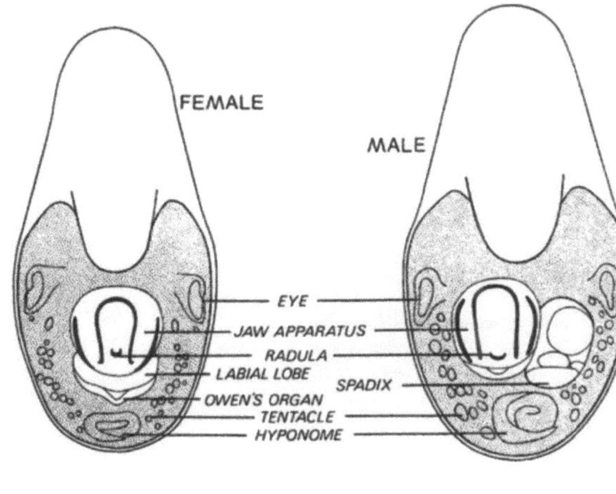

Figure 9. Cross sections of mature male and female *N. belauensis*, showing sexual differences. Note the offset jaw apparatus in the male, in response to the development of the spadix. The figures are based on xerographic images of sectioned, frozen carcasses. Reprinted with permission from Saunders and Spinosa (1978).

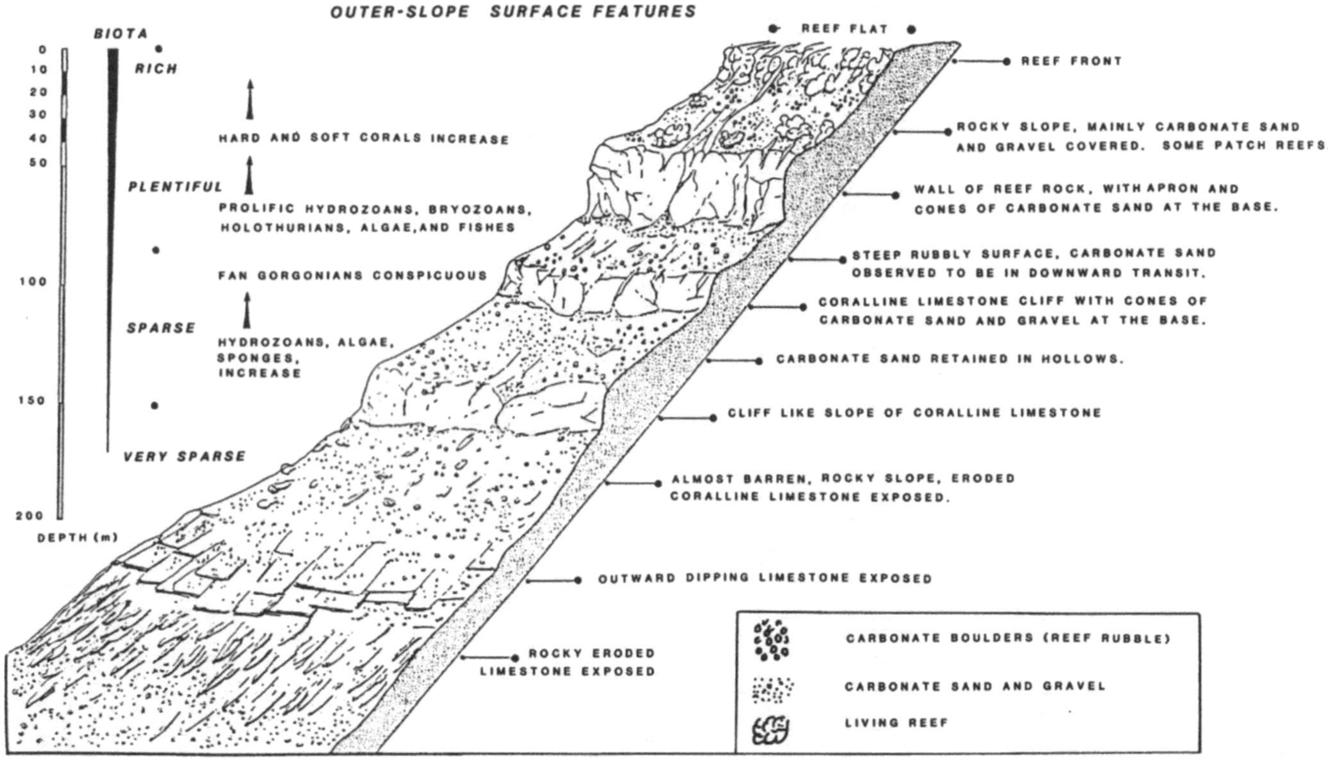

Figure 10. Observer's impression of upper 300 m of forereef slope biota and topography, off Flinders Reefs, Great Barrier Reef, Australia, based on observations from a submersible. This portrayal is probably representative of the *Nautilus* habitat in general. Although no *Nautilus* were observed during submersible transects in this region, the animals are probably present, as indicated by the presence of *N. pompilius* and *N. stenomphalus* off Lizard Island, approximately 400 km to the northwest. After Orme (1977); published with the author's permission and with permission of Academic Press.

year-round, the eggs are extremely large for invertebrates, and the clutch size is small—usually one or two eggs (Arnold, 1984; Carlson, 1979; Saunders, personal observations) (see also Chapters 26 and 35). Indeed, if these generalizations for deep-water animals are valid, it would be anomalous to see size classes in *Nautilus*.

7. Summary: A Profile of *Nautilus* in Its Natural Habitat

Nautilus can be summarily characterized as a warm-water (≈ 10–25°C), fore-reef slope inhabitant (Fig. 10), the overall depth range of which extends from near-surface waters to approximately 600 m, but there is a preferred, or optimal, depth range of approximately 150–300 or 400 m depending on locale (Fig. 11). This is a remarkably large range, considering that it may be traversed by an individual animal in a day or less. Depth limits involve a number of physiological, physical, and biological factors. Shell implosion and siphuncle rupture depths determine absolute lower limits, at about 700–800 m. Cameral flooding may limit long-term habitat depths to less than 300 m, although this limitation would not prevent short-term incursions into deeper water. The upper limit, which seems to be determined by water temperatures above 25°C, varies both geographically and seasonally. The role of biological interactions, e.g., predation by teleost fishes, is probably also important. Within the reef-slope habitat, *Nautilus* is a bottom-dwelling scavenger and opportunistic predator, ranging widely in search of food. Its diet is catholic, but features crustaceans supplemented by windfalls. The animal is active both nocturnally and diurnally within its optimal depth range, but appears to make shallow-water forays on a near-nightly basis.

Population characteristics of *Nautilus*, which are typical of deep-water or-

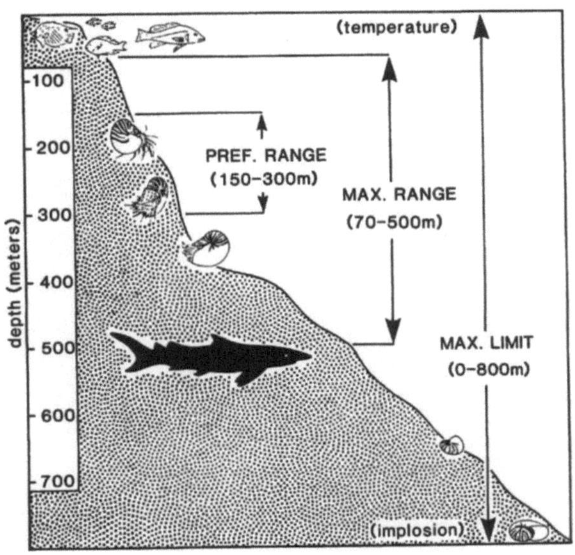

Figure 11. Composite showing the vertical distribution, range, and depth limits of *N. belauensis* in Palau. The maximum depth limit determined by shell implosion and siphuncle rupture is approximately 800 m, but photosequences and trapping data indicate that the actual maximum depth range is approximately 500 m. Cameral flooding may limit long-term habitat depths to approximately 300 m maximum, which is supported by the higher frequency with which animals are recorded in photosequences above 300 m. The upper long-term limit is determined by water temperature (≈ 25°C) and possibly by teleost predation; shallow occurrences (≈ 70 m) appear to be limited to nocturnal forays. After Saunders (1984b).

ganisms, include a large proportion of adults (≈ 80%); juveniles are rare, and size classes are lacking. Males characteristically comprise approximately 75% of the population, and there is no evidence that this is an artifact of sex/depth segregation or trapping.

ACKNOWLEDGMENTS. The data synthesized in this chapter were derived primarily from field work, and it is therefore difficult to acknowledge the many individuals who made this seemingly straightforward chapter possible. If experience in Palau and New Guinea is any measure, few traps for *Nautilus* have been set and retrieved without some degree of misadventure. The efforts and assistance of fisheries personnel are due particular acknowledgment: Toshiro Paulis and his team in Palau; Trevor Bell and the Manus Coastal Fisheries group in Manus; A. Wright, A. Richards, and P. Dalzell, of the Kavieng Fisheries Research Station; J. Locke and Ursula Kolkolo, Konedobu; D. Itano, R. Buckley, R. Tulafono, and others at the Office of Marine Resources, Pago Pago, American Samoa; Lui Bell, Fisheries Division, Apia, Western Samoa; and V. and S. Longi, Office of Fisheries, Nuku alofa, Tonga. Others who assisted include Claude Spinosa, Boise State University; the late Michael Weekley, Waikiki Aquarium; George Monaco, Los Angeles, California; the R. L. Knight family, Manus; N. Quinn, B. Kojis, and R. Johns, Papua New Guinea University of Technology; N. Polunin, P. Colin, and L. Colin, Motupore Island Research Station; B. Goldman, L. Goldman, P. Pini, and G. Pini, Lizard Island Research Station; Larry Davis, Washington State University; and Tom Landry, Seattle Washington. Special thanks are due to personnel of Operation Raleigh, particularly Lee Hastie, Byron White, David King, and the crew of SES *Sir Walter Raleigh*. This study was supported by National Science Foundation Grant EAR 83 19832, BSR 86-08065, and by the National Geographic Society.

Chapter 10

Incidence and Kinds of Epizoans on the Shells of Live *Nautilus*

NEIL H. LANDMAN, W. BRUCE SAUNDERS,
JUDITH E. WINSTON, and PETER J. HARRIES

1. Introduction

Like the shells of other mollusks, the shells of *Nautilus* are subject to marine fouling. Seilacher (1982) observed epizoans on four live specimens of *N. pompilius* and *N. macromphalus* and one drift shell of *N. pompilius*. Landman (1983a) documented the occurrence of a large barnacle, *Chirona tenuis*, that grew on a juvenile specimen of *N. pompilius* while the *Nautilus* was still alive. Hamada (1964, 1983) reported epizoans on drift shells and live specimens of *N. pompilius*.

Although no systematic survey of the kind and extent of encrustation on *Nautilus* has been made, such a study is particularly interesting because of the life history of *Nautilus* and their evolutionary background. First, *Nautilus* are mobile animals and do not provide a stationary substrate for epizoan settlement. In this respect, they more closely resemble other pelagic animals such as sea snakes, whales, and turtles, rather than sedentary mollusks. Second, *Nautilus* maintain near-neutral buoyancy, and the growth of epizoans may therefore become a serious encumbrance. However, antifouling adaptations such as periostracum may inhibit epizoan settlement (Bottjer, 1981, 1982). Third, *Nautilus* live as deep as 600 m and seldom ascend to depths less than 50 m except in New Caledonia (Chapter 9). Epizoans at these depths may differ from those in shallow water, which are more likely to encrust the drift shells of dead animals. Fourth,

NEIL H. LANDMAN and JUDITH E. WINSTON • Department of Invertebrates, American Museum of Natural History, New York, New York 10024. W. BRUCE SAUNDERS • Department of Geology, Bryn Mawr College, Bryn Mawr, Pennsylvania 19010. PETER J. HARRIES • Department of Geology, University of Colorado, Boulder, Colorado 80309.

Nautilus is related to many fossil forms, and we may compare their incidence of encrustation to that of *Nautilus* to uncover differences in their ecology.

In this chapter, we investigate the kinds and incidence of epizoans on shells of live *Nautilus* to answer the following questions: (1) Do the same kinds of epizoans occur on all species of *Nautilus*, or are certain kinds species-specific? (2) What is the variation in the incidence of encrustation among different species of *Nautilus* and among dissimilar geographic and ecological settings? (3) What is the distribution of epizoans on the shell's surface?

2. Material and Methods

A total of 2029 shells of live *Nautilus* were used in this study, representing three different species, *N. belauensis* Saunders, 1981, *N. pompilius* Linnaeus, 1758, and *N. scrobiculatus* [Lightfoot, 1786].

Nautilus belauensis has a large, relatively smooth shell with a shallow umbilicus and little or no periostracum at maturity. Data were compiled from 1372 specimens of this species caught in steep forereef environments off Palau (Chapter 3) and from several drift shells collected on Palau [reposited in the American Museum of Natural History (AMNH)].

Nautilus pompilius is typically smaller than *N. belauensis*, but it is similar in shape, with little or no periostracum at maturity. Specimens of *N. pompilius* were obtained from two sites in the tropical Indo-Pacific, although this species occurs over a much wider area: (1) Tañon Strait (350 animals), which is a relatively narrow basin between Negros and Cebu Islands (Chapter 11), and (2) Papua New Guinea (281 animals), which represents several environments including steeply dipping forereefs (Port Moresby, Central Province; Manus), broadly dipping forereefs (Kavieng, New Ireland Province), and river delta (Lae, Huon Gulf) (see Chapter 3 for details on all these collecting sites). Several drift shells collected on the beaches of Indonesia were also studied (reposited in the AMNH).

Nautilus scrobiculatus has a deeply umbilicate shell with a unique, thick periostracal covering. This periostracum is best developed on the bottom and flanks of the shell; it is generally absent in the umbilicus, even in live-caught animals. A total of 26 animals were caught in steep forereef environments off Manus, Papua New Guinea, at a site at which *N. pompilius* also occurs. The shells of several dead animals were also inspected (reposited in the AMNH).

Shells of freshly caught animals were inspected for epizoans, which were classified as serpulids, bryozoans, barnacles, and scyphozoans. On the basis of these data, the percentage of encrusted shells and the percentage of shells encrusted by each kind of epizoan were computed. Subsamples of encrusted shells of all three species were chosen for closer study, including 36 shells of *N. belauensis* from steep forereef environments off Palau (reposited in the AMNH), 12 shells of *N. pompilius* from three steep forereef environments off Papua New Guinea (Port Moresby, Central Province; Ndrova and Komuli, off Manus, reposited in the AMNH), and 12 shells of *N. scrobiculatus* from one of these same sites (Ndrova off Manus; reposited in the AMNH). All these specimens were dried shells, but the specimens of *N. scrobiculatus* retained most of their periostracum.

Epizoans were identified using light and scanning electron microscopy. The bryozoan colonies were also counted, and the percentage of them alive at the time of capture was inferred from the presence of chitinous opercula and cuticles that were preserved on dried specimens. Because the embryos of most bryozoan species are brooded in calcified brood chambers (ovicells), we also calculated the percentage of colonies that were able to reproduce on the *Nautilus* shell.

Shells were drawn or photographed from each side and normalized to the same size. Epizoans were described as inside or outside the umbilical area, defined by a circle with a diameter equal to one quarter that of the shell and centered on the umbilicus. In *N. scrobiculatus*, this circle circumscribed the deep umbilicus, whereas in the other two species, it corresponded to a central depression. The surface area of the shell and the part covered by epizoans were measured on the drawings or photographs. These areal measurements are underestimates, because the three-dimensional shell and its epizoans are represented in only two dimensions. On the basis of these measurements, the percentage of the total shell surface covered by epizoans was computed.

The percentage composition of the encrusted surface area represented by each kind of epizoan (e.g., serpulid, bryozoan) was also computed. This percentage differs from the percentage of shells encrusted by each kind of epizoan previously calculated on the basis of the complete sample of shells, for three reasons: First, foraminifers were not included in calculating the first percentage. Second, epizoans were not treated equally in calculating the two percentages. For example, five colonies of bryozoans may have comprised more surface area than a single serpulid, but each kind of epizoan was counted only once in calculating the first percentage. Third, the sample of shells used to calculate the percentage of encrusted shell surface was small and may have underrepresented or even excluded epizoans that occur at low frequencies.

3. Results

3.1. Incidence of Encrustation

The incidence of encrustation varied among the three *Nautilus* species and among their geographic and environmental settings (Table I). In *N. pompilius* from the Philippines, 12% of the shells were encrusted, whereas 49% of the shells of this species were encrusted in the Papua New Guinea sample. Within this sample, however, 63% of the shells from steep forereef environments were encrusted, compared to only 10–20% from broadly sloping forereef and river delta environments. In *N. belauensis* from steep forereefs off Palau, the incidence was 68%. The highest incidence of encrustation, 92%, occurred in *N. scrobiculatus* from steep forereef environments off Papua New Guinea.

3.2. Incidence According to Epizoan

The percentage of shells encrusted by each kind of epizoan is shown in Fig. 1. In *N. pompilius*, the greatest diversity of epizoans occurred in steep forereefs

Table I. Percentage of Epizoan-Encrusted Shells for Three Species of Live *Nautilus*

Species	Locality	Environment	Number of shells	Shells encrusted
N. pompilius	Tañon Strait, Philippines	Restricted basin	350	12%
	Papua New Guinea	Various	281	49%
	Port Moresby, Central Province; Manus	Steep forereef	196	63%
	Kavieng, New Ireland Province	Broad forereef	49	10%
	Lae, Huon Gulf	River delta	36	22%
N. scrobiculatus	Manus, Papua New Guinea	Steep forereef	26	92%
N. belauensis	Palau	Steep forereef	1372	68%

off Papua New Guinea. A third of all shells were encrusted by serpulids, a third by barnacles (almost all stalked), and 3% by bryozoans. In other environments, both in the Philippines and in Papua New Guinea, *N. pompilius* was encrusted by, at most, two kinds of epizoans. In *N. scrobiculatus*, almost all the shells were encrusted by serpulids (92%), approximately half by scyphozoans (46%), and 12% by barnacles (acorn barnacles 4%, stalked barnacles 8%). In *N. belauensis*, more than half the shells were encrusted by serpulids (62%), approximately one third by bryozoans (36%), and 5% by barnacles (acorn barnacles 4%, stalked barnacles 1%).

3.3. Species Composition of Epizoans and Their Distribution on the Shell Surface

3.3.1. *Nautilus belauensis*

The encrusted area averaged 2.6% of the shell surface and occurred equally on the left and right sides of the shell, both inside and outside the umbilicus (Figs. 2 and 3A). The maximum encrustation on any specimen was 16%.

Bryozoans comprised 56% of the encrusted area and extended over large parts of the flanks and venter (Figs. 3A and 4A and B). A total of 95 bryozoan colonies were observed, and about two thirds of them were alive at the time of collection. Almost one third of these colonies (27%) were sexually reproductive, indicating that *Nautilus* act as dispersal agents. One species, *Celleporina excisa* (Fig. 4A), comprised 54% of the total number of colonies. Of these colonies, 82% were alive at the time of collection, though their skeletons showed considerable abrasion; 16% of them were ovicelled. Of all bryozoan colonies, 12% belonged to an undescribed species of *Stomachetosella*. Of these, 90% were alive, and 64% were ovicelled. Of all bryozoan colonies, 6% belonged to *Hippothoa flagellum* (Fig. 4B); most of these colonies appeared to have been alive and reproductive at the time of collection. The remaining colonies belonged to four other species (Table II).

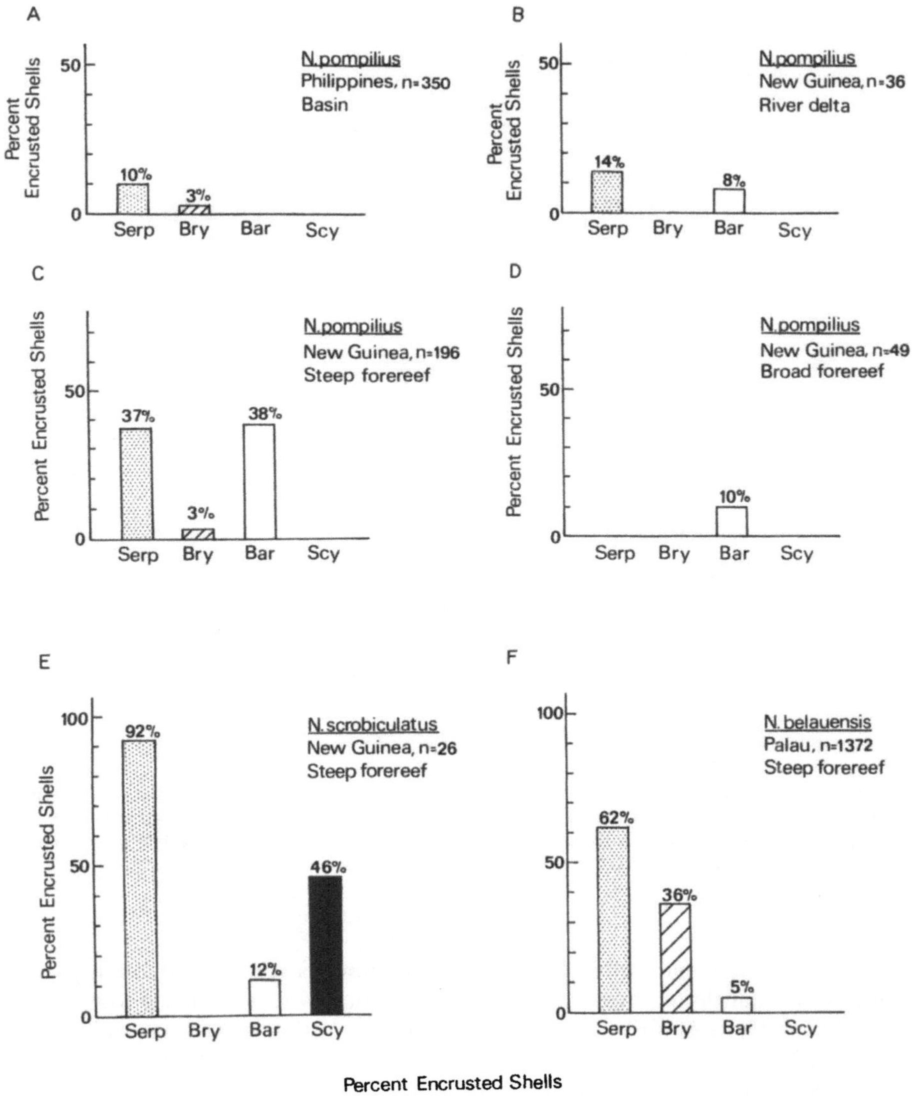

Figure 1. Percentage of *Nautilus* shells encrusted with each kind of epizoan. (A–D) Percentages for *N. pompilius* from four different environments: (A) Tañon Strait, the Philippines; (B) Lae, Huon Gulf, Papua New Guinea; (C) Port Moresby, Central Province and Manus, Papua New Guinea; and (D) Kavieng, New Ireland Province, Papua New Guinea. (F) Percentages for *N. scrobiculatus* from steep forereefs off Manus, Papua New Guinea. (G) Percentages for *N. belauensis* from steep forereefs off Palau.

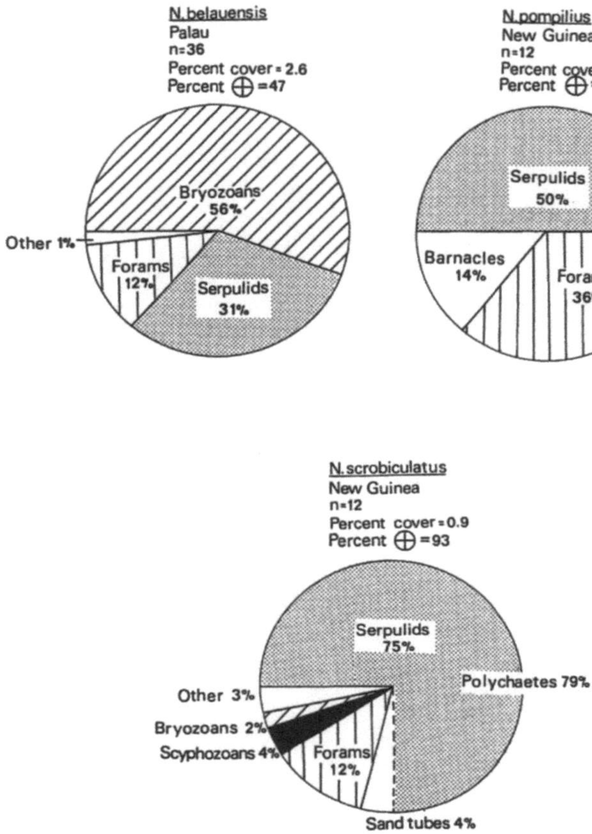

Figure 2. Pie charts depicting the average percentage composition of the encrusted surface for each species of *Nautilus*, based on the component area of each kind of epizoan. "Percent cover" is the average percentage of the total surface area of the shell covered by epizoans. "Percent ⊕" is the average percentage of the encrusted area that occurs within the umbilical region.

The other epizoans on *N. belauensis* consisted of serpulids and foraminifera; these were distributed over the entire shell surface, with a tendency toward the umbilicus. The serpulids consisted of small, encrusting tubes, many of which were remnants; at least six species were present (Fig. 3B and Table II). The foraminifera consisted of encrusting types that formed dishlike excavations in the shell. They belonged to the genus *Cibicides* or the family Cibicididae (Fig. 4D and Table II). Rare occurrences included acorn and stalked barnacles, ostreid bivalves, sponge spicules, diatoms, and scyphozoans. These scyphozoans belonged to the genus *Stephanoscyphus* and were more common on *N. scrobiculatus*, but occurred on four specimens of *N. belauensis* (Fig. 4C). Minute borings were also detected on the shell surface of some specimens. These borings mea-

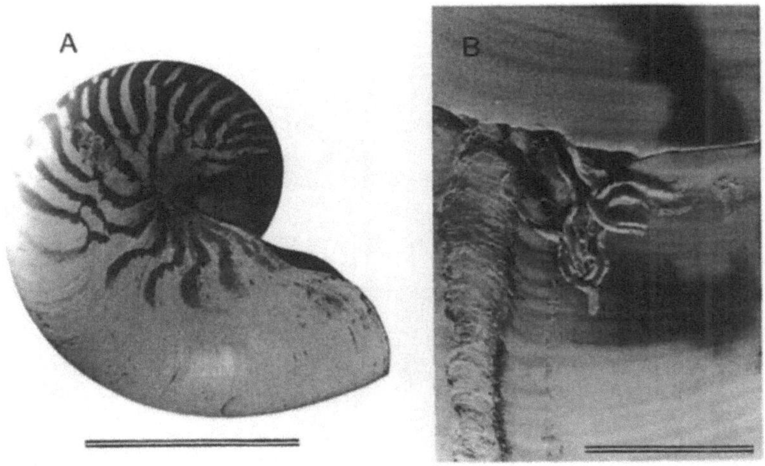

Figure 3. (A) *Nautilus belauensis* (AMNH 43026), mature female, caught May 28, 1977, by W. B. Saunders and C. Spinosa at a depth of approximately 240 m, Mautremdiu Point, Palau. The shell is encrusted with serpulids and several colonies of bryozoans and serpulids. Scale bar: 10 cm. (B) Close-up of *N. belauensis* (AMNH 43027), mature male, caught July 31, 1982, by W. B. Saunders and M. Weekley at a depth of approximately 330 m, Mautremdiu Point, Palau. Serpulids settled and grew in a repaired shell injury. Adapertural direction is toward the top of the photo. Scale bar: 1 cm.

sured approximately 30 μm in diameter and may have been produced by algae, fungi, or dissolution (Fig. 4F).

3.3.2. *Nautilus pompilius*

The encrusted area averaged 0.5% of the shell surface, or less than one fifth that in *N. belauensis* (Fig. 5A). The left and right sides of the shell were approximately equally covered by epizoans, and the maximum encrustation on any specimen was 2.2%. Most of the encrusted area (59%) occurred near the umbilicus (see Fig. 2).

Unlike the case of *N. belauensis*, bryozoans were rare on *N. pompilius* and were represented by only a single species, *Hippothoa flagellum*, which also occurred on *N. belauensis* (see Fig. 4B). In lieu of bryozoans, serpulids composed half the encrusted area (Fig. 2). They belonged to three species (Table II) and were distributed over the shell surface with a tendency toward the umbilicus. Foraminifera represented 36% of the encrusted area and included species similar to those on *N. belauensis* (Fig. 4D and Table II). They were equally distributed inside and outside the umbilical region. Barnacles comprised 14% of the encrusted area and consisted mostly of small, stalked forms; these epizoans occurred almost exclusively in the umbilicus (Figs. 4E and 5B). Rare occurrences included venerid bivalves.

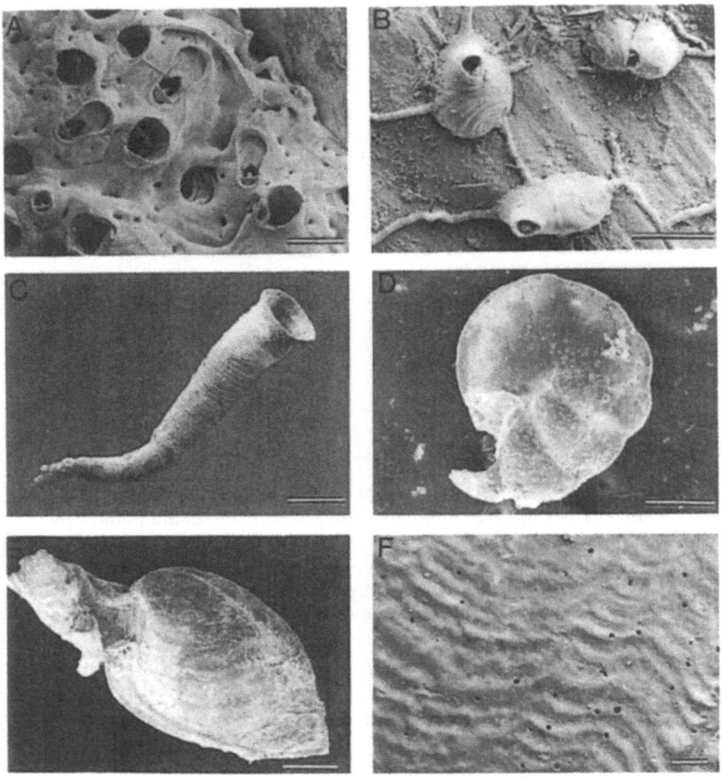

Figure 4. Examples of encrusting organisms. (A) Close-up of *Celleporina excisa*, the most common bryozoan on *N. belauensis*. Scale bar: 200 μm. (B) Close-up of *Hippothoa flagellum*, the bryozoan that occurs on all three species of *Nautilus*. Scale bar: 200 μm. (C) *Stephanoscyphus* sp., the scyphozoan that commonly occurs on *N. scrobiculatus*. Scale bar: 1 mm. (D) *Cibicides* sp. cf. *C. refulgens*, a common foraminifer. Scale bar: 200 μm. (E) Small, stalked barnacle from the umbilicus of a specimen of *N. pompilius*. Scale bar: 1 mm. (F) Bore holes approximately 30 μm in diameter on *N. belauensis* that may have been produced by algae, fungi, or dissolution. Scale bar: 200 μm.

3.3.3. *Nautilus scrobiculatus*

The encrusted area averaged 0.9% of the shell surface (Fig. 6A), which is comparable to that in *N. pompilius* (Fig. 6A). The left and right sides of the shell were equally covered by epizoans, and the maximum encrustation on any shell was 2.9%. However, the epizoans occurred almost exclusively in the umbilicus; they were virtually absent elsewhere, except on the top surface of the shell and at the sites of repaired injuries.

The umbilicus supported a lush growth of serpulids (75%), other polychaetes with agglutinated tubes (4%), and scyphozoans (4%) (Figs. 2 and 6B). The serpulids included at least eight species (Table II), many of which exhibited large, convoluted tubes extending out from the umbilicus. The scyphozoans belonged

Table II. Epizoans on the Shells of Live *Nautilus*

Epizoan	*N. belauensis* Palau	*N. pompilius* New Guinea	*N. scrobiculatus* New Guinea
Bryozoans			
Celleporina excisa	X	—	—
Electra angulata	X	—	—
Exochella tricuspis	X	—	—
Hippothoa flagellum	X	X	X
Iodictyum sp.	X	—	—
Membranipora undescribed sp. 1	X	—	—
Microporella orientalis	X	—	—
Stomachetosella sp.	X	—	—
Foraminiferans			
Carpenteria cf. *C. monticularis*	—	—	X
Carpenteria sp.	—	X	—
Cibicides refulgens	X	—	—
Cibicides cf. *C. refulgens*	—	X	X
Cibicides cf. *C. lobatulus*	—	X	X
Cibicides spp.	X	X	X
Cymbaloporetta sp.	—	—	X
Globigerinoides spp.	—	X	—
?*Hemisphaerammina* sp.	X	—	—
Homotrema spp.	X	—	—
Planorbulina spp.	X	—	—
Rectocibicides spp.	—	—	X
?*Rectocibicides* spp.	X	X	—
Remaneica spp.	—	—	X
Polychaetes			
Hydroides bifidus	X	—	—
Hydroides bifidus?	—	—	X
Hydroides longispinosa?	X	—	—
Hydroides spp.	X	X	X
Serpula concharum?	—	—	X
Serpula cf. *S. kaempfari*	X	—	—
Serpula vermicularis?	—	—	X
Serpula vittata?	—	—	X
Spirobranchus giganteus giganteus	X	—	—
Spirobranchus tetraceros	—	X	X
Spirobranchus cf. *S. tricornigerus decoratus*?	X	X	X
Polychaetes spp. (agglutinated tubes)	—	—	X
Barnacles			
Acorn	X	X	X
Stalked	X	X	X
Corals	—	—	X
Scyphozoans			
Stephanoscyphus sp.	X	—	X
Bivalve mollusks			
Ostreids	X	—	X
Venerids	—	X	—
Others	—	—	X
Sponge	X	—	X
Diatoms	X	—	X
Coccoliths	—	—	X
Pollen (on barnacles)	—	X	X

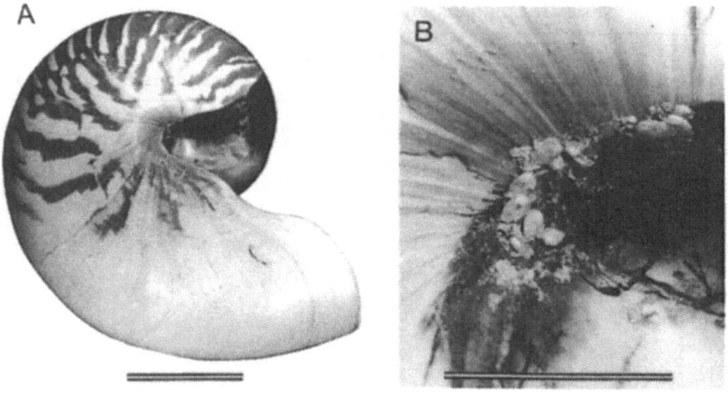

Figure 5. (A) *Nautilus pompilius* (AMNH 43028), mature male, caught June 17, 1984, by W. B. Saunders and L. E. Davis at a depth of approximately 250 m, Ndrova Island, Manus Province, Papua New Guinea. Scale bar: 5 cm. (B) Close-up of the umbilicus showing numerous stalked barnacles. Scale bar: 1 cm.

to the genus *Stephanoscyphus* (Fig. 4C) (see Werner, 1966, 1967, 1974) and consisted of erect chitinous tubes up to 1 cm in height that extended perpendicularly out from the umbilicus (Fig. 6B). Foraminifera and bryozoans consisted of many of the same species previously observed on *N. belauensis* and *N. pompilius* (Fig. 4B and D and Table II). Bryozoans were represented by *Hippothoa flagellum*, which also occurred on *N. belauensis* and *N. pompilius*. Rare occurrences in-

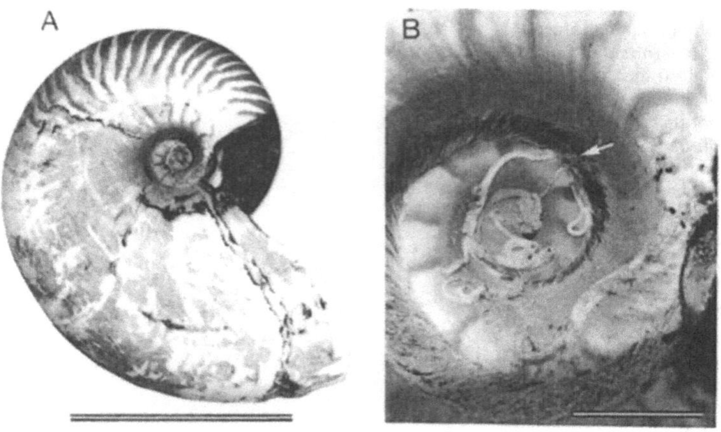

Figure 6. (A) *Nautilus scrobiculatus* (AMNH 43029), mature male, caught May 28, 1985, by W. B. Saunders and L. E. Davis at a depth of approximately 370 m, Ndrova Island, Manus Province, Papua New Guinea. Scale bar: 10 cm. (B) Close-up of the umbilicus showing a variety of serpulids and a scyphozoan (⇐). Scale bar: 1 cm.

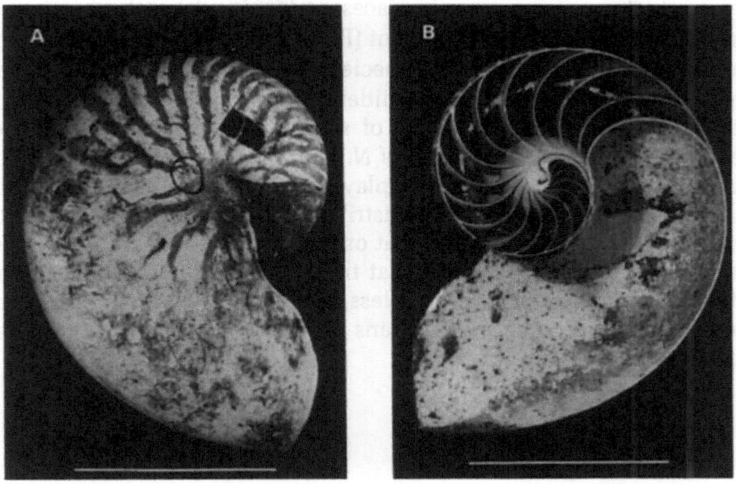

Figure 7. *Nautilus belauensis* (AMNH 43030) showing epizoans that grew on the shell during 46 days in shallow water (6 m depth) in Malakal Harbor, Palau. This specimen was captured alive June 26, 1982, and placed in a shallow holding cage; it was observed to be dead 3 days later, and 2 days afterward, its body had been removed from the shell by scavengers. The epizoans occur over the entire shell surface, including the black layer (A) and inside the body chamber (B). Scale bars: 10 cm.

cluded bivalve mollusks, acorn and stalked barnacles, corals, diatoms, coccolithophores, sponges, and pollen.

3.3.4. Drift Shells

Epizoans were randomly distributed over the entire shell surfaces of the two drift specimens of *N. belauensis* examined. The epizoans occurred on top of the

Table III. Bryozoans on Selected Drift Shells of *Nautilus belauensis* and *Nautilus pompilius*

Species	*N. belauensis* Palau	*N. pompilius* Indonesia
Celloporaria vagans?	X	—
Celloporaria sp.	X	—
Electra angulata	X	X
Fenenstrulina catastichos	X	—
Mebranipora undescribed sp. 1	X	X
Mebranipora undescribed sp. 2	—	X
Microporella undescribed sp.	X	—
Parasmittina hastingsae	X	—
Parasmittina parsevalii	X	—
Parasmittina tropica	X	—
Smittoidea pacifica	X	—

black layer, which was worn away in places, and inside the body chamber, except in the areas of former muscle attachment (Fig. 7A and B). The serpulids on these two shells consisted of at least two species of *Hydroides*, and the foraminifera consisted of *Planorbulina* sp. and an unidentified species belonging to the family Cibicididae. The bryozoans consisted of several species absent on live-caught specimens but present on drift shells of *N. pompilius* from Indonesia (Table III). Other drift shells of *N. belauensis* displayed broken body chambers with large oysters attached inside. However, the distribution of epizoans on three drift shells of *N. scrobiculatus* was identical to that on the shells of live animals. This similarity may be explained by the fact that these shells were obtained by rewards offered for perfect specimens. Nevertheless, the periostracum was already missing from these shells, and no or few epizoans occurred on that part of the shell that previously had been covered.

4. Discussion

Epizoans occur on more than half the *Nautilus* that inhabit steep forereef environments, even in widely separated areas (see Table I). In *N. scrobiculatus*, this incidence approaches 100%. By comparison, the percentage is lower in *N. pompilius* from broadly sloping forereef (Kavieng, New Ireland Province, Papua New Guinea), river delta (Lae, Huon Gulf, Papua New Guinea), and restricted basin (Tañon Strait, the Philippines). An extremely muddy bottom may account for the low incidence of encrustation in the river delta (22%), but the low incidence in the other two environments (10–12%) is more difficult to explain.

In general, the epizoans consist of serpulids, bryozoans, barnacles, foraminifera, and scyphozoans. All these epizoans, except scyphozoans, have previously been reported on *N. pompilius* from the Philippines (Seilacher, 1982; Hamada, 1983). Stalked barnacles and serpulids have also been reported on *N. macromphalus* from New Caledonia (Seilacher, 1982; Hamada, 1983).

The distribution of the various kinds of epizoans among the three species of *Nautilus* suggests that the epizoans may be species-specific. On the basis of our data, bryozoans prefer *N. belauensis*. The highest percentage of encrusted shells (36%), the highest percentage of encrusted shell surface (56%), and the greatest species diversity (8) occur in this species. In *N. scrobiculatus* and *N. pompilius*, on the other hand, bryozoans occur on fewer than 3% of all shells and comprise less than 2% of their encrusted shell surface. They display only one species in common with *N. belauensis*, although Hamada (1983) reported another species on *N. pompilius* from the Philippines.

Serpulids prefer *N. scrobiculatus*. This species is marked by the highest percentage of encrusted shells (92%), the highest percentage of encrusted shell surface (75%), and the greatest species diversity, including polychaetes with agglutinated tubes. In *N. belauensis* and *N. pompilius*, serpulids are less abundant but still common and occur on approximately 40–60% of all shells and compose 30–50% of their encrusted shell surface. *Nautilus belauensis* exhibits the second highest species diversity, but has few species in common with *N. scrobiculatus*. *Nautilus pompilius* shows the lowest diversity, but Hamada (1983) reported three additional species on *N. pompilius* from the Philippines, one of which occurs on *N. belauensis*.

Barnacles prefer *N. pompilius*, as indicated by the high percentage of encrusted shells in this species (38%). Barnacles comprise 14% of the encrusted shell surface and include stalked and acorn forms. Both these forms also occur on *N. pompilius* from the Philippines (Hamada, 1983) and on *N. belauensis* and *N. scrobiculatus*. Foraminifera, on the other hand, are distributed with almost equal diversity among the three species of *Nautilus* and mainly consist of species of *Cibicides*. *Nautilus scrobiculatus* and *N. pompilius* share the most species in common, although Hamada (1983) reported a species of *Homotrema* on *N. pompilius* from the Philippines that is present in our samples on *N. belauensis*. The highest percentage of surface area encrusted occurs in *N. pompilius* (36%).

These preferential patterns may reflect partly the desirability of the various *Nautilus* species as substrates for the different kinds of epizoans. For example, the relatively smooth shell surface of *N. belauensis* and *N. pompilius*, with little or no periostracum at maturity, provides a more attractive substrate for bryozoans than the periostracum-covered surface of *N. scrobiculatus*. The deep umbilicus in this species favors serpulids, but may inhibit barnacles. For example, *N. pompilius* and *N. scrobiculatus* co-occur, but the percentage of shells encrusted by serpulids is almost three times higher in *N. scrobiculatus*, whereas the percentage of shells encrusted by barnacles in this same species is less than one third lower. In general, however, the umbilicus is a preferred site of encrustation. This same pattern was observed in *N. macromphalus* (Hamada, 1983) and may be the result of a hydrological effect produced by water flow into the umbilicus when *Nautilus* is in motion (see Chamberlain, 1976). This circulation could transport epizoan larvae as well as food particles into the umbilicus. Repaired shell breaks may also be preferred sites of epizoan settlement due to increased water turbulence.

In contrast, epizoans are generally excluded from the black area and apertural margin of the shell, although Hamada (1983) noted exceptions in *N. pompilius* from the Philippines. These parts of the shell are in contact with the soft body and tentacles, which actively move across the apertural region during life. The black layer itself serves to cover epizoans and provide a more uniform surface for deposition of the dorsal wall (Landman, 1983a).

The shaggy periostracum in *N. scrobiculatus* appears to discourage epizoans, as indicated by their low incidence outside the umbilicus, where the periostracum is present. Exceptions occur on the top (dorsal) surface of the shell, where the periostracum is thinner, and in repaired shell breaks. The average area of encrustation on the part of the shell covered by periostracum is less than one quarter that on similar periostracum-free shell areas in *N. pompilius* and *N. belauensis*. These two species do not show periostracum at maturity, although it is well developed in juveniles.

In all species, the encrusted surface comprises less than 3% of the total shell surface and less than 1% of the shell weight in air, a negligible factor in the overall buoyancy of the animal. This low percentage of encrustation may suggest that *Nautilus* employs other means of inhibiting epizoan settlement, e.g., biochemical substances or a thin covering of mucus on the shell surface.

The sparseness of epizoans on the shells of live *Nautilus* contrasts with the condition in drift specimens, in which epizoans are densely and randomly distributed on the shell surface. On drift shells of *N. belauensis* and *N. pompilius*, the epizoans consist of several species of serpulids, bryozoans, oysters, and algae that are absent on live animals (Table III). Stalked barnacles have also been reported on drift shells of *N. pompilius* (Hamada, 1964; Seilacher, 1982). Epizoans are generally absent, however, on beached specimens in which the outer shell layer has eroded away.

The geographic ranges of the epizoan species on live specimens also differ from those on drift shells. In the bryozoans, for example, the two most abundant species on live specimens have a restricted distribution. *Celloporina excisa* is known only from two stations in eastern Indonesia, ranging from 73 to 113 m in depth (Harmer, 1957). *Stomachetosella* sp. is undescribed at present and is known only from our samples. On the drift shells, however, the individual species have a broader geographic distribution; an exception is *Fenestrulina cotastichos*, which was recently described from New Zealand (D. P. Gordon, 1984). *Electra angulata* and the two undescribed species of *Membranipora* are widely distributed on surface drift objects (seeds, wood, plastic trash). They occur throughout the tropical Indo-Pacific and possibly even the Caribbean. Surprisingly, however, two colonies of *Electra angulata* were also observed on live specimens of *N. belauensis*. Their occurrence may suggest that these *Nautilus* approached the surface at some time in their lives, and indeed, *N. belauensis* has been caught in overnight traps as shallow as 66 m (Saunders, 1984b).

The low percentage of the total surface area encrusted on live specimens is similar to the pattern in other pelagic animals, such as sea snakes, whales, and turtles. Sea snakes are encrusted with diatoms, foraminifera, hydrozoans, serpulids, bivalves, bryozoans, and stalked barnacles. In general, the extent of encrustation on sea snakes is low, which may be due to periodic shedding of the skin and to twisting and knotting behavior (Zann *et al.*, 1975). Whales are commonly encrusted with acorn and stalked barnacles on the snout and elsewhere on the body (C. Potter, personal communication, 1985). Sea turtles are encrusted with a variety of epizoans, including algae, barnacles, bryozoans, isopods, crabs, amphipods, hydrozoans, and bivalves. The extent of encrustation is low; this may be due to a flaking away of the scutes or to the presence of a grease-impregnated integument (Frazier *et al.*, 1984).

How do the patterns of encrustation on *Nautilus* compare to those on fossil nautilids and ammonoids? In the Late Cretaceous of the Western Interior of North America, the nautilid species *Eutrephoceras dekayi* is often encrusted during early growth with bryozoans and serpulids (Fig. 8). In contrast, associated specimens of scaphitid ammonoids, which have comparatively deeper umbilici and coarser ornament, show no indication of encrustation during life. Possibly, the absence of epizoans on the scaphites may be due to the former presence of a heavy periostracum or mucuslike covering. Alternatively, the two forms may have inhabited different depths, and their co-occurrence in the same fossil bed may simply be an artifact of preservation or postmortem transportation.

Seilacher (1982, Figs. 3B and 6) figured two specimens of nautilids from the Jurassic Posidonia Shales of Holzmaden that were also encrusted with serpulids and bryozoans. Many of the co-occurring ammonoids, however, were much more heavily encrusted with serpulids, pelecypods (*Ostrea, Exogyra, Gervillia, Inoceramus*), bryozoans, and inarticulate brachipods. Although no figures were pro-

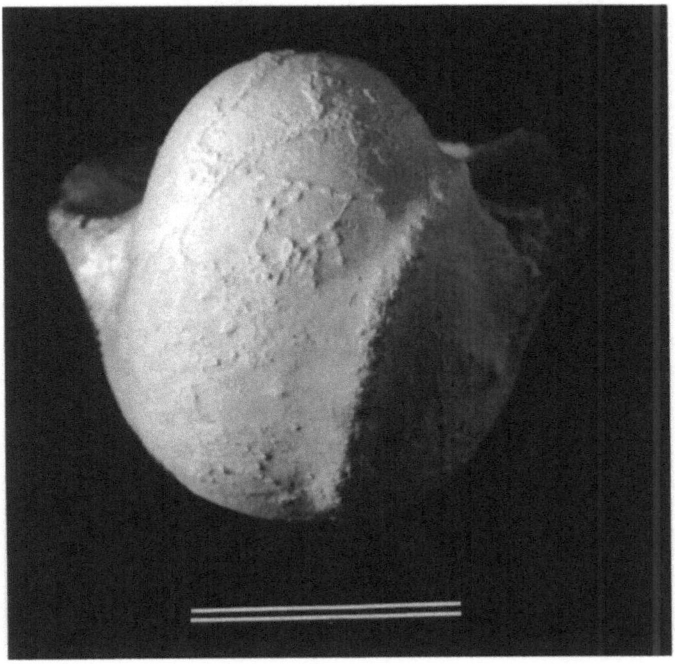

Figure 8. Specimen of *Eutrephoceras dekayi* (AMNH 38944) from the Late Cretaceous of North America. Adult whorls were peeled back to expose juvenile encrustation by serpulids and bryozoans. Scale bar: 1 cm.

vided as to the percentage of shells encrusted, Seilacher concluded, on the basis of the distribution of epizoans on the shell surface, that encrustation occurred while the shell was still in the water column, above the bottom. In specimens in which epizoans occur just above the aperture and along the peristome, some encrustation may have occurred after death (Seilacher, 1982, Figs. 5B and 8B and F). However, in other specimens, the distribution and orientation of epizoans suggest that they grew during the life of the ammonoid (Seilacher, 1960; 1982, Figs. 7C and D, 8C, 9, and 10B and C). If this suggestion is true, it would seem that these ammonoids were much less successful at discouraging epizoan settlement than are the species of modern *Nautilus*.

Acknowledgments. We thank Drs. John Lee and Ernst Kirsteuer, American Museum of Natural History, who helped identify the foraminifera and scyphozoans, respectively, and Amy Adelson, Beverly Heimberg, and Susan Klofak, who assisted in statistics and preparation of the figures and plates. This research was partly funded by NSF Grant No. EAR 83-18932 to W. Bruce Saunders.

Chapter 11

On the Habitat of *Nautilus pompilius* in Tañon Strait (Philippines) and the Fiji Islands

SHOZO HAYASAKA, KIMIHIKO ŌKI, KAZUSHIGE TANABE, TOSHIO SAISHO, and AKIHIKO SHINOMIYA

1. Introduction

Since 1981, we have been engaged in field studies of the habitat of *Nautilus pompilius* in the Philippines (1981 and 1982) and in the Fiji Islands (1982 and 1983) (Hayasaka *et al.*, 1982; Hayasaka, 1983, 1985). The main purpose of these studies was to obtain basic data on the habitat of *N. pompilius* in the Philippines

SHOZO HAYASAKA and KIMIHIKO ŌKI • Institute of Earth Sciences, Faculty of Science, Kagoshima University, Kagoshima 890, Japan. KAZUSHIGE TANABE • Geological Institute, Faculty of Science, University of Tokyo, Tokyo 113, Japan. TOSHIO SAISHO and AKIHIKO SHINOMIYA • Laboratory of Marine Biology, Faculty of Fisheries, Kagoshima University, Kagoshima 890, Japan.

and in Fiji, which are at opposite ends of the vast distribution range of this species. Although the overall project is still in progress, the results of study to date are summarized in this chapter.

2. *Nautilus* Distribution and Fishery Techniques in the Philippines

It is well known that *Nautilus* is ubiquitous in the Philippines; thousands of specimens are captured and exported each year (Talavera and Faustino, 1931). Our reconnaissance survey in 1980 in the central part of the islands (Panai, Negros, Cebu, and Bohol Islands) resulted in the following conclusions:

1. *Nautilus* usually ranges in depths of approximately 150–500 m.
2. Water temperatures at depths of 150–200 m are approximately 20°C.
3. Organisms associated with *Nautilus* include several kinds of crustaceans (crabs, spiny lobsters, prawns, and shrimps), deep-water bivalves, gastropods, sponges, and, commonly, the echinoid *Malepia cordata*.
4. Most fishermen trap *Nautilus* at about 200 m, but the maximum known depth is more than 525 m.
5. Diverse types of bait are used successfully, including toad meat (along the west coast of Panai Island), chicken and fish (along the east coast of Negros Island), and eel (along the south coast of Bohol Island).
6. *Nautilus* fishing is usually done during the night, and several types of traps are used successfully.

3. Southern Tañon Strait: A Case Study

3.1. Setting and Background

Tañon Strait is one of the best-known areas for *Nautilus* studies, beginning with the accounts of Dean (1901). Located between the islands of Cebu and Negros, the strait is rather narrow (15–27 km across) and long (220 km from north to south). According to Dean (1901) and to hydrographic charts (Hydrographic Office of Japan, 1962), the deepest point (555 m) is situated in the center, and depths greater than 500 m extend widely from north to south (Fig. 1). Both sides of the strait are rather steep, except where the submarine terraces are developed. The strait becomes gradually shallower northward and is separated from the Visayan Sea by a shallow region with many small islands (including Bayantan, Don, and Guintan). The narrow (≈ 6 km) southern outlet of the strait is approximately 125 m deep, off Liloan Point, Cebu Island. Farther south, outside the strait, the bottom deepens toward Bohol Strait. Physiographically, and in terms of seawater characteristics, the area within Tañon Strait is relatively closed and isolated.

To obtain basic data regarding the habitat of *Nautilus*, a systematic survey of seawater characteristics, plankton, and bottom sediments was carried out at nine stations deployed along two traverses in the southern part of the Strait (Fig. 1). Bottom topography was surveyed by means of an echo-sounder (Honda-HE

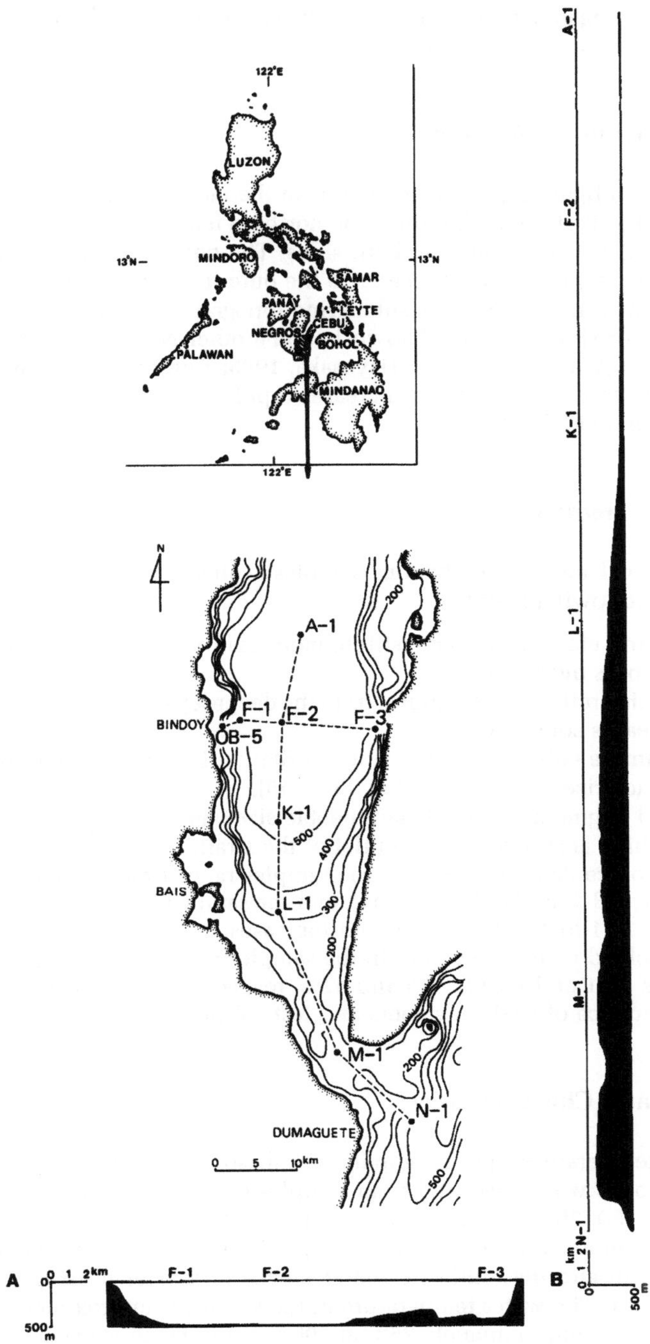

Figure 1. Bathymetric profiles, transverse (A) and longitudinal (B), of Tañon Strait (Philippines), with a map showing the lines of the profiles. Reprinted with permission from Hayasaka (1983).

315), and trapping for *Nautilus* and its associated fauna was undertaken off Bindoy on the east coast of Negros Island.

3.2. Submarine Topography

Bottom configuration seems to be one of the most fundamental features that control the distribution of *Nautilus*. The northern half of the basin is the deepest (≈ 500 m), with steep slopes on both sides. The bottom is almost flat, and the sides ascend steeply to the surface or to the outer margin of stepped submarine terraces (Figs. 1 and 2). Similar submarine topography (deep water adjacent to steep scarps or slopes) is also known in some other *Nautilus* habitats, such as Fiji and Palau (Ward *et al.*, 1977; Hayasaka, 1985; Saunders and Spinosa, 1978). South of the strait, bottom depth increases rapidly to more than 600 m between Negros and Bohol Islands.

3.3. Bottom Sediments

Mechanical analysis of bottom samples (collected at 11 stations) (Fig. 3) yielded the following results:

1. Silt and clay fractions constitute more than 80% of the samples obtained at depths over 400 m.
2. Silt/clay ratios of the samples from the deepest part and the marginal slopes decreases southward.
3. A sample collected on the slope at Station L-1 from a depth of 307 m is characterized by poor sorting (So: 8.45).
4. Coral fragments of cobble and pebble size were collected from Station M-1, which is situated at the outlet of the strait.
5. The percentages of clay in three samples range from 3.5 to 4.4% (stations OB-1, OB-2, and OB-3, Bindoy), and the sand content increases uniformly from 20.1 to 49.8% as the depth increases.
6. Dissolved carbonate values in the samples collected from the deepest stations (A-1 and F-3) are 30 and 47%, respectively, and seem to reflect the abundance of shell fragments and tests of planktonic foraminifera.

3.4. Seawater Characteristics

Water temperature, specific gravity, pH, dissolved oxygen (DO), salinity, and oxygen isotopes were measured from samples collected at depths of 10, 20, 30, 50, 75, 100, 150, 200, 300, and 400 m at nine stations (shown in Fig. 1).

The maximum value for dissolved oxygen is usually observed between depths of 0 and 30 m and may indicate the existence of a layer of oxygen-producing phytoplankton. The water temperature at the surface is approximately 30°C and falls gradually to approximately 25°C at 100 m; a thermocline occurs consistently at 100–150 m. Immediately below the thermocline, the temperature is about 20°C.

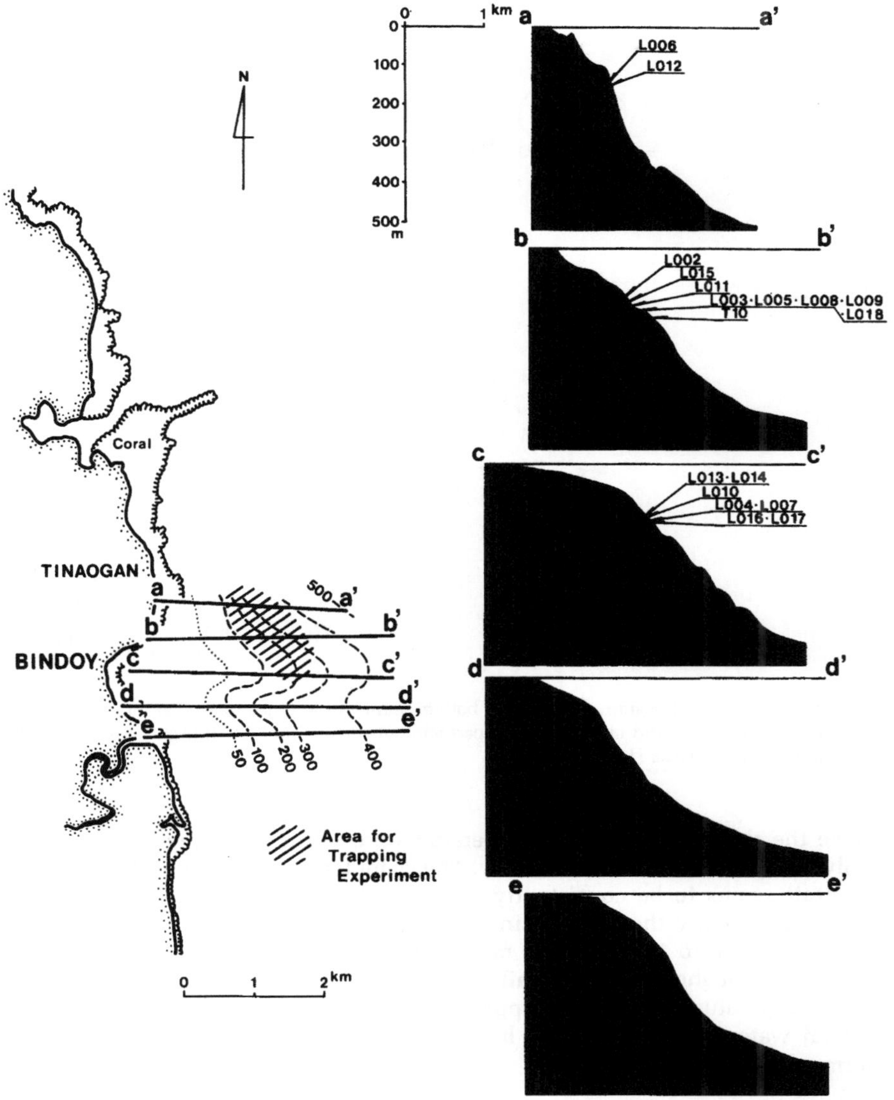

Figure 2. Map showing the lines of echo-sounding and the trapping area (*left*) and the positions of trapping *Nautilus* on the bathymetric cross sections representing rapid and steplike deepening of the bottom topography off Bindoy (*right*). Reprinted with permission from Hayasaka *et al.* (1982).

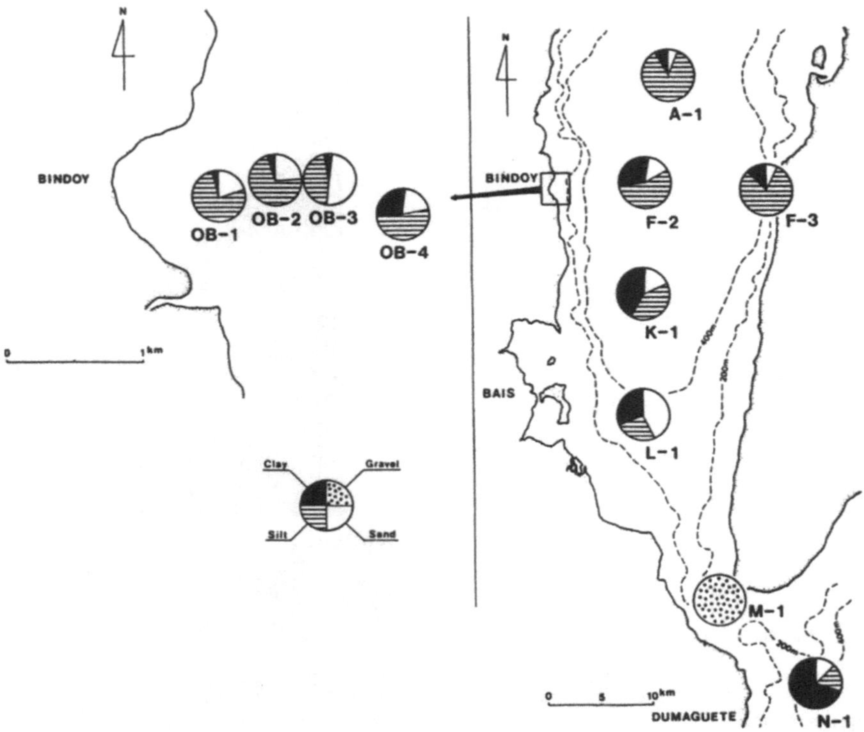

Figure 3. Map showing the grain-size ratios of bottom sediments at 11 stations other than F-1 and OB-5 (Fig. 1) off Binoy (*left*) and in the southern part and outside of Tañon Strait (*right*). Reprinted with permission from Hayasaka (1983).

Above the thermocline, water temperatures are stratified throughout the strait, implying a remarkable water stability. The water temperature below 200 m within the strait seems to be consistently about 17°C, whereas outside the strait, the temperature below the thermocline decreases rapidly from 14.6°C at 200 m to 13.2°C at 300 m to 12.6°C at 400 m (Fig. 4). These figures show that water temperatures throughout Tañon Strait are suitable for *Nautilus* (the lower optimal water temperature of which is approximately 15°C), whereas outside the strait, optimal water temperature may be restricted to a narrow zone just below the thermocline.

3.5. Biota

3.5.1. Plankton

Sampling was carried out at eight stations within the strait and at one station outside the strait, off Dumaguete City (shown in Fig. 1), using a closing-type net

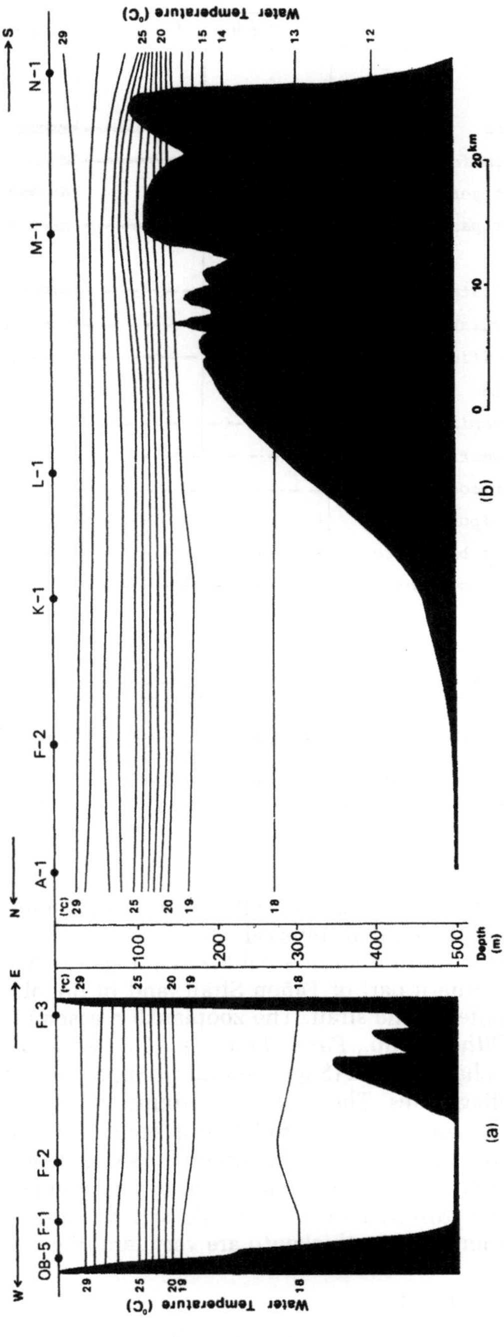

Figure 4. Distribution of water temperature in the transverse (a) and the longitudinal (b) profiles of the southern part of Tañon Strait. The lines of the profiles are shown in Fig. 1. Reprinted with permission from Hayasaka (1983).

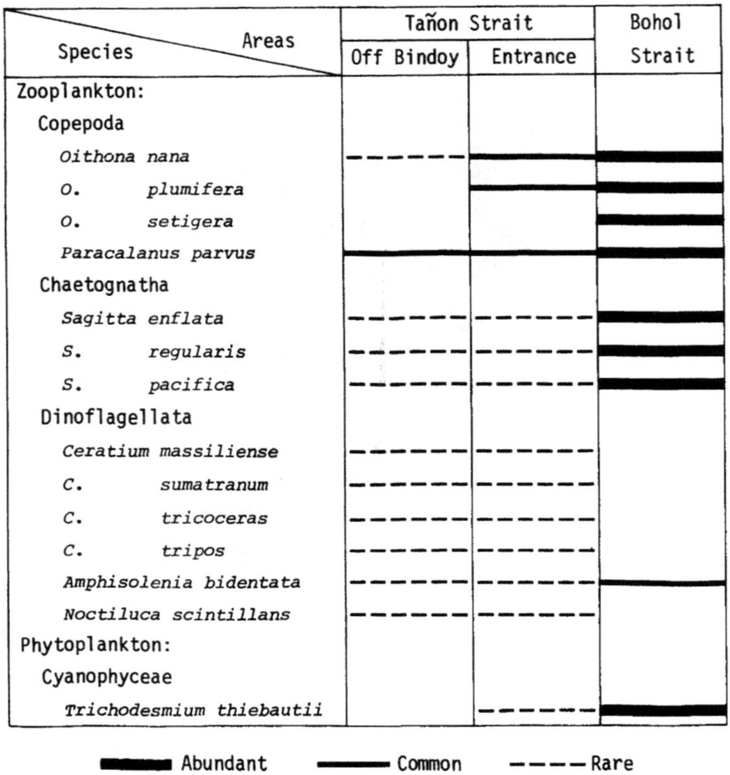

Figure 5. Horizontal distribution of representative species of plankton collected from the surface layer (0–50 m) in the areas inside and outside Tañon Strait. "Off Bindoy": stations OB-5, F-1, F-2, F-3, and A-1. "Entrance": L-1 and M-1. "Bohol Strait": N-1.

(30 cm diameter, with XX13 bolting silk Müller gause; 0.097 mm mesh); it was done during daylight, at three depth intervals (0–50, 50–100, and 100–200 m).

The average plankton settling volume was 3.2 ml (range 1.8–5.6 ml). Plankton are abundant in the southern part of Tañon Strait (and in Bohol Strait), but are rather scarce in the center of the strait. The zooplankton assemblage was dominated by copepods (*Oithona* spp., *Paracalanus* sp., *Oncaea* spp., *Acartia* spp., and copepod nauplii), chaetognatha (*Sagitta enflata*), Appendicularia, pteropods, amphipods, and dinoflagellates. The first three groups are especially abundant (Fig. 5). As a whole, the zooplankton population is small in Tañon Strait, while it is rather large in Bohol Strait. Phytoplankton include Diatomaceae and Cyanophyceae (*Rhizosolenia styliformis*, *Rh. calca avis*, *Biddulphia sinensis*, *Chaetoceros coarctatus*, *Coscinodiscus granii*, *Nitzschia seriata*, and *Trichodesmium thiebautii*). Species other than *Tr. thiebautii* are very rare (Fig. 5).

3.5.2. Benthonic Formaninifera

Bottom samples were collected with a small dredge from 10 stations within Tañon Strait and at one station outside the strait (see Fig. 3). All samples were preserved in buffered formalin (5%), and a portion were stained with Rose Bengal. The total number (per 10 cc of sample) of benthonic foraminifera, planktonic foraminifera, and radiolaria, and the respective proportions of each, were recorded (see Table 2 in Ōki, 1983) and provide the basis for the following conclusions:

1. There are more than 5000 benthonic foraminifera per 10 cc in the nearshore samples. In the central part of the strait and at one station outside the strait (N-1), the numbers range from 803 to 1947 per 10 cc of sample. The total number of planktonic foraminifera and radiolaria at Stations OB-3 and F-3, where rather large numbers of individual benthonic foraminifera occur, is apparently larger than those at the four stations in the central part of the strait.
2. At every station within the strait, the frequency of calcareous hyaline tests exceeds 80%; calcareous porcellaneous tests range from 2.0 to 5.0%, and agglutinated tests comprise less than 14%. Outside the strait (Station N-1), the frequency of agglutinated tests is 24.6%.
3. The ratio of planktonic foraminifera to the total foraminifera in Tañon Strait increases with depth.
4. The ratio of numbers of individual planktonic foraminifera and radiolaria contained in the bottom sediments of Tañon Strait does not correlate with water depth. The number of planktonic foraminifera as a percentage of the total number of planktonic foraminifera and radiolaria is about 70% at the three near-shore stations (OB-1, OB-2, and F-3), whereas at the four stations in the central part of the strait, it increases to 85–95%, suggesting a good correlation with distance from shore. At Station N-1, outside the strait, this ratio is smaller (58.7%); i.e., the relative frequency of radiolaria is higher outside the strait.

3.5.3. Fishes

Long-line fishing, beach seine nets, and two types of *Nautilus* traps were used to obtain data on the associated fish fauna. The fishes collected with traps and long lines (see Table 6 in Hayasaka *et al.*, 1982) typically live along island or continental slopes, and some of them are known to occur from tropical to temperate zones in the Indo-Pacific region (e.g., *Halaelurus buergeri*, *Brotula multibarbata*, *Saurida undosquamis*, and *Rexea prometheoides*). The shallow-water fishes collected with the beach seine nets included 74 species belonging to 41 families. By contrast, the fauna collected around the coral reef shoal about 1 km off the Tinaogan Reef included species belonging to only four families (Pomacentridae, Labridae, Scaridae, and Chaetodontidae). Fishes of the families Mullidae, Sillaginidae, Gerreidae, Leiognathidae, Uranoscopidae, and Pleuronectidae were frequently collected. Among them, *Leiognathus equulus* and *L. leuciscus* are regarded as the best bait for *Nautilus* (W. Vailoces, personal communication).

3.6. *Nautilus* Trapping and Associated Fauna

As a result of trapping during September 7–24, 1981, 52 live *N. pompilius* were obtained from Tañon Strait [see Table 10 in Hayasaka *et al.* (1982) and Table

1 in Tanabe *et al.* (1983)]. Trapping was undertaken at 27 sites in the area south of Tinaogan Reef, about 3 km ENE of Bindoy (see Fig. 2). The submarine topography of this area changes from a wide intertidal flat, through a scarp that drops from 50 to 400 m in depth in a distance of approximately 2–3 km to a gently sloping floor 400–500 m deep. The bottom substrate at the trap sites is primarily dark gray silt.

Two types of traps were used: (1) a large trap (known locally as a "bobo") made of bamboo and fishnet (2 × 1.5 × 1 m) and (2) a small, double-entry "crab pot" (0.7 × 0.5 × 0.3 m in size). The former design is that used by native fishermen and is typical of traditional Indo-Pacific fish traps; it has probably been in use for centuries (see Dean, 1901). The latter was devised by Japanese fishermen. The traps were set overnight at depths of 120–312 m off Bindoy.

Bait included various meats such as chicken, shrimp, fish, and tortoise. A total of 52 *Nautilus* were caught in 27 traps, and only 2 traps were empty.

The associated fauna obtained in the traps included crabs, shrimps, fish, snagged octocoral branches, gastropods, and sea urchins. The number of specimens of each group and its depth range were as follows:

Sea urchins	88 (120–150 m)
Crabs	22 (130–132 m)
Shrimps	8 (170 m)
Fishes	8 (135–312 m)
Octocorals	4 (120–150 m)
Gastropods	1 (312 m)

Sea urchins (*Malepia cordata*) were most abundant, and crabs were common. The records of the highest catches of *Nautilus* (4–6 individuals per trap) were concentrated at depths of 130–150 m, where all the sea urchins were obtained. Four *Nautilus* were trapped at a depth of 312 m, and three adult specimens were trapped at a depth of 525 m, in the central part of the strait. The results of our trapping are similar to those reported by Haven (1972, 1977a). Observations by underwater television showed sea urchins at depths of 80 and 150 m, with a density of about 160 individuals m^2 (Hayasaka *et al.*, 1982), suggesting a close association between sea urchins and *Nautilus*; a comparable relationship has been recorded in Fijian waters (Hayasaka, 1985). Epifauna on the surface of *Nautilus* shells included barnacles, bryozoans, foraminifera, and serpulid worm-tubes (Hayasaka, 1983).

Following capture, each *Nautilus* was labeled, weighed, and sexed, and maximum shell diameter and aperture width were measured using a slide caliper (±0.05 mm precision). Volumes of cameral liquid in the last chamber were also estimated, in selected specimens, by extracting the liquid with a hypodermic syringe (±0.05 ml precision) through a small hole drilled into the chamber.

3.7. Sexual Dimorphism

A number of authors have previously reported on various aspects of sexual dimorphism in *Nautilus* (e.g., Willey, 1902; Haven, 1977a; Saunders and Spinosa, 1978; JECOLN, 1980b). Mature males possess a spadix on either the left or the right side of the buccal mass, whereas in females, the buccal mass is centrally

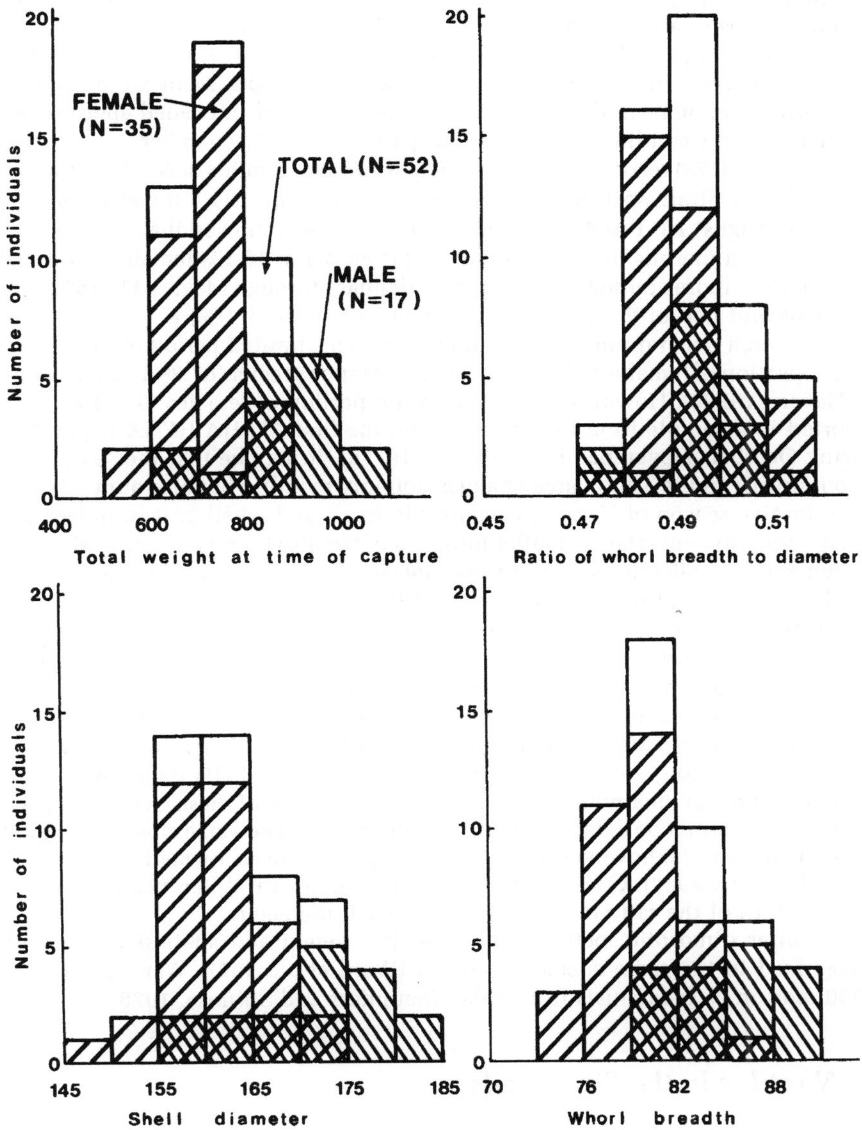

Figure 6. Frequency distributions of total live weight, shell diameter (D), whorl breadth (B), and the ratio B/D for *N. pompilius* from the Tañon area. The labeling in the top left graph applies to the other three as well. After Hayasaka *et al.* (1982).

located, along with accessory organs [labial lobe and Owen's organ (see Saunders and Spinosa, 1978, Fig. 3)]. Females also possess a yellow-brown, kidney-shaped nidamental gland in the mantle cavity. In combination, these differences permit accurate determination of sex without removing the animal from its shell. Of the 52 specimens captured, 17 (32.7%) were males and 35 (67.3%) were females. Except for several immature males and females, most specimens were regarded as mature, judging from the presence of a blackened and thickened aperture and the lack of color bandings on the ventral part of the body chamber.

Haven (1977a) recorded seasonal fluctuations in sex ratios of *Nautilus* trapped in the Tañon Strait area, based on data from 3000 animals trapped during the period of August 1971 to August 1972. Her results also showed that females were always less numerous than males (varying from 2.9 to 13.1% females), although females were more abundant in the spring (January through May; ≈ 11–15%) than in the second half of the year (June through December; 4–9%).

Although our trapping was conducted during a limited period in September, the proportion of females in the total catch was much greater (67.3%) than reported by Haven (1977a). However, a fairly large proportion of females (33%) was also reported in a sample from the same area on the first *ALPHA HELIX* Expedition, during October–November 1975 (Haven, 1977a). It may be that the sex ratio of *N. pompilius* in the Tañon area changes not only monthly but also annually. The reproductive season of *N. pompilius* in this area—and whether it is, in fact, seasonal—remains uncertain. Smaller numbers of females than males have also been recognized in collections of the same species from Blanche Bay, New Britain (Willey, 1902), and from Fiji (Ward *et al.*, 1977; Ward and Martin, 1980; Hayasaka, 1985); similar results were found in *N. belauensis* from Palau by Saunders and Spinosa (1978) and Saunders (1981a) and in *N. macromphalus* from New Caledonia by Ward and Martin (1980), but not by Willey (1902).

Mature specimens show dimorphic differences in total live weight and shell dimensions (Fig. 6). Mature males are generally heavier (881 g mean total weight) than females (720 g), and males ordinarily have larger shells than females (171 vs. 161 mm mean maximum diameter). Since the ranges of these characters partly overlap, sex cannot be determined by the difference in shell size alone. The results of Student's *t*-tests, however, indicate that the differences are statistically significant (95% accuracy). Conversely, there is no significant difference between the mean values of the ratio of whorl width to shell diameter.

Similar dimorphic patterns for total live weight and shell size have been recognized in the same species from Fiji (Ward *et al.*, 1977; Ward and Martin, 1980) and in *N. belauensis* from Palau (Saunders and Spinosa, 1978).

4. *Nautilus* in the Fiji Islands

In Fiji, unlike the Philippines, commercial fishing for *Nautilus* has scarcely been practiced, and the *Nautilus* population is therefore relatively undisturbed. Through numerous investigations of deep-sea fishes by the Institute of Marine Resources (IMR), the University of the South Pacific, the area off the southeast coast of Viti Levu, the largest island in Fiji, has been known for some time to be densely populated by *Nautilus*.

5. Area off the Southeast Coast of Viti Levu: A Case Study

5.1. Setting

Viti Levu is surrounded by rather shallow seas with many small islands, except for its southeastern and southwestern corners, where the water deepens rapidly to more than 1000 m. The field studies for our project were carried out in two areas: off the south coast of Suva and off Pacific Harbor (Fig. 7).

5.2. Submarine Topography

The area off Suva has a barrier reef, interrupted by a few narrow passages, outside which is a steep slope (5–9°) that extends from the shore to the bottom, 1000 m deep. Slope profiles perpendicular to the coast show small-scale step- or valleylike topography at different depths along the slope (Fig. 8). It is interesting that the highest *Nautilus* yields (at stations SV-11–14) were concentrated around the steplike promontories (i' in Fig. 8).

By contrast, the area studied off Pacific Harbor is a narrow ($\approx$ 2 km) passage (Mbengga Passage) between the barrier reefs along the coast of Viti Levu and surrounding Mbengga Island (Fig. 7C). The reef slopes of the passage are steep and the bottom is almost flat, at about 260 m. Trapping close to the barrier reef around Mbengga Island was quite productive.

5.3. Bottom Sediments

Mechanical analysis of 23 bottom samples collected at 19 stations (see Table 1 in Hayasaka, 1985) yielded the following results:

1. Median diameters of bottom sediments from all stations deeper than 180 m fall within a narrow range from 5.18 to 6.41 (Mean: 5.19; standard deviation: 0.40). This suggests that both areas deeper than 180 m off Suva and off Pacific Harbor are the sites of silt deposition, irrespective of water depth and topography.
2. Bottom samples are divided into two groups: (a) sandy silt from two stations off Suva (SV-9F and SV-11L) and three off Pacific Harbor (PH-1L, PH-2L, and PH-4F) and (b) silt from all the other stations. The samples of the first group are poorly sorted (So = 1.7 $\approx$ 2.6) and show bimodal grain size distribution, with peaks around 6 and 3, and negative skewnesses. The median diameters of these samples, and the presence of shallow water foraminifera, suggest supply of coarse materials from the shallower bottom area, e.g., by submarine sliding.
3. Dissolved carbonate values in the 19 bottom samples show considerable variation, from 1.69 to 35.49% (mean and standard deviation: 14.06 $\pm$ 8.92%) irrespective of water depth. This variation may reflect different rates of deposition of calcareous organic skeletons, perhaps controlled by the relief of bottom topography.

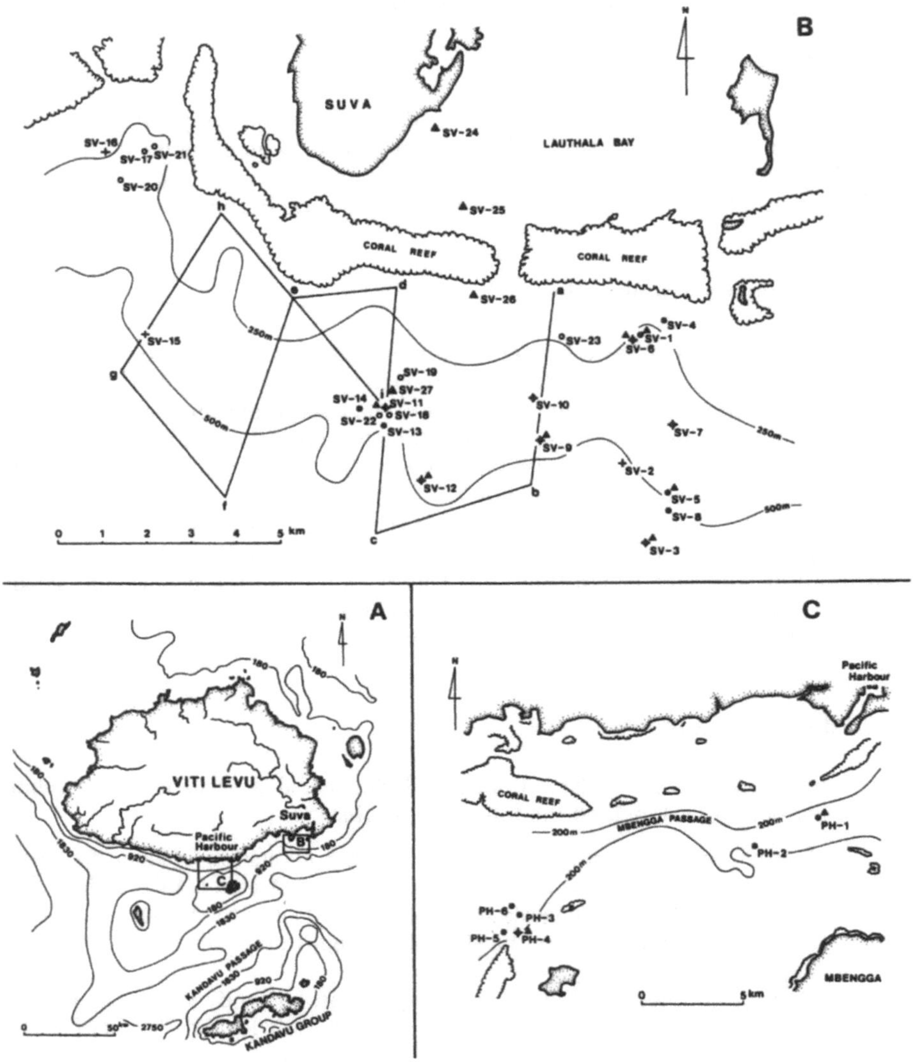

Figure 7. Maps of the studied areas. (A) Index map; (B) off Suva; (C) off Pacific Harbor. Activities: (●) trapping; (◆) oceanographic survey; (▲) plankton sampling; (○) underwater TV and still camera work. Reprinted with permission from Hayasaka (1985).

5.4. Seawater Characteristics

Water temperature, salinity, and dissolved oxygen (DO) were measured from samples collected at depths of 10, 20, 30, 50, 75, 100, 200, 300, 400, 500, and 600 m, at ten stations off Suva and one off Pacific Harbor (see Fig. 7B and C).

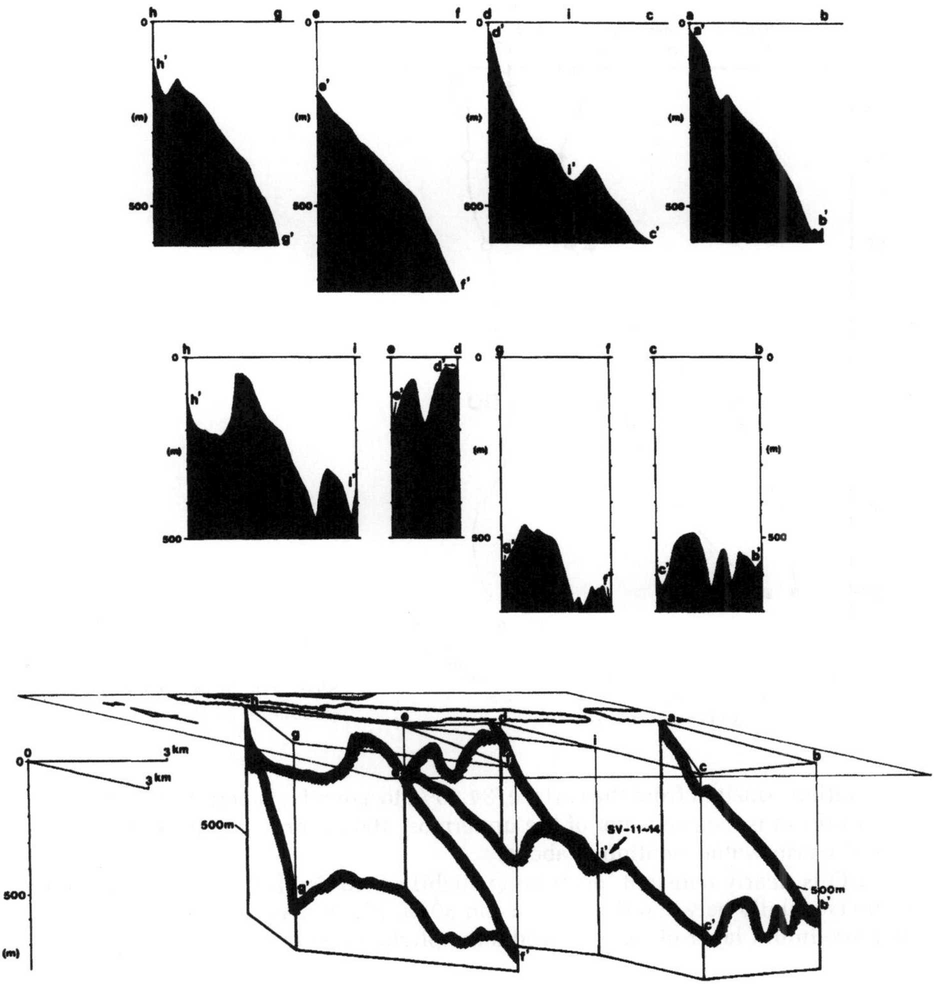

Figure 8. Bathymetry of the studied area off Suva. Bathymetric cross sections along the lines perpendicular to (A) and parallel to (B) the coastline. The lines of cross sections are shown in Fig. 7-B. (C) Panel diagram showing submarine topographic features. Reprinted with permission from Hayasaka (1985).

Changes of water temperature with depth are gradual, without any conspicuous change (thermocline). Generally, the surface-mixed water is constant (about 25°C) to a depth of about 100 m and then falls rapidly to about 12°C, from a depth of 100 to 400 m; the temperature at 600 m was 7.4°C (Fig. 9).

In a profile of water temperature extending from coast to offshore (Fig. 10), water temperature between depths of 240 and 500 m, from which *Nautilus* specimens were collected, ranges from 20 to 8°C.

Salinity variation in the area, represented by the record of station SV-3 (Fig.

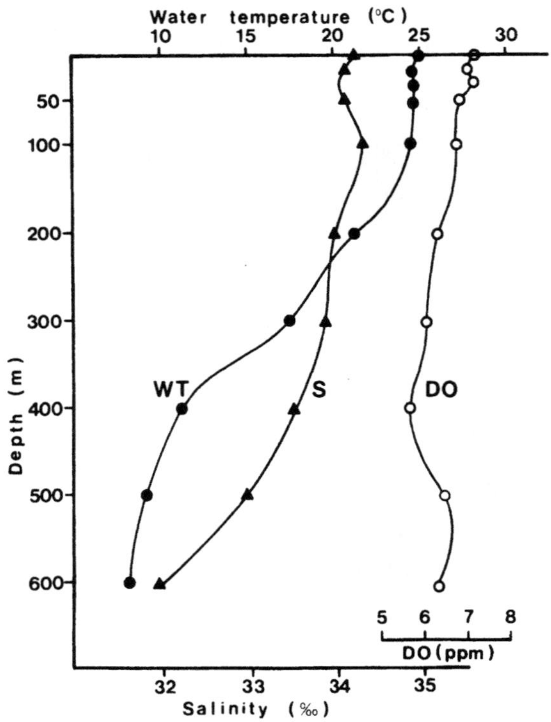

Figure 9. Vertical distribution of water temperature (WT), salinity (S), and dissolved oxygen (DO) at station SV-3 off Suva. After Hayasaka (1985).

9), is rather constant from the surface (34.30%) to a depth of about 100 m (34.38%). The uniform isohaline water of the uppermost 100 m corresponds to the surface-mixed warm water mentioned above.

DO is nearly constant, decreasing slightly with depth (Fig. 9). Its maximum value is usually between the surface and 30 m, which may indicate the existence of a maximum layer of oxygen-producing phytoplankton.

5.5. Biota

5.5.1. Plankton

Plankton sampling was carried out at 15 stations off Suva (inside and outside the barrier reef) and Pacific Harbor (see Fig. 7B and C), using a closing-type net (the same as used in the Philippines) during daytime, at a depth between 0 and 30 m, by vertical towing. The quantity of plankton was determined by settling volume (ml/m^3) in a test tube after 24 hr.

The plankton volumes collected from 11 stations off Suva (Fig. 7B) varied from 1.2 to 2.1 ml/m^3 and averaged 1.6 ml/m^3. Phytoplankton were dominated by *Trichodesmium hildebrantii* and *Pelagothrix clevei* (Cyanophyceae), *Rhizosolenia alata* and *Thalassiothrix longissima* (Bacillariophyceae), and *Ceratium macroceros, C. inflatum, C. atium pullchellum, C. trichoceros,* and *Ornithocercus*

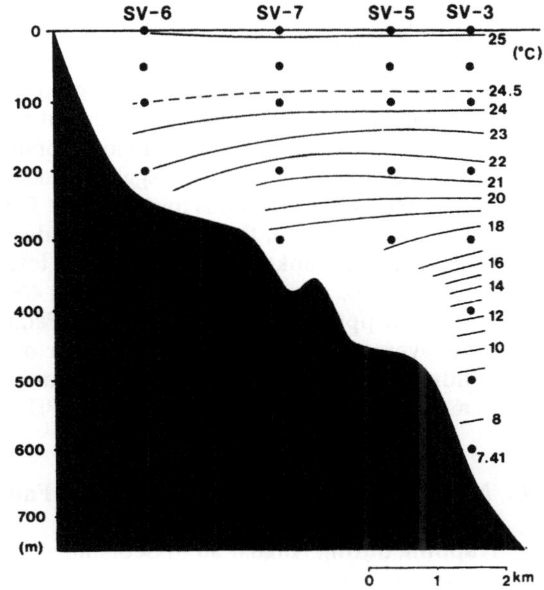

Figure 10. Distribution of water temperature in the profile perpendicular to the coastline off Suva. Reprinted with permission from Hayasaka (1985).

serratus (Dinophyceae). Zooplankton included *Sagitta* (Chaetognatha) and *Paracalanus*, *Oithona*, *Acartia*, *Oncaea*, *Corycaeus*, and *Oikopleura* (Appendicularia). In general, the plankton community was poor in volume but rich in variety.

In Lauthala Bay (Fig. 7B), inside the barrier reef off Suva, phytoplankton included the neritic plankton *Rhizosolenia alata* and *Navicula* sp. (Bacillariophyceae), *Ceratium macroceros* and *C. pullchellum* (Dinophyceae), and *Acartia* spp. and *Oncaea* spp. (Copepoda).

Settling volumes of plankton at the two stations off Pacific Harbor (Fig. 7C) were 0.8 ml/m^3 (PH-1) and 0.9 ml/m^3 (PH-4) (the lowest values recorded). Phytoplankton included *Trichodesmium hildebrantii* and *Pelagothrix clevei* (Cyanophyceae), *Thalassiothrix longissima* (Bacillariophyceae), and *Ceratium macroceros*, *C. pullchellum*, and *Ornithocercus serratus* (Dinophyceae). Zookplankton assemblages were dominated by *Acanthometron* (Radiolaria), *Siphonophora* (Coelenterata) and *Sagitta* (Chaetognatha), *Oithona* spp. and *Oncaea* sp. (Copepoda), and *Oikopleura* sp. (Appendicularia).

5.5.2. Foraminifera

Bottom samples were collected from ten stations off Suva (Fig. 7B) and three stations off Pacific Harbor (Fig. 7C). The total number (per 10 cc of sample) of benthonic and planktonic foraminifera and the respective ratios between them [Table 1 in Ōki (1985)] led to the following conclusions:

1. The total number of benthonic foraminifera per sample ranges from 5678 to 54714 in 10 cc of sediment, but does not correlate with water depth.

The same may be said of the total number of planktonic foraminifera and radiolaria.

2. The assemblages of benthonic foraminifera belong to two groups: the calcareous porcellaneous and the agglutinated foraminifera. One group is characterized by less than 7.4% agglutinated foraminifera and from 11.7 to 16.7% calcareous porcellaneous foraminifera; another group is characterized by more than 14% agglutinated foraminifera and less than 12.6% calcareous porcellaneous foraminifera. The former is present in water shallower than 330 m, the latter in water deeper than 365 m.
3. The ratio of planktonic foraminifera to total foraminifera off Suva increases with depth; at every depth, it is about 40% less than that in Tañon Strait, the Philippines. At three stations off Pacific Harbor, it is less than 17.4%.
4. At every station, the ratio of number of individuals of planktonic foraminifera to the total number of planktonic foraminifera and radiolaria is more than 87.2% except for stations PH-3 and PH-5.

5.6. *Nautilus* Trapping and Associated Fauna

Trapping during August 30 to September 27, 1983 yielded 163 live *N. pompilius* from off Suva and Pacific Harbor, Viti Levu Island, Fiji (see Tables 1 and 2 in Hayasaka, 1985). Trap locations included 13 sites on an abruptly dropping slope at depths of 180–640 m off Suva and 6 sites at depths of 255–465 m in a narrow submarine channel off Pacific Harbor, both outside the barrier reef (see Fig. 7B and C). The bottom substrates at the trap sites are primarily light gray silts with a few sandy silts.

Traps of three different sizes (ranging from approximately 1 to 2 m^3) were used; they were made of iron frame covered with 15 mm wire netting, originally devised for deep-sea fishing by the IMR and the University of the South Pacific. The traps were set overnight, baited with whole bodies of sardine or small tuna.

In the study areas, *Nautilus* occurred rather abundantly between depths of 360 and 470 m, but were less numerous at above 300 m and below 470 m. Furthermore, no *Nautilus* were obtained at 180 or 640 m. Our data match those of Ward *et al.* (1977) during July–August 1976; no *Nautilus* were obtained deeper than 550 m or shallower than 75 m.

Following capture, each *Nautilus* was labeled and processed as in the Philippines (see Section 3.6). Cameral liquid was extracted from the last chamber of 22 specimens, using a hypodermic syringe (± 0.05 ml accuracy), to analyze the oxygen isotope ratio and the relationship between cameral liquid volume and last-septum thickness (Tanabe *et al.*, 1985).

The associated fauna in the traps included shrimps (13 species), crabs (6 species), and fishes (14 species) (see Table 1 in Shinomiya *et al.*, 1985). The number of individuals of each fish species was small except for species of *Pristipomoides*. There seemed to be no correlation in catch between *Nautilus* and the species of fish. Scientists at the IMR have the impression that there is a close correlation between *Nautilus* and shrimps in the area, and this idea is supported by our trapping data. In photographs taken with an underwater still camera by a member of our group (Hattori *et al.*, 1985), a young *Nautilus* can be seen along

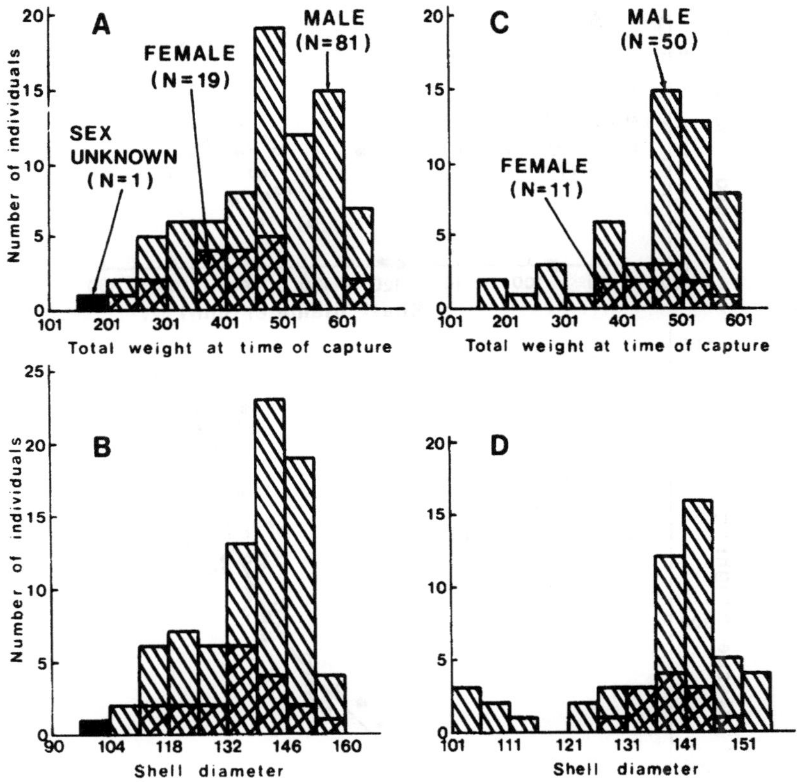

Figure 11. Weight and size distributions of the specimens of *N. pompilius* captured from the Suva area during August 29 to September 9 (A, B) and from the Pacific Harbor area during September 19–22 (C, D), 1983. Reprinted with permission from Hayasaka (1985).

with several shrimps gathered around the bait. Analysis of esophagus and stomach contents (Saisho and Tanabe, 1985), and aquarium maintenance (Kakinuma and Tsukahara, 1985), confirms a crustacean diet. The coexistence of *Nautilus* with the shrimp *Heterocarpus sibogae* is common in the study area (Shinomiya *et al.*, 1985).

5.7. Sexual Dimorphism

Of the 100 *Nautilus* captured off Suva, males comprised 81 (80.2%) of the total. Similarly, of the 61 specimens obtained off Pacific Harbor, males ($N = 50$) represented 82% of the total. Only one unsexed young animal was collected from the Suva area. No correlation between sex and either location or depth was observed (see Tables 1–4 in Hayasaka, 1985). Size and weight distributions of the sample from the Suva area (Fig. 11A and B) show that males are larger and heavier than females. Alternatively, in the sample from off Pacific Harbor, mean values

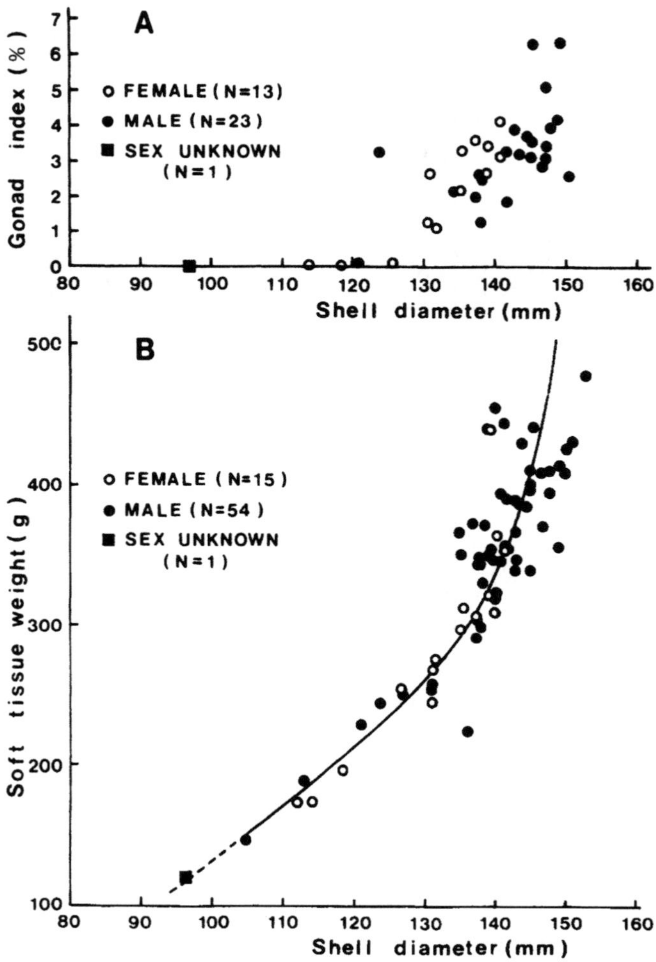

Figure 12. Scatter plots of gonad index (A) and soft tissue weight (B) vs. shell diameter for selected specimens captured from the Suva and the Pacific Harbor areas. Reprinted with permission from Hayasaka (1985).

of total animal weight and shell size in males are slightly larger than those in females (Fig. 11C and D).

On the basis of the measurement data, allometric relationships of shell and soft tissue were examined in relation to sexual maturity. Scatter plots of gonad index (ovary or testis weight/soft tissue weight) vs. shell diameter in selected specimens from the Suva and Pacific Harbor areas (Fig. 12A) shows that the female gonad index increases exponentially when shell diameter exceeds 130 mm in diameter. In contrast, full development of the testis in males is prolonged, to between 140 and 150 mm shell diameter. Therefore, between 130 and 140 mm shell diameter, the gonad index in females is slightly larger than that in similar-size males. Most specimens larger than 140 mm in diameter are males, and the gonad index in full-grown males appears to be much larger than that in mature females. If a gonad index of 3.0 is regarded as a provisional standard of sexual maturity, mature males are mostly larger than mature females. In specimens larger than 140 mm in shell diameter, males usually possess heavier soft tissue than females (Fig. 12B).

6. Summary

In Tañon Strait, *Nautilus* seems to prefer the waters below the thermocline (a depth of about 150 m, ≈ 20°C, down to about 500 m, ≈ 17°C). However, their vertical distribution is variable, probably depending on submarine topography and faunal associations.

Owing to the basinlike submarine topography of Tañon Strait, the population of *Nautilus* within the strait should to some extent be isolated from those outside the strait. Our observations show that water temperature profiles inside and outside the strait show a striking contrast (see Fig. 4b). Outside Tañon Strait, namely, in Bohol Strait, the water temperature below the thermocline declines more rapidly than inside. Data on *Nautilus* from outside the strait are quite limited, but five mature shells of *N. pompilius* obtained from a fisherman and reported to be from off Siquijor Island, in Bohol Strait, are much smaller than those from the Tañon area (≈ 130 vs. ≈ 165 mm in average diameter). This difference suggests a fairly large range of intraspecific variation, although the distance between the two habitats is less than 50 km. It is also notable that color bandings on the Siquijor shells are rose-colored, whereas those of the Tañon shells are scarlet. While the uncertain reliability of source data and the small sample size of the Siquijor specimens are a handicap, the fact that the water temperatures below 150 m at station N-1 (southern margin of Bohol Strait near Siquijor Island) are much lower than those in Tañon Strait (Fig. 4b) suggests a possible ecological difference that might relate to these morphological differences.

Off the southeast coast of Viti Levu Island, Fiji, distribution of *N. pompilius* on the steep slope between depths of 240 and 500 m seems to be rather sporadic, probably depending on the irregular topography. The vertical distribution ranges from a depth of 240 m (water temperature ≈ 20°C) to 500 m (≈ 8°C), and the highest density is between 360 m (≈ 14°C) and 470 m (≈ 10°C). Trapping data suggest a relationship between *Nautilus* and shrimps, in particular the species *Heterocarpus sibogae*.

ACKNOWLEDGMENTS. Our deepest thanks are due Dr. Alfredo E. Evangelista, Assistant Director of the Philippine National Museum, Manila, Professor Angel C. Alcala, Vice-President of Silliman University, Dumaguete City, Philippines, and Professor Uday Raj, Director of Marine Resources, University of the South Pacific, Suva City, Fiji, for their invaluable help in planning and conducting the project. We are also indebted to Mr. and Mrs. Wilson Vailoces, Bindoy, Philippines, for their assistance in trapping and for accommodations of the crew of research vessels of the Institute of Marine Resources, Suva. Other members of the project in the Philippines and Japan are due particular thanks for their help in field operation and discussions. This study was supported by Grants-in-Aid for Oversea Scientific Research from the Ministry of Education, Science, and Culture of the Japanese Government for five years (1980 and 1981, No. 504207; 1982, No. 57043059; 1983, No. 58041055; 1984, No. 59043050).

Chapter 12

Predation on *Nautilus*

W. BRUCE SAUNDERS, CLAUDE SPINOSA, and
LARRY E. DAVIS

1. Introduction

Little has been written regarding predation on *Nautilus*, because information on the subject has largely been limited to inferences based on indirect evidence, such as healed injuries, shell breaks, and borings. For example, Willey (1902) remarked on the presence of healed severe injuries to the hoods of several males and speculated that they might be due either to attacks by fishes or to fighting between sexes. Haven (1972) regarded shell breakage and hood injuries in *N. pompilius* as a fairly common product of fighting within the species and demonstrated that the distinctive, V-shaped breaks so common in *Nautilus* shells are a product of biting (Haven, 1972, Figs. 4 and 5). Arnold (1985) described and categorized a wide variety of shell abnormalities in *N. pompilius* from the same region. In addition, evidence indicates that *Nautilus* is regularly preyed upon by *Octopus*, and attacks on *Nautilus* by fishes have now been documented. This chapter briefly summarizes the results of previous studies and presents new information regarding predation on *Nautilus*.

2. *Octopus* Predation

Shell borings, consisting of small (1–4 mm diameter), subcircular to elliptical perforations or pits are frequently observed in drifted *Nautilus* shells and, less commonly, in the shells of live-caught animals. Tucker and Mapes (1978) were

W. BRUCE SAUNDERS • Department of Geology, Bryn Mawr College, Bryn Mawr, Pennsylvania 19010. CLAUDE SPINOSA • Department of Geology and Geophysics, Boise State University, Boise, Idaho 83725. LARRY E. DAVIS • Department of Geology, Washington State University, Pullman, Washington 99164.

the first to report these borings. They noted that 43 of 150 shells (28.7%) of *N. pompilius* (obtained from shell dealers in the Philippines) had been bored, and attributed this to either naticid gastropods or coleoid cephalopods. Evidence for the octopod origin of the borings is their close resemblance to *Octopus* borings in other mollusks, which have now been well documented (Arnold and Okerlund-Arnold, 1969; Nixon, 1979 a,b). In addition, the report by Arnold (1985) of a trap containing a specimen of *Octopus cyanea* along with six *N. pompilius*, one of which had been bored and partially eaten, seems to provide fairly unequivocal evidence of *Octopus* predation on *Nautilus*.

2.1. Frequency of *Octopus* Predation

The frequency of *Octopus* attacks on *Nautilus*, and their effectiveness, is not well known because of the small numbers of shells on which previous studies were based and because of uncertainty as to their origin (i.e., it was not known whether they had been live-caught or were drift shells, the causes of death of which were unknown). Thus, the 28.7% boring frequency for Philippine *Nautilus* reported by Tucker and Mapes (1978) probably represents an unknown combination of both lethal and sublethal attacks. Some sublethal frequency data are available for Philippine *N. pompilius*, observed during the 1979 ALPHA HELIX Expedition. Only 4 (1.1%) of the 353 live-caught specimens showed evidence of boring (W. B. Saunders, unpublished observations; Arnold, 1985). Additionally,

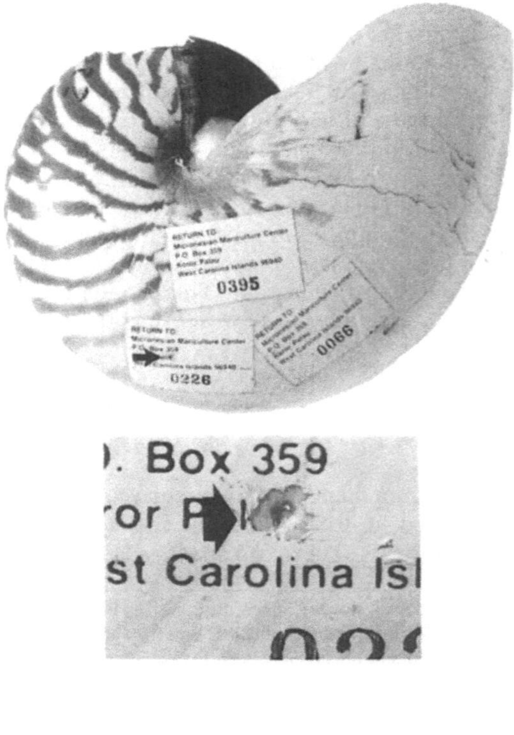

Figure 1. Mature *N. belauensis* trapped three times in 1977, off Mutremdiu Point, Palau (shell diameter 216 mm). The arrows point to the *Octopus* boring on tag 0226.

Table I. Occurrences of Borings in *Nautilus belauensis* from Palau and *Nautilus pompilius* from the Philippines

	Total		Number and percent (%) of total bored				
Year	Catch	Bored	Males	Females	Immature	Mature	Multiple
			1977–1982 Palau				
1977	407	42 (10.3%)	24 (57.1%)	18 (42.9%)	4 (9.5%)	38 (90.5%)	11 (26.2%)
1978	460	58 (12.6%)	44 (75.9%)	14 (24.1%)	7 (12.1%)	51 (87.9%)	6 (10.3%)
1979	264	28 (10.6%)	$\frac{16}{26^a}$ (61.5%)	$\frac{10}{26^a}$ (38.5%)	2 (7.1%)	26 (92.9%)	6 (21.4%)
1981	489	28 (5.7%)	15 (53.6%)	13 (46.4%)	$\frac{7}{27^b}$ (25.9%)	$\frac{20}{27^b}$ (74.1%)	2 (7.1%)
1982	1100	55 (5.0%)	$\frac{18}{50^a}$ (36.0%)	$\frac{32}{50^a}$ (64.0%)	$\frac{6}{51^b}$ (11.8%)	$\frac{45}{51^b}$ (88.2%)	12 (21.8%)
Total	2720	211 (7.8%)	$\frac{117}{204^a}$ (57.4%)	$\frac{87}{204^a}$ (42.6%)	$\frac{26}{206^b}$ (12.6%)	$\frac{180}{206^b}$ (87.4%)	37 (17.5%)
			1979 Philippines ALPHA HELIX Expedition				
1979	353	4 (1.1%)	—	2 (50%)	2 (50%)	2 (50%)	—

[a] Not all bored specimens were sexed. The figures 26, 50, and 204 represent the number of specimens caught (1977, 1982, and totals, respectively) for which sex could be determined; these figures were utilized to determine percentages [e.g., 16/26 (61.5%)]. Figures for total bored specimens (42, 58, 28) were utilized where total catch could be sexed.

[b] It could not be determined for all specimens whether they were mature or immature. Only the number of specimens determined to be mature or immature are utilized to obtain percentages.

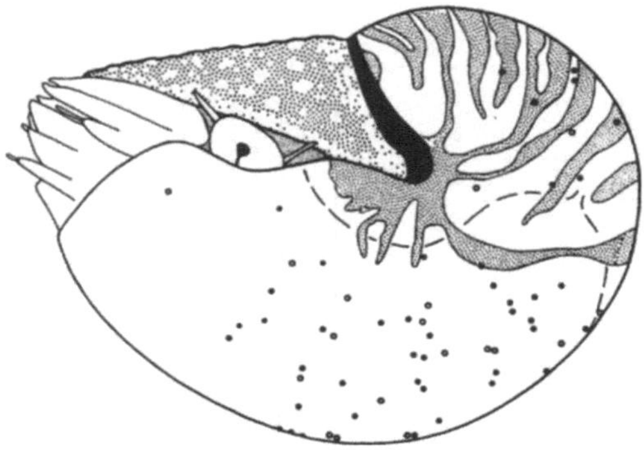

Figure 2. Diagrammatic composite of sublethal boring locations in live-caught *N. belauensis*. Orientation: (○) left side of shell; (●) right side. Note the concentration of borings in the posterior portion of the body chamber, corresponding approximately to the location of retractor muscles and viscera. The data are based on the locations of 61 borings in 39 specimens (measured prior to release). (– – –) Approximate body outline.

of 270 live-caught specimens of *N. pompilius* from various localities in Papua New Guinea (Saunders and Davis, 1985), only 5 (1.9%) possessed *Octopus* borings.

Considerably more data are available regarding live-captured *N. belauensis* from Palau. In the course of mark–recapture studies during 1977–1982, 2720 live-caught specimens were examined for borings prior to release (Table I); 211 specimens (7.8%) exhibited sublethal borings, and 37 of these specimens (17.5%) had multiple borings. One specimen, trapped and tagged three times during the 1977 field season, had sustained boring between recaptures (Fig. 1). It was first trapped on June 19 and tagged as number 0066 (no borings recorded); it was recaptured on July 3 (0226; no borings recorded) and recaptured again on July 13 (0395), by which time an attack by *Octopus* had occurred, as evidenced by a boring through tag 0226. *Octopus* borings appear to be concentrated in the region of the retractor muscle (Arnold, 1985); sublethal borings also occur on the ventral and posterior portion of the shell (Fig. 2).

2.2. Characterization of *Octopus* Borings

Octopus borings are characteristically somewhat elliptical in outline (Figs. 3 and 4), approximately 1–4 mm in diameter, and conical in cross section. Generally, they comprise several elongate marks (presumably caused by radular rasping in different directions or at different times), imparting a scalloped appearance to the boring (Fig. 3A and B). Some borings are surrounded by smooth or finely etched surfaces (Figs. 3A and 4A); others are surrounded by coarser, scratchlike striations (Figs. 3B and 4B).

Superficially similar rounded marks are seen less commonly on the shells of *Nautilus* (Fig. 5). However, the latter lack the polished, etched surfaces of *Octopus* borings; they seem to have been caused by impact or breakage and may represent either teleost teeth marks or *Nautilus* bites. Most sublethal borings do not perforate the shell (Figs. 3A and 4); of those that do (e.g., Fig. 3B), most show an organic, black layer, in some cases covered by a nacreous layer. Presumably this deposition represents an effort to repair the perforation and may play a role in the animal's survival of *Octopus* attacks.

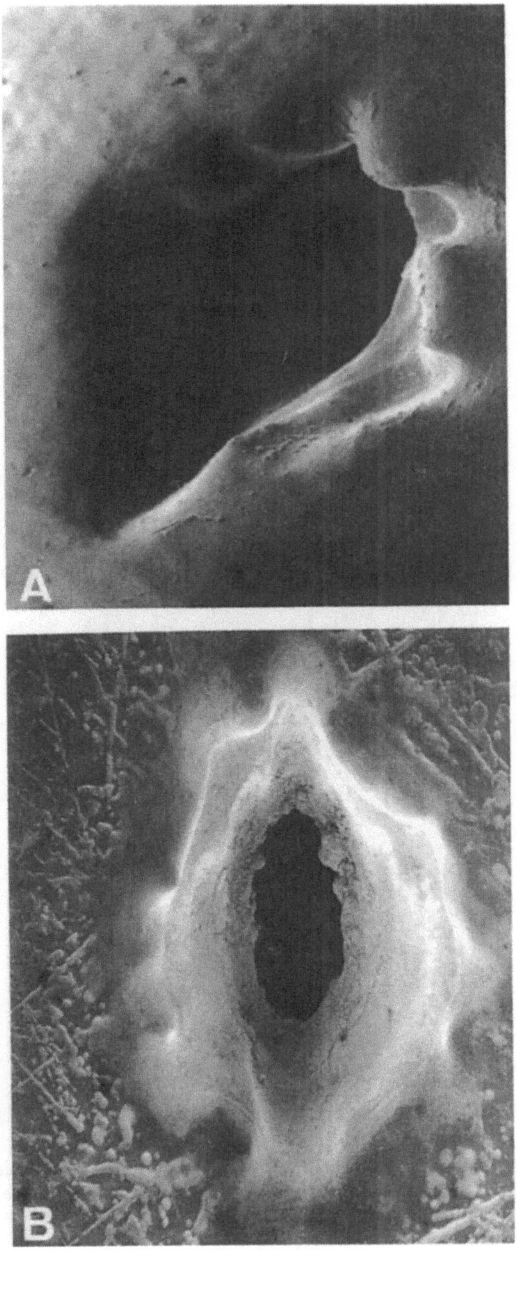

Figure 3. Scanning electron photomicrographs of *Octopus* borings in shells of *N. belauensis*. Note the scalloped edges produced by radular rasping (A) and the scratched shell surface around borings (B). The boring in (B) perforated the shell, and a black layer and a nacreous layer had been secreted inside the body chamber. Both borings are approximately 1.5 mm in diameter.

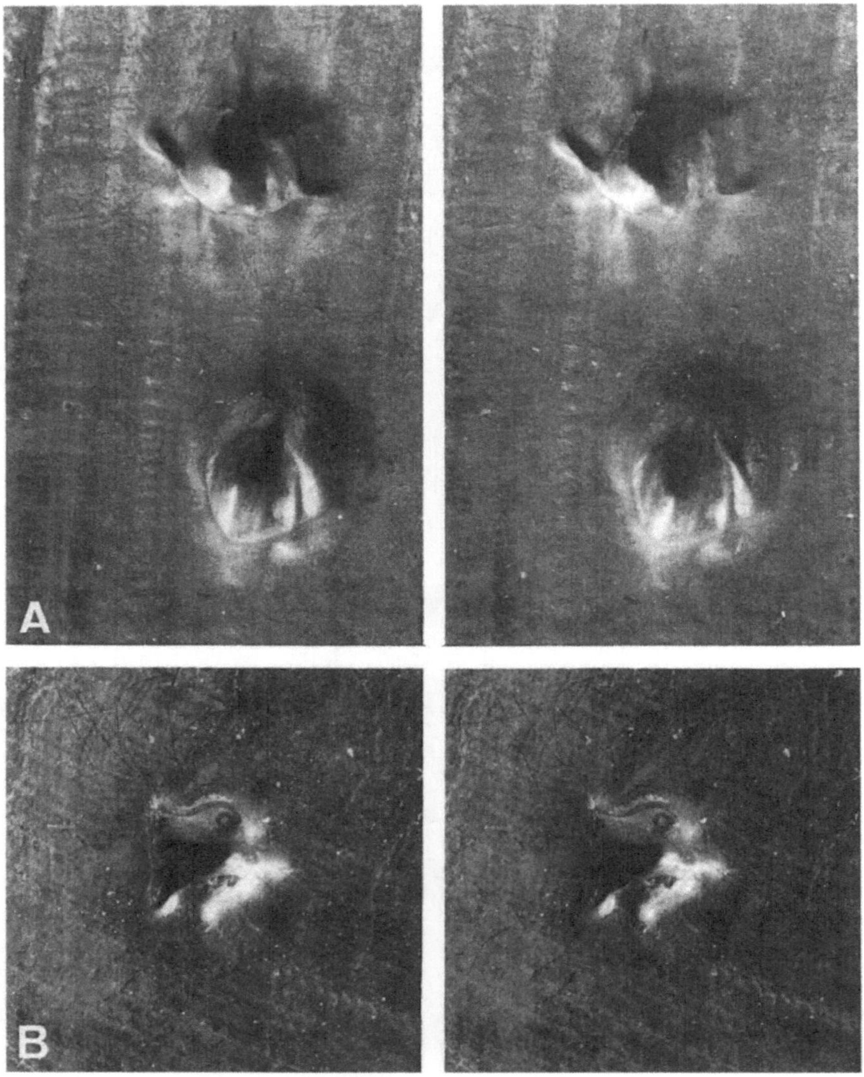

Figure 4. Stereo pairs of *Octopus* borings in *N. belauensis*. (A) Two adjacent borings (diameter ≈ 1.5 mm), possibly representing a single predatory attack (the upper boring appears to record several different boring attempts). Note the highly irregular outline of the boring in (B).

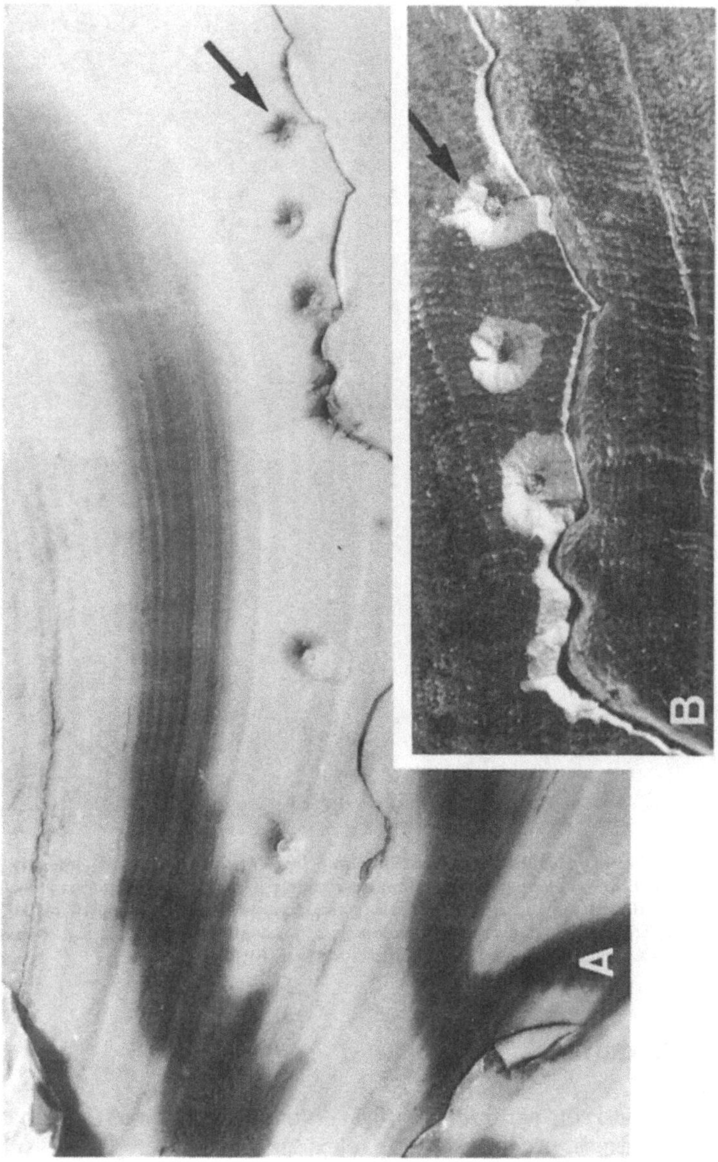

Figure 5. (A) Presumed bite marks on *N. belauensis*, probably inflicted by teleost or by other *Nautilus*, near the apertural margin. Note the sharp, angular surfaces, in contrast to *Octopus* borings (Figs. 3 and 4). (B) Enlarged view (diameter of pits ≈ 1 mm).

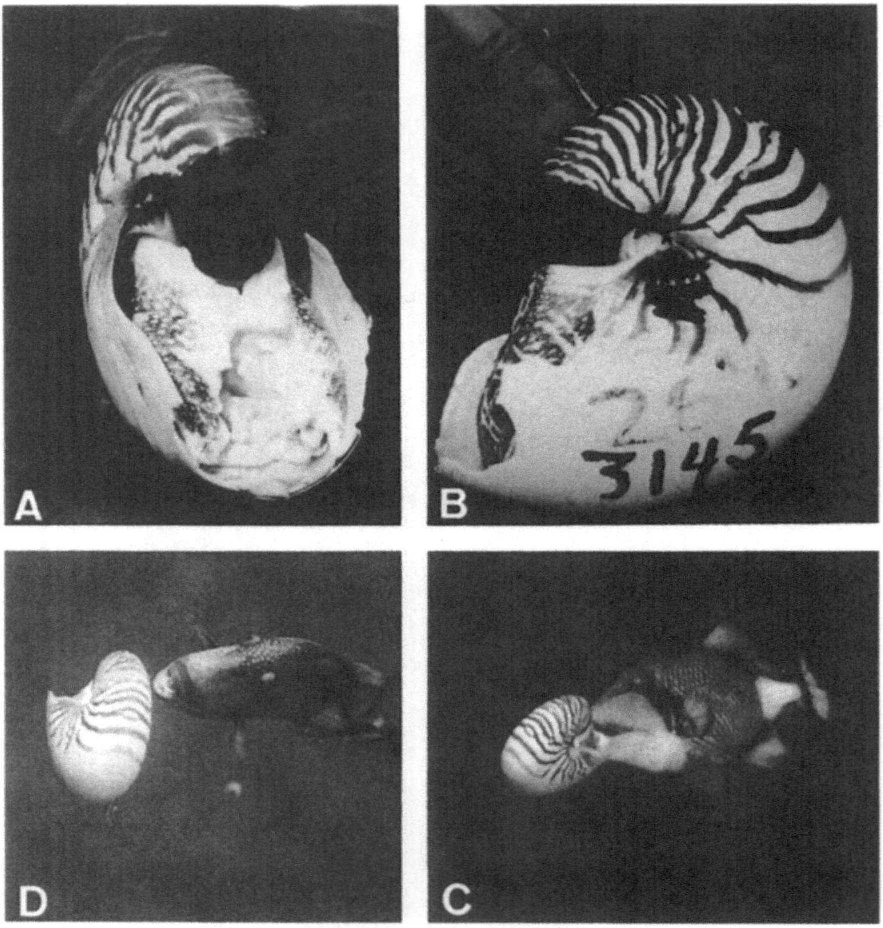

Figure 6. Teleost predation on *Nautilus*. (A) *Nautilus pompilius* showing shell, hood, and body damage inflicted by grouper (*Epinephelus*) at a depth of approximately 27 m off Manus, Papua New Guinea. (B) Shell breakage and teeth marks on *N. belauensis* recovered after attack by a triggerfish (*Balistoides viridescens*). (C, D) Triggerfish attacking on *N. belauensis* at a depth of approximately 25 m off Ngemelis, Palau [this is a separate incident from that recorded in (B)].

3. Teleost Predation

Shell breaks and soft-part injuries (Figs. 6A and B and 7B and D) are common in *Nautilus* and suggest predatory efforts in excess of what *Octopus* (or *Nautilus*) could inflict. The origin of such injuries had been uncertain; thus, it was instructive to witness, in 1982, a series of attacks on *Nautilus* in shallow water (Fig. 6C and D). A group of 16 animals, which had been retrieved from a depth of ap-

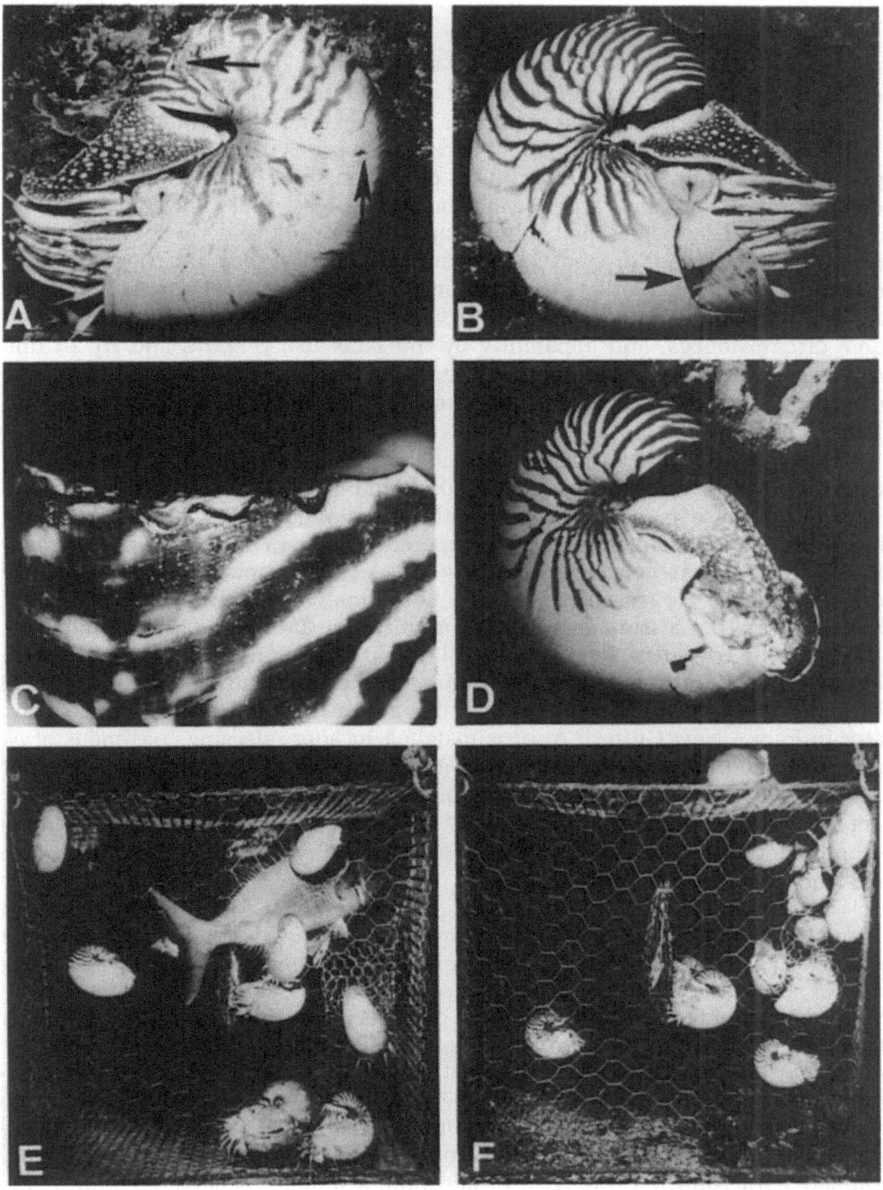

Figure 7. (A) Lateral view of *N. belauensis* in Palau, showing characteristic shell breakage–repair. (B) Major repaired break showing mantle damage at arrow. (C) *Nautilus* bites inflicted on juvenile *N. belauensis* while in an aquarium with mature specimens. (D) Severe damage to the shell and hood of *N. belauensis* inflicted simultaneously by bite. (E) Deep-water remote photograph (depth, 270 m, night), off Manus, Papua New Guinea, showing a large teleost (*Etelis carbunculus*) attracted to a baited trap with *N. pompilius* and *N. scrobiculatus*. (F) Different photo sequence at the same locale as (E) (depth, 325 m, night) showing *Octopus* sp. on top of the trap (trap width 1 m).

proximately 300 m, were being released at a depth of 20 m, when a large (≈ 0.5 m) triggerfish (*Balistoides viridescens*) began bumping aggressively at the specimens, ignoring the accompanying divers (M. Weekley and B. Saunders). The fish pinned one *Nautilus* against the reef and made a series of slashing attacks at the aperture and hood, spitting out chunks of broken shell. When the divers intervened, the fish retreated and began attacking at least four other animals. The injured *Nautilus* was rescued, photographed (Fig. 6B), and rereleased, but its condition offered poor prognosis for recovery. When last seen, the triggerfish was repeatedly attacking another specimen on a ledge at a depth of approximately 45 m. This scenario was repeated several weeks later, at the same site, under the same circumstances, and apparently by the same fish (Fig. 6C and D). In this instance, the fish began circling and attacking the *Nautilus* in open water, but inflicted little damage, because it was unable to obtain purchase on the smooth, slippery shell.

These events were revealing, in that they showed the relative helplessness of *Nautilus* against such attacks, particularly in open water, as well as indicating a lack of a defense or escape response in *Nautilus*. When attacked, *Nautilus* withdrew into its shell, made no effort to escape, and drifted in midwater until rescued. At the time, it was concluded that the attacks were probably anomalous and unique, because (1) triggerfish may be notoriously aggressive when defending their territory (B. Carlson, personal communication) and (2) the near-anticipatory actions of the triggerfish suggested that it might be responding to habit—indeed, the release of hundreds of animals at this site may have turned it into a feeding station. Thus, it was surprising to encounter a near-replay of this situation 3 years later in Papua New Guinea. Approximately 1 hr after releasing specimens of *N. pompilius* from the surface, divers (L. Davis and B. Saunders) observed a grouper (*Epinephelus* sp., ≈ 0.5 m long) vigorously attacking one specimen against the reef face at a depth of approximately 27 m. Considerable damage had been done to both shell and hood, and the specimen was dead when brought to the surface and photographed (Fig. 6A).

These records show that the action and the responses observed in Palau and Manus were neither unique to a particular species of fish nor just an isolated instance. They also provide a measure of the damage that can be inflicted by fish attacks and of the relative defenselessness of *Nautilus*; it is clearly vulnerable in head-on encounters with fishes. This discovery may be relevant to an earlier suggestion (Saunders, 1984b) that nocturnal migrations of *Nautilus* into shallow water may be coordinated with the greatly reduced nocturnal activity of shallow-water teleosts (Goldman and Talbot, 1976). This does not imply that there are no deep-water fishes capable of preying on *Nautilus*. To the contrary, large deep-water snappers (*Etelis carbunculus*) approaching 1 m in length are commonly trapped with *Nautilus* (Fig. 7E). In addition, a number of large (≈ 2 m) deep-water sharks (including *Echinorhinus cookei* and *Hexanchus griseus*) have been photographed by remote camera with *Nautilus* (Saunders, 1984b, Fig. 10a and b).

3.1. Frequency of Inferred Teleost Predation

Records of the incidence of shell breakage and injury are available for 270 live-caught *N. pompilius* from Papua New Guinea (Table II). Categorization of

Table II. Incidence of Injuries in Live-Caught *Nautilus pompilius* and *Nautilus scrobiculatus* from Papua New Guinea

Locality	Number of specimens	Injury				
		Boring	Major	Minor	Mantle	Hood
N. pompilius						
Ndrova	91	3 (3.3%)	31 (34.1%)	41 (45.1%)	17 (18.7%)	7 (7.7%)
Komuli	40	1 (2.5%)	5 (12.5%)	22 (55.0%)	8 (20.0%)	3 (7.5%)
Lorengau	5	—	1 (20.0%)	4 (80.0%)	—	—
Port Moresby	49	1 (2.0%)	3 (6.1%)	27 (55.1%)	5 (10.2%)	1 (2.0%)
Kavieng	49	—	9 (18.4%)	32 (65.3%)	14 (28.6%)	1 (2.0%)
Lae	36	—	—	27 (75.0%)	2 (5.6%)	—
TOTALS:	270	5 (1.9%)	49 (18.1%)	153 (56.7%)	46 (17.0%)	12 (4.4%)
N. scrobiculatus						
Ndrova	29	1 (3.4%)	14 (48.3%)	17 (58.6%)	25 (86.2%)	2 (6.9%)
Komuli	1	—	—	—	—	—
TOTALS:	30	1 (3.3%)	14 (46.7%)	17 (56.7%)	25 (83.3%)	2 (6.7%)

these injuries is somewhat subjective, inasmuch as it is difficult to define precisely what constitutes a major vs. a minor injury. However, shell breaks less than about 10 mm deep were recorded as minor; major breaks were those that spanned as much as 45° of the body chamber length (≈ 150°). Injuries that affected subsequent shell secretion, usually in the form of longitudinal grooves accompanied by black material, were recorded as mantle injuries. *Nautilus* heals, but appears unable to regenerate, injuries to the hood (Fig. 7D). The incidence of this type of injury was noted in *N. pompilius* and *N. scrobiculatus* from Papua New Guinea (Table II).

The overall incidence of major shell breaks was 18.1, and 56.7% of the shells showed minor breakage. Since these figures represent nonlethal injuries, they indicate that the animals sustained a surprisingly large number of apparently traumatic events. The irregularly crescentic shape of the major breaks is consistent with those observed to have been produced by fish attacks, although other causes may have been involved. The high incidence of minor breaks is not surprising, considering the thin, delicate nature of the immature peristome. Many were probably caused by other *Nautilus*, as has been noted earlier, particularly the distinctive V-shaped breaks (Fig. 7C). Injuries that affected the mantle were also fairly common (17%), but relatively few animals showed healed hood injuries (4.4%).

One interesting statistic emerges from the Papua New Guinea data; the Lae sample of *N. pompilius* (N = 36) shows a remarkably low incidence of major shell breaks (0%), compared to the overall incidence of such breaks in the other Papua New Guinea populations (20.9%). In this regard, it may be relevant that the Lae setting is quite different from most other sites, being located within the Markham River delta (see Chapter 3), while all other populations occur on forereef slopes. In terms of associated macrofauna, the trapping at Lae yielded very few teleosts, but produced the highest number of sharks of any *Nautilus* site studied to date. It would be of particular interest to see whether other deltaic settings (e.g., off Suva, Fiji) show similarly low frequencies of shell breakage.

ACKNOWLEDGMENTS. The assistance of the late M. W. Weekley in Palau in 1982, in the work that led to documenting the teleost attacks on *Nautilus*, was deeply appreciated. This research was supported by NSF grants EAR 81-00629 (Palau) and EAR 83-18923 (Papua New Guinea) and by a Boise State University research grant (Palau).

V

Physiology

Chapter 13

The Central Nervous System

J. Z. YOUNG

1. Introduction

The nervous system of *Nautilus* contains a much larger number of neurons than is found in any noncephalopod mollusk. Many of the neurons lie in peripheral ganglia, especially those of the tentacles and the brachial nerve cords in the arms. On the other hand, there is a great concentration of neurons to form a "brain," which is composed of cords around the esophagus. The nervous organization thus consists of peripheral reflexes for detailed actions subordinated to a brain devoted to control of the whole animal. Cephalopods have probably used this system since the Cambrian Period (≈ 500 Ma), and it was probably in large part responsible for their early success as dominant carnivores of the seas.

The nervous organization also differs from that of other mollusks, such as that of the gastropods with their few large cells, in that it consists not of compact ganglia, but of cords with cell bodies around the outside and a neuropil at the center. This arrangement probably allows for complex patterns of interaction between neurons and thereby for elaborate behavior patterns.

2. Nervous System of *Nautilus*

The nerve cords of *Nautilus* seem at first sight to be very different from the brain of coleoid cephalopods (Fig. 1). They are, in fact, arranged on the same plan, with a supraesophageal and two subesophageal parts and a magnocellular lobe at the side (Fig. 2). The different appearance of nautiloid and coleoid brains is due to the wide separation of the nerve cords in *Nautilus*. This separation is

J. Z. YOUNG • The Department of Experimental Psychology, Oxford OX1 3UD, England.

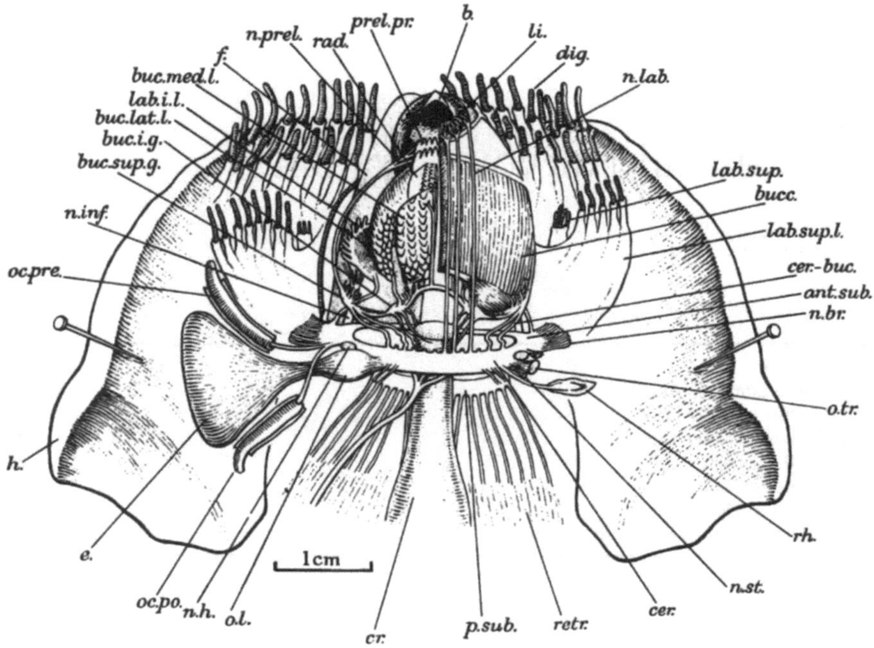

Figure 1. Drawing of a dissection of the nervous system and buccal mass. The dissection has been carried deeper on the left. The tentacles are somewhat diagrammatic. Anatomy: (*ant.sub.*) anterior subesophageal mass; (*b*) beak; (*bucc.*) buccal mass; (*buc.i.g.*) inferior buccal ganglion; (*buc.lat.l.*) lateral buccal lobe; (*buc.med.l.*) medial buccal lobe; (*buc.sup.g.*) superior buccal ganglion; (*cer.*) cerebral (supraesophageal) cord; (*cer.-buc.*) cerebrobuccal connective; (*cr.*) crop; (*dig.*) digital tentacles; (*e.*) eye; (*f.*) funnel; (*h.*) hood; (*lab.i.l.*) inferior labial lobe; (*lab.sup.*) inner tentacles of superior labial lobe; (*lab.sup.l.*) superior labial lobe; (*li.*) lips; (*n.br.*) brachial nerves; (*n.h.*) hood nerve; (*n.inf.*) infundibular nerve; (*n.lab.*) labial nerve; (*n.prel.*) prelingual nerve; (*n.st.*) static nerve; (*oc.po.*) postocular tentacle; (*oc.pre.*) preocular tentacle; (*o.l.*) optic lobe; (*o.tr.*) optic tract; (*prel.pr.*) prelingual process; (*p.sub.*) posterior subesophageal mass; (*rad.*) radula; (*retr.*) retractor muscles; (*rh.*) rhinophore. Reprinted with permission from Young (1965a).

probably made necessary by the need for a wide esophagus to receive large pieces of food. This need, in turn, is connected with the absence of posterior salivary glands and salivary papilla, which means that there is no external digestion.

There are superior and inferior buccal ganglia, joined to the supraesophageal cord by two pairs of connectives (Fig. 1). These ganglia send nerves to the muscles of the jaws and other parts of the buccal mass, and to the buccal and labial palps and two pairs of prelingual processes, which are probably gustatory (Griffin, 1900). Sympathetic nerves proceed from the inferior buccal ganglia to the esophagus.

2.1. Anterior Nerve Cord

The anterior nerve cord is often called "pedal," but in fact it consists of two distinct parts (Young, 1965a). A more anterior region gives rise to nerves to the

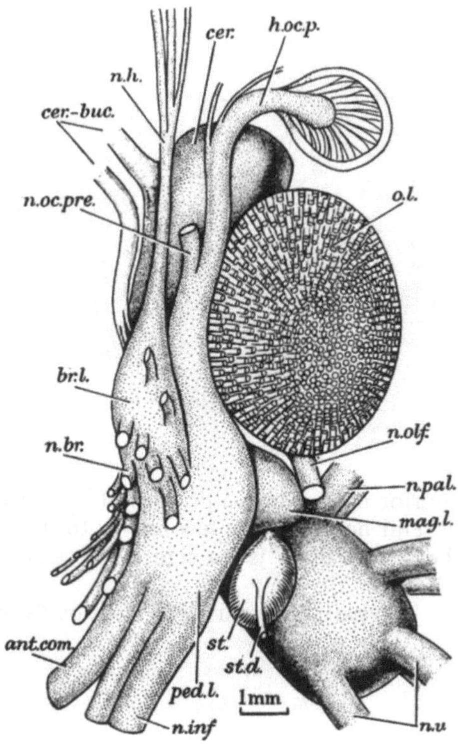

Figure 2. Reconstruction of the outlines of the main parts of the nervous system from tracings of serial sagittal sections. Lateral view. Anatomy: (*ant.com.*) anterior subesophageal commissure; (*br.l.*) brachial lobe; (*cer.*) cerebral cord; (*cer.-buc.*) cerebrobuccal connective; (*mag.l.*) magnocellular lobe; (*n.br.*) brachial nerves; (*n.h.*) hood nerve; (*n.inf.*) infundibular nerve; (*n.oc.p.*) nerve of postocular tentacle; (*n.oc.pre.*) nerve of preocular tentacle; (*n.olf.*) olfactory nerve; (*n.pal.*) palatine nerve; (*n.v.*) visceral nerve; (*o.l.*) optic lobe; (*ped.l.*) pedal lobe; (*st.*) statocyst; (*st.d.*) Kolliker's duct. Reprinted with permission from Young (1965a).

arms and hood and to the pre- and postoptic tentacles; it may be called the "brachial lobe." The nerves to the funnel arise from the more posterior part; hence, it is called the "infundibular lobe" (Fig. 2).

The brachial lobe is a straplike structure, narrowing ventrally to form the anterior subesophageal commissure. It consists of islands of large and small cells and bundles of fibers that are directly continuous with the cerebral cord (Fig. 3). The infundibular part of the anterior cord consists mainly of large cells and fibers running dorsoventrally. It is continuous with the magnocellular lobe. The funnel is undoubtedly homologous with the foot of other mollusks; this "infundibular" part of the anterior subesophageal cord can therefore be called "pedal." It is argued that the brachial part of the anterior cord is "cerebral" in origin and that the arms, hood, and tentacles are not morphologically parts of the foot. Other support for this view is given by Young (1965a).

2.2. Posterior Nerve Cord

The posterior cerebral cord corresponds to the palliovisceral region of the brain of coleoids. It consists of an outer layer of nerve cell bodies and a central neuropil (Fig. 4). In this morphology, it resembles the infundibular part of the

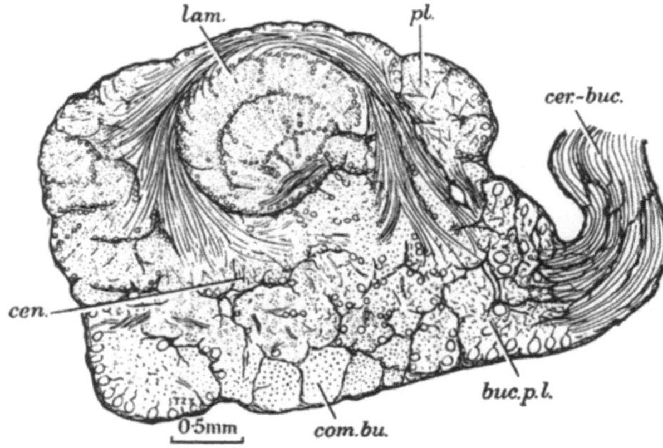

Figure 3. Drawing of a sagittal section of the supraesophageal cord. The cells are shown larger than life-size. Anatomy: (*buc.p.l.*) posterior buccal lobe; (*cen.*) central zone; (*cer.-buc.*) cerebrobuccal connective; (*com.bu.*) commissural bundles; (*lam.*) laminated zone; (*pl.*) plexiform zone. Reprinted with permission from Young (1965a).

anterior cord. The pallioviseral lobe gives rise to the numerous retractor and visceral nerves and a few nerves to the mantle. The latter contains few muscle fibers, and there is no sign of the stellate ganglia that are so characteristic of coleoids.

The palliovisceral cords are broadly joined across the midline. They are con-

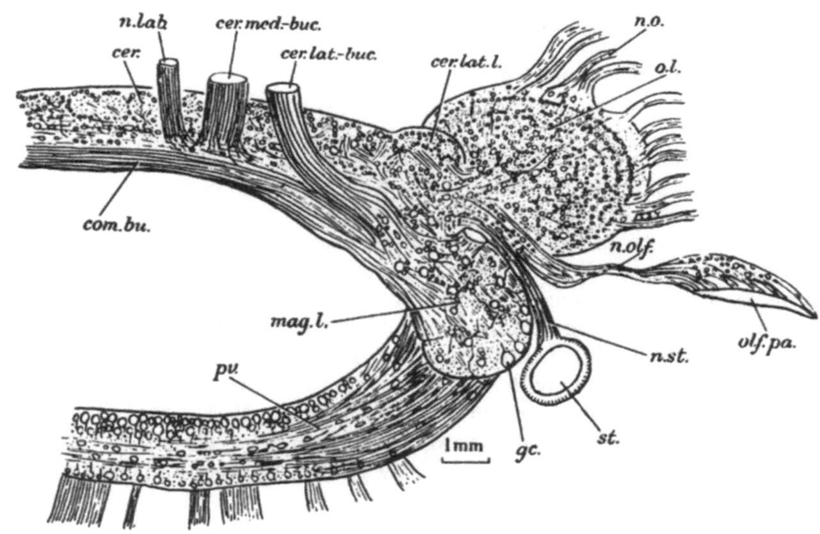

Figure 4. Drawing of a transverse section of the central part of the nervous system reconstructed from photographs. The nerve cells are shown larger than life-size. Anatomy: (cer.) cerebral cord; (*cer.lat.l.*) lateral cerebral lobe; (*cer.lat.-buc.*) lateral cerebrobuccal connective; (*cer.med.-buc.*) medial cerebrobuccal connective; (*com.bu.*) commissural bundles; (*gc.*) giant cell; (*mag.l.*) magnocellular lobe; (*n.lab.*) labial nerve; (*n.o.*) optic nerves; (*n.olf.*) olfactory nerve; (*n.st.*) static nerve; (*o.l.*) optic lobe; (*olf.pa.*) olfactory papilla; (*p.v.*) palliovisceral cord; (*st.*) statocyst. Reprinted with permission from Young (1965a).

tinuous above with the magnocellular region, where all the cords join. This is also the region of entry of the nerves from the statocyst, and here there are the largest cells in the brain (Fig. 4). This lobe corresponds to the magnocellular lobe of coleoids and is probably an important center in coordinating movements of the whole animal.

2.3. Optic Lobes

The magnocellular lobe is also a region in which influences from the eyes and rhinophore converge. The optic lobes have a general similarity to those of coleoids, but they are smaller and simpler. They have a volume only 0.14 times that of all the rest of the brain, whereas in *Octopus*, they are at least twice as large as the rest, and in *Sepia* three times or more.

The optic nerves probably do not pass through a dorsoventral chiasm as they do in all other living cephalopods (except *Vampyroteuthis*). This lack of chiasmatic interconnection suggests that information provided by the pinhole eye is not projected onto the optic lobe in a topographically systematic manner. The layers in the outer part of the lobe are much less differentiated than in coleoids (Fig. 4). There are only irregular inner and outer layers of small cells, and the plexiform zone shows no sign of the separate layers that constitute the system for coding the input in higher cephalopods (Young, 1971). The center of the optic lobe contains a laminated zone and central islands of cells directly continuous with those of the cerebral cord. Thus, the organization of the optic lobe seems not to be suited for the detailed analysis of visual forms; however, experimental data about the visual capacity of *Nautilus* are needed and would be most valuable (see Chapters 15 and 16). The retina and optic lobes are more fully differentiated than those of any noncephalopod mollusk, but less so than those of coleoids. What do they allow the animal to see?

2.4. Olfactory Lobes

The olfactory lobes are of interest because they are relatively larger than in any coleoids, but unfortunately little is known of their organization or function. They are continuous with the cerebral cords and have the same organization as the latter. An outer plexiform layer receives some of the olfactory fibers; others pass to a laminated zone or to the larger ventral region of islands of cells. This plan of organization is similar to that of the optic lobes and suggests that there may be considerable analysis of the signals from the olfactory organ. Tracts of fibers join the olfactory lobe to the optic, cerebral, magnocellular, infundibular, and palliovisceral regions. It is clear that information from the rhinophore plays a large part in cerebral functioning.

2.5. Cerebral Nerve Cord

This cord is essentially the part of the brain that controls the process of eating. Because of the importance of obtaining food, it probably becomes responsible for directing the behavior of the whole animal. In *Nautilus*, it is not divided into a series of distinct parts as it is in coleoids, but its plan of organization is similar in some respects to that of their supraesophageal lobes (Figs. 3 and 4).

The anterior part of the cord is comprised of large cells with axons that proceed to the superior buccal lobes through the cerebrobuccal connectives, lying at the sides. This region of large cells may be called the "posterior buccal lobes," by analogy with *Octopus*. The labial nerves enter from the lips along the whole anterior face, presumably carrying signals of taste (Fig. 1).

The remainder of the cerebral cord is covered by an outer plexiform zone. Within this is a laminated zone dorsally, and below that a large central mass of islands of cells and neuropil. Large commissural bundles occupy the ventral part of the cord. The plexiform zone allows for interweaving of fibers coming from the lips in the labial nerves with those entering at the sides from the eyes, arms, and other receptors. There is unfortunately no evidence as to how these fiber systems interact or of their relationship to the layered systems of the laminated zone. Further detailed study of the connections in this cerebral region would be very rewarding.

At the sides of the cerebral cord, the plexiform zone is enlarged to form a lateral cerebral lobe (Fig. 4). This lobe receives large bundles of fibers from each of the arms and tentacles and from the optic and other lobes. Its interweaving bundles of fibers recall the inferior frontal lobes of octopods. This lobe lies above the point at which all the various afferent input channels enter the cerebral cord and from which the efferent channels proceed outward to the magnocellular lobes and elsewhere.

The cerebral cord does not contain any distinct region with many small cells comparable to the vertical and subfrontal lobes, which in coleoids are associated with visual and tactile memory systems (Young, 1965c, 1983). The laminated zone of the cerebral cord occupies approximately the position of the vertical lobes. This area receives fibers from the plexiform zone, which corresponds in structure and connections with frontal lobes of coleoids. It would be of great interest to investigate the capacity for memory storage in *Nautilus* and, if possible, whether it is associated with these parts of the cerebral cord.

A major difference between the cerebral cord of *Nautilus* and the supraesophageal lobes of coleoids is the absence of any sign of the basal lobes or peduncle lobes. In the higher cephalopods, these lobes serve as higher motor centers analogous to the vertebrate cerebellum. They contain systems of fine parallel fibers, which perhaps serve as timing devices to ensure precisely timed ballistic movements. That *Nautilus* lacks these systems is interesting, in conjunction with its lack of fins and a muscular mantle. The development of these more precise nervous and muscular agents of locomotion may have been important factors in ensuring survival of the coleoids and near extinction of the nautiloids and ammonoids in the face of competition with the fishes.

Nautilus has no distinct optic gland similar to that of coleoids, but groups of cells in the position of that gland at the junction of the optic and olfactory lobes may have a similar endocrine function. These cells have abundant cytoplasm and argentophil rodlike inclusions. They probably carry axons and may be neurosecretory cells concerned with reproduction.

3. Conclusion: Adaptive Strategy of *Nautilus*

The organization of the nervous system of *Nautilus* is suited to a mode of life different from that of coleoids: *Nautilus* appears to live by "smelling and groping" rather than by "visual spotting and attacking," as pointed out to me by W. B. Saunders (1985). Correspondingly, the retina shows poor visual discrimination, there is no chiasm of the optic nerves, and the optic lobes lack the regular layers and columns that are so conspicuous in many coleoids. Conversely, the olfactory organ and the associated lobes are well developed. On the motor side, there is the small size of the magnocellular lobe and the absence of any of the higher motor centers with "cerebellar" organization, which are associated with ballistic attacks on prey. The statocyst lacks the specializations with maculae and cristae to give information about linear and angular acceleration. Most interesting of all is the absence of the small-celled regions of the vertical and subfrontal lobes that are associated with memory formation in coleoids. Perhaps a largely opportunistic scavenging habit would not be assisted by a memory record of the details of the positive and negative features of prey that have been attacked in the past, though such inutility is not obvious. Maybe experiments will show that *Nautilus* has its own specific form of memory storage. Certainly its nervous organization is markedly different from that of all other living cephalopods.

Chapter 14

The Sense Organs of *Nautilus*

VERNON C. BARBER

1. Introduction

The soft parts of *Nautilus* differ from those of coleoid cephalopods in numerous ways; one of the most obvious is that there is an array of tens of tentacles in place of the eight or ten arms of octopuses or squids. The eyes, although large, have no lens; rather, each has a small "pinhole" opening to allow the entry of light. In front of and behind the eyes are two sensory tentacles, the pre- and postocular tentacles (Willey, 1898a). There are also pairs of digital tentacles which are used for prehension and have a sensory function, and buccal (labial) tentacles that surround the buccal region (Willey, 1898b). The pre- and postocular tentacles and the digital tentacles are accommodated in sheaths, but the buccal tentacles are not. Below the eye is a sensory sac, the rhinophore, and there is also a pair of simple statocysts.

There have been a number of studies of the anatomy of *Nautilus* since the first anatomical description by Richard Owen (1832). There have been several light-microscopic descriptions of the structure of the sense organs, e.g., the retina (Hensen, 1865; Merton, 1905), the tentacles (Fernandez, 1907; Griffin, 1900; Vaysièrre, 1896; Willey, 1898a,b), the rhinophore (Fernandez, 1907; Zernoff, 1869), and the statocyst and nervous system (Bassot and Gabe, 1966; Young, 1965a). In contrast, there have been few studies of the ultrastructure of the sense organs and none of the structure of the nervous system. However, the structures of the retina, rhinophore, digital tentacles, and postocular tentacles were studied by the author (Barber, 1967; Barber and Wright, 1969), while a more recent study has presented

VERNON C. BARBER • Department of Sciences, Roehampton Institute, Whitelands College, West Hill, London SW15 3SN, England.

ultrastructural information on the iris groove (Muntz and Raj, 1984) (see also Chapter 16). In addition, Japanese researchers have presented further information on the structure of the tentacles (Fukuda, 1980; Fukuda *et al.*, 1977a,b) (see also Chapter 17). Other recent studies have centered on the physiology of the visual system (Hurley *et al.*, 1978; G. D. Lange *et al.*, 1979; Muntz and Raj, 1984) and on the statocyst-mediated oculomotor reflex (Hartline *et al.*, 1979). These studies present information that can be compared with our extensive knowledge of coleoid cephalopods as presented in several comprehensive accounts (see Nixon and Messenger, 1977; Wells, 1978; Young, 1971).

2. Sense Organs

2.1. Eye and Retina

There are several obvious differences between the structure of the eye of *Nautilus* and that of the eye of coleoid cephalopods. For example, there are no ocular muscles that would allow the eye to move or accommodate in the way described for other cephalopod eyes (Budelmann and Young, 1984). There is no lens or cornea, only a "pinhole" aperture to admit light; this aperture can open and close in response to changes in light intensity.

Apart from these gross differences, there are many ultrastructural ones. As in other cephalopods, there are two types of cells present in the retina: the retinula cells, or receptor cells, and the supporting cells (epithelial cells). Both these cell types form an epithelium that rests on a basal lamina (Fig. 1). This arrangement is in contrast to that in other cephalopods, in which the supporting cells are present only distal to this lamina, which separates the nuclear portion of the

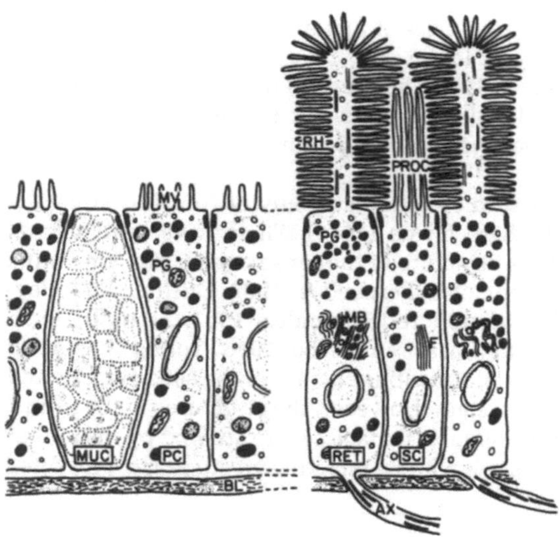

Figure 1. Diagram of part of the retinal structure. The retina consists of a rhabdomal portion and, near the pupil, a nonrhabdomal portion (at left in the diagram). Anatomy: (AX) axon; (BL) basal lamina; (MB) myeloid body (phaosome); (MUC) mucous cell; (MV) microvilli; (PC) pigmented cell; (PG) pigment granule; (PROC) processes of supporting cells; (RH) rhabdom; (RET) retinula cell; (SC) supporting cell. Abbreviations and diagram based on Barber and Wright (1969).

retinula cells from the rhabdomeric portion (Cohen, 1973a; Yamamoto *et al.*, 1965). No blood vessels have been observed in the *Nautilus* retina, in contrast to the situation in the retina of coleoid cephalopods, in which blood vessels are located just below the basal lamina.

The retinula cells can be distinguished easily. The myeloid body found in other cephalopods is a complex tubular array in *Nautilus*, located distal to the nucleus and oriented approximately at right angles to the path of the incident light. This structure was recognized in earlier light-microscopic studies and was termed the "phaosome" (Merton, 1905). It is formed by the fusion of numerous membranes and forms a structure reminiscent of the myelin sheath that envelops vertebrate axons. The membranes of the phaosome are connected to the general cell-membrane system. The function of the phaosome is unknown, although it may be concerned in the metabolism of retinochrome, as suggested by studies of the squid retina (Hara and Hara, 1976). Retinochrome is one of the two photosensitive chromoproteins with retinaldehyde as the prosthetic group, rhodopsin being the other one. The retinula cells and the supporting cells contain pigment granules, plus other organelles, such as mitochondria. The microvillar processes that make up the rhabdoms of the retinula cells are not arranged in organized rectangular arrays as in other cephalopods, but there is, nevertheless, an extensive array in each cell. However, the packing density of retinula cells is similar to that found in other cephalopods (Muntz and Raj, 1984).

As far as is known, the retinula cells are primary receptors and bear axons that pass to the optic lobe of the animal. The axons contain numerous microtubules. As was found in the optic nerves of other cephalopods (Dilly *et al.*, 1963), many axons are enclosed by one glial cell. Muscle fibers and collagen fibers are also present in this region. Synaptic endings, as have been found in other cephalopod retinae (Cohen, 1973b; Lund, 1966; Tonosaki, 1965), have not been found at the bases of the retinula cells in *Nautilus*, although I do not rule out the possibility that some may exist.

Although the retinula and supporting cells take up much of the internal surface of the eye, the retinula cells are absent near the pinhole. There are pigmented and mucous cells in this region. The pigmented cells bear microvilli at their distal ends and contain numerous pigment granules.

Observations of live *Nautilus* have shown that the pinhole aperture of the eye can vary in diameter (Barber and Wright, 1969; Hurley *et al.*, 1978). The latter study has shown that the pupil dilates or constricts in response to changes in light intensity. In response to a sudden increase in light intensity, the closing times for the pupil were on the order of 90 sec, whereas opening times in response to a decrease in light intensity were approximately 50 sec. Stimulation of one eye caused a constriction of the pupil in the unstimulated eye. The dynamics of this contralateral response differed from those of the ipsilateral response already described (presumably because of central nervous system involvement), the major difference being that the constriction response usually exhibited slow adaptations to higher light levels. Hurley *et al.* (1978) showed that other factors could cause a change in pupil opening. For example, struggling behavior or anesthesia with urethane caused pupil dilation. The exact diameter of the pupil is difficult to determine, because of the pigment that surrounds the pinhole, but it is oval in shape, being longer along the horizontal axis. The anatomical feature that causes the pupillary response has not been determined, but because muscle fibers have been noted in the pupillary region, it is presumed that the response is muscular in origin (Muntz and Raj, 1984).

Earlier descriptions of the eye of *Nautilus* suggested that the retina is in direct contact with the seawater. I found that the interior of the eye was filled with a jellylike substance that could have been produced by the mucous cells that occur near the edge of the retina. This may not be the case, since an exchange of fluid between the ocean and the interior of the eye may occur. It has also been determined that the refractive index of the fluid extracted from the chamber of the eye and the refractive index of seawater are very similar (Hurley *et al.*, 1978). Also, there are ciliated cells in the vertically oriented iris groove, and these cells generate a strand of mucus that is directed downward, away from the pupil (Muntz and Raj, 1984) (also see Chapter 16). This presumably provides some protection against the entry of foreign material into the eye chamber.

Another interesting physiological response in *Nautilus* is the ability of the animal to rotate its eyes relative to its body, so as to compensate for changes in its body orientation and maintain its eye fixed with respect to gravity. Such a response is particularly necessary because of the rocking motion that occurs as the animal swims by jet propulsion. The response is mediated by the statocyst, as has been shown by statocyst-ablation experiments (Hartline *et al.*, 1979). This eye movement is thought to be made via the stalk of the eye, which contains a complicated array of muscular and connective tissue elements. It need hardly be noted that the nervous system of the animal must control this relatively complex reflex.

The exact function of the eye is yet to be determined. Earlier studies with model eyes showed that a fairly sharp image could be produced on the retina (Mugglin, 1937). This fact, and the large number of retinula cells, suggest that the eyes are at least capable of simple visual discriminations and can function as more than simple light-intensity monitors. Recent visual acuity tests, using the optomotor response, showed that the minimum separable angle discrimination discernible by the animal lies between 5.5 and 11.25° (Muntz and Raj, 1984). This value agrees with what would be expected on the basis of the gross dimensions of the eye and pupil. However, it is considerably worse than would be expected from the packing density of the retinula cells. Other studies, of the electroretinogram of *Nautilus*, show that it is of a complexity comparable to that found in other cephalopods (G. D. Lange *et al.*, 1979). How this complexity is related to the functioning of the eye is uncertain at this time.

The use that *Nautilus* makes of vision in its day-to-day life is uncertain, particularly in view of its deep-water habitat. The presence of a well-developed retina, a statocyst-mediated eye stabilization, and a pupillary response suggests that vision is of some importance. Recent studies have also shown that there is a positive phototactic response in *Nautilus* (Chapter 15). So, the function of the eye in the way of life of *Nautilus* must be said to be uncertain at this time (see Chapter 15 for a more detailed discussion).

2.2. Rhinophore

The rhinophore consists of a sac of cells that is located just below the eye. This sac opens to the exterior via a pore. There are three types of cells present

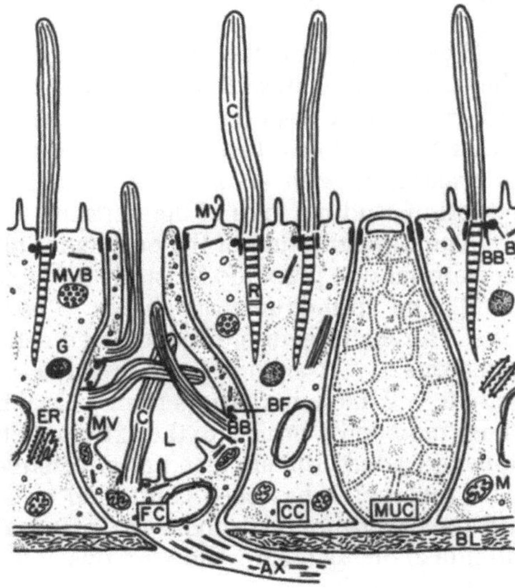

Figure 2. Diagram of part of the rhinophore structure. Anatomy: (AX) axon; (BB) basal body; (BF) basal foot; (BL) basal lamina; (C) cilium; (CC) ciliated cell; (ER) endoplasmic reticulum; (FC) flask-shaped, ciliated receptor cell; (G) Golgi body; (L) lumen of flask-shaped cell; (M) mitochondrion; (MUC) mucous cell; (MV) microvilli; (MVB) multivesicular body; (R) root. Abbreviations and diagram based on Barber and Wright (1969).

in the epithelium of the rhinophore: flask-shaped ciliated cells, other ciliated cells, and mucous cells (Fernandez, 1907; Young, 1965a; Zernoff, 1869). These cells form an epithelium that rests on a basal lamina (Fig. 2).

The flask-shaped cells each open to the lumen of the sac by a narrow pore. The lumen of the flask contains numerous cilia with a 9 + 2 axonemal structure; the cilia arise from basal bodies that have basal feet but no roots. One or two cilia pass up the narrow neck of the flask and reach the lumen of the sac via the pore. Some microvilli are also present. The nucleus of the cell lies at the base of the cell, and it is presumed that an axon arises in this region. However, the presence of an axon was not confirmed in ultrastructural studies (Barber and Wright, 1969).

Cells of the second type bear numerous cilia and microvilli. The cilia have a 9 + 2 axonemal structure and arise from basal bodies that bear basal feet and long striated roots. It is presumed that these cilia are motile and move a current of mucus over the epithelium. The third type of cell is a mucus-producing cell with usual mucus granules. Below the basal lamina are bundles of axons, muscle fibers, and collagen fibers.

The structure of the rhinophore in *Nautilus* is similar to the structure of the so-called olfactory organ in other cephalopods (Emery, 1975c, 1976; Woodhams and Messenger, 1974). For example, flask-shaped cells, presumed to be receptor cells, have been found in the olfactory organs of *Octopus* and *Lolliguncula*. The presumed chemoreceptor cells in the sucker of *Octopus* and the lip of *Sepia* also have some similarities to the flask-shaped receptor cells in *Nautilus* (Graziadei, 1964, 1965b). It seems likely, therefore, that the flask-shaped, ciliated cells are receptors. However, since the function of the olfactory organ in cephalopods has not been determined, the modality of response of the *Nautilus* rhinophore is uncertain. A chemoreceptor function is most likely, with the organ acting as a food detector or as a means of sampling respiratory water.

2.3. Preocular and Postocular Tentacles

The ocular tentacles are shorter and more slender than the digital tentacles. They can be as small as 2 mm in diameter, half the diameter of the digital tentacles, and are circular in cross section (Fukuda, 1980). Along the inner surface of the ocular tentacles are closely spaced, tonguelike processes that are large enough to be discernible with the naked eye. These processes are arranged at right angles to the longitudinal axes of the tentacles.

The ultrastructure of the postocular tentacles was studied by the author, but no material was prepared for study of the preocular tentacles (Barber and Wright, 1969). The epithelium was shown to be composed of ciliated and nonciliated cells. The main feature of the ciliated cells is that they bear macrocilia, namely, numerous 9 + 2 axonemes, each enclosed in a single outer membrane. Each of these 9 + 2 axonemes arises from a basal body that bears a basal foot and a root. These macrocilia project to the exterior between the various folds of the tentacle. Nonaggregated cilia are also present (Barber and Wright, 1969). The presence of these macrocilia has been confirmed by scanning-electron-microscopic studies, and they were found in both the pre- and postocular tentacles (Fukuda, 1980; Fukuda *et al.*, 1977a,b). The reactions of the pre- and postocular tentacles suggest that they may bear mechanoreceptors (Bidder, 1962). It is therefore possible that the cells that bear the macrocilia are mechanoreceptors, perhaps used to help protect the large eyes from injury.

2.4. Digital Tentacles

Behavioral observations of live *Nautilus* presented with a distant food source showed that the digital tentacles were differential in their response, some functioning as "alert," some as "search," and the rest as "feeding" groups (Bidder, 1962). Unlike the case in most other cephalopods, there are no suckers on the tentacles, but they do have a series of adhesive, transverse ridges.

The epithelium of the digital tentacles consists of a single layer of cells that rests on a basal lamina (Barber and Wright, 1969). Two cell types were found, pigmented cells (the more common) and mucus-producing cells, similar to those found in the rhinophore. No recognizable receptor cells were seen in the material studied by Barber and Wright (1969). However, more recent scanning-electron-microscopic and light-microscopic studies show that there are "taste-bud-like" cells in the depressions between the ridges (Fukuda, 1980; Fukuda *et al.*, 1977a, b). It should also be noted that these latter authors report that there are fewer of these proposed sensory cells, or mucus-producing cells, on the outer (nonadhesive) surface of the tentacle. The variety of reactions of the digital tentacles suggests that they may have more than one type of receptor cell. It is also likely that the mucus-producing cells secrete the adhesive substance used to attach the tentacles to the substratum.

2.5. Buccal Tentacles and Labial Margin

As far as is known, the ultrastructure of the buccal (labial) tentacles has not been studied. However, they probably also contain sensory receptor cells. Possible sensory cells have been reported in the labial margin, but no studies of the ultrastructure of these putative receptor cells have been reported (Fukuda, 1980).

2.6. Statocysts

The statocyst consists of a single sac of cells, probably composed of a large number of hair cells and supporting cells. The cavity of the statocyst is almost completely filled with statolith crystals. There is an excretory duct, Kolliker's canal, that opens directly to the exterior (Young, 1965a). In general, it can be said that the statocyst of *Nautilus* is similar to that found in most noncephalopod mollusks, and it does not have the more complex structure (and presumably functions) found in coleoid cephalopods (Barber, 1968).

Physiological experiments on the *Nautilus* statocyst have not yet been carried out, although the organ has been shown to be involved in the eye stabilization response (Hartline *et al.*, 1979). However, the similarity in structure to that of the statocysts found in gastropod mollusks suggests a similarity in function. So, it is presumed that the central nervous system receives information about the animal's position in space by the spatial orientation of the excitation pattern of the responding hair cells. The direction of rotation may be detected by the sequence in which different hair cells are stimulated (Budelmann, 1975).

3. Discussion

The ultrastructures of the retina and rhinophore of *Nautilus* are known in reasonable detail, and there is some information on receptor cells in the tentacles (Barber, 1967; Barber and Wright, 1969; Fukuda, 1980; Fukuda *et al.*, 1977a,b) (see also Chapter 17). There have been some physiological investigations on the optics of the eye and one study of the eye stabilization response that involves the statocyst (Hartline *et al.*, 1979; Hurley *et al.*, 1978; Muntz and Raj, 1984) (see also Chapter 15). However, only one electrophysiological study has been made, and we know effectively nothing concerning the mode of response of the receptor cells in any of the sense organs. Also, it is not known whether there are species, age, or gender differences in the structures and functions of the sense organs.

In such circumstances, all one can say is that *Nautilus* has a full complement of sense organs that are comparable with, although in some cases more poorly organized than, those of coleoid cephalopods. Although the general anatomy of the retina is less complex in nautiloids, the retinula cells are similar in basic structure in nautiloid and coleoid cephalopods (albeit the myeloid body, or phaosome, is more complex in nautiloids!). It is probable that few or no synaptic interactions are present in the *Nautilus* retina, although this latter point has to be clarified by further anatomical study. The rhinophore of nautiloids is similar in structure to the olfactory organ of coleoid cephalopods (Emery, 1975c, 1976; Woodhams and Messenger, 1974) and is presumed to have a chemosensory function. The tentacles of *Nautilus* contain a variety of sensory cells, but more study is needed to determine the structure and likely modality of response of these cells. The statocyst of *Nautilus* is simpler than in coleoid cephalopods, since the sensory parts of the statocyst are not divided into a crista and macula, and so the statocyst has a structure similar to that found in noncephalopod mollusks (Barber, 1968; Budelmann, 1975; Young, 1965a).

The simplicity of the sense organs, together with the less centralized nervous system in *Nautilus*, compared with other living cephalopods, suggests that the sense organs of ancient nautiloids were similar to those found in modern *Nautilus*. However, it cannot be ruled out that secondary simplification of the various sense organs has taken place.

ACKNOWLEDGMENTS. This chapter is dedicated to Professor J. Z. Young, who has done so much to further our understanding of the nervous system and sense organ of cephalopods. Thanks are due to Ms. G. Campbell for drawing the figures.

Chapter 15

Visual Behavior and Visual Sensitivity of *Nautilus pompilius*

W. R. A. MUNTZ

1. Introduction

The various species of *Nautilus* found today show many features that are probably primitive. Of them, the eye has attracted attention because it has no lens and appears to operate on the principle of a pinhole camera; consequently, it should have a performance considerably inferior to that of the lens-bearing eyes of other cephalopods. This inferiority was confirmed for visual acuity in a previous paper (Muntz and Raj, 1984). It was found, using the optomotor response, that the minimum separable angle lay between 5.5 and 11.25°, which agreed well with the value expected on the basis of the gross dimensions of the eye and also with expectations based on photographing a visual test chart with a scale model of the eye. In *Octopus vulgaris*, on the other hand, the minimum separable visual acuity is 17° or better (Sutherland, 1963).

This chapter reports some additional behavioral observations on the optomotor response of *Nautilus* and also on the positive phototactic response, both types of visual behavior being well developed in this animal. Estimates of sensitivity were also made by use of these responses. Finally, the absorbance spectrum of the rhodopsin of *Nautilus* was determined. Assuming that the animal's spectral sensitivity agrees with the spectral absorbance of its rhodopsin allows

W. R. A. MUNTZ • Faculty of Science, Monash University, Clayton, Victoria 3168, Australia.

for a calculation of the maximum depth at which the animal would be able to see.

2. Material

Nautilus were trapped off the main reef at Suva, Fiji, at depths of about 500 m. They were kept in a holding tank with a closed circulation system containing a sand filter and a heat exchanger that maintained the temperature at about 16°C. They were fed occasionally on prawns or fish. Under these conditions, the animals remained alive for several weeks, but behavioral data were always collected within the first 7–10 days, to minimize the chances that deterioration in the animals' condition was affecting the results. Individual animals were marked with a waterproof felt-tipped pen for indentification.

3. Optomotor Response

3.1. Methods

The optomotor response was tested using the apparatus described in Muntz and Raj (1984) and shown in Fig. 1. The animal was placed in the center of a glass vessel filled with water to a depth such that the bottom of the shell just

Figure 1. Optomotor apparatus. The inner glass vessel containing the animal is 24 cm in diameter. From Muntz and Raj (1984).

rested on the bottom of the vessel. In this way, the animal could be positioned so that the eyes were near the center of the apparatus—this to reduce, as much as possible, distortions caused by the curved sides of the glass vessel. The outer drum was constructed from thin sheet metal, and the stripes were made on cylinders of thick paper, which could be inserted within the drum. The drum was rotated by hand at a speed such that one complete revolution took approximately 14 sec. The initial response of the animal was usually a simple rotation, achieved using the funnel; this response could be seen clearly before the animal left its central position.

Sensitivity was determined with this apparatus using eight each of alternating black and white stripes, each subtending 22.5°. The stripes were illuminated by a 15 W tungsten bulb, suspended 77 cm directly above the upper surface of the animal container. This bulb was enclosed in a light-tight box with a single opening, over which neutral-density filters could be placed. As is usually the case with gelatin filters, their spectral absorbance curves were not strictly neutral, and the values given in the figures refer to intensities at 468 nm, the wavelength at which it was found that *Nautilus* rhodopsin absorbed maximally. The brightness of the white stripes, at the highest intensity available, was 23.9 candelas per square meter (cd m^{-2}) as measured with a PIN silicone photodiode placed in the position of the animal container; the photodiode itself was calibrated against a UDT 40X optometer. The color temperature of the bulb, from the manufacturer's data, was 2750 ± 200°K.

In any given trial, the animal was placed in the center of the apparatus. A positive response was scored if the animal rotated 90° or more in the same direction as the stripes and a negative response if it rotated 90° or more counter to the direction of stripe movement. A trial was complete when the drum had rotated through 360° or when a response occurred, whichever came first. Failure of the animal to rotate 90° in either direction during a trial was scored as a failure to respond.

Nine animals were used, and they were tested over three experimental sessions held in the morning and evening of one day and the morning of the next day. Before each session, the animals were dark-adapted for at least 30 min. Seven intensity levels were tested during each session, except for the first session, in which the lowest intensity was not tested. The individual animals did not all receive exactly the same tests during each session, but the trials were arranged so that by the end of the third session, they all had received four trials at each intensity, two for each direction of drum rotation. Thus, there were 36 tests overall at each intensity level. Animals were tested in turn, receiving two consecutive trials (one for each direction of rotation), and then returned to the holding tank while the other animals were tested.

3.2. Results

The results for each of the three sessions and their mean are shown in Fig. 2, along with the number of failures to respond, which increased as the light level decreased. In the absence of stimulation, the animals tended to show a preference for rotating in one direction, which accounts for the clustering of points on the

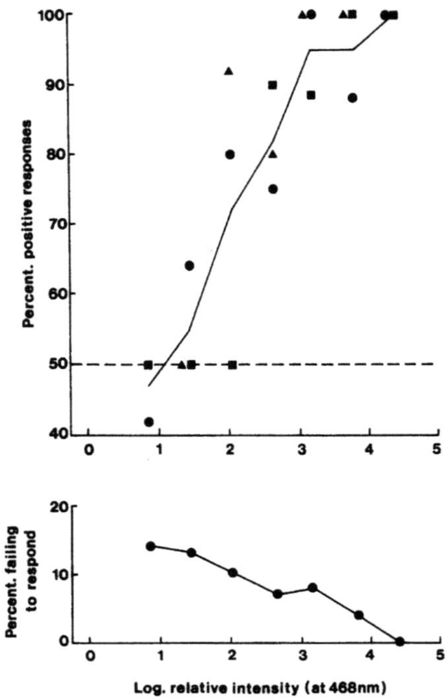

Figure 2. Sensitivity results for the optomotor situation. (A) Percentage of positive responses for each of three testing sessions at the different intensity levels. The different testing sessions are shown by different symbols, and only two testing sessions were given at the lowest intensity level used. Some of the points have been displaced slightly to the right of their true positions for clarity. The line is drawn through the mean for the three sessions. (B) Percentage of failures to respond at the different intensity levels. A value of 5 on the abscissa represents 23.9 cd m^{-2}.

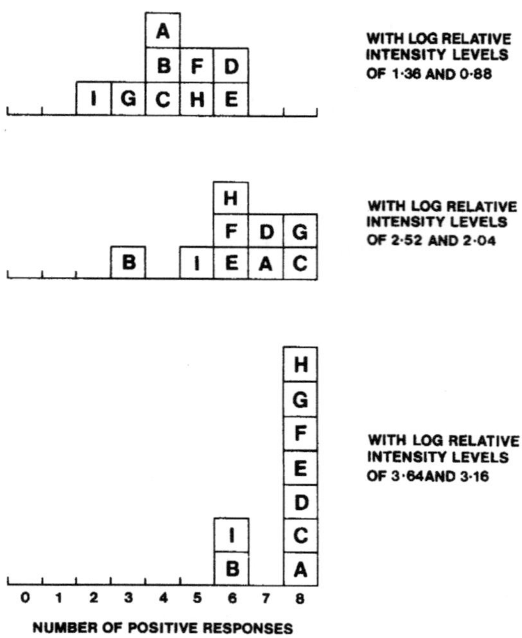

Figure 3. Number of positive responses, of a possible maximum of eight, made by individual animals A–I at different intensity levels.

50% level at low intensities. Figure 3 shows the results for individual animals, pooled over adjacent intensity levels.

4. Phototactic Response

4.1. Methods

Two different arrangements were used to study phototactic behavior. In the first, all the walls of the apparatus were lined with black polythene (Fig. 4). Illumination was provided by general room lighting (5 Matsushita Daylight FL40 SD fluorescent tubes). In any given trial, six animals were placed in one or the other end of the apparatus. This end of the apparatus was then covered with black polythene, and the number of animals in either end of the tank was then noted at various time intervals.

The second apparatus (Fig. 5) consisted of two wooden partitions (AA and BB) inserted into a glass aquarium so as to form a T-maze, with two 60-cm-long arms and a starting chamber (S). Most of the experiments involved a single light source, which illuminated a tracing paper screen (T) at the end of one of the arms. The light source and neutral-density filters used were the same as those used to test the optomotor response. The rest of the apparatus was lined with black polythene. To guard against side preferences, the position of the wooden partitions could be reversed, so that the starting chamber would be positioned at the top in the plan view (Fig. 5) and the position of the illuminated screen would be at T′, as opposed to T, and the animal would have to turn right instead of left, to move toward the stimulus. The size of the illuminated screen was 20 cm wide by 40 cm high in either case.

Some experiments were also conducted with two light sources illuminating two screens, one at the end of each arm of the maze. A fixed light source illuminated one screen, whereas the other arm was illuminated by the same variable light source as before. The fixed light source was a 25 W bulb, in front of which was a diffuser and two layers of neutral-density filter; the bulb was placed 57.5

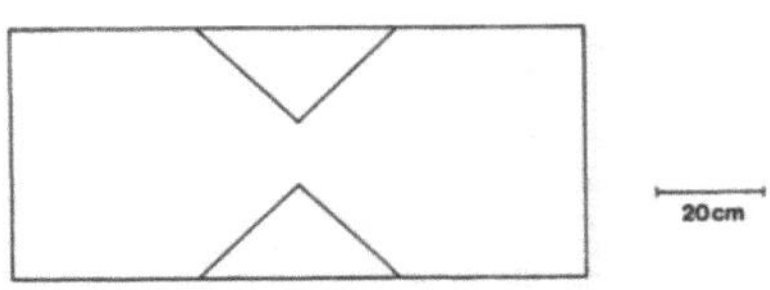

Figure 4. Plan view of the first phototactic situation. The water depth was 33 cm.

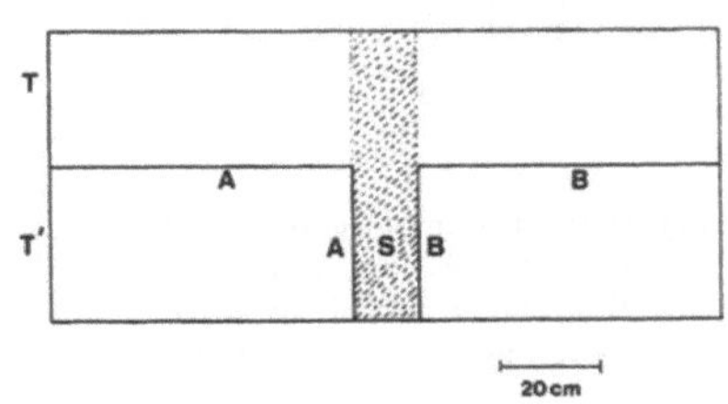

Figure 5. Plan view of the second phototactic situation. The water depth was 40 cm. Key: (AA, BB) wooden partitions; (T, T′) alternate positions for the stimulus; (S) starting chamber. The shaded area was raised 25 cm, and the space beneath it blocked off, in the two-light experiment.

cm from the screen at the end of the arm of the T-maze. The resulting luminance of the screen was about 2.4 cd m^{-2}. In the apparatus as so far described, some of the light from one screen was transmitted across the apparatus and affected the intensity of the second screen. To avoid this, the floor of the starting chamber and central part of the T-maze (shaded area in Fig. 5) was raised by 25 cm using a wooden platform inserted into the maze. The space beneath the platform was blocked off, and the size of the tracing paper screens was reduced, so that they occupied only the bottom 20 cm of the ends of each arm. This arrangement eliminated any direct light path from one screen to the other.

The positive phototactic response, with very few exceptions, involved the animal swimming backward (i.e., with the shell leading) toward the light. In any given trial, the animal was placed in the starting chamber with the shell oriented toward the opening, so that normal swimming would take the animal into the body of the maze. The trial was terminated if the animal did not leave the starting chamber within 30 sec. Once the animal had left the starting chamber, it was allowed a further 45 sec to respond; failure to reach the end of one or the other arm of the maze within this time was recorded as no response. A rough sketch of the track followed by the animal was made on each occasion, and a record was also kept of those rare occasions when the animal swam forward rather than backward. The animals were tested in turn, being returned to the holding tanks between trials.

4.2. Results

4.2.1. With the "Hourglass" Apparatus

Most of the animals moved rapidly out of the dark into the light and remained there (Fig. 6). The general response was to swim backward and only occasionally forward. In either case, the response appeared to be directed and not the result of random movement. A few animals did not respond, but remained in the dark end of the tank for as long as 5 hr.

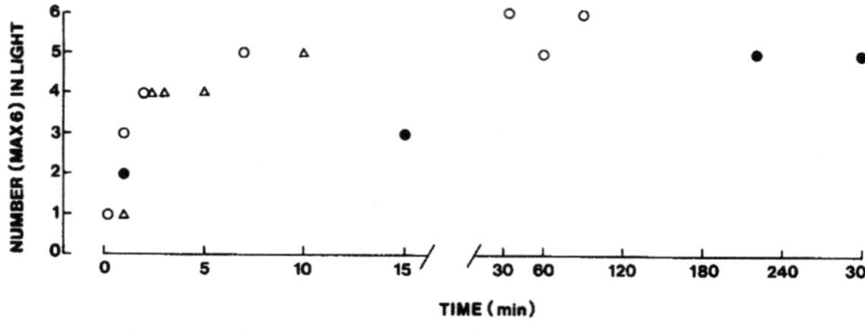

Figure 6. Results of the first phototactic situation. The different symbols denote the results for three experimental sessions. Six animals were used in each session, and the points show the number in the illuminated half of the apparatus at different times after the beginning of the session. Note the change in the scale of the abscissa between 15 and 30 min.

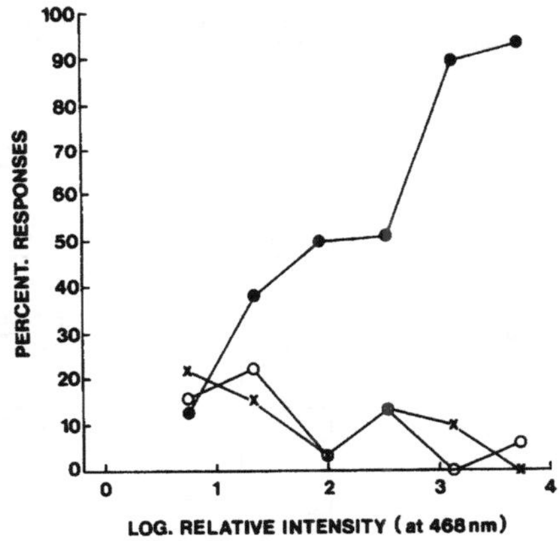

Figure 7. Results of the second phototactic experiment. Key: (●) positive responses; (○) negative responses; (×) failures to leave the starting chamber. At the lower intensities, these three types of responses do not add up to 100%; the remaining trials were occasions on which the animals left the starting chamber, but failed to complete either a positive or a negative response within the prescribed time. The abscissa is directly comparable to that of Fig. 2.

4.2.2. With the T-Maze Apparatus

The results of testing with the single stimulus when the intensity was varied involved the use of 20 animals (Fig. 7). Of these animals, 9 were tested twice at each intensity and the remaining 11 once (each point in Fig. 7 thus representing the result of 29 trials). It can be seen that as the intensity was reduced, the number of positive responses decreased, whereas the number of negative responses and failures to leave the starting chamber increased. There were also more failures to complete either a positive or a negative response by reaching the end of one or the other arm of the maze within 45 sec. For example, at the lowest intensity used, there were 13% positive responses, 16% negative responses, and 21% failures to leave the starting chamber; therefore, on 50% of the trials, the animals left the starting chamber but failed to complete a response (Fig. 7). Apart from this, there were also qualitative changes in the animals' behavior. At the higher intensities, the animals generally swam directly toward the light. At the lower intensities, the animals' movements were much less obviously directed toward the light, and even if overall they showed positively phototactic behavior, the tracks followed during individual trials showed many circling movements and changes in direction (Fig. 8).

An experiment was also carried out to see whether the behavior would reverse, if sufficiently intense stimuli were used. The stimulus chosen was a 300 W reflector flood lamp placed 65 cm from the screen. The luminance of this stimulus was about 32 cd m^{-2}, which is much greater than anything the animals would be exposed to in their normal environment. Ten animals were tested; all swam directly to the stimulus.

In the two-light experiments, 12 animals were used, with each stimulus combination being shown twice to each animal (Fig. 9). An intensity change of 2.4 log units was sufficient to reverse the animals' behavior from preferring the fixed light source to preferring the variable stimulus. The behavior of the animals was also clearly directed toward one or the other of the stimuli, and circling behavior (such as that shown in Fig. 8) was seldom seen. In fact, in over 48 trials, there

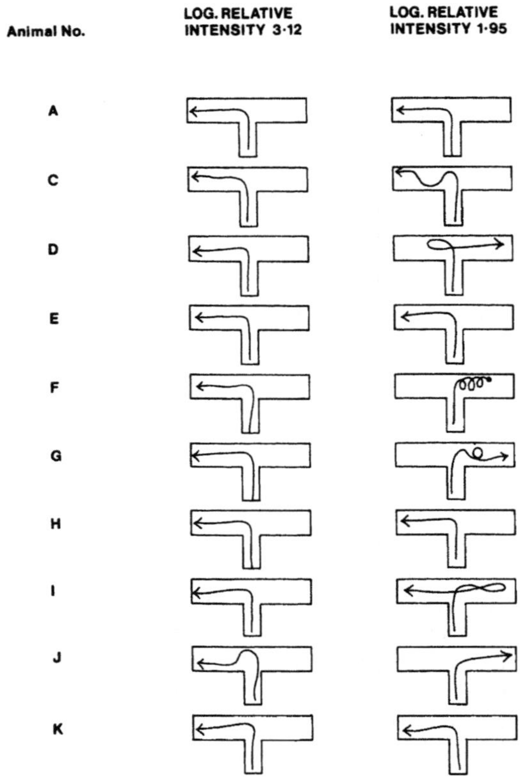

Figure 8. Freehand sketches of the paths followed by ten individual animals at two intensity levels in the second phototactic experiment. The stimulus in all cases was presented on the left.

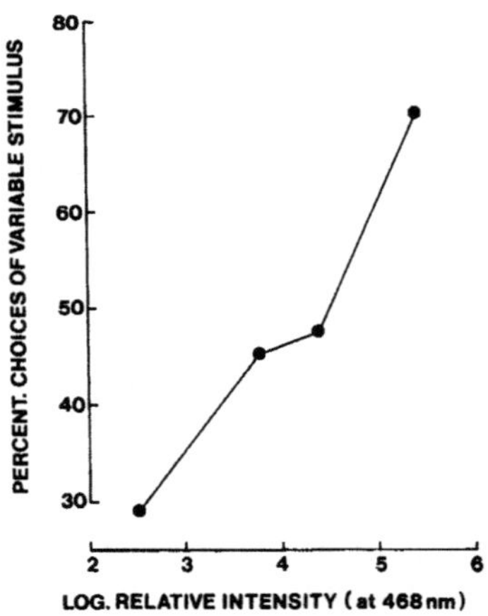

Figure 9. Log relative intensity of the variable stimulus against the percentage choices of that stimulus in the two-light experiment. The abscissa is directly comparable to those of Figs. 2 and 7.

were only 2 failures to leave the starting chamber within the prescribed 30 sec and only 3 examples of circling behavior (Fig. 9).

5. Visual Pigment

5.1. Methods

Ten eyes from freshly killed, dark-adapted animals were hemisected, and the outer segments of the receptors were brushed off into seawater with a soft brush. The outer segments were washed in five changes of acid (pH 4.6) buffer and then extracted at room temperature overnight with 1.5 ml 3% digitonin solution. Hydroxylamine, 0.2 M, 10% by volume, and the same amount of saturated sodium borate solution were added to the extract before the absorption spectrum was measured. The extract was then bleached by being immersed in a water bath at 70°C for about 90 sec, thus allowing a difference spectrum to be constructed. The addition of hydroxylamine should have prevented the results from being affected by the presence of retinochrome (Hara and Hara, 1976), and all operations were carried out under deep-red light to minimize the formation of metarhodopsin. These methods are the same as those used by Muntz and Johnson (1978) and allow comparison of the *Nautilus* data to those obtained with other cephalopods.

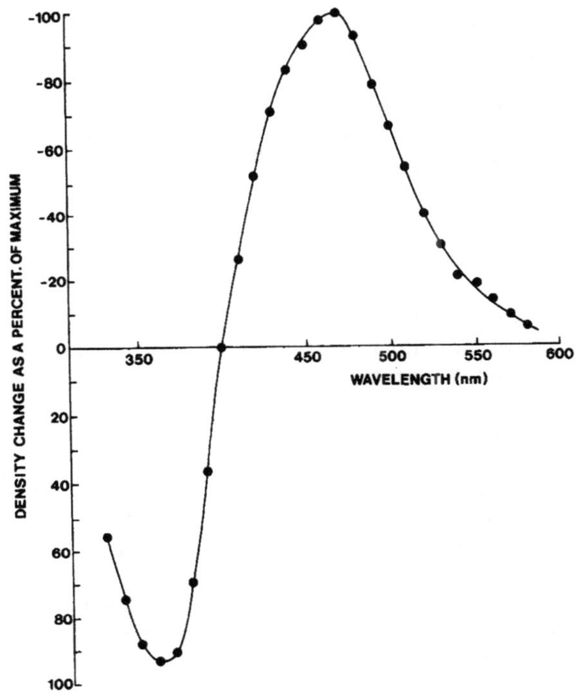

Figure 10. Difference spectrum for *Nautilus* rhodopsin as a percentage of the maximum.

5.2. Results

The difference spectrum of the retinal pigment obtained on thermal bleaching shows that the maximum density loss occurred at about 467 nm, and the position of the maximum density gain on bleaching, in the presence of hydroxylamine, indicates that it is based on vitamin A_1 (Fig. 10).

6. Discussion

Nautilus rhodopsin has its maximum absorption (λ_{max}) at shorter wavelengths than do the rhodopsins of other cephalopods (e.g., Muntz and Johnson, 1978), although the rhodopsin of *Eledone moschata*, with its λ_{max} at 470 nm (Hamdorf *et al.*, 1968), approaches *Nautilus* rhodopsin closely. Some other deep-sea animals have rhodopsins that absorb at the same or even shorter wavelengths [e.g., *Bathylagus microphthalumus* with λ_{max} at 467 nm (Dartnall and Lythgoe, 1965) and *Euphasia pacifica* with λ_{max} at 462 nm (Kampa, 1955)], so we may conclude that the rhodopsin of *Nautilus* is typical for a deep-sea animal.

The sensitivity of *Nautilus* was determined using two different behavioral methods. The same light source and filters were used on each occasion. The distances of the source from the screen of the phototactic apparatus and from the optomotor drum at the level opposite the animal's eyes are also known, as is the angle at which the light hits the drum in the latter case. Finally, the fraction of light transmitted by the tracing paper screen was measured, as was the fraction reflected by the white paper used for the optomotor drum. With this information, it is possible to compare the stimulus intensities used in the two situations. The probability that a positive response will occur by chance also differs for the two methods, being 0.5 for the optomotor response and (from Fig. 7) about 0.15 in the phototactic situation. This possibility can be allowed for by applying to the data Abbott's correction (Finney, 1962):

$$P_d = \frac{P_e - P_c}{1 - P_c} \tag{1}$$

where P_d is the corrected probability of detecting the stimulus, P_e is the experimentally obtained probability of a correct response, and P_c is the probability of obtaining a correct response by chance. Comparison of the two sets of results after this correction has been applied (Fig. 11) shows that the agreement is good, suggesting that the absolute threshold is at about 1 on the log relative intensity axis, which represents 2.4×10^{-5} cd m^{-2}. This value is, however, not very useful, because candelas per square meter are photometric units and thus relevant only to human vision and also because the tungsten light source used in the experiments presented herein has a spectral output very different from that to which *Nautilus* is exposed in its natural habitat. With certain assumptions, it is possible, however, to use the result to estimate the greatest depth at which vision would be possible for *Nautilus*, for at this depth the product of the animal's spectral sensitivity, the surface daylight, and the transmission characteristics of the sea-

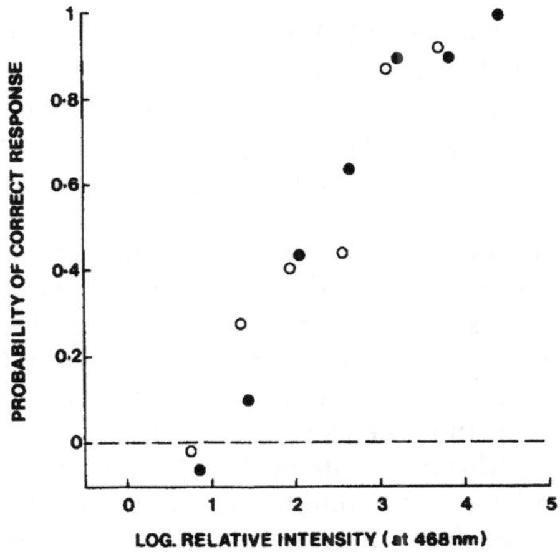

Figure 11. Comparison of the optomotor and phototactic results after applying Abbott's correction. Key: (●) positive responses in the optomotor situation; (○) positive responses in the phototactic situation. It is assumed that the probability that a correct response will occur by chance is 0.5 in the optomotor situation and 0.15 in the phototactic situation.

water will be equal to the product of the animal's spectral sensitivity, the energy output of the tungsten bulb, and the transmission characteristics of the neutral-density filters that defined the animal's absolute threshold. That is, at the maximum depth at which vision is possible, the following relationship holds:

$$K \int_0^\infty V_\lambda \cdot E_\lambda \cdot T_\lambda \cdot d_\lambda = K \int_0^\infty V_\lambda \cdot S_\lambda \cdot N_\lambda \cdot d_\lambda \tag{2}$$

where, for the wavelength λ, V_λ is the animal's relative sensitivity, E_λ is the energy of light reaching the surface of the earth, T_λ is the proportion of light transmitted by the water at the greatest depth at which the animal can see, S_λ is the energy of the tungsten light source, N_λ is the proportion of light transmitted by the neutral-density filters at threshold, and K is a constant needed because V_λ is in relative, not absolute, terms.

Estimates may be made of all these factors, except K, which cancels out. In this instance, V_λ was assumed to be the same as the absorbance of the rhodopsin of *Nautilus*, a reasonable assumption because cephalopods apparently lack color vision (Messenger *et al.*, 1973; Flores *et al.*, 1978) and have only one type of receptor; V_λ was thus taken from the nomogram of Dartnall (1953) for a visual pigment with λ_{max} at 467 nm. There is some uncertainty about using such a nomogram at very short wavelengths, but any such uncertainty is unimportant, because there is very little light in either the sea or the laboratory situation at such wavelengths.

E_λ was taken from the data of Tyler and Smith (1970) for the light reaching the surface of Crater Lake at 11:00 hr on May 8, 1965, and T_λ from the values of Jerlov (1968) for J1 oceanic water, and N_λ was measured directly on the filters actually used. S_λ was estimated from the color temperature of the source provided by the manufacturers, the known brightness of the stimulus in cd m^{-2}, and the absolute luminosities of different wavelengths in lumens/W, taken from Weaver (1949).

The value of the right-hand side of equation (2) was then calculated by plotting ($V_\lambda \cdot S_\lambda \cdot N_\lambda$) against wavelength and measuring the area under the resulting curve. The value of the left-hand side of the equation was also calculated in the same way for various depths, which then allows the depth at which equation (2) holds to be found by interpolation. In equation (2), the limits of the integrals are 0 and ∞. In fact, the value of the right-hand expression is effectively 0 below 400 nm and above 570 nm, as is that of the left-hand expression below 420 nm and 570 nm, so only the wavelengths between these limits need be considered. The final estimate of the maximum visual depth obtained in this way was 770 m. This may be compared with the estimate by Clarke and Denton (1962) of 1000 m for the maximum depth at which vision should be possible for deep-sea fish.

It thus seems likely that some vision of downwelling daylight is possible for *Nautilus* to a depth of about 770 m. This estimate is for optimal weather conditions and for light coming vertically downward in very clear seawater. Clearly, the visual depth limit will be very much less during the night and for lines of sight other than upward. Thus, Dartnall (1975) has calculated that in clear oceanic seawater, at the depths at which *Nautilus* lives, the light intensity decreases by about 1 log unit for every 115 m increase in depth; furthermore, Tyler (1960) has shown that the light reaching the animal horizontally will be about 1.5 log units less than the downwelling light (equivalent to a depth of about 173 m) and the upwelling light 2.5 log units less (equivalent to a depth of about 288 m). Also, there is a range of about 3 log units of intensity over which, although the animals respond to the light on average, they do not respond consistently on every trial (Fig. 11). If this inconsistency also holds true under natural conditions, it represents a very large depth range over which their responses to light are uncertain. Finally, at depth in the ocean, bioluminescent light can be as great as the natural light penetrating from the surface (Clarke and Backus, 1956; Kampa and Boden, 1957), so it seems clear that bioluminescence will be detectable by *Nautilus*.

Comparison of the maximum depth at which vision should be possible for *Nautilus* and deep-sea fish with various findings on the depths at which *Nautilus* is found and the depth at which the shell implodes is shown in Fig. 12. This last factor clearly represents an ultimate limit to the depth at which *Nautilus* can live. In life, *Nautilus* probably does not approach within 100 m of its implosion depth, and some limited vision of natural daylight may be possible throughout its depth range.

The use that *Nautilus* makes of vision during its natural life remains obscure. The facts that the eye shows statocyst-mediated stabilization (Hartline *et al.*, 1979) and that there is a well-developed retina and some forms of well-developed visual behavior suggest that vision is important. The animals have usually been considered to be nocturnal; they are trapped mostly at night (Haven, 1972) and are also more active at this time (Hayasaka *et al.*, 1983; Zann, 1984). Even during daylight hours, the animals live at considerable depths where light levels are low, which means that pupil size will be large (Hurley *et al.*, 1978) and visual acuity, consequently, extremely poor (Muntz and Raj, 1984). Under such conditions, their ability to discriminate form must be very limited, and it seems more likely that it is the detection of light sources and some form of orientation to them that is important.

The positive phototaxis shown by *Nautilus* is an example of such orientation

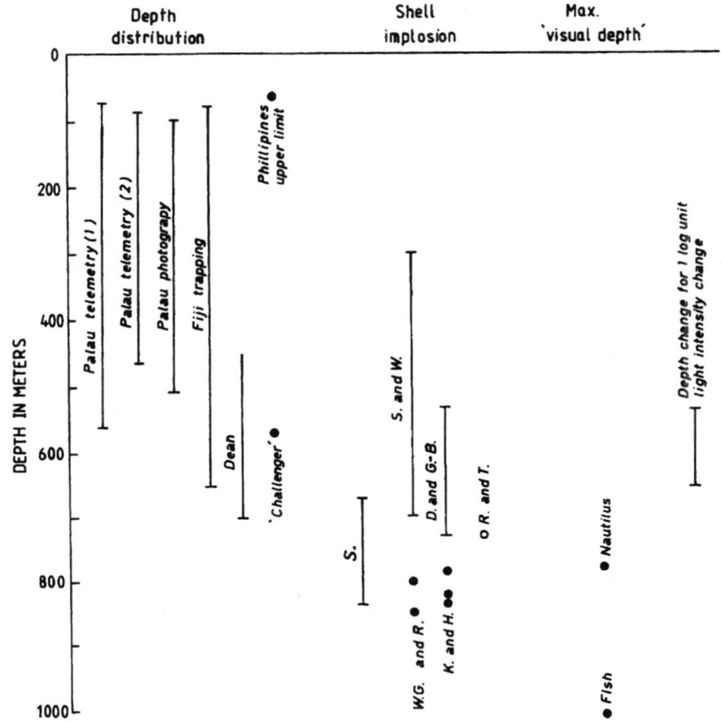

Figure 12. Depth distribution, shell implosion depths, and the maximum depth at which surface light should be visible for *Nautilus* and a typical deep-sea fish. The data for the depth distribution are from Carlson *et al.* (1984), Saunders (1984b), and Ward *et al.* (1984) (Palau telemetry and photography), Ward and Martin (1980) (Fiji trapping), Dean (1901) (Dean), Haven (1972) (Philippines upper limit), and Moseley (1879) ("Challenger"). The data for implosion depths for shells are from Saunders (1984b) (S.), Saunders and Wehman (1977) (S. and W.), Denton and Gilpin-Brown (1966) (D. and G.-B.), and Raup and Takahashi (1966) (R. and T.). The data for implosion depths of living animals are from Ward *et al.* (1980b) (W.G. and R.) and Kanie and Hattori (1983) (K. and H.). The maximum "visual depth" for *Nautilus* is from this chapter, and that for a typical deep-sea fish is from Clarke and Denton (1962). The estimate of depth change necessary for a 1 log unit light intensity change is from Dartnall (1975).

to light sources. The change from directed behavior to circling behavior and failures to respond as light intensity is reduced, as well as the directed nature of the response in the two-light situation, show that this is a true phototaxis in the terminology of Fraenkel and Gunn (1961). Finding positive phototaxis in a nocturnal animal is unexpected. However, in Palau, *Nautilus* has been shown to undergo a daily vertical migration, moving toward the surface at night (Ward et al., 1984; Carlson *et al.*, 1984). Moving toward the light could contribute to this behavior, for the animals are known to be benthic (Ward *et al.*, 1984; Zann, 1984), and if the slope of the bottom is sufficiently steep, there will be more light in the uphill direction. Provided positive phototaxis occurs predominantly at dusk, these circumstances would result in the animals moving upward over the slope at this time. The first set of phototactic experiments showed that animals can remain inactive, in which case positive phototaxis is not manifest, and the data of Zann (1984) showed a particular burst of activity in the evening. Evening activity, together with positive phototaxis, thus could result in upward movement at this time. Downward movement in the morning would remain unexplained.

With shallow slopes, such as those found off Fiji, there is no appreciable difference in light intensity facing up or down the slope, because the light distribution at depth is symmetrical. It is unlikely, however, that any appreciable daily vertical migration occurs in the Fijian population [Zahn (1984) has calculated that the animals cannot swim fast enough to achieve, in the time available, an appreciable change in depth moving up or down a shallow slope, if they remain in contact with the bottom].

Finally, it should not be forgotten that daylight is not the only source of light to which *Nautilus* is exposed: Bioluminescence may be equally important. Positive phototaxis could possibly take *Nautilus* toward areas of bioluminescent activity. It is clear, however, that at our present state of knowledge, efforts to explain the significance of the positive phototaxis of *Nautilus* must remain very speculative.

ACKNOWLEDGMENTS. I am very grateful to the Director and staff of the Institute of Marine Resources at the University of the South Pacific, Suva, Fiji, for facilities, assistance, and much helpful discussion; to J. H. S. Blaxter for reading and commenting on the manuscript; and to the Science and Engineering Research Council for financial assistance.

Chapter 16

A Possible Function of the Iris Groove of *Nautilus*

W. R. A. MUNTZ

1. Introduction

The eye of *Nautilus* has no lens or cornea. The interior of the eye is apparently in direct communication with the surrounding water by way of the open pupil. Running ventrally from the pupil, on the outside of the iris, is a groove. The groove is wide and shallow immediately beneath the pupil and becomes deeper toward the ventral edge of the iris (Fig. 1). It is lined with ciliated epithelium (Muntz and Raj, 1984). In 1865, Hensen suggested that the function of the groove might be to direct a current of water across the pupil, to prevent foreign objects from entering the eye. The observations reported in this chapter suggest that the iris groove may indeed provide some protection for the eye, though not in the manner suggested by Hensen.

2. Discussion and Results

Barber and Wright (1969) stated that the interior of the *Nautilus* eye is filled with a jellylike substance, which might reduce the need for the pupil to be protected. However, Hurley *et al.* (1978) could find no detectable difference in the refractive index of the eye contents and that of the surrounding water, and measurements made with an American Optics portable refractometer showed that this was also true in Fijian specimens. Furthermore, the eye contents of animals placed in seawater, diluted to a salinity of about 32% and having a refractive index of 1.3389, came to have the same refractive index as the surrounding water within 2 hr. This change in the eye contents may have occurred in part by mixing through the pupil, but also probably through changes in the general body fluids, since the one euryhaline cephalopod known is an osmoconformer (Hendrix et *al.*, 1981). Many specimens of *Nautilus* carry a parasitic copepod, *Anchicaligus nautili* (Wil-

W. R. A. MUNTZ • Faculty of Science, Monash University, Clayton, Victoria 3168, Australia.

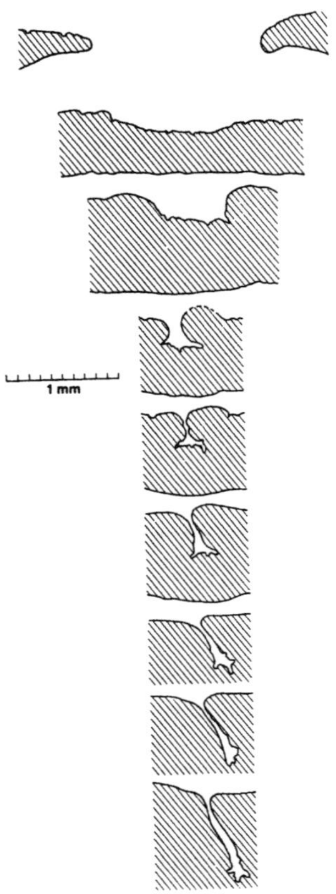

Figure 1. Sections through the iris groove taken at regular intervals between the level A (topmost section) and the level B (bottom section) shown in Fig. 2. Anterior is toward the left.

ley) (see Ho, 1980), and the need for some pupil protection is illustrated further by the finding, on one occasion, of a specimen of this copepod actually within the eye of a *Nautilus*.

To find out whether the ciliated epithelium of the iris groove behaves as Hensen suggested, the front of the eye was removed from freshly killed animals and placed in a dish of seawater, with the groove facing upward. Small amounts of diluted India ink were then pipetted onto the external surface of the iris, around the pupil. The external surface of the iris consists of columnar epithelial cells bearing microvilli, among which are many mucous cells (Muntz and Raj, 1984), and it was found that the particles of India ink became trapped in mucus and were then transported ventrally down the groove away from the pupil and eventually appeared at the ventral end of the groove. On one occasion, some isolated particles were timed moving in the groove at a speed of about 0.9 mm/min. Particles of India ink could often be seen lying over the pupil. The movement of mucus into and down the groove in this way results in a sheet of mucus being drawn across the pupil as indicated in Fig. 2.

Sheets of mucus moving over the surface are common among mollusks and are associated with epithelia that contain columnar cells bearing microvilli, mu-

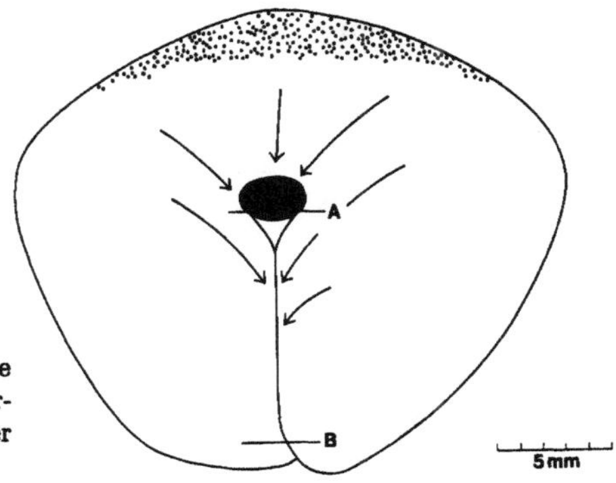

Figure 2. General appearance of the front of the eye of *Nautilus*, anterior being toward the left. The arrows show the direction of mucus movement over the surface of the iris.

cous cells, and ciliated cells, the different types not necessarily being uniformly spread (e.g., Simkiss and Wilbur, 1977; Cook and Shirbhate, 1983). Such sheets of mucus are used, for example, for feeding, cleaning, and locomotion. It seems possible that in *Nautilus*, a standard molluscan mechanism has become adapted so as to provide some protection for the eye.

Chapter 17

Histology of the Long Digital Tentacles

YOSHIO FUKUDA

1. Introduction

Nautilus differs from the coleoid cephalopods in possessing approximately 90 tentacles that lack both sucker disks and arm hooks. Although the gross anatomy and functional morphology have been described previously (Owen, 1832; Griffin, 1900; Dean, 1901; Willey, 1902; Bidder, 1962; Stenzel, 1964), ultramicroscopic structures and functional properties have been studied in detail only by Barber and Wright (1969). This chapter is a summary of the results of a histological examination of the paired long digital tentacles in two species of *Nautilus* (*N. macromphalus* Sowerby and *N. pompilius* Linnaeus).

2. Materials and Methods

This study is based on one specimen each of *N. macromphalus* and *N. pompilius*. The former was a mature female captured in 1976 off Noumea, New Caledonia, and kept in Yomiuri Marine Aquarium, Tokyo (Specimen No. 5, 640 g total body weight). The latter was a fully grown, mature female captured from Tañon Strait, the Philippines, in 1979 (Specimen No. 8-50, 830 g total body weight).

The tentacles were removed immediately after death (*N. macromphalus*) or while alive (*N. pompilius*) and were fixed in Bouin's solution. Selected portions of the paired sensory digital tentacles (Fig. 1) were dissected, embedded in paraffin, cut into thin sections, stained for histochemical examination with periodic acid–Schiff (PAS), Alcian-Blue, and Millon's technique, and studied with a light microscope. Scanning (SEM) and transmission electron-microscopic (TEM) stud-

YOSHIO FUKUDA • Laboratory of Pathology and Animal Physiology, Chiba Prefectural Institute of Public Health, Chiba 280, Japan.

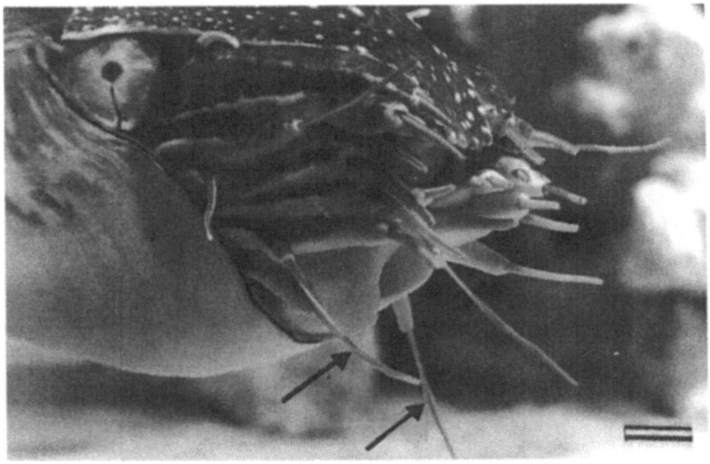

Figure 1. *Nautilus macromphalus* in an aquarium, showing food-seeking behavior with extended tentacles near the bottom. (→) Pair of long digital tentacles. Scale bar: 5 mm.

ies were also performed on selected portions with a Hitachi S-450 and a Jeol-100S electron microscope, respectively.

3. Observations

3.1. General

Nautilus has approximately 90 tentacles, classified in three groups, following Owen (1832): (1) labial tentacles, surrounding the buccal region; (2) 19 pairs of digital tentacles surrounding the labial tentacles [a pair of long, trailing tentacles belong to this group (Fig. 1)]; and (3) two thick pre- and postocular tentacles, one of each situated in front of and behind the eye, respectively (Griffin, 1900; Dean, 1901). The tentacles can be used to identify the sex of an animal: The female has 11 pairs of well-developed labial tentacles that are not accommodated in tentacular sheaths; the labial tentacles of the male are poorly developed. The labial tentacles are considered to participate mainly in reproductive behavior. The other tentacles can be retracted into tentacular sheaths. Most of the tentacles are pale brown on their outer (aboral) surface and white on their inner (oral) surface, and their outer surface exhibits many thin transverse ridges.

Each tentacle is subtriangular in cross section (Fig. 2B). The oral side is specialized as an adhesive tactile surface with many transverse ridges. Each tentacle ends in a blunt or sharp tip, where the transverse ridges are generally indistinct. Although Bidder (1962) suggested that the digital tentacles show no morphological differences, the long digital tentacles appear to serve primarily a sensory

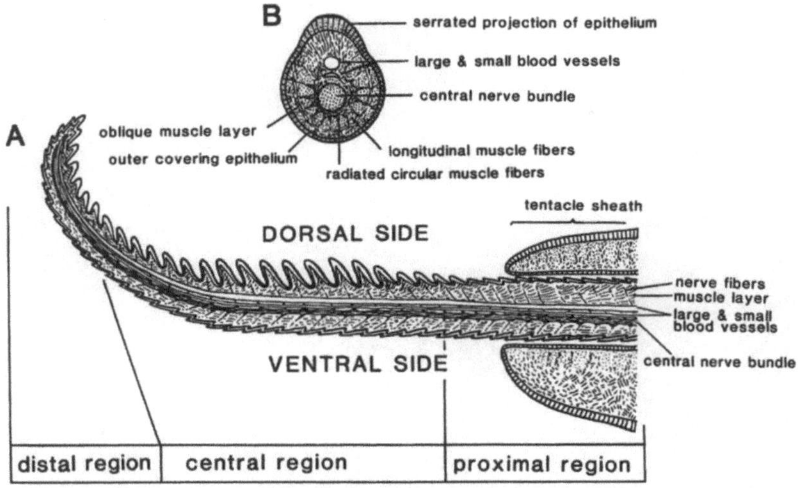

Figure 2. Structure of a long digital tentacle of *Nautilus*. Sections: (A) longitudinal; (B) transverse.

function, whereas the others seem to be used for both tactile searching and grasping.

3.2. Long Digital Tentacles

The same microstructure is observed in cross sections of the long digital tentacles in *N. macromphalus* and *N. pompilius* (Fig. 2). Many transverse ridges are present on the inner (oral) surface of the tentacles, and the ridges tend to be wider toward the distal end. The tentacles are roughly divided into distal, central, and proximal regions on the basis of the degree of development of the transverse ridges (Fig. 2A).

Scanning electron micrographs of a long digital tentacle of *N. pompilius* in cross section reveal a bundle of thick, nonmyelinated nerve fibers (cn in Fig. 3A and B) in the center of the tentacle. Two blood vessels (b_1 and b_2 in Fig. 3A and B) are present on the side of the oral surface; the one nearer the oral surface is the thicker. Light microscopy of the long digital tentacles shows that the central nerve bundle and blood vessels parallel the long axis of the tentacle (Fig. 2A). Longitudinal and radial muscle fibers (lm and rm in Figs. 3D and E) surround the central nerve bundle and blood vessels, forming the main body of the tentacle, which measures about 4 mm in diameter. The radial muscle fibers divide the longitudinal muscle fibers into many bundles (Fig. 3E). There are about four times as many longitudinal muscle fibers as radial ones; this may be advantageous for quick retraction of the tentacle into the tentacular sheath.

A longitudinal section of the oral surface of the tentacle of *N. macromphalus* shows an evenly spaced row of transverse ridges and grooves (Figs. 4 and 5). The ridges and grooves are low and shallow in the distal region, but are well developed

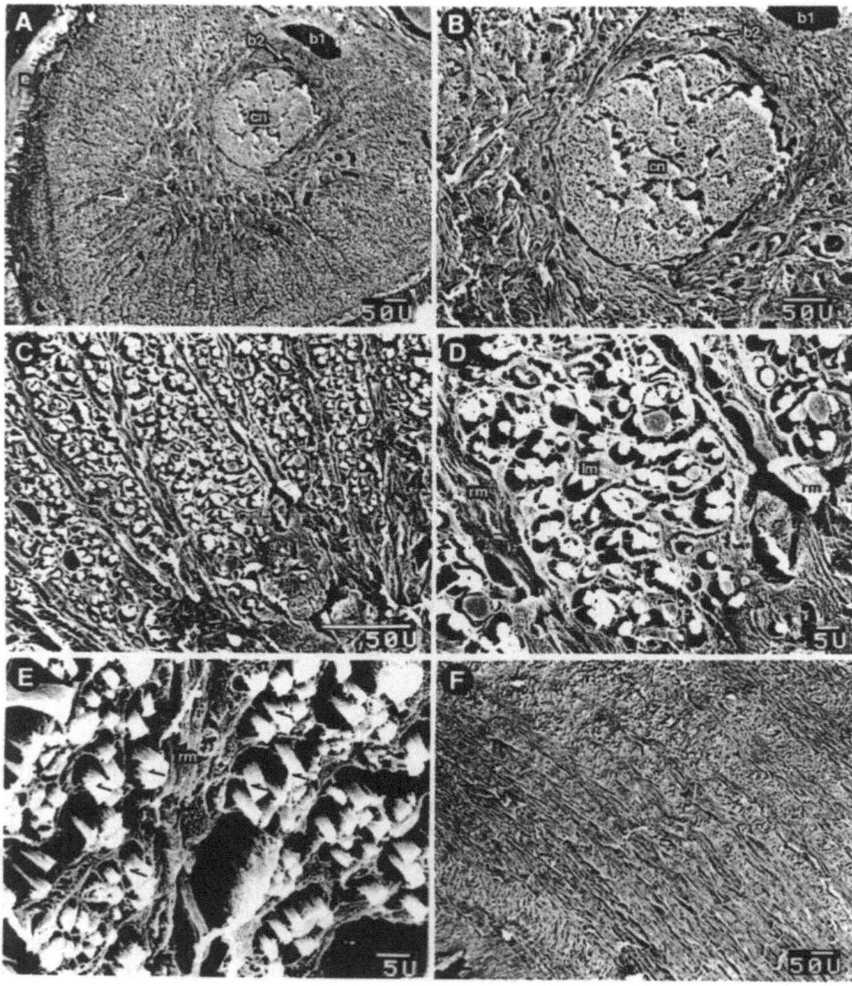

Figure 3. A long digital tentacle in *N. pompilius* under SEM. (A) Cross section in the distal region (see also Fig. 2B); (B) Part of (A), showing the broad central nerve bundle; (C,D) Network of longitudinal and radial muscle fibers; (E) Part of (C), showing many bundles of longitudinal muscle fibers (←), which are divided by radial muscle fibers (F). Key: (cn) central nerve bundle; (b1, b2) large and small blood vessels; (lm) longitudinal muscle fibers; (rm) radiated circular muscle fibers; (e) outer covering epithelium.

in the central region, where they are about 500–700 μm in height. The ridges are serrated and are slightly concave anteriorly (Fig. 4A and B). The transverse ridges gradually decrease in height proximally near or within the tentacular sheath (Fig. 2A). By contrast, the ridges on the aboral surface are low and relatively flat, showing little regional variation (Fig. 5). On the oral surface of the tentacle, the

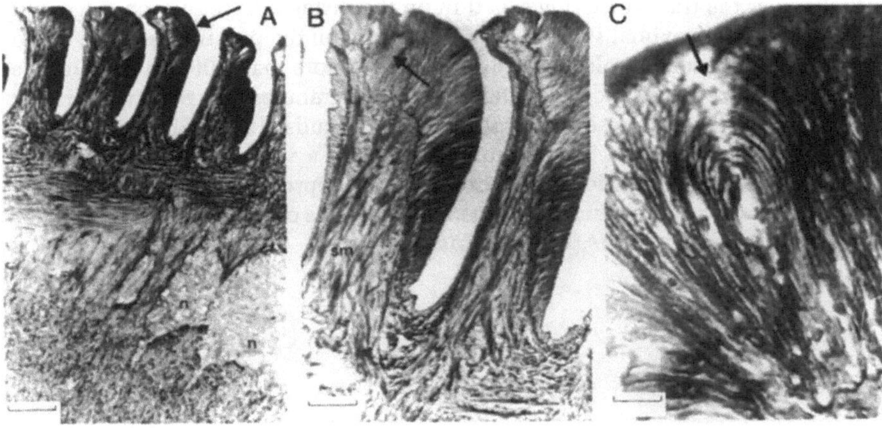

Figure 4. Transmitted light micrographs of a long digital tentacle in *N. macromphalus*. Longitudinal section; Azan stain. (A) Row of serrated transverse ridges on the oral surface of the tentacle in the central region [(n) nerve fiber bundles]; (B) enlarged view of serrated transverse ridges, consisting of basal smooth muscle fibers (sm) and high epithelial cells (e) [(←) position of a taste-bud-like putative chemoreceptor]; (C) whole view of the taste-bud-like putative chemoreceptor on the top of the serrated ridge [(→) an opening of the receptor, which is usually associated with a crypt or cavity]. Scale bars: (A) 200 μm; (B) 100 μm; (C) 25 μm.

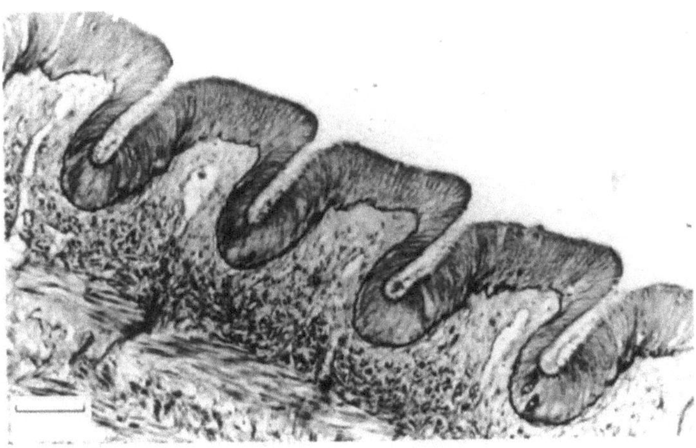

Figure 5. Transmitted light micrograph of the tilelike epithelium on the aboral (outer) surface of the long digital tentacle in *N. macromphalus*. Longitudinal section; Azan stain. No muscular element is observed in the triangular processes. Scale bar: 100 μm.

epithelium of the transverse ridges is thin on the distal surface of each ridge, but it is thick on the proximal surface. The epithelium on the proximal surface of the tentacle is composed of columnar epithelial cells more than 100 μm in height (e in Fig. 4B). The epithelial cells contain numerous granules that react moderately to PAS, suggesting the presence of glycomucus (Fukuda, 1980). Thus, these cells are likely to be secretory. The granules also show a weak positive reaction to protein stain (Millon reaction), suggesting that glycoprotein is also present. Glycoprotein has also been found in the special cutaneous organ, known as Kolliker's organ, of a juvenile octopus (Brocco *et al.*, 1974).

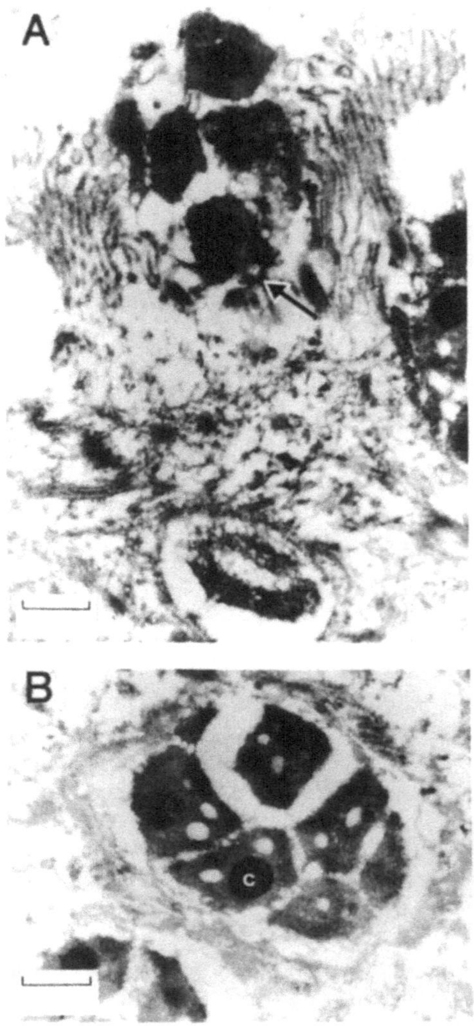

Figure 6. Free surface of the tall columnar epithelial cells, which constitute the proximal region of serrated ridges of a tentacle of *N. macromphalus*. TEM; longitudinal section. (A) Epithelial cell, which has a secretory function [electron-dense granular materials during discharge are seen among microvilli (←)]; (B) secretory granules within the epithelial cell, consisting of electron-dense central core (c) and surrounding electron-lucent vacuole. Scale bars: (A,B) 1 μm.

Under TEM observation, the outer surface of the epithelium shows a mass of granules interpreted to be in the process of discharge among the microvilli [Fig. 6A(←)]. Each granule is about 1 μm in diameter and polygonal in shape and contains an electron-dense central core (c in Fig. 6B) surrounded by an electron-lucent vacuole. Histochemical observation shows that the central core and surrounding vacuole react to PAS and protein stain. This evidence, as well as the presence of glycoprotein in the granule, indicates that the core and vacuole may correspond to protein and saccharide, respectively. Other types of granular cells containing acid glycomucus, which shows a positive reaction to Alcian-Blue, are not abundant in the epithelium on the oral surface of the tentacle. This reaction also applies to the other tentacles.

In the distal region of each long digital tentacle, a remarkable mass of taste-bud-like cells is associated with cavities in the tall columnar epithelium of the serrated transverse ridge [Fig. 4B and C(←)]. On the basis of their morphology, these cells are tentatively classified as chemoreceptors. They occur in groups in *N. macromphalus*, whereas single receptor cells are found in the labial region of octopods and squids (Graziadei, 1965a, b; Emery, 1975a,b). The cells also resemble those found on the sensory epithelium of the osphradium of the gastropod *Buccinum undatum* (Welsch and Storch, 1969, Fig. 10). They are rare on the labial tentacles around the buccal region and on other digital tentacles of *Nautilus*.

4. Discussion

In a holding tank, living *Nautilus* uses the tentacles to capture and hold food and to grasp the partner's shell during mating. During copulation, a mucous mass appears to envelop the shell. It is apparent that such abundant mucus could not have been supplied by a small number of the granular cells or by the Alcian-Blue-positive secretory cells within the epithelium on the dorsal tactile surface of the tentacles. It is likely that adhesion of the tentacles of *Nautilus* is aided by secretion of glycoprotein from the tall columnar epithelium of the oral ridges. The digital tentacles are usually retracted in the tentacular sheaths, but they are extended when necessary. The sheaths may serve to limit the waste of glycoprotein, as well as to prevent physical damage to the tentacles (Fig. 7A-B). Muscle fibers below the tall columnar epithelium on the oral tentacle surface are probably responsible for moving the transverse ridges during copulation and food-seeking. Movement of the ridges would also expose the putative chemoreceptors on the concave surface of the transverse ridges. *Nautilus* commonly swims with the pair of long digital tentacles protruding. In its natural habitat, *Nautilus* generally feeds on crustaceans, as shown by analysis of gut contents (Ward and Wicksten, 1980; Saisho and Tanabe, 1985). The putative chemoreceptor cells may aid in location of bottom-dwelling prey in dark environments. In contrast, there are few kinds of epithelial cells on the transverse ridges on the aboral surface (see Fig. 5). The density of mucus-secreting cells positive to Alcian-Blue is higher on the aboral side than on the oral side. Moreover, no muscle fibers are found on the aboral side of the tentacle, suggesting that the epithelium there is mainly protective. The

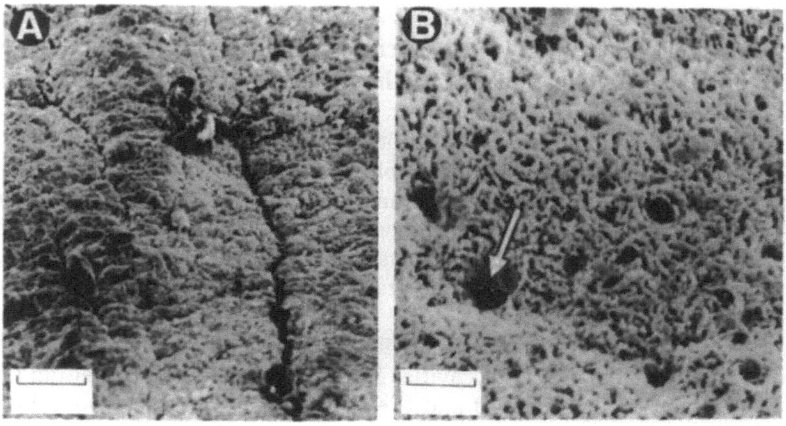

Figure 7. SEM view of the inner surface of a tentacular sheath in *N. macromphalus*. (A) Rough aboral surface of the sheath, consisting of columnar epithelium and secretory (mucous) cells; (B) Oral surface of the sheath, which faces the adhesive surface of the tentacle. This region consists of low cuboidal epithelium and secretory (mucous) cells. (⇐) An opening of a secretory gland. Scale bars: (A) 25 μm; (B) 5 μm.

triangular processes on the aboral surface of the long digital tentacles are low and stationary.

ACKNOWLEDGMENTS. I thank members of JECOLN (Japanese Expert Consultation on Living *Nautilus*) and Prof. Shozo Hayasaka of Kagoshima University for providing soft tissues of *Nautilus* for this study.

Chapter 18

The Functional Morphology of the Tentacle Musculature of *Nautilus pompilius*

WILLIAM M. KIER

1. Introduction

The morphology of the musculature of cephalopods, and indeed that of many mollusks, is characterized by a tightly packed three-dimensional arrangement of muscle fibers that lack extensive fluid-filled cavities or hardened skeletal elements. Previous research on the arms and tentacles of squids (Kier, 1982) and the arms of octopuses (Kier, 1987) suggests that the skeletal support of these appendages is provided by a type of hydrostatic skeleton that differs from the classic conception of a hydrostatic skeleton (e.g., Chapman, 1958, 1975; Clark, 1964, 1981; Clark and Cowey, 1958; Wainwright, 1970, 1982) in that the musculature both creates movement and provides skeletal support. These appendages, termed *muscular-hydrostats*, are capable of diverse, complex, and highly controlled movements (Kier and Smith, 1985). This study of the functional morphology of *Nautilus* tentacles was undertaken to explore further the diversity of muscular arrangement and function in cephalopods.

2. Materials and Methods

This study of the morphology and behavior of tentacles of *N. pompilius* Linneaus, 1758 was based on specimens maintained in captivity in the New York

WILLIAM M. KIER • Department of Zoology, Duke University, Durham, North Carolina 27706. *Present address:* Department of Biology, University of North Carolina, Chapel Hill, North Carolina 27514.

Aquarium, Brooklyn, New York, by Dr. J. Chamberlain, Brooklyn College of the City University of New York. The animals were transferred from the exhibit area to an aquarium suitable for photography, where 16 mm cine films and still photographs were made of tentacle movements during foraging and feeding. The cine films were analyzed with an NAC Inc. DF-16C 16 mm projector.

Whole tentacles from specimens that had recently died were fixed for 24–48 hr in either Bouin's fixative or phosphate-buffered 2.5% glutaraldehyde and 4% paraformaldehyde. Excellent fixation and staining were achieved, especially with the buffered glutaraldehyde and paraformaldehyde fixative. The tissue was dehydrated through a graded series of ethanols, cleared in xylene, and embedded in paraffin (melting point 56–58°C). Transverse, parasagittal, and frontal serial sections were cut at 7–10 μm with a rotary microtome. Two staining procedures were used: (1) Milligan's Trichrome Stain, with aniline blue substituted for fast green, and (2) Picro-Ponceau with Weigert hematoxylin (Humason, 1979). Both stains show strong contrast between muscle and connective tissue. The sections were studied with direct, phase-contrast; and polarized-light microscopy.

3. Results

3.1. Gross Morphology and Movements of the Tentacles

Nautilus possesses numerous tentacles, classified by Owen (1832) into three groups: (1) one pre- and one postocular tentacle located in front of and in back of each eye, respectively; (2) a variable number of labial tentacles, arrayed on lobes surrounding the buccal mass (some of which are modified to form secondary sexual structures in both male and female animals); and (3) 19 pairs of digital tentacles, surrounding and extending beyond the labial tentacles (Fig. 1). Each digital tentacle consists of an extensible, muscular cirrus enclosed in a protective sheath. The cirri of the labial and digital tentacles are equipped on one side with adhesive annular ridges. The cirri of the ocular tentacles are not adhesive and are thought to be sensory (Griffin, 1900; Willey, 1898a). The morphology and function of the digital tentacles of the chambered nautilus are described in this chapter (Fernandez, 1907; Griffin, 1900; Hamada *et al.*, 1980a; Willey, 1898b) (see also Chapter 17). Descriptions of the structure of the labial tentacles and associated secondary sexual structures of *Nautilus* are provided by Griffin (1898b), Haswell (1896), and Owen (1843); the structure of the ocular tentacles has been described by Griffin (1900) and Willey (1898a, 1902).

The sheaths of the digital tentacles (19 per side) are fused in a mass to form the cephalic sheath. Four of the digital tentacles per side are small and are directed laterally, with sheaths that open closer to the apertural margin of the shell than do the sheaths of the other, larger, and more medial digital tentacles (Fig. 1). The terminal portions ($\approx$ 1 cm) of the sheaths of the digital tentacles are not fused to the cephalic mass and are triangular to quadrangular in cross section. The sheaths taper to a blunt point, where a slitlike or oval opening is located.

The muscular cirri of the digital tentacles can be retracted completely into

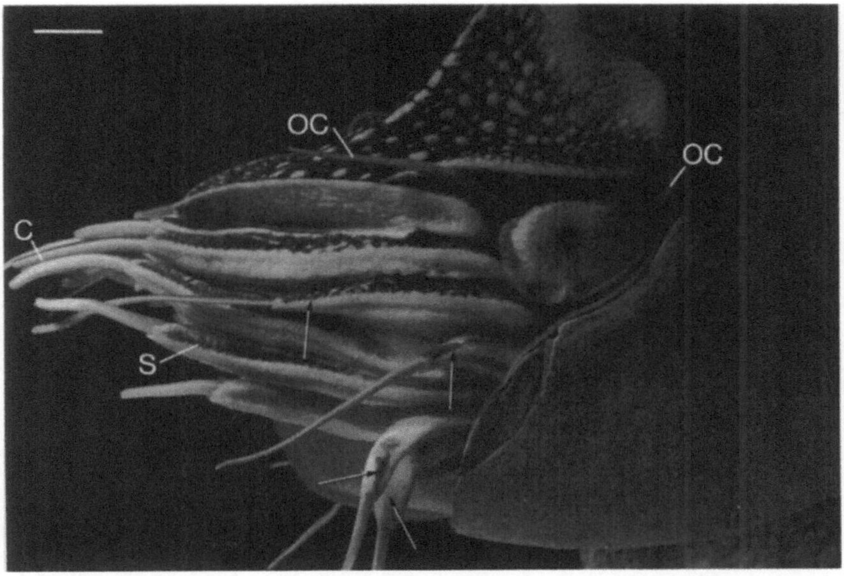

Figure 1. Photograph of live *N. pompilius*. Many of the cirri (C) of the digital tentacles are extended beyond sheaths (S). The labial tentacles are not visible in the photograph, but the ocular tentacles (OC) can be seen in front of and in back of the eye. (→) Four smaller digital tentacles. Scale bar: 1 cm.

the sheaths. The cirri are slender and slightly tapered, with a blunt, rounded tip. The distal portion of the cirrus is approximately triangular in cross section, but the proximal portion is circular. The cirri are encircled with a series of narrow, annular grooves and ridges. The ridges on the oral side of the cirrus (the side facing the mouth) are adhesive and are more pronounced than those on the two aboral faces. The annular grooves and ridges become more closely spaced and less pronounced proximally. The base of each cirrus tapers slightly to insert at the base of the sheath and at this point the tissues are continuous.

The digital tentacles are used to seize and manipulate food and to attach to surfaces. Measurements from cine films of *N. pompilius* show the cirri to be capable of extensions of 90–110% of their fully retracted length. Extension of a cirrus occurs quite slowly, often requiring 5–10 sec or longer. Once extended, the cirri are capable of bending movements in any plane. Bending can occur either sharply at one point on the cirrus or more gently, over the entire extended length. Lateral bending of the free tips of the sheaths was observed during bending of extended cirri. Twisting of the cirri along their long axes (torsion) was also observed (Fig. 2). The total amount of torsion was usually small; the maximum observed was approximately 90° (at the distal tip, relative to the base). Torsion was observed to occur in both directions and commonly occurred during bending. For additional descriptions of the use of the tentacles, see Bidder (1962).

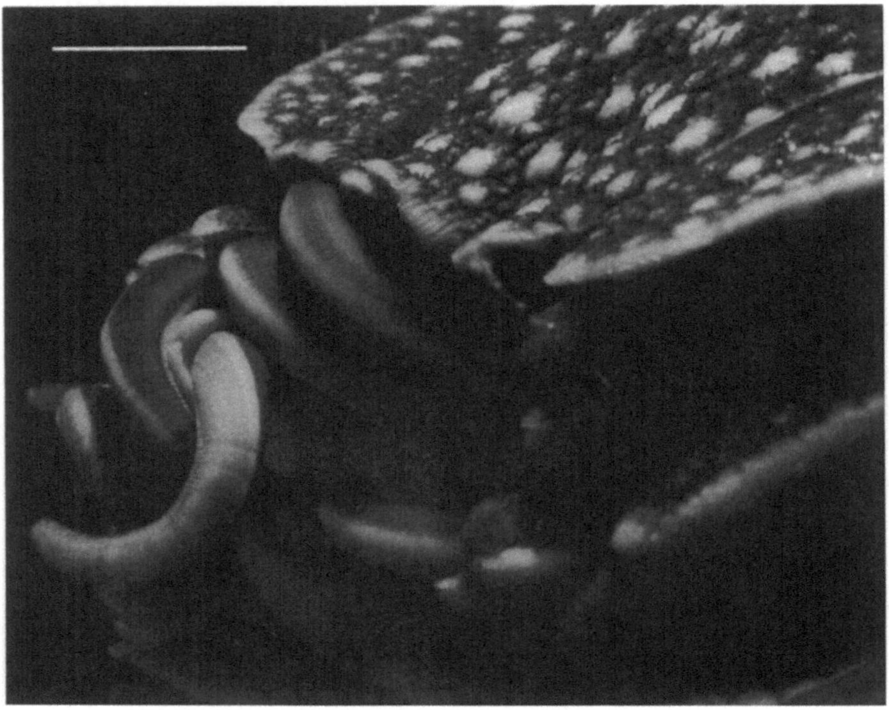

Figure 2. Close-up photograph of torsion and bending in the cirrus of a digital tentacle of a live *N. pompilius*. Scale bar: 1 cm.

3.2. Microanatomy of the Digital Tentacles

The axial nerve cord lies within the musculature and runs longitudinally down the cirrus in a subcentral location on the oral side (Figs. 3 and 4). The nerve cord is enlarged at the location of each oral adhesive ridge along the length of the cirrus (Fig. 5a). Nerves branch off laterally and orally from the axial nerve cord at the location of each oral adhesive ridge and extend within the trabeculae of the radial muscles (Fig. 5b) to the oral side of the cirrus, where they connect with a peripheral network of nervous tissue (described below). The nerve cord itself is enveloped by a thin sheath of fibrous connective tissue. Adjacent to the nerve cord on the oral side is a thick-walled artery. A larger and thinner-walled vein is located toward the outer surface on the oral side of the cirrus. Surrounding the axial nerve and occupying the core of the cirrus is a mass of radially arranged muscle fibers. The radial muscle fibers are arranged perpendicularly to the long axis of the cirrus. Radial muscle fibers extend from the central muscle mass of the cirrus to the periphery in longitudinally oriented sheets. Similar extensions of the central muscle mass of the arms of *Octopus* were termed *trabeculae* by Graziadei (1965a), and the term is therefore adopted for extensions of the radial

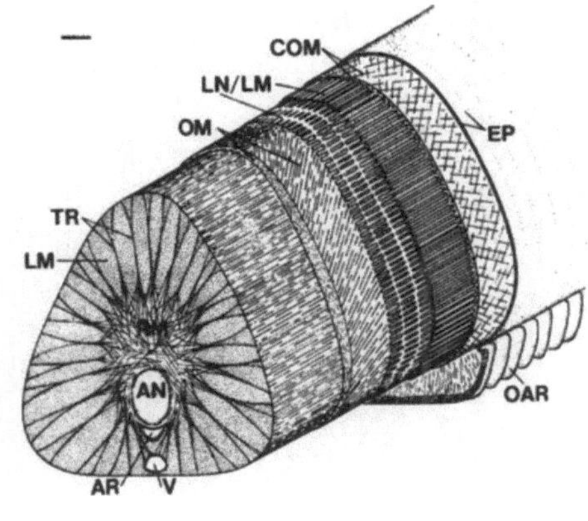

Figure 3. Schematic cutaway view of the cirrus of a digital tentacle of *N. pompilius*. Key: (AN) axial nerve cord; (AR) artery; (COM) crossed oblique muscle; (EP) epidermis; (LM) longitudinal muscle; (LN/LM) longitudinal nerve/LM network; (OAR) oral adhesive ridge; (OM) oblique muscle; (RM) radial muscle; (TR) trabeculae of radial muscle; (V) vein. Scale bar: 0.5 mm.

muscle in the cirrus of *Nautilus* (Figs. 3, 4, and 5). The trabeculae divide the surrounding longitudinal muscle into bundles and branch as they approach the periphery of the cirrus, where they insert on a layer of connective tissue immediately beneath the epidermis (see below). Both the longitudinal muscle and radial muscle are divided by and associated with a fine network of connective tissue (Figs. 5a and 6). The muscle fibers that make up the radial musculature in the center of the cirrus are not tightly packed, even when the shrinkage resulting from

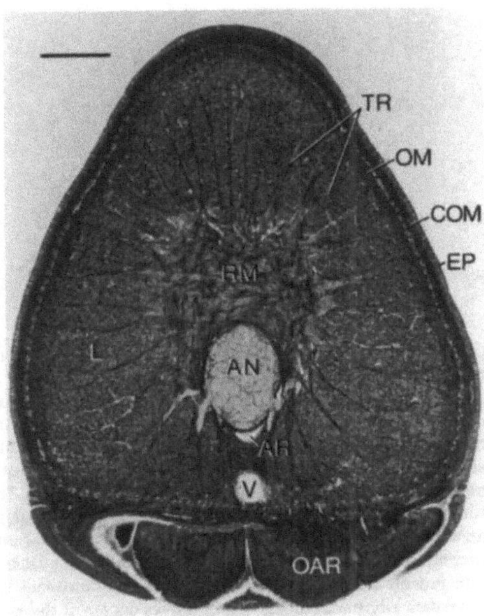

Figure 4. Micrograph of a transverse section of the cirrus of a digital tentacle of *N. pompilius*. See the Fig. 3 caption for details. Direct microscopy of paraffin section stained with Milligan's Trichrome. Scale bar: 0.5 mm.

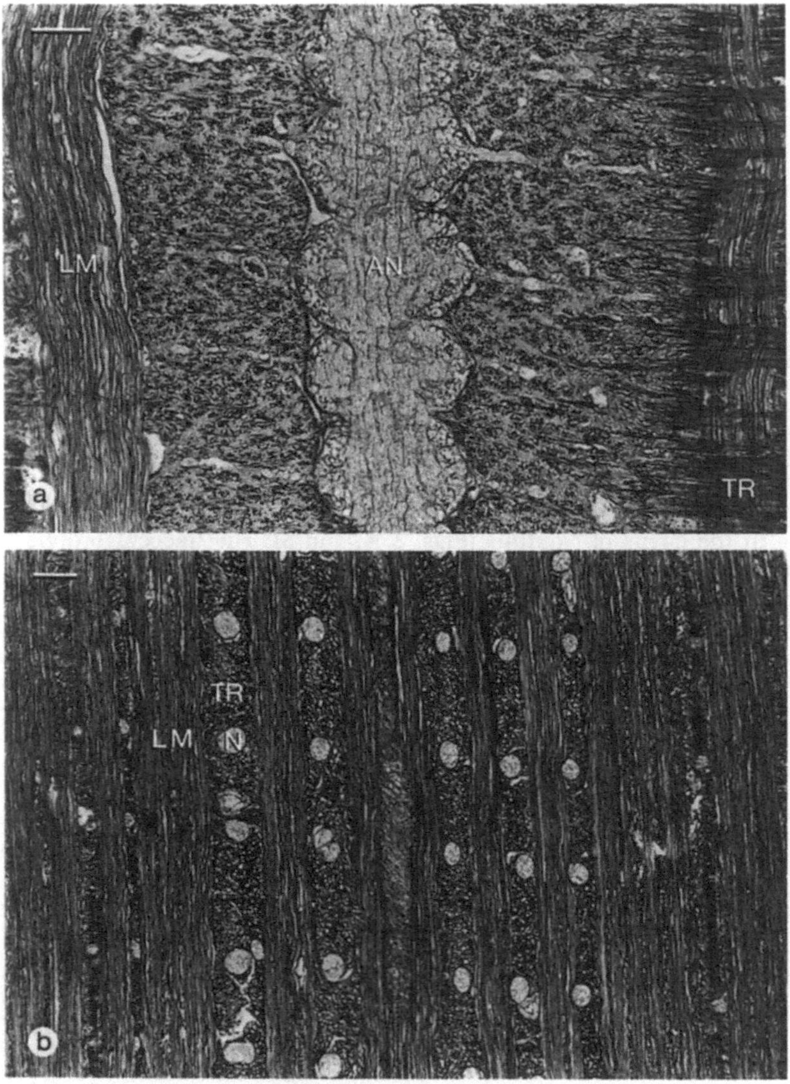

Figure 5. (a) Micrograph of a frontal section of the cirrus of a digital tentacle of *N. pompilius* at the level of the axial nerve cord. The nerve cord (AN) runs vertically in the micrograph and has greater diameter at the location of each adhesive ridge along the length of the cirrus. Surrounding the nerve cord is the radial muscle mass. Note that the radial muscle fibers, many of which are cut in cross section, are not closely packed but, instead, are separated by connective tissue (less densely stained). Longitudinal muscle (LM) is visible at the far right and far left of the micrograph. Trabeculae (TR) of the radial muscle mass are visible at right. Direct microscopy of paraffin section stained with Milligan's Trichrome. Scale bar: 100 μm. (b) Micrograph of a frontal section of the cirrus of a digital tentacle of *N. pompilius*. This section is closer to the oral side of the cirrus than the section in (a). Bundles of longitudinal muscle (LM) can be seen running vertically in the micrograph. Between the bundles of longitudinal muscle are the trabeculae (TR) of the radial muscle mass with muscle fibers in cross section. Note the numerous nerves (N) in cross section in the trabeculae. Each transverse alignment of a group of nerves occurs at the location of an oral adhesive ridge. Direct microscopy of paraffin section stained with Milligan's Trichrome. Scale bar: 100 μm.

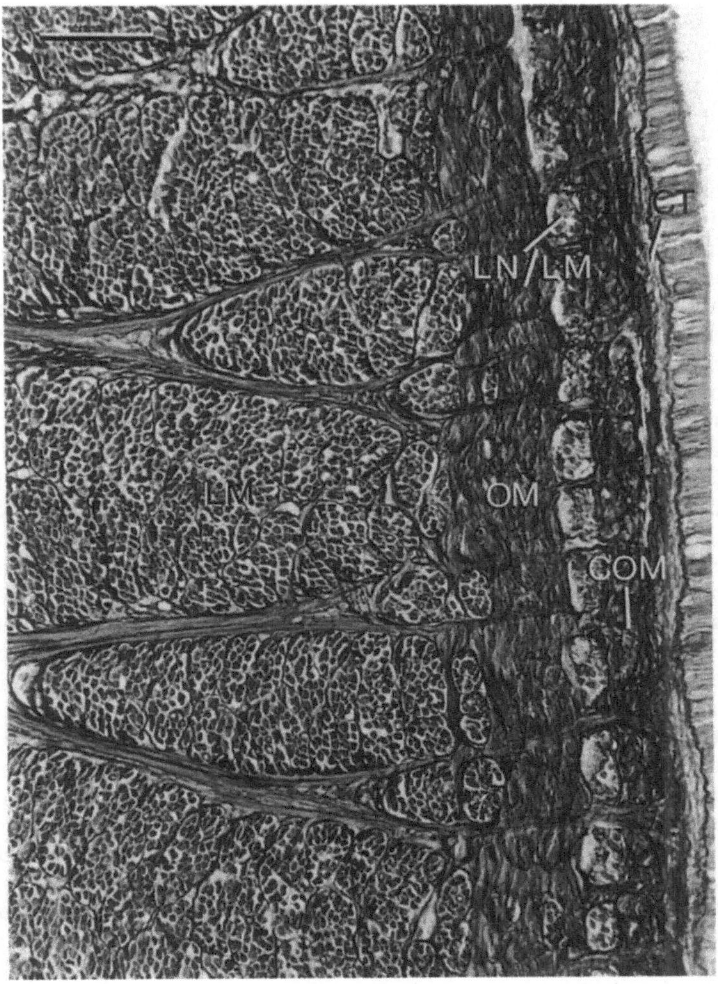

Figure 6. Micrograph of a transverse section of the cirrus of a digital tentacle of *N. pompilius*. The trabeculae of the radial muscle mass extend from the central radial muscle mass and branch to insert on connective tissue (CT) immediately beneath the epidermis. The trabeculae pass through (from left to right in the micrograph) the longitudinal muscle [(LM) cut in cross section], the oblique muscle layer [(OM) cut in oblique section], the longitudinal nerve/longitudinal muscle (LN/LM) bundle network, and finally the thin crossed oblique muscle layer (COM) underneath the connective tissue and epidermis. Note the connective tissue network in the longitudinal muscle. Direct microscopy of paraffin section stained with Milligan's Trichrome. Scale bar: 100 μm.

histological techniques is taken into account. Instead, individual muscle fibers and bundles of muscle fibers are separated from one another by the fine connective tissue network. The connective tissue is especially apparent in frontal sections such as Fig. 5a, but can also be seen in transverse sections (Fig. 6).

Enclosing the longitudinal muscle on each side of the cirrus is a layer of oblique muscle fibers (Figs. 3, 4, and 6). The oblique muscle layer is thickest laterally and thins aborally and especially orally. Parasagittal serial sections of the cirrus show that the handedness of the oblique muscle fibers on one side of the cirrus is opposite to that on the other side. If an oblique muscle fiber from either oblique muscle layer is traced from the oral side to the aboral side of the cirrus, it follows an oblique course from distal to proximal (Fig. 7a). The muscle fibers from the oblique layer on each side of the cirrus cross and interdigitate orally and aborally. No distinct connective tissue structure is present at their origin and insertion. The fiber angle of the oblique muscle layers (the angle that a muscle fiber makes with the long axis of the cirrus) ranged from 40 to 50° in the cirri examined. (However, the state of extension or retraction of the cirri was not known.)

Just outside the perimeter formed by the oblique muscle layers is an unusual array of longitudinal muscle bundles that are associated with a network of nervous tissue (Figs. 6 and 7). The longitudinal muscles of this network lie adjacent to and outside the longitudinally arrayed nervous tissue. Muscle fibers from the trabeculae of the radial muscle mass extend between the longitudinal nerve/muscle bundles. In addition to the longitudinally arrayed nerve bundles, circumferential nerve bundles connect each longitudinal nerve bundle around the periphery (Fig. 7). The circumferential nerve bundles have the same periodicity as the oral adhesive ridges. Nerve branches from the axial nerve cord connect to this peripheral array of nervous tissue at the positions of the circumferential nerve bundles.

A thin, interlaced layer of crossed oblique muscle fibers wraps the longitudinal bundles of muscle and nervous tissue (Figs. 3 and 7b). This layer was incorrectly identified by Griffin (1900) as a circular muscle layer. The crossed oblique muscle layer is wrapped by a thin layer of fibrous connective tissue that is in turn covered by a simple columnar epithelium on the aboral side of the cirrus. A large number of goblet cells are found in the epithelium. The epithelium covering the adhesive ridges on the oral side of the cirrus is different from that on the aboral surface. The epithelial cells that line the grooves between adhesive ridges are cuboidal, with large nuclei. There is an abrupt transition on the proximal side of each ridge from cuboidal to tall, columnar epithelial cells that comprise the surface of the adhesive ridge (Fig. 8). The nuclei of these cells are basal, and the terminal one third of each columnar cell is filled with intensely staining granules (red in Milligan's, black in Picro-Ponceau with hematoxylin) (see Chapter 17). The musculature of the adhesive ridges inserts on a thick basement membrane underlying the tall columnar epithelium (Fig. 8). Muscle fibers project from this insertion in trajectories that are perpendicular to the oral surface of the cirrus and converge to pass between the longitudinal nerve/muscle bundles. After passing between the longitudinal nerve/muscle bundles, the musculature of the adhesive ridge becomes incorporated into the musculature of the cirrus and is commonly observed to be continuous with the trabeculae of the radial muscle mass.

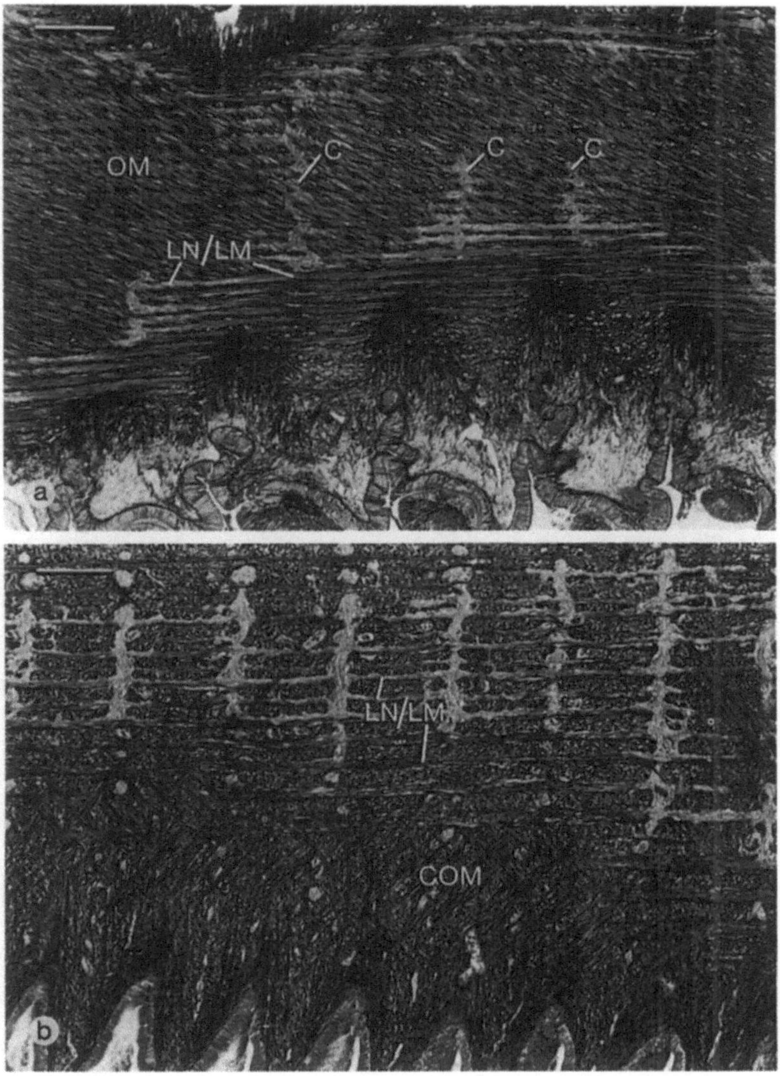

Figure 7. (a) Micrograph of a parasagittal section of the cirrus of a digital tentacle of *N. pompilius*. The long axis of the cirrus is horizontal. The oral adhesive ridges can be seen at the bottom of the micrograph. The oblique muscle layer (OM) runs across the upper part of the micrograph. The longitudinal nerve/longitudinal muscle (LN/LM) network is also visible. Circumferential nerve bundles (C) of this network have the same periodicity as the oral adhesive ridges. Direct microscopy of paraffin section stained with Milligan's Trichrome. Scale bar: 200 μm. (b) Micrograph of a frontal section of the cirrus of a digital tentacle of *N. pompilius*. The long axis of the cirrus is horizontal. The oral adhesive ridges are visible at the bottom of the micrograph. The upper portion of the micrograph is deeper into the cirrus than the lower portion. The crossed oblique muscle layer (COM) is visible. The longitudinal nerve/longitudinal muscle (LN/LM) network is visible in the upper portion of the micrograph. Note the circumferential nerve bundles of this network (oriented vertically in the micrograph). Direct microscopy of paraffin section stained with Milligan's Trichrome. Scale bar: 200 μm.

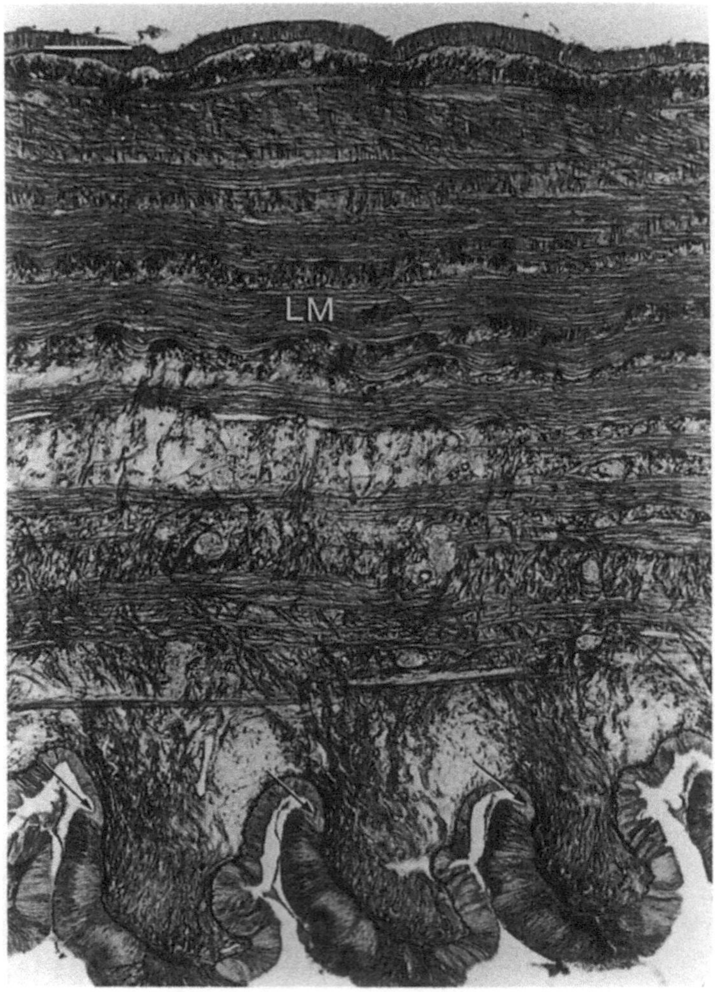

Figure 8. Micrograph of a parasagittal section of the cirrus of a digital tentacle of *N. pompilius*. The long axis of the cirrus is horizontal. The oral adhesive ridges are at the bottom of the micrograph. Note the abrupt transition in the epithelium from groove to ridge (→). The musculature of the adhesive ridges inserts on a thick basement membrane underlying the epithelium and extends up into the musculature of the cirrus. Longitudinal muscle (LM) bundles are also visible. Direct microscopy of paraffin section stained with Milligan's Trichrome. Scale bar: 200 μm.

In section, the muscle cells of the cirrus have the appearance of regular, obliquely striated muscle (terminology of Millman, 1967). Transverse sections of the muscle fibers show the cells to be made up of a central core, presumably of mitochondria, surrounded by a mass of myofilaments that show a banding pattern characteristic of obliquely striated muscle (see Kier, 1985) [regular, obliquely striated muscle is distinguished from irregular, obliquely striated muscle by the presence of a mitochondrial core in the former and the absence of such a core in the latter (Millman, 1967)]. In an ultrastructural study of *Nautilus*, Hochachka et *al*. (1978) found regular, obliquely striated muscle in cirrus number two of the spadix (a modified labial tentacle of the male), in the funnel musculature, and in the retractor muscles. Obliquely striated muscle is the most common type in cephalopods (Kawaguti and Ikemoto, 1957; Kawaguti, 1962; Cloney and Florey, 1968; Florey, 1969; Socastro, 1969; Gonzalez-Santander and Socastro Garcia-Blanco, 1972; D. V. Ward and Wainwright, 1972; Moon and Hulbert, 1975; Amsellem and Nicaise, 1980; Mommsen *et al*., 1981; Bone *et al*., 1981; Chantler, 1983; Nicaise and Amsellem, 1983; Kier, 1985).

4. Discussion

The morphology of the musculature of the cirri of *Nautilus* is similar in many respects to that of other muscular cephalopod appendages, such as the arms and tentacles of many squids (Kier, 1982) and the arms of octopuses (Kier, 1987). Therefore, discussion of the functional role of the muscular morphology of the cirri is allied with the proposals outlined previously for squids and octopuses. The most important biomechanical feature of these "muscular-hydrostats" is that they are constant in volume. The musculature and body fluids of these appendages are composed primarily of an aqueous liquid that is practically incompressible at physiological pressures. In addition, there is no evidence of flow of fluid into or out of the cirri, and no gas-filled spaces are present. In a structure of constant volume, a change in one dimension will cause a compensatory change in at least one other dimension. The following analysis of the function of the musculature of the cirri of *Nautilus* is based on this principle.

Elongation of the cirri probably results from the contraction of the radial musculature. Radial muscle contraction decreases the cross-sectional area of the cirrus, and because the cirrus is constant in volume, a decrease in cross section results in an increase in length. The displacement resulting from radial muscle contraction is amplified mechanically, because of the relationship between the diameter and the length of initially elongate, constant-volume structures like the cirri (see Kier and Smith, 1985). To illustrate the effects of this mechanical amplification, consider a constant-volume cylinder 4 mm in diameter and 80 mm long, the approximate dimensions of a retracted cirrus of one of the specimens studied. (Because the cirri can be completely retracted within the sheaths, measurements of the dimensions of retracted cirri could not be taken from the cine films of live animals and were determined by dissection of preserved specimens instead.) An increase in the length of such a cylinder by 100%, or 80 mm (a typical extension observed), is caused by a decrease in diameter of slightly less than 1.2 mm. The displacement produced by radial muscle contraction is thus amplified mechanically and is analogous to the mechanical amplification observed in support systems that use hardened skeletal elements to provide leverage.

Branching of the radial muscle bundles as they extend outward from the central, radial muscle mass of the cirrus (see Figs. 4 and 6) may be important in distributing the load produced by radial muscle contraction. Distributed loads rather than point loads are common in support systems that lack rigid skeletal elements, such as the cirrus (Wainwright *et al.*, 1976).

The longitudinal muscle is probably responsible for retraction of the cirrus. During elongation and shortening of the cirrus, the longitudinal muscle operates over a range of extension and contraction of approximately 100% of its length at rest. The operating range of retractor muscles is discussed in Kier and Smith (1985) and Kier (1987).

During elongation and shortening, the radial and longitudinal muscles operate independently, but bending movements of the cirrus probably require simultaneous activity of portions of the radial and longitudinal musculature. Bending movements are produced by contraction of longitudinal muscle on one side of the cirrus. This unilateral longitudinal muscle contraction will cause bending only if the longitudinal compressional force tending to shorten the cirrus is resisted (Kier, 1982; Kier and Smith, 1985). Because shortening increases the diameter of the cirrus, shortening due to longitudinal muscle contraction can be prevented by resisting increase in diameter. Constant diameter may be maintained by contractile activity of the radial muscles of the cirrus. Thus, bending requires simultaneous, coordinated activity of the radial and longitudinal musculature. Note that with appropriate nervous control, the same arrangement of musculature used for creating bending movements also can be used for extension and retraction.

The cirri were observed to be capable of bending movements in any plane. Such bending requires longitudinal muscle arrayed around the entire circumference, and as Fig. 4 shows, such a situation obtains in the cirrus. In addition, much of the longitudinal muscle is situated peripherally, away from the central axis of the cirrus. This location is of interest, because a greater moment arm for bending the cirrus is provided by a peripheral rather than a central arrangement of longitudinal muscle.

Torsional movements of the cirri are probably caused by contraction of the oblique musculature (Kier, 1982; Kier and Smith, 1985). The direction of torsion depends on the handedness of the contracting oblique muscle. For example, contraction of oblique muscle arranged as a right-hand helix will create counterclockwise torsion of the distal portion of the cirrus relative to the proximal portion when viewed from the proximal end of the cirrus, looking distally. Both right- and left-handed oblique muscle is present, and torsion of the cirri in either direction was observed. Figure 4 shows that the oblique musculature is situated peripherally in the cirrus, away from the central axis. This arrangement provides a larger torsional moment through which to apply torque than a more central location would.

In addition to creating a torsional force, helical or oblique muscle fibers may also create force for elongation or shortening, depending on their fiber angle. A theoretical model of a constant-volume cylinder wrapped by helically arranged, extensible fibers (Kier and Smith, 1985) predicts that in addition to creating a torsional force, oblique muscle fibers with a fiber angle greater than 54°44′ create a force for elongation, and oblique muscle fibers with a fiber angle less than 54°44′ create a force for shortening. Oblique muscle fibers with a fiber angle equal to 54°44′ create a torsional force and do not create force for elongation or shortening.

The fiber angles measured in oblique muscles of the cirri approximate 50°, suggesting that the oblique muscles contribute primarily to torsional movements and play a minor role, if any, in creating changes in length of the cirri. It should be noted, however, that the state of extension or retraction of the cirri sampled was not known. Because the fiber angle increases and decreases as the cirri shorten and elongate, respectively, it is possible that the oblique muscle could contribute to length changes of the cirri at the extremes of extension and retraction.

The arrangement of the musculature of *Nautilus* cirri is similar to that of the arms and tentacles of many squids and octopuses. The details of the arrangement are different, however, and may be correlated with differences in function. The radial musculature of the cirrus, suggested here to be responsible for elongation, is not as extensive or tightly packed as the transverse musculature of the tentacles of loliginid squid, which may be responsible for elongation (Kier, 1982). The difference in extent and packing of muscle is perhaps significant, because squid tentacles are extended rapidly and forcefully (15–35 msec), whereas the cirri of *Nautilus* are extended comparatively slowly (5–10 sec). In both appendage types, the displacement created by the radial and transverse musculature is mechanically amplified, due to the geometric relationships of these initially elongate, constant-volume structures.

The oblique musculature of the cirri is less extensive and complex than the oblique musculature of many squid and octopus arms (Kier, 1982; Kier, 1987). Squid and octopus arms are wrapped by two layers of oblique muscle, each forming complete right- and left-handed helical muscle/connective tissue systems. The oblique musculature of the cirri primarily consists of a single oblique layer in which the handedness of one side is opposite that of the other. The relatively small amount of torsion observed in the cirri compared with that in octopus and squid arms may be correlated with the less extensive oblique musculature of the cirri.

In *Nautilus* cirri, as in octopus arms, the same arrangement of muscle is capable of creating changes in length and also bending movements, with appropriate nervous control. The pattern and form of bending observed are variable, suggesting complex motor control and a complex, subdivided neural system. Little is known about sensory input and motor control of the cirri, and in this regard, the elaborate longitudinal nerve/muscle and associated circumferential connections described here deserve further study.

ACKNOWLEDGMENTS. J. A. Chamberlain of Brooklyn College, New York City, arranged for me to film live *Nautilus* and provided the specimens used for histological examination. I am grateful for his time and assistance. I thank K. K. Smith and S. A. Wainwright for discussion and advice. This work was supported in part by a Cocos Foundation Morphology Traineeship and a grant from Sigma Xi.

Chapter 19

The Circulatory System

GEORGE B. BOURNE

1. Introduction

Ever since Owen (1832) published descriptions of the soft parts of *Nautilus*, zoologists have hoped that studies of living *Nautilus* would shed light on possible modes of existence of the once-dominant nautiloids and ammonoids, as well as aid in understanding cephalopod evolution. Dating back to Willey (1895), studies of the living animal in its natural habitat have been undertaken during infrequent expeditions, and information about the animal has been gathered at scattered intervals. In this chapter, an attempt is made to discuss some of the morphological and functional features of the *Nautilus* circulatory system by drawing on several sources and by introducing some new information.

2. Update on the Anatomy of the *Nautilus* Circulatory System

Lawrence E. Griffin (1900) produced the definitive anatomical study of *N. pompilius* in 1900. He had at his disposal all the earlier anatomical studies, and in describing the circulatory system, he relied heavily on the earlier work by Arthur Willey (see Chapter 1 for a bibliography of the early works). One feature that baffled Griffin was the apparent lack of valves guarding the entrance to and exit from the ventricle; he described this state of affairs thus (Griffin, 1900, p. 180):

> I have not been able to find any valves at the openings of the vessels leading into or from the heart, except possibly the dorsal aorta. The openings are, however, tightly closed, and it is possible that at the commencement of systole the walls of the heart contract first around the openings of the branchial veins, and thus the regurgitation of blood is prevented.

Prior to their physiological experimentation, Bourne *et al.* (1978) failed to locate valves in the hearts of freshly killed *Nautilus*, but did note the constrictions

GEORGE B. BOURNE • Department of Biological Sciences, The University of Calgary, Calgary, Alberta T2N 1N4, Canada.

that Griffin had observed earlier (Bourne, unpublished observations). The published data on pressure (Bourne *et al.*, 1978) strongly suggested that a valve guards the entrance into the dorsal aorta; therefore, a new attempt was made to locate valves in the heart of *Nautilus*. Results of this effort are presented herein, along with more anatomical information based on a study of corrosion models of the *Nautilus* arterial system. Also included is a summary of isolated bits of information on circulatory anatomy that have been published since the first R/V ALPHA HELIX *Nautilus* Expedition to the Philippines in 1975 (results published in Bourne *et al.*, 1978).

3. Materials and Methods

The animals used in the dissections were captured in the Philippines in 1975 and 1979 (for details, see Bourne *et al.*, 1978; Redmond and Bourne, 1982). Freshly killed animals, which had been used for hemodynamic studies, were perfused with 10% formalin in seawater through an efferent branchial vein and then preserved in the formalin solution.

Corrosion casts were made with Batson's No. 17 corrosion compound (Polysciences, Inc., P.O. Box 4, Rydal, Pennsylvania 19046), following the directions of the manufacturer. The liquid polymer was introduced into the circulatory system via an efferent branchial cannula. After the polymer hardened, the soft parts were macerated in 34% potassium hydroxide solution, and the casts were stored in a 50% glycerin solution.

4. Results and Discussion

Photographs of the heart valves of *Nautilus* are shown in Fig. 1. Griffin (1900) found some indication of the dorsal aortic valve, but not of the others. The single large cusps that guard the entrances to the dorsal and lesser aortae are reminiscent of the aortic pocket valve in the prosobranch *Haliotis corrugata* (see Bourne, 1974). However, the valve to the *Nautilus* genital artery is bicuspid, with one very small and one very large flap. All these valves tend to be filmy in appearance, with small clumps of dead cells (probably amebocytes) accumulated around them. Due to their fragile nature, it is understandable that earlier workers missed these valves.

The four auriculo-ventricular (A–V) valves are all single, platelike cusps attached to the A-V opening for about two thirds of their circumference. Unlike the valves that guard the outflow from the ventricle, these valves are more robust, with interlacing muscle fibers continuing into the ventricular myocardium. It is possible that earlier workers simply assumed that the A-V valves were part of the ventricular myocardium.

The ultrastructure of the ventricular myocardium has been examined by Hochachka *et al.* (1978) and Dykens *et al.* (1982). Because the two groups of workers had different end purposes in mind, their transmission electron micrographs reveal slightly divergent features of *Nautilus* myocardium. Hochachka *et al.* (1978)

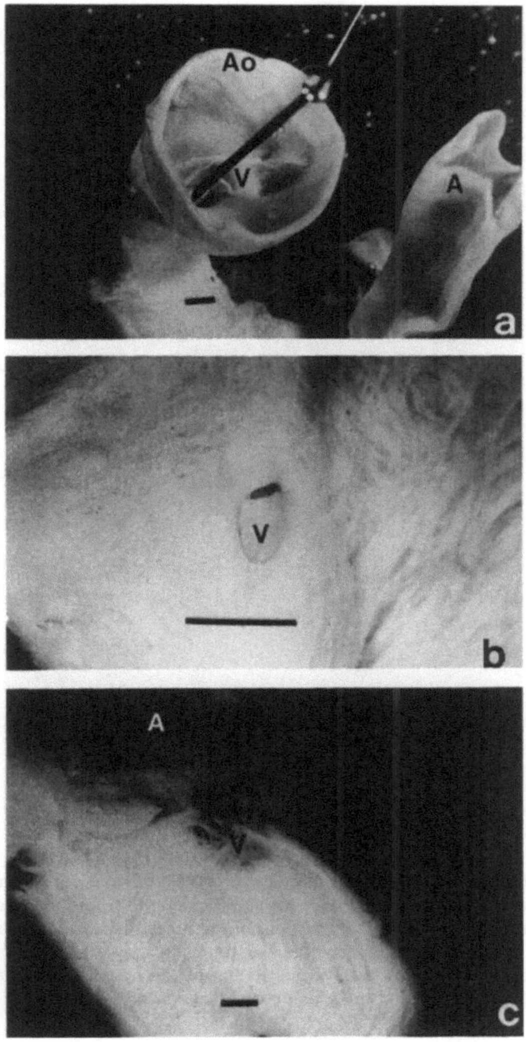

Figure 1. Photographs of valves from the heart of *N. pompilius*. (a) Dorsal aortic valve, from the cut end of the aorta; (b) valve of the gonadal artery, from the ventricular lumen; (c) A-V valve, from the ventricular lumen. Key: (A) auricle; (Ao) dorsal aorta; (v) valve. Scale bars: 5 mm.

found that the ventricle was covered with an epicardium 7–13 μm thick, which was underlain with a 14-μm connective tissue layer consisting of small muscle filaments and axons embedded in an acellular collagenous matrix. Both groups of workers found the myocardium to be composed of muscle fibers (25–55 μm in diameter) that ran in several directions, thus explaining the "almost" trabeculated appearance referred to by Griffin (1900). Dykens *et al.* (1982) felt that the overall appearance of the myocardium was strongly reminiscent of that of bivalves rather than that of decapods. They based this assessment on the ratio of myofil-

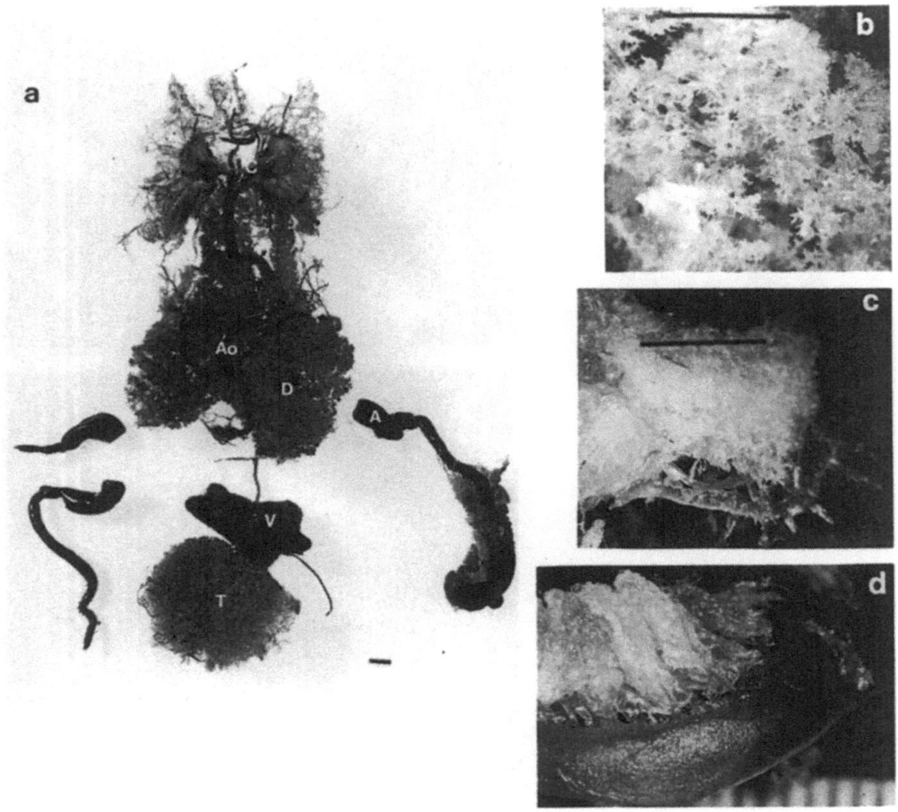

Figure 2. Photographs of a corrosion cast of the *Nautilus* arterial system. (a) Overall view—note the good penetration by the corrosion compound from the visceral region through to the head, the hood, and the tentacles; (b) close-up of the digestive gland region; (c) close-up of the cerebral cord region; (d) close-up of gill lamellae. See the Fig. 1 caption for the key; additional key: (D) digestive gland; (T) testis. Although the leaflike nature of the lamellae is apparent, the arterially introduced corrosion compound filled the gills poorly. Scale bars: 5 mm.

aments to mitochondria (60–80% myofilaments, 10–30% mitochondria), the scattering of Z bodies throughout the sarcoplasm, and the serpentine nature of the sarcolemma. Dykens *et al.* (1982) felt that neither their study nor that of Hochachka *et al.* (1978) unequivocally demonstrated a sarcoplasmic reticulum.

A second reason for the reexamination herein of the anatomy of the *Nautilus* circulatory system was to gain an understanding of the extent of the microcirculation. The *Nautilus* arterial system is now seen to terminate in a reticulated series of minute channels (Fig. 2). The torturous nature of the dorsal aorta is evident in Fig. 2. Presumably, this vessel straightens out and lengthens to some degree when the animal protracts from its shell. Gosline and Shadwick (1982) found elastic fibers arranged around the dorsal aorta that permitted considerable extension in the longitudinal as well as the radial direction. These same workers found that the fiber arrangement of decapods allowed only radial extension, a feature that has considerable importance in the mechanics of decapod circulation.

Griffin (1900) wrote of capillaries in reference to some of the small channels in the circulatory system. However, the available evidence does not reveal any minute vessels that have a structure comparable to those seen in squid (Barber and Graziadei, 1965, 1967; Kawaguti, 1970). Published transmission electron micrographs from a number of sources have been examined (Arnold and Williams-Arnold, 1978; Hochachka *et al.*, 1978; Mangum and Towle, 1982; Schipp and Martin, 1981), and in those instances in which obvious blood spaces occur, the spaces all appear to be lacunae. Thus, without further study, the circulatory system of *Nautilus* cannot be defined as either "open" or "closed."

5. Functional Attributes of the Circulatory System

The relatively deep-water habitat of *Nautilus* in the Indo-Pacific has rendered it inaccessible to a systematic physiological study. However, two scientific expeditions to the Philippines aboard the R/V ALPHA HELIX in 1975 and 1979 form the basis for much of what we know about the physiology of this organism. Most of the information presented in this section was obtained during those expeditions (for results, see Bourne *et al.*, 1978; Redmond and Bourne, 1982).

Bourne *et al.* (1978) pointed out that despite some differences in the circulatory system of *Nautilus* compared with those of coleoids, there are enough similarities to adopt the same scheme devised to discuss the functional properties of the *Octopus dofleini* circulatory system (Johansen and Martin, 1962). Seven functional elements were proposed:

1. Two types of active circulatory pumps: the systemic heart and the branchial hearts
2. A passive pump: the respiratory apparatus conveying pressure changes to the large, thin-walled elements of the vascular system
3. Propulsive vessels: vessels that actively propel blood, e.g., the branchial vessels
4. "Windkessel" vessels: the aortae and their major branches
5. Resistance vessels: the terminal portions of the arterial system
6. Exchange vessels: the capillaries and sinuses
7. Capacitance vessels: the distensible, thin-walled vessels, such as the vena cava

The systemic ventricle is the major propulsive organ in the circulatory system of *Nautilus*. Although ultrastructural evidence suggests that the *Nautilus* systemic myocardium is not as well developed as that of squid (Dykens *et al.*, 1982), a study of the enzyme profiles of the *Nautilus* ventricle indicates that its metabolism is mostly aerobic (Hochachka *et al.*, 1978). This finding implies that the ventricle beats continuously to deliver oxygen and other nutrients to the tissues. There are no reports of the *Nautilus* heart stopping spontaneously like that of several gas-

Table I. Comparison of Some Cardiovascular Data from Different Molluscan Species

Class	Temperature (°C)	Heart rate (beats/min)[a]	Aortic systolic pressure (kPa)[a]	Source
Gastropoda				
Haliotis corrugata	15	21 (16–24)	0.88 (0.74–1.05)	Bourne (1974)
Bivalvia				
Anodonta anatina	Room temperature	—	— (0.20–0.98)	Brand (1972)
Cephalopoda				
Nautilus pompilius	17	12.5 (5–18)	3.43 (0.78–7.85)	Bourne *et al.* (1978)
	—	—	— (1.60–5.89)	Gosline and Shadwick (1982)
Loligo pealei	19–22	102 (72–131)	7.16 (5.23–10.07)	Bourne (1982)
Nototodarus sloani	—	—	— (9.81–19.62)	Gosline and Shadwick (1982)
Octopus dofleini	7–9	— (8–18)	— (4.41–6.87)	Johansen and Martin (1962)
Octopus vulgaris	22	— (40–50)	3.92 (1.96–3.92)	Wells (1979)

[a] The ranges in parentheses are those found in the study reported herein.

tropods (Bourne, 1983); this phenomenon is indicative of a higher anaerobic capacity of the heart. Such a capacity has been found in the heart of the gastropod *Busycon* (Ellington, 1981).

The *Nautilus* ventricle produces a systolic pressure that averages 3.43 kPa (range 0.78–7.85 kPa) in the dorsal aorta (Bourne *et al.*, 1978). Gosline and Shadwick (1982), using stress extension curves, determined that the normal physiological pressure range in the dorsal aorta was from 1.96 to 5.89 kPa. Table I compares some aspects of ventricular performance of *Nautilus* to that of several other mollusks and demonstrates that although *Nautilus* is at the lower end of the cephalopod range, the performance of the *Nautilus* ventricle clearly exceeds that of the ventricles of mollusks from other classes.

The other active pump drives the blood through the branchial part of the circulation. Although Willey (1902) first described the pulsations of the renal appendages and pericardial glands in *Nautilus* and Naef (1913) referred to their heartlike function, textbooks and other secondary and tertiary sources view the active branchial circulation as a coleoid and not a nautiloid feature.

Bourne *et al.* (1977, 1978) clearly demonstrated that there is a significant increase in pressure of the blood as it passes from the vena cava to the afferent branchial vessels (Fig. 3b). Bourne and his co-workers found that the blood pressure in afferent branchial vessels possessed a mean value of 0.42 kPa, whereas that of both the vena cava and the efferent branchial vessel had mean values close to 0 kPa and occasionally became negative. This confirms that the pressure assist through the branchial circulation is a characteristic of cephalopods. The arrangements for pumping branchial blood and for excretion are somewhat different between nautiloids and coleoids, but clear homologies exist for these various structures (see A. W. Martin, 1975; Schipp and Martin, 1981).

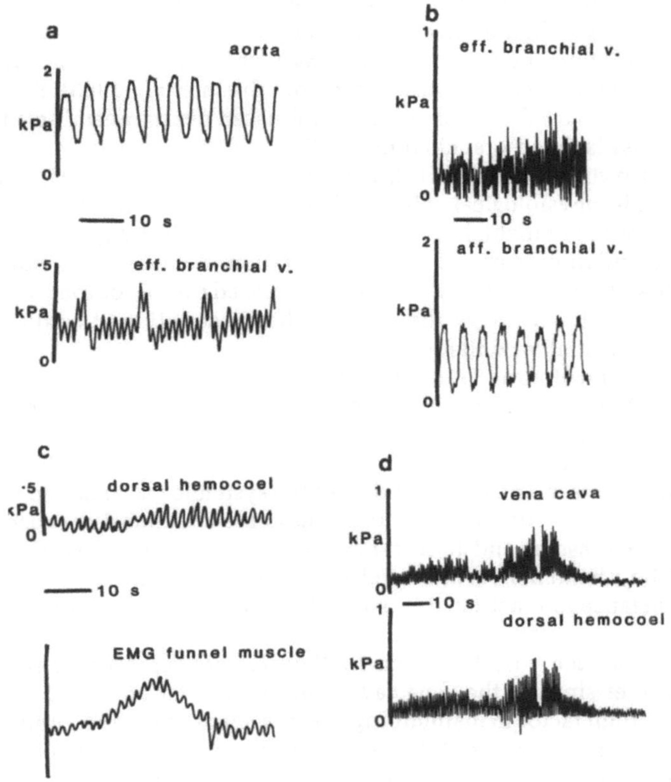

Figure 3. Pressure measurements from a number of regions of the *Nautilus* circulatory system. (a) Simultaneous recordings of blood pressure from the aorta and efferent branchial vessel; (b) simultaneous recording of blood pressure from the efferent branchial vessel and the afferent branchial vessel; (c) simultaneous recording of blood pressure from the dorsal hemocoel and an electromyogram of the right funnel muscle; (d) simultaneous recording of blood pressure from the vena cava and the dorsal hemocoel. Modified from Bourne *et al.* (1978).

Another element in the Johansen and Martin (1962) model is the "windkessel" vessels. In this model, the aorta and its major branches, coupled with their higher-resistance terminals, act as resistance-capacitance filters and convert the pressure output of the ventricle into the smoother output detected in the more distal parts of the circulatory system (for a discussion of various types of vascular system models, see McDonald, 1974). The work of Bourne *et al.* (1978) and Gosline and Shadwick (1982) supports the application of the "windkessel" model in discussing the dynamics of the *Nautilus* circulatory system.

Once blood leaves the arterial system, it enters the exchange vessels, which in *Nautilus* consist mostly of lacunae and sinuses of small diameter (as in Fig. 2; see also Griffin, 1900). There are no studies that directly investigate whether *Nautilus* can regulate flow distribution in the systemic circuit, but Redmond and

Bourne (1982) explored vascular control in the branchial circuit of *Nautilus*. They used excised gills to find a linear relationship between pressure and flow in the branchial circulation and found that this vascular bed is a "nonreactive" or "passive" bed (for a discussion of these terms, see Green *et al.*, 1963). Redmond and Bourne (1982) also tried several putative molluscan neurotransmitters to alter the response of the isolated branchial circulation. Of the four [acetylcholine, dopamine, 5-hydroxytryptamine (5-HT), and noradrenaline], only 5-HT appeared to have a possible physiological role. 5-HT, at a concentration of 10^{-9} M, caused vasodilation of the branchial circulation. The major role assigned to 5-HT in the molluscan circulation is one of cardioacceleration. Redmond and Bourne (1982) suggested that the same factors that call for increased cardiac output would require increased blood flow through the gills and hence coupling of cardiac output to the state of branchial vasodilation.

Redmond and Bourne (1982) found that dopamine mimicked the effects of 5-HT on the *Nautilus* gill, but that the gill was less sensitive to dopamine than to 5-HT. These workers also discovered that *Octopus* gill responded in a fashion similar to the gill of *Nautilus*.

Although there are no direct studies of the systemic vascular bed in *Nautilus*, Bourne *et al.* (1978) found that *Nautilus* possesses an extensive capacity to alter aortic diastolic pressure and at the same time keep the return pressure to the heart constant. The phenomenon indicated a change in the cardiac output–total peripheral resistance interaction, thus suggesting some capacity to alter the systemic vascular bed.

The capacitance vessels, which in *Nautilus* are represented by the dorsal hemocoel, other sinuses, the vena cava, and the efferent branchial veins, appear to be an important factor in maintaining effective circulation by forming a variable-volume reservoir. Bourne *et al.* (1978) found the action of the ventilatory pump on the capacitance vessels to be of great importance in maintaining effective circulation. In *Nautilus*, the ventilatory movements and a portion of the waterjet for locomotion are produced by undulating contractions of the funnel base (Bidder, 1962). More vigorous ventilation and rapid locomotion are brought about by the protraction and retraction of the head, utilizing the head muscles (Ward *et al.*, 1977; Packard *et al.*, 1980). This ventilatory action interacts rhythmically with the capacitance vessels (Fig. 3c and d) to physically couple venous return to respiratory activity and locomotion.

Although operationally different, the interaction of ventilation and circulation is similar to that seen in coleoids (Johansen and Martin, 1962; Wells, 1979; Bourne, 1982) and is analogous to the "venous pump" described in any textbook of mammalian physiology.

Ultimately, the performance of a circulatory system is measured by the ability to deliver oxygen and other metabolites to tissues. One means of evaluating this performance is to measure the cardiac output; in essence, this means measuring the amount of blood put out by the heart in unit time. Johansen *et al.* (1978), using the Fick method, calculated a cardiac output of 51.6 ml kg^{-1} min^{-1} for *Nautilus* at 17°C, a value that is at the low end of the cephalopod scale. Furthermore, Bourne *et al.* (1978) found that the contour of the flow pulse was much more mammal-like than it was gastropodlike (Bourne and Redmond, 1977).

In summing up the functional attributes of the *Nautilus* circulatory system, one is struck more by similarity to than by differences from the coleoid circulatory system. Thus, the morphological differences in the circulatory systems of these two classes of cephalopods probably represent the added requirement for the coleoid system to operate at higher levels necessary to sustain the more active life style of the latter group.

Chapter 20

The Excretory System of *Nautilus*

R. SCHIPP and A. W. MARTIN

1. Introduction

Paleontologists trace the nautiloids to an origin from septate monoplacophorans in the Late Cambrian, whereas the coleoids appear to have had a more recent origin, from belemnites. The divergence of these cephalopod lines is so far back in antiquity that major differences are to be expected in both structure and habits. Nevertheless, the groups show functional homology in their excretory apparatus; therefore, a uniform nomenclature for the pericardial and renal appendages has been adopted (Naef, 1913). The existing knowledge of excretion in *Nautilus* is based entirely on studies of two species: *N. pompilius* Linnaeus and *N. macromphalus* Sowerby, with few differences having been noted between the two and every likelihood that excretion in the other few known species will prove to function in much the same fashion.

1.1. Topography

The excretory organs of *Nautilus*, like their homologues in the coleoids (Schipp and von Boletzky, 1976), may be conceived as outfoldings of the paired

R. SCHIPP • Institute for General and Special Zoology, Justus Liebig University, D 6300 Giessen, Federal Republic of Germany. A. W. MARTIN • Department of Zoology, University of Washington, Seattle, Washington 98105.

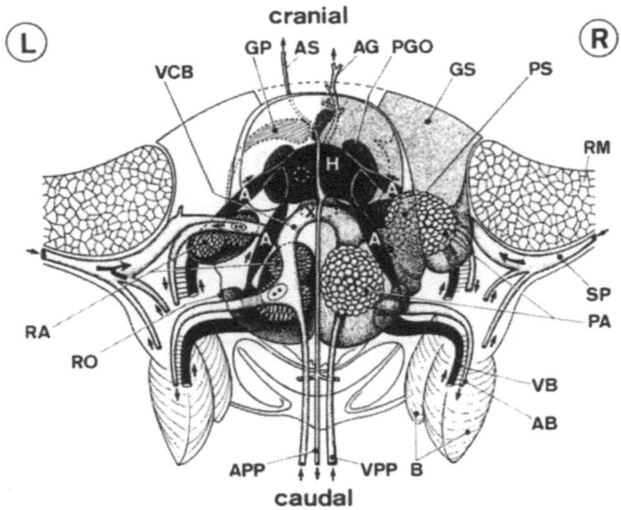

Figure 1. Dorsal view of the central circulation system and the position of the excretory organs of *Nautilus*. Anatomy: (A) atria; (AB) a. branchialis; (AG) a. genitalis; (APP, VPP) a. and v. pallialis posterior; (GS) genital septum; (H) heart with the opening of aorta ant. sin.; (PA) pericardial appendages within the pericardial coelom removed on the left side; (PS) pericardial septum with openings to the genital coelom (PGO); (RA) renal appendages; (RM) retractor muscle; (RO) left caudal opening of the renal sac; (SP) sinus pallialis; (VB) v. branchialis; (VCB) bifurcation of the v. cava with the opening of the v. cava ant. (X) branchia (B); (AS) septal artery; (GP) glandula pyriformis. Modified after Naef (1913); reprinted with permission from Schipp *et al.* (1985).

venae cavae. The pericardial appendages insert dorsally, and the renal appendages insert ventrally, on the bases of the four afferent branchial vessels (Fig. 1) (Keferstein, 1866; Vigelius, 1880; Haller, 1895; Griffin, 1900; Naef, 1913). The morphological and functional subunits of both appendages are contractile, the wave of contraction being propagated along the veins. The beat frequency is low, 9–12/min in *N. macromphalus*, in this respect conforming with that of many of the coleoids. Much of the blood in the lacunae of the organs is emptied into the veins at each beat, or it is moved back and forth between the apex and base, and in the case of the pericardial appendages, the contraction helps to move the blood through the venous system. The branchial hearts of the coleoids have considerably more muscle in proportion to the amount of glandular tissue than do the pericardial glands of *Nautilus*, and in the coleoids, a specialized filter, the branchial heart appendage, has been developed.

The specialized development of the coelomic system in *Nautilus* is of great functional importance (Haller, 1895; Naef, 1913). The four renal appendages lie in four isolated ventral coelomic cavities (renal sacs), each with its own short duct and opening to the mantle cavity. The dorsal pericardial appendages project into four coelomic areas that have open connections to each other and to the central pericardial coelom, which in turn communicates with the widespread gonadal–midgut coelomic space through paired openings in the genital septum

(Fig. 1). From the pericardial coelom, two ducts on each side drain the coelomic fluid into the seawater in the mantle cavity. Only one of each pair communicates directly with the mantle cavity; the other one joins the duct of a renal appendage to emit urine through a common pore (Haller, 1895; Griffin, 1900; Schipp et al., 1975). In *Octopus* (Harrison and Martin, 1965) and *Sepia* (Schipp and von Boletzky, 1975), a filtrate passes from the branchial heart appendages (pericardial glands) through a renopericardial canal, in which reabsorption may take place, and into renal sacs for further processing; such a circulatory route is not possible in *Nautilus*, in which, it must be emphasized, the pericardial appendages and the renal appendages are isolated into two systems. Their methods of functioning are therefore examined independently in the discussion that follows.

2. Pericardial Appendages

2.1. Structural and Histochemical Aspects

The pericardial appendages, often referred to in the older literature as pericardial glands (Haller, 1895), consist of numerous fingerlike villi (3–5 mm long and 1–1.5 mm in diameter in adult animals), giving the organs a powder-puff appearance (Keferstein, 1866). As described by Haller (1895) and Griffin (1900), each villus consists of two morphologically and functionally different parts (Fig. 2). The apex, the surface relief of which is smooth, joins a basomedial segment bordered by irregular protrusions and cryptlike invaginations; the apex generally measures one quarter to one third of the total villus length and has a terminal pore, from which deep tubular invaginations of the epithelium reach internally almost to the basomedial region. The central area of each villus contains much of the contractile substrate, a network of unmyelinated nerve fibers that serve obliquely striated muscle cells, some of which reach all the way to the apex (Figs. 2 and 3B).

In the epithelium of the basomedial segment (Figs. 2 and 3) are cuboidal cells, which are tightly interdigitated and possess S-shaped desmosomes. The folds are lined with the epithelium and probably serve to permit the considerable extension and contraction of each villus. The basal labyrinth of each cell is low and relatively free of mitochondria. The mitochondria in the epithelial cells are more numerous toward the medial region, where large groups can be found together with scattered lysosomal and peroxisomal dense bodies (Figs. 3 and 4).

The bulk of the central area of the basomedial portion of each villus is occupied by large ovoid–polygonal cells (Fig. 3), each nearly surrounded by hemolymph in the blood lacunae (Schipp and Schafer, 1969; Schipp et al., 1971; Witmer and Martin, 1973; Sundermann, 1980). The cells contain numerous dictyosomes, lysosomes, coated and uncoated vesicles, and vacuoles of various sizes, and they give evidence of high endocytotic and exocytotic activity (Schipp et al., 1975). A complete catalogue of the functions of these cells must be left for the future. Their homology to the branchial heart cells of coleoids and their similarity to the "pore cells" of lower mollusks suggest that their functions are similar to those so far described for both these kinds of cells: concentration of toxic metals

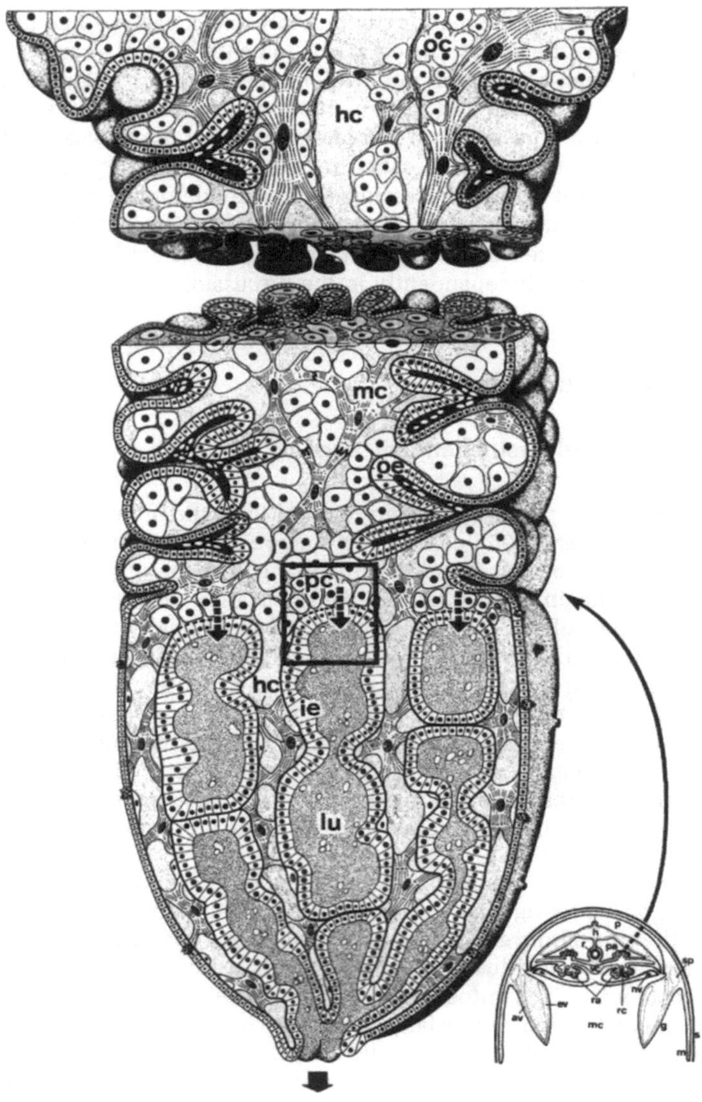

Figure 2. Schematic diagram of a pericardial villus. Anatomy: infolded epithelium in the apical (ie) and basomedial (oe) area; (hc) hemocyanin in blood lacunae; (oc) ovoid cells; (mc) muscle cells; (pc) podocytes; (lu) lumen; possible direction of ultrafiltration (--->) and secretion processes (⟶). The boxed section is seen in close-up in Fig. 5. *Inset:* Transverse section of the posterior body of *Nautilus*. Anatomy: (s) shell; (m) mantle; (mc) mantle cavity; (g) gill; (r) rectum; (h) heart; (p) pericardium; (pa, ra) pericardial and renal appendages; (rc) renal cavity; (vc) vena cava; (sp) sinus pallialis; (av, ev) afferent and efferent branchial vessel; (nv) nervus visceralis. Modified after Naef (1913); reprinted with permission from Schipp *et al.* (1985).

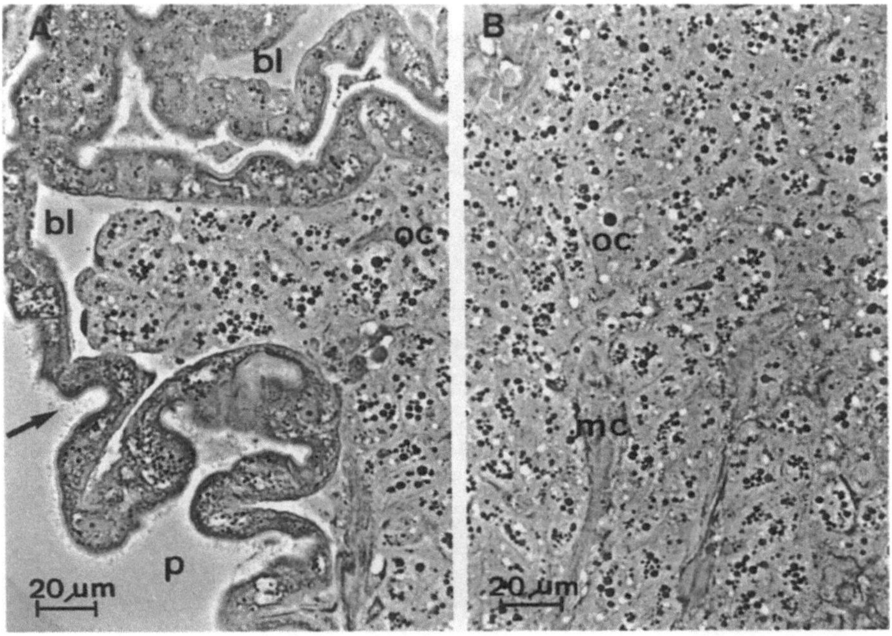

Figure 3. Infolded epithelium of the basomedial segment of a villus in *N. macromphalus* (A) and the basocentral area (B) with a large mass of ovoid–polygonal cells (oc) and single muscle cells (mc). (bl) Blood lacuna; (p) pericardium; (⟶) commensal bacterial within the brush border.

(Guary and Fowler, 1982; Simkiss and Mason, 1983), concentration of copper and iron in the dense bodies (Muzii and de Marca, 1970; J. H. Martin and Flegal, 1975; Schipp and Hevert, 1978), synthesis of adenochrome (Palumbo *et al.*, 1977) and of a dopa FeIII complex (Regneri, 1981), and phagocytotic and immunological functions (Sminia, 1980; Sundermann, 1980). Detoxification by formation of hippuric acid appears likely (A. W. Martin and Meenakshi, 1974), as does the excretion of ammonia (Potts, 1965). Storage of catabolites would be expected. The enzymatic components catalase, MAO, acid phosphatase, and β-glucuronidase, already demonstrated in the lysosomes and peroxisomes of *N. pompilius*, may be of importance (Table I).

Structures that probably initiate an important filtration process include the ovoid cells; these cells are located on the boundary between the medial and apical portions of each villus, and each possesses large foot processes similar to those of the ovoid cells that occupy the bulk of the basomedial region. The foot processes are increased on the basal surface, and the cells have become specialized as podocytes (Fig. 5). The basal lamina constitutes a fairly effective filter for hemocyanin, as in the channel cells of a gastropod (Luchtel *et al.*, 1984), because of the large size of these molecules [(mol. wt. 3.51×10^6 (Bonaventura *et al.*, 1981)]. It may be assumed that an ultrafiltrate of the blood is carried through the laby-

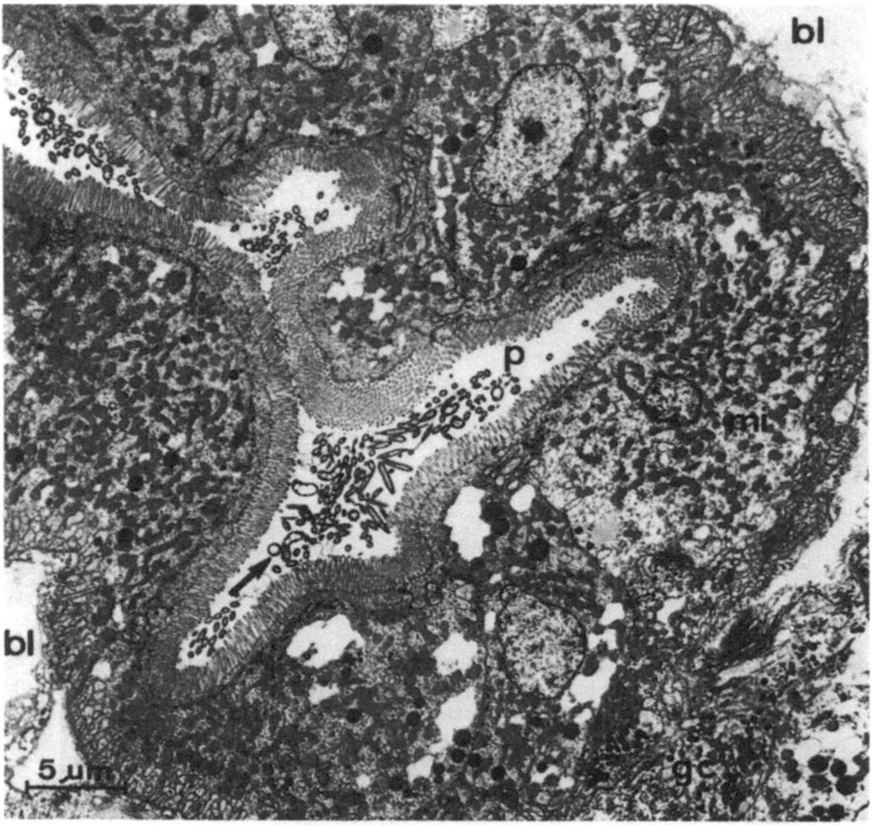

Figure 4. Transmission electron micrograph (TEM) of the infolded epithelium of the lateromedial area of a villus in *N. macromphalus*. Anatomy: (bl) blood lacunae; (p) pericardium with numerous commensal bacteria (→); (mi) mitochondria concentrated within the medioapical area of the cells; (gc) granulocyte.

rinthine canaliculi to the hemocyanin-free intercellular space between the ovoid cells and the epithelial cells of the apical infoldings.

The tip of each villus has a pore, from which deep, tubular invaginations of the epithelium reach all the way to the boundary between basomedial and apical portions. This provides a set of channels through which the filtrate formed by the podocytes can reach the perivisceral coelom. In a sense, each villus can be compared to the branchial heart–branchial heart appendage–renal coelom system of coleoids (Fig. 6). The apical portion of the villus serves not only as a conduit, but also the secretory and reabsorptive functions of the renopericardial canal or renal coelom of coleoids. To make these functions possible, each tubular invagination (Fig. 2) is bathed in blood. The epithelium of the basomedial segment

Table I. Enzymatic Activities in the Excretory Organs of *Nautilus pompilius*

		Pericardial appendages					Renal appendages	
Enzyme[a]	EC No.	Peripheral epithelium	Epithelium of the basomedial infoldings	Epithelium of the apical infoldings	Polygonal cells	Muscle cells	Epithelium of the tubules	Authors of the methods
MDH	1.1.1.37	++	++	–	(+)	+	++	Lojda (1965)
SDH	1.3.99.1	+++	+++	+/(+)	(+)	+++	++	Lojda (1965)
D-AAO	1.4.3.3	+/(+)	+	+	(+)?	0?	+	Shnitka and Talibi (1971)
MAO	1.4.3.4	+	++	–	+	(+)	+++	Glenner *et al.* (1957)
Catalase	1.11.1.6	(+)/+	(+)/+	(+)/+	+/++	0	+	Graham and Karnovsky (1966)
GOT	2.6.1.1	(+)	(+)	(+)	0	0	(+)	Kishino (1968)
AChE	3.1.1.7	+	+	(+)	(+)	++ Nerve fiber +++	+?(+)	Karnovsky and Roots (1964)
Al.phos.	3.1.3.1	+	+	(+)	0	0	+	Pearse (1961)
Ac.phos.	3.1.3.2	++	+	++	+	+	+++	Pearse (1961)
β-Gluc.	3.2.1.31	(+)/+	+	(+)/+	(+)	Blood cells 0	++	Seligmann *et al.* (1954)
LAP	3.4.11.2	0	0	0	0	0	0	Nachlas *et al.* (1960)
Mg^{2+}-ATPase	3.6.1.4	++	++	++	(+)	++	++	Ernst (1972b)
Na^+/K^+-ATPase	3.6.1.3	+	+	+	(+)	–	+	
CAH	4.2.1.1	+++	+++	–	–	–	+++	Meijer and Bloem (1966)

[a] Enzymes: (MDH) malate dehydrogenase; (SDH) succinate dehydrogenase; (D-AAO) D-amino acid oxidase; (MAO) monoamine oxidase; (GOT) glutamic oxalacetic transaminase; (AChE) acetylcholinesterase; (Al.phos.) alkaline phosphatase; (Ac.phos.) acid phosphatase; (β-Gluc.) β-glucuronidase; (LAP) lactic acid peroxidase; (CAH) carbonic anhydrase.

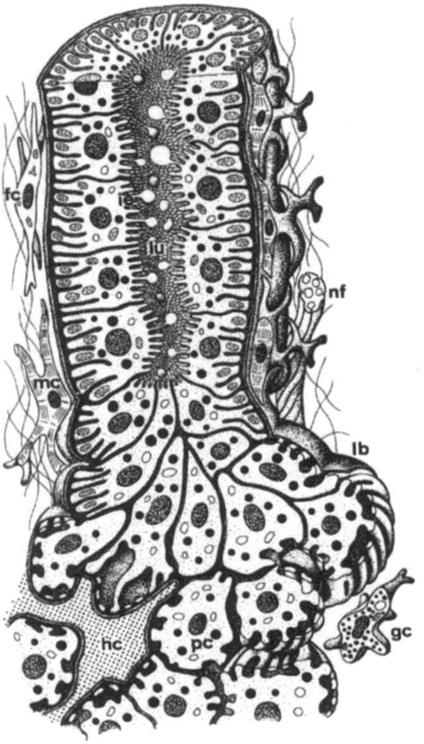

Figure 5. Diagrammatic reconstruction of the contact area between podocytes (pc) and the epithelium (ie) of the apical infoldings. Anatomy: (lb) lamina basalis of the right side partly removed; (hc) hemocyanin; (fc) fibrocyte; (mc) myocyte; (gc) granulocyte; (nf) nerve fiber; (lu) lumen. Reprinted with permission from Schipp *et al.* (1985).

has been characterized as low and relatively free of mitochondria. In contrast, the cells of the epithelial of the apical infoldings are high (Figs. 2, 5, and 7), and numerous mitochondria lie in the deeply interdigitated basal labyrinth. Beneath a high microvillous border are numerous dense bodies. The enzymes of these cells, demonstrated histochemically, are listed in Table I. The pattern of distribution differs from that of the enzymes in the basomedial epithelium cells, in accordance with the different distribution of the mitochondria. In the infolded apical epithelium, the oxidoreductases (MDH and SDH), as well as the Mg^{2+}-ATPase, occupy a basal position in the cells, whereas in the basomedial epithelium, they are more concentrated in the apical cell area. MAO and catalase, as well as the hydrolases (acid phosphatase and β-glucuronidase), reflect the positions of the dense bodies. Both basomedial and apical epithelial cells show the presence of Na^{+}/K^{+}-ATPase (Ernst, 1972a,b), as well as CAH, in the basal labyrinth (Tables I and II). These can also be demonstrated cytologically, as was CAH, in the microvillous region of both epithelia.

It is to be expected that an excretory product will emerge as a result of filtration and intense chemical activity. Indeed, the cells show terminally enlarged vacuolar areas of microvilli and apical exocytosis of vesicles. It has long been known that a secretion forms in the lumen and is transported via the apical pore

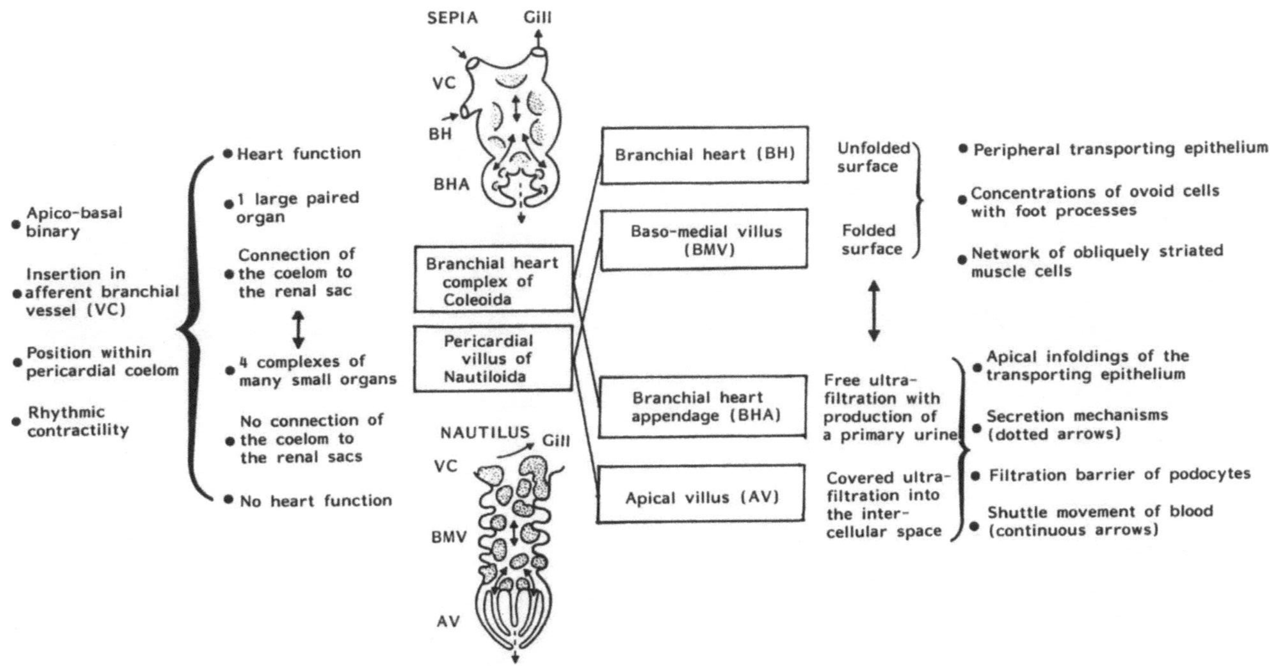

Figure 6. Structural and functional comparison of the branchial heart complex of Coleoida and the villus of the pericardial appendages in Nautiloida.

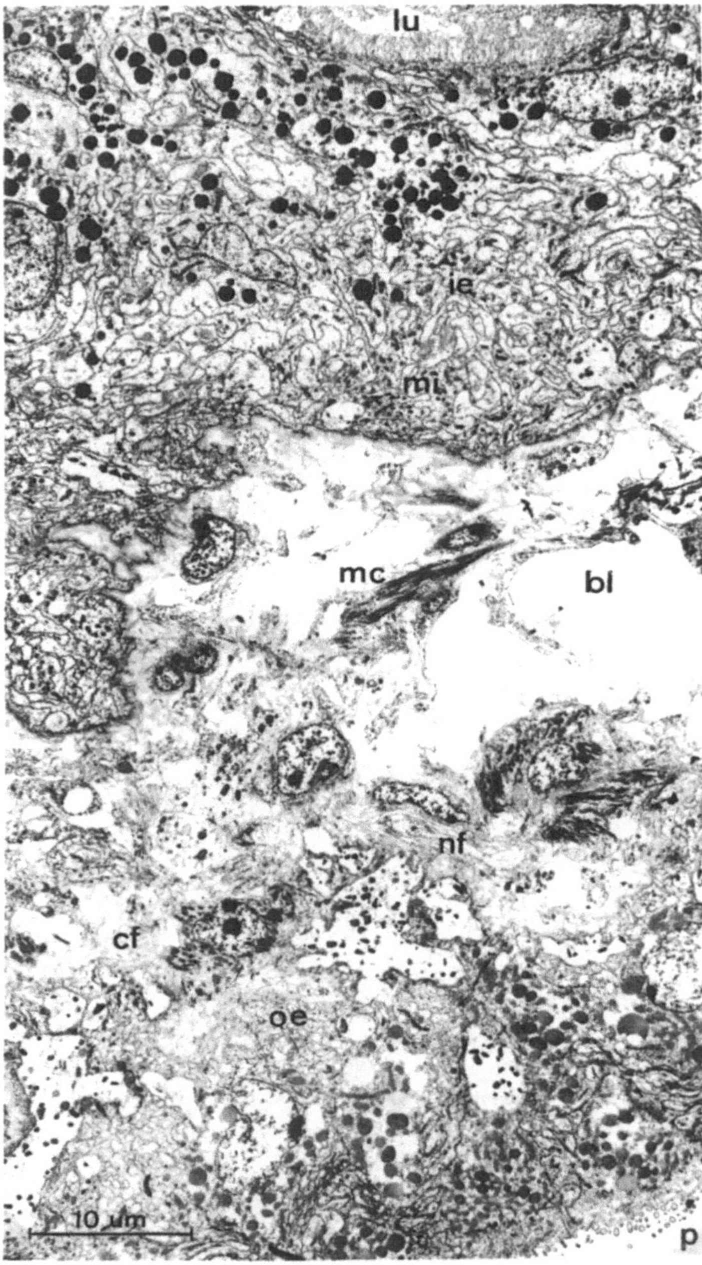

Figure 7. TEM of the lateroapical area of a villus in *N. macromphalus*. Anatomy: (lu) lumen of an apical infolding; (p) pericardium; (ie, oe) inner and outer epithelium; (bl) blood lacuna with muscle cells (mc); (cf, nf) collagenous fiber and nerve fiber; (mi) mitochondria.

Table II. Cytochemical Reaction of Na^+/K^+-ATPase (Method of Ernst) in the Pericardial Appendages (Middle Area of Villus) of *Nautilus macromphalus*

Incubation media (mM)	a	b	c	d	e	f
Tris-HCl buffer (pH 9)	100	100	100	100	100	100
Nitrophenyl phosphate	5	—	5	5	5	—
KCl	10	10	—	10	10	10
$MgCl_2$	10	10	10	—	10	10
$SrCl_2$	20	20	20	20	20	20
Na-β-glycerophosphate	—	—	—	—	—	20
Ouabain	—	—	—	—	0	—
Localization	Relative intensity of precipitations					
Folded epithelium						
Microvilli	(+)	(+)	(+)	0	(+)	+
Lateral interdigitations	+	(+)	(+)	0	0	(+)
Outer mitochondrial membrane	+	(+)	(+)	(+)	(+)	(+)
Basal invaginations of plasma membrane	+	0	(+)	(+)	0	+
Lamina basalis	++	++	++	++	+	++
Ovoid cells						
Plasma membrane	+	(+)	+	(+)	(+)	+
Vesicles	(+)	(+)	(+)	(+)	(+)	+
Dense bodies	0	0	0	0	0	0
Plasma membrane of muscle cells	+	(+)	(+)	(+)	(+)	+

into the pericardial coelom. Keferstein (1866, p. 1391), following a study of fixed specimens, described the product as a "fatty-like mass" and Haller (1895, p. 196), as a "hard, putty-like mass." Fixation and dehydration may have been responsible for the consistency reported.

In completion of the analogy to the entire coleoid system, we may visualize filtration across the podocytes followed by passage of the fluid either intercellularly or through the epithelial cells that line the tubules, through the tubules to the coelom, and thence to the exterior. Because the columnar infoldings are bathed in hemolymph, every opportunity is available for reabsorption of metabolites from the filtrate and for addition of metabolic wastes to the fluid. In this way, the pericardial appendages have been enabled to take on many of the excretory duties ordinarily fulfilled by the renal appendages of coleoids and leave to the renal appendages of *Nautilus* a prime involvement in calcium metabolism.

2.2. Commensal Bacteria

It is of interest to note the presence of flagellated bacteria along the pericardial epithelia of both *N. macromphalus* and *N. pompilius*, except in the tubules of the apical portion of each villus. They may have close contact with the glycocalyx, apparently anchored to the microvilli with their flagellae, but may lie as free groups in the lateral invaginations (see Figs. 3A and 4). Like the dicyemid mesozoans in the renal sacs of coleoids, these microorganisms occupy a niche relatively high in ammonia concentration. The question remains unsolved, as it does

Table III. Osmotic Measurements in Body Fluids of *Nautilus macromphalus*

Animal no.	Medium	Osmolality (measured by freezing point depression)				Oncotic pressure (measured with a membrane of 70,000 daltons)				Protein content (g/liter)
		N	Mean (mOsmole)	SD	Variance (%)	N	Mean (mmH_2O)	SD	Variance (%)	
	Seawater	6	1026.7	+ 5	0.5	—	—	—	—	—
V	Blood BW[b] 560 g	5	1204 (1092)[a]	± 12.2	1	5	19.9	± 0.7	3.7	67.6
VI	Blood BW 480 g	4	1188.33 (1078)[a]	± 2.5	0.21	5	22.8	± 1.3	5.7	67.4
VII	Blood BW 423 g	4	1196.8 (1086)[a]	± 13.1	1.0	6	21.9	± 2.5	11.5	69.0
VI	Renal sac fluid	4	1014.8	± 11.1	1.1	—	—	—	—	—
II	Chamber fluid BW 512 g	9	2236.9	± 106.7	4.8	—	—	—	—	—
VII	Chamber fluid	6	698.3	± 23.2	3.3	—	—	—	—	—

[a] Mathematically corrected values, expected using vapor pressure osmometry.
[b] BW, body weight.

for the mesozoans (Lapan and Morowitz, 1972; Lapan, 1975), as to whether they are simply commensals or whether they serve some function in the tract. In any case, it is instructive that they are not in the renal appendages, as in coleoids, but in the pericardial glands.

2.3. Physiological Data

Experimental studies to elucidate the mechanisms of urine formation in the pericardial appendages of *Nautilus* are only beginning. Recent osmotic measurements (Table III) show that the blood is isotonic to ambient seawater. Exact measurements of total osmolality of the blood by the freezing point method are hindered by the high protein content (Mangum and Johansen, 1975). The blood oncotic pressure approximates 20 mm of water, slightly under that of *Octopus* and *Sepia*. However, if one compares this value with the mean hydrostatic pressure in the afferent branchial vessels, 42 mm of water (Bourne *et al.*, 1978), it may be seen that pressure filtration for urine formation is possible.

The function of ammonia excretion in *Nautilus* is centered in the pericardial appendages, in contradistinction to the coleoids, where it is a function of the renal appendages. As shown in Table IV, the total ammonia concentration (by the Berthelot method) in pericardial fluid is significantly higher than in blood or in the contents of the renal sacs. Enzymes are available for deamination and ammonia production. As shown in Table I, MAO and D-AAO are available, as is GOT, located in the epithelial cells. Lee (1970) has suggested an active process of exocytosis of vesicles to transport the ammonia; Potts (1965) has shown that in *Octopus dofleini*, the transport is pH-dependent. In accordance with the Henderson–Hasselbach equation, in the tissues at pH 7.4 and a pK′ of $NH_3/NH^+{}_4$ of 9.25, nearly 1% of the ammonia is present as NH_3, which diffuses across cell membranes much more freely than the charged cation. At an acid pH, established normally or artificially in the renal sac of *O. dofleini*, ammonia was trapped as the ammonium ion up to 100 times the blood level of ammonia. If the contents

Table IV. Ammonia Concentrations (ppm) of Some Body Fluid Samples of *Nautilus macromphalus*

Animal no.	Seawater	Blood	Fluid: Renal sac	Fluid: Pericardial coelom
I	—	14.8	46.4	248.0
II	5.2	8.0	5.4	14.0
II	—	—	6.8	13.2
IV	13.6	14.0	15.2	270.0
V	2.7	32.4	9.6	22.0
V	—	—	—	20.8
VI	1.8	9.2	48.4	19.2
VI	—	—	—	23.0
VII	0.6	16.0	—	—
$\bar{X}$ ± S.E.M.	4.78 ± 5.2	15.7 ± 8.8	21.8 ± 20.2	78.8 ± 111.4

of the renal sac were made alkaline, only a small amount of ammonia was excreted by the kidneys. We do not know the pH of the fluid in the perivisceral coelom of *Nautilus*, and the structural differences between coleoids and *Nautilus* are such that a choice between these processes cannot be made at present. An Na^+–$NH^+{}_4$ exchange, mediated by the basolaterally localized Na^+/K^+-ATPase, cannot be excluded (Mangum and Towle, 1977; Mallery, 1983).

3. Renal Appendages

3.1. Structural and Histochemical Aspects

As shown above, the four renal appendage complexes originate from the afferent branchial vessels close to the pericardial appendages, but each in its own, isolated coelomic derivative, a renal sac. Each sac then opens only into the mantle cavity (Keferstein, 1866; Griffin, 1900; Naef, 1913). If any fluid is to pass out through these short channels, it must originate from the renal appendages or be seawater that has penetrated into the sacs. Morphologically, the appendages are vein-wall-derived lacunal extrusions into the coelomic sacs that surround them (Figs. 8 and 9) (Schipp and Martin, 1981). From a relatively smooth surface epithelium, deep tubular invaginations extend inward to fill the structure, but do not include the lumen of the vein. There is no endothelium, so the blood bathes the basement membrane and internal faces of the epithelial cells (Figs. 8 and 10). There are no podocytes, nor even any of the ovoid–polygonal cells so characteristic of the pericardial appendages. The appendage must be able to empty itself of blood, and to meet this need, there is a wide-meshed network of obliquely striated muscle cells accompanied by single nerve fibers and, finally, a very sparse network of collagenous fibrils that make contact with the basal lamina of the tubule epithelium (Fig. 11). It may be assumed that the muscular contractions coordinated with the contractions of the veins not only exchange the blood, but also promote the movement of the spherical concrements along and out of the tubules (Figs. 9B, 11, and 12).

The epithelial cells alone, then, must be responsible for all the secretory processes. They are of an active cell type, with basal infoldings and basolateral interdigitations, a microvillous border, and a high content of mitochondria and partly periodic acid–Schiff (PAS)-positive dense bodies (Figs. 8, 10, and 11). The cells of the surface epithelium appear to be slightly flatter and have somewhat less content of the organelles mentioned, but they cannot be considered to be of a different cell type.

There may be functional significance in the cytological resemblance of these cells to the siphuncular epithelial cells in the same species. Both epithelia show exocytotic processes and blebs at the apices of the enlarged microvilli. The mitochondria are not accumulated along basal or apical boundaries, but are spread throughout the entire cell. The cells possess very long, straight, basoapical endoplasmic reticulum channels and relatively broad intercellular spaces in the area of the lateral and basal interdigitations (Fig. 10). Apparently these spaces can be widened or narrowed, as can those in the emptying phase of the siphuncle

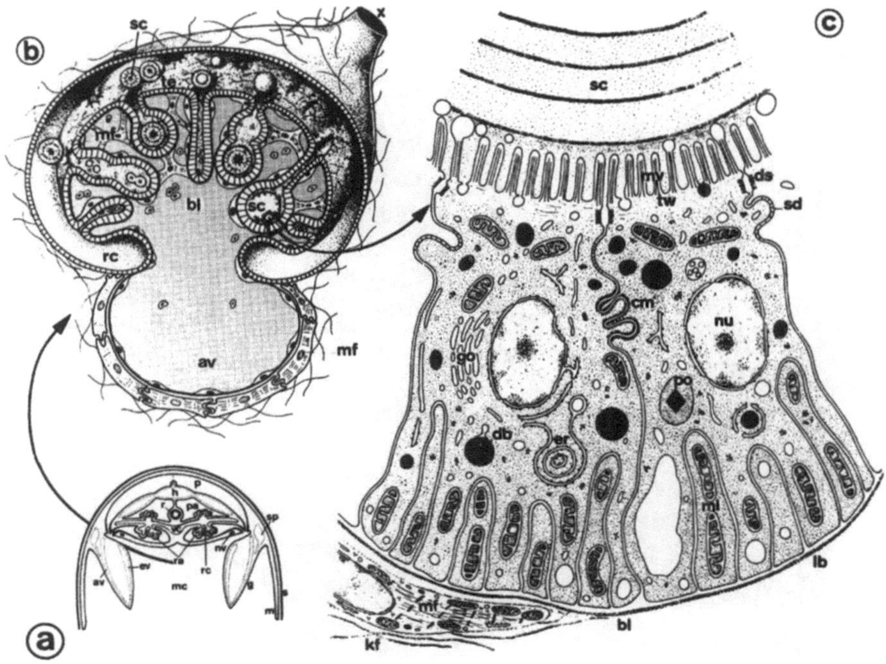

Figure 8. (a) Topography of the renal appendages (ra) in a schematic cross section through the posterior portion of the body of *Nautilus*. Anatomy: (s) shell; (m) mantle; (g) gill; (r) rectum; (h) heart; (p) pericardium; (pa) pericardial appendages; (pc) renal cavity with opening into the mantle cavity (mc); (vc) vena cava; (sp) sinus pallialis; (av) afferent branchial vessel; (ev) efferent branchial vessel; (nv) nervus visceralis. Modified after Naef (1913). (b) Section of (a), showing renal appendages in a three-dimensional reconstruction [turned 180° and sectioned perpendicular to the plane of (a)]. Anatomy: (av) afferent branchial vessel; (rc) renal cavity; see the key in (c) for the remaining abbreviations. Section of (b), showing the transporting epithelium (te), spherical concrements (sc), muscle fibers (mf), collagenous fibers (kf), nucleus (nu), microvilli (mv), basal lamina (lb), terminal web (tw), desmosome (ds), septate desmosome (sd), coated membrane (cm), dense bodies (db), peroxisome (po), endoplasmic reticulum (er), Golgi system (go), mitochondria (mi), blood lacunae (bl), and opening of the v. cava ant. (x). Modified after Schipp and Martin (1981).

epithelium (Greenwald *et al.*, 1982). Variations in the width of the intercellular spaces occur from specimen to specimen, despite the use of the same method of fixation, an observation similar to the findings on the kidney sac epithelium of the pulmonate *Helisoma*, after the animal was treated with extracts of the visceral ganglion (H. R. Kahn and Saleuddin, 1979). According to the classification of this type of epithelium (Berridge and Oschman, 1972), the epithelium may be capable of ion transport not only from cell base to apex, but also in the reverse direction. Such bidirectional transport has been shown for the siphuncular epithelium of *Nautilus* (Ward and Greenwald, 1982).

In the intensity and distribution of enzymatic activities as studied by cytochemistry, the renal epithelium does not show marked differences from the epithelia of pericardial appendages (Table I). There is high MAO, catalase, acid phosphatase, and β-glucuronidase activity, which can be assigned to the numer-

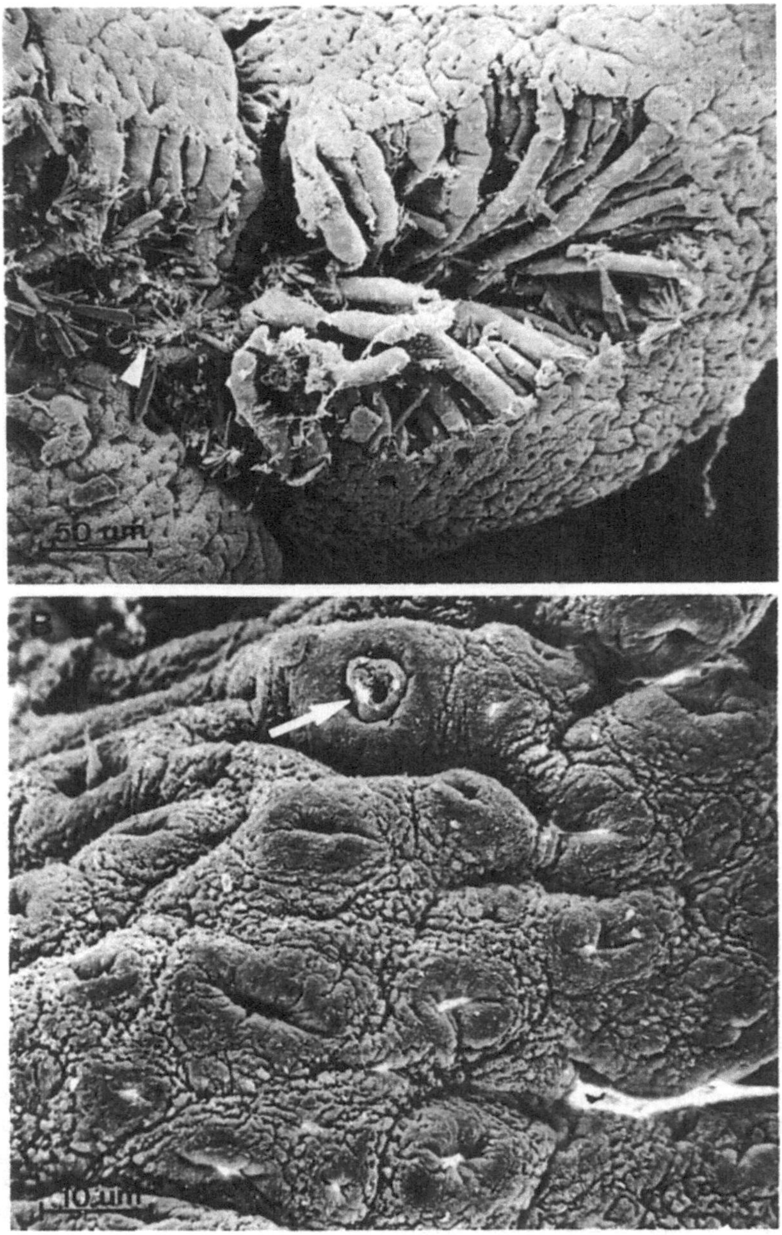

Figure 9. Scanning electron micrograph (SEM) of a partially opened lobus of the renal appendages of *N. macromphalus*, showing the rough surface of the openings of the tubules (A), needles of aragonite (▷), and (B) protruding concrement (⟹).

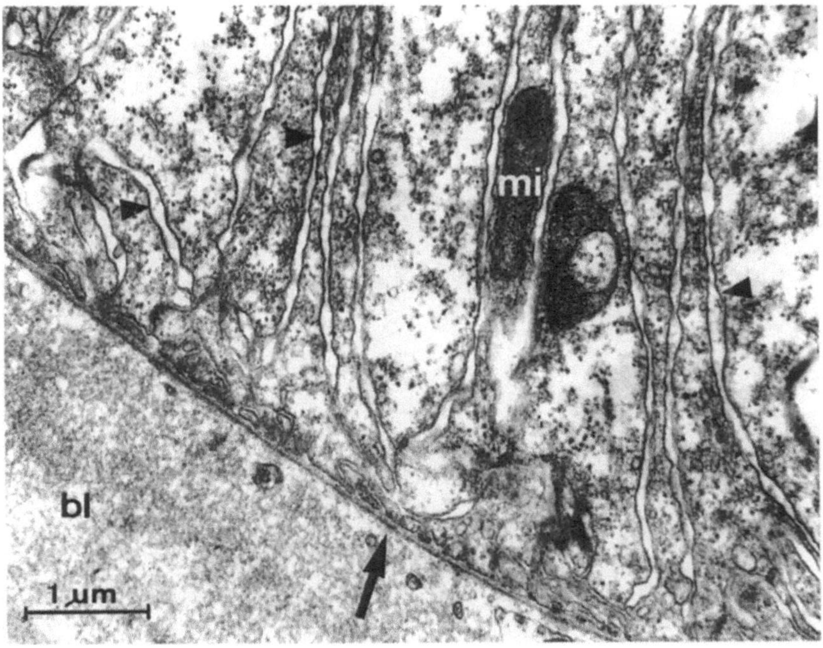

Figure 10. TEM of the basal area of renal epithelium. Anatomy: (bl) blood lacunae; (→) lamina basalis; (▶) enlarged channels of the intercellular spaces; (mi) mitochondria.

ous lysosomes and peroxisomes (Schipp and Martin, 1981), as well as high activity of CAH located in the plasmalemma of the cell base and apex (Fig. 13), and high activity of Na+/K+ ATPase located in the area of the basolateral infoldings (Schipp, 1983, unpublished data).

3.2. Morphogenesis and Structure of *Nautilus* Concrements

Griffin (1900, p. 165) stated: "The renal sacs are often completely filled with a gritty substance, like fine sand. Sometimes it is white, sometimes a faint rose-pink in color. It is composed of rounded grains, formed by numerous concentric layers . . ." (Figs. 11, 12, 14). Electron micrographs show terminal blebs or vacuoles at the tips of the microvilli of epthelial cells (Figs. 14a and b). These blebs appear to grow rapidly in size, and in larger, neighboring concrements, mineral and organic substances are layered alternately at the PAS- and ruthenium-red-positive glycocalyces (Schipp and Martin, 1981). As the granules pass down the tubules, they increase in size as layers of proteoglycan and mineral continue to be deposited until they are expelled from the tubules. The process of movement

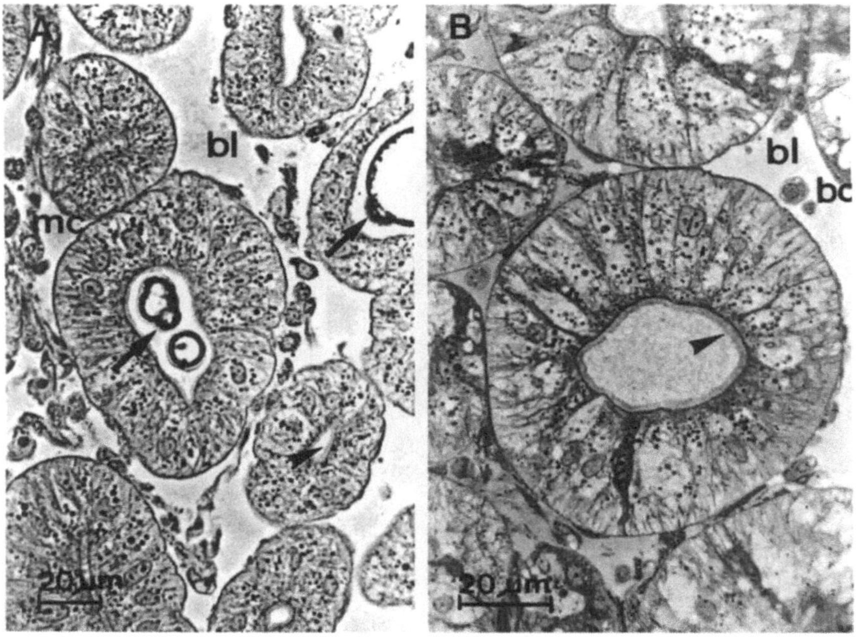

Figure 11. Sections of renal tubules of *N. macromphalus* (A) and *N. pompilius* (B) in phase-contrast. Anatomy: (bl) blood lacunae with blood cells (bc) and muscle cells (mc); (⟶) spherical concrements; (▶) brush border of the epithelium.

and expulsion (Fig. 9b) is probably assisted by the regular contractions of each villus, which may help to retain a spherical form in the concrements.

Qualitative X-ray microanalysis confirms the presence of Ca, Mg, and P in the concrements of *N. macromphalus* (Fig. 15) and *N. pompilius* (McConnell and Ward, 1978). The S and Na demonstrated in similar concentrically layered structures of the renal appendages of *Sepia* (Schipp *et al.*, 1975) are not present. A small amount of calicum carbonate may also be present and, under certain conditions, form needles of aragonite, as seen in scanning electron micrographs (Figs. 9 and 16). Such crystals have been described in the calcification process of growing and regenerating shells of the mussel *Amblema* (Petit, 1977; Petit *et al.*, 1980; Watabe, 1981).

3.3. Quantity and Inorganic Composition of the Concrements

Keferstein (1866) drew attention to the high content of concrements in the renal sacs of *N. pompilius* in 1866. Griffin (1900) reported 3.28 g in one specimen, and Schipp and Martin (1981) found more than 5 g in the renal sacs of one *N. pompilius*. The shell and calcitic jaw surface in *Nautilus* are almost pure calcium

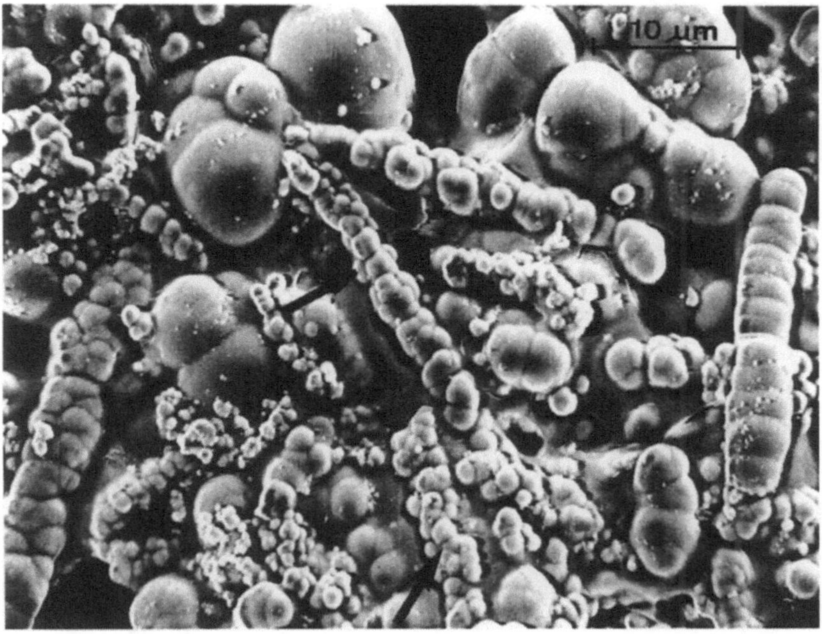

Figure 12. Spherical concrements of *N. macromphalus* in an SEM. The chains of concrements (→) represent positive copies of the lumina of the tubules.

carbonate (Lowenstam *et al.*, 1984). The concrements, as noted by Keferstein (1866), contain mostly Ca and Mg phosphate. R. E. Crick *et al.* (1985), reporting on concrements from 4 specimens of *N. pompilius* analyzed by X-ray, found magnesium oxalate dihydrate and hydroxyapatite, with a preponderance of Mg over Ca. The amounts were (ppm): Sr, 3200; Mg, 188,000; and Ca, 51,000. We report here on concrement samples taken from 5 *N. macromphalus* in New Caledonia. There was a small predominance of Mg over Ca in 2 of the 5 animals; the other 3 showed a slight predominance of Ca. The means and standard deviations were (moles/kg): Ca, 2.58 ± 0.16; Mg, 3.51 ± 1.33; and P, 6.14 ± 0.10. Some oxalate, ammonium, sulfate, and carbonate were present. In 15 specimens of *N. pompilius*, freshly caught in the Philippine Islands, the concrements were recovered quantitatively, except for those in the renal appendage tubules. The amount per animal ranged from 0.081 to 3.70 g with a mean of 0.79 ± 1.09 g. Despite the diversity in amount, there was a striking similarity in composition, the means (where N = 15) being Ca, 5.20 ± 0.25; Mg, 3.05 ± 0.24; and P, 5.89 ± 0.30 moles/kg. This computes roughly to a mixture of $MgNH_4PO_4$ and $Ca_3(PO_4)_2$, but with insufficient P for both salts, confirming the presence of some other anions. Lowenstam *et al.* (1984) reported values for the uroliths of *N. belauensis* that suggest a predominance of Ca. Values for Sr, Mg, and P were, respectively (wt.%): about 0.4, 5.6, and 54.7, and the assumption may be made that of the P, the remainder was bound

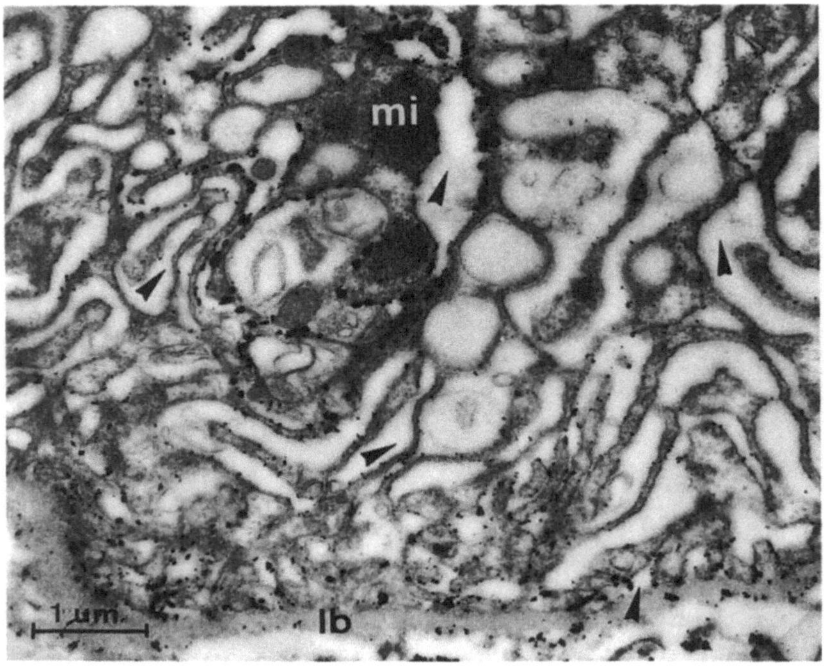

Figure 13. Cytochemical localization of the CAH in the basal infoldings of the renal epithelium of *N. macromphalus*. Anatomy: (lb) lamina basalis; (mi) mitochondria; (▶) enlarged intercellular spaces.

to Ca. From these reports, we suggest, as a middle ground, that the ordinary concrements contain about the same amounts of Ca and Mg. This is true of some other mollusks with large accumulations of nonshell minerals, e.g., the nudibranch *Archidoris brittanica* (McCance and Masters, 1937).

3.4. Function Hypotheses

The earliest hypothesis of the function of the renal appendages, and the simplest, viewed these organs as merely eliminating excess mineral substances. This view was expressed both by Haller (1895) and by Griffin (1900). *Nautilus* ingests large amounts of Ca. Growth time to maturity has been reported as being from about 2 years to nearly 30 years (Chapter 28). Animals need not move far, and they grow more quickly, if there is sufficient food (Ward, 1985), but they may move long distances and grow very slowly (Saunders, 1983; Saunders and Spinosa, 1979). Perhaps, in the absence of sufficient food, they will eat crustacean exuviae, or coral or bite off pieces of neighbors' shell margins (Ward and Wicksten, 1980; Magnier, personal communication). The large amounts of Ca, Mg, and perhaps P ingested might necessitate a provision for some efficient form of excretion.

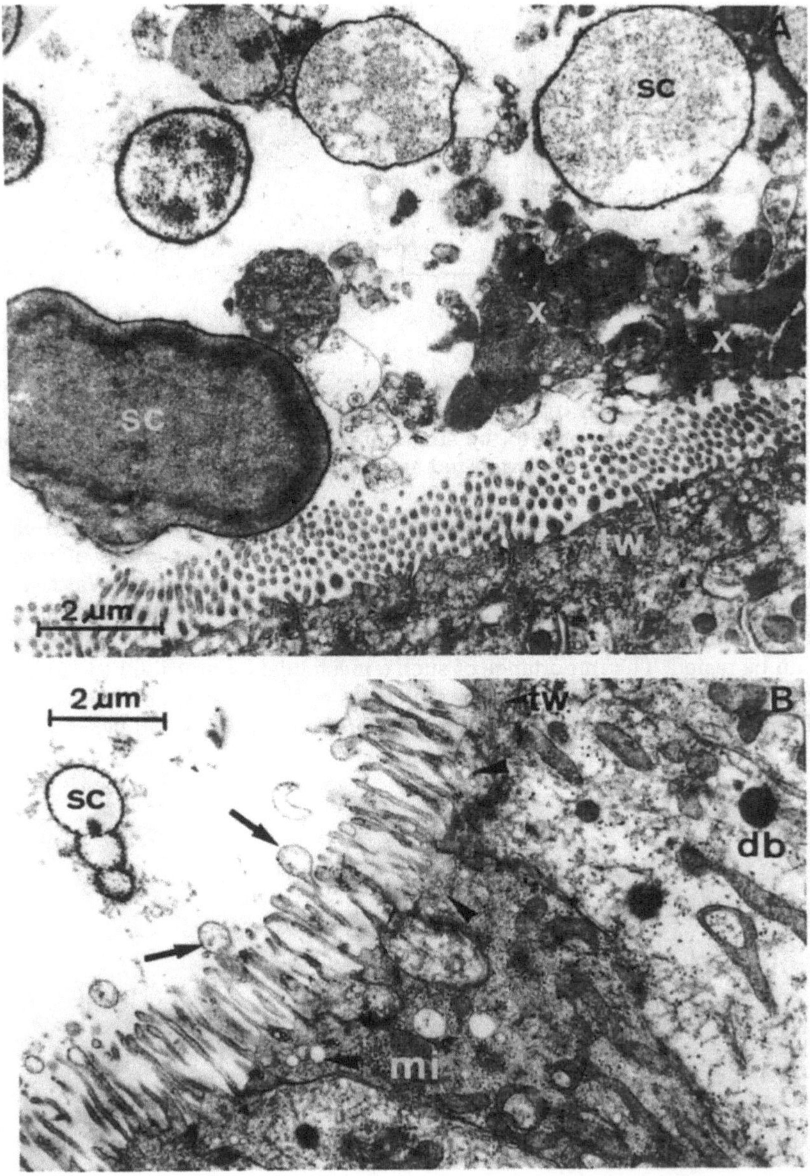

Figure 14. (A, B) Apical area of the ephithelium of the renal appendages of *N. macromphalus* in a TEM. Anatomy: (→) exocytotic mechanisms at the apices of enlarged microvilli; (◀) exocytotic vesicles; (sc) spherical concrements *in stadio nascendi*; (tw) terminal web; (mi) mitochondria; (x) cell detritus; (db) dense bodies.

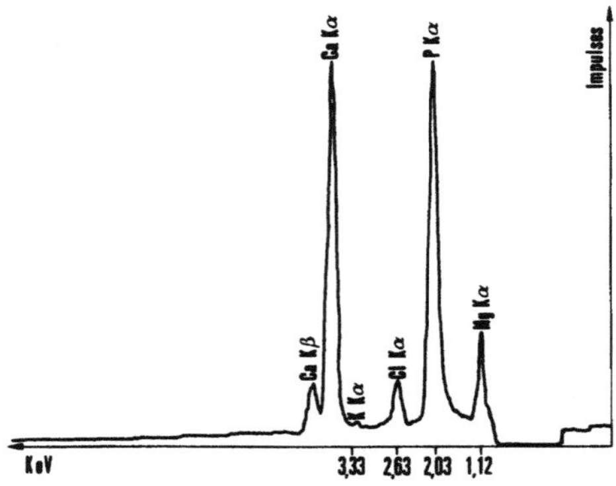

Figure 15. X-ray microanalysis of the spherical concrements from the renal sac of an adult *N. macromphalus*.

This hypothesis is not very satisfying as an explanation of why the important renal appendages, which in other cephalopods carry out a variety of excretory functions, have been devoted to the single problem of mineral excretion in *Nautilus*. Nor does it explain why well-fed animals in an aquarium may accumulate large quantities of concrements.

Another possibility is that in the cycle of changing buoyancy (Ward *et al.*, 1981; Ward and Chamberlain, 1983), the jettisoning of a few grams of mineral might enable the animal to maintain its nearly neutral buoyancy. Ward (1987) has shown regular changes in the X-ray density of the concrement accumulations. If concrements are discharged regularly by animals held in aquaria, this hypothesis can be tested. The importance of such a mechanism seems too little to justify the effort of maintaining such active structures as the renal appendages.

Perhaps as a result of their finding of a preponderance of Mg over Ca in the

Figure 16. SEM of needles of aragonite in formaldehyde-fixed renal appendages of *N. macromphalus*. Compare with Fig. 9.

concrements of four specimens of *N. pompilius*, R. E. Crick *et al.* (1985) proposed another hypothesis, namely, that the renal appendages are responsible for removing Mg from the blood. The animals deposit Ca in the shell very rapidly, and because the blood contains about three times more Mg than Ca, the result would be an increase in the blood Mg level. This increase could be prevented by deposition of magnesium oxalate dihydrate around existing nuclei of hydroxyapatite.

An alternative proposition by A. W. Martin (1975) was based at first on the observation that randomly dissected animals varied widely in the amount of concrement that had been accumulated. He hypothesized that living animals might make use of the Ca in some cyclic pattern, e.g., in septum formation. So important an activity might justify the maintenance of these large and active organs in present-day *Nautilus*. Some support for this view may be found in the extraordinarily high level of phosphatase in the renal appendages of *N. pompilius* (Schipp and Martin, 1981), where it could help in the mobilization of the calcium phosphate of the concrements. An observation by R. E. Crick *et al.* (1985) can be interpreted to fit this view. The uroliths in their low-Ca specimens were stained Mg-purple with alizarin red S, but only in the outer layers (of magnesium oxalate), leaving unstained the centers, which contained the hydroxyapatite. By the Ca-storage hypothesis, this result would suggest that most of the calcium phosphate had been removed biologically before the samples were taken. Perhaps more important is the afore-cited observation of Ward (1987), based on X-ray observation of living animals, that showed a regular sequence of deposition and removal of concrements from the renal appendage sacs. The removal was, however, not coordinated with rapid calcification of new septa. It may be noted, finally, that only cephalopods that possess calcareous structures show mineralized spherical concrements in the renal sacs (Schipp *et al.*, 1975; Lowenstam *et al.*, 1984).

We suggest a reevaluation of the roles of Mg and P in the *Nautilus* concrements. Many other mollusks use intracellular or extracellular spherules of calcium carbonate or phosphate. Of an extensive literature on their use for pH regulation, shell growth and repair, heavy metal detoxification, and mucus formation, we need cite only a few. Burton (1976), for example, pointed out the role of phosphate in preventing precipitation of crystals from supersaturated hemolymph of *Helix pomatia;* he has pursued these problems in a number of subsequent publications and has reviewed the subject (Burton, 1983). Simkiss (1976) pointed out the role of phosphate in calcium mobilization and subsequently reviewed the field (Simkiss and Mason, 1983). From experimental reports on the role of Mg, we cite only that of Zorkendorfer (1930), who showed that the shaking of a magnesium chloride solution with freshly precipitated calcium carbonate solubilized increasing amounts of calcium with increasing concentrations of magnesium chloride. This result he attributed to formation of magnesium calcium complexes. Approaching the problem from the opposite point of attack, Bachra *et al.* (1965) tested the effects of Mg on precipitation of calcium carbonates and phosphates under conditions that favored a slow spontaneous precipitation; Mg ions stabilized the amorphous precipitates of calcium carbonate and phosphate and disturbed the crystallization of apatite. These experiments were done without the added complication of organic ions. The difficulties of describing these interactions in biological solutions in strictly physicochemical terms has been summarized nicely by Robertson (1982). It seems possible that different mollusks have adapted their capacity for fine discrimination between Ca and Mg in different ways to utilize mixtures of these minerals in optimal proportions for the immediate task. In the case of *Nautilus*, the mechanism might serve for an unusually large reservoir of stored calcium. It seems likely that the hypothesis could be tested by the use of radioactive tracers.

4. Summary

4.1. Pericardial Appendages

In *Nautilus*, the pericardial appendages carry out the greater part of the tasks of excretion. Each villus may be compared, in a way, with a vertebrate nephron. The villus develops its own filtration pressure, which forces a filtrate through podocytes about two thirds of the way down the villus. The filtrate is processed by very active cells that line the apical end of each villus, reabsorbing useful materials and excreting others. The epithelial cells of the basomedial portion of each villus also appear to be active cells and play some part in secretory and reabsorptive processes and to participate in keeping the urine isotonic to the blood and ambient seawater.

4.2. Renal Appendages

The renal appendages regulate either the excretion of calcium or its conservation; a definitive judgment cannot be made at this time. At the very least, these organs protect the pericardial appendages from the formation of deposits of calcium carbonate or phosphate, which, in some other molluscan groups, appear likely to be a burden on the kidneys.

ACKNOWLEDGMENTS. The authors thank Dr. P. de Boissezon and Dr. Y. Magnier, who provided the facilities to work in the laboratories of ORSTOM and the Aquarium de Noumea, as well as Dr. G. Doell, A. Bleichner, S. Beckermann, A. Hudel, N. Kopch, J. Luh, H. Schmidt, and R. Vomrath, J. L. U. Giessen, for their technical help. We also thank Dr. Doris Stewart, University of Baltimore, for chemical analysis of the concrements. The research was supported by the Deutsche Forschungsgemeinschaft.

Chapter 21

Respiratory Physiology

JAMES R. REDMOND

1. Introduction

Like the coleoid cephalopods, *Nautilus* has a well-developed respiratory system, but there are numerous anatomical differences between the two groups. *Nautilus* possesses four gills instead of two; the hyponome (funnel) is formed by two overlapping muscular flaps instead of being fused distally into a continuous tube; most important, the presence of an external shell with a thin, lining mantle requires that ventilation be accomplished by a means other than mantle contraction and expansion. This anatomy and the process of ventilation are described in detail in Chapter 25. In this chapter, respiratory physiology will be discussed primarily from the standpoint of oxygen uptake and utilization.

2. Hemocyanin

2.1. Oxygen-Carrying Capacity

The blood respiratory pigment of all cephalopods is hemocyanin. This copper-based oxygen transporter occurs in high concentrations in the blood of *Nautilus*, but because of the large size of the molecules, the total oxygen-carrying capacity of the blood is low. Johansen *et al.* (1978) reported an oxygen-carrying capacity of 2.30 ± 0.57 vol.% for frozen blood samples from ten specimens of *N. pompilius* from the Philippines, and Redmond (1978) reported values of 1.63–2.64 vol.% (mean 2.28) for blood samples from five specimens of *N. macromphalus*

JAMES R. REDMOND • Department of Zoology, Iowa State University, Ames, Iowa 50011.

from New Caledonia. These figures refer to oxygen bound by the hemocyanin and do not include oxygen that may physically dissolve in the blood. These values are somewhat lower than the 3–4.5 vol.% oxygen-carrying capacities commonly reported for the coleoid cephalopods (Mangum, 1980).

2.2. Oxygen Equilibrium Curves

One of the very interesting aspects of respiratory physiology in the coleoid mollusks is the extraordinary sensitivity of the hemocyanin–oxygen bond to pH. This extreme Bohr effect was first reported by Redfield and Goodkind (1929) in their classic studies on the squid *Loligo pealei*. They found that at 23°C, an increase in pCO_2 of 4 mm Hg was enough to shift the half-saturation pressure of the squid hemocyanin from about 50 to 100 mm Hg oxygen. The resulting pH change, estimated to be 0.13 unit, was responsible for the release of one fourth to one third of the oxygen bound to the hemocyanin. It was then of great interest to determine the characteristics of oxygen binding by the hemocyanin of *Nautilus*, both to help understand its respiratory role and possibly to shed light on the unusual Bohr effect found in its modern coleoid relatives.

Oxygen equilibrium curves have been obtained for the hemocyanin of *N. pompilius* (Johansen *et al.*, 1978) and *N. macromphalus* (Redmond, 1978). These curves are illustrated in Fig. 1. and the effect of pH on the half-saturation pressure of *N. macromphalus* hemocyanin in Fig. 2. These results show the oxygen binding curves to be sigmoid, to exhibit moderately low oxygen affinities, and to be affected only slightly by changing pH. Data from these curves (summarized in Table I) show the magnitude of the Bohr effect in *Nautilus* hemocyanin to be strikingly

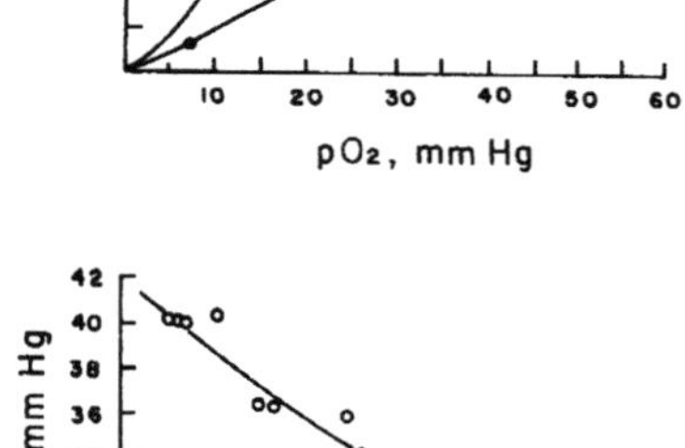

Figure 1. Oxygen equilibrium curves of *N. pompilius* and *N. macromphalus* hemocyanins. (○) *N. pompilius*, 18°C, pH 7.45 (data from Johansen *et al.*, 1978); (●) *N. macromphalus*, 25°C, pH 7.53 (data from Redmond, 1978).

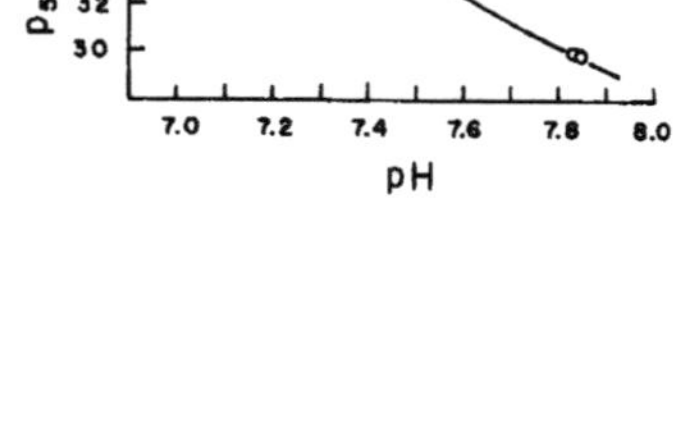

Figure 2. Half-saturation pressures for the hemocyanin of *N. macromphalus*, 25°C. Reprinted with permission from Redmond (1978).

Table I. Characteristics of *Nautilus* Hemocyanin

Property	*N. pompilius*[a]	*N. macromphalus*[b]
O_2 capacity (vol. %)	2.30 ± 0.57	2.28 ± 0.43[c]
P_{50} (mm Hg)	25.2 (25°C, pH 7.4)	34 (25°C, pH 7.45)
P_{50} (mm Hg)	16.9 (18°C, pH 7.45)	—
Bohr effect (ϕ)	−0.20 (18°C)	−0.16 (25°C)
Hill coefficient (n)	2.7 (18°C, pH 7.45)	3.5 (25°C, pH 7.24)

[a] Johansen *et al.* (1978). [b] Redmond (1978). [c] The values are estimates, based on the copper content of blood.

less than the effects reported for the hemocyanins of the coleoid cephalopods. Expressed as phi values, where $\phi = \log P_{50}/\Delta$ pH, *Nautilus* hemocyanin has a very small Bohr effect of −0.16 to −0.20, whereas the Bohr effects of the coleoids range from −0.80 to −1.90 for *Octopus dofleini* and for the squid *L. pelaei*, respectively (Lenfant and Johansen, 1965; Redfield and Goodkind, 1929). A possible significance of this difference is discussed later.

3. Oxygen Uptake

3.1. Oxygen Consumption

Oxygen uptake data are available for *N. pompilius* from the Philippines (Redmond *et al.*, 1978) and from Papua New Guinea (Wells and Wells, 1985). Measurements were made at 17 and 25°C, the former temperature corresponding to the water temperature at depths of 200–400 m where the Philippine specimens were trapped. Table II summarizes some of these data.

It should be kept in mind that *Nautilus* normally lives at about 17°C and that 25°C is close to lethal temperature, at least for *N. pompilius*. The measurements at 17°C should represent reasonably normal oxygen consumption rates. Taking into account the size differences of the specimens used, the two studies found similar rates of oxygen uptake. The rates found for active specimens (those ob-

Table II. Oxygen Uptake by *Nautilus pompilius*[a]

Conditions	Shell weight (g)	Soft weight (g)	ml O_2 kg^{-1} hr^{-1} (soft wt.)	N
Calm				
25°C	168.5	543.8	65.2	6
17°C	157.0	463.7	26.5	4
15–17°C	117.7	256.7	30.0	3[b]
Active				
25°C	162.8	557.3	109.3	4
17°C	127	378	52	1

[a] The data are from Redmond *et al.* (1978) unless otherwise indicated. [b] Wells and Wells (1985).

Table III. Oxygen Consumption by Cephalopods

Species	Temperature (°C)	ml O_2 kg^{-1} hr^{-1}	Sources
Nautilus pompilius	17	27	Redmond *et al.* (1978)
	17	30	Wells and Wells (1985)
	25	66	Redmond *et al.* (1978)
Octopus vulgaris	16	67	Jolyet and Regnard (1877)
	20	117	Vernon (1896)
	24	84	Montuori (1913)
	21	72	Wells *et al.* (1983a)
Octopus cyanea	26	94	Maginniss and Wells (1969)
Octopus dofleini[a]	11	28	Johansen (1965)
Sepia officinalis	24	233	Montuori (1913)
Eledone moschata	16	181	Cohnheim (1912)
	24	33	Montuori (1913)
Loligo pealei[b]	24	600	Redfield and Goodkind (1929)
Loligo opalescens	14	254	O'Dor (1982)

[a] Low oxygen uptake is a reflection of low temperature and very large body size. The average weight of the specimens in this study was 9.59 kg.
[b] Experimental conditions probably gave oxygen uptakes approaching maximal values.

served to show continuous forceful ventilatory activity) are a little more than twice the resting rates.

If the values cited above are compared with oxygen utilization of the coleoid cephalopods (Table III), it can be seen that *Nautilus* respires at a somewhat lower rate than its modern coleoid relatives. Although differences in body size and temperature make close comparisons difficult, it appears that *Nautilus* respires at about half the rate of *Octopus*, but at a much lower rate than a squid.

3.2. Effect of Oxygen Partial Pressure on Oxygen Uptake

The effect of varying the external oxygen partial pressure on the oxygen uptake of *N. pompilius* was examined in two studies. Wells and Wells (1985) found that at 15–17°C, *Nautilus* maintained its rate of oxygen consumption down to an ambient pO_2 of approximately 75 mm Hg. At 25°C, oxygen consumption varied with external pO_2, and below a pO_2 of approximately 50 mm Hg, *Nautilus* could not obtain sufficient oxygen (Redmond *et al.*, 1978) (Fig. 3). Both groups of investigators noted a tendency for ventilation rate to increase as the ambient pO_2 decreased.

The inability of *Nautilus* to regulate its oxygen uptake at 25°C is presumably an aspect of temperature stress. At this temperature, these animals apparently can obtain barely enough oxygen to meet their needs, and any decrease in oxygen availability will result in decreased uptake.

Whether *Nautilus* normally ever experiences hypoxia is unknown. Although the oxygen content of its environmental waters has not been measured, it seems unlikely that *Nautilus* encounter water of low oxygen content. It is possible that its ability to tolerate considerable periods of hypoxia is of use when it is defen-

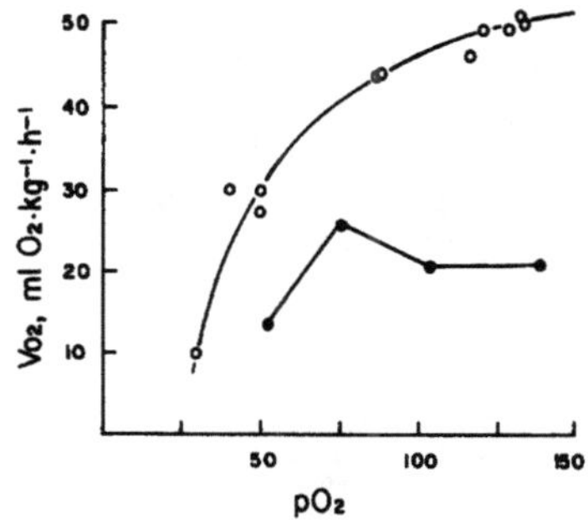

Figure 3. Influence of oxygen partial pressure on the oxygen uptake of *N. pompilius*. All calculations are based on soft-tissue weights. (○) 25°C (from Redmond *et al.*, 1978); (●) 17 ± 0.5°C (data from Wells and Wells, 1985).

sively withdrawn into its shell. Even while it is withdrawn, some ventilation may continue, but the rate is presumably reduced.

3.3. Oxygen Transport by the Blood

Blood gas transport was examined in *N. pompilius* by implanting catheters in various parts of the circulatory system and withdrawing blood from unrestrained specimens after their recovery from surgery (Johansen *et al.*, 1978). Mixed venous blood was taken from the cephalic vena cava, and oxygenated blood was taken from one of the efferent branchial vessels near a gill tip. In two experiments, arterial blood was sampled from the cephalic aorta. Blood from the latter gave blood gas and pH values similar to those from the efferent branchial vessel. The results of these experiments, conducted at 18°C, can be summarized as follows (and see Figs. 4 and 5):

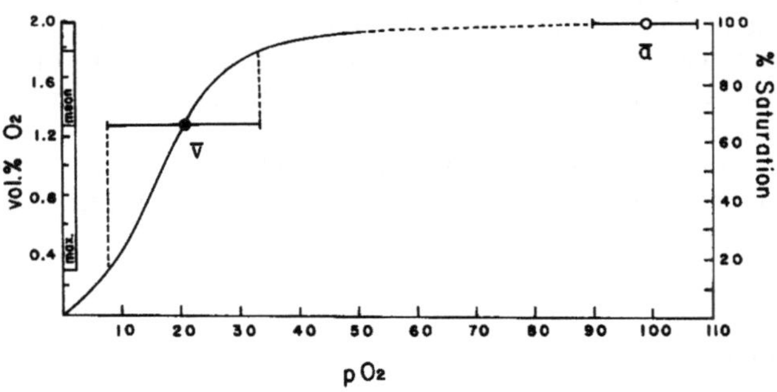

Figure 4. Oxygen content of the postbranchial and venous hemocyanin of *N. pompilius*. The hemocyanin becomes nearly saturated with oxygen in the gills (ā) and, on average, returns from the tissues ($\bar{v}$) still carrying 63% of its oxygen. The bars through the circles indicate one standard deviation about the means. Based on an oxygen-carrying capacity of 2.0 vol.%, the open bar on the vertical axis indicates the oxygen given up to the tissues by the hemocyanin in 100 ml of blood. The large variation in oxygen delivery is probably a result of different levels of activity among the specimens. Redrawn from Johansen *et al.* (1978).

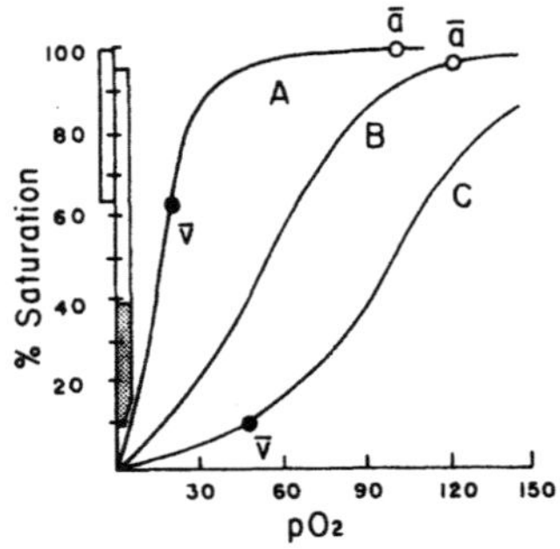

Figure 5. Influence of the Bohr effect on oxygen transport by the hemocyanins of *N. pompilius* and the squid *Loligo pealei.* (A) Oxygen equilibrium curve of *N. pompilius.* Under resting conditions, the shift of P_{50} in the tissues would be less than 0.5 mm Hg oxygen and is too small to show on this scale. The Bohr effect would have no significant influence on overall oxygen exchange. (B) Oxygen equilibrium curve of *L. pealei* hemocyanin in the presence of 2 mm Hg carbon dioxide (postbranchial conditions). (C) Oxygen equilibrium curve of *L. pealei* hemocyanin in the presence of 6 mm Hg carbon dioxide (venous conditions). The large Bohr shift is responsible for the release of an additional 20% of the oxygen being carried by the hemocyanin. Note also that the oxygen is released at high partial pressures. Key: ($\bar{a}$) mean postbranchial pO_2 and percentage saturation; ($\bar{v}$) mean venous pO_2 and percentage saturation. The open bar outside the vertical axis indicates average oxygen released to tissues by the hemocyanin of *N. pompilius;* the open bar inside the vertical axis indicates average oxygen that would be released to tissues by *L. pealei* hemocyanin if no Bohr effect were present; the stippled portion of the bar represents additional oxygen released by *L. pealei* hemocyanin as a result of the Bohr effect. (Note: *Nautilus* is in resting condition; *Loligo* is in active condition. Comparable measurements for resting squid are not available.) Data from Johansen *et al.* (1978) and Redfield and Goodkind (1929).

1. Blood pH variation among normal specimens was great, ranging from 7.20 to 7.80.
2. Under normal conditions, the pH difference between arterial and venous blood was very small, usually less than 0.05 unit. Because of the very small Bohr effect, this pH change should not play a significant role in oxygen delivery during resting conditions. The possibility remains, however, that larger pH shifts in local tissues during periods of greater activity could facilitate the release there of additional oxygen from the hemocyanin. Under conditions of hypoxia, the A-V difference could exceed 1.0 pH unit.
3. Arterial pO_2 was high. The mean value was 99.2 mm Hg, with a maximum measurement of 125 mm Hg.
4. Venous pO_2 varied from 10 to 50 and averaged 20 mm Hg.
5. The pO_2 values cited above indicate that, on average, the hemocyanin becomes essentially saturated with oxygen in the gills and returns from the tissues still 65% oxygen-saturated.
6. Based on a measured oxygen uptake of 0.5 ml $O_2 \cdot kg^{-1} \cdot min^{-1}$ and an oxygen-carrying capacity of 2.0 vol.%, a cardiac output of 54 ml $kg^{-1} \cdot min^{-1}$ was estimated.

4. Discussion

As the only surviving tetrabranchiate cephalopod, *Nautilus* is of great interest as a key to many features of the extinct forms and as a base from which to speculate on the evolution of the modern coleoid cephalopods. Many aspects of these relationships are discussed elsewhere in this volume. The following comments are restricted to facets of respiratory physiology.

From the data available, the *Nautilus* respiratory system appears well developed and, together with an excellent circulatory system (Bourne *et al.*, 1978), appears to be well capable of delivering oxygen adequate to meet all the animal's needs. One can speculate that if *Nautilus* is representative of the extinct shelled cephalopods, it seems unlikely that oxygen delivery to tissues was a limiting factor in their evolution or physiology. For the coleoid cephalopods, however, the reverse may well be true.

The respiratory pigment found in the blood of all living cephalopods is, and in all extinct cephalopods very likely was, hemocyanin. This protein occurs dissolved in the blood, and although it may be present in high concentrations, it severely limits the oxygen-carrying capacity of the blood. This limitation is a result of the large protein component that is associated with each oxygen combining site. For every two copper atoms that can bind an oxygen molecule, there are approximately 50,000 daltons of protein (hemoglobins commonly have approximately 16,000 daltons of protein per oxygen combining site). An oxygen-carrying capacity of about 4.5 vol.% is probably close to the maximum for a hemocyanin blood. Concentrations of hemocyanin greater than this may actually reduce oxygen delivery per unit time due to increasing blood viscosity (Snyder and Mangum, 1982).

Because the modern coleoid cephalopods commonly have blood with oxygen-carrying capacities of 3–4 vol.%, and *Nautilus* blood can carry 2–2.6 vol.%, it does not appear that there has been a strong selective pressure to increase hemocyanin concentration in the blood of *Nautilus*. This, together with the very small Bohr effect shown by *Nautilus* hemocyanin, suggests that oxygen delivery to the tissues is very adequately serviced by the circulatory system.

Again assuming that the modern coleoid cephalopods evolved from species that possessed characteristics similar to those of *Nautilus*, some interesting speculations may be made concerning the respiratory system of the coleoids. It appears that the major changes introduced by the coleoids were associated with greater mobility and higher metabolic rates. Among these changes were the loss of the shell, thickening and muscularization of the mantle, increased concentrations of hemocyanin in the blood, the appearance of a very large Bohr effect, and the development of branchial hearts from the pericardial glands (A. W. Martin, 1975; Bourne *et al.*, 1977). Metabolic rates, particularly of the squids, may be very high. This increased metabolic rate placed great demands on the oxygen delivery system. One of the responses to this demand was to increase hemocyanin concentrations in the blood. This response was limited by the total amount of dissolved protein that the blood could contain and still be circulated effectively. Selective pressures continued on the respiratory system, and the sensitivity of the hemocyanin–oxygen bond to pH increased until a very small pH drop in the tissues was sufficient to drive off a considerable proportion of the oxygen carried by the hemocyanin. This large Bohr effect maximized oxygen delivery at high pO_2 by blood of low oxygen-carrying capacity. Associated with these changes, the muscular mantle has become a very effective ventilatory device, and the new branchial hearts now help speed the blood by boosting blood pressure at the base of the gills.

The highly developed respiratory and circulatory systems of the shelled cephalopods provided an excellent base, perhaps an essential one, from which the evolution of the very successful modern cephalopods could proceed. The coleoids, in turn, may be taxing their blood oxygen transport to the extreme. The presence of hemocyanin in the blood of these animals, more than adequate for oxygen delivery in the ancient cephalopods, may now be a serious impediment to the development of even higher metabolic rates by their modern descendants.

Chapter 22

Mouth Part Histology and Morphology

KAZUSHIGE TANABE and YOSHIO FUKUDA

1. Introduction

The primary digestive organ of *Nautilus* is a powerful, remarkably well developed buccal mass; it is distinguished from those of modern dibranchiate cephalopods by the presence of prominent calcified deposits and by the shorter inner lamellae of the lower jaw (Teichert *et al.*, 1964a; Gasiorowski, 1973; Müller, 1974; Okutani and Mikami, 1977; Saunders *et al.*, 1978). The mouth-part anatomy of *Nautilus* has been described in some detail previously (Owen, 1832; Vayssière, 1896; Griffin, 1900; Naef, 1923; Stenzel, 1964; Solem and Richardson, 1975; Okutani and Mikami, 1977). These works demonstrated that the buccal mass of *Nautilus* differs from those of coleoids in the nature of both jaw and radular structures. Detailed relationships of the soft and hard tissues, however, have not been examined fully. In this chapter, we describe such relationships on the basis of investigations using a scanning electron microscope (SEM) and discuss their functional meanings as they pertain to feeding.

2. Material and Methods

This report is based on detailed study of a mature specimen of *N. pompilius* Linnaeus captured in Tañon Strait, the Philippines, in December 1981. Immediately after capture, the buccal portion was removed from the body and was

KAZUSHIGE TANABE • Geological Institute, Faculty of Science, University of Tokyo, Tokyo 113, Japan. YOSHIO FUKUDA • Laboratory of Pathology and Animal Physiology, Chiba Prefectural Institute of Public Health, Chiba 280, Japan.

preserved in 10% formalin in seawater buffered with carbonate. It was photographed 1 month later by soft X-ray and then transferred to 70% ethanol buffered with borax. Cross sections were prepared for SEM study by being bathed in 90 and 100% ethanol and then in 100% amyl acetate, followed by critical-point drying. SEM observations were made using a Hitachi Model S-450 scanning electron microscope, with 20 kV acceleration voltage.

Comparison of the mouth-part microstructure of *N. pompilius* with that of *N. macromphalus* is based on an earlier study of the latter species by Fukuda (1980). However, it should be stressed that, at the histological level, there is no significant difference in the buccal structure of these two species.

3. Microstructural Observations

The buccal mass of *Nautilus* is relatively large (≈5 cm in maximum length in mature *N. pompilius*) and globular. Except for the anterior portion, it is covered by a thin buccal membrane (Fig. 1). Internally, the framework of the buccal mass consists mainly of semiflexible chitinous jaws that support muscles and a prominent radula. The oral opening is surrounded by the lateral buccal palp, salivary papillae, and labial margin. There follows an account of the details of microstructure and general morphology of the various jaw components.

3.1. Labial Margin

The anterior opening of the buccal mass, called the *labial margin*, includes a distal area of epithelium covering the mass and consists of four rows of dense, triangular projections of mucosa, each of which is 2–3 mm in length and 0.2–0.3 mm in width (Fig. 2A and C). Light-microscopic observations indicate that the labial margin consists of a simple epithelial layer, in which there are some tall,

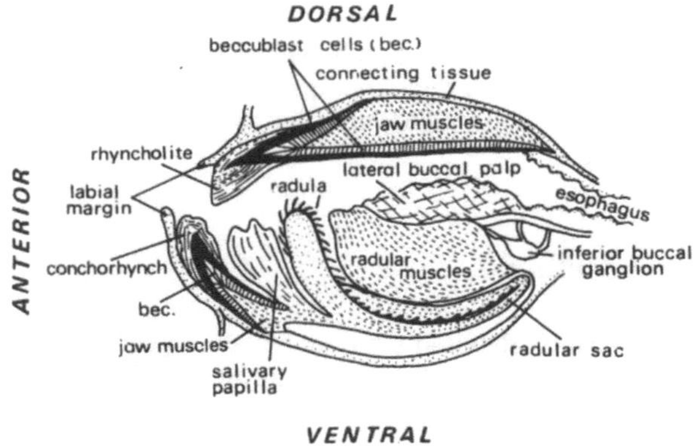

Figure 1. Diagram of the buccal mass of *Nautilus* (median section). The terminology is mostly from Young (1965b). Modified from Tanabe and Fukuda (1983).

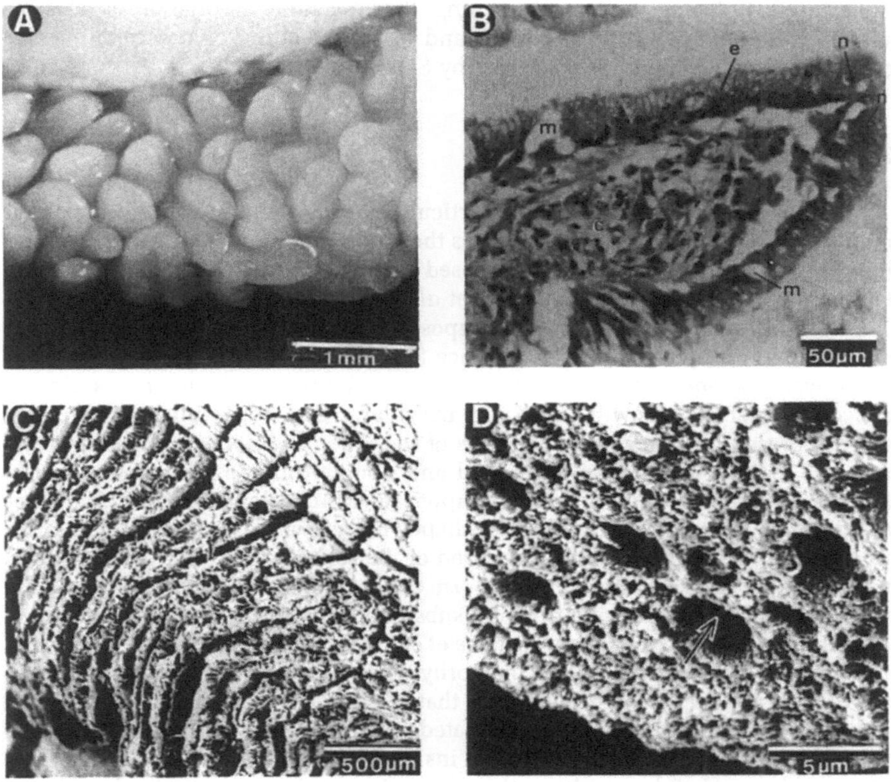

Figure 2. Labial margin of *N. pompilius.* (A) Optical micrograph of a frontal view of the labial margin, consisting of four rows of projecting mucosa. (B) Optical micrograph of a sagittally sectioned triangular mucosal projection. Anatomy: (e) columnar epithelium; (n) nerve bundles; (c) connecting tissue of submucosa; (m) mucous cells. (C) SEM of sagittally sectioned rows of projections of mucosa surrounded by columnar epithelial cells. (D) SEM of the columnar epithelial surface, showing the opening of a mucous-secreting cell (→).

columnar epithelial cells, mucous cells, and isolated sensory cells (Fukuda, 1980). The muscular system is poorly developed in the labial margin, except for several longitudinal muscle bundles in the connecting tissue beneath the epithelium and a simple circular muscle bundle under the outermost epithelium of the buccal mass (Fukuda, 1980). In each triangular projection of mucosa, the nerve bundles are crowded (Fig. 2B), suggesting the presence of sensory receptors, as in coleoids (Young, 1965b; Emery, 1975a,b). The outermost part of the mucosa is made of a layer of columnar epithelial cells about 50 μm thick (e in Fig. 2B). On the surface of the columnar epithelium, cells are densely distributed (Fig. 2D), and they secrete ellipsoidal granules (Fukuda, 1980).

From these observations, it is evident that the labial margin does not function as a "lip" to hold or grasp food; rather, it probably serves a number of functions, including (1) closing the buccal opening, (2) preventing exfoliation of jaws by secretion of mucus, (3) feeling food, and (4) regenerating or precipitating the calcareous jaw elements, as suggested by Saunders (1981b).

3.2. Jaws

The jaws of *Nautilus* consist of articulated upper and lower chitinous elements (Fig. 3A). The upper jaw encloses the larger lower jaw. The jaws are strong but flexible during life and are composed mainly of black chitin (Griffin, 1900; Okutani and Mikami, 1977; Saunders *et al.*, 1978; Lowenstam *et al.*, 1984).

Microstructurally, the jaws are composed of layered membranes of chitin that are oriented obliquely to the jaw surface (Fig. 4A and C). The outer surfaces of the lower jaw and the upper jaw are both sculptured with numerous closely spaced, concentric "growth lines" with irregular, radial ridges. Such growth lines are not present on the interior surfaces of the jaws. The jaws are thickest in the anterior region (where they are folded and double-walled) and support a conspicuous calcareous covering. In the upper jaw, this calcite element [the rhyncholite (r in Fig. 3A)] is an arrowhead-shaped object, with a distinct median ridge in the anterior portion, and is composed of densely stacked, thin layers of prismatic calcite, each measuring about 5 μm thick (Fig. 3C and D). The rhyncholite is firmly attached to the protein–chitin substrate, with a thin insertion of aragonite (≈100 μm thick) between them (Tanabe *et al.*, 1980; Lowenstam *et al.*, 1984). The calcified lower jaw element [the conchorhynch (c in Fig. 3A)] is divided into three areas: (1) irregular calcareous deposits that cover the exterior surface of the wide chitinous jaw, (2) a distinctly denticulated occlusal or oral surface (Fig. 3B), and (3) a thick calcareous layer lining the inside of the chitinous jaw. Both (1) and (2) are composed of dense, alternating, thin prismatic layers, as in the rhyncholite of the upper jaw, but the crenulated denticles of the oral surface are characterized by a radiating, prismatic microstructure (Fig. 3E and F). These structural and mineralogical characteristics are well designed for a cutting and shearing function (Saunders *et al.*, 1978; Tanabe *et al.*, 1980).

3.3. Jaw Muscles and Beccublast Cells

The outer surfaces of the upper and lower jaws are mostly covered by a thin buccal membrane that comprises connecting tissue and epithelium. In these areas, the chitinous jaw plates are free of muscle. The inner side of the lower jaw and outer side of the upper jaw, by contrast, are connected by a thick, muscular complex, with a thin layer of tall columnar cells lying between them (bec. in Fig. 4A). These cells each measure about 20 μm in height and 3 μm in width. Because of the fundamental similarity in the position and arrangement, they are regarded as homologous with beccublast cells, the chitin-secreting cells in modern coleoids (Dilly and Nixon, 1976). On the side of the jaw plate, the beccublast cells are branched into many fine trabeculae, each of which is about 0.3 μm in diameter (Fig. 4B), and the ends of the trabeculae are deeply inserted within the chitinous tissue, forming numerous micropores (Fig. 4D). This is different from the anchor-

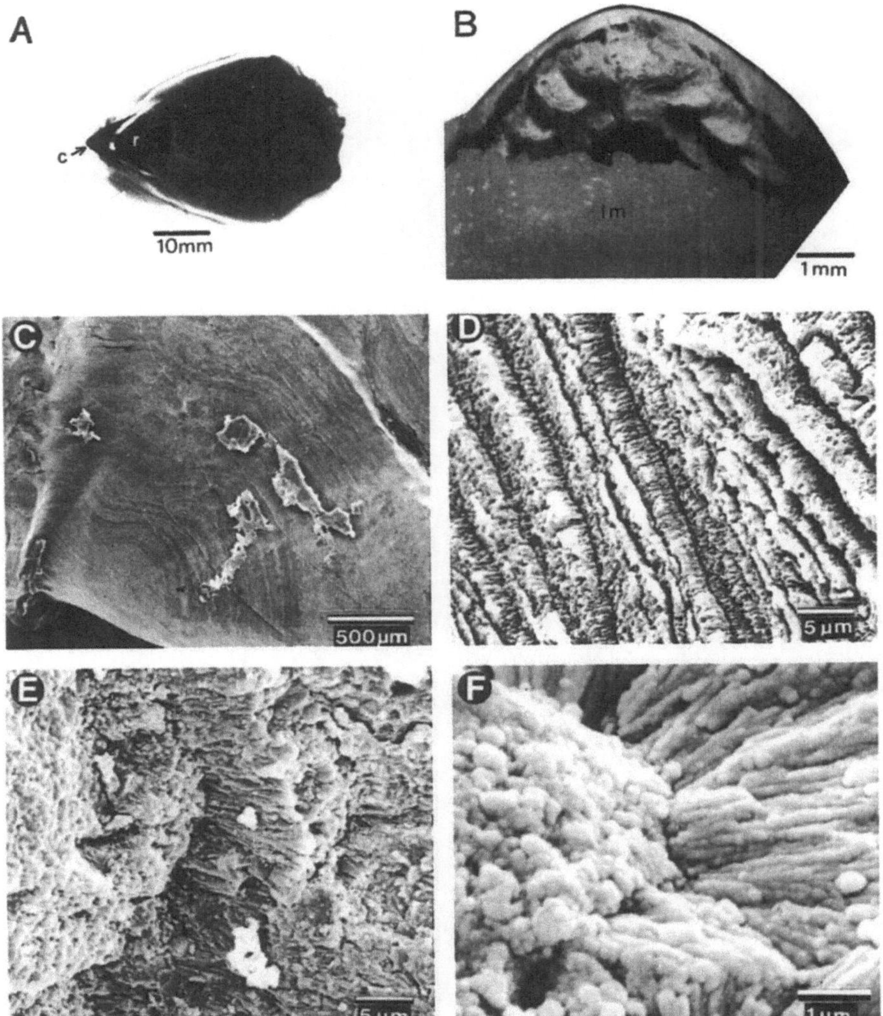

Figure 3. Calcareous deposits on the anterior region of the jaws of *N. pompilius*. (A) X-ray radiograph of the buccal mass (dorsal view). Anatomy: (r) rhyncholite on the upper jaw; (c) conchorhynch on the lower jaw. (B) Optical micrograph of the denticulated portion of the conchorhynch (dorsal view) shown protruding from beneath the labial margin (lm). (C) SEM of the anterior portion of a rhyncholite, showing the multilayered structure (lateral view). (D) Enlarged SEM of the area shown in (D), showing the densely layered prismatic structure. (E, F) SEM of the denticulated portion of the conchorhynch in Fig. 2B, showing the radiating prismatic structure.

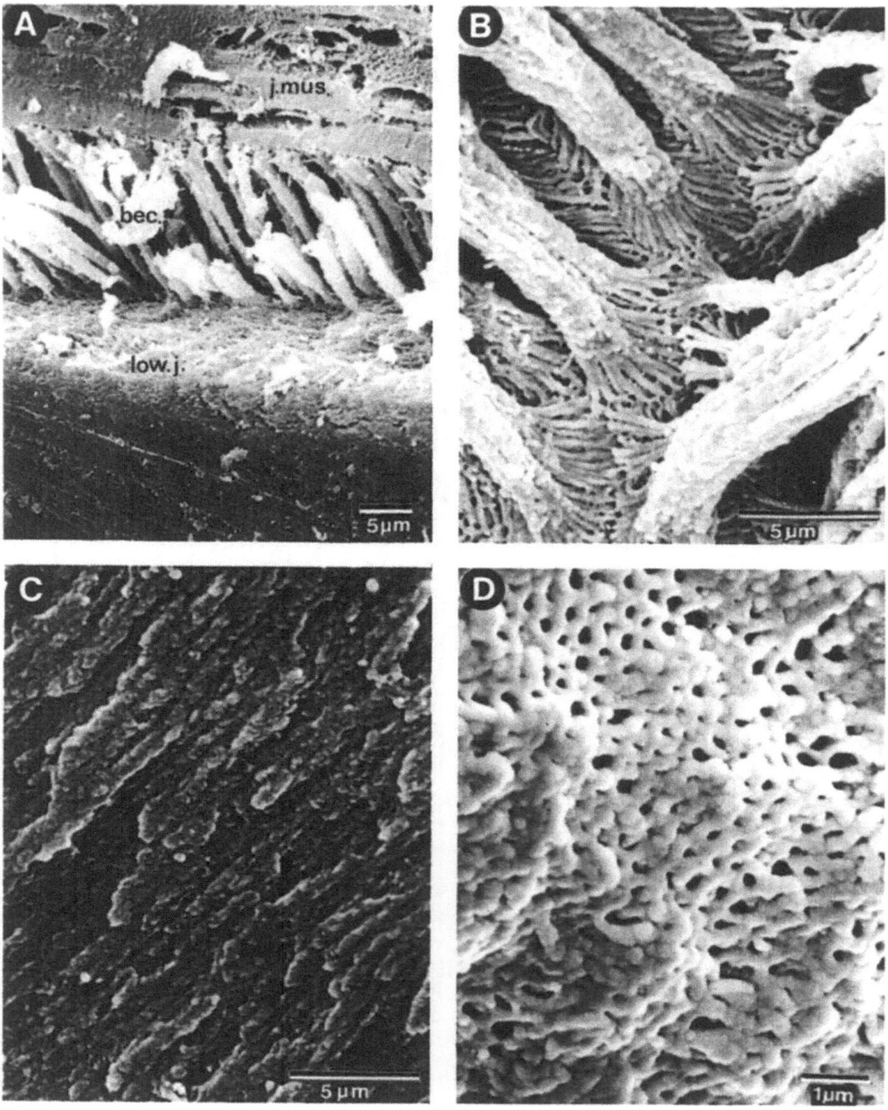

Figure 4. SEM of the jaw, beccublast cells, and jaw muscles in *N. pompilius*. (A) Part of a median-sectioned buccal mass in the posterior region, showing the beccublast cells (bec.), jaw (low j.), and jaw muscles (j. mus.). (B) Enlarged section of (A), showing the branching trabeculae of the beccublast cells attached to the jaw surface. (C) Microstructure of the chitinous jaw plate. (D) Micropores on the inner surface of the lower jaw, into which the branching ends of beccublasts are deeply inserted. Reprinted with permission from Tanabe and Fukuda (1983).

type weak attachment of the beccublasts on the jaw surface in modern coleoids (Dilly and Nixon, 1976). The branching ends of the beccublasts in *Nautilus* therefore provide a special function; in addition to secreting chitin, they serve as a firm attachment site for the jaw muscles on the hard tissue.

3.4. Radula

The radula of *Nautilus* is more complex than that of coleoids. In mature specimens, it measures approximately 4 mm by 20 mm. Its morphological features have previously been described in detail by a number of authors (e.g., Vayssière, 1896; Griffin, 1900; Naef, 1923; Solem and Richardson, 1975; Okutani and Mikami, 1980; Fukuda, 1980; Saunders, 1981a,b). The radula consists of nine primary teeth (one central and two laterals and two marginals on each side) and two pairs of marginal support plates. This arrangement differs substantially from that of the radula of dibranchiate cephalopods, which has a total of seven elements (Solem and Richardson, 1975). Our observations show that, in the posterior portion of the radular sac, the radular teeth are wholly surrounded by radular muscles (Fig. 5A). A thick ($\approx$30 μm) epithelium is present on the outer surface of the radular teeth (Fig. 5B); this epithelium probably consists of chitin-secreting, columnar cells. The posteriormost radular teeth are not fully mineralized, but are composed of a network of radially arranged connecting tissues (ct in Fig. 5B). In the central part of the radula, the teeth are free of muscle and are fully chitinized (Fig. 5C and D), but they are covered by a thick epithelial cell layer (rc in Fig. 5C). In the anterior portion of the radula, the epithelial cell layer is completely exfoliated from the radular teeth (Fig. 5E). The central and lateral teeth in the posterior (and presumably nonfunctional) area are much shorter than the marginal teeth (Solem and Richardson, 1975), but they tend to be projected anteriorly in the distal region near the buccal cavity (Fig. 5F). The distal radular elements are probably used for grasping and conveying food and have been observed to operate in a synchronous reflex with opening and closing of the mouth (Saunders, 1981a, Figs. 13 and 14). Data from energy-dispersion X-ray analysis have shown considerable amounts of calcium and silica in the radula of *N. macromphalus* (Fukuda, 1980); this mineralization has been interpreted as important in increasing the hardness of the radula.

3.5. Esophagus and Crop

The esophagus of *Nautilus* is a muscular tube about 5 mm in diameter, originating from the hindsection of the buccal mass. The inner wall of the esophagus consists of columnar epithelium and folded mucous cells, filled with granules of the acid mucopolysaccharide (Fig. 6) (Fukuda, 1980). The esophagus is continuous with an ovoid swelling called the *crop*. The crop is an elastic, bag-shaped organ, in which predigested pieces of food are stored temporarily. It is capable of enormous enlargement; it may measure as much as 8 cm long by 5 cm in diameter when filled with food (Tanabe *et al.*, 1980). Except for the presence of a network of arteries accompanied by nerve fiber bundles, the wall structure of

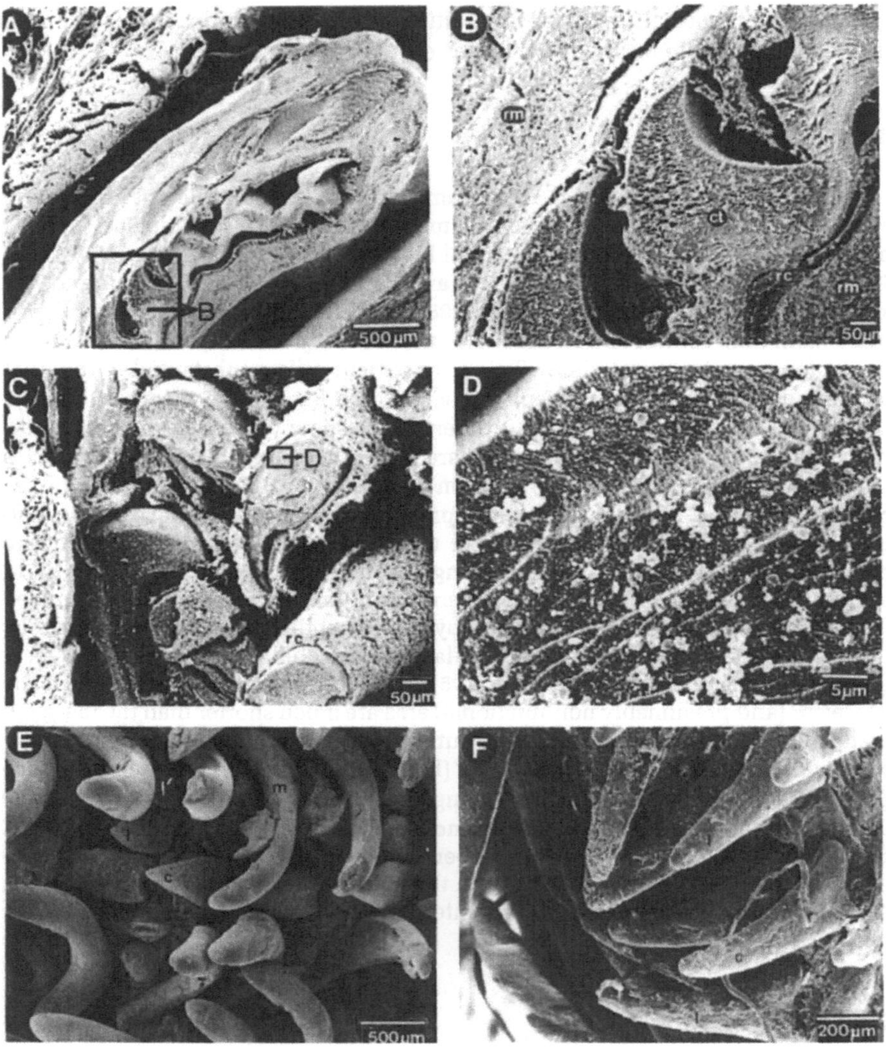

Figure 5. SEM of the radula of *N. pompilius*. (A) Posterior portion of the radular sac (median section). (B) Enlarged area of (A), showing unmineralized radular teeth (ct) surrounded by thick radula-secreting columnar epithelial cells (rc) and radular muscles (rm). (C) Central part of the radular sac, showing the exposed radular teeth with a covering of epithelial cells (rc). (D) Enlarged area of (C), showing the radially arranged chitinous layers within the radular tooth. (E) Dorsal view of the radular teeth in the anterior region. Anatomy: (c) central tooth; (l,l′) inner and outer lateral teeth; (m) inner marginal tooth. (F) Part of radular teeth in the distal area, showing the strongly projected central (c) and inner lateral (l) teeth.

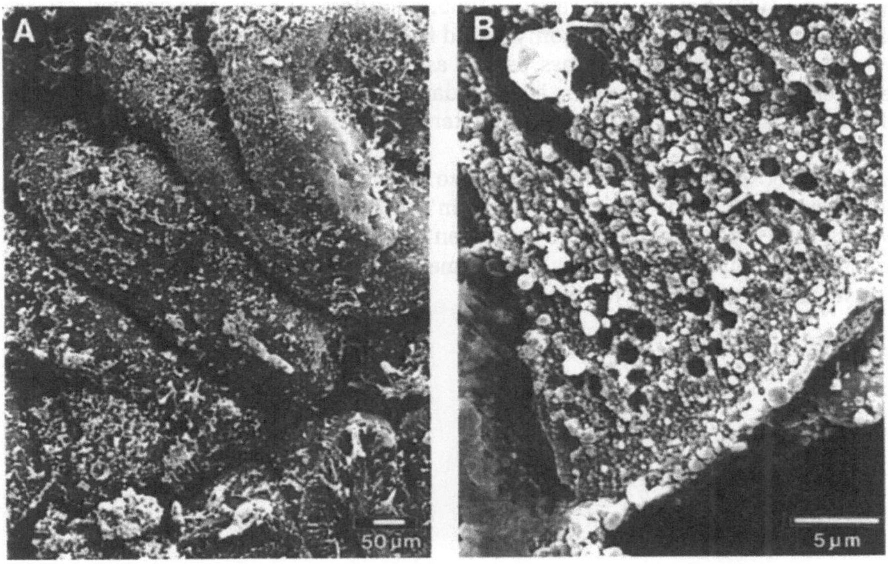

Figure 6. SEM of the inner wall of the esophagus of *N. pompilius*. (A) Longitudinal folds of mucosa. (B) Enlarged section of (A), showing the aggregation of granules of mucopolysaccharide secreted by mucous cells.

the crop is essentially the same as that of the esophagus (Fukuda, 1980). It is assumed that the arteries and nerve fiber bundles in the inner crop wall may function to monitor the tension of the crop wall and, hence, the amount of food (Fukuda, 1980).

4. Discussion

As described above, the structure of the buccal mass of *Nautilus* differs considerably from that of modern coleoids in having anterior calcified elements with denticles, firmer attachment of beccublast cells on the surface of the jaw plate, and a more complex radula. These structural features correlate well with the mineralogical and trace-element data of the jaws (Lowenstam *et al.*, 1984), and both provide an unusually strong biting and shearing ability (Saunders *et al.*, 1978; Tanabe and Fukuda, 1983). In contrast to coleoids, however, *Nautilus* lacks a submandibular gland within the buccal mass and a projective chitinous membrane in the wall of the esophagus and crop (Fukuda, 1980; Tanabe and Fukuda, 1983); therefore, chewed pieces of food found within the crop are poorly digested (Tanabe *et al.*, 1980). Perhaps in compensation, thick mucous cells are developed within the inner wall of the esophagus and crop and serve to protect the inner surface from predigested food. In its natural habitats, *Nautilus* regularly feeds on crustaceans of various sizes, including hermit crabs and carid shrimps (Ward and

Wicksten, 1980; Saisho and Tanabe, 1985; Saunders, 1984b), and chewed pieces of these crustaceans are commonly found in the crops of trapped specimens. The overall design of the buccal mass, as well as the microstructural and mineralogical features of the digestive organs, is well adapted to a scavenging–predacious mode of feeding, such as is thought to characterize *Nautilus*.

ACKNOWLEDGMENTS. We thank Prof. Shozo Hayasaka, Kagoshima University, and other members of the Oversea Project on the Biology of *Nautilus* in Philippine waters (1981) for providing the specimen of *N. pompilius* utilized. We also appreciate the helpful discussions with former members of JECOLN (Japanese Expert Consultation on Living *Nautilus*).

VI

Metabolism

IV

Chapter 23

Energy Metabolism of *Nautilus* Swimming Muscles

JOHN BALDWIN

1. Introduction

Increasing attention is being paid to the energy metabolism of cephalopods, particularly to the metabolic pathways used to provide ATP for muscle contraction during locomotion (for recent reviews, see Hochachka *et al.*, 1983; Storey and Storey, 1983). Extending these studies to *Nautilus* has generated considerable interest among comparative biochemists for a variety of reasons. As the only extant genus of the once abundant and highly diverse externally shelled cephalopods, *Nautilus* offers a unique opportunity for gaining information about metabolic organization in an ancient lineage that has been separated since Paleozoic times from those that lead to other modern cephalopods. In addition, most studies of energy metabolism have been carried out with the highly active squids and cuttlefishes. *Nautilus*, with lower oxygen demands and a less efficient oxygen delivery system (Johansen *et al.*, 1978; Redmond *et al.*, 1978), provides an interesting contrast of cephalopod metabolism in "slow motion." Further, because correlations exist between the intensity and duration of locomotion and biochemical properties of the muscles involved in propulsion, comparative studies utilizing a range of cephalopods enable predictions to be made about locomotory behavior in animals such as *Nautilus* for which direct observations of swimming in nature are limited.

The relationships between metabolic organization and locomotory behavior are seen most clearly in those properties of muscle that reflect the relative contributions of aerobic and anaerobic pathways to ATP production during muscle work. This relationship arises from limitations imposed by the oxygen delivery system. Oxygen levels in muscle may become limiting during short-term bursts

JOHN BALDWIN • Zoology Department, Monash University, Clayton, Victoria 3168, Australia.

of maximum work. When this occurs, ATP can be provided from the rapid hydrolysis of muscle phosphagens and from the anaerobic catabolism of carbohydrates following activation of glycolysis. Under longer-term submaximal work loads, adequate oxygen delivery can be maintained for ATP to be generated by oxidative phosphorylation following the complete oxidation through the Kreb's cycle of fuels such as lipids, carbohydrates, and amino acids.

Properties of muscle that have been used for assessing the contributions of these aerobic and anaerobic ATP-generating pathways include muscle ultrastructure, particularly the distribution of fiber types and abundance of mitochondria, endogenous fuel stores and phosphagen content, maximum activities of enzymes unique to the different pathways, metabolite changes associated with muscle work, and the pH buffering capacity of muscle homogenates. All these approaches have been applied, although to differing degrees, to *Nautilus*. When considered together, the results of these studies provide a coherent picture of energy metabolism in the muscles used to power swimming, which in turn makes possible comparisons with the metabolic organization of muscles from other cephalopods and provides insights into the locomotory behavior of *Nautilus* in nature.

2. Muscles Used to Power Swimming

Cephalopods display a wide range of locomotory behavior, varying both in the muscles used for propulsion and in the intensity and duration of the work performed. However, although they differ greatly in general morphology, all groups share the common feature of powering swimming movements by rapidly expelling water from the mantle cavity. *Nautilus* is the only extant cephalopod in which the body is confined within a rigid shell, an arrangement that places considerable constraints on swimming behavior. In this animal, the mantle is represented by a thin muscle sheet that lines the outermost shell chamber and, unlike other cephalopods, plays no part in jetting water from the mantle cavity. Instead, the volume of the mantle cavity is controlled by two separate sets of muscles: the funnel and retractor (or shell) muscles. The funnel is formed from a triangular muscle sheet that is rolled up to make a tube that can be extended beyond the shell edge. When the two halves of the funnel separate, the internal volume increases, and water is drawn into the mantle cavity. Water is expelled through the funnel tube when the funnel flaps come together. Swimming direction during jetting is controlled by the orientation of the funnel aperture. The retractor muscles attach the body of the animal to the inside walls of the body chamber. Contraction of these retractors pulls the body back into the shell, thereby reducing mantle cavity volume and forcing water out through the funnel and the top edge of the shell aperture. Although the funnel muscle is used during both gentle respiratory movements and powerful jetting associated with swimming, the retractor muscles are reported to operate only during the latter (Bidder, 1962; Chamberlain, 1980a; Packard *et al.*, 1980).

3. Ultrastructure of Funnel and Retractor Muscles

Vertebrate skeletal muscle contains a range of fiber types specialized for different kinds of muscle work. At the extremes, red fibers containing large numbers of mitochondria are used aerobically for sustained submaximal work, whereas white fibers with fewer mitochondria are used anaerobically for short bursts of maximal work (Holloszy and Booth, 1976). Muscle fiber specialization also occurs in cephalopods. For example, the circular mantle muscle of squid is composed of inner and outer zones that are rich in mitochondria and function during gentle sustained respiratory movements. These zones are separated by a middle zone with fewer mitochondria that is used to power rapid bursts of jetting during swimming (Bone *et al.*, 1981; Mommsen *et al.*, 1981). Distinct fiber types that can be distinguished on the basis of mitochondrial abundance also occur in muscles of *Nautilus* (Hochachka *et al.*, 1978). In the funnel muscle, type A fibers and type B fibers occur in equal numbers, with mitochondria occupying about 1 and 22%, respectively, of the cross-sectional area. This distribution differs markedly from the retractor muscle, in which 90% of the muscle fibers are of type A, with mitochondria accounting for less than 2% of the cross-sectional area. These differences in fiber distribution correlate well with muscle usage. Presumably, gentle respiratory movements of the funnel are supported by oxidative pathways in type B fibers, whereas bursts of jetting during locomotion utilize anaerobic glycolysis in the type A fibers of both funnel and retractor muscles.

4. Pathways of ATP Production in Swimming Muscles of *Nautilus*

Carbohydrates and amino acids appear to be the major fuels used during muscle work in cephalopods, with little contribution from lipids. Evidence for this has been reviewed by Storey and Storey (1983) and is based on (1) concentrations of glycogen, free amino acids, and lipids in muscle; (2) activities of catabolic enzymes; and (3) the ability of isolated mitochondria to oxidize pyruvate and amino acids, but not lipid substrates. High concentrations of glycogen, the free amino acids proline and arginine, and taurine have been detected in *Nautilus* retractor muscle (Hochachka *et al.*, 1978; Fields and Hochachka, 1982). Tissue-slice experiments show much higher rates of [^{14}C]glucose than of [^{14}C]proline incorporation into [^{14}C]O_2 by *Nautilus* retractor muscle, a result opposite that observed with muscles of squid and octopus (Fields and Hochachka, 1982; Hochachka and Fields, 1982).

When carbohydrates are catabolyzed aerobically by cephalopod muscle, both the glycerol phosphate and malate–aspartate shuttles may be involved in the cytoplasmic reoxidation of glycolytically derived NADH (Hochachka *et al.*, 1975; Zammit and Newsholme, 1976; Baldwin, 1982). Significant activities of glycerol phosphate dehydrogenase and glutamate-oxaloacetate aminotransferase in *Nautilus* funnel and retractor muscles indicate the potential for use of both shuttle systems during aerobic glycolysis (Hochachka *et al.*, 1978; Baldwin, 1982).

During anaerobic glycolysis, cytoplasmic NAD^+ is regenerated at the terminal step in the pathway by the reaction

$$\text{Pyruvate} + \text{arginine} + \text{NADH} = \text{octopine} + \text{NAD}^+$$

which is catalyzed by octopine dehydrogenase. *Nautilus* funnel and retractor muscles contain moderate activities of this enzyme, and octopine has been shown to accumulate in the retractor muscle during removal of the animal from its shell or following electrical stimulation of isolated muscle strips (Hochachka *et al.*, 1977, 1978). The arginine that condenses with pyruvate to form octopine is derived from the high concentrations of free arginine in muscle, and this concentration increases during anaerobic work as the muscle phosphagen arginine phosphate is hydrolyzed. While arginine phosphate concentrations do not appear to have been determined directly, an increase in arginine concentration from about 24 to 70 μmoles g^{-1} wet weight retractor muscle following electrical stimulation (Hochachka *et al.*, 1978) suggests high concentrations of the phosphagen in resting muscle. This muscle also contains high activities of arginine kinase, the enzyme involved in phosphagen synthesis and hydrolysis (Morris and Baldwin, 1984). High arginine kinase activities are associated with high concentrations of arginine phosphate in resting muscles of other mollusks (Beis and Newsholme, 1975; Zammit and Newsholme, 1976).

5. Relationship between Metabolic Organization and Swimming Behavior

The relative contributions of aerobic and anaerobic muscle work during locomotion in free-ranging animals can be assessed by comparing the maximum activities of enzymes unique to the different ATP-generating pathways. The results of such a comparison carried out with fresh muscle samples from *Nautilus*, octopods, cuttlefishes, and squids that were assayed under identical conditions are shown in Table I. In this study, the activity of octopine dehydrogenase, an enzyme unique to anaerobic as opposed to aerobic glycolysis, was used as an index of the use of a muscle for short bursts of anaerobic swimming. Citrate synthase (Kreb's cycle), glycerophosphate dehydrogenase (glycerol phosphate

Table I. Maximum Activities of Enzymes in Cephalopod Muscles[a]

Animal	Muscle	ODH	CS	GPDH	MDH	GOT	ODH/GPDH + GOT
Octopods							
Octopus macropus	Mantle	321	4.4	6.8	55.4	23.2	10.7
	Arm	251	0.2	5.1	20.1	7.3	20.2
Octopus horridus	Mantle	337	3.3	12.7	73.0	25.3	8.9
	Arm	339	1.5	10.9	36.5	11.8	14.9
Squids							
Sepioteuthis lessoniana	Mantle	520	5.9	10.3	59.4	11.2	24.2
	Fin	353	10.0	8.4	112	15.6	14.7
Thysanoteuthis rhombus	Mantle	952	15.6	4.7	42.3	4.7	101.3
	Fin	391	62.8	2.0	263.0	50.6	7.4
Cuttlefishes							
Sepia bandensis	Mantle	251	2.2	3.3	38.0	8.8	20.7
	Fin	394	5.7	6.8	148	37.2	9.0
Sepia latimanus	Mantle	646	3.0	14.8	32.5	9.1	27.0
	Fin	468	6.4	10.2	97.7	32.3	11.0
Nautiloid							
Nautilus pompilius	Funnel	56	2.3	1.8	69.6	7.1	6.2
	Retractor	88	6.2	6.2	93.4	9.2	5.7

[a] Adapted from Baldwin (1982). Data are expressed in micromoles per minute per gram wet weight muscle. Temperature 25°C. Enzymes: (ODH) octopine dehydrogenase; (CS) citrate synthase; (GPDH) α-glycerol phosphate dehydrogenase; (MDH) malate dehydrogenase; (GOT) glutamate-oxaloacetate transaminase.

shuttle), malate dehydrogenase (Kreb's cycle and malate–aspartate shuttle), and glutamate-oxaloacetate transaminase (malate–aspartate shuttle) were selected as enzymes unique to aerobic metabolism and associated with sustained swimming at submaximal speeds. The ratio octopine dehydrogenase/glycerol phosphate dehydrogenase + glutamate-oxaloacetate transaminase activities provides a useful index of the relative importance of these two types of muscle usage.

For *Nautilus*, low activities of the aerobic enzymes in both funnel and retractor muscle, compared with highly aerobic muscles such as squid fin, suggest low levels of aerobic energy metabolism during slow, sustained swimming. This interpretation is supported by the lower oxygen demands and less efficient oxygen delivery system of *Nautilus*, relative to other cephalopods (Johansen *et al.*, 1978; Redmond, 1978). Octopine dehydrogenase activity is also very low in funnel and retractor muscle, relative to the other cephalopod muscles, while the low octopine dehydrogenase/glycerol phosphate dehydrogenase + glutamate-oxaloacetate transaminase ratios imply basically aerobic rather than anaerobic muscle work. The low pH buffering capacity of 38 slykes obtained for the *Nautilus* retractor is similar to that of the aerobic squid fin (Morris and Baldwin, 1984), again indicating that *Nautilus* does not rely heavily on short bursts of rapid anaerobic swimming.

Information on the swimming behavior of *Nautilus* in their natural habitat is very limited. Although there are anecdotal accounts of animals swimming "as fast as fish" (see Bidder, 1962), observations made by divers and with remote cameras suggest that they are slow swimmers that feed by foraging near the sea floor (Ward *et al.*, 1980a; Saunders, 1984b). Ward *et al.* (1977) reported a maximum speed of 0.25 m sec^{-1} for free-swimming animals released at sea, and this rate is in keeping with the predictions made from the metabolic organizations of the muscles used to power swimming.

Chapter 24

Oxygen Conformity and Metabolic Arrest in *Nautilus*

Analysis of Mechanisms and Functions

P. W. HOCHACHKA

1. Introduction

In 1976 and 1979, two R/V ALPHA HELIX Expeditions to Tañon Strait, Republic of the Philippines, were launched for the purpose of exploring the biochemistry and physiology of *Nautilus pompilius*. In these studies, only preliminary attempts were made to integrate biochemical information with data obtained from whole organisms (Hochachka *et al.*, 1978; Baldwin, 1982). As a result, the biochemical basis and significance of two rather important metabolic responses in *Nautilus* to varying O_2 availability were overlooked. One of these responses is O_2 conformity: As O_2 availability in inspired water declines, the rate of O_2 uptake ($\dot{V}O_2$) also declines. M. J. Wells (personal communication) has obtained data suggesting that at lower temperatures, O_2 conformity sets in at lower O_2 concentrations, which is perhaps not too surprising. A second, more remarkable response is metabolic arrest: As O_2 availability declines to some critical level, $\dot{V}O_2$ rather abruptly stops (Redmond *et al.*, 1978). Three problems arise that apply not only to *Nautilus* but also to all systems that show similar metabolic behavior. First, as O_2 availability declines, how is metabolism suppressed and, at the limit, arrested? Second, how does the organism avoid an energy shortfall? Third, how are cell maintenance functions balanced with the reduced rates of ATP synthesis? As a result of developments that occurred subsequent to the 1976 and 1979 *Nautilus* expeditions (Hochachka, 1986), this species now can be used to illustrate (1) the nature of

P. W. HOCHACHKA • Department of Zoology and The Sports Medicine Clinic, University of British Columbia, Vancouver, British Columbia V6T 2A9, Canada.

these problems, (2) how O_2 conformity and metabolic arrest are achievable, and (3) what they may mean to organisms that rely on them.

2. Oxygen Conformers versus Oxygen Regulators

Cells, tissues, and organisms the $\dot{V}O_2$ of which varies with O_2 availability over broad ranges are termed *O_2 conformers*, to contrast their behavior with that of *O_2 regulators*, the $\dot{V}O_2$ of which is independent of O_2 availability down to very low values. The metabolic response of the mammalian brain to changes in pO_2 fits the pattern of O_2 regulators (Kinter *et al.*, 1984) and represents an extreme end of a spectrum of responses observable in mammalian tissues. Liver expresses an intermediate pattern (Edelstone *et al.*, 1984), whereas skeletal muscle is perhaps the most O_2 conforming of all mammalian tissues. At least in some mammals, no plateau in muscle O_2 consumption is reached even at very high O_2 tensions or very high O_2 delivery rates (Whalen *et al.*, 1973). Similar O_2 conforming patterns are common among numerous invertebrate groups and among ectothermic vertebrates (Mangum and Van Winkle, 1973).

In contrast, isolated mitochondria are universally found to display very high O_2 affinities and respiration rates that are largely independent of O_2 concentration. Their respiratory patterns are clearly those of O_2 regulators, as they should be, since the K_m for O_2 is usually in the 10^{-6}–10^{-7} M range (Sugano *et al.*, 1974). Thus, O_2 conforming patterns displayed by a wide diversity of organisms present us with an interesting metabolic paradox: Why should $\dot{V}O_2$ be declining at O_2 concentrations that are fully saturating for isolated mitochondria? Our recent analysis of this problem indicates that there are three ways to look at the problem (Hochachka and Guppy, 1987).

One possible explanation assumes O_2 sensing. In this model, O_2 concentration serves as a signal for a respiratory set point, but it is unknown how an O_2 receptor might work, if one exists (Mulligan and Lahiri, 1981). A second possibility is that the electron-transfer system (ETS), per se, involves a reversible O_2 binding step that is close to equilibrium with all earlier steps in the chain. If so, this could explain how mitochondrial O_2 uptake rates can respond sensitively to O_2 concentration (D. F. Wilson *et al.*, 1979). Although this model is mechanistic and heuristic, it is not, at this point, widely accepted, partly because cytochrome oxidase catalytic mechanisms are still poorly understood. A third alternative simply assumes true O_2 limitation under *in vivo* conditions due to O_2 sinks coupled with diffusion barriers (D. P. Jones *et al.*, 1985).

One of the most convincing lines of evidence in favor of this interpretation arises from comparisons of the O_2 dependence of different intracellular O_2-requiring functions, such as myoglobin O_2 binding, reduction of ETS components, and urate oxidase O_2 binding. *In vivo*, as O_2 availability declines, all three functions decline in parallel despite very different O_2 dependencies (very different *in vitro* O_2 affinities). Over a decade ago, Chance (1976) took such data to indicate very steep $[O_2]$ gradients with $[O_2]$ at the mitochondria being much lower than would be anticipated from $[O_2]$ in plasma, and that seems still to be a favored interpretation today (D. P. Jones *et al.*, 1985; Chance and Leigh, 1985). True O_2

limitation of mitochondrial oxidative metabolism as O_2 availability declines seems the only possible explanation for conformity in the lungless salamander (Hochachka and Guppy, 1987) and is the alternative here favored for O_2 conformity in *Nautilus* as well, because it is consistent with a limited O_2 delivery capacity—limited by capacities of ventilation, cardiac output, and blood-to-tissue transport (Johansen *et al.*, 1978; Lykkeboe and Johansen, 1982). Limited O_2 delivery capacities are found in most O_2 conforming animals (Mangum and Van Winkle, 1973), so it may be universally applicable that plots of $\dot{V}O_2$ vs. $[O_2]$, although right-shifted on the $[O_2]$ axis, nevertheless reflect the form of true O_2 saturation curves for working mitochondria. However, even if this interpretation can adequately explain O_2 conformity down to low pO_2, it does little to account for $\dot{V}O_2$ being fully arrested at this O_2 availability. A new and different mechanism must be set in motion at this point.

3. Arresting Oxidative Metabolism

Although O_2 limitation may be the most probable explanation for O_2 conformity, the most plausible of potential mechanisms of metabolic arrest appears to involve adenylate translocation, for it is known that in a variety of systems during normoxia–hypoxia transitions, the steady-state distribution of adenylates is shifted in favor of the cytosolic pool (Aprille and Brennan, 1985; Austin and Aprille, 1984; Brennan and Aprille, 1985; Tagawa *et al.*, 1985); there is no reason to believe that the same process does not occur generally in O_2 conforming animals as O_2 availability declines. In effect, the steady state in the equation

$$\text{Cytosolic adenylates} \underset{\text{Normoxia}}{\overset{\text{Hypoxia}}{\rightleftharpoons}} \text{Mitochondrial adenylates}$$

is shifted to the left during hypoxia, and various ADP- and ATP-linked mitochondrial processes, including ATP synthesis rates, decline accordingly (Aprille and Brennan, 1985; Brennan and Aprille, 1985). These shifts in adenylate distribution, presumably coupled with O_2 limitation in all forms of O_2 conforming behavior, are fully and rapidly reversible *in vitro* (Tagawa *et al.*, 1985) and *in vivo* (Brennan and Aprille, 1985). In forms like *Nautilus*, we suggest that O_2 conformity becomes metabolic arrest when adenylate translocation to the cytosol is complete, for at that point, most, and probably all, mitochondrial functions linked to adenylate metabolism become necessarily, if reversibly, arrested (Aprille and Brennan, 1985; Tagawa *et al.*, 1985).

In this view, then, metabolic arrest is O_2 conformity carried to completion (mitochondrial adenylate pools fully depleted), an interpretation that is helpful in explaining why metabolic arrest is most commonly observed in animals that are well adapted to hypoxia (Hochachka, 1985, 1986; Mangum and Van Winkle, 1973). Tissues in these organisms are poor in mitochondria, and the bulk of the adenylate pool is already cytosolic even in normoxia. In such systems, O_2 conforming responses to declining $[O_2]$ presumably would deplete mitochondrial adenylate pools more rapidly than in homologous, mitochondria-rich tissues;

thus, metabolic arrest (i.e., complete depletion of mitochondrial adenylates) may be expected to cut in before O_2 is fully depleted; indeed, this may occur at different $[O_2]$ in different tissues and in different species, a prediction consistent with observations (Mangum and Van Winkle, 1973). In some systems, of course, the process may not reach completion until very low $[O_2]$ (O_2 conformity without metabolic arrest), a situation again commonly observed (Whalen *et al.*, 1973; Mangum and Van Winkle, 1973). For several reasons, then, the adenylate translocation hypothesis (Aprille and Brennan, 1985) seems particularly useful in explaining the various characteristics of metabolic arrest and of O_2 conformity. The next question that arises is how the organism gets by when oxidative metabolism is fully blocked.

4. Minimizing Anaerobic Problems

When O_2 is limiting to any cell, tissue, organ, or organism, two metabolic problems must be resolved. The depletion of fermentable substrate reserves in inefficient anaerobic metabolism must be avoided, and the accumulation of undesirable anaerobic end products must be minimized. Animals that are so hypoxia-tolerant that they are commonly termed facultative anaerobes typically harness three mechanisms for resolving these problems (Hochachka, 1986). These mechanisms include: (1) storing high amounts of fermentable substrate, (2) utilizing more efficient fermentation pathways, and (3) reducing rates of ATP synthesis (Hochachka, 1986). Although *Nautilus* is not so anoxia-tolerant, it shares with facultatively anaerobic animals a notably high ratio of anaerobic/aerobic metabolic capacity. Some tissues, such as the spadix, display very low mitochondrial volume densities and appear to rely almost solely on anaerobic ATP-generating pathways; even tissues such as the heart retain oxidative capacities that are only about one fifth to one fourth those found in more active, O_2-dependent cephalopods (Hochachka *et al.*, 1978; Baldwin, 1982; Storey *et al.*, 1978). In general, all tissues so far examined, including heart, funnel muscle, retractor muscle, gill, kidney, and pericardial gland, tend to have relatively low capacities for oxidative metabolism. Not surprisingly, cardiac output and sustainable swimming speeds are, as a result, only a fraction of those achievable by more active cephalopods (Johansen *et al.*, 1978). The myocardium may be the most O_2-dependent organ in the body of more active cephalopods. In *Nautilus*, however, the ultrastructure of its myocardium more closely resembles that of the myocardium of bivalves (Dykens *et al.*, 1982); the latter are commonly hypoxia-tolerant and capable of metabolic arrest when exposed to hypoxia stress (Hochachka, 1985). Like *Mytilus* and other bivalves, *Nautilus* also expresses a greater dependence on glucose metabolism than is found in other cephalopods. However, no tissues store unusual amounts of glycogen (Fields and Hochachka, 1982), which in all facultative anaerobes is a major fuel for surviving O_2-free episodes. Thus, although *Nautilus* shares many bivalve characteristics, strategy (1) above is not available to it as a defense measure against hypoxia. Enzyme studies indicate that strategy (2) of using more efficient fermentations is also unavailable. Anaerobic metabolism in this genus seems to rely mainly on phosphagen hydrolysis and glycolysis.

On the basis of relative enzyme activities, the main pathway apparently involves the fermentation of glycogen (glucose) to octopine (Hochachka *et al.*, 1977, 1978; Baldwin, 1982). Alanopine and strombine dehydrogenases are absent (Baldwin, 1982), which is presumably why neither alanine nor glycine pools are depleted during periods of O_2-limited work (Hochachka *et al.*, 1978). Some tissues do possess lactate dehydrogenase, implying the presence of an additional pathway of glycolysis, generating lactate as an end product (Fields and Hochachka, 1982), but the energy yield of this process, like that of octopine fermentation, is only 2 moles ATP per mole glucose, so this does not help much. In marine invertebrates, the main means for increasing ATP yield in anoxia is to couple the fermentation of glucose with aspartate, generating succinate or propionate as anaerobic end products (Hochachka and Somero, 1984). In *Nautilus*, however, any significant contribution of aspartate fermentation to succinate or propionate is considered unlikely because aspartate levels are low and do not decline further during hypoxia (Hochachka *et al.*, 1978), unlike the situation in facultatively anaerobic bivalves, in which both the aforementioned conditions prevail (Collicut and Hochachka, 1977). Thus, it seems safe to assume that *Nautilus* cannot utilize energetically improved pathways of fermentation during hypoxia stress [alternative (2) above]. Of the three strategies theoretically available, therefore, the third, reducing ATP synthesis rates, seems most critical for surviving hypoxic episodes. Does this then mean that *Nautilus* must sustain an energy shortfall during hypoxia?

5. Positive and Reversed Pasteur Effects

The question just posed can be rephrased in terms of how *Nautilus* sustains "maintenance" functions when oxidative metabolism is arrested. One possibility, common in many organisms, is to make up the energy deficit by activating anaerobic metabolism (a process that is termed the *Pasteur effect* and requires the rate of accumulation of anaerobic end product and of substrate utilization to increase greatly). With glucose as a substrate, for example, to maintain fully normoxic ATP turnover rates by anaerobic glycolysis would require up to 18-fold-elevated rates of substrate utilization. In *Nautilus* at 17°C and normoxic ATP turnover rates (Redmond *et al.*, 1978), glucosyl consumption would have to rise to about 2.2 μmoles glucosyl units g^{-1} min^{-1}. On known reserves (Fields and Hochachka, 1982), some tissues could sustain this rate of anaerobic metabolism for 15–20 min, but most could not; included in the latter are the retractor muscles, heart, kidney, and gill. If we assume that *Nautilus* at 16–17°C can sustain 60 min of anoxia, a not unreasonable assumption, then it is clear that a Pasteur effect large enough to fully compensate for the energy shortfall is unlikely, a situation also observed in facultative anaerobes. In the latter, usually only minor Pasteur effects are observed, accounting for ATP turnover rates equivalent to about one fifth of normoxic (Hochachka, 1986; Hochachka and Guppy, 1987). In contrast, anoxia-tolerant genera (e.g., *Mytilus*) sustain a reversed Pasteur effect, and anoxic ATP turnover rates fall to about 1/20th of normoxic rates (Hochachka, 1985, 1986; Hochachka and Guppy, 1987; Hochachka and Somero, 1984). Although hypothermic, anoxic turtles so strongly reverse the Pasteur effect that ATP turnover rates are reduced by two orders of magnitude (Hochachka and Guppy, 1987; Hochachka and Somero, 1984), in postanoxic *Nautilus*, the O_2 debt repaid is only about 25–50% of that which would be required if anoxic ATP turnover rates were sustained at normoxic levels (Redmond *et al.*, 1978); i.e., in anoxic *Nautilus*, rates of anaerobic metabolism are suppressed to perhaps one fourth the value required to fully make up the energy shortfall caused by O_2 lack.

The question of how the reversed Pasteur effect is achieved has been examined closely elsewhere (Storey, 1985; Hochachka, 1985) and will not be repeated in detail here. Suffice it to point out that at least some of the metabolite signals required for a standard Pasteur effect, such as adjustment in [adenylate] and [P_i], appear to occur universally on transition from normoxia to hypoxia. Animals that display such reversed Pasteur effects (or only minor positive Pasteur effects) require mechanisms for overriding the coarse controls described above *so as to maintain low glycolytic fluxes despite O_2 lack.* It appears that there are three or four such overriding mechanisms available to organisms capable of such metabolic arrests, and these mechanisms may well be species- or even tissue-specific (Hochachka, 1985).

Thus, we are faced with an interesting situation in *Nautilus:* As O_2 availability declines, $\dot{V}O_2$ declines (O_2 conformity with partial loss of mitochondrial adenylates), and at a critical pO_2 (which appears to be temperature-dependent), $\dot{V}O_2$ ceases altogether (oxidative metabolism is fully arrested, presumably because the mitochondrial adenylate pool is fully depleted). At this time, not all (perhaps about 25%) of the consequent energy shortfall is made up by anaerobic metabolism, indicating that a potentially 18-fold glycolytic activation is also suppressed to about one fourth of the expected value. Similar metabolic behavior appears in numerous facultative anaerobes under hypoxic stress, and in both cases the same question arises: *How are maintenance functions sustained as ATP synthesis rates decline in hypoxia or anoxia?* Since the possibility available to many organisms of making up the energy shortfall with powerfully activated glycolysis (Kinter *et al.*, 1984) is ruled out, in principle the only alternative left for O_2 conforming organisms is to cut the costs of maintenance activities in proportion to declining O_2 availability (more specifically, in proportion to declining rates of ATP generation). Obviously, numerous processes contribute to tissue maintenance metabolisms, yet one of the most energetically costly of such processes, common to all tissues, is the maintenance of electrochemical gradients across cell membranes. Our analysis implies that the cost of maintaining electrochemical gradients across cell membranes must decline as O_2 availability declines; otherwise, metabolic and membrane functions in effect would become decoupled.

6. Stabilizing Membrane Functions during Anoxia

A good way to illustrate the critical role of maintaining membrane electrochemical gradients is to consider hypoxia-sensitive systems, such as the mammalian brain. Within 60 sec of ischemia, decreased rates of ATP generation are unable to keep ion pumping in balance with ion leaks; K^+ begins to flood out of cells into the extracellular fluid, and Na^+ moves in the opposite direction. When the change in voltage potential becomes large enough, Ca^{2+} floods into the intracellular fluid via voltage-activated Ca^{2+} channels or pores. Simultaneously, anoxic mitochondria appear to lose their capacities for maintaining ion and electrical gradients, and low levels of intramitochondrial Ca^{2+} pools are depleted, thus also contributing to cytosolic accumulation of Ca^{2+}, a potential cellular toxin. If uncontrolled, such processes may lead to cell damage or cell death by mechanisms that are gradually being unraveled (for recent reviews, see Hochachka, 1986; Tagawa *et al.*, 1985). It is an instructive observation that such a breakdown of electrochemical gradients does not occur in homologous organs of anoxia-tolerant animals (Sick *et al.*, 1982) and cannot occur during the metabolic arrest of anoxic *Nautilus* (otherwise, the organism clearly could not survive the episode). In both instances, membrane stabilization is achievable (i.e., electrochemical gradients are largely preservable), not because anaerobic glycolysis can be activated to keep ion pumping in pace with ion leaks, but because the latter can be slowed down. Since such leaks are due to ion movements through ion-specific channels and leak rates depend on the functional density of ion-specific protein pores, we arrive at a potentially useful generalization: *Metabolic arrest and oxygen conformity appear to be possible, if—and only if—the densities of functional ion channels are reduced in proportion to declining rates of ATP synthesis* (for expansion of this concept, see Hochachka and Guppy, 1987). This should be true not only for *Nautilus*, but also for all systems that show capacities for oxygen conformity, for metabolic arrest, or for both processes (Mangum and Van Winkle, 1973; Hochachka and Guppy, 1987).

Although this analysis satisfactorily accounts for current data, if we turn the situation around (move up the O_2 saturation curve rather than down), we are confronted with the question of why increased densities of functional ion channels develop as O_2 availability increases (as ATP synthesis rates increase). Although we do not know the answer to this question at this time, an interesting possibility is that cells and membrane-bound organelles in effect become more reactive at higher channel densities. Ion fluxes through channels usually subserve specific biochemical or physiological functions and, presumably, these would be favored at higher channel densities; hence, high channel densities and high ATP turnover may codevelop and be coadaptive. Conversely, we anticipate that a metabolically arrested organism, such as *Nautilus* below a critical pO_2, may be in a relatively nonreactive state, analogous to that in anesthesia. We do not know whether this adequately describes an anoxic *Nautilus*; however, the description empirically fits other anoxic organisms, such as anoxic turtles, very well indeed. In this view, then, an additional cost of O_2 conformity or of metabolic arrest capacities may be reduced reactivities under O_2-limiting conditions. That this is an acceptable price for survival is indicated by the wide variety of organisms that share with *Nautilus* these strategies of adaptation to O_2 lack (Mangum and Van Winkle, 1973).

ACKNOWLEDGMENTS. Much of my own work on this exotic organism was supported by NSERC (Canada), while the 1976 and 1979 ALPHA HELIX Expeditions were supported by the NSF (U.S.A.). K. Johansen and C. P. Mangum deserve special thanks for making me aware of these problems in the first place. I would like to dedicate this paper to the memory of K. Johansen, who shared with us a fascination in the metabolism and physiology of *Nautilus*.

Chapter 25

Ventilation and Oxygen Extraction by *Nautilus*

M. J. WELLS

1. Introduction

Because of the shell, *Nautilus* cannot ventilate in the same manner as other cephalopods, which use the mantle musculature to drive the respiratory stream. In *Nautilus*, the respiratory stream is propelled by the funnel and its associated structures, aided at times by contraction and extension of the head retractor muscles. The account that follows shows how these structures generate the ventilatory flow. It is based largely on experiments and observations made by the author during a visit to the Motupore Island Research Department of the University of Papua New Guinea during February to May 1984. A more detailed report is given in Wells and Wells (1985).

2. Anatomy

An outline of the structures concerned with ventilation in *Nautilus* is given in Fig. 1, which also shows the direction of flow of the water through the mantle cavity. The position of the gills is critical to any understanding of the ventilatory arrangements. Classically (e.g., in the monograph by Owen, 1832), the gills are illustrated as hanging loose in the mantle cavity. This depiction is a result of

M. J. WELLS • Department of Zoology, University of Cambridge, Cambridge CB2 3EJ, United Kingdom.

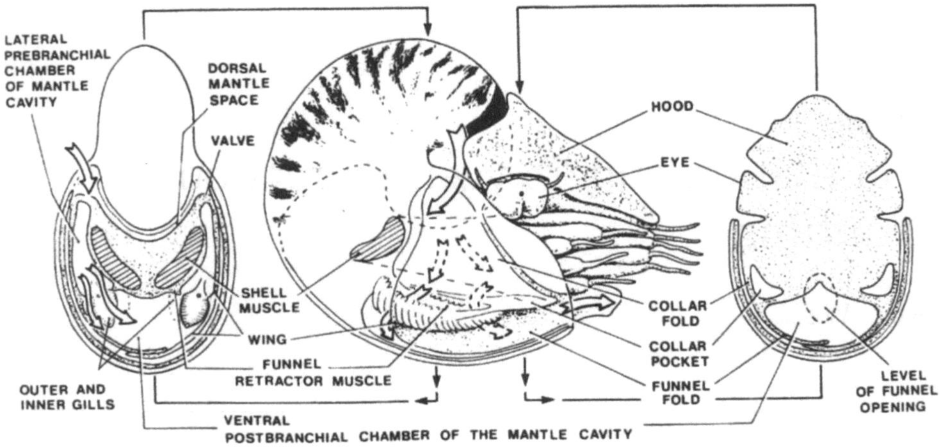

Figure 1. Structures within the mantle cavity of *Nautilus*. Arrows show the direction of water flow. Reprinted with permission from Wells and Wells (1985).

figures being drawn from fixed material, with the gills, funnel, wings, and mantle all more or less contracted. A more realistic appraisal of the position of these structures in the living animal can be made by breaking away parts of the shell from lightly anesthetized or freshly dead specimens held in their normal orientation, with the soft structures supported by water.

Dissected from the side in this manner, the mantle must first be cut away to reveal one of the wings (Fig. 2a). They are presumably derived from the backward extension of the funnel and collar folds, which fuse to form a pair of flaps, overlapping below and joined to the head above (Fig. 1). The wings are thin and muscular and in life always active. They propel the ventilatory stream in a manner explained in Section 4.

Folding back the wing exposes the outer gill on the side concerned (Fig. 2b). The gill extends forward horizontally from the viscera, which form the hind wall of the mantle cavity. It is supported in this position by a rib of connective tissue that forms a bracket that holds the upper surface of the gill against a shelf of tissue lying along the side of the head and digestive gland. Toward the front end of the gill, the shelf extends laterally to form the floor of the collar pocket (Fig. 1) and thence joins the wing, which lies along the outer edge of the gill. Seen from above, the lamellae that form the exchange surfaces of the gill are held out by a series of bars extending at right angles to the central rib (Fig. 2b). Despite the apparent cocurrent layout of the afferent and efferent branchial vessels, the blood in the exchange surfaces runs countercurrent to the water flow over them [Joubin (1890) includes a description of the anatomy of the gills]. If the first visible gill is pulled aside, a second, smaller gill is revealed, again bracketed to the viscera; this gill blocks a gap between the posterior inner margins of the outer gill and the central body mass that joins head to viscera (Fig. 2c).

The gills thus divide the mantle cavity into three chambers, two lateral and prebranchial, the third ventral and postbranchial. Because the gill lamellae are

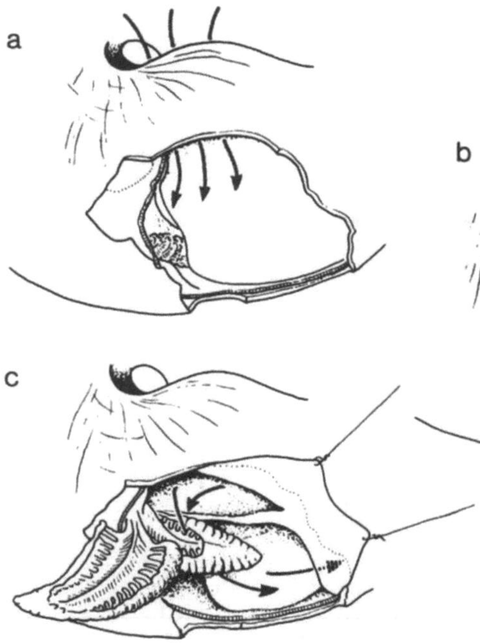

Figure 2. Three stages in the dissection of a freshly killed *Nautilus*, showing the position of the gills in life. Part of the mantle has been cut away to show the wing on the right side. Reprinted with permission from Wells and Wells (1985).

soft, they will fold against one another to block any tendency for a return flow upward from the postbranchial to the prebranchial compartments.

3. Pressures That Drive the Ventilatory Flow

Colored markers (skimmed milk or methylene blue) injected close to the lateroposterior margins of the hood are sucked in close to the umbilicus at each side. The upper margin of each wing acts as a valve that opens and closes at a rate that depends mainly on temperature (Fig. 3). If the animal is active and using its jet for propulsion, the rate (about 0.5 Hz at 16°C at rest) can more than double, and there is generally some backflow through the valve. Outflow through the funnel begins soon after the inlet valves open and continues after the valves have shut. In ventilation at rest, there is again no sign of a reverse flow.

The pressures that drive the ventilation stream are evidently quite low. Packard *et al.* (1980) held a cannula in front of the funnel of a restrained *N. macromphalus* that weighed 687 g and found peak pressures of less than 0.1 kPa. Similar low pressures were observed in a somewhat smaller (387 g), free-moving *N. pom-*

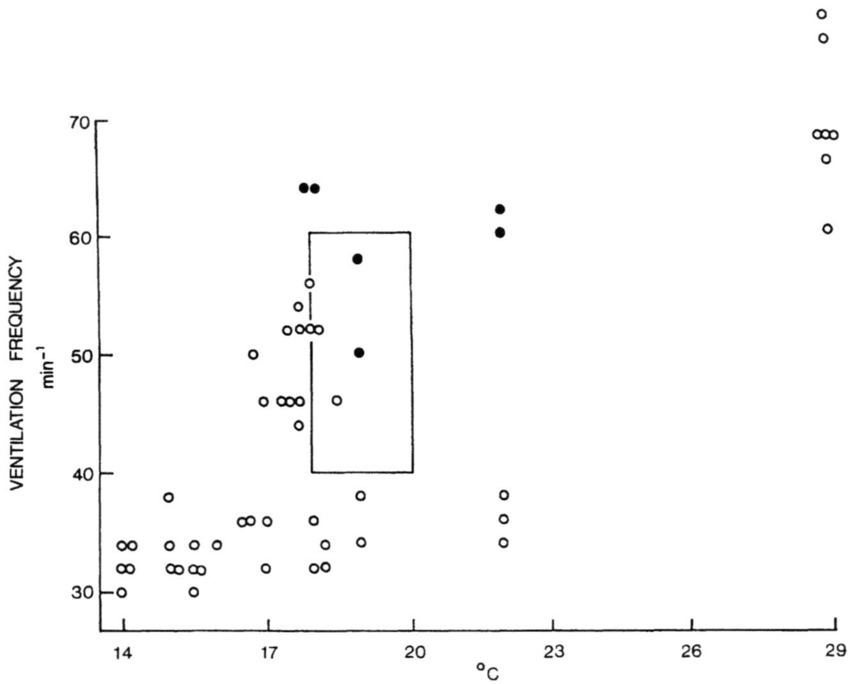

Figure 3. Ventilation frequency and temperature. Key: (○) animals at rest; (●) active animals. The rectangle shows the range of frequencies reported by Johansen *et al.* (1978). Reprinted with permission from Wells and Wells (1985).

pilius, using a cannula running through the shell and mantle into the ventral mantle cavity (Wells and Wells, 1985).

The cycle of pressure changes that occur within the mantle cavity has been observed in two ways. The least intrusive technique was to cut small windows (1–2 cm in diameter) in the shell. Changes in the convexity of the mantle blocking the gap showed pressure differentials between the inside and the outside. Similarly, and with little more interference, holes were bored in the shell, and cannulas, penetrating through punctures in the mantle, were used to observe pressures manometrically. These experiments showed that pressure within the mantle is generally positive, with a brief negative phase that immediately precedes the opening of the valves. Changes below the gills lagged slightly behind those above.

In a third type of experiment, a cannula formed a loop running from a lateral prebranchial to the ventral postbranchial chamber, bypassing the gills on one side (Fig. 4a). The movement of colored markers injected into the loop showed the direction of flow (and therefore the pressure gradient) between the two compartments. This sort of experiment proved particularly revealing. The markers always moved from prebranchial to postbranchial, never in the opposite direction. Cycles of acceleration and deceleration were superimposed on a slow, steady flow. Whenever the loop was cut, marker was ejected from both ends.

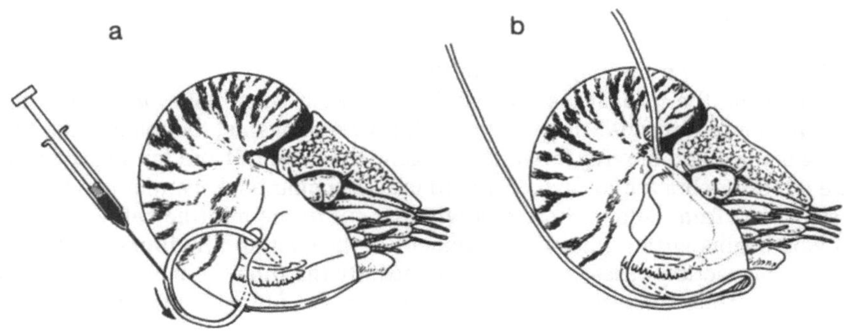

Figure 4. Positions of cannulas used to study ventilation in *Nautilus*. (a) The cannula penetrated shell and mantle. Injected markers showed flow from above to below the gills at all stages in the ventilation cycle. (b) The cannulas were attached to the shell. The upper cannula collected samples from a point outside the mantle cavity close to the inlet valve on one side; the lower cannula looped round and ran into the mantle cavity between the mantle and funnel folds, to emerge between the latter just below the gills. The two were used to collect inhalant and exhalant samples when assessing oxygen extraction from the ventilatory stream. Reprinted with permission from Wells and Wells (1985).

From the pressure observations, it would seem that there is always, or very nearly always, a pressure gradient across the gills, whatever the stage of the ventilatory cycle. In the absence of measurements of gill resistance and reliable figures for the absolute pressures in different parts of the mantle cavity, it is impossible to be certain that flow across the gills is continuous at all stages in the ventilation cycle. It is evident, however, that flow across the exchange surfaces will proceed more or less steadily throughout the greater part of the ventilatory cycle, so that the stagnant period, if any, is very short.

A continuous or nearly continuous flow across the gills would be expected in any case, because there is no clear arithmetic relationship between heartbeat frequency and ventilation frequency. In *N. pompilius*, the heart beats at about 13 times/min at 17°C (Bourne *et al.*, 1978). Ventilation is two or three times as fast (see Fig. 3). The mismatch between the frequencies would lead to fluctuations in the oxygen supply to the blood passing through the exchange vessels if ventilatory flow across the gills were other than constant. A similar situation exists in coleoids, in which the heartbeat rate is commonly faster than ventilation and in which, again, the problem appears to be resolved by a pumping pattern that results in a near-continuous flow across the gills (Wells and Smith, 1985).

4. Wing Movement and the Ventilatory Flow

In life, the wings move continuously. This movement can be seen by observing the valves from above or by looking into the funnel from the front. Further details of the movement of the wings are hidden by the shell. To take matters further, it is necessary to break windows in the shell and sometimes also to cut away the mantle, so that the movements of the wings can be observed. Transparent polystyrene sheeting can then be glued to the shell to prevent the wings and the remaining mantle from bulging out of the holes.

Seen from the side, at the level of the gills, each wing moves in a cycle, first extending backward with the posterior edge pressed against the mantle and then folding inward and moving forward, caressing the outer gill and forcing the water in the lateral prebranchial cavity forward into the collar pocket and down through the gill. Seen from below, through a window cut in the midline below the gills, the cycle begins with an inward movement of the overlapping funnel folds. The wave of contraction passes rapidly backward. By the time the lateral parts of the wings reach their maximum forward excursion, the funnel folds have begun to expand again. The inward movement of the funnel folds coincides with an inward and downward movement of the gills, which can be seen from below or by watching from in front, into the aperture of the funnel. As the funnel folds move apart, the gills rise again. Further details of experimental techniques are given in Wells and Wells (1985).

5. Ventilation Cycle

Putting the pressure and movement observations together, it is possible to reconstruct the sequence of events that drive the ventilation stream as follows:

1. The margins of the funnel folds move inward. The wave of movement spreads rapidly back along the wings, which start to move forward, stroking the gills.

2. An early effect of this forward movement is to pull down the upper margins of the collar folds, behind the eyes. This opens the region between the wings and the mantle to the outside, and water flows in from above as the wing margins move inward and forward.

3. As the forward movement continues, pressure builds up in front of the wings, expanding the collar pockets and forcing water down through the gills. The gills move downward and toward the midline of the animal. The total volume of the prebranchial cavities—water in front of the wings plus water drawn in behind—increases, whereas that of the postbranchial cavity decreases.

4. Water flows out in a jet through the funnel.

5. As the water trapped in front of the wings escapes through the gills, the volume of the prebranchial cavities decreases and the gills draw apart and move upward. The upper margins of the collar folds relax to close the inlet valves.

6. As the wings reach the limit of their forward movement, a wave of expansion begins in the funnel folds and spreads backward. Expansion draws water down through the gills and maintains the pressure differential across the gills.

7. The wings begin to return backward, sweeping along the sides of the mantle cavity to enclose the water that flowed in behind them on the previous forward movement. Flow out through the funnel stops.

8. Flow across the gills nevertheless continues while the funnel folds are still expanding; the postbranchial cavity is still increasing in volume, while the volume of the prebranchial cavities continues to decrease.

9. The wings are now fully extended, almost touching the hind wall of the mantle cavity. The fully expanded folds of the funnel press against the ventral wall of the mantle. There must be a brief period during which there is no muscular contraction to generate a pressure differential across the gills. This does not necessarily mean that flow stops. Inertia would tend to maintain flow despite gill resistance.

10. Contraction begins with the funnel folds and spreads backward. The cycle has begun again.

6. Oxygen Extraction

An account of the respiratory physiology is presented by Redmond (Chapter 21). Compared with the coleoids that have been studied, *Nautilus* has a rather low metabolic rate. Oxygen uptakes in the region of 30 ml kg^{-1} hr^{-1} (flesh weight) have been found at 17°C, about half the consumption that we would expect for *Octopus* and rather less than one third of the uptake of *Loligo* (Redmond *et al.*, 1978; Wells and Wells, 1985). It should perhaps be pointed out that the coleoids studied have tended to be active pelagic or littoral animals, living in habitats very different from the deep-water haunts of *Nautilus*. When figures for deep-water coleoids become available, the view that *Nautilus* is a rather slow sort of creature by cephalopod standards may need revision.

Oxygen extraction from the ventilatory stream has been studied by samples taken through cannulas attached to the shells of free-moving animals, as shown in Fig. 4b. Some typical results are summarized in Fig. 5. Extraction is highly variable. If the animal is resting quietly on the bottom or attached by a few tentacles to the side of the respirometer, extraction is commonly in the region of 5–10% (Fig. 5a). At rest, following a period of activity, it may rise to 20% or more

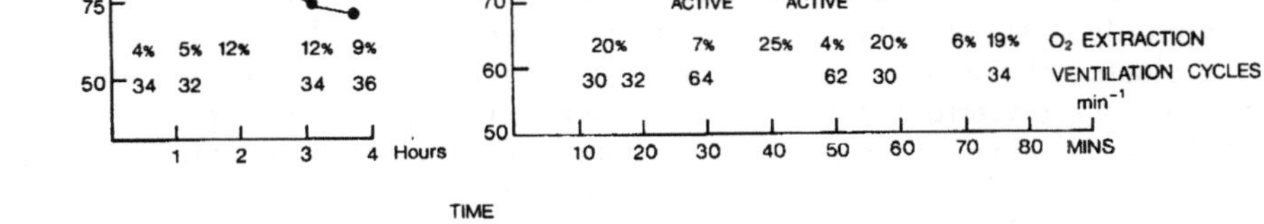

Figure 5. (a) Oxygen uptake and extraction in a closed respirometer, showing the inhalant (O) and exhalant pO_2 (●) collected through cannulas mounted as in Fig. 4b. The percentages are the proportions of the available oxygen removed. (b) Similar, but in an open respirometer, showing the effect of activity on oxygen extraction. The experiments were made at 14–16°C. Modified from Wells and Wells (1985).

(Fig. 5b). The highest extraction rate recorded was 43% at a tank-water pO_2 of 130 mm Hg at 19°C.

These values are similar to those found by Johansen *et al.* (1978) and Redmond *et al.* (1978): 7.2 ± 4% S.D. measured directly from the exhalant flow, 35% estimated from values for O_2 uptake and ventilation flow rate found in separate experiments. Compared with coleoids, these are low rates of extraction. *Octopus vulgaris* and *Sepia officinalis* regularly remove more than 50% of the oxygen from the ventilatory stream (Hazelhoff, 1938; Wells and Wells, 1984; Winterstein, 1909), and even the cold-water species *O. dofleini* averaged 27%, considerably higher than most of the values found with *Nautilus* (Johansen and Lenfant, 1966).

7. Ventilation Stroke Volume

Direct measurements of the exhaled volume from *Nautilus* have been made by Johansen *et al.* (1978). They cut a plastic bottle to fit over the shell and hood (leaving the inhalant openings free) and then collected the exhaled volume in a condom placed over the neck of the bottle. Stroke volumes averaged 2.9 ml at 18–20°C for four animals having a mean flesh weight of 488 g. A single 378-g specimen, tested both when quiet and when active, showed a fivefold increase in ventilation volume per minute, with ventilation frequency rising from 40 to 70 cycles/min and stroke volume increasing from 0.74 to 2.1 ml.

Estimates of the stroke volume based on Fick calculations (oxygen uptake, percentage extraction, and ventilation frequency being known, as in Fig. 5a) yield values three or four times (exceptionally ten times) greater than the 3 ml obtained by direct collection of the ventilation stream. The stroke volume of 3 ml would seem, intuitively, a rather small throughput from a system with a total volume of around 75 ml [mantle volume from a fixed specimen of 470 g with the retractor muscles contracted (Packard *et al.*, 1980)]. In view of the very low pressures now known to drive the ventilation stream, the most likely explanation of the differences found is that even the inflation of a collapsed condom creates a backpressure sufficient to impede the ventilation stream.

8. Retractor Muscles, Ventilation, and Jet Propulsion

Jet propulsion by *Nautilus* is discussed by Chamberlain (Chapter 32). When the animal is at rest and has not recently been active, ventilatory movements are commonly so slight that nothing can be seen from outside other than the opening and closing of the valves behind the eyes and a minuscule expansion and contraction of the aperture of the funnel. There is generally no perceptible movement of the hood. There is nevertheless commonly a sufficient jet to drift the animal slowly backward, if it is neutrally buoyant, or to rock it gently to and fro, if it is resting on the bottom. In aquaria, resting *Nautilus* generally anchor themselves to the side of the tank by a few of the tentacles.

Bidder (1962) reported this and was of the opinion that the animal could develop a sufficient jet for effective locomotion without contraction of the head retractor muscles. This may or may not be so; what is certain is that the head retractor ("shell") muscles soon come into play if the animal is handled and are certainly used in rapid jetting. The pressures produced are then an order of magnitude greater than those found in quiet ventilation. If the animal is breathing deeply after a period of activity, the hood rises and falls rhythmically in time with the ventilation cycle. Packard *et al.* (1980) have pointed out that *Nautilus* can change from quiet ventilation to jet propulsion without interrupting the rhythm, which would suggest that elements of the same musculature are involved. The very large retractors include transverse as well as longitudinal muscle fibers. When the transverse fibers contract, the retractors will be extended, raising the hood and drawing water into the mantle. Oxygen extraction declines during periods of activity (Fig. 5b), which would suggest that the gills are displaced during vigorous jet propulsion and then no longer form a complete barrier between the lateral and ventral chambers of the mantle cavity.

9. Oxygen Debt

At present, we have far too few data to make reliable statements about oxygen debt in *Nautilus*. The animal appears to run up an oxygen debt in extreme hypoxia. If it is allowed to deplete the oxygen in a closed respirometer, *Nautilus* regulates well down to pO_2's of 75 mm Hg (Chapter 21) and in some cases well beyond this (Fig. 6). Somewhere in the region of 20 mm Hg, *Nautilus* ceases to ventilate.

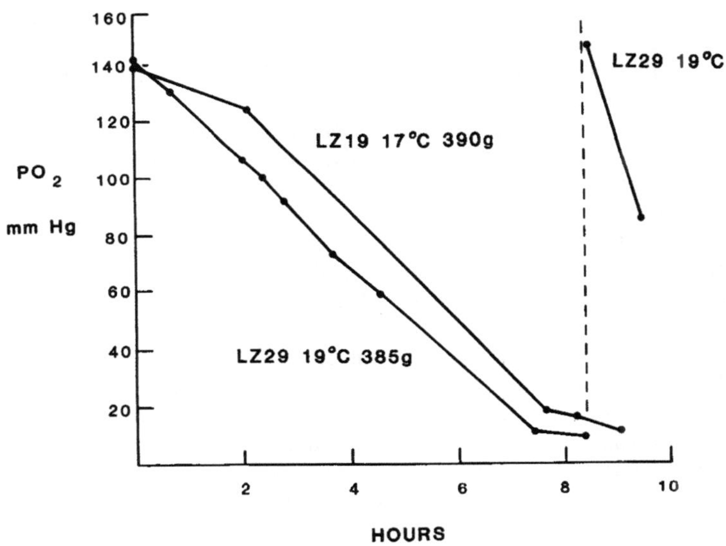

Figure 6. Oxygen uptake in extreme hypoxia in *N. pompilius* (Nos. LZ19 and LZ29). Specimen LZ19 was still ventilating at 40 min^{-1} at the end of its run. Specimen LZ29 had ceased, but revived within a few minutes of being placed in a respirometer with well-aerated seawater; it then consumed oxygen at the accelerated rate shown.

Animals in water depleted to this extent are typically somewhat extended, with the tentacles limp, but they still respond to touch and may withdraw into the shell or produce repeated jets from the funnel. If their tanks are aerated, they very soon recover. Oxygen uptake is then well in excess of previous resting levels (Fig. 6).

The oxygen debt that can be carried by other cephalopods is very limited; figures of 73 ml kg^{-1} have been found for *Illex* (O'Dor, 1982) and 22 ml kg^{-1} for *Octopus* (Wells and Wells, 1984). The titer of octopine dehydrogenase in *Nautilus*, and the balance between this enzyme and enzymes concerned in aerobic metabolism (see Chapter 23), indicate that *Nautilus* is no exception; it is overwhelmingly an aerobic organism, with little capacity for sustained anaerobiosis.

10. Ventilation, Oxygen Uptake, and Exercise

Activity can double oxygen uptake (Chapter 21). Ventilation frequency also doubles in activity, but oxygen extraction from the ventilation stream falls when the animal jets (Wells and Wells, 1985). If the animal is to avoid accumulating an oxygen debt, the ventilation stroke volume must increase to compensate for the lowered rate of extraction. If the rates of extraction shown in Fig. 5b are typical, the ventilation stroke volume would have to triple from about 10 to about 30 ml in an animal of 500 g. A value of 30 ml is in close agreement with the only value so far observed directly (Chapter 32).

We do not know, of course, how activity in a respirometer would relate to movement over the ground in the sea. Ward *et al.* (1977) estimated 0.25 m sec^{-1} as the maximum speed that they observed in the sea. Chamberlain (1976) measured the coefficient of drag as 0.43 (using vol $^{0.67}$ as the characteristic area), and on this basis the mechanical power required to drive a 500-g *Nautilus* at 0.25 m sec^{-1} would be 0.023 W (Fig. 7).

The extra oxygen uptake of an active *Nautilus* of this size is around 0.25 ml min^{-1}, which at 1 ml = 20 J would yield 0.08 W. But the wings, funnel, and retractor muscles are unlikely to be more than 25% efficient, and a rough estimate of the Froude efficiency suggests that this cannot be greater than 40% and is more probably in the region of 25% (see Fig. 7). So the flat-out power requirement is about three times what is continuously available.

Plainly, the animal cannot travel at full speed without running up an oxygen debt, but it could manage about 15 cm sec^{-1} almost indefinitely. This speed would give it a maximum range of about 14 km in 24 hr, if it remained continuously active. In fact, telemetered animals generally moved less than 2.0 km in 24 hr (Ward *et al.*, 1984). This distance was their horizontal movement, usually associated with diurnal inshore–offshore migrations up and down the reef slopes, to shallower water at night (vertical movement adds little to the power requirement, since the animals are neutrally buoyant). *Nautilus* is probably only intermittently active in the sea, with a few minutes of activity every 40 min or so, as Zann (1984) found in aquaria. As he points out, it would make good sense for a chemosensitive scavenger to shift its location once in a while, and it would be more economical to sit and sample the current from time to time rather than remain continuously on the move.

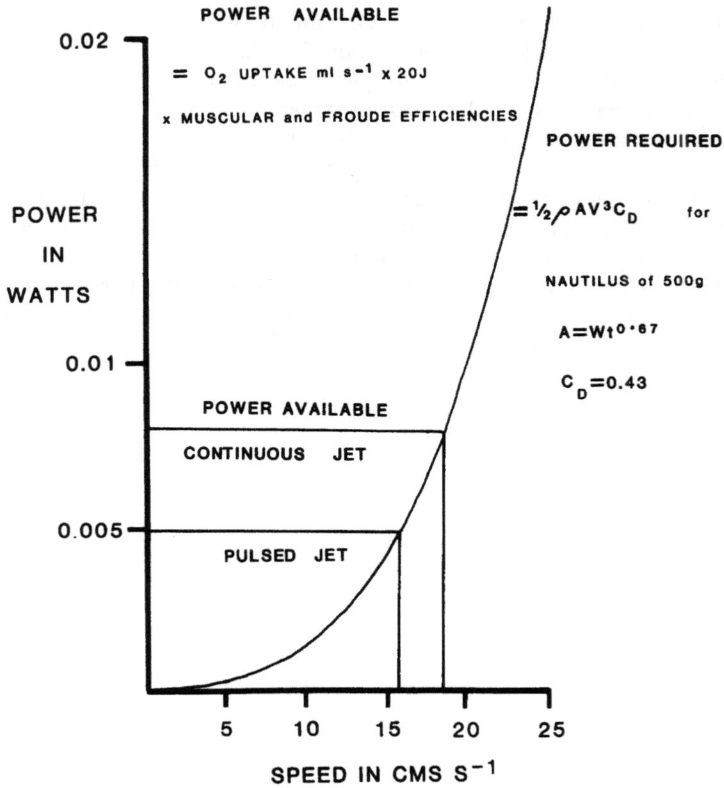

Figure 7. Sustainable rate of locomotion in *Nautilus*. Power required = drag × velocity. Power available assumes a largely protein diet with breakdown to ammonia yielding 20 J/ml O_2 consumed. Muscular efficiency is taken as 25% (it is likely to be less rather than more). Froude efficiency (F.E.) of the jet is estimated from animal velocity/(animal velocity + ½ jet velocity) or

$$\frac{2}{1 + \sqrt{\dfrac{1 + AC_D}{A_F}}}$$

where A_F is the cross-sectional area of the funnel aperture. This gives a value of F.E. = 40% for a continuous jet. The jet is of course pulsed, and this would reduce F.E. to around 25%.

The cost of transport, at around 56 ml O_2 kg^{-1} km^{-1} at 0.54 km hr^{-1} (15 cm sec^{-1}), is low compared with that of *Octopus* walking (250 ml kg^{-1} km^{-1} at 0.34 km hr^{-1}) or squid (*Loligo*) swimming by jet propulsion (170 ml kg^{-1} km^{-1} at 3.6 km hr^{-1}) (O'Dor, 1982; Wells *et al.*, 1983b; Wells and Wells, 1984). Neutral buoyancy has distinct advantages in fuel economy, but the bulky, high-drag means of achieving it restricts *Nautilus* to slow speeds. Fishes, like the squids, tend to have very low drag coefficients, and the teleosts at least have neutral buoyancy, but they trade the advantage of low drag for very much higher cruising speeds, ending up with costs of transport similar to that of *Nautilus*.

When one considers the locomotor performance of *Nautilus* in relation to its life style, the advantages of a low cost of transport are evident. The animal can keep going for a long time between infrequent meals. The potential disadvantage of relatively slow progress toward the food source depends very much on the nature of the competition. In the photographic sequences in Chapter 3, Saunders shows that in *Nautilus*'s present habitat, scavenging fish do not appear to represent a serious threat, whereas scavenging crustaceans arrive first only at night. The tradeoff between speed and economy is nicely balanced and must be one factor that determines the relatively small differences in size and shape of extant species, as compared to the much larger variability in extinct forms.

VII

Reproduction and Growth

Chapter 26

Reproduction and Embryology of *Nautilus*

JOHN M. ARNOLD

1. Introduction

In the past 15 years, there has been tremendous progress in our understanding of many aspects of *Nautilus* biology. This progress is due in part to an increase in the number of investigators now interested in this genus, but is largely due to successful collection and aquarium culture techniques (Carlson, 1979; Mikami et *al.*, 1976) (see also Chapter 35). When I began writing this chapter, I jokingly offered to leave four blank pages to cover what was then known about the embryonic development of *Nautilus*. That same afternoon, I opened some neglected egg capsules and found the first living *Nautilus* embryo (Arnold and Carlson, 1986). What follows is more a progress report than an attempt to present definitive data and meaningful interpretations.

2. Anatomy of the Reproductive System

Like all other Cephalopoda, *Nautilus* is in essence bilaterally symmetrical with secondary modifications of the body plan that make it asymmetrical. The reproductive system reflects this generalization, but until the internal anatomy of the embryo has been thoroughly studied, the origins of these modifications

JOHN M. ARNOLD • Pacific Biomedical Research Center, University of Hawaii, Honolulu, Hawaii 96822; Marine Biological Laboratory, Woods Hole, Massachusetts 02543.

will not be known. There is still much to be learned about the histology of all the organ systems of *Nautilus*, particularly the reproductive tract. The following description of the sexual anatomy is based primarily on the work of Griffin (1900) and Willey (1902) and on the author's personal observations.

2.1. Female Reproductive Tract

In the female, a large unpaired ovary lies at the posterior end of the visceral mass and conforms in shape to the contour of the last septum. Anteriorly, the ovary is rounded, so the entire structure is roughly ovate. It is enclosed in a translucent membrane and supported by ligaments attached to the mantle and surrounding viscera. It opens anteriorly by a short, thick-walled tube that inserts into the thin-walled portion of the oviduct. These two structures are not connected, so, anatomically, the oocytes are shed into the coelom, but functionally they pass into the proximal portion of the oviduct (Willey, 1902). The distal end of the oviduct is quite muscular, wide, and short. The surface is covered with deep circumferential folds, which suggest that it is capable of great distension as the oocytes are passed through it. As they are shed, the oocytes must pass by the openings of the nidamental glands and the glandular oviductal tissue. It is presumably here that the egg jelly and two layers of egg capsule must be laid down.

The ovary is large and filled with several sizes of oocytes. The largest oocytes are ovate in shape and measure about 2.5 cm in length. There appear to be several size classes of oocytes in any given gonad, the number in each class varying roughly from 6 to 10. The exact number of oocyte size classes is, as yet, unknown. This suggests that unlike other cephalopods, *Nautilus* females lay a clutch of eggs, wait for another batch of oocytes to mature, then lay a second clutch and repeat this cycle. Carlson (personal communication) confirms this reproductive activity, with a period of a few months for *N. belauensis* and *N. pompilius* in the Waikiki Aquarium in Hawaii. A. W. Martin *et al.* (1978) reported a yearly cycle of copulation (June to September) for *N. macromphalus* in the Aquarium de Nouméa in New Caledonia. Accurate data are sparse, but it is obvious that *Nautilus*, unlike most other cephalopods, is not semelparous.

The largest and (by analogy to other cephalopods) most mature oocytes (Selman and Arnold, 1977) are approximately 2.5 × 1.6 mm, ovate masses of thick, liquid yolk. In the ovary, they conform to the shape of the gonad on the outside and are compressed into shapes that fit the adjoining oocytes and tissues on their inner surfaces. At the apex of each oocyte at the point where the arterial supply contacts the oocyte, there is an accumulation of protein-rich cytoplasm that is grayish in color, in contrast to the yellow-brownish yolk of the rest of the oocyte. In this area, Willey (1897a) saw a large "germinal tract" that, according to my observations, must be the sizable germinal vesicle. When the oocytes are moved from the ovary into a dish of seawater, they gradually assume a smooth contour but retain the ovate shape. The younger oocytes can be seen to retain folds of follicular epithelium that penetrate down into the yolk and form the "follicular meshes" noted by Willey (1897a). Near what will develop into the future animal pole the vascular supply is obvious, as it is in other cephalopod embryos. Associated with one side of this area is a cellular nonoocyte structure, which is pale yellow to white in life. Preliminary histological sections suggest that it may possibly in some way be involved in yolk production in an extraoocyte fashion, similar to that seen in some insect eggs. Younger size classes are progressively smaller, with more and more of their volume occupied by the follicular epithelium. The very youngest oocytes are surrounded by only a relatively few follicular cells, which are not syncytial as in the older follicular epithelium. The follicular epithelium not only is involved in the secretion of the yolk in the developing oocyte, but also, as the oocyte is reaching maximum size, produces the thin chorion that encompasses the entire cell and is penetrated by the micropyle. The very smallest (= youngest) oocytes appear to have only one follicle cell next to them. Except for the possible accessory yolk organ and the extreme size of the oocyte, oogenesis is quite analogous to oogenesis in *Loligo* (Selman and Arnold, 1977). I have not seen any suggestion of an albuminiparous gland in the ovary as reported by Keferstein (1865).

The secondary sexual organs of the female include the "squashed-W-shaped" nidamental glands in the mantle cavity and the organ of Valenciennes below the mouth. As Willey (1902) and Haven (1977a) observed, when the nidamental glands mature, they change from cream-colored to yellowish, to yellowish green, to greenish, and, finally, to an olive green color at maturity. This color can be observed in the living female, using the technique of Haven (1977a) of removing the animal from the water and holding it upside down so that the muscles tire and relax. By looking over the gills, one can clearly see the nidamental glands inside the shell. This procedure does not harm the animal, but it accurately identifies its sex and has the potential for providing a maturity index.

The organ of Valenciennes lies below the buccal apparatus as an area of folded glandular tissue on which spermatophores are occasionally found coiled. It therefore appears to be analogous with the seminal receptacle in squid, except the spermatophore of *Nautilus* has not been ejaculated (Arnold, 1984). The details of spermatophore transfer are as yet unknown, but Mikami and Okutani (1977) have reported on copulation in captive *Nautilus*. According to their observations, the spadix of the male can be clearly seen to be inserted into the subbuccal region of the female (for an illustration, see also Arnold, 1984). Presumably, when the eggs are laid, the massive spermatophore breaks, and the sperm are concentrated around the oocyte's micropyle before or as the egg jelly or egg capsules or both are forming. The teardrop shape of the inner egg capsule suggests that it might possibly be formed in the oviduct, whereas the convoluted shape of the outer egg capsule strongly suggests that it is manipulated into shape by the arms. It is well known that decapod cephalopods form egg capsules and manipulate them into place with their arms (Drew, 1911; Waller and Wicklund, 1968; Arnold, 1962).

2.2. Male Reproductive Tract and Accessory Organs

As in the female, the male reproductive tract has a single, large gonad ($\approx 2 \times 4$ cm) at the extreme posterior end of the animal just below the proximal end of the siphuncle. Its posterior surface fits against the septum and conforms to the shape of the surrounding organs. The testis is covered by a thin membrane and is divided into an indefinite number of lobes, which are compressed by the covering tunic, so that it has a smooth outline. Several ligaments attach it to other viscera in the surrounding coelom. The testis has a central cavity and is attached to the vas deferens by a closely applied funnel-shaped tube of the covering tunic. It opens by a slit into the vas deferens, which is embedded in the accessory organ. Because of the close application of the tunic-derived funnel to the vas deferens, the sperm are shed into the coelom anatomically, but functionally are released directly into the vas deferens in a fashion similar to that found in the female for release of eggs.

The testis is composed of intertwining, cordlike seminiferous tubules that form the lobes and contain many different stages of spermatogenesis at any given location. Spermiogenesis has been described in detail by Arnold and Williams-Arnold (1978) and Arnold (1978), and only unusual features, uncommon to other Mollusca, will be mentioned here. The spermatids are linked by large cytoplasmic bridges, which interconnect even the residual cytoplasmic blobs when they are shed in final sperm maturation. The nucleus becomes extremely elongated, and in *N. pompilius*, the sperm head becomes approximately 35 μm in length, making it one of the largest known sperm. The spermatids have specialized "ball-and-socket-like" junctions with branches of the Sertoli cells, and multivesiculate vesicles suggest that junctions may be nutritional in nature. During development, the nucleus goes from spherical to teardrop shaped, to spindle-shaped, to elongate, and finally to ropelike. This progression is accompanied by the posterior-to-anterior development of an elaborate manchette of microtubules that are extensively cross-linked. Colchicine prevents the assembly of these microtubules. If the spermatids are treated with colchicine when the head is only partially elongated, the region of the nucleus surrounded by the microtubules is compressed into a cylindrical shape, but the nucleus beyond the manchette becomes an indefinite "blob-form" (Arnold, 1984). Interestingly, the chromatin beyond the region in which the microtubules are organized into a manchette does not condense, thus supporting the suggestion of Bergstrom and Arnold (1974) that microtubules are possibly associated with condensation of the DNA in the cephalopod sperm. There is a massive acrosome derived from the Golgi apparatus, and the mitochrondia, instead of being terminal, or in a "spur," lie in two parallel grooves that run the length of the nucleus. The two parallel centrioles are composed of nine sets of doublet microtubules, unlike the vast majority of centrioles that are made of triplet centrioles. There are massive amounts of glycogen in the head surrounding the nucleus and therefore in proximity to the mitochondria. Arnold and Williams-Arnold (1978) suggested that the unique features of *Nautilus* sperm structure may be due to their long evolutionary history rather than correlated with the mode of sperm release and fertilization. The sperm is elongate, and movement, if any is observed, is extremely slow. There is some suggestion that the sperm head may flex, but this impression may be due to the slow rotation of a slightly curved head or the passive bending in response to tail movement. The *Nautilus* sperm is unique in structure and form, and this uniqueness may possibly reflect either the phyletic history or some as yet unknown specialized features necessary for the mode of life in these primitive animals. The sperm certainly does not fall in the "primitive" category in the classification of molluscan spermatozoa by Franzen (1955).

The testis empties directly into the vas deferens, which is embedded in the "accessory organ," where the spermatozoa are collected and formed into the spermatophores. The details of this procedure are unknown, but the sperm are precisely aligned and covered with tunics and membranes. Before the spermatophore is deposited on the female, it is covered with a mucous coat, but where this coat is formed is unknown. The accessory organ is about one half to two thirds the size of the testis, and the vas deferens is highly convoluted. There are three regions of the vas deferens, each of which has a different function. The distal portion of the vas deferens is thin-walled and expands into a seminal vesicle. It leads directly into the thick-walled spermatophoric sac, or Needham's sac, where highly coiled spermatophores can often be found. Needham's sac is attached directly to the right side of the penis, which is quite muscular and thick-walled with a rounded tip and slitlike opening. The penis is 4–5 mm in diameter and Y-shaped, with the major portion on the midline. The right side of the penis is the aforementioned spermatophoric sac while the left side ends as a blind pouch 10–12 mm long. The right side is divided by a longitudinal membrane. Unlike those of other cephalopods, the spermatophore of *Nautilus* is a relatively simple elongate sac without an elaborate ejaculatory apparatus. In *N. pompilius*, it averages 14 cm in length and 5 mm in diameter. One end is a simple blunt tip, while the other end tapers to about one third the diameter of the rest of the spermatophore and then ends bluntly. The spermatozoa are all contained within a "sperm rope," which is twisted and coiled inside the tunics of the spermatophore. The sperm heads are attached to the wall of the sperm rope and all point in the same direction, while the tails project into the central region of the sperm rope. There appears to be no cement gland, no spiral filaments, and no cap thread, and when the spermatophore is punctured or even cut into pieces, the sperm are not forcefully ejaculated as would be the case in "higher" cephalopods. The whole spermatophore is in tight, irregular coils and appears to be folded over, possibly by the longitudinal membrane in Needham's sac.

The secondary male sexual organs are the spadix (Hoeven, 1857), with its associated modified cirri, the antispadix, and Hoeven's organ. The spadix, usually on the right side, is a large mass of conical erectile tissue ($\approx 2.5 \times 4.5$ cm) surrounded by three lappets. All are highly modified cirri and have slime glands in yellowish-gray patches slightly back from their tips. The spadix is a solid intromittent organ that becomes erect in copulation when it is pushed between the arms of the female toward the organ of Valenciennes. The details of how the spadix functions in spermatophore transfer are unknown (Mikami and Okutani, 1977, 1981; Arnold, 1984), but it can be seen from above that the hoods of the copulating animals are distorted as they pull together. Griffin (1900) provides a detailed description of the structure of the spadix sheath and lappets. The spadix and lappets are so large they displace the buccal mass to one side in the male. Hochachka *et al.* (1977) stimulated the spadix for prolonged periods and found it capable of extensive protrusions and shape changes for a considerable period, even after being excised.

The antispadix is a group of slightly modified cirri in the contralateral position of the spadix of mature males. It may represent repressed or vestigial bilaterality. The organ of Hoeven is a glandular area below the buccal mass of the males. Most investigators have assumed that it is a glandular organ that is in some way associated with reproduction, because it is found only in the male. The organ is oval ($2.5 \times 1.5 \times 1$ cm) and flattened dorsoventrally. Running toward that ventral cleft are laminae that are apparently secretory (Griffin, 1900). The organ is located in a depression formed by the labial ridges. Obviously, a modern histological investigation would contribute significantly to our understanding of this organ.

2.3. Bilaterality of the Reproductive Tract

In both the male and the female, there are suggestions of vestigial sexual organs that would indicate a primitive bilaterally symmetrical body plan. Both the male and the female have structures referred to as *pryoform sacs* or *glands* [Owen (1832) first used the term, and Lankester and Bourne (1883) expanded on it]. This organ occupies the same position on the left side of the animal as do the genital ducts on the right side. Willey (1902) found that the arteries that supply these structures are symmetrical with those that supply the genital ducts, and Kerr (1895) previously had found that in the young female, the genital duct and pyroform sac have a similar appearance. In the young male, there is also a strong resemblance between the left saclike extension from the penis and the future Needham's sac. In all instances, however, it appears that the gonad itself is unpaired. Only embryological evidence can resolve this question of bilaterality in the reproductive tract.

3. Reproduction

Unlike that of octopods and teuthoids, "courtship" and copulation by *Nautilus* are disappointingly passive. Exactly how sexual recognition is accomplished is unclear, but it must be rudimentary when compared with the signaling that occurs, for example, in *Sepia officinalis* (Grimpe, 1926) or *Sepioteuthis sepioidea* (Arnold, 1965). This description of reproductive behavior in *Nautilus* is based mainly on personal observations, discussions with Bruce Carlson of the Waikiki Aquarium, and the papers of Mikami and Okutani (1977, 1981). Basically, it appears that the males are aggressors and will attempt copulation with anything of the general shape and size of another *Nautilus*. If a new *Nautilus* of the same or a different species is introduced to an aquarium containing male animals, males will attempt copulation with it, regardless of its sex. In fact, if an animal is removed from the aquarium and its shell is wiped off before it is returned, undisturbed males will attempt copulation with it, regardless of its sex (Carlson, personal communication). The factors that stimulate copulation thus may be chemical or tactile, but the cues are certainly subtle and as yet unknown. If a male attempts to copulate with another male, it holds the shell in the copulatory position for minutes to hours, but eventually releases it and swims away. If, somehow, the aggressor male determines that the approached animal is a female *Nautilus*, copulation usually follows.

The male grasps the female with several arms and pulls the apertures into approximation. The hoods are forced together so that they are distorted, and generally the female's hood is covered at the distal edge by the male's hood. Occasionally, the female may grasp the male with a few tentacles, but typically it appears to attempt to retract into the shell slightly. Because the spadix of the male is displaced to one side, the male holds the female at an angle to its midline. The position of the arms is displaced, and this angle makes it possible to identify the male *in copulo*. The spadix may be visible between the arms on one side of the female's hood. It is erected and mobile, and because of its ivory color, it can be distinguished easily from the pigmented sheaths of the cirri. The male will continue to grasp the female for several minutes to several hours. During that time, the male may frequently bite the female's shell or mantle or both. The breaks in the shell may reflect the number of times the female has been in "copulatory embrace" and hence may be an indication of the number of egg-layings (Arnold, 1985). At best, the breaks may provide a poor approximation of reproductive activity, offering some insight into the reproductive history of individual females.

Precisely how the spermatophores are transferred is unknown (Arnold, 1984; Mikami and Okutani, 1977, 1981), but observations by Haven (1977a,b) and my own investigations indicate that in *N. pompilius*, only one spermatophore is ever present on the organ of Valenciennes. How copulation is related to the timing of egg deposition is also unknown at this time. Because the records on captive *Nautilus* have been made mainly on animals on display in public aquaria, precise data-keeping has been difficult.

The methods of egg deposition have been observed only once (Hamada *et al.*, 1980a) (see also Chapter 36). When a female is in position to lay an egg capsule, the animal pulls its aperture and hood against the substrate and thereby blocks observation with its body. We have not observed the egg-laying on a glass aquarium surface. The egg capsules are generally deposited singly, but they are also found in groups of two or three with their outer capsules confluent at their bases. In the Waikiki Aquarium, the egg capsules generally appear in batches of five to eight, which would correlate with the number of eggs in the size classes discussed previously.

4. Embryology

4.1. Structure of the Egg Capsule and Egg

Willey (1897d) described the first egg capsule and egg of *N. macromphalus* in 1897, and his description applies to all known species if allowances are made for size differences. As stated previously, the shape of the egg capsule is apparently determined while it is laid. The inner capsule is probably formed in the oviduct, whereas the outer capsule is manipulated to conform with the substrate to which it is attached. When freshly laid and in seawater, the capsule wall has the toughness of a human fingernail, but when dried it tends to be fairly brittle. As seen with the scanning electron microscope (SEM), the walls of the inner and outer capsule appear to be uniformly dense, with vacuoles, voids, pores, and tunnels throughout. Just at laying, the egg capsules are flexible, but harden in sea water within a few hours.

The inner capsule is less variable in shape and contour than is the outer capsule; the former is teardrop-shaped, with the blunt end attached to the inside of the outer capsule over about one quarter or one third of its surface area. There are distinct sutures in the inner capsule that run from its tip toward the base and eventually disappear where the inner and outer capsules fuse. These sutures are quite tightly appressed and divide the inner capsule equally into four quadrants. They seem to be mechanical junctions, rather than folds or lines in a continuous sheet of the egg capsule material. It is not known whether these sutures represent future areas of dehiscence when the juvenile *Nautilus* hatches, but it seems likely that this is the case. How the sperm might penetrate the inner capsule is unknown, and it may be that fertilization is accomplished before the inner capsule is formed. As far as I know, no oocyte has ever been observed in the oviduct, and until that is accomplished, we can only speculate about the mechanism of encapsulation.

The outer capsule is saclike in general structure, and where it is attached to the substrate, a base area commonly extends beyond the capsule proper. Willey (1897b, p. 468) made the apt analogy that the form of the outer capsule is like "folds of drapery, giving it a graceful appearance." Between these undulations there are about eight to ten large slits a few millimeters wide and several millimeters long; these slits allow seawater to circulate freely between the two parts of the egg capsule. At the edges of these slits, the egg capsule material thins from its normal thickness of 0.8 mm and consequently becomes more flexible. There is a strong tendency for the outer egg capsule to be bilaterally symmetrical, and there is commonly a line of demarcation at the midline. In many of the egg capsules, the outer capsule is not completely formed, and gaps occur so that the inner capsule can be seen easily. These gaps tend to occur along the midline, further suggesting that the outer capsule is formed from two halves (see Willey, 1897b, Fig. 4). The opening of the outer capsule always seems to be oriented upward and away from the attachment area. The size varies with the species and size of the female, but the entire capsule in *N. belauensis* is about 4 cm long.

The oocyte or zygote is situated inside the inner capsule, like a hen's egg in a small paper sac. The lower part of the capsule more or less conforms to the shape of the oocyte, so it has a rather smooth outline. The egg itself is ovoid and is contained in a clear chorion that is enclosed in a whitish egg jelly that probably functions as padding. There is a micropyle visible as a depression at the apex of the egg. In the many infertile eggs and the two blastoderm stages I have examined, there is a slight space in the area between the egg and chorion. A description of the oocyte is given in Section 2.1.

How the juvenile *Nautilus* hatches out of this capsule is an open question at this time. In the oldest embryos I have seen (three chambers), the shell and periostracum cover the entire mantle where the Hoyle organ would be found in other cephalopod embryos. It is possible that the sutures of the inner capsule open so the juvenile can escape into the somewhat larger space within the outer capsule. Of course, at the presumed stage of hatching [seven chambers (see Chapter 27)], the juvenile has a well-developed beak and buccal apparatus, so it is entirely possible that the animal simply bites its way out of the capsule.

4.2. Embryonic Development

Until recently, this section on embryology could not have been written. The first embryos of *Nautilus* were not discovered until March 4, 1985 (Arnold and

Carlson, 1986), despite repeated previous attempts, beginning with that of Willey (1897b). What follows here is not a definitive description, but rather a progress report based on the seven living embryos and nine shells of disintegrating embryos thus far obtained. The complete description of a hitherto unknown embryo is a large undertaking, complicated by the rarity of material and the necessity of caution to gain as much information as possible before proceeding on to other techniques that obviate certain observations. In writing this description, I have assumed that the reader has little knowledge of other cephalopod embryos and have attempted not to make too many comparisons to "modern" cephalopod embryos, with which I have a quarter of a century of familiarity. I have tried to avoid unwarranted speculation, but some "informed opinion" will be necessary. Much of this information is preliminary, so a number of photographs are included, and future readers will be able to make their own interpretations.

As stated above, the zygote of *Nautilus* is one of the largest known invertebrate eggs (20 × 14 mm, for *N. belauensis*). It is ovate, with the animal pole in the space below the pointed end of the inner egg capsule. In infertile eggs and early stages of fertile eggs, there is a distinct chorion, which is somewhat opaque, but it only slightly restricts observation. In the later stages, when the embryonic body stands out from the contour of the yolk mass, the chorion is not apparent, but by then the yolk mass is completely covered by the external yolk sac. However, the whitish egg jelly is still present. Thus, the early cleavage stages have not been found, which is unfortunate, because even a quick glance at the 8- or 16-cell stage would provide important information about the origin of the nonspiral cleavage pattern in cephalopods.

The earliest stages found to date have both been blastoderm stages, one of which was known to be 14 days old (Fig. 1A and B). Both measured 6.7 mm in diameter. The blastoderm at this stage is divided into three regions: a central organogenetic region, a surrounding uniform region, and a marginal region of large, opaque cells.

In the central region, there are symmetrically placed areas of cell concentrations arranged on either side of the midlines (Fig. 1B). At this early stage, it is not possible to identify which organ primordia these thickenings represent, but serial sections should provide that information. Surrounding this organogenetic area is a ring of uniformly dense tissue that represents (by analogy to other cephalopod embryos) the future external yolk sac. At the margin of the blastoderm is a ring of what appears to be large whitish triangles or trapezoids. These structures are comparable in position and appearance to the blastocones seen in other cephalopod embryos (Vialleton, 1888; Naef, 1923). Obvious cytoplasmic processes that extend over the "uncellulated" yolk surface, although common in other large yolked cephalopod embryos, were not observed. The structure of this blastoderm is quite comparable to that of the blastoderm of other cephalopods, when the large yolk mass is considered.

The description of the development of the surface organs is based on seven embryos of the two stages obtained thus far. Fortunately, these two stages provide important organogenetic events that give insight into some of the significant developmental events. The earlier embryo is roughly equivalent in development to stage 22 or 23 of *Loligo* (Arnold, 1965) or stage IX or X of *Sepia* (Naef, 1923). The major organ primordia are forming, but still retain the position of their origins,

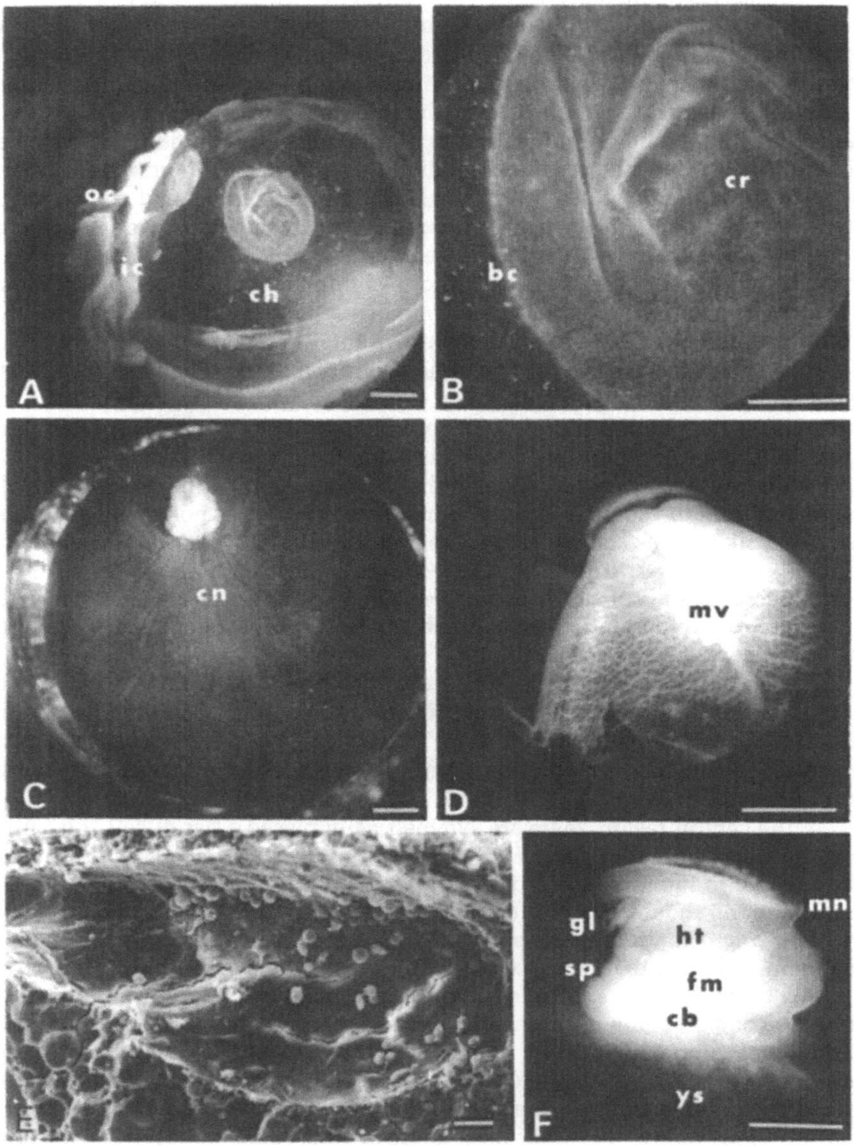

Figure 1. (A) Blastoderm of 14-day-old embryo *in situ*. Notice that the outer capsule (oc) and inner capsule (ic) have been trimmed away and the egg jelly partially removed. The chorion (ch) has been torn open, and the blastoderm is clearly visible. Scale bar: 2 mm. (B) Central region (cr) of the blastoderm showing the beginnings of organ differentiation. At the edge of the blastoderm, there are large white cells that apparently are blastocones (bc). Between these two regions is an area that becomes the external yolk. The folds are artifacts of yolk loss elsewhere in the embryo. Scale bar: 1 mm. (C) Embryo in early organogenesis showing the extraembryonic circulatory system (cn) in the future dorsal surface. Scale bar: 2 mm. (D) Extraembryonic circulatory system on the future ventral side of the one-

and the mantle has not yet covered the visceral primordia. The three older embryos are in the same stage of development, with the shell being well developed as a cone and the first chamber being formed. Although the chamber is present, a calcified septum has not yet been formed, and the siphuncle does not cross the space between the septum and the protoseptum. Gill filaments have begun to form in the proximal pair of gills and are well advanced in the distal pair. While these older embryos were alive, their respiratory movements were active and regular. The hearts were active, and other well-coordinated movements occurred. This stage of development is roughly equivalent to stage 27 in *Loligo* or stage XVII in *Sepia* (Naef, 1923).

The external yolk sac apparently originates in the blastoderm peripheral to the central organogenetic area. This sac is similar to the case in other cephalopod embryos. As the organ primordia begin to develop, the external yolk sac expands at its margin. At the early organogenetic stage, it has completely engulfed the yolk mass. In the living embryo, it appears to be ciliated; at least particles are transported across its surface. Under the SEM, the surface of the early organogenic, external yolk sac was seen to be covered with a contaminating precipitate, and cilia cannot be seen. In the older embryos, the surface appears bare, but this appearance may be due to preparation techniques and surface contamination. Plastic sections prepared for the light microscope indicate cilia and basal bodies, but no pattern is evident. Beneath the surface epithelium are two layers of muscle cells that run at cross angles to each other. Below these muscle layers lie either the yolk syncytium or a system of blood vessels that form the extraembryonic circulatory system similar to that found in large-egged vertebrate embryos (Fig. 1C–E). As far as I know, this extraembryonic circulatory system is unique among the invertebrates and certainly is unique among the cephalopods for which embryogenesis has been described. From the area beneath the hood and eyes, at least five large blood vessels emerge and branch almost immediately to ramify over the surface of the external yolk sac, where they anastomose and spread over the yolk surface. Farther away from the embryonic body, the vessels form finer and finer branches until they can no longer be seen with a dissecting microscope at 100×. At the midline below the future siphon region, a single large vessel runs between two prominent muscle bands (Fig. 1C and E). It is joined a short distance from the embryonic body by numerous smaller vessels that tend to be arranged perpendicular to its axis. On the basis of the anatomical position and the pattern of these two vessel groups, it seems likely that the vessels that emerge near the hood are the arterial supply and that the single medial vessel is the venous return.

---

chambered (older) embryo. Note that there is a single median vessel (mv) that is joined by multiple anastomosing branches. This network is assumed to be the venous return from the yolk sac to the embryonic body. Scale bar: 2 mm. (E) Fractured blood vessel of the extraembryonic yolk sac. The vessel wall is smooth and has frequent blood cells fixed to its wall. There seems to be no elaborate lining to this vessel. Scale bar: 20 μm. (F) Living early organogenetic stage seen from the side. The future ventral surface is to the left, with the mantle (mn) on the top. The embryonic body develops on the massive external yolk sac (ys). The siphon primordium (sp), the future retractor muscle (fm), and valves develop above the cirral buds (cb). The ventral pair of gills (gl) have a distinct hollow in them that is probably a developing blood vessel. The dorsal gills are smaller than the ventral gills and are partially covered by the mantle (cf. Fig. 3B). The heart region (ht) is closely associated with the gills (cf. Fig. 2A and B). Scale bar: 1 mm.

This pattern would return blood near the gills. It was not possible to observe the route of blood flow directly because the blood is transparent, and the hemocrit is low.

In the literature, the terms "arm," "tentacle," and "cirrus" are used more or less interchangeably, but it should be kept in mind that *Nautilus* appendages are distinctly different from the arms of octopods and from the arms and tentacles of squids and cuttlefish. In addition to being more numerous (≈90 vs. 8 or 10) and to being contained in a rather firm sheath, they lack suckers and instead have transverse ridges that function in grasping. The cirri arise in two bilateral groups above the yolk sac and below the eyes and siphon–retractor muscle complex (Figs. 1F, 2A, 3A, and 4A–C). In the early embryo, they arise above the margin of the external yolk sac as a row of about seven protrusions of tissue (= cirral buds) that begin below the eye region and extend to the siphon region (Figs. 1F, 2A, and 3A). Two are definitely associated with each eye and most likely become the optic cirri (Griffin, 1900; Willey, 1902). On the future ventral surface, the outlines of the arms are vague, and at the midline, there is a smooth area of tissue directly above the median extraembryonic blood vessel that lacks the cirral buds (Fig. 2A). In the living embryo, each cirral bud can be seen to have a dark center, which may indicate that it has either a hollow core or a median blood vessel. Examination of the tips of the arms shows a smooth surface with no indication of an indentation or invagination (Figs. 3A, 4A, and 5B). Some of the cirral buds tend to be spindle-shaped. In the older embryo, 17 cirral primordia can be counted on each side, but because they are packed so closely together and are partially covered by the developing mantle and siphon, this count can be considered only a minimal number. It is not possible to distinguish an inner or outer group of cirri, but the cirral buds nearest the yolk sac appear to be smallest (Fig. 5A). It appears that arms develop as bilateral groups that must unite secondarily at the midline. The exact grouping and placement of arms require further study.

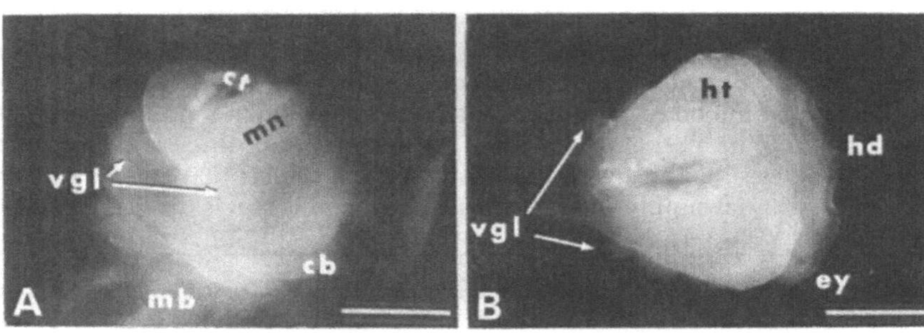

Figure 2. (A) Ventral view of the early organogenetic stage. Two muscle bands (mb) run over the surface of the external yolk sac, and the median blood vessel (seen in Fig. 1D) runs between them. The developing cirral buds (cb) are quite prominent between the siphon primordium and the yolk sac. Both pairs of gills lie just below the mantle (mn). Between the ventral gills (vgl) is a mass of tissue that is assumed to be hindgut primordium. The cicatrix (ct) is prominent in the mantle. Scale bar: 1 mm. (B) Viewed from above, the outline of the ventral gills (vgl) is conspicuous, as are the eye complex primordia (ey), the hood primordium (hd), and the midline. The edges of the valves border the heart (ht). Scale bar: 1 mm.

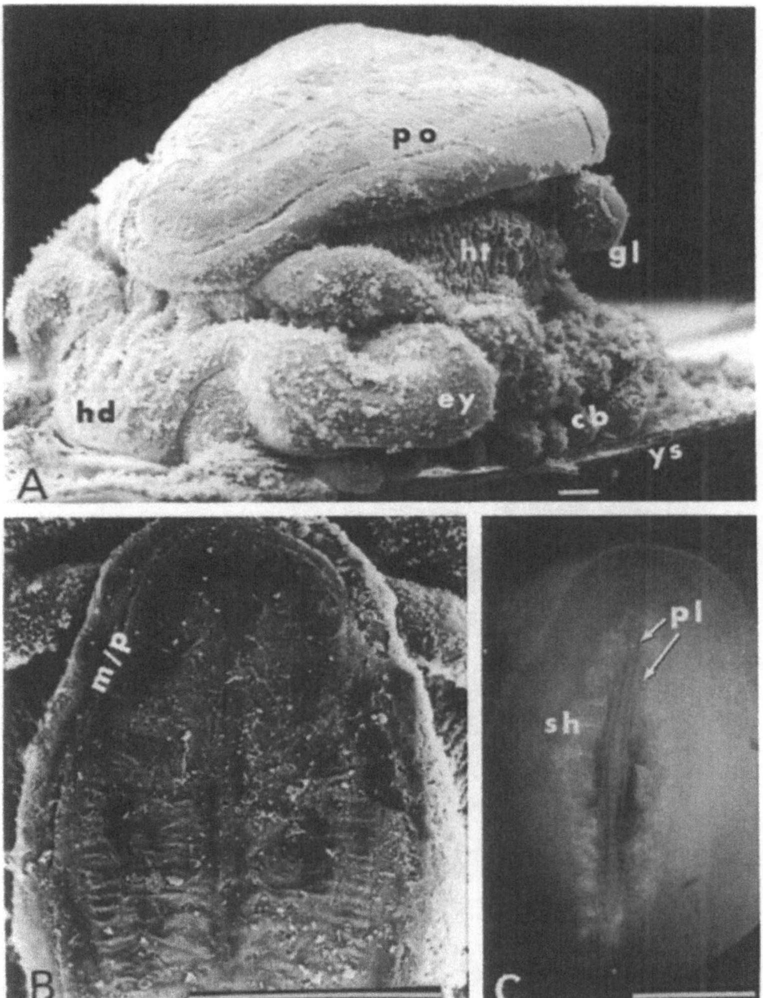

Figure 3. (A) SEM of the same embryo as shown in Figs. 1C–F and 2A and B. The mantle can be seen to be covered by a layer of periostracum (po). Other anatomy: (cb) cirral buds; (ey) eye complex primordia; (gl) ventral gills; (hd) hood primordium; (ht) heart; (mn) mantle; (ys) yolk sac. Scale bar: 100 μm. (B) Mantle of the living early organogenetic stage viewed from above. The mantle is covered with periostracum, which has uniform undulations radiating outward from the central line in the cicatrix. There is a sharp demarcation between the mantle edge and the periostracum (m/p). Scale bar: 500 μm. (C) Cicatrix and developing shell, as seen through the periostracum. At this stage, there are four pigmented lines (pl) paralleling the midline, and the shell prisms (sh) are accumulating in roughly perpendicular lines on either side of these lines. There is no evidence of a protoconch. Scale bar: 500 μm.

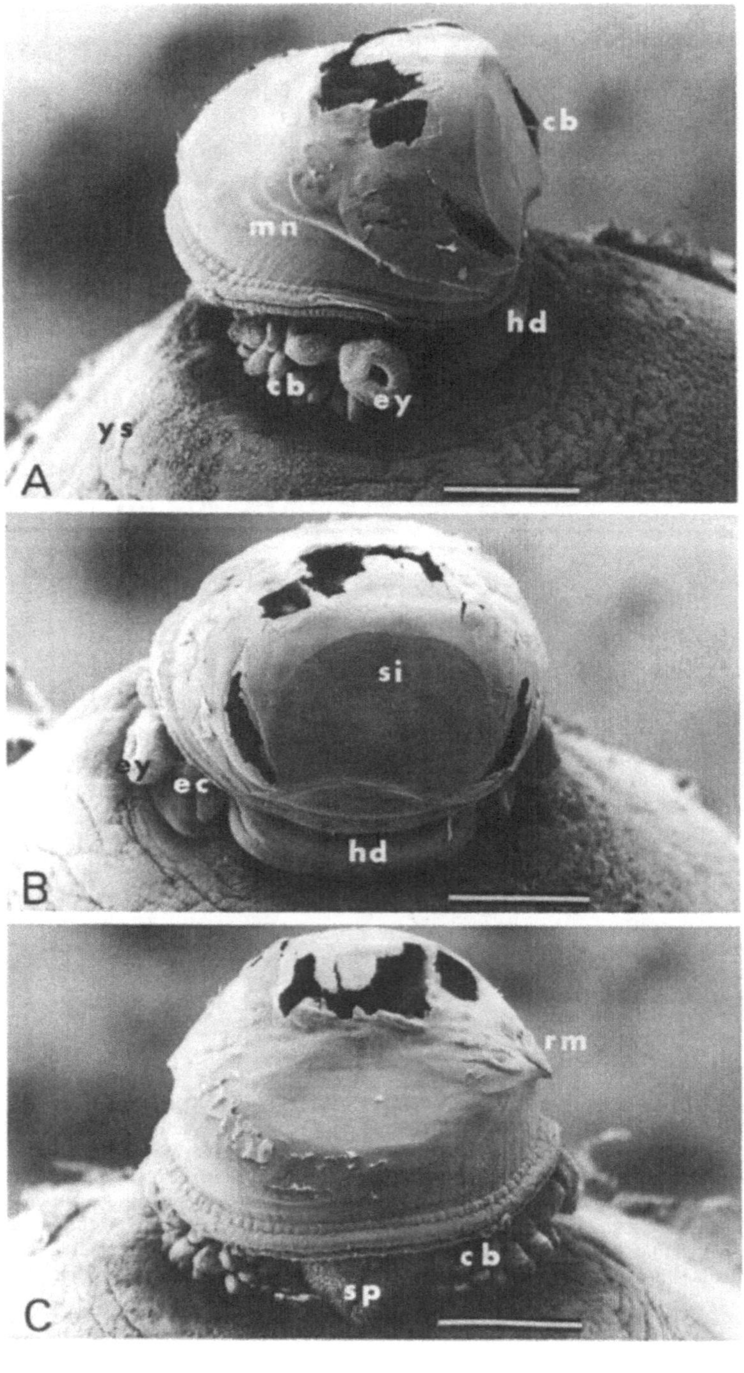
cb
mn
hd
cb
ey
ys
A
si
ey
ec
hd
B
rm
cb
sp
C

The origin of the hood and mouth is obscure at this writing. The mouth primordium is apparently covered by the developing hood, but a slit between the external yolk sac and hood does exist, and this slit probably communicates with the developing mouth and buccal mass. The hood appears to develop as a fold of tissue on the midline between the eye complex primordia (Figs. 2B, 3A, 5B, and 6). Because there is no counterpart in other known extant cephalopod embryos, the origin of the hood and the development of its innervation are interesting questions that must be approached using sectioned material. In the adult, there are two modified cirri, the sheaths of which form the lateral margins of the hood; these cirri possibly arise from the mass of tissue below and behind the eye complex (Figs. 4B and 5B).

The siphon and retractor muscles will be considered together here, although in the adult they are anatomically distinct. Above the eye complex and developing arms in the early organogenetic stage, two bands of tissue nearly encircle the embryonic body (Figs. 1F, 2A and B, and 3A). In the future ventral region, the lower band meets at the midline (Fig. 2A) and extends back toward the eye, where it meets with the second band in the future dorsal region. This is the siphon complex. In the living embryo, there is a distinct groove above the siphon complex, and it projects out from the body. It forms a tube with overlapping edges that slide past each other during the inhalant and exhalant phases. It can be protruded for a considerable distance from the shell and retracted, if the embryo is jarred (Fig. 5A). Above the eye, there is an elongate tissue mass at an angle to (and in contact with) the lower edge of the dorsal mantle (Figs. 1F, 2B, 3B and C). This mass apparently gives rise to the retractor muscle and the valves of the mantle. In the older embryos, the retractor muscle is quite dense, and it is attached via the mantle to the inside of the shell so tightly that when the shell is removed, a thin layer of shell pulls free and remains attached to the muscles. The structure of the shell is modified in this area and is not covered with nacre.

The eyes and associated structures arise on either side of the embryonic body, just ventral to the hood (Figs. 1F, 2B, and 3A). In the early organogenetic stage, invagination has been completed, and the optic vesicle communicates to the outside by a small, barely visible opening, associated with a mass of tissue, that might be the primordium of the rhinophore or the optic lobe (final identification will require sectioning). In the older embryos, the eye is pigmented (bright red) and the opening prominent. Three of four eyes examined have a prominent line that runs ventrally from the pupil toward the cirral buds. This position would seem to correspond with the band of pigmentation seen in the adult (Griffin, 1900).

←

Figure 4. (A) Side view of the embryo with the first chamber forming. The folds in the large external yolk sac (ys) are probably due to muscular contractions that occurred when the embryo was fixed. Most of the embryonic organs are covered by the mantle, but the eye (ey), cirral buds (cb), and hood (hd) are evident. The hood appears to arise as a folded structure. The apex of the embryo is truncated, indicating that the first chamber was forming but that the first septum had not yet calcified. Scale bar: 1 mm. (B) Viewed from the future dorsal surface, the hood (hd), eye complexes (ec), and optic cirri are evident. The septum was apparently just beginning to form shell, and at the midline, slightly above center, there is a slight depression that probably corresponds to the future site of the siphuncle (si). Scale bar: 1 mm. (C) The future ventral surface shows the cirral buds (cb) and the siphon (sp). The two lateral projections are immediately over the retractor muscles (rm). In this region, a thin layer of shell remains attached to the mantle. Compare with Fig. 5B. Scale bar: 1 mm.

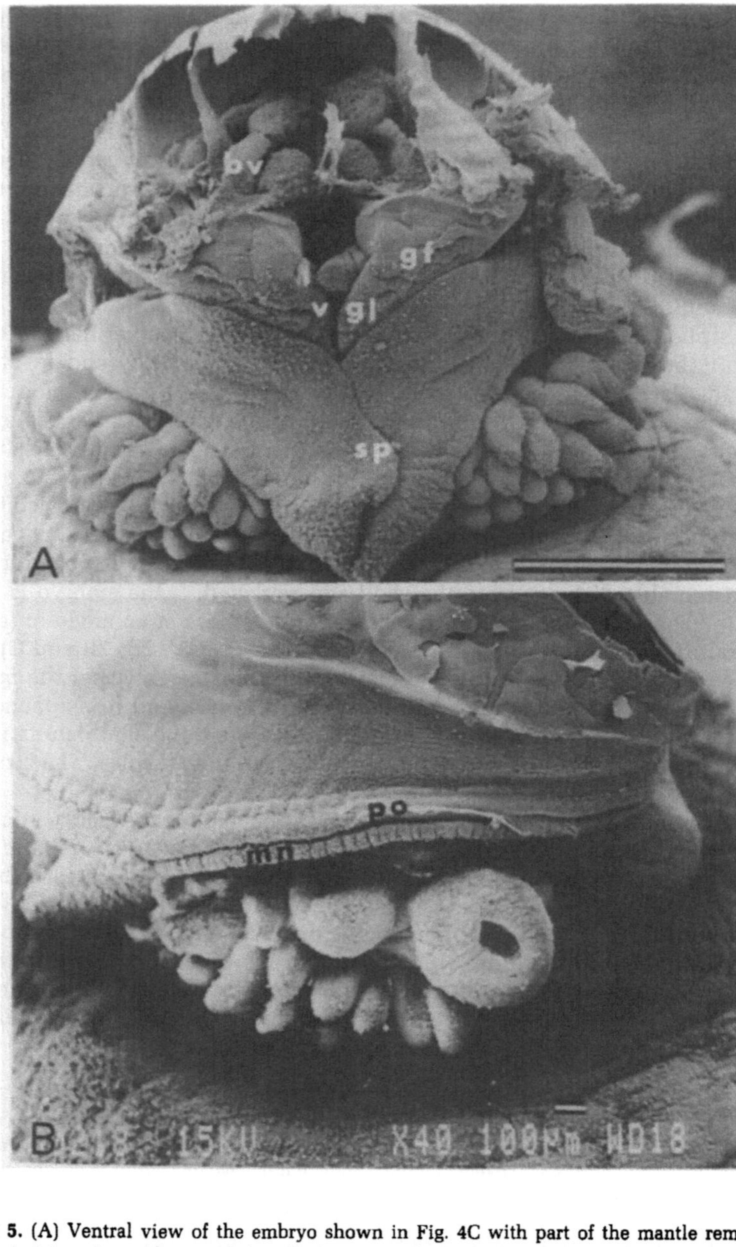

Figure 5. (A) Ventral view of the embryo shown in Fig. 4C with part of the mantle removed. The siphon (sp) is quite evident, and the gills insert inside the inner margins of the siphon. The ventral gills (vgl) have progressed further in development than the dorsal pair. One tip of a dorsal gill can be seen projecting from behind the right ventral gill. It has begun a series of folds that will become the gill filaments (gf). The blood vessels (bv) and hearts leading from the gills are evident in the visceral cavity. Note that some of the cirral buds are spindle-shaped. Scale bar: 1 mm. (B) At higher magni-

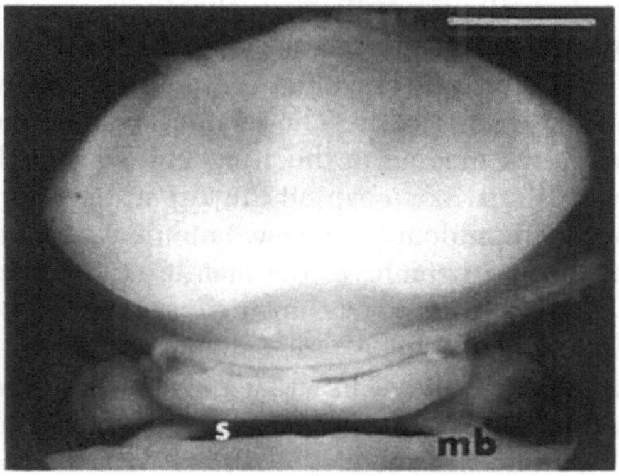

Figure 6. Dorsal view of a gluteraldehyde-fixed embryo after the shell has been removed. Note below the hood the prominent slit (s) that probably communicates with the mouth. On either side of it are two muscle bands (mb), extending over the surface of the external yolk sac. On the midline of the slit is a slight notch that corresponds to the position of the median blood vessel. Scale bar = 1 mm.

The eye is subspherical at this stage of development, rather than ellipsoidal, with the outer surface flattened, as in the adult.

The gills are quite prominent on the future ventral surface just below the mantle of the younger embryo (Figs. 1F, 2A and B, and 3A). There are two pairs, the more ventral of which appears to develop earlier or faster (Figs. 1F and 3A). The ventral gills appear to have a hollow central core that most likely is a developing blood vessel. The more dorsal pair is separated from the ventral pair by a slight depression, and at the early organogenetic stage, they are considerably smaller and barely project from under the mantle edge (Figs. 1F and 3B). In the older embryos, the gill filaments arise as a series of folds on the ventral gill surface (Fig. 5A). At this stage, the ventral pair is developmentally advanced and much larger than the dorsal pair. The gill primordia project into the space formed within the siphon, so that as water is pumped through the mantle cavity, the gills are bathed. As development proceeds, the dorsal pair of gills must accelerate in development, because they are all the same size in the adult. Between the gills on the midline is a small mass of tissue that is probably the developing hindgut (Fig. 2A).

Dorsal to the gills are two large areas that undergo rhythmic expansions and contractions (Arnold and Carlson, 1986) with an average cycle duration of about 2.4 sec. When the specimen is fixed and examined under the SEM, parallel folds can be seen covering this area (Fig. 3A). These areas comprise the developing branchial hearts, and they circulate the blood within the embryonic body and the external yolk sac. As the slow, rhythmic contractions of the hearts occur, the

---

fication, the mantle (mn) has a beaded appearance. A thin layer of what may be the periostracum (po) lies in the groove proximal to the mantle edge. The ventral portion of the mantle also has a beaded line that approximates the edge of the shell. These periodic repeating structures may correlate with radiating lines on the shell. There is a line running from the pupil toward the edge of the optic cup. The fold of the hood is evident, and behind the eye is a mass of tissue that probably becomes incorporated into the hood at its distal margins. In the central top portion of the picture are thin layers of shell attached over the retractor muscles; these layers remained when the shell was removed. Scale bar: 100 μm.

ventral gill primordia move slightly in synchrony. In the older embryos, there are obvious vessels and tubelike hearts associated with each of the gills (Fig. 5A).

The mantle and shell complex are of considerable interest to both biologists and paleontologists. The mantle arises at the animal pole of the embryo as an oval area median to the arms and eyes. In the early organogenetic embryo, the mantle sits like a cap on the top of the embryonic body, and it projects slightly outward and curves somewhat downward, particularly in the future dorsal area. In the living embryo, the mantle is transparent enough so that the developing shell can be seen within it (Figs. 2A and B and 3C). There is a clear layer of periostracum that extends almost to the edge of the mantle where it is demarked by a sharp line. The surface of the periostracum has uniform undulations that radiate out from a central linear depression. Arnold (1987) has suggested that these undulations may correlate with the radial lines of ornamentation seen in the older shell. Below the periostracum, the early development of the cicatrix is evident, and four lines of black pigmentation lie beside the exact midline (Fig. 3C). Lateral to the outer pair of lines is more pigment clustered in the center of the mantle. Perpendicular to these lines of pigment, crystals of shell are arranged in linear arrays. There is no evidence of any precursor chamber, structure, or protoconch (Hyatt, 1884b).

Later in development, the pigmented region of the cicatrix expands, laterally, but not longitudinally, so it is lentoid in shape (see Fig. 1B in Chapter 27). The shell has become cone-shaped, and a network of lines is evident on the outer surface. Clear periostracum extends beyond the calcified shell, and the same lines of ornamentation can be seen in it. Where these lines intersect, crystals of shell can commonly be seen (see Chapter 27 for details).

A tentative five-step model of early shell development in *Nautilus* has been proposed as follows (Arnold, 1987):

1. The cicatrix is the first site of shell deposition. The crystals of shell are laid down in association with the black pigment that may be the organic matrix of the shell.
2. At the midline, prismatic shell fills in the area between the original black lines. This forms a central bar that terminates in two circular areas where the prisms are irregularly arranged.
3. Shell prisms accumulate beneath the tops of the undulations of the periostracum, and these prisms act as centers of further shell accumulation.
4. Cycles of peripheral outgrowth form the concentric lines of ornamentation parallel to the aperture.
5. Later layers of shell (e.g., nacreous and inner prismatic) are added by regional secretion from the mantle onto the inner surface of the original shell, as a secondary developmental event that is not related to the cicatrix.

Thus, shell growth is initiated at the cicatrix, which serves as a growth center of the embryonic shell, rather than being the remnant of any earlier structure that disappeared in early embryonic life. This view is contrary to the suggestion of Hyatt (1884b).

5. Conclusions

Although there have been considerable advances in our knowledge of the life history and development of *Nautilus* in the past several years, there is still much to be learned about this primitive and interesting animal. With renewed interest and modern technical advances, even more new data will emerge soon. What little we know about the reproductive biology of *Nautilus* is mainly inferred from their anatomy and from comparisons with other, better-known cephalopods [see Wells (1978) or Arnold (1984)] and from limited observations of living animals (e.g., A. W. Martin *et al.*, 1978; Mikami and Okutani, 1977; Haven, 1977a). Basically, *Nautilus* is much simpler than other cephalopods but development is, not surprisingly, cephalopodlike rather than gastropodlike.

One striking feature of *Nautilus* development is the large yolk mass, which indicates a long developmental period and imposes some limitations on developmental strategies. Because *Nautilus* seems to be well adapted to its environment as an adult, there would seem to be obvious advantages in using the same life style and habitat for the newly hatched forms. This would dictate that the newly hatched animals would be miniature adults, and this appears to be the case. There are certain structural and hydrodynamic advantages to having a completed whorl at hatching, when the animal first faces its environment (Chamberlain, 1980a). This would necessitate a large yolk mass and therefore an extraembryonic circulatory system. The extraembryonic circulatory system would not only enhance yolk utilization, but would also offer respiratory advantages not found in simple hemal spaces. The reasons other large-egged cephalopods, such as *Sepia*, do not have a network of vessels in their external yolk sac may possibly be related to the separate evolutionary histories of the coleoids and the nautiloids. It is interesting that these "primitive" cephalopods have independently evolved an efficient mechanism of yolk utilization in a manner parallel to the presumably "advanced" vertebrates.

The branchial hearts are large and develop relatively early, because they must function in propelling the blood across the external yolk sac. In other cephalopod embryos, there are muscle bands that traverse the hemal space and have the circulatory function. Since the external embryonic circulatory system of *Nautilus* does not have these single-celled muscles, the embryonic branchial hearts must have this function. The asynchrony of development of the ventral and dorsal gills is puzzling. It does not seem that there should be a gradient of developmental sequence, such as occurs in vertebrate embryos, and except for the cirral bands, development of the other organs seems to be synchronous. Unlike the otherwise insightful hypothetical model of Naef (1923), the cirri develop in two restricted bilateral patches, rather than circumferentially.

Obviously, only a beginning of the understanding of the reproductive biology and the embryonic development of *Nautilus* has been made. Enough work remains to be done to occupy more than one lifetime, and embryonic material is scarce, so advances will occur slowly. *Nautilus* has guarded its secrets well for eons and will not yield answers easily. Perhaps Bashford Dean (1901, p. 819) explained it best when he said: ". . . Nautilus [is] a form which one usually associates with remote and cannibal islands. . . ." By its remoteness and insular nature, *Nautilus* remains a fascinating question to those of us fortunate enough to study its mysteries.

ACKNOWLEDGMENTS. This research was sponsored by the Grass Foundation, the *Nautilus* Research Corporation, the Hawaiian Malacological Society, the Friends Of The Waikiki Aquarium, and the *Teuthis Obscura* Fund. The author thanks Bruce A. Carlson for his collaboration with this research. The staff of the Waikiki Aquarium have been particularly helpful in all phases of maintaining the adult animals and providing support facilities. Dr. Barbara Boyer read and improved this manuscript.

Chapter 27

Development of the Embryonic Shell of *Nautilus*

JOHN M. ARNOLD, NEIL H. LANDMAN, and HARRY MUTVEI

1. Introduction

The organization of this chapter follows the development of the shell of *Nautilus* from the formation of the cicatrix through hatching. The description of the cicatrix is based mainly on study of one embryo of *N. belauensis* in early organogenesis, in which periostracum and crystals of aragonite were just beginning to form (Arnold and Carlson, 1986). Development through the one-chambered stage is based largely on a second embryo of *N. belauensis*, in which the first chamber was in the process of forming, but a calcified septum had not yet developed (Arnold and Carlson, 1986). Subsequent to these initial observations, our findings were con-

JOHN M. ARNOLD • Pacific Biomedical Research Center, University of Hawaii, Honolulu, Hawaii 96822; Marine Biological Laboratory, Woods Hole, Massachusetts 02543. NEIL H. LANDMAN • Department of Invertebrates, American Museum of Natural History, New York, New York 10024. HARRY MUTVEI • Swedish Museum of Natural History, 104 05 Stockholm, Sweden.

firmed by study of a second early embryo and two additional first-chamber embryos. Later embryonic growth is based on the embryonic whorls of juvenile and adult specimens of *N. belauensis*, *N. pompilius*, *N. macromphalus*, and *N. scrobiculatus*. These latter specimens are in the collections of the American Museum of Natural History (AMNH), Bryn Mawr College (BMC), the Delaware Museum of Natural History (DMNH), the Museum of Comparative Zoology (MCZ), the National Museum of Natural History (NMNH), the University of Iowa (SUI), and the Yale Peabody Museum (YPM).

The sequence of early shell development can be divided into three stages: (1) initial shell secretion at the cicatrix, including the formation of the periostracum, lines of black material, and initial calcification; (2) marginal growth of the initial shell wall and the appearance of structurally differentiated shell layers, e.g., the nacreous and inner prismatic layers; and (3) formation of the protoseptum, the septa proper, and the siphuncle.

2. Initial Shell Formation: Cicatrix

The initial shell develops on the mantle of the embryo in early organogenesis (Chapter 26). The shape of the mantle is probably determined by the placement of the bilateral organ primordia, such as the eye complex, siphon, retractor muscles, and arms. The mantle thus takes on a lentoid shape, with the future dorsal region narrower than the future ventral region (Fig. 1A).

The initial shell has a shiny outer coating, presumably the periostracum. The periostracum extends to a conspicuous groove that occurs a short distance above the mantle edge (Fig. 1C). Parallel to the midline, four lines and two lateral areas of black material are visible through the periostracum (Fig. 1A). Crystals of aragonite form a lentoid area surrounding this black material. These crystals appear to develop in association with the black material and form lines perpendicular to the midline. At this stage, no crystals are visible between the inner black lines. The lentoid area is obviously the initial stage of the cicatrix, the so-called "scar" on the apex of the shell (for a complete description of this feature, see Erben and Flajs, 1975; Blind, 1976). Shell formation begins here and proceeds laterally and medially.

A central, shallow depression occurs over the pigmented area. Laterally, the periostracum exhibits a series of low undulations that radiate outward from the central depression toward the periostracal groove (Fig. 1C). The crystals of aragonite are deposited in relation to these undulations and form the initial shell sculpture. The shell bears no sign of a precursor structure such as an organic protoconch as suggested by Hyatt (1872, 1884b).

The black organic material is absent when the periostracum is removed and hence is part of the periostracum. It consists of distinct, small granules, although these granules may represent clumps of even smaller particles. This black material, which may contain melanin, is resistant to acidic conditions that will dissolve the embryonic shell. Similar black material is associated commonly, if not always, with areas of stress and shell repair (Landman, 1983a; Arnold, 1985).

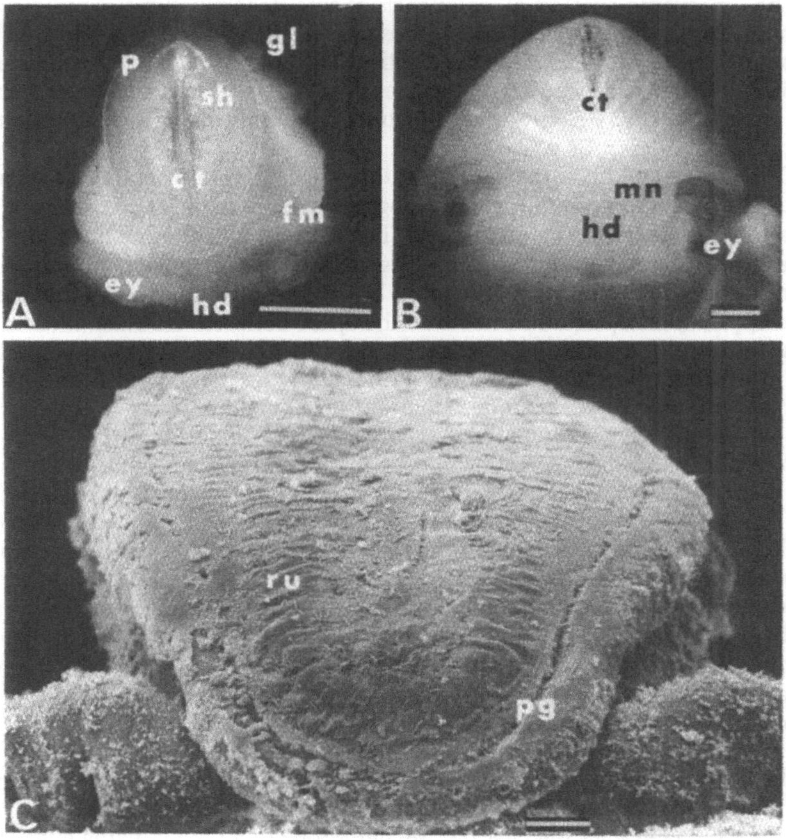

Figure 1. (A) Developing shell of the embryo in early organogenesis. The four black lines parallel to the midline represent the beginning of the cicatrix (ct). Crystals of aragonite (sh) develop in arrays perpendicular to the black material. The lentoid shell primordium is covered with clear periostracum (p) visible as a reflective area on the future ventral side (near the top of the photograph). Other organs: (gl) gill; (fm) funnel musculature; (ey) eye complex; (hd) future hood primordium. Scale bar: 1 mm. (B) Dorsal area and cicatrix (ct) of the embryo at the one-chambered stage. The dark pigmentation of the cicatrix appears as a lentoid area similar in length to the cicatrix in the earlier embryo. The mantle (mn) is visible at the dorsal margin of the shell, but the periostracum is transparent. Scale bar: 1 mm. (C) Scanning electron micrograph of the mantle of the embryo in early organogenesis. The periostracum exhibits a series of low undulations (ru) that radiate outward from a central depression. The periostracal groove (pg) is prominent near the edge of the mantle. Scale bar: 100 μm.

3. One-Chambered Stage

3.1. External Morphology

The embryonic shell of *N. belauensis* at this stage measures approximately 3 mm in maximum diameter and represents about ⅛ whorl. The shell wall is approximately 75 μm thick near the cicatrix. The apex of the shell is cap-shaped and similar to that of other *Nautilus* species (Fig. 1B) (see also Hyatt, 1894; Stenzel, 1964; Stumbur, 1975). When the periostracum is removed, it exhibits a pearly luster.

The cicatrix region of the embryo at this stage appears to be the same length as the cicatrix region of the early organogenetic embryo, but the amount of pigmentation has increased and expanded (Fig. 1A). Thus, although the amount of pigmentation may vary in *N. belauensis*, at least the length of the cicatrix is remarkably constant. An elongate central bar occurs at the midline of the cicatrix and is approximately equal in width to the space between the black lines in the younger embryo (compare Fig. 1A and B). Dorsally and ventrally, this central bar ends in a circular terminus. Removal of the overlying periostracum reveals that the aragonitic crystals that compose the central bar are uniformly oriented, while those of the circular termini are irregularly packed (Fig. 2A and B).

A slight ovoid depression surrounds the central bar and approximates the limit of pigmentation in the cicatrix region. This depression is surrounded, in turn, by a second, deeper, subcircular depression enclosed in an outer ridge (Fig. 2C). This region marks the beginning of a reticulate pattern of shell sculpture composed of fine, raised lirae. Spiral lirae diverge from the apex of the shell and intersect the radial lirae that parallel the aperture (Figs. 2C and 3A). The parallel lirae bear no signs of an ocular or hyponomic sinus (the indentations by the eye and hyponome, respectively). They originate in the periostracum and therefore indicate the position of the mantle edge. They may reflect growth cycles and exhibit an outward flexure where they intersect the spiral lirae (Fig. 3B).

The embryonic shell is composed of several layers including the *periostracum* and the *outer prismatic, nacreous*, and *inner prismatic* layers (for definitions of these terms, see Chapter 31). In describing the structure of the shell, the *apex* refers to the region near the cicatrix, the *margin* refers to the growing edge at the aperture, and the *protoseptum* (which is described in more detail later) refers to a calcareous deposit that invests the interior of the shell apex. The dorsal and ventral directions of the embryonic shell are indicated in Figs. 1B, 4, 8, 10, and 13.

3.2. Outer Prismatic Layer

The periostracum is secreted in the periostracal groove at the mantle edge. A zone of the mantle surface adjacent to this groove secretes the outer prismatic layer (Wilbur and Saleudin, 1983). On the dorsal side of the embryonic shell, the outer prismatic layer formed by this zone is approximately 120 μm wide (Fig. 4). At high magnification, its inner surface appears granular (Fig. 5D).

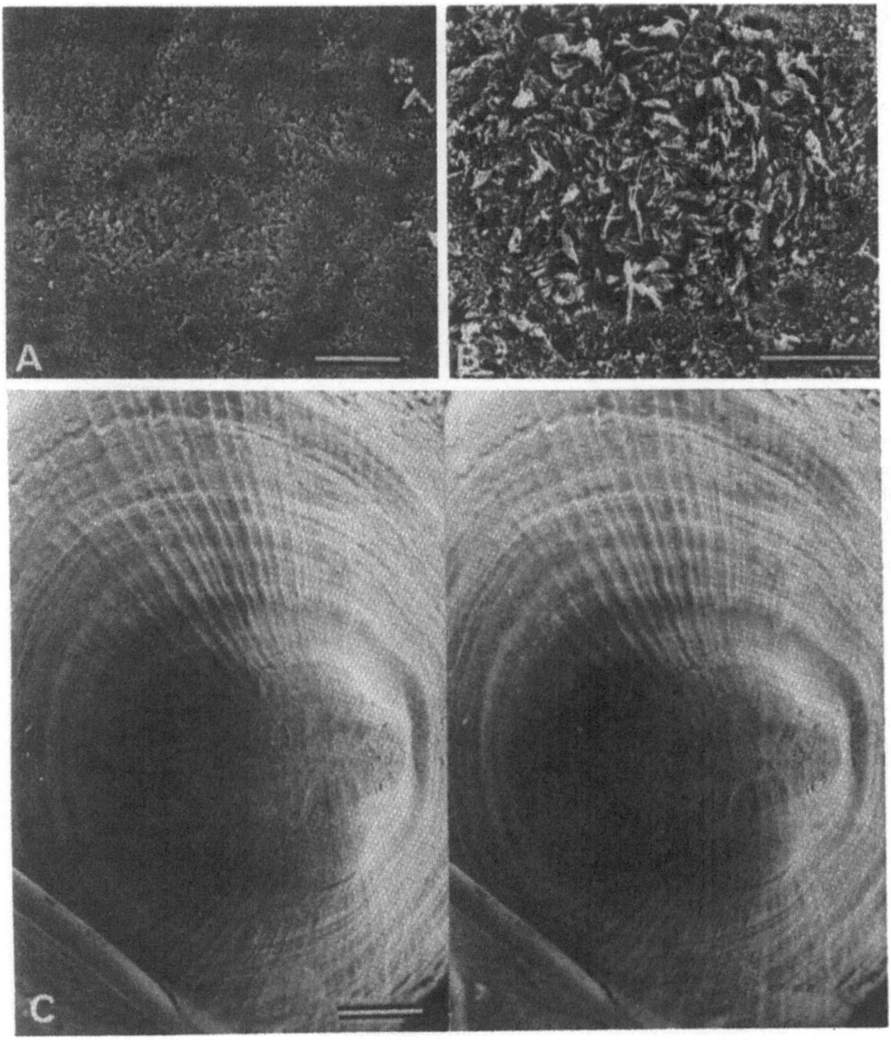

Figure 2. Cicatrix and apex of the shell at the one-chambered stage, showing the central bar (A) and circular terminus (B) of the cicatrix. Note that in the central bar, crystals are uniformly packed, whereas in the circular terminus, they are clumped irregularly. Scale bars: 20 μm. (C) Stereo view of the cicatrix and apex of the shell at the one-chambered stage. Note the two depressions that surround the cicatrix and the general contour of the shell. In this micrograph, periostracum still covers the cicatrix and has developed cracks near the cicatrix margin, parallel to the midline. The shell sculpture originates in the periostracum. Scale bar: 500 μm.

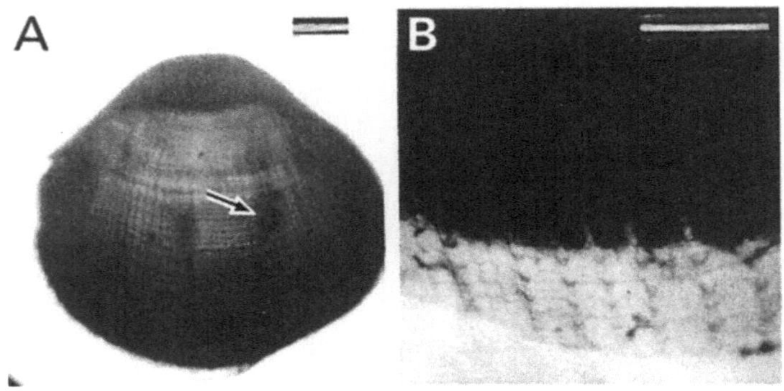

Figure 3. Shell at the one-chambered stage photographed with transmitted light. (A) The reticulate pattern of shell sculpture is conspicuous and originates in the periostracum. (→) A patch of thicker shell. Scale bar: 1 mm. (B) Close-up of the growing edge of the shell and the periostracum. The radial lirae (parallel to the aperture) exhibit an outward flexure where they intersect the spiral lirae (diverging from the apex of the shell). Calcification occurs more slowly along the spiral lirae than in the spaces between them. Scale bar: 500 μm.

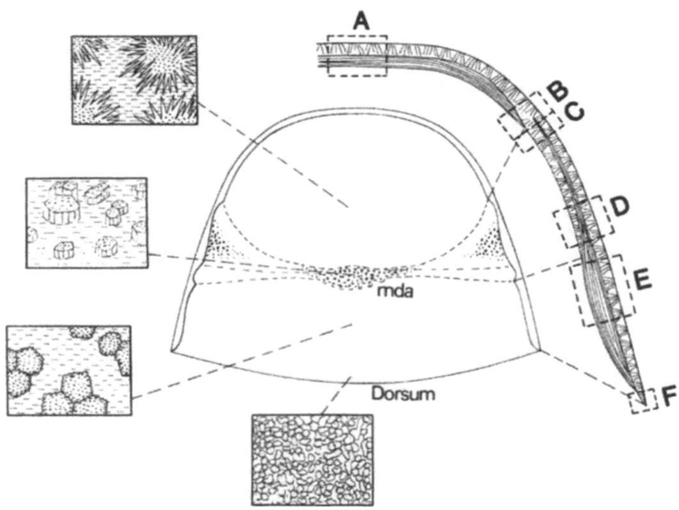

Figure 4. Diagrammatic view of the interior of part of the apical portion of the shell at the one-chambered stage of development. On the exterior (not shown), the cicatrix extends dorsally, along the median plane of symmetry, to the region that corresponds approximately to the middorsal area (mda). The shell structure is illustrated in a diagrammatic cross section along the shell edge. (A–F) Refer to Fig. 6A–F. Cross section A transects the ventralmost portion of the cicatrix. The regions of shell secretion are shown on the interior surface with close-ups of representative crystals. From the top counterclockwise, these close-ups represent nacreous crystals of the protoseptum (the innermost layer in cross sections A and B), prismatic crystals of the inner prismatic layer (the innermost layer in cross sections C and D), nacreous crystals of the medial nacreous layer (the innermost layer in cross section E), and prismatic crystals of the outer prismatic layer (the layer in cross section F). These close-ups appear as micrographs in Fig. 5.

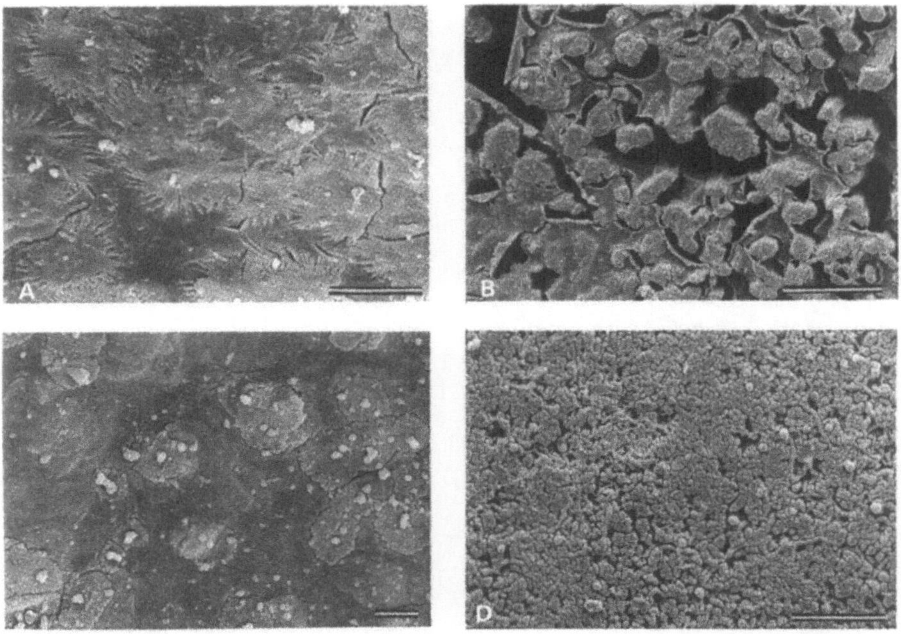

Figure 5. Close-ups of the crystals developing on the interior surface of the apical part of the shell at the one-chambered stage of development. (A) Nacreous crystals of the protoseptum. Scale bar: 10 μm. (B) Prismatic crystals of the inner prismatic layer in the middorsal area. Scale bar: 10 μm. (C) Nacreous crystals of the medial nacreous layer. Scale bar: 4 μm. (D) Prismatic crystals of the outer prismatic layer. Scale bar: 10 μm.

In section, the outer prismatic layer consists of numerous, more or less distinct, prismatic sectors of spherules (Figs. 6A–F and 7) The acicular (needlelike) crystalline elements in each sector radiate from the periostracum toward the shell interior. As the sectors grow inward, new acicular elements are added, making the resultant shell denser. Commonly, horizontal organic sheets interconnect the crystalline elements at their bases. Clumps of aragonitic crystals develop at the intersection of the spiral and radial lirae (Fig. 7).

In later ontogeny, the main portion of the outer prismatic layer is spherulitic, and only a thin, inner portion in contact with the succeeding nacreous layer is prismatic. In the embryonic shell, however, the spherulitic portion has not yet developed, and the entire layer is prismatic (Erben *et al.*, 1969b; Blind, 1976).

3.3. Nacreous Layer

The nacreous layer is secreted by an epithelial zone on the mantle surface behind the zone that secretes the outer prismatic layer. On the interior surface of the embryonic shell, this layer is exposed as a wide band between the inner and

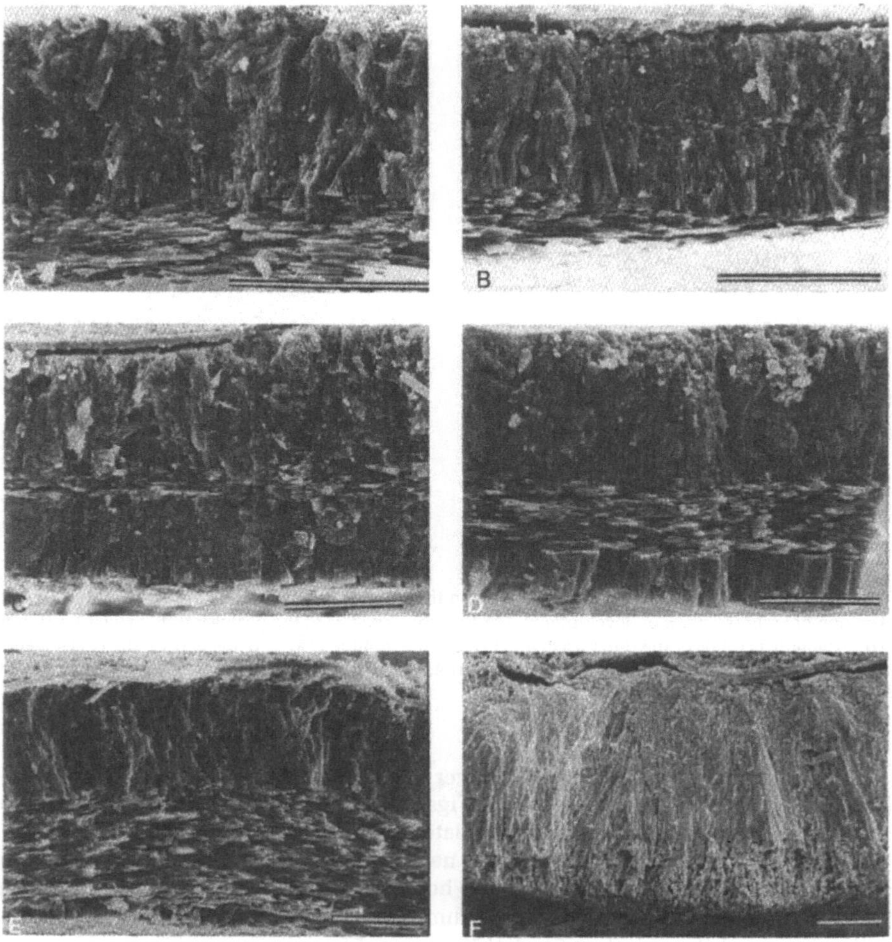

Figure 6. Cross sections of the outer wall of the apical part of the shell at the one-chambered stage of development, as indicated in Fig. 4. (A) Outer and inner (?) prismatic layers of the shell wall and outer prismatic (?) and nacreous layers of the protoseptum in the region of the cicatrix. Scale bar: 40 μm. (B) Periostracum, outer prismatic, beginning of the medial nacreous, and inner prismatic layers of the shell wall; and nacreous layer of the protoseptum. Scale bar: 40 μm. (C) Periostracum and outer prismatic, medial nacreous, and inner prismatic layers of the shell wall. Scale bar: 20 μm. (D) Outer prismatic, medial nacreous, and inner prismatic layers of the shell wall. Scale bar: 20 μm. (E) Periostracum and outer prismatic and medial nacreous layers of the shell wall. Scale bar: 20 μm. (F) Periostracum and outer prismatic layer of the shell wall. Scale bar: 10 μm.

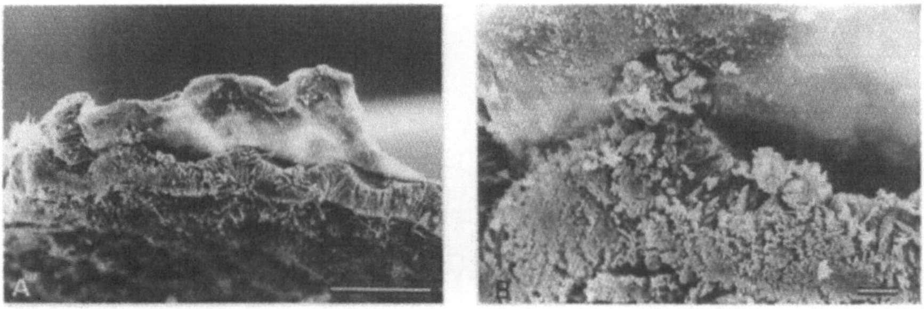

Figure 7. Growing edge of the shell at the one-chambered stage of development. (A) The shell is prismatic and reflects the ornamentation in the periostracum. Scale bar: 100 μm. (B) Close-up of a spherical bundle of crystals formed at the intersection between spiral and radial lirae. Scale bar: 10 μm.

outer prismatic layers. On the dorsal side of the embryonic shell, this band measures approximately 450 μm in width and ends approximately 120 μm from the apertural margin (Fig. 4). At low magnification, the surface of this layer exhibits a granular appearance, but closer inspection reveals that the nacreous crystals are indistinct and embedded in an organic matrix (Fig. 5C).

The nacreous layer comprises the medial layer of the completed three-layered shell wall [not counting the periostracum (Figs. 4 and 6B–E)]. It is thin near the dorsal edge of the embryonic shell, but rapidly increases in thickness toward the apex. Its maximum thickness is approximately 30 μm, about 6 μm thicker than the overlying outer prismatic layer (Fig. 6E). Farther toward the apex, it decreases in thickness, and it wedges out completely at the protoseptal margin (Fig. 6B). It is absent in the entire apical region.

Measurements of the thickness of the nacreous layer in the early whorls of adult shells indicate that, on the ventral side, it is equal in thickness to that of the outer prismatic layer at the site of the first septum and double the thickness of the outer prismatic layer at the site of the second septum (Fig. 8). In contrast, on the dorsal side, it is four times the thickness of the outer prismatic layer at the site of the second septum.

3.4. Inner Prismatic Layer

This layer is secreted by the adhesive epithelium at the site of its attachment to the shell wall. In the embryonic shell, it forms a girdle that invests the apical region of the shell adapertural to the protoseptum. It is narrow middorsally, but broadens laterally (Fig. 4). At high magnification, its surface consists of the ends of hexagonal prisms embedded in an organic matrix (Fig. 5B).

In section, the prisms of the inner prismatic layer generally show a more regular vertical orientation than do those in the outer prismatic layer (Fig. 6D). In the apical region of the shell wall where the medial nacreous layer wedges out, the inner prismatic layer is in contact with the outer prismatic layer (Fig. 6A

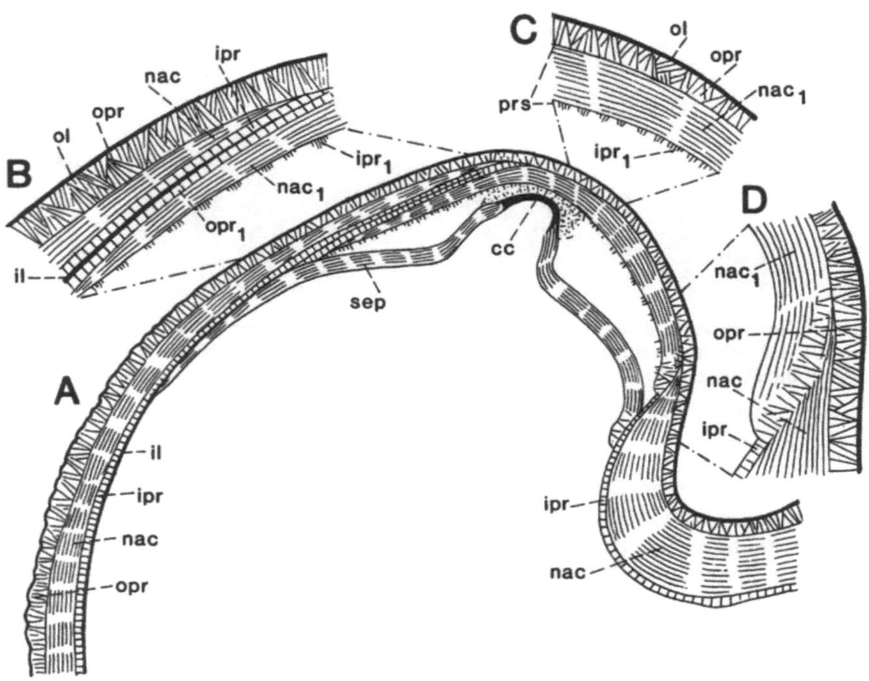

Figure 8. Diagrammatic median cross section through the apex of the shell of an adult specimen of *N. pompilius*. The dorsal direction is toward the right of the diagram. (A) Reconstructed growth stage after the formation of the first septum and cecum. (B, C, D) Details of the structure of the outer wall and protoseptum. Anatomy: (cc) cecum; (il) inner organic film of the outer wall; (ipr, ipr_1) inner prismatic layer of the outer wall and protoseptum, respectively; (nac, nac_1) nacreous layer of the outer wall and protoseptum, respectively; (ol) outer organic film of the outer wall; (opr, opr_1) outer prismatic layer of the outer wall and protoseptum, respectively; (prs) protoseptum; (sep) first septum.

and B). Near the dorsal edge of the shell, the inner prismatic layer is distinct and measures 20 μm in maximum thickness (Fig. 6C).

On the ventral side of the early whorls of adult shells, the inner prismatic layer consists of two sublayers, separated by a conspicuous organic layer. The outer sublayer is probably secreted at the shell aperture while the inner sublayer is secreted in front of the first septum. On the dorsal side of the shell, the inner prismatic layer does not show a clear separation into sublayers.

3.5. Protoseptum

After the secretion of the three-layered shell wall surrounding the cicatrix, the protoseptum is deposited in the area that will eventually be enclosed by the first chamber. It should be emphasized that the protoseptum is a secondary deposit on the interior of the apical region of the embryonic shell and not a true septum that delineates a chamber. In other words, it forms the apical surface of the first chamber and is not penetrated by the siphuncle. In median sections the protoseptum consists of an outer prismatic layer, a median nacreous layer, and an inner prismatic layer. The inner prismatic layer is diffuse and hard to differentiate. In the dorsal part of the protoseptum, the outer prismatic layer is also inconspicous while the nacreous layer is relatively thick. The maximum thickness of the protoseptum occurs in the unilayered dorsoapical region of the shell where the outer wall is thinnest (Fig. 8C and D).

In the embryonic shell, the protoseptum is not fully developed in that it lacks the inner porous prismatic layer. Its inner surface instead consists of nacreous tablets that exhibit a modified spherulitic shape (Fig. 5A). Similar structural modifications are common in nacre rich in organic matrix (Mutvei, 1972). The nacreous layer is thickest on the dorsalmost part of the protoseptum (Fig. 6A). The layer that overlies it is probably the outer prismatic layer of the protoseptum, but the boundary between this layer and the shell wall is indistinct. In the ventral direction, the nacreous layer of the protoseptum decreases in thickness (Fig. 8).

The morphology of the protoseptum reflects the structure and position of the cicatrix (Figs. 9B and 10A). The cicatrix appears as a shallow median depression in the protoseptum with ovoid expansions at each end. This depression is rimmed by a low and broad elevation. The ovoid dorsal area corresponds to the dorsal circular terminus while the deep circular ventral portion corresponds to the ventral circular terminus (Fig. 10A). Only the more conspicuous dorsal area, referred to here as the *dorsal septal depression*, is retained in subsequent septa ("hyposeptal fossa") (Fig. 10B and C) (Mutvei, 1957). It always occurs above the middorsal area. During ontogeny, this depression becomes more conspicuous and attains a maximum depth in the septa of the second whorl. These septa bend adapically to form a prominent flexure (Fig. 10C.) The dorsal septal depression diminishes in the septa of the third whorl; in the last few septa of the mature shell, it becomes the septal furrow (Fig. 10D).

3.6. Epithelial Attachment to the Shell

The retractor muscles occur as a pair on the dorsolateral sides of the soft body of the embryo and form scars on the interior of the shell (Fig. 9A) (see also Fig. 4C in Chapter 26). These muscles maintain their position throughout growth. In addition to the retractor muscles, two myoadhesive epithelial bands also develop, but only in later ontogeny (see, for example, Mutvei, 1957; Bandel and Spaeth, 1983). One of these bands extends between the retractor muscles on the ventral side of the body; the other follows the outline of the septal epithelium.

Inspection of the inner surface of the embryonic shell reveals another possible site of muscle attachment. A shallow groove, approximately 100 μm wide, occurs in front of the protoseptum (Figs. 9 and 10A). This groove broadens middorsally to form a well-defined region, called the *middorsal area*. This area extends to the protoseptal margin, where it ends in a slightly elevated ridge. The adapertural boundary of the middorsal area is less pronounced and more irregular. The middorsal area and the dorsal groove exhibit a prismatic structure distinct from that of the nacre of the protoseptum and that of the shell wall (Fig. 5B). On the basis of its shape, prismatic structure, and ontogenetic development, the middorsal area may be the site of muscle attachment.

The mantle surface of the embryo also bears bundles of longitudinal pallial muscles (Chapter 26). These muscles form a ring approximately 150 μm above

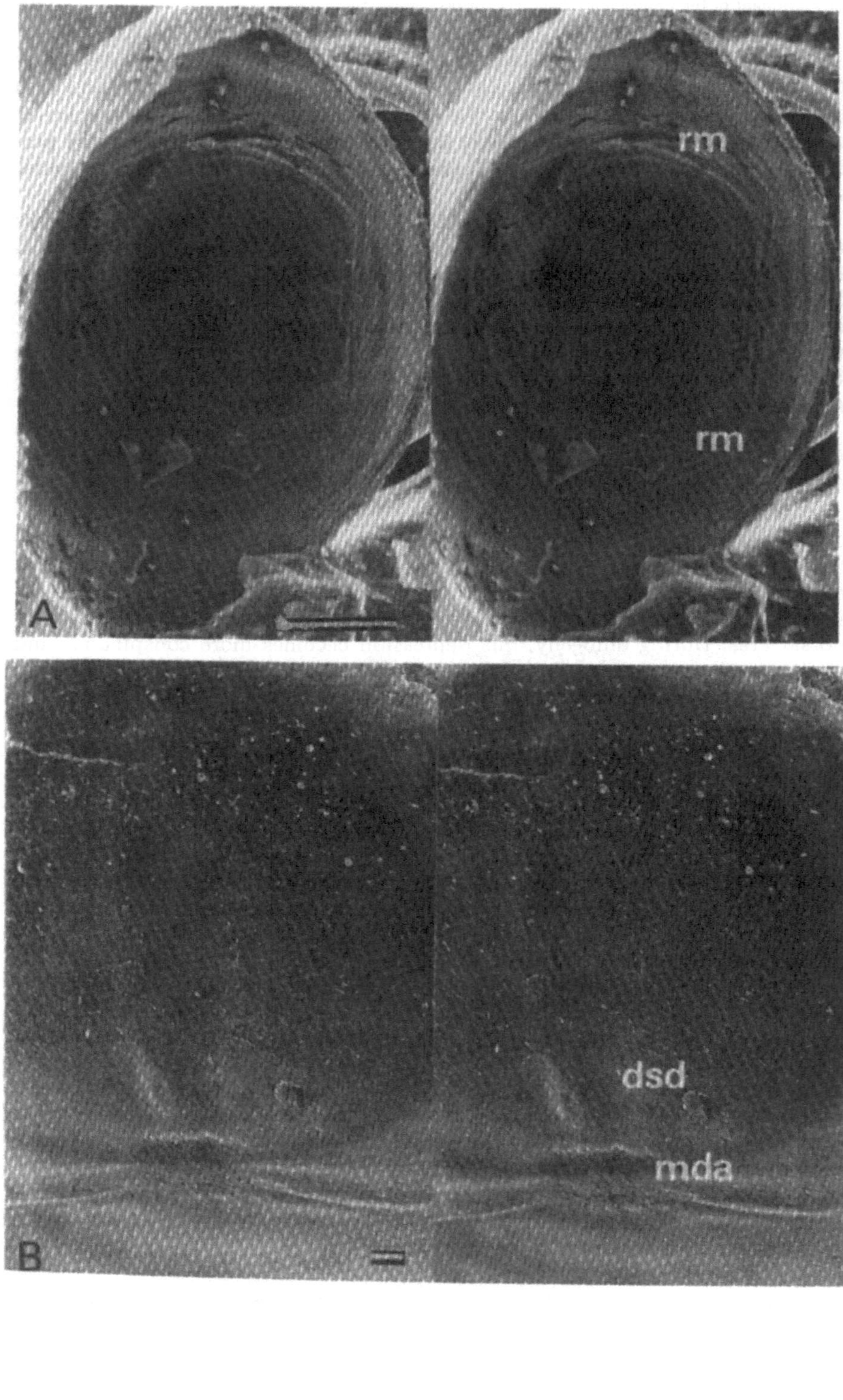
rm
rm
A
dsd
mda
B

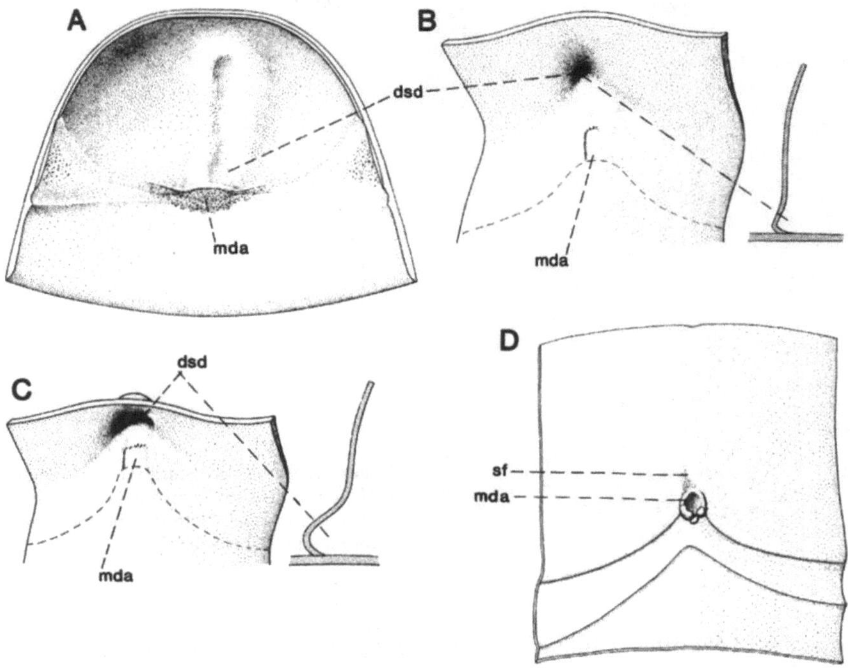

Figure 10. Development of the middorsal area (mda), dorsal septal depression (dsd), and septal furrow (sf). The dorsal direction is toward the bottom of the diagrams. Septa: (A) protoseptum; (B) third septum; (C) septum in the second whorl; (D) septum in the adult.

the periostracal groove and are similar to the pallial muscles in monoplacophorans, most bivalves, and patellid gastropods (see Fig. 5B in Chapter 26). Each bundle terminates downward in a myoadhesive epithelial area that has a circular or quadrangular outline. In later ontogeny, the mode of attachment of the mantle to the shell aperture changes radically and consists of epithelial extensions attached to pores in the shell wall (Mutvei and Doguzhaeva, manuscript in preparation).

4. Multiple-Chambered Stage

4.1. External Morphology

In this stage, the outer shell displays gradual changes in color, curvature, ornamentation, and thickness. These changes occur in the embryonic shells of

←

Figure 9. (A) Stereo view of the interior of the shell apex at the one-chambered stage of development showing the protoseptum and the sites of retractor muscle (rm) attachment. The dorsal direction is toward the right of the photograph. Scale bar: 1 mm. (B) Stereo view of the protoseptum, middorsal area (mda), and dorsal septal depression (dsd). The dorsal direction is toward the bottom of the photograph. Scale bar: 100 μm.

Figure 11. Median cross section of a specimen of *N. pompilius* showing the axis of coiling (⇨) near the umbilicus. Note the reduced spacing (approximation) between septa 7 and 8.

all species of *Nautilus*, although developmental differences may appear (Dauphin, 1979). In describing the external morphology of the embryonic shell, we will refer to the number of whorls from the apex, with the axis of coiling located near the umbilicus (Fig. 11), or from the nepionic constriction. This feature consists of a conspicuous groove near the umbilicus of the shell (Fig. 12). It occurs at $1\frac{1}{4}$–$1\frac{1}{2}$ whorls from the apex in all species of *Nautilus* and has been interpreted as marking the end of embryonic development (Willey, 1897c; Naef, 1923; Stenzel, 1964; Stumbur, 1975; Cochran *et al.*, 1981; Ward, 1983a).

At approximately $\frac{1}{6}$ whorl from the apex (about 3 mm maximum diameter), the shell curvature may change slightly. A more marked change in curvature occurs at $\frac{1}{3}$ whorl (4–6 mm maximum diameter); this change was interpreted by Hyatt (1894) as the end of his "metanepionic stage." The dorsal part of the shell changes at this point from a flat to a curved shape, and a slight constriction may appear near the umbilicus (Fig. 13A). Thereafter, the shell coils as a spiral with little modification (Fig. 13B) (Eichler and Ristedt, 1966b; Hirano and Obata, 1979; Hirano *et al.*, 1980; Tanabe *et al.*, 1985). This developmental pattern is supported by measurements of the radius of the shell spiral versus whorl number. The radius is measured from the axis of coiling to the ventral margin of the shell, along the median plane. These measurements indicate an almost uniform growth of the spiral radius (Fig. 14) [note that oscillations in such graphs may be artifacts due to mislocation of the axis of coiling (for a method to pinpoint the coiling axis, see Landman, 1982a)]. At approximately $\frac{1}{2}$ whorl from the apex, the dorsal part of the shell begins to turn inward, and at 1 whorl (16–20 mm maximum diameter), the shell becomes involute, producing a narrow and deep umbilical perforation in *N. pompilius*, *N. belauensis*, and *N. scrobiculatus*, but only a shallow pit in

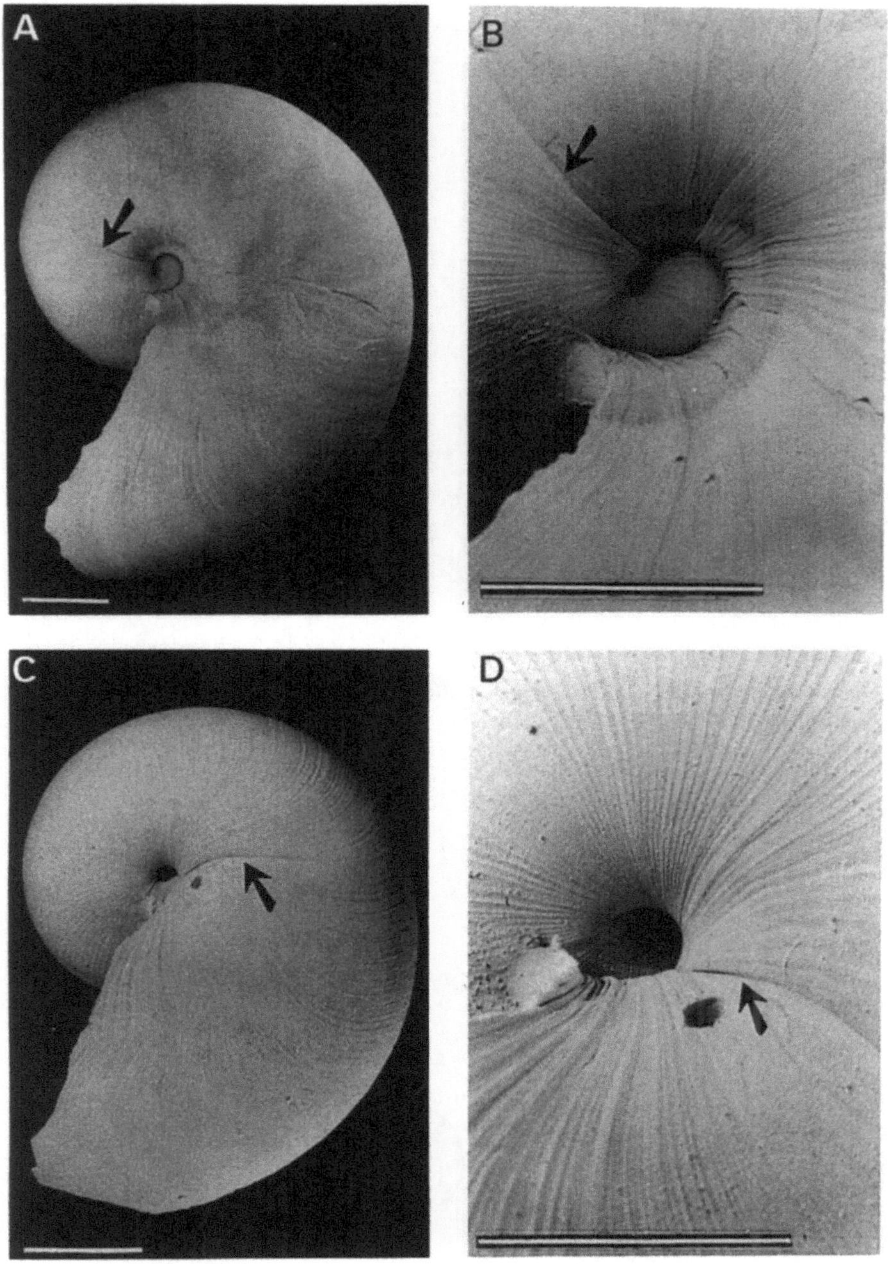

Figure 12. Nepionic constriction (⬅) in three species of *Nautilus*, showing the change in the umbilicus at this point. Species: (A) *N. macromphalus* from New Caledonia (MCZ 230061); (B) enlargement; (C) *N. belauensis* from Palau (AMNH 43031); (D) enlargement; (E) *N. scrobiculatus* from Papua New Guinea (DMNH 19941); (F) enlargement. Scale bars: 1 cm.

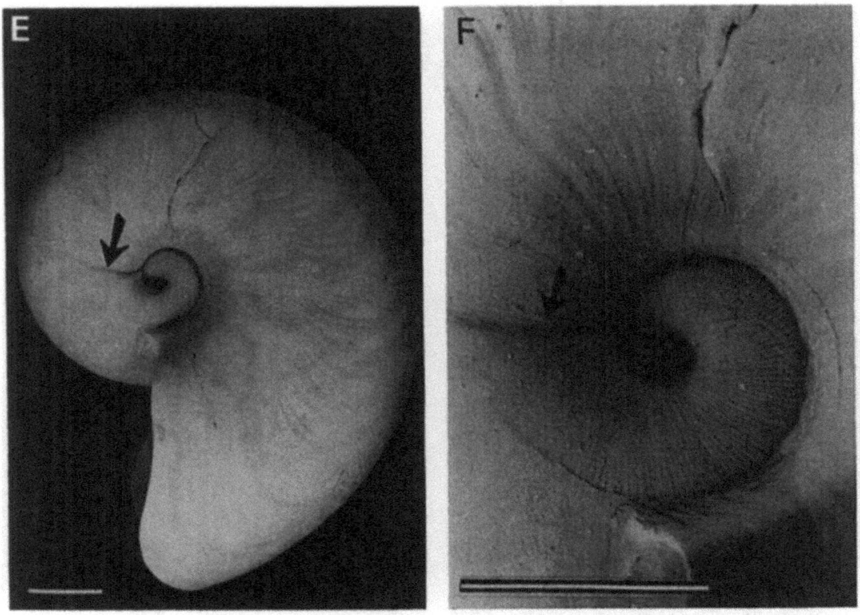

Figure 12. *(Continued)*

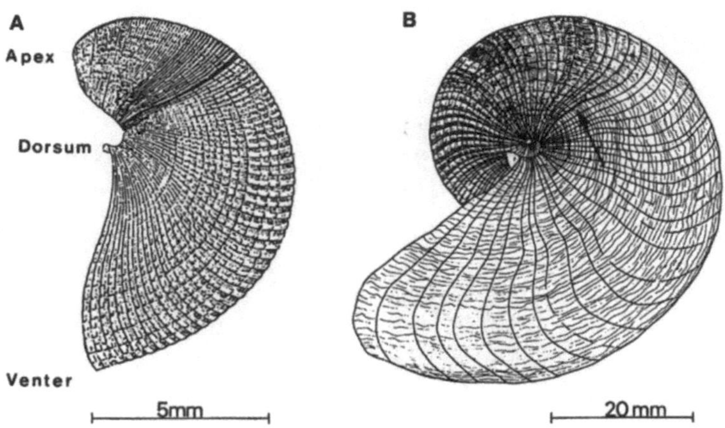

Figure 13. Sculpture and shape of the embryonic shell of *N. pompilius*. (A) Drawing of a specimen of *N. pompilius* approximately ½ whorl long, modified from Hyatt (1894). A slight constriction occurs at approximately ⅓ whorl from the apex. (B) Drawing of a specimen of *N. pompilius* from Papua New Guinea (AMNH 43032) showing the nepionic constriction (←).

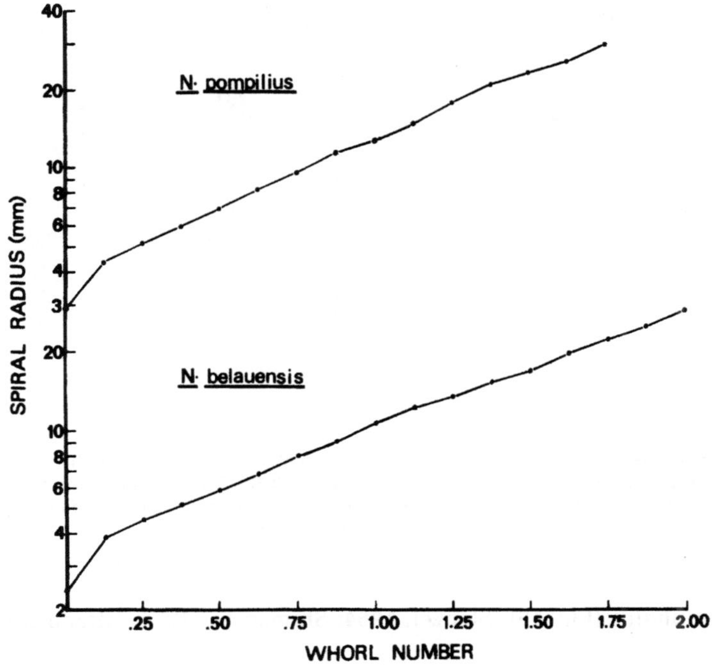

Figure 14. Measurements of the radius of the shell spiral vs. whorl number in a specimen of *N. pompilius* and a specimen of *N. belauensis*. These measurements indicate an almost uniform growth of the spiral radius after the apex.

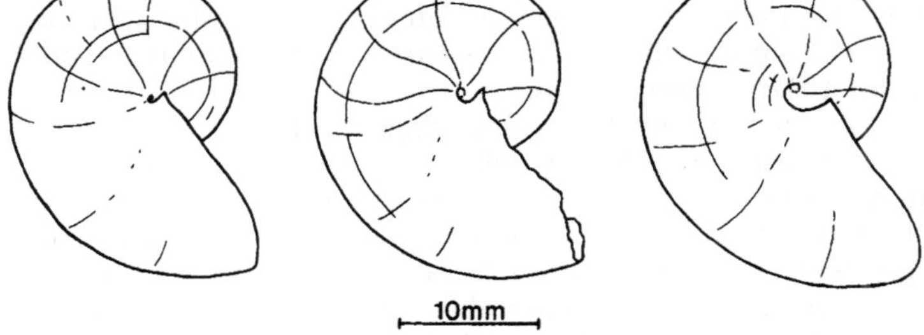

Figure 15. Drawings of the embryonic shells of three species of *Nautilus* prior to the formation of the nepionic constriction. Note the differences in the sizes of the umbilici. (A) Specimen of *N. macromphalus* (YPM 6387, 12530), prepared from a larger shell. (B) Specimen of *N. pompilius* (YPM 12530) preserved intact. This specimen may have died at hatching. (C) Specimen of *N. scrobiculatus* (YPM 6067), prepared from a larger shell.

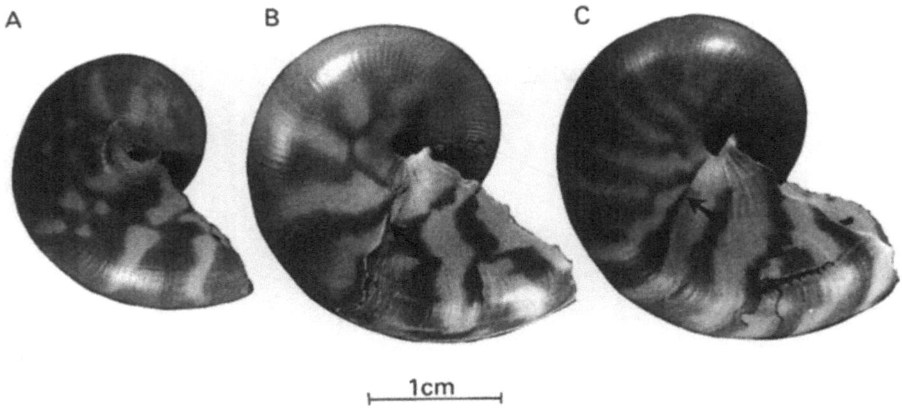

Figure 16. Ontogenetic development of the protective coloration. Nepionic constriction (←). (A) Specimen of *Nautilus* sp. from the Loyalty Islands (DMNH 117070). This specimen lacks the nepionic constriction and may have died at hatching. (B) Specimen of *N. pompilius* from the Philippines (SUI 40044). (C) Specimen of *N. pompilius* from the Philippines (SUI 40064).

N. macromphalus (Fig. 15) (Stenzel, 1964; Shimansky, 1974; Hirano and Obata, 1979).

The reticulate pattern of shell sculpture gradually attenuates during embryonic development and generally disappears altogether before the nepionic constriction. At approximately $\frac{1}{3}$ whorl from the apex, coincident with the change in shell curvature, the radial lirae start to become progressively broader and later develop a hyponomic sinus (Fig. 13A) (Stenzel, 1964). After approximately $\frac{1}{2}$ whorl from the apex (9–11 mm maximum diameter), they become smooth, although they may persist near the umbilicus (Fig. 13B). They are replaced by growth lines that exhibit a shallow hyponomic sinus. The spiral lirae become weak after $\frac{1}{2}$ whorl from the apex in *N. pompilius* and gradually fade away, first on the venter and later on the flanks (Hyatt, 1894; Eichler and Ristedt, 1966a,b; Stumbur, 1975; Dauphin, 1979; Hirano and Obata, 1979). However, in *N. belauensis*, *N. macromphalus*, *N. scrobiculatus*, and some specimens of *N. pompilius*, a weak crenulate sculpture may persist past the nepionic constriction (Fig. 13B) (Willey, 1897c; Stenzel, 1964).

The color pattern generally develops slightly before or immediately after the nepionic constriction (Fig. 16). Initially, the shell is ivory colored, but at approximately $\frac{1}{2}$ whorl from the apex, it develops a pale orange color that subsequently becomes darker. In *N. pompilius*, the white and brown color pattern appears gradually, first on the flanks and later on the venter. It becomes visible on the flanks about $\frac{1}{4}$–$\frac{1}{2}$ whorl before the constriction. Initially, the boundaries between the color bands are indefinite, but later become better defined (Stenzel, 1964). Species-specific differences in the developmental pattern of the coloration occur, but require further study.

The whorl section of the embryonic shell changes from a depressed to a more

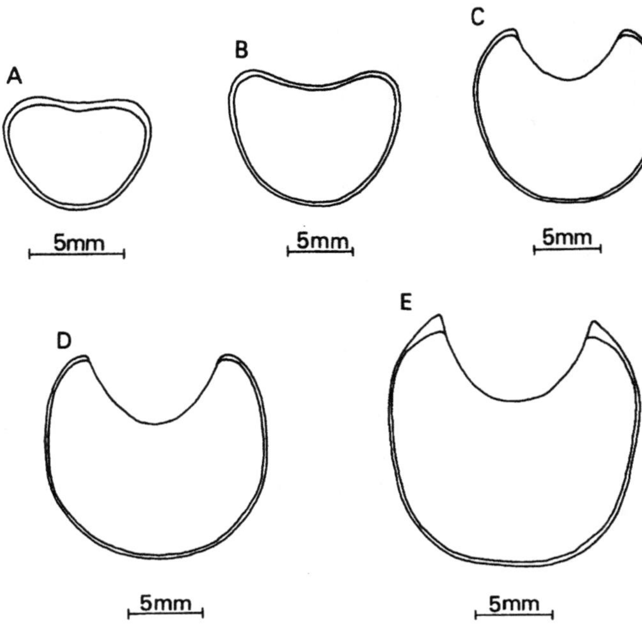

Figure 17. Changes in the whorl section during the embryonic development of *N. pompilius*, drawn from five different specimens. (A) At 0.4 whorl (≈8 mm diameter), from a specimen prepared from a larger shell (YPM 12530). The whorl width/whorl height ratio is 1.3. (B) At 0.8 whorl (15.1 mm diameter), from a specimen prepared from a larger shell (YPM 12530). The whorl width/whorl height ratio is 1.3. (C) At 21.9 mm diameter, from a specimen preserved intact at this size (SUI 40041). The whorl width/whorl height ratio is 1.2. (D) At 24.7 mm diameter, from a specimen preserved intact at this size (YPM 12530, illustrated in Fig. 15B). The whorl width/whorl height ratio is 1.1. (E) At 29.8 mm diameter (0.1 whorl past the nepionic constriction), from a specimen preserved intact at this size (SUI 42473). The whorl width/whorl height ratio is 1.0.

compressed shape throughout growth (Fig. 17). At approximately ⅓ whorl from the apex, the whorl section is subtriangular, with a slight dorsal impression (Fig. 17A). At approximately ½ whorl from the apex, the whorl section is still depressed, with an average width/height ratio of 1.5 in *N. pompilius* and 1.6 in *N. macromphalus* (Hirano and Obata, 1979). At the nepionic constriction, the average width/height ratio has decreased to 1.1 in *N. pompilius* and 1.0 in *N. macromphalus*, and the dorsal impression is relatively deep (Fig. 17C and D) (Stumbur, 1975; Hirano and Obata, 1979). Thereafter, the width/height ratio remains nearly the same (Fig. 17E) (Chapter 7).

The embryonic shell surface is glossy and rarely displays any signs of shell breakage and repair. As a result, the pattern of shell sculpture and growth lines is generally uninterrupted. However, shell abnormalities sometimes occur during embryonic development, as illustrated in Fig. 18 (see also Eichler and Ristedt, 1966b).

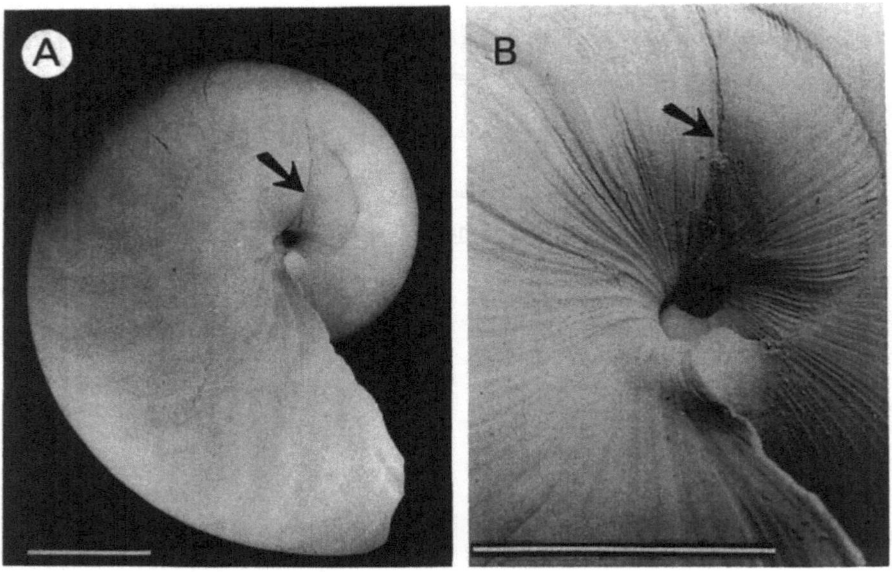

Figure 18. (A) Specimen of *N. pompilius* (AMNH 170788) from the Philippines, showing abnormal shell development just adapical of the nepionic constriction (→). (B) Close-up. Scale bars: 1 cm.

4.2. Septa

Several septa and the adjoining length of siphuncle are inferred to have formed in embryonic development (Willey, 1897c, 1902; Naef, 1923; Stenzel, 1964; Cochran *et al.*, 1981). These septa resemble those of the postembryonic stage in their microstructure and placement against the outer shell wall. A common pattern of septal spacing occurs in the embryonic whorls of all *Nautilus* species; it consists of closer spacing, or approximation, between septa 7 and 8 (see Fig. 11 and Table I) (Stenzel, 1964; Davis and Mohorter, 1973; Stumbur, 1975; Dauphin, 1979; Cochran *et al.*, 1981). However, this approximation may occur as early as between septa 4 and 5 or as late as between septa 8 and 9 (Stenzel, 1964). The thickness of the embryonic septa also shows a more or less consistent pattern (Blind, 1976; Hirano and Obata, 1979, Fig. 8; Hirano *et al.*, 1980). As illustrated in three specimens of *N. pompilius*, this pattern consists of a variable thickness averaging approximately 100 μm for the first 7 septa, generally followed by a steeper pattern of increase beginning at the 8th septum (Fig. 19).

4.3. Cecum

The saclike initial portion of the siphuncle, termed the cecum, occurs on the inner surface of the protoseptum and extends to the first septum. Its structure is variable and may be the same as that of later septal necks and connecting rings

Table I. Average Angular Length of the First Ten Chambers in Four Species of *Nautilus* Based on 15 Specimens That Display the Common Pattern of Reduced Septal Spacing Between Septa 7 and 8

Chamber no.	*Nautilus pompilius* (N = 4)	*Nautilus macromphalus* (N = 2)	*Nautilus belauensis* (N = 4)	*Nautilus scrobiculatus* (N = 5)
2	33	36	35	29
3	52	53	54	45
4	60	59	60	55
5	60	58	67	60
6	56	60	60	58
7	46	48	47	48
8	24	30	26	26
9	30	28	30	32
10	28	30	24	28

(see Mutvei, 1972). For example, the cecum illustrated in Fig. 20F is structurally unmodified. The septal neck of the first septum consists of an outer spherulitic–prismatic layer, a nacreous layer that is structurally modified distally, and an inner prismatic layer. The connecting ring of the first septum consists of an outer spherulitic–prismatic layer that extends to the inner prismatic layer of the protoseptum and an inner conchiolin layer that invests the inner surface of the cecum. The connecting ring of the second septum is fused to the septal neck of the first septum by means of an auxiliary deposit. This structurally unmodified cecum is porous.

More commonly, however, the cecum exhibits a modified multilayered structure (Erben *et al.*, 1969b; Blind, 1976; Bandel and Boletzky, 1979). The number of layers that compose the cecum is variable. One or several of these layers may

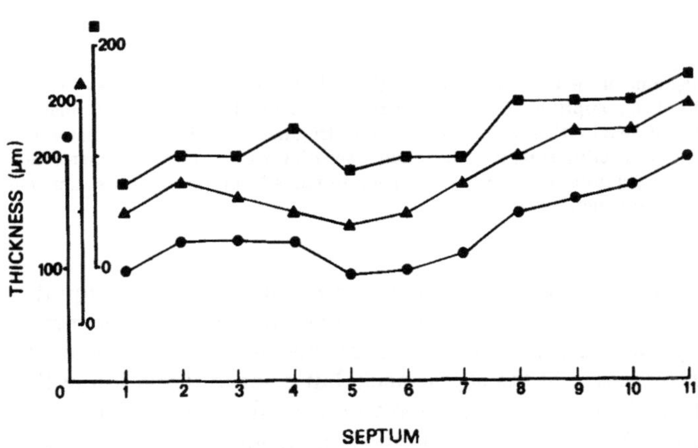

Figure 19. Septal thickness vs. septal number for the first 11 septa in three specimens of *N. pompilius* that show reduced spacing between septa 7 and 8.

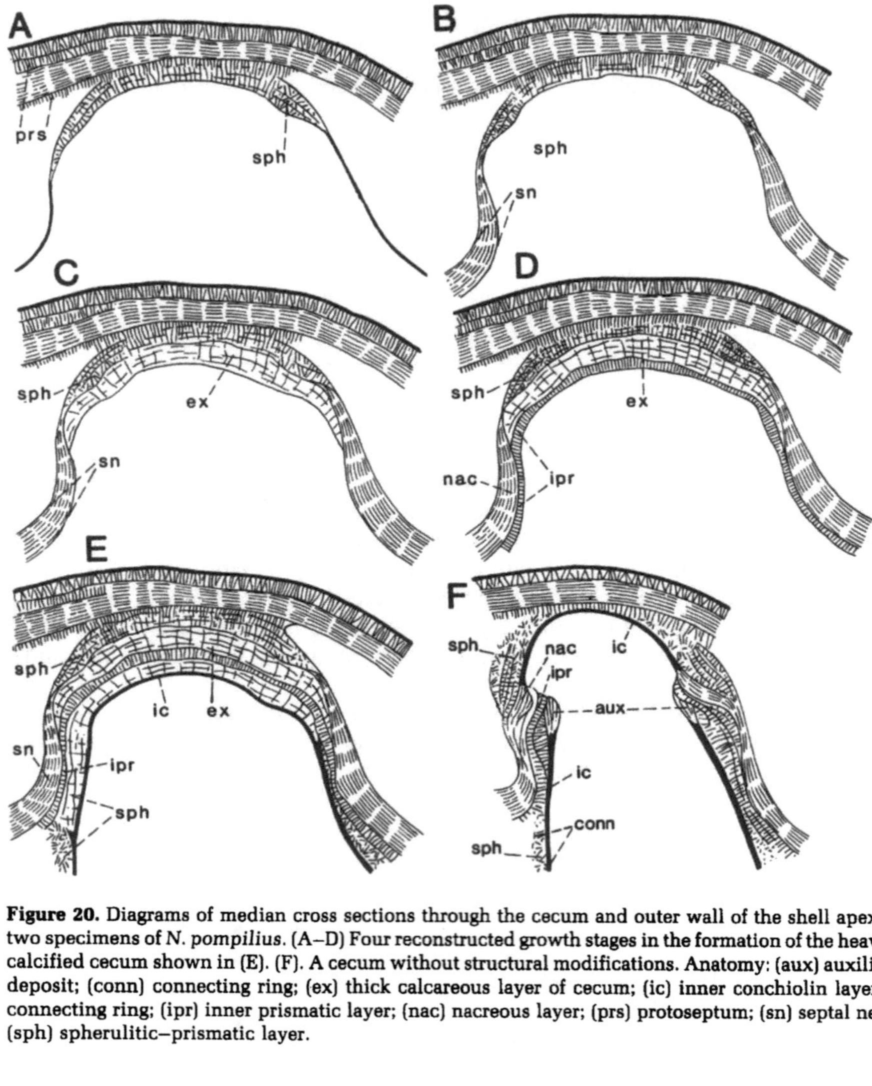

Figure 20. Diagrams of median cross sections through the cecum and outer wall of the shell apex in two specimens of *N. pompilius*. (A–D) Four reconstructed growth stages in the formation of the heavily calcified cecum shown in (E). (F). A cecum without structural modifications. Anatomy: (aux) auxiliary deposit; (conn) connecting ring; (ex) thick calcareous layer of cecum; (ic) inner conchiolin layer of connecting ring; (ipr) inner prismatic layer; (nac) nacreous layer; (prs) protoseptum; (sn) septal neck; (sph) spherulitic–prismatic layer.

be heavily calcified and therefore impermeable to gas and liquid. The cecum illustrated in Fig. 20E exhibits a thick, calcified layer (ex). Reconstruction of the successive growth stages of this layer indicates that it was secreted after formation of the nacreous layer of the septal neck, but prior to formation of the inner prismatic layer, and therefore would have prevented removal of liquid from the first chamber (Fig. 20A–E). Another calcified, compact layer occurs in this cecum and probably represents an extension of the spherulitic–prismatic layer of the connecting ring of the second septum (Fig. 20E).

5. Hatching

5.1. External Morphology

The nepionic constriction has been interpreted as marking the end of embryonic development (Willey, 1897c; Naef, 1923; Stenzel, 1964; Stumbur, 1975; Cochran *et al.*, 1981; Ward, 1983a). It occurs approximately $1\frac{1}{4}$–$1\frac{1}{2}$ whorls from the apex of the shell in all species of *Nautilus*. This whorl number corresponds to an average diameter of 26 mm and an average whorl width of 16 mm. The constriction itself consists of a narrow angular groove that extends from the umbilical shoulder onto the flanks and eventually dies out on the venter [see Fig. 12 (the "nepionic line" of Willey, 1897c)]. On steinkerns, it also appears as a groove, because it is a flexure in the actual shell wall. The constriction is accentuated *a posteriori* by a postembryonic expansion in whorl width in both *N. pompilius* and *N. macromphalus* (Hirano and Obata, 1979).

In cross section, the constriction involves a thickening and subsequent thinning of the outer prismatic layer (Fig. 21). Unlike a repaired shell break, the increase in thickness of the outer prismatic layer does not reflect a withdrawal of the mantle margin, but suggests, instead, a pause in the forward growth of the aperture. Along the venter, however, secretion of the shell is apparently unaffected.

Commonly, instead of the nepionic constriction, a break in the shell is present. As a result, the growth lines on the surface of the shell are unconformable. In cross section, the outer prismatic layer repeats, indicating a withdrawal of the mantle margin. In such specimens, breakage occurred immediately after hatching, and the break extended adapical of the original position of the constriction. The break therefore represents only the approximate point of hatching.

The number of whorls and the diameter of the shell at the constriction show some variation within and among species (Tables II and III). The number of whorls ranges from 1.20 to approximately 1.49; the diameter of the shell ranges from 22.6 to 32.6 mm. The largest embryonic shells occur in *N. pompilius* from Fiji and *N. belauensis* from Palau; the smallest occur in *N. pompilius* from the Philippines. The diameter of the shell at the constriction approximately matches the maximum size of the inner egg capsule, which measures approximately 25 mm in both *N. macromphalus* and *N. belauensis* (Willey, 1897b) (see also Chapter 26).

After formation of the nepionic constriction, the shell displays several changes indicative of postembryonic life outside the egg capsule (Stenzel, 1964). The shell surface loses its luster, and the growth lines become coarser. They form a shallow ocular sinus and a deeper hyponomic sinus. The color pattern of the shell becomes a darker reddish brown, with sharply defined white bands. Most important, the surface of the shell bears numerous scars from repaired shell injuries. The umbilical seam also changes from one spiral to another; as a result, the umbilicus immediately after the constriction is much wider (see Fig. 12) (Stenzel, 1964).

5.2. Septa

The approximation between septa 7 and 8 coincides with the formation of the nepionic constriction on the outer shell. This was observed in three live spec-

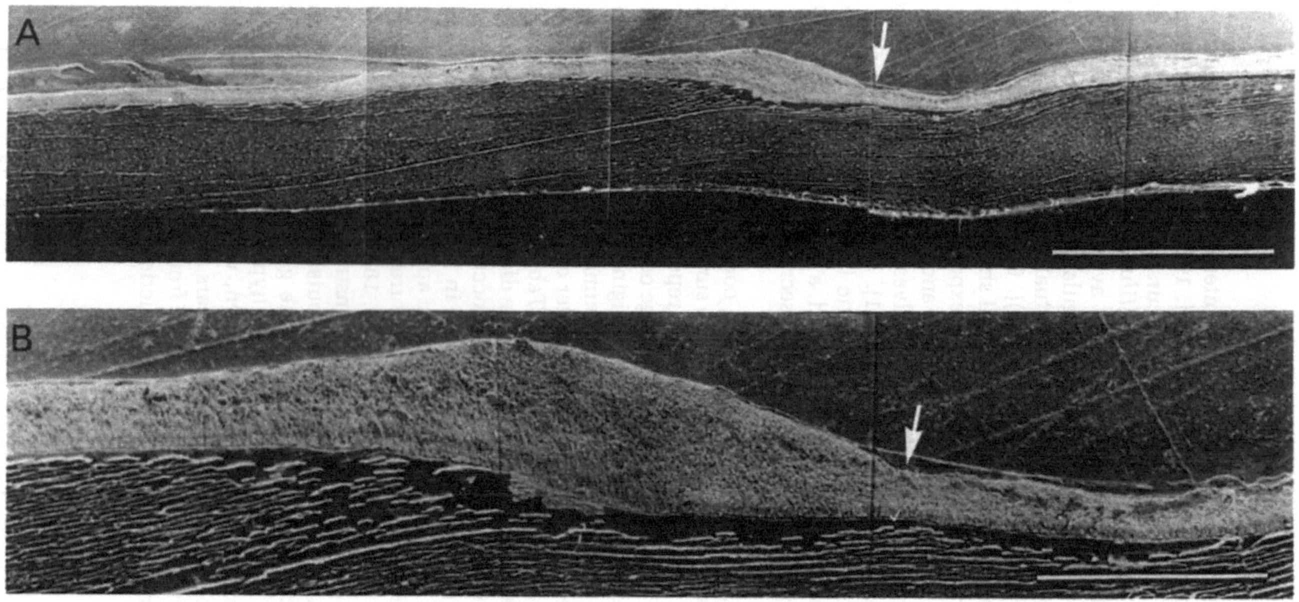

Figure 21. (A) Cross section through the nepionic constriction (⇦) near the umbilicus in a specimen of *N. pompilius*. The outer prismatic layer initially thickens and then thins. Scale bar: 1 mm. (B) Close-up. Scale bar: 300 μm.

Table II. Number of Whorls at the Nepionic Constriction in Three Species of *Nautilus*[a]

Species	Locality	Collection(s)[b] or reference	N	Number of whorls		
				Mean	SD	Range
N. macromphalus	New Caledonia	AMNH	2	1.30	0.020	1.28–1.32
N. macromphalus	New Caledonia	Hirano and Obata (1979)	1	1.48	—	—
N. scrobiculatus	Papua New Guinea	AMNH, BMC	5	1.27	0.046	1.20–1.32
N. pompilius	Philippines	Hirano and Obata (1979)	11	1.49	0.119	—
N. pompilius	Philippines	AMNH, SUI	4	1.33	0.024	1.30–1.36

[a] The number of whorls is measured from the apex of a shell on a median section. A discrepancy of approximately 0.08 whorl may occur between our data and those of Hirano and Obata (1979) due to a difference in the choice of the center of the spiral and the point at which the number of whorls equals 0.0.
[b] Collections: (AMNH) American Museum of Natural History; (BMC) Bryn Mawr College; (SUI) University of Iowa.

Table III. Diameter of the Shell at the Nepionic Constriction in Four Species of *Nautilus*

Species	Locality	Collection(s)[a] or reference	N	Diameter (mm)		
				Mean	SD	Range
N. belauensis	Palau	AMNH	2	31.1	0.70	30.4–31.8
N. macromphalus	New Caledonia	MCZ, SUI, YPM	4	26.2	0.73	25.4–27.4
N. scrobiculatus	Papua New Guinea	BMC	4	25.6	0.62	24.5–26.0
N. pompilius	Tañon Str., P.I.	Tanabe *et al.* (1985)	6	25.7	0.90	24.7–27.1
	Bohol Str., P.I.	MCZ, SUI, Tanabe *et al.* (1985)	7	25.4	1.56	22.7–27.1
	Mindoro, P.I.	AMNH, MCZ, NMNH	4	26.5	0.98	25.3–28.0
	Jolo Isl., P.I.	NMNH	2	29.0	0.85	28.1–29.8
	Philippines	AMNH, NMNH, SUI	4	25.9	0.72	25.2–27.1
	Philippines?	AMNH, SUI, YPM	50	26.2	1.38	22.6–30.0
	Amboina, Indonesia	AMNH, MCZ	2	27.1	0.30	26.8–27.4
	Papua New Guinea	AMNH	1	23.1	—	—
	Solomon Isl.	NMNH	1	24.5	—	—
	New Hebrides	NMNH	3	26.8	0.79	25.7–27.6
	Fiji	MCZ, Tanabe and Tsukahara (Chapter 7), Davis and Mohorter (1973)	11	27.1	2.11	24.3–32.6

[a] Collections: (AMNH) American Museum of Natural History; (BMC) Bryn Mawr College; (MCZ) Museum of Comparative Zoology; (NMNH) National Museum of Natural History; (SUI) University of Iowa; (YPM) Yale Peabody Museum.

imens of N. *pompilius* from Fiji in which the constriction had just formed (Davis and Mohorter, 1973). This reduction in septal spacing may reflect stress immediately after emergence from the egg capsule. The corresponding increase in septal thickness may imply a postembryonic increase in septal strength (Westermann, 1973) (see also Chapter 30).

Changes in the oxygen isotope composition of the septa provide some of the strongest evidence for interpreting the septal approximation and constriction in the outer shell as marking the end of embryonic development. Analyses of 13 specimens (8 *N. pompilius* and 5 *N. macromphalus*) indicated a shift of approximately 1 ‰ in the ratio of oxygen isotopes from light values characteristic of the first 7 septa to heavier values for septum 8 forward (Eichler and Ristedt, 1966a,b; Cochran *et al.*, 1981; B. E. Taylor and Ward, 1983; Oba and Tanabe, 1983). This shift (see Fig. 5 in Cochran *et al.*, 1981) appeared in most specimens analyzed, although Eichler and Ristedt (1966a,b) and B. E. Taylor and Ward (1983) also reported wide fluctuations among the first few septa. In one specimen of *N. pompilius*, the septal approximation occurred between septa 6 and 7 and coincided with the shift in the oxygen isotope ratio (Oba and Tanabe, 1983).

The light isotope values characteristic of the first 7 septa reflect the isotope composition of the fluid within the inner egg capsule (Cochran *et al.*, 1981; B. E. Taylor and Ward, 1983) (see also Chapter 9). This fluid is depleted in $\delta^{18}O$ by 1 ‰ relative to ambient seawater due to fractionation across the egg membrane (Crocker *et al.*, 1985). The shift to heavier values at septum 8 reflects equilibration with seawater on emergence of the animal from the egg capsule.

The pattern of carbon isotopes has also been analyzed in these same specimens. It generally shows a similar shift from light to heavier values, but at 2 or 3 septa after the increase in $\delta^{18}O$ (Eichler and Ristedt, 1966a,b; Cochran *et al.*, 1981; B. E. Taylor and Ward, 1983; Crocker *et al.*, 1985). This delay may suggest that a kinetic effect also occurs at hatching (Cochran *et al.*, 1981; B. E. Taylor and Ward, 1983; Crocker *et al.*, 1985) (see also Chapter 9).

6. Speculation on the Mode of Life and Environment at Hatching

On the basis of the morphological and isotopic data presented in Section 5, *Nautilus* presumably hatches at a shell diameter of approximately 26 mm. The shape of the embryonic shell is nearly the same as that of the adult. Its phragmocone measures about 14.5–18.5 mm in diameter and commonly consists of 7 buoyancy chambers (Stumbur, 1975). The angular length of the body chamber measures approximately ⅓ whorl (Davis and Mohorter, 1973).

The smallest animals caught alive were collected in Fiji and were described as active swimmers (Davis and Mohorter, 1973). They measured approximately 25 mm in shell diameter and had just formed the constriction. Most other small shells known consist of drift specimens that probably represent animals that died immediately after hatching.

The depth at which *Nautilus* hatches has not yet been determined. The young animals collected in Fiji by Mohorter occurred in a shallow lagoon [at a depth of 1 m (see Davis and Mohorter, 1973)]. Similarly, Willey (1902) observed egg-laying on the sides of his trap in shallow water (depth of 6 m) in the Loyalty Islands, but all the eggs were infertile. However, turbulence in very shallow water could pose problems for young animals (Chamberlain, 1978). Ward and Martin (1980) have also argued that the high temperatures in these environments are generally lethal to adults. They believe that hatching more probably occurs at water depths of at least 100 m (in Fiji and New Caledonia), and these estimates correlate well with those based on the oxygen isotope composition of the first few postembryonic septa (B. E. Taylor and Ward, 1983). In Palau, juvenile animals are rare, but occur at water depths of 160–300 m, along with adults (Saunders, 1984b).

7. Comparison with Other Cephalopods

As in all other recent cephalopods, development in *Nautilus* is direct. Hatching occurs at a relatively large size and, unlike the hatching of some squids and cuttlefish, is not followed by a planktonic phase (Arnold, 1984). The embryonic shells of fossil nautilids are also relatively large and similar in appearance to those of modern *Nautilus* (Landman *et al.*, 1983). This large size may represent an adaptation to a deep-water habitat (Saunders, 1984b) and contrasts sharply with hatching size in the co-occurring group of fossil shelled cephalopods, the ammonoids (Landman, 1982a, 1987). Embryonic shells in these animals average 1 mm in diameter and exhibit a distinctive microornamentation, a uniform shell structure, and a single large buoyancy chamber called the *protoconch* (Landman, 1987). Immediately after hatching, the ammonoids may have spent some time as plankton, which may have contributed to their differential extinction at the end of the Cretaceous (Emiliani *et al.*, 1981; Ward, 1983b; Landman, 1984).

8. Conclusions

The preceding discussion cannot be considered a definitive description of the embryonic development of any structure as complex as the *Nautilus* shell. We have attempted, instead, to present all available data and to lay the groundwork for further investigation. We still do not have the ontogenetic stages that show the formation of the cecum and its transformation into the siphuncular tube, the developmental changes in the attachment of the soft body to the shell, or the formation of all the layers in the shell wall. Nor do we yet know the events in the mantle that activate shell secretion or their basis at the cellular and biochemical levels. However, now that it appears possible to obtain *Nautilus* embryos on a consistent basis, some of these goals may be attainable.

ACKNOWLEDGMENTS. John Arnold gratefully acknowledges the support of the Grass Foundation, the *Nautilus* Research Corporation, the Hawaiian Malacological Society, the Friends of the Waikiki Aquarium, and the *Teuthis obscura* Fund. Neil Landman acknowledges the support of the American Museum of Natural History.

Harry Mutvei thanks the Swedish Natural Science Research Council for support (Grant No. 287–12). Dr. Barbara Boyer read and improved the manuscript. Miss Amy Adelson (AMNH) prepared Figs. 12, 13B, and 14–19. Miss Susan Klofak (AMNH) and Miss Lauren Duffy (AMNH) assisted in specimen preparation and scanning electron microscopy, respectively.

Chapter 28

Growth and Longevity of *Nautilus*

NEIL H. LANDMAN and J. KIRK COCHRAN

1. Introduction

Like many other aspects of the life history of *Nautilus*, its rate of growth and longevity have until recently been unknown. Growth rate is defined as the increase in dimensions of the shell or number of septa (chambers) with respect to time. The growth of the shell involves two processes: (1) Shell is accreted at the aperture in conjunction with growth of the soft body; as traced along the venter, the shell forms a logarithmic spiral in which radial increase is an exponential function of rotational angle. (2) Periodically, the animal moves forward to begin formation of a new septum (chamber) at the rear of the body chamber (see Chapter 34, this volume). Except near maturity and immediately after hatching, septa are usually secreted at equal angular intervals and the ratio of chamber volumes between adjacent chambers is constant. These two growth rhythms are interrelated, and as the animal adds body and shell weight, it must initiate the formation of a new chamber to maintain neutral buoyancy. The angular length of shell added to the aperture must be sufficient to compensate for the sector the new chamber subtends. The timing of these two processes and their variation over ontogeny are the subjects of this chapter.

Apertural growth occurs in the form of irregular increments called *growth lines*. However, the rate of apertural growth is commonly expressed as the number of millimeters of shell added to the aperture per day (measured along the venter). It is usually reported as an average over the entire time of observation, because

NEIL H. LANDMAN • Department of Invertebrates, American Museum of Natural History, New York, New York 10024. J. KIRK COCHRAN • Marine Sciences Research Center, State University of New York at Stony Brook, Stony Brook, New York 11794.

it may vary widely over shorter intervals. The reported value may also represent an average over several developmental stages. Relative growth is measured with respect to the total ventral circumference, measured in longitudinal cross section, from the apex of the shell to the aperture. The septal or chamber period is defined as the time between successive points in septal or chamber formation. Another metric expression of growth, although infrequently used, is overall size increase, measured as a function of diameter, radius, or weight. Like other animals, *Nautilus* progresses in its ontogeny from immaturity through maturity, the point at which growth ceases. This stage is marked by the development of distinctive features, such as thickening of the aperture and deepening of the ocular sinus (see Chapter 29).

The orderly arrangement of septa and the spiral shape of the *Nautilus* shell have led to the impression that *Nautilus* grows in a similarly even fashion. In his celebrated poem, Oliver Wendell Holmes (1858) speculated that a new chamber forms every year. More recently, P. G. K. Kahn and Pompea (1978) assumed that chambers are formed every lunar month and that growth lines are deposited daily. These assumptions were based on analogy with growth-line development in other mollusks, such as clams, and on their interpretation that approximately 30 growth lines occurred on the outer shell between adjacent septa. Using these assumptions, they recorded the number of growth lines per chamber in fossil nautiloids in order to determine the number of days per lunar month and by implication the distance between the earth and the moon throughout the Phanerozoic. However, these assumptions were false, and this approach has been widely discounted (Hughes, 1979, 1985; D. S. Jones and Thompson, 1979; Runcorn, 1979; Saunders and Ward, 1979).

Two general methods have been used to investigate the rate of growth of *Nautilus*: (1) Direct methods, such as release–recapture and aquarium study, record the rate of growth over the period of observation. (2) Indirect methods, principally radionuclides, reconstruct the rate of growth over some time interval prior to capture. We will consider each method in turn and at the end of the chapter attempt a synthesis. We hope to answer the following questions: (1) How similar are the rates of growth derived by the various methods? (2) How does the rate of growth vary over ontogeny? (3) How similar are the rates of growth among the various species of *Nautilus*? (4) Can we construct a general curve to describe the rate of growth? (5) How does the rate of growth of *Nautilus* compare to that of other cephalopods?

2. Direct Methods of Growth Measurement

2.1. Release–Recapture

The release–recapture studies of Saunders (1983) yielded the first direct measurements of *in situ* growth. Over 2000 specimens of *N. belauensis* were caught, measured, tagged, and released during five field seasons in Palau, West Caroline Islands. Growth rates were obtained for seven animals that were either immature or submature at release and grew for periods of 45–355 days before recapture.

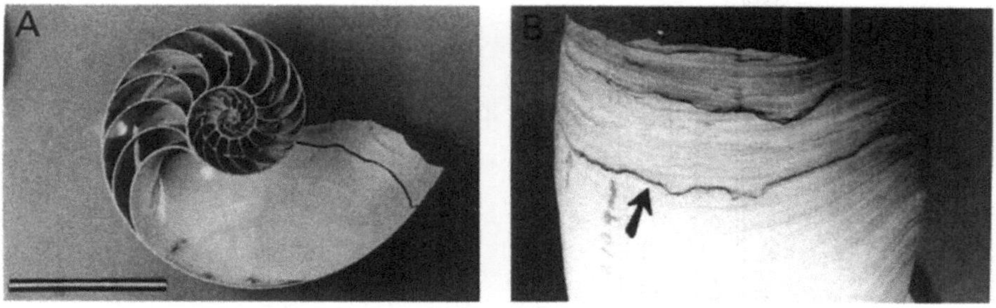

Figure 1. *Nautilus belauensis* (specimen F), immature at release, showing the growth increment formed after 353 days. (A) The inked line on the interior of the body chamber indicates the position of the aperture at the time of release. Scale bar: 10 cm. (B) In a ventral view, the position of the aperture at the time of release is marked by a "shock line" (→) Data and photographs from Saunders (1983).

Several of these animals reached maturity before recapture (E, B, and G); therefore, the resultant growth rates are averages over several developmental stages (Table I). Four animals were immature at release (A, C, E, and F) and ranged in size from 177 to 192 mm with 34–36 septa (Fig. 1). Three animals were submature at release (B, G, and D) and ranged in size from 202 to 208 mm with 34–36 septa. Five other submature or mature animals (J, I, H, K, and L), ranging in size from 192 to 210 mm, were recaptured after periods of 1054–1496 days and, except in one specimen (J), exhibited no apparent growth.

2.1.1. Rate of Apertural Growth

Despite the variation in the intervals of time elapsed between release and recapture, the four immature specimens that grew to submaturity or maturity (A, C, E, and F) displayed similar growth rates, ranging from 0.08 to 0.12 mm/day [average = 0.10 mm/day (Fig. 2 and Table I)]. The three submature animals (B, G, and D) grew more slowly (0.02–0.07 mm/day), although two of them reached maturity before recapture and may have experienced a time interval during which no growth occurred (Table I). Thus, their rates of apertural growth may have been depressed and do not necessarily imply a gradual deceleration through ontogeny. Animal J, which grew from submature to mature, exhibited a 7% increase in diameter, although its apertural growth rate was not reported. Growth ceased at maturity, as indicated by animals I, H, K, and L, which showed no growth in the time interval between release and recapture.

The growth lines added to the aperture between release and recapture were difficult to count. The computed number of days per growth line was lowest (4.0–9.0 days) in the four immature specimens that grew to submaturity or maturity (A, C, E, and F) and the single submature specimen that remained submature (D) (Table I). The longest estimates (13.8 and 25.4 days) occurred in the two submature specimens that grew to maturity (B and G), although they may have experienced an interval of nongrowth.

Table I. Rate of Apertural Growth and Period of Septal Formation in *Nautilus belauensis*[a]

Specimen	At release			Time elapsed (days)	At recapture			Rate of apertural growth		Period of septal formation (days)
	Stage[b]	Diameter (mm)	Septa		Stage[b]	Diameter (mm)	Septa	mm/day	Days/growth line	
A	I	192	36	82	SM	196	36	0.12	8.2	—
C	I	177	35	46	SM	178	35	0.11	7.7	—
E	I	179	35	342	M	190	36	0.09	4.0	≲342
F	I	192	34	353	SM	204	35	0.08	7.1	≈353
B	SM	208	36	330	M	213	36	0.04	13.8	—
G	SM	203	34	355	M	205	34	0.02	25.4	—
D	SM	202	35	45	SM	203	35	0.07	9.0	—
J	SM	192	—	1496	M	205	—	—	—	—
I	SM	210	—	1448	M	209	—	—	—	—
H	M	194	—	1108	M	194	—	—	—	—
K	M	203	—	1054	M	204	—	—	—	—
L	M	199	—	1083	M	198	—	—	—	—

[a] The table is based on release–recapture date of Saunders (1983). The apparent changes in diameter in specimens I, K, and L are due to approximating to the nearest integer.

[b] Stages: (I) immature; (SM) submature; (M) mature.

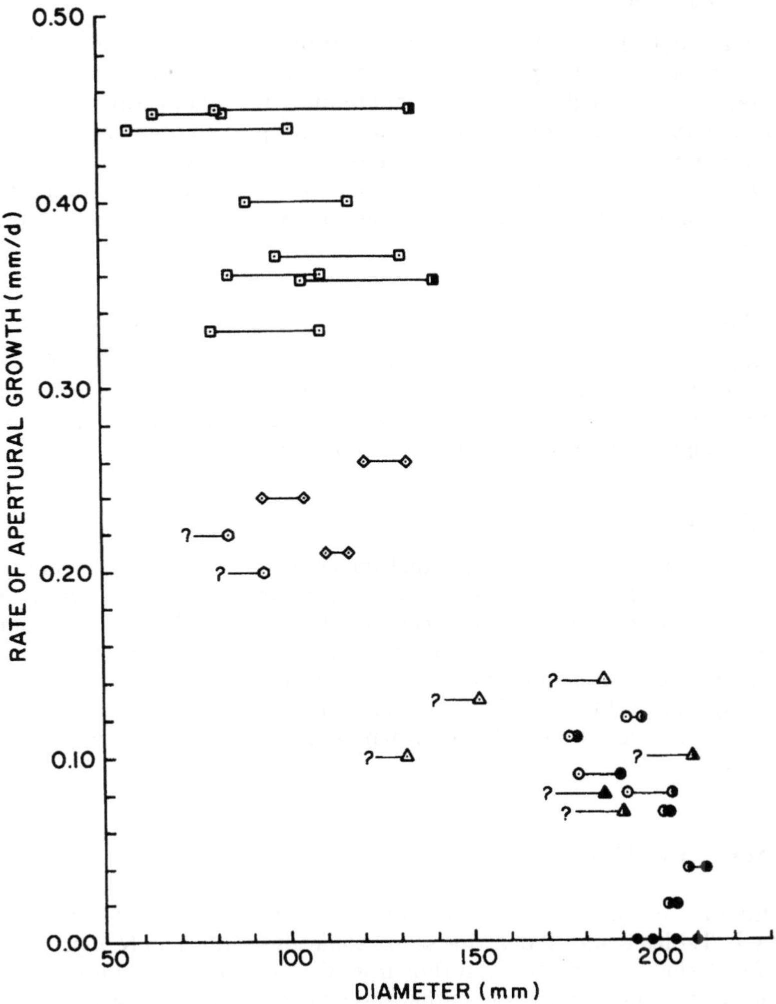

Figure 2. Rate of apertural growth vs. shell diameter for two species of *Nautilus:* (1) *N. belauensis* from the release–recapture study of Saunders (1983) (⊙, ◑, ●) and radiometric measurements [(△, ▲, ▲) specimen 4 omitted (Cochran and Landman, 1984)] and (2) *N. pompilius* from radiometric measurements [⊙ (Cochran *et al.*, 1981)] and two different aquarium studies [⊡, ◨ (Ward, 1985); ◇ (Ward and Chamberlain, 1983)]. Shell diameters for specimens of *N. pompilius* from Ward (1985) were estimated from his Fig. 2. Stages: (⊙, △, ⊙ , ⊡, ◇) immature; (◑, ▲, ◨) submature; (●, ▲) fully mature. The rate of apertural growth for any particular specimen may represent an average of more than one growth stage. (?) Indicates that the initial developmental stage and diameter are unknown. The specimens of *N. pompilius* in aquaria are smaller and grow more rapidly than the specimens of *N. belauensis* in nature, suggesting an inverse relationship between the rate of apertural growth and shell diameter over their combined range. The discrepancy in the rates of growth may be due to actual differences in the two species or to differences in aquarium vs. natural conditions. However, specimens must be compared at the same stage of development because of variation in the rate of growth over ontogeny, especially at maturity.

2.1.2. Period of Septal (Chamber) Formation

Two animals (E and F) each secreted their final septa in the interval between release and recapture. In specimen E, which grew from an immature to a mature stage, 342 days elapsed. This figure represents a maximum estimate for septal formation, because the animal had stopped growing some time before recapture. In specimen F, which grew from an immature to a submature stage, 353 days elapsed. The final septum in this specimen was thickened and approximated relative to the preceding septum, but the animal was still growing at the aperture when recaptured.

2.1.3. Rate of Weight Increase

Animal weight was an unreliable indicator of growth (Saunders, 1983, Tables 2 and 3). It showed irregular fluctuations over time due to variation in the amount of cameral liquid occluded and the weight of crop contents (as much as 20% of the total weight can be accommodated in the crop).

2.1.4. Critique

No growth rate data were obtained for very small immature specimens because (1) only 20% of the animals released were immature or submature and (2) none of the very small animals was recaptured, suggesting that they may not have survived the initial capture and release. Recaptured animals exhibited a "shock line" on the shell, probably produced by breakage and abrasion of the aperture during trap retrieval (Fig. 1). This shock may have caused a reduction in the rate of growth, although shell material grown subsequent to release was normal in appearance.

2.2. Aquarium Maintenance

This method has been used widely and has yielded the most data on rates of growth. Tables II and III describe seven aquarium studies on *N. macromphalus* from New Caledonia and *N. pompilius* from the Philippines, representing a total of about 60 specimens for which either the rate of apertural growth or the period of septal formation, or both, was reported. To facilitate comparisons with other studies, whenever possible, we classified the initial and final stages of development as immature, submature, or mature and expressed the initial and final sizes in terms of diameter.

The smallest immature specimens were reported in the study of Ward (1985), in which both species were grown under the same conditions. Nevertheless, several of these immature specimens attained maturity by the end of the study, although at a smaller size than that in nature. The number of septa ranged from 17–29 at the beginning of this study to 23–34 at the end (Ward, 1985, Table 1). In other studies in which the number of septa was reported, it ranged from 24–32 at the start to 25–34 at the end (Ward *et al.*, 1981, Table 1; Ward and Chamberlain, 1983, Fig. 1).

Table II. Rate of Apertural Growth and Period of Septal Formation in *Nautilus macromphalus* Based on Aquarium Studies

Study	Temperature (°C)	Number of specimens	Stage (St)[a] and diameter (D): Initial St	Initial D (mm)	Final St	Final D (mm)	Time elapsed (days)	X-ray?	Rate of apertural growth (mm/day): Mean ± SD	Range	Period of septal formation (days)
Hamada et al. (1978)[b]	14–17	1	—	—	I	129	100	No	0.05	—	—
		3	—	—	SM	142–155	100	No	0.11 ± 0.02	0.08–0.12	—
		2	—	—	M	163–165	100	No	0.00 ± 0.00	0.00–0.01	—
A. W. Martin et al. (1978)[c]	22–29	7	I	—	I	—	90	Yes	0.25	0.17–0.30	—
		3	I	—	I	—	85–100	No	—	—	≈30
Kanie et al. (1979)[b]	14–19	1	—	—	I	128	100	No	0.04	—	—
		4	—	—	SM	142–155	57–100	No	0.08 ± 0.05	0.00–0.13	—
		2	—	—	M	163–166	100	No	0.00 ± 0.00	0.00–0.01	—
Ward et al. (1981)[d]	16	19	I	—	I	—	12–77	Yes	0.09 ± 0.04	0.05–0.18	70 (composite), 80–120
		2	I	—	I	—	105–135	No	0.15 ± 0.00	0.15–0.15	
Ward (1985)[e]	18–20	2	I	125–134	M	153–159	300	Yes	(I) 0.54 ± 0.08	0.49 ± 0.04–0.60 ± 0.10	49–98

[a] Stages: (I) immature; (SM) submature; (M) mature. [b] The stage of development was inferred.
[c] Longer estimates (60–70 days) for the period of septal formation were obtained by dividing the calculated age to maturity by 30 chambers.
[d] The estimate of 80–120 days was based on one specimen. The rate of apertural growth (0.09 ± 0.04 mm/day) was based on seven specimens.
[e] Diameters were calculated from radial measurements (Ward, 1985, Fig. 2). The initial diameters mark the start of the period of infrequent radiography (≈300 days), during which time growth appeared normal. The range in the period of septal formation from 49 to 98 days reflects an ontogenetic increase through maturity over a time interval of approximately 390 days.

Table III. Rate of Apertural Growth and Period of Septal Formation in *Nautilus pompilius* Based on Aquarium Studies

Study	Temperature (°C)	Number of specimens	Stage (St)[a] and diameter (D): Initial St	Initial D (mm)	Final St	Final D (mm)	Time elapsed (days)	X-ray?	Rate of apertural growth (mm/day): Mean ± S.D.	Range	Period of septal formation (days)
Ward and Chamberlain (1983)[b]	18–20	4	I	—	I	—	300	Yes	0.23 ± 0.02	0.21–0.26	85 ± 3–132 ± 7
Ward (1985)[c]	18–20	9	I	58–105	SM	84–141	300	Yes	0.40 ± 0.11	0.33 ± 0.09–0.45 ± 0.11	14–105

[a] Stages: (I) immature; (SM) submature.
[b] The rates of apertural growth are based on original data supplied by the authors.
[c] Diameters were calculated from radial measurements (Ward, 1985, Fig. 2). The initial diameters mark the start of the period of infrequent radiography (≈300 days), during which time growth appeared normal. The range in the period of septal formation from 14 to 105 days reflects an ontogenetic increase over a time interval of approximately 390 days.

The water temperatures ranged from 14 to 20°C in all studies except the study of A. W. Martin *et al.* (1978), in which temperatures ranged from 22 to 29°C. The periods of observation ranged from 12 days to about a year. In four of the studies, animals were radiographed weekly or biweekly, although in Ward (1985), radiography was carried out progressively less frequently (about once a month) during the course of the study.

2.2.1. Rate of Apertural Growth

In *N. macromphalus*, the rate of apertural growth varied widely among individual specimens at the same developmental stage in different studies and, to a lesser extent, within the same study as well (Table II). The slowest rates averaged less than 0.15 mm/day (Hamada *et al.*, 1978; Kanie *et al.*, 1979; Ward *et al.*, 1981). A. W. Martin *et al.* (1978) obtained an average rate of 0.25 mm/day, although the weekly rate fluctuated from 0 to 0.60 mm/day, and shell growth appeared abnormal. Nevertheless, growth line counts on ten specimens yielded a consistent period of 2 days/growth line (Table IV). Ward (1985) obtained the highest average rate of 0.54 mm/day for two specimens observed over approximately 300 days (during which time radiography was performed with relative infrequency). Rates varied widely over shorter intervals of time (1–3 months), but apertural growth was continuous during septal formation. At the onset of maturity, rates of apertural growth declined abruptly to zero (Fig. 3) (Hamada *et al.*, 1978; Kanie *et al.*, 1979; Ward, 1985).

In *N. pompilius*, Ward and Chamberlain (1983) obtained an average rate of 0.23 mm/day (Fig. 2 and Table III). Apertural growth was continuous during the course of septal formation. Ward (1985) obtained a higher average rate of 0.40 mm/day, comparable to that of *N. macromphalus* in the same study (Table III). Averages of individual specimens ranged from 0.33 ± 0.09 to 0.45 ± 0.11 mm/day (Fig. 2). Rates varied widely over shorter intervals of time (1–3 months) and displayed no obvious trend, except a decline at maturity (Fig. 3).

Table IV. Periodicity of Growth Line Formation in 10 Specimens of *Nautilus macromphalus* in Aquaria[a]

Specimen	Period of observation (days)	Number of growth lines (mean ± S.D.)	Days/growth line
A	44	21.4 ± 1.63	2.0
B	28	14.4 ± 1.17	1.9
C	136	51.4 ± 3.66	2.6
D	101	37.7 ± 3.23	2.7
E	88	41.2 ± 1.86	2.1
F	78	33.7 ± 2.55	2.3
G	149	70.7 ± 3.26	2.1
H	113	58.2 ± 2.78	1.9
I	105	52.9 ± 4.27	2.0
J	84	38.9 ± 5.03	2.2

[a] From A. W. Martin *et al.* (1978).

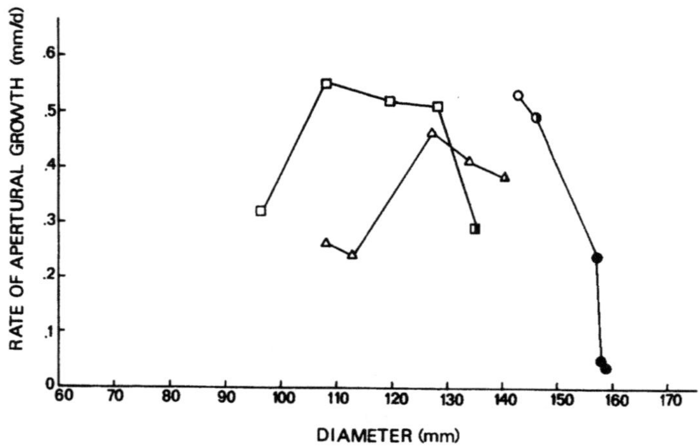

Figure 3. Rate of apertural growth vs. shell diameter in two specimens of *N. pompilius* (□, ◨, △, ▲) and one specimen of *N. macromphalus* (○, ◑, ●) as they approach maturity. The three specimens were grown in the same aquarium and show variation in the rate of growth with a decline at the onset of maturity. The symbols are explained in the Fig. 2 caption. Data from Ward (1985, Fig. 2); shell diameters were estimated from radial measurements.

2.2.2. Period of Septal (Chamber) Formation

On the basis of cyclic fluctuations in weight, A. W. Martin *et al.* (1978) suggested that *N. macromphalus* forms a new chamber every month (Table II). However, longer estimates of 60–70 days were obtained by dividing the estimated age to maturity (5–6 years, based on extrapolated rates of apertural growth and weight increase) by 30 chambers. Ward *et al.* (1981) also suggested a period of 70 days on the basis of composite data for 19 specimens [no single animal was followed through a complete chamber formation cycle (Table II)]. Longer estimates of 80–120 days were obtained, on the basis of rates of septal calcification and apertural growth. In two *N. macromphalus* observed by Ward (1985), the period of chamber formation increased from 49 to 98 days through ontogeny (Table II).

In *N. pompilius*, the period of chamber formation also increased ontogenetically. In Ward and Chamberlain (1983), the period ranged from 85 ± 3 to 132 ± 7 days, with increased duration in larger, later-formed chambers (Table III). In Ward (1985), the period ranged from a minimum of 14–28 days in a small, immature specimen (chamber 23) to a maximum of 105 days in a larger, submature specimen [chamber 30 (Table III)]. A semilog plot of septal number vs. period of septal formation suggests that this increase was exponential (Fig. 4). For example, the pattern of ontogenetic increase in one specimen may be represented by the equation

$$y = 2.8e^{0.12x}$$

where y is the period of septal formation in days and x is the septal number. However, the period of chamber formation for any particular chamber (e.g., chamber 30) varied widely among individual specimens (Fig. 4).

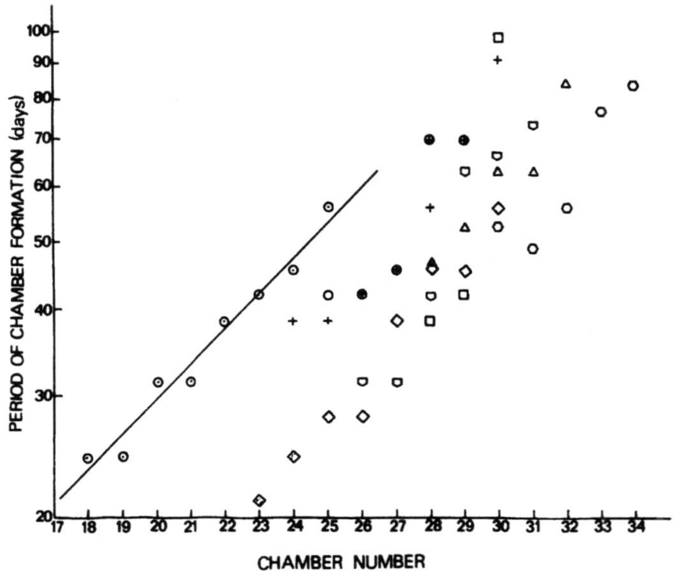

Figure 4. Semilog plot of the period of chamber formation vs. chamber number (smaller numbers refer to earlier chambers) for seven specimens of *N. pompilius* (various symbols) grown in the same aquarium. Specimens included in the plot consisted of at least four points (four chamber formations). The increase in the period of chamber formation through ontogeny appears to be exponential; the best-fit line is shown for one specimen. Data from Ward (1985, Fig. 4); the value at any point represents the average of the reported maximum and minimum estimates.

2.2.3. Rate of Weight Increase

On the basis of weekly measurements, A. W. Martin *et al.* (1978) reported a linear increase in weight (discounting fluctuations) for immature specimens of *N. macromphalus* over periods of 3–4 months. Instantaneous growth rates averaged 0.35 ± 0.046% weight increase per day, with no indication of a deceleration with increasing weight.

2.2.4. Critique

Although the effects of confinement are unknown, growth rates in aquaria may differ from those in nature for several reasons (Hirano, 1981). Higher than normal growth rates may be due to regular feedings and an abundance of food. Surface conditions (1 atm pressure) may also contribute to faster rates and promote rapid cameral liquid emptying (see Ward, 1985). Lower than normal rates may be due to high temperatures (A. W. Martin *et al.*, 1978) and poor water quality (Kanie *et al.*, 1979). The frequency of observation (weighing, measuring, and radiography out of the water) may also play a role (see Ward, 1985). These processes may stress the animal and cause breakage of the thin apertural lip. Similar breakage may also result from chance collisions against the sides of the aquarium. These breaks retard the rates of apertural growth and septal formation (Ward, 1985).

Figure 5. Specimen of *N. pompilius* showing abnormal shell growth in an aquarium. Scale bar: 5 cm.

Shells in aquaria bear numerous such scars, and the shape of the shell may even differ from that of a normal log spiral (Fig. 5) (Arnold, 1985) (see also Chapter 35).

3. Indirect Methods of Growth Measurement

3.1. Radionuclides

The radiometric method was used to determine the rate of chamber formation in nine specimens of *N. pompilius* and *N. belauensis* (Cochran *et al.*, 1981; Cochran and Landman, 1984). This technique is inferential and relies on the radioactive decay of nuclides incorporated into the shell from seawater during growth in the animal's natural environment. Radioactive decay proceeds at definite, well-known rates and therefore provides a natural chronometer for growth.

The nuclides that were used are ^{210}Pb (half-life of 22.3 years) and ^{210}Po (half-life of 138 days). ^{210}Pb occurs in seawater as a result of *in situ* radioactive decay of its grandparent ^{226}Ra and through input from the atmosphere into surface waters. It decays through a short-lived intermediate daughter, ^{210}Bi (half-life of 5 days), to its granddaughter, ^{210}Po:

$$^{210}\mathrm{Pb} \xrightarrow{22.3\ \mathrm{yr}} {}^{210}\mathrm{Bi} \xrightarrow{5\ \mathrm{days}} {}^{210}\mathrm{Po} \xrightarrow{138\ \mathrm{days}}$$

To circumvent the problem of variable concentrations of ^{210}Pb in seawater, the ingrowth of the daughter nuclide is followed with respect to the parent, i.e.,

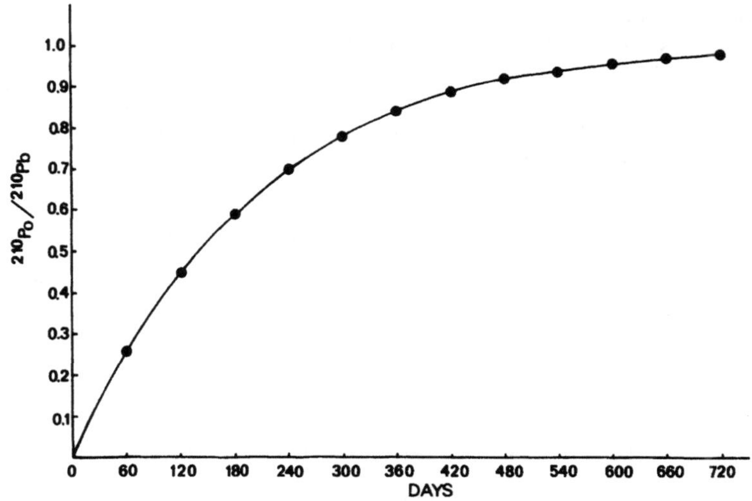

Figure 6. Plot of the $^{210}Po/^{210}Pb$ activity ratio. The trend shows a predictable increase with time and reaches an equilibrium value of approximately 1.0 in about 2 years.

the ratio of the radioactivity of ^{210}Po to that of ^{210}Pb. This ratio follows a predictable increase with time, based only on the half-lives of the two nuclides, and reaches an equilibrium value of approximately 1.0 in about 2 years (Fig. 6). On the basis of this relationship, an age can be assigned for any measured value of the activity ratio.

The validity of this age determination for dating septa is based on the fact, established in recently formed shell material, that ^{210}Pb is incorporated into the shell but its granddaughter ^{210}Po is excluded. The relevant equation is

$$A_{Po}/A_{Pb} = 1.0179[1 - e^{(-\lambda_{Po} + \lambda_{Pb})t}]$$

where A_{Po} is the activity of ^{210}Po, A_{Pb} is the activity of ^{210}Pb, λ_{Po} is the decay constant for ^{210}Po (5.01×10^{-3} day^{-1}), λ_{Pb} is the decay constant for ^{210}Pb (8.51×10^{-5} day^{-1}), and t is the septal age (days).
The calculated value of t is a mean over the length of septal formation, and differences in t between adjacent, completely formed septa represent times between successive points in septal formation (Cochran *et al.*, 1981).

The four to nine most recently formed septa were analyzed in the shells of nine animals representing immature to mature stages of growth (Table V). These septa were analyzed 26–120 days after capture and reveal the rate of growth for some time interval prior to capture. This rate may represent an average over several developmental stages. Two of the specimens were immature *N. pompilius* from the Philippines, with 26 and 24 septa and shell diameters of 94 and 84 mm, respectively. The other seven specimens were *N. belauensis* captured during the tag–release study of Saunders (1983) in Palau (although they were not released

Table V. Rate of Apertural Growth and Period of Septal Formation in *Nautilus pompilius* and *Nautilus belauensis* Based on Radiometric Measurements[a]

Specimen	Stage[b]	Septa	Diameter (mm)	Age of septum 1 (days)	Period of septal formation (days)	Rate of apertural growth (mm/day)
N. pompilius						
Np1	I	26	94	0	$\gtrsim 92^{+45}_{-37}$	$\lesssim 0.20^{+0.13}_{-0.07}$
Np2	I	24	84	4^{+15}_{-4}	$\gtrsim 67^{+41}_{-46}$	$\lesssim 0.22^{+0.49}_{-0.08}$
N. belauensis						
Nb1	I	28	132	28^{+4}_{-5}	$\approx 228^{+44}_{-36}$	$\approx 0.10 \pm 0.02$
Nb2	I	31	152	52 ± 8	$>168^{+31}_{-32}$	$<0.13^{+0.03}_{-0.02}$
Nb7	I	36	186	10 ± 2	$>120^{+13}_{-21}$	$<0.14^{+0.03}_{-0.01}$
Nb5	SM	33	210	110^{+12}_{-15}	$>290^{+15}_{-12}$	$<0.10 \pm 0.00$
Nb4	BM	33	199	240^{+47}_{-34}	100^{+115}_{-100}	$0.28^{+0.00}_{-0.15}$
Nb6	BM	36	191	120^{+17}_{-19}	$>280^{+19}_{-17}$	$<0.07^{+0.00}_{-0.01}$
Nb3	M	33	186	180^{+32}_{-26}	$\approx 220^{+26}_{-32}$	$\approx 0.08 \pm 0.01$

[a] Data from Cochran *et al.* (1981) and Cochran and Landman (1984). Errors were calculated using the counting uncertainties on the activity ratios to set upper and lower limits. With the exception of Nb4, the periods of septal formation are minima (or approximations); correspondingly, the rates of apertural growth are maxima (or approximations), because either the last septum was still incomplete (Np1, Np2, Nb1, Nb2, Nb7) or the second septum was already at radioactive equilibrium (Nb5, Nb6, Nb3). (Septa with $^{210}Po/^{210}Pb$ activity ratios ≥ 0.90 were assigned a minimum age of 400 days, or about 3 half-lives of ^{210}Po.) On the other hand, Nb3 was already mature, and its period of septal formation may represent a maximum, because growth may have stopped some time prior to capture.
[b] Stages: (I) immature; (SM) submature; (BM) barely mature; (M) mature.

and subsequently recaptured). These animals ranged from immature to mature, were 132–210 mm in diameter, and had 28–36 septa.

3.1.1. Period of Septal (Chamber) Formation

The septa were assigned ages based on the measured values of their activity ratios (Figs. 7–9 and Table V). The period of septal formation in *N. belauensis* ranged from >120 to ≈230 days in immature specimens [1, 2, and 7 (Figs. 7 and 8)] to 100 to >290 days in specimens near or at maturity [3, 4, 5, and 6 (Fig. 9)]. The error was approximately ±20%, except in specimen 4, in which it was ± 100%. The shortest periods occurred in the two immature *N. pompilius*. Calculations based on the age difference between the two most recent septa in these specimens yielded periods of ≳92 and ≳67 days, respectively, with substantial percentage errors. However, equilibrum was not reached until at least septum 8 in specimen 1, and not at all in specimen 2. In view of this, a best-fit time curve may be drawn through all the values to obtain an average time interval over the

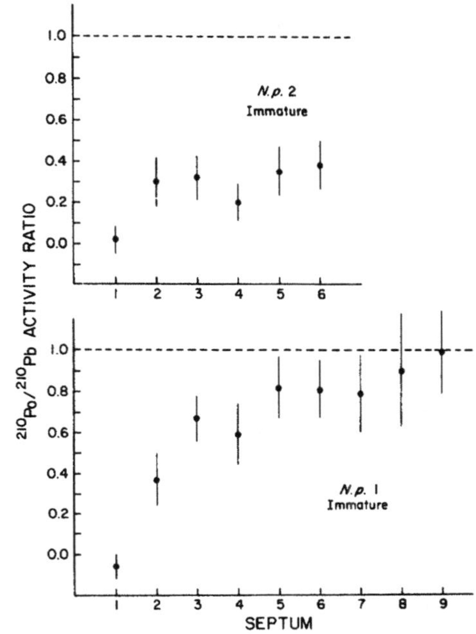

Figure 7. $^{210}Po/^{210}Pb$ activity ratios for the six to nine most recent septa in two specimens of *N. pompilius* from the Philippines (1, 2). The activity ratio is 0 in the most recently formed septum. In specimen 1, it increases regularly to equilibrium values by septum 9. In specimen 2, the second septum reaches a value of 0.4, and the next four septa have similar values. (The vertical bars represent the 1σ counting uncertainty associated with the measured $^{210}Po/^{210}Pb$ activity ratio. Because a correction is required to account for the production of ^{210}Po from ^{210}Pb decay between collection and analysis, specimens that were analyzed rapidly after collection have somewhat lower uncertainties than specimens for which a long time elapsed.) Data from Cochran *et al.* (1981).

last 6–9 septa (Cochran *et al.*, 1981). This method yields an average period of septal formation of 75 days (range 30–90 days) in specimen 1 and 23 days (range 10–60 days) in specimen 2. These values fall within the range of values based on the measurements between the two most recent septa, but are lower because they average over earlier ontogeny. In specimen 2, the near similarity in activity ratios for several consecutive septa may indicate times of more rapid growth (≈20 days/septum) followed by slower growth (Fig. 7); such a deceleration may also have occurred in specimen 1.

3.1.2. Rate of Apertural Growth

The rate of apertural growth is an approximation, inferred by dividing the distance along the venter between the last two septa by the rate of septal formation based on the age difference between the last two septa (Table V). In the immature specimens of *N. belauensis*, the rates ranged from ≈0.10 to <0.14 mm/day. In specimens near or at maturity, rates were slightly lower, ranging from <0.07 to <0.10 mm/day, with the exception of specimen 4, which was subject to large error. The highest estimates (≤0.20 to ≤0.22 mm/day) occurred in the smaller immature specimens of *N. pompilius*.

3.1.3. Critique

The radiometric approach is potentially the most valuable method, because growth is not interrupted for observation. It is based solely on the radioactive

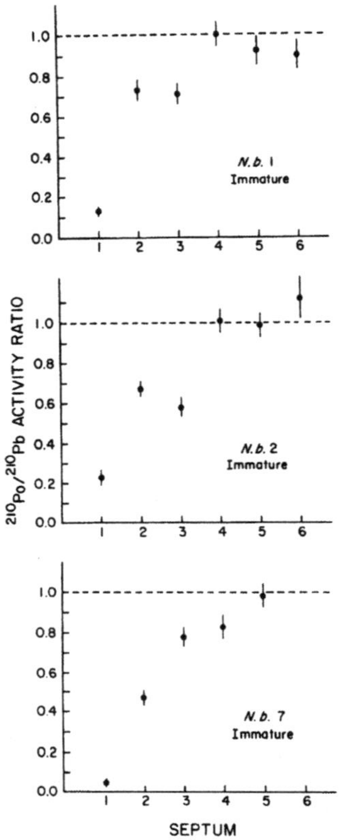

Figure 8. $^{210}Po/^{210}Pb$ activity ratios for the five or six most recent septa in three immature specimens of *N. belauensis* (1, 2, 7) from Palau, West Caroline Islands. Radioactive disequilibrium persists through the three or four most recently deposited septa. (For an explanation of the error bars, refer to the Fig. 7 caption.) Data from Cochran and Landman (1984).

decay of incorporated nuclides. However, it requires the assumption that no process other than radioactive decay affects the activity ratio after formation of a septum (closed system) and entails substantial error bars around the measurements if the activity ratios are low or if a long time has elapsed between collection and analysis.

3.2. Partial Pressures

On the basis of partial pressures and rate of diffusion, Denton and Gilpin-Brown (1966) calculated a constant fortnightly (13-day) period of chamber formation in *N. macromphalus*. The volume of chambers increases exponentially through ontogeny; therefore, if the period of chamber formation were constant, the rate of apertural growth would be exponential (Denton and Gilpin-Brown, 1966). However, to calculate the period of chamber formation, Denton and Gilpin-Brown (1966) assumed that liquid in newly formed chambers is pumped out immediately. In fact, the removal of cameral liquid is gradual (to compensate for gradual growth at the aperture); therefore, their estimate of 13 days is probably a minimum (Collins *et al.*, 1980).

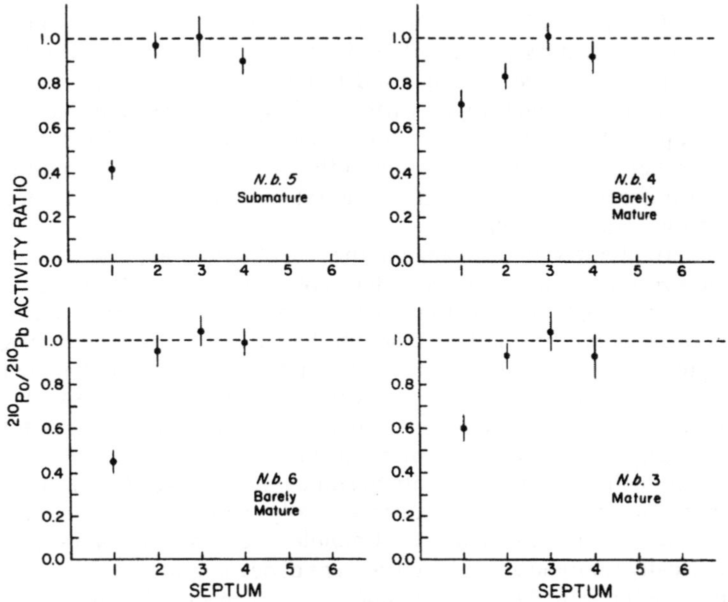

Figure 9. $^{210}Po/^{210}Pb$ activity ratios for the four most recent septa in four specimens of *N. belauensis* near or at maturity (3, 4, 5, 6). In specimens 3, 5, and 6, the activity ratio approaches equilibrium by the second septum. (For an explanation of the error bars, refer to the Fig. 7 caption.) Data from Cochran and Landman (1984).

3.3. Epizoans

The growth of a barnacle, *Chirona tenuis* (1.8 cm in height), onto the midventer of a juvenile *N. pompilius* (10.7 cm in diameter) during its life provides an estimate of the growth rate of *Nautilus* in nature (Landman, 1983a). The barnacle settled almost exactly on the midline of the shell, and the *Nautilus* attempted to overgrow it less than one whorl later. Therefore, the age of the barnacle (calculated at 340 days) provides the basis for a minimum estimate for the rate of growth of the last *Nautilus* whorl. Judging from the spacing of chambers, about 16 chambers formed during this time interval, yielding a minimum average of about 21 days/chamber.

4. Discussion

Synthesizing a general picture of the rate of growth in *Nautilus* is difficult, because the available data are derived from at least two different methods and three different species and reflect ontogenetic and intraspecific variation. The fastest rates and most variation occurred in *N. macromphalus* in aquaria (see Table II). Rates of apertural growth ranged from 0 to 0.60 mm/day, and periods of chamber formation averaged less than 100 days. Growth rates were also high in aquarium studies of *N. pompilius:* 0.21–0.45 mm/day with periods of chamber formation less than 130 days (see Fig. 2 and Table III). The rate of growth in *N. pompilius* in nature was similar (Fig. 2 and Table V), but it was consistently lower in *N. belauensis;* rates of apertural growth in this species averaged less than 0.14 mm/day, and periods of chamber formation exceeded 100 days (Fig. 2 and Tables I and V). To what extent is this variation in the rate of growth due to differences in species or differences in aquarium vs. natural conditions? To what extent does it reflect ontogenetic variation?

The pattern of ontogenetic change in the rate of apertural growth was similar in all species regardless of the method. Rates of apertural growth fluctuated randomly or declined slightly, with a marked deceleration at maturity. In immature specimens of *N. pompilius* and *N. macromphalus* in aquaria, apertural growth fluctuated from approximately 0.20 to 0.60 mm/day throughout ontogeny (Ward, 1985) and was continuous during septal formation (Ward *et al.*, 1981; Ward and Chamberlain, 1983). In *N. belauensis* in nature, no single animal was followed through ontogeny, but several specimens at different developmental stages suggested a slight decline from immaturity to submaturity. Rates of apertural growth in animals that were recaptured or radiometrically sampled at an immature stage of development ranged from 0.10 to 0.14 mm/day. Animals that were recaptured or radiometrically sampled at a submature stage of development displayed slightly lower rates, ranging from 0.07 to 0.12 mm/day. In all species, however, regardless of the method, the rate of growth decelerated at maturity; forward growth of the aperture ceased, and the aperture thickened (Hamada *et al.*, 1978; Kanie *et al.*, 1979; Cochran and Landman, 1984; Saunders, 1983; Ward, 1985) (see also Chapter 29).

On the other hand, the period of chamber formation increased uniformly over the course of ontogeny. This pattern was predicted on the basis of the ontogenetic decrease in the ratio of the surface area of the siphuncle to chamber volume (Chamberlain, 1978). The period of chamber formation in specimens of *N. pompilius* in aquaria increased from a minimum of 14 to a maximum of 132 days (Ward and Chamberlain, 1983; Ward, 1985). Similarly, the period of chamber formation in specimens of *N. macromphalus* increased from 49 to 98 days (Ward, 1985). This pattern of increase in aquaria appears exponential, although the period of formation for any particular chamber varies widely among individuals (see Fig. 4). In immature specimens of *N. pompilius* in nature, the period of chamber formation averaged 60–90 days, but shorter periods may have occurred in earlier ontogeny (see Fig. 7). In *N. belauensis* in nature, the period of chamber formation averaged 120–230 days in immature animals and approximately 1 year in submature–mature animals.

These patterns of ontogenetic change in the rate of growth suggest that individual specimens must be compared at the same developmental stage. Even then, however, differences in growth rate persist and may relate to the differences in aquaria vs. natural habitats. The high rates of growth obtained for *N. pompilius* and *N. macromphalus* in aquaria may be due to a steady food supply and surface water conditions (1 atm pressure). Variations in aquarium conditions themselves may, in turn, account for the discrepancies in growth rates among aquarium studies [compare Hamada *et al.* (1978) and Ward (1985)]. On the other hand, the rate of growth in nature, as shown by *N. belauensis*, is slower and may be uneven due to variation in food supply, predation, and other parameters of the natural habitat.

The differences in the rates of growth may also be species-specific. Immature specimens of *N. pompilius* and *N. macromphalus* at comparable sizes in the same aquarium exhibited similarly rapid rates, suggesting that under the same conditions, these two species grow alike (Ward, 1985). Immature specimens of *N. pompilius* at the same size showed similar rates in nature, although maturity occurs at larger sizes (Cochran *et al.*, 1981; Ward, 1985). However, the rates of growth in immature and submature specimens of *N. belauensis* were much slower. These specimens were larger than *N. pompilius* and *N. macromphalus* at a comparable stage of development (although they had a similar or slightly greater number of septa) and indicate that developmental stage and size may not correlate between species. *Nautilus belauensis* reaches maturity at a larger size than *N. pompilius*, which it otherwise resembles. In fact, the evolutionary change from *N. pompilius* to *N. belauensis* may have been facilitated through a prolongation of ontogeny with a subsequent delay in the onset of sexual maturation. *Nautilus belauensis* may attain larger sizes at slower rates, although small immature specimens of this species, comparable in size to *N. pompilius* and *N. macromphalus*, may grow at a similarly rapid rate. However, until more data are obtained about the natural rates of growth for all species throughout ontogeny, the question of species-specific differences will be unresolved.

Nevertheless, on the basis of available data, we may construct a general growth curve for *Nautilus* and estimate the time to maturity. A minimum estimate of 2.5 years for *N. macromphalus* was based on cyclic weight changes in aquarium-reared animals, which were interpreted as marking monthly periods of chamber formation (A. W. Martin *et al.*, 1978). Longer estimates of 5–6 years for this species were based on: (1) instantaneous rates of weight increase of 0.35%/day, (2) apertural growth rates of 0.25 mm/day, and (3) the number of growth lines counted on composite sections of several shells [estimated at 2 days/growth line (A. W. Martin *et al.*, 1978)]. In contrast, a time interval of approximately 15 years was estimated for *N. belauensis* on the basis of the highest rate of apertural growth (0.12 mm/day) measured in recaptured animals (Saunders, 1983). Another estimate of 10 years for this species was based on a simple combination of two different periods of septal formation throughout ontogeny for an animal with 36 septa (Cochran and Landman, 1984): 120 days/septum for the first 25 postembryonic septa and 220 days/septum as maturity approached through septum 36.

This last technique can be refined further by assuming a smooth exponential increase in the period of chamber formation. Such a pattern of exponential increase appeared in the growth rate data for *N. pompilius* in aquaria (see Fig. 4) (Ward, 1985). Using this pattern as a model, we can construct a growth curve for *N. belauensis* based on its rate of growth in nature. For example, the period of formation of the last chamber (e.g., 36) in *N. belauensis* may be assigned 350 days, the average time for the formation of the last chamber in recaptured specimens of this species (Saunders, 1983). Chamber 28 may be assigned a period of 230 days, based on an actual value from radiometric measurements (Cochran and Landman, 1984). No natural growth rate data are available for the first few postembryonic chambers, but a period of 24 days will be assigned to the first postembryonic chamber (8), which was the value reported for the smallest chamber

(18) in *N. pompilius* raised in aquaria (Ward, 1985). The resultant exponential equation is

$$y = 11.6e^{0.10x}$$

for chambers 8–36, where *y* is the duration of chamber formation and x is the chamber number. The integral of this equation is

$$A = 0.32[e^{0.10x} - 2.3]$$

where *A* is the age in years and x is the chamber number. The posthatching time to maturity for this hypothetical specimen of *N. belauensis* is computed at 10.9 years. (A similar equation can be derived for *N. pompilius*, basing it on the two radiometric measurements of this species and assigning the same period of 24 days for the first postembryonic chamber. The resultant time to maturity is 5.6 years.) On the basis of one specimen of *N. belauensis* recaptured 4 years after being released as mature, this species may live for several years after maturity is attained (Saunders, 1983).

These growth rates and great longevity contrast markedly with those of most other modern cephalopods. Octopus, squid, and cuttlefish reach maturity rapidly (within 1–2 years), reproduce, and die (Packard, 1972; Wells, 1983). The slow rates of growth in *Nautilus*, on the other hand, may represent the "costs" of growing a large external shell (Ward, 1985). Slow growth rates may also represent one aspect of a life history strategy, including small clutch size, large yolky eggs, and generalized scavenging behavior, that is typical of life in a deep-sea habitat (Saunders, 1984b).

Comparisons with extinct shelled cephalopods, especially ammonoids, are tenuous (Hirano, 1981; Landman, 1983b). Nevertheless, on the basis of the ratio of siphuncular area to chamber volume, the period of chamber formation in the extinct forms also probably increased through ontogeny, as it does in *Nautilus* (Chamberlain, 1978; Ward, 1985). Therefore, within one species, we may compare the relative ages of individual specimens at very different diameters and numbers of septa. However, determination of absolute growth rate and time to maturity is more difficult (Westermann, 1971; Kennedy and Cobban, 1976). Several attempts have been based on analogy with *Nautilus* and interpretations of ontogenetic patterns of growth, including (1) growth lines and lirae (Kulicki, 1974; P. G. K. Kahn and Pompea, 1978; Doguzhaeva, 1982), (2) septal spacing (Rieber, 1963; Oechsle, 1958), (3) whorl section and coiling (Currie, 1944; Raup, 1967), (4) shell constrictions (Hirano, 1981), and (5) oxygen isotopes in the septa (Jordan and Stahl, 1970). Growth rates have also been based on the known growth rates of encrusting organisms that grew while the ammonoid was alive (Schindewolf, 1934; Seilacher, 1960, 1982; Merkt, 1966; Meischner, 1968; Westermann, 1971; Hirano, 1981) and the identification of size or year classes within a preserved assemblage (Trueman, 1941). These methods yield diverse estimates, but generally suggest a slow rate of growth more similar to that of *Nautilus* than to that of most modern dibranchiates.

ACKNOWLEDGMENTS. We thank P. D. Ward (University of Washington, Seattle) for sending us a preprint of his 1985 paper and Amy Adelson (American Museum of Natural History) for preparing the illustrations.

Chapter 29

Adolescent Growth and Maturity in *Nautilus*

DESMOND COLLINS and PETER D. WARD

1. Introduction

The shell of *Nautilus* is commonly thought to be fully mature when it has certain characteristics, namely, a white venter, a thickened apertural edge, a black band along the inside of the aperture, a pronounced ocular sinus on each side, a marked thickening of the final septum, and approximation of the final septa (Stenzel, 1964). This chapter will examine and evaluate these and other putative indicators of maturity and will endeavor to show how mature modifications develop in the ontogeny of the animal.

Late ontogenetic morphological criteria were examined in more than 200 mature or almost mature *Nautilus* shells from the Philippines and Fiji (*N. pompilius* Linnaeus, 1758), from New Caledonia (*N. macromphalus* Sowerby, 1849), and from Palau [*N. belauensis* Saunders, 1981 as used by Saunders (Chapter 3)]. The shell and soft parts of 26 mature and 14 immature *N. macromphalus* from New Caledonia were studied in order to relate sexual maturation, as shown by the size of the reproductive organs, to the characteristics of the shell. To establish a time frame for the growth of the shell and reproductive organs, we studied six *Nautilus* from Palau, the growth rates of which had been determined by the mark–recapture work of Saunders (1983, 1984a). The relationships thus determined enable us to present, in order and time of development, the events involved in the growth of *Nautilus* from the late juvenile stage, through adolescence, and on to full maturity. Detailed tables of data on which this analysis was based are provided elsewhere (Collins and Ward, 1987).

DESMOND COLLINS • Department of Invertebrate Palaeontology, Royal Ontario Museum, Toronto, Ontario M5S 2C6, Canada. PETER D. WARD • Department of Geological Sciences, University of Washington, Seattle, Washington 98195.

2. Fully Mature Shell

A number of characteristics of the shell were examined to determine whether they provide reliable indicators of completion of shell growth.

1. *White ventral area:* In fully mature *N. pompilius*, the exterior of the body chamber is white, except near the umbilicus. This coloration results because all animals cease secretion of the red-brown pigment at about the same stage of immaturity, i.e., about ⅓ whorl adapical of the fully mature aperture.

2. *Thickened apertural edge:* The apertural edge of an immature shell is thin and fragile (Fig. 1A). The featheredge is totally within the outer, spherulitic–prismatic shell layer (Mutvei, 1964). The apertural margin thickens progressively in an adapical direction as more and more of the nacreous shell layer is added. Growth increments are secreted in thin sheets on the inside of the apertural margin at an acute angle to the external surface of the shell (Fig. 1). These increments form a band up to 25 mm wide around the inside of the aperture, called the *shell growth band* (Fig. 2) (see Collins and Ward, 1987). The apertural edge in fully mature shells is thick (>1 mm) and blunt (Fig. 1B). About one quarter of its thickness is spherulitic–prismatic layer, and the rest is made up mostly or wholly of layers of nacre. A thickened apertural edge is a reliable indicator of full maturity.

3. *Black band at the aperture:* Immature shells have no black band at the aperture. Such a band does occur in most shells with a thickened apertural margin (90% of fully mature shells examined) (Fig. 3). Because it does not occur in all

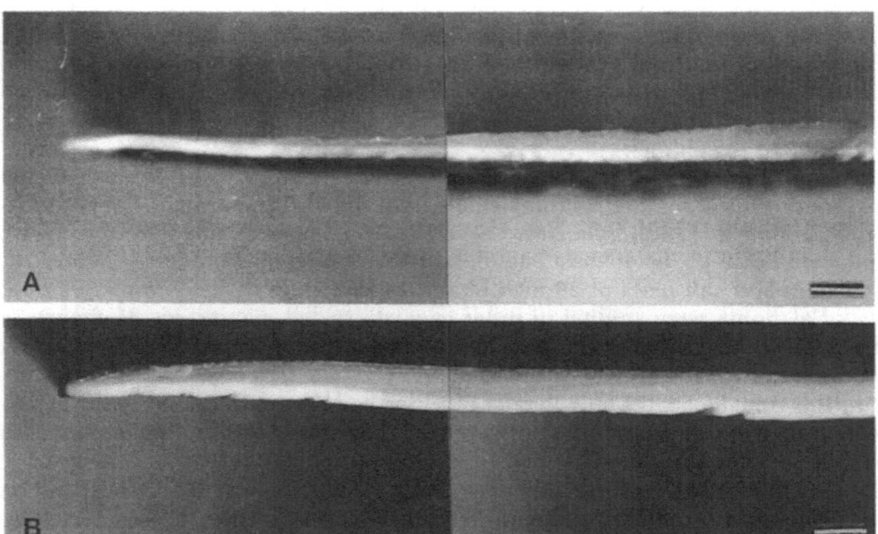

Figure 1. Apertural edges of *N. pompilius* from Fiji: (A) immature edge; (B) mature edge. The outer, white band is the spherulitic–prismatic layer; the inner, gray band is the nacreous layer. Scale bars: 1 mm.

Figure 2. Shell growth band inside the aperture of a juvenile *N. pompilius* from Fiji. The limits of the band are indicated by lines.

shells, its absence alone should not be used as an indicator that the animal has not reached full maturity.

4. *Pronounced ocular sinuses:* The ocular sinuses deepen in the final stage of shell secretion (Fig. 4). This deepening occurs because the space between growth lines increases more rapidly dorsally and ventrally from each sinus than it does at the sinus itself. Growth lines more than a few millimeters adapical of

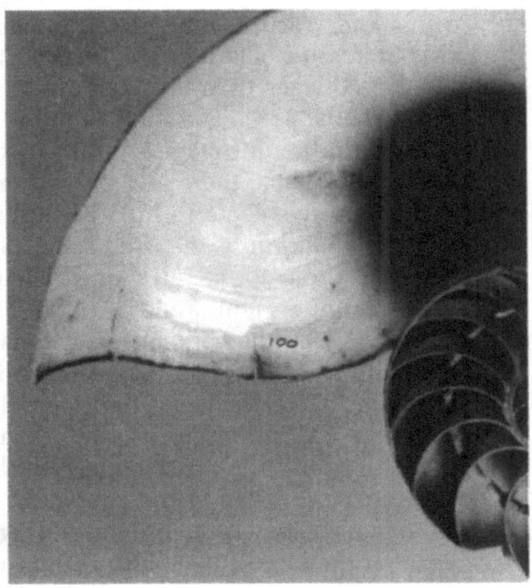

Figure 3. Inner surface of a mature aperture of an *N. pompilius* from the Philippines, showing the black band.

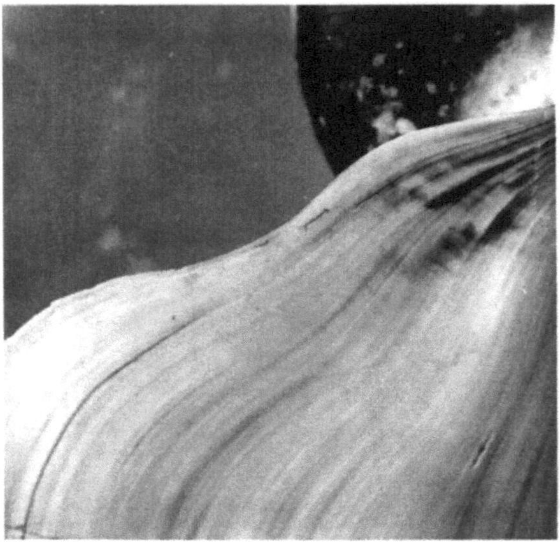

Figure 4. Mature ocular sinus of an *N. pompilius* from the Philippines showing deepening of the growth lines in the final 5 mm of shell growth.

the fully mature peristome have much shallower ocular sinuses than does the peristome itself.

5. *Thickening of the final septum:* To test the observation that an indicator of full maturity in *Nautilus* is a final septum that is markedly thicker than the preceding septa, the thicknesses of the last five septa of 68 mature *Nautilus* from the Philippines and Fiji (*N. pompilius*) and from New Caledonia (*N. macromphalus*) were measured. They show a surprisingly consistent pattern (for data, see Collins and Ward, 1987). Up to the penultimate septum, there is an average thickness increase of about 12%. The penultimate septum in the New Caledonian shells is 13% thicker than those before it; in the Fijian shells, 18%; and in the Philippine shells, 24%. The final septum in the New Caledonian shells is 19% thicker than the penultimate septum; in the Fijian and Philippine shells, the increase is 30%. Thus, there is a marked increase in septum thickness in one or both of the final *two* septa in fully mature shells.

6. *Approximation of the final septa:* Septal crowding, or approximation, is a widely used indicator of full maturity (e.g., Willey, 1902; Stenzel, 1964; Ward *et al.*, 1977; Saunders and Spinosa, 1978). The cameral angle (the angle subtended at the umbilicus by the contacts of consecutive septa with the outer shell wall) can be used to measure septal approximation, and it is independent of shell size. Among the last five chambers, the cameral angle up to the penultimate septum averages about 24° and varies no more than 2° in 80% of the shells examined. In contrast, the cameral angle of the final chamber is highly variable and is commonly much smaller than those of the preceding four chambers. The cameral angle of the final chamber may be as little as 9° (Fig. 5). Weak septal approximation may also occur in the penultimate chamber, particularly in shells in which the final

Figure 5. Dorsoventral view of an *N. pompilius* from the Philippines showing the final two septa thickened and strongly approximated.

two septa are strongly approximated. Thus, septal approximation may occur over the final two chambers as full maturity is approached.

Of 76 mature shells from the Philippines and Fiji (*N. pompilius*) and New Caledonia (*N. macromphalus*) measured, 11 show strong approximation, 52 exhibit moderate to weak approximation, and 13 show none at all. It is evident, therefore, that septal approximation occurs in most, but not all, fully mature shells, an observation first made by Willey (1902).

Approximate septa are also present in many large immature shells. Their presence indicates that final septa are secreted before completion of the body chamber. Thus, septal approximation, when present, is a reliable indicator of approaching, rather than full, maturity.

7. *Diameter of the shell:* The wide variation in the diameter of fully mature shells from population to population (Chapter 3) indicates that size alone is not a reliable indicator of maturity. However, the small variation in mature shell diameter of individual populations may allow this measurement to be used as an index to maturity in single populations (Collins and Ward, 1987).

8. *Rounded broadening of the fully mature body chamber:* Just after the time that secretion of red-brown pigment ceases on the venter (Fig. 6), the body chamber begins to expand laterally and to become more rounded at the venter (Fig. 7). This morphological change results in a more capacious body chamber, presumably to house the large reproductive organs of the mature animal.

9. *Apertural contraction* ("contracted" defined by Teichert, 1964, p. K55): Shortly before the shell reaches its fully mature size, there is a slight contraction of apertural width. This contraction, combined with the expansion discussed in characteristic (8), results in a fully mature body chamber that is egg-shaped (Fig. 7B).

10. *Change in shell coiling:* In almost all fully mature shells examined, the

Figure 6. White ventral area. Comparative views of a juvenile *N. pompilius* from the Philippines showing the final color bands and oriented to demonstrate growth of the shell still to come (A) and a mature *N. pompilius* from the Philippines with a large white ventral area (B).

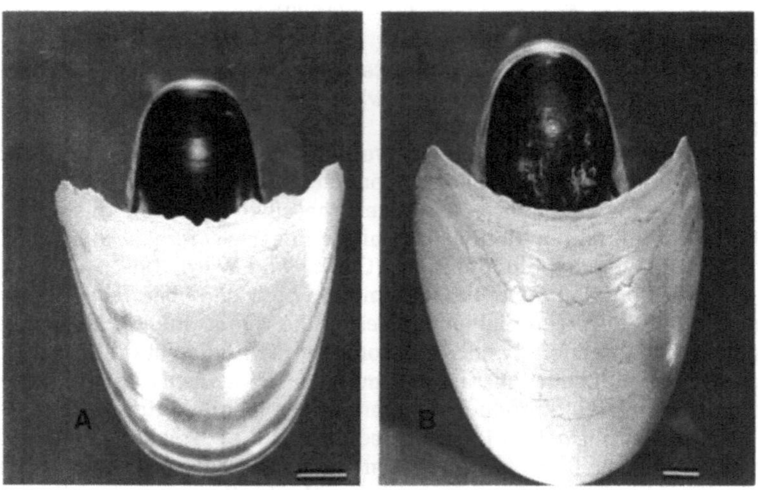

Figure 7. Ventral view of shells: (A) juvenile *N. pompilius* from the Philippines showing round-bottomed V-shaped (25° angle) cross section; (B) adult *N. pompilius* from the Philippines with an egg-shaped cross section. Scale bars: 1 cm.

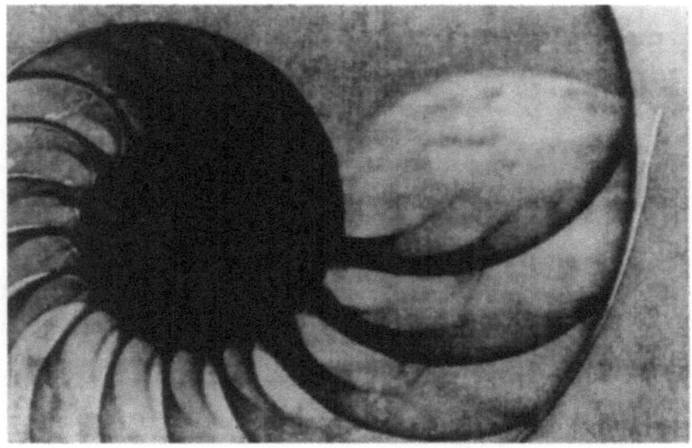

Figure 8. Divergence of the mature aperture from the phragmocone spiral, shown by superimposing an enlarged X-radiograph of a juvenile *N. pompilius* shell on a photocopy of a bisected mature *N. pompilius* shell.

curvature of the ventral surface of the body chamber coincides with that of juvenile shells until the last 20–30° of the body chamber, where it flattens out (Fig. 8). The change in coiling is a characteristic of nearly all fully mature *Nautilus* shells.

11. *Increase in the length of the mature body chamber:* There is significant growth of the body chamber after the final septum is secreted (Collins *et al.*, 1980; Tanabe *et al.*, 1985). This growth can be measured by means of the change in the body chamber angle [the angle subtended by (a) a line from the axis of coiling to the point where the last septum touches the venter and (b) the line from the axis of coiling to the aperture at the venter].

The body chamber angle of 9 immature shells from Fiji (*N. pompilius*) varies from 93 to 113°; the body chamber angle of 25 fully mature shells from Fiji averages 127° (Fig. 9). Immature shells of the same species from the Philippines have a similar immature body chamber angle, but in 25 fully mature shells, the angle averages 129°. The body chamber angle of 6 immature shells from New Caledonia (*N. macromphalus*) varies from 86 to 106°; in 19 fully mature shells, the angle averages 122° (for full details, see Collins and Ward, 1987).

It is clear, then, that the body chamber increases in length about 15° more than the maximum length of the immature body chamber. This increase is the most significant mature modification in the *Nautilus* shell.

12. *Reduction in cameral liquid:* The length of the immature body chamber fluctuates cyclically with the addition of new septa. The amount of cameral liquid in the phragmocone also varies in the same cycle as new chambers are added and the liquid in them is removed; however, even at its lowest volume, there is a significant amount of liquid present in the most recently formed chambers. For example, in three juvenile *N. pompilius* shells from Fiji, with body chamber angles averaging 110°, there was an average of 7.2 ml of liquid in the most recently formed chambers (Collins *et al.*, 1980, Table 2). After the final septum is secreted, cameral liquid is removed to compensate for the increase in the length of the fully mature body chamber; hence, at full maturity, there is virtually no cameral liquid in the phragmocone (Collins *et al.*, 1980).

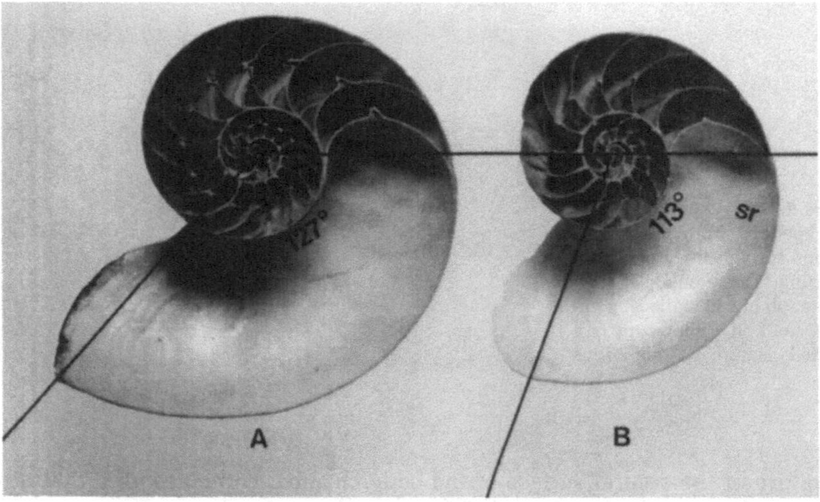

Figure 9. Comparison in *N. pompilius* shells from Fiji of the mature body chamber length (A) and the maximum juvenile body chamber length [B (sr) mural ridge].

3. Sequence of Mature Modifications in the *Nautilus* Shell

Mature modifications develop in the following order (modified from Collins *et al.*, 1978):

1. Cessation of secretion of color bands on the venter.
2. Rounded broadening of the aperture.
3. Secretion of the final two septa, characterized by septal approximation and increased septal thickness, especially of the final septum.
4. Change in coiling.
5. Contraction of the aperture.
6. Deepening of the ocular sinuses.
7. Full growth of the fully mature body chamber.
8. Thickening of the apertural edge.
9. Secretion of the black band inside the aperture.
10. Final extraction of cameral liquid.

Note that there is some overlap among these modifications; e.g., the deepening of the ocular sinuses and contraction of the aperture are coincident with the last part of the growth of the body chamber to its fully mature size. Note also that mature modifications such as septal approximation and secretion of the black band at the aperture do not always occur.

4. Mature Modifications in the Shell versus Sexual Maturity

Willey (1902, p. 746) stated that propagation by *Nautilus* takes place only after the final septum has been formed, implying that *Nautilus* attains sexual maturity after the formation of the final septum. This statement was tested in 40 *N. macromphalus* from New Caledonia by relating their sexual maturity, as expressed in the size of their reproductive organs, to mature modifications of their shells. The 24 males and 16 females were sorted into mature and immature groups, and their soft-tissue weights, gonad weights (and spadix weights, for males), and body chamber angles were measured. For the purpose of comparison, gonad weight as a percentage of the total soft-tissue weight was used, rather than the actual gonad weight.

We consider that adolescence in *Nautilus* begins when the gonads and accessory reproductive organs start to increase significantly in size, as compared to total body size. In juveniles, even large ones, the average gonad percentage is less than 0.06%; in several adults, it averaged 6%. Thus, there is a 100-fold increase in the relative size of the gonads during adolescence.

Other reproductive organs are likewise tiny in juveniles. The spadix was 0.55% of the body weight in the smallest male examined and averaged 7.7% in mature males—a 14-fold increase in size. The spadix grows so large that it pushes the buccal mass off center (see Saunders and Spinosa, 1978, Fig. 4A and B).

In the sample of 14 immature *N. macromphalus* examined by Collins and Ward (1987), five distinct clusters were recognized, based on size, gonad size and degree of sexual maturation, presence or absence of final septa, and body chamber angle. They considered these clusters to represent five stages in the growth of *Nautilus:*

1. Juvenile—small size, tiny gonads.
2. Late juvenile—moderate size, tiny gonads.
3. Early adolescence—gonad percentage is three times that of juveniles; the penultimate septum is secreted.
4. Moderate adolescence—gonad percentage is two to three times larger again; secretion of the final septum is begun; the body chamber angle averages 99°.
5. Late adolescence—gonad percentage is five times larger again; secretion of the final septum has been completed; the body chamber angle averages 119°.

In contrast, in the 26 fully mature specimens examined, the gonad percentage was twice that of late adolescents and the body chamber angle was 123°. A similar pattern of growth of the gonads and the spadix was recorded by Tsukahara (1985) for *N. pompilius* from Fiji.

Sexual maturity, at least in males, seems to be attained in very late adolescence. This timing is shown by the presence of a full-grown spermatophore in only the latest adolescents from the New Caledonia and Fiji samples. In both cases, the body chamber angle is within a degree or two of the average at full maturity, indicating that the shell is very close to full growth. Willey's statement that propagation takes place only after the secretion of the final septum is thus borne out by the pattern of maturation of the reproductive organs.

5. Duration of Adolescence

The best data on natural growth of *Nautilus* come from six immature animals caught, measured, released, and recaptured off Palau in 1981 and 1982 (Saunders,

1983, 1984a). From photocopies of dorsoventral sections of these shells provided by Saunders, we measured the body chamber angles at the time of the initial capture of each animal and at the time of recapture (Collins and Ward, 1987).

In each of the shells, the final septum had been formed. Thus, the six shells provide 12 significant body chamber angles, with associated calendar dates. We found three remarkably distinct clusters in the 12 body chamber angles measured. There were 3 between 103 and 105° and 6 between 117 and 120°; the 2 mature shells had body chamber angles of 132°. One body chamber angle (126°) fitted none of the clusters.

By combining the data on shell changes and gonad growth in *N. macromphalus* from New Caledonia with the data on natural growth of the animals off Palau, it is possible to construct a timetable for the growth of *Nautilus* from juvenility to maturity. For example, one animal (designated F in Saunders, 1983, Table 3) was first captured on June 20, 1981, and, at that time, had a body chamber angle of 104°; this angle would assign it to middle adolescence. When animal F was recaptured on June 8, 1982, it had a body chamber angle of 117°, characteristic of late adolescence. Animal E, when captured on July 5, 1981, had a body chamber angle of 117°, which puts it in late adolescence; when it was recaptured on June 12, 1982, it had a body chamber angle of 132° and was fully mature. Thus, it would appear that it takes *Nautilus* a year to go from middle adolescence to late adolescence and another year to go from late adolescence to full maturity. In short, the three Palauan shell clusters apparently represent year classes.

If one extrapolates from the Palau specimens to those from New Caledonia and numbers from full maturity: fully mature animals are year class 1; late adolescents comprise year class 2; middle adolescents, year class 3; early adolescents, year class 4; late juveniles, year class 5; and the smaller juveniles, year class 6 or 7.

Note that this timetable is based on the known growth of the shell of one species of *Nautilus*, whereas the soft-part data are derived from another species. Are we justified in relating the clustering in one species to that in another? We believe so, because the pattern of changes in the shell is so similar in all four populations examined. Moreover, the pattern of growth of the gonads and other soft tissues is the same in *N. pompilius* from Fiji (Tsukahara, 1985) as it is in *N. macromphalus* from New Caledonia. The biggest difference is among the average body chamber angles of fully mature individuals of the four populations: New Caledonia, 123°; Fiji, 127°; Philippines, 129°; and Palau, 132°. Such differences, however, are well within the range of variation for each group of shells.

A. W. Martin *et al.* (1978) recognized the possibility of two year classes in *Nautilus* from New Caledonia. Their classification was based on live weight only. Our work indicates that the sizes of the gonads, spadix, and body chamber angle are more sensitive indicators for year-class assignment.

We draw two significant conclusions: First, ontogenetic changes in *Nautilus* from the beginning to the end of the sexual maturation process take at least 3 years. Second, the clustering of body chamber angles and gonad weight percentages in the Palau and New Caledonia samples indicates that these changes follow yearly cycles, i.e., that growth is seasonal.

The notion that sexual growth and activity are seasonal is not a new one. Haven (1977a, p. 180) recorded that mature female *Nautilus* in the Philippines produced the largest oocytes in July (3.6 g) and the smallest in January (1.4 g). In their study of reproductive behavior in *N. macromphalus* from New Caledonia, A. W. Martin *et al.* (1978, p. 219) reported that from June to September 1972, "first caught animals began copulation even while being transported to the aquarium in the surging water of plastic containers." Ward (1983a) showed that egg-laying by this species in aquaria was limited to September through January.

From the size of their gonads and spadix relative to those of mature animals, and from their rate of growth, it is apparent that the late adolescents from New Caledonia in this study would probably have reached sexual maturity in about 6 months, in August or September. Palau animal E had probably achieved full growth about a month before recapture, i.e., in May 1982. At the same rate of growth, Palau animals A, C, D, and F would have reached the same state at about the same time of year. Apparently, northern *Nautilus* populations reach sexual maturity some 2–3 months before southern ones.

6. The Two Programs of Growth to Maturity

Two characteristics of the shell development of *Nautilus* demonstrate that growth to maturity is programmed:

1. Cessation of secretion of brown pigment on the venter
2. The rounded broadening of the body chamber (to accommodate the mature reproductive organs)

Both modifications begin when the animal is only about half-grown and still has tiny gonads.

For an animal to reach full maturity in the May–September period of a given year, the beginning of adolescence must take place 3 years earlier, triggered, perhaps, by an event related to the time of year. The first visible sign of approaching reproductive capability is a slow increase in the size of the gonads. If the animal is a male, the spadix begins to get bigger, relative to the antispadix and other tentacles. During this first year of adolescence, the penultimate septum is secreted, the gonads increase about five times in size, and, in males, the spadix doubles in size.

During the second year of adolescence, secretion of the final septum begins. For the animal to attain sexual maturity at a particular time of year, the secretion of the final septum must occur at a predetermined time beforehand. The formation of the final septum will thus begin regardless of where the shell is in its regular cycle of septal formation. Since formation of a chamber late in the shell's growth takes most of a year (Cochran and Landman, 1984), it is probable that secretion of the final septum will begin within a chamber formation cycle, rather than at its end, producing a narrower chamber. In other words, in most shells, the final septum will be approximated. The apparent randomness of occurrence and degree of septal approximation can thus be explained as the effect of the imposition of a more powerful growth program, that of adolescence, interrupting the regular, cyclical septal formation of the juvenile growth program.

By the end of the second year of adolescence, the gonads have quintupled in size over that at the beginning of the year, and in males, the spadix is two to three times larger. The reproductive organs now comprise a significant proportion of the soft-tissue weight of the animal: the gonads 1–2% and the spadix 3%.

In the final year of growth, the gonads quadruple in size, the spadix doubles in size, and another 15° is added to the body chamber angle. Most of the mature modifications to the shell occur in this final year. First, continuation of rounded shell growth at the aperture causes it to contract slightly. At the same time, the ventral surface of the body chamber becomes flatter, causing it to unwind slightly from the whorl. During secretion of the final few millimeters of shell at the aperture, the ocular sinuses become deeper and the reproductive organs reach sexual maturity, shown in males by the growth of a full-sized spermatophore. The shell is finally completed as shell secretion brings the apertural edge to its full thickness, usually marked by a black band, and balanced by the final extraction of cameral water. Thus, step by programmed step, *Nautilus* achieves sexual maturity and full growth.

Acknowledgments. We thank Bruce Saunders, Bryn Mawr College, for the photocopies of the Palau *Nautilus* shells and Dr. Kazushige Tanabe, Ehime University, Japan, and Drs. Shozo Hayasaka and Junzo Tsukahara, Kagoshima University, Japan, for promptly providing, at our request, data and reprints on their studies on *Nautilus* from Fiji and the Philippines. Extensive editorial assistance was kindly provided by R. A. Davis, Cincinnati Museum of Natural History. Of the people at the Royal Ontario Museum, we are particularly grateful to Bill Robertson and Allan McColl for the photographs and Peter Fenton for the figures. Most of all, we thank Joan Burke for producing the many drafts of the manuscript.

VIII

The Shell and Its Architecture

Chapter 30

Nautilus Shell Architecture

ROGER A. HEWITT and G. E. G. WESTERMANN

1. Introduction

The familiar shape of the shell of *Nautilus pompilius* has been described thoroughly (e.g., Stenzel, 1964) and needs little introduction. We wish to consider to what extent the shell's design resists water pressures as great as 8.34 megapascals (MPa) [830 m depth equivalent (Kanie and Hattori, 1983)]. The loading of the last septum and phragmocone wall is due to the difference in pressure between the partial vacuum within the gas chambers (<1 atm) and the ambient hydrostatic head of seawater. The strength and stiffness of the aragonite shell permit it to act as a constant-volume hydrostatic apparatus that enables *Nautilus* to maintain neutral buoyancy. The gas diffuses into the chambers via the siphuncular tube as the cameral water is removed from behind the recently completed last septum of the growing shell. Because the equilibrium gas pressures in the ocean are largely independent of water depth, there is no difference in the final gas pressure of chambers grown at different depths.

Ambient water pressure is a linear function of water depth, approximating an increase of 1 MPa/100 m. The differential pressure between seawater and the phragmocone interior acts as an added constant once the gas chambers have reached equilibrium by diffusion through the siphuncular tube. The internal pressure is nearly equal to the partial pressure of nitrogen in the atmosphere (0.78 atm), and the differential pressure on the phragmocone is therefore $0.1013 - 0.0780 = 0.0233$ MPa. Partially dry chambers are in equilibrium only with the more mobile oxygen component of seawater, with a vacuum pressure of about 0.23 atm, which yields a differential pressure of 0.080 MPa.

The obvious strength and buoyancy requirements of *Nautilus* shells can be

ROGER A. HEWITT and G. E. G. WESTERMANN • Department of Geology, McMaster University, Hamilton, Ontario L8S 4M1, Canada.

understood by making analogies with submarines. The submarine paradigm provides a model optimum design (Westermann, 1977) and allows estimates of original loads via mechanical analogues for orthoconic nautiloid septa and siphuncles (Westermann, 1973, 1985a). However, this analogy may be faulty even though the equations of elementary elasticity used to define the septal strength index are valid (S.S.I. = 1000 d_s/R, where R is curvature radius and d_s is minimum thickness).

Nevertheless, our adaptationist approach is justified on the grounds that *Nautilus* and other cephalopod shells, such as *Sepia*, must resist water pressure as long as they contain gas for buoyancy (Birchall and Thomas, 1983; Ward and von Boletzky, 1984). The small safety factor of 1.3 implied by the difference in implosion and habitat depth of *Nautilus* (Ward and Martin, 1980) may result from the shallow-water origin of the ancestral Tertiary nautilids. Eocene nautilids have thinner septa, indicative of implosion depths of only 100–350 m (Hewitt, 1987). The contrary implication of Saunders (1984b), that ancestral nautilids could have lived at depths similar to those at which *Nautilus* lives, can be maintained only by regarding the tensile strength of *Nautilus* septal nacre as being seriously underestimated by Westermann (1973) (see also Chamberlain and Chamberlain, 1985). We will investigate the mechanical properties of the *Nautilus* shell and consider the problem of hydrostatic loading using finite element analysis and a mechanical experimental approach.

2. Experimental Results from *Nautilus*

Shells that were kept moist continuously after death are termed "fresh shells" and those that had been allowed to dry are termed "dry shells." Pressures in megapascals are simply multiplied by 100 to give an approximate water depth in meters, accurate enough for most purposes.

2.1. Mechanical Properties of Nacre

The relatively small and uniform hydrostatic load on the *Nautilus* shell becomes concentrated within the carbonate and chitinous shell material. The body tissue and cameral water are incompressible and merely transfer the hydrostatic load, whereas the gas space cannot bear any load. Thus, the critical strength of the shell may be defined as a force per unit cross-sectional area equivalent to a local pressure in megapascals (e.g., 1 MPa = MN/m^2). These stresses are defined by three orthogonal stress axes (S_1, S_2, S_3) at any point within the *Nautilus* shell and may reach critical values defined by uniaxial or bending tests on samples loaded in one direction within the laboratory. Most of the deformation produced by these stresses is a recoverable and linearly proportional expansion or contraction termed the *linear elastic strain*. The strain is simply the deformation divided by the initial size of the sample (e.g., as a percentage), and the relationship between the three orthogonal stresses and the strain ellipsoid is defined by the elastic constants of the material. In an isotropic substance, the same constants apply in all directions and consist of a number of simple stress–strain relationships in both "fresh" and "dry" shells (Currey, 1979).

Bulk modulus (K) = change in volume under hydrostatic pressure ($S_1 = S_2 = S_3$)/initial volume
Young's modulus (E) = change in length under uniaxial loading (S_1, $S_2 = 0$; $S_3 = 0$)/initial length
Shear modulus (G) = change in shape due to shear stress/initial shape
Poisson's ratio (ν) = change in cross-sectional area and length due to flow normal to uniaxial stress

The actual strength of shell material is determined by the strain energy (the area under a stress–strain line or curve), the size and frequency of holes concentrating stress within the structure (Griffith cracks or flaws), and the ability of the material to reduce strain energy by nondestructive plastic deformation. The mechanical testing of tensile and compressive strengths is merely an empirical method of determining these effects under simple loading conditions. Similarly, the elastic constants are just an expression of the gross stiffness of the shell under one set of loading conditions and will vary on a microscopic scale and with stress orientations. The relationship between the mean crystallographic orientation of the shell carbonate and the principal stress direction (S_1) has a major influence on shell stiffness. It is therefore important to measure the elastic constants in the direction of critical loading of the shell material.

Studies of bivalve nacre by Srinivasan (1941) and Currey (1977) have confirmed that both the stiffness (Young's modulus E) and the work of fracture are reduced very considerably when the aragonite c-axis is compressed or when the organic layers of the nacre are pulled apart. The shell is "designed" with the c-axis normal to the nacreous layers of the shell and with all or most of the stress due to water pressure applied within the a/b plane defined by these organic membranes. In *Nautilus*, these membranes are composed of chitin fibrils aligned at right angles to structural protein fibrils, so as to form a plywoodlike organic sheet (Lowenstam *et al.*, 1984). It is unlikely to show the extensive viscoelastic effects known in uniaxial fabrics (Wainwright *et al.*, 1976). Even rapidly submerging *Nautilus* has a maximum strain rate of only 2×10^{-7} sec^{-1} (based on Ward *et al.*, 1984, Fig. 2), some three or more orders of magnitude less than that of mechanical testing data.

Lowenstam *et al.* (1984) and Grégoire (1962) reported that the chitin fibrils and the aragonite a-axes are aligned normal to the growth lines on the body chamber, whereas the chitin fibrils are normal to the median line of the septum. Srinivasan (1941) claimed that the body chamber stiffness (E) increased from 19.3 ± 0.5 GPa normal to the growth lines to 45.4 GPa at an angle of 52.5° and 43.8 GPa parallel to b, due to twinning of the aragonite tablets. A random aragonite fabric should have a mean stiffness of 92 GPa (Srinivasan, 1941). There is evidence for some reduction in stiffness with the volume of organic matrix, but these results have not been confirmed by similar bending deflection measurements taken on the body chamber by Currey and Taylor (1974) and Currey (1976). These and other data given below refer to the a/b crystallographic plane in the body chamber

Table I. Acoustic Velocities Measured from Transverse Sections of the Umbilical Plug Region of *Nautilus pompilius*[a]

	Callus site (prismatic aragonite)[b]					
	1	2	3	4	5	Nacre
P km sec^{-1}	6.345	6.253	6.297	6.530	6.359	6.309
S km sec^{-1}	3.559	3.517	3.552	3.560	3.347	3.283
ν	0.270	0.269	0.267	0.289	0.308	0.314
K′ GPa	63.11	61.03	61.66	69.51	68.86	68.64
E GPa	86.89	84.74	86.30	88.18	79.15	76.50
P wave path mm	1.637	1.632	1.609	1.577	2.194	2.208
Deviation of c axes	5°	5–10°	? 10°	5–45°	15–30°	5–10°

[a] Thickness variation and local deviation of the aragonite *c*-axis from normal to the P wave direction shown for adjacent sites which overlapped. Apparent elastic constants explained in the text.
[b] Callus sites 1–4 are from thin specimens; site 5 is from a thick specimen next to the nacre site.

nacre unless otherwise stated (probably a, since that yields the longest strips of shell).

The 13 bending tests of Currey yielded a mean static Young's modulus (stiffness *E*) of 45.8 GPa [compare with Birch, (1966, pp. 164–165)]. Acoustic P- and S-wave velocities were measured for us by Dr. Patel of the McMaster University Ceramics Department. These results were converted to moduli by assuming elastic isotropy and a density of 2700 kg mm^{-3} (Table I) (for density measurements, see Reyment, 1958, p. 117; Currey, 1975).

The dynamic shear modulus (*G*) is not increased by thermal anelasticity, and *G* shows little variation between static and dynamic measurements from limestones. Birch (1966) cited a mean modulus of 23.5 GPa, less than that of the nacreous shell wall (29.1 GPa) and callus (33.2 GPa). The apparent bulk modulus (*K*′) obtained from the aligned crystallites had a uniform mean of 65.47 GPa (Table I) and is the reciprocal of the dynamic compressibility in an isotropic medium.

The Poisson's ratio (ν) was calculated without having to assume the density and was 0.314 in the nacre (Table I). This value appears reasonable for use in the interpretation of static strain gauge data from essentially dry shells. None of the available strain data is from "fresh shells," and the larger strains recorded by Kanie *et al.* (1981) reflect amplifier problems (Kanie, personal communication, 1982).

The compressive or tensile straining of *Nautilus* shell generally shows a linear trend (the Young's modulus) and is therefore proportional to the square root of strain energy. This relationship is apparent from strain gauges attached to imploded shells by Saunders and Wehman (1977), Kanie *et al.* (1981), and Kanie and Hattori (1983, Fig. 2), as well as from our own more precise data from low-pressure deformation (Figs. 1, 2, and 3). Our data show hysteresis loops and relaxation times, but they appear not to be related to the strain rate or the load and may even be a correctable phenomenon of the gauges and electronic wire resistance under water (despite rubber coatings) (Figs. 1, 2, and 3). But some viscoelastic hysteresis has been reported during more rapid loading of bivalve nacre (Currey, 1977).

The compressive (crushing) stress was measured in the a/b plane by cutting

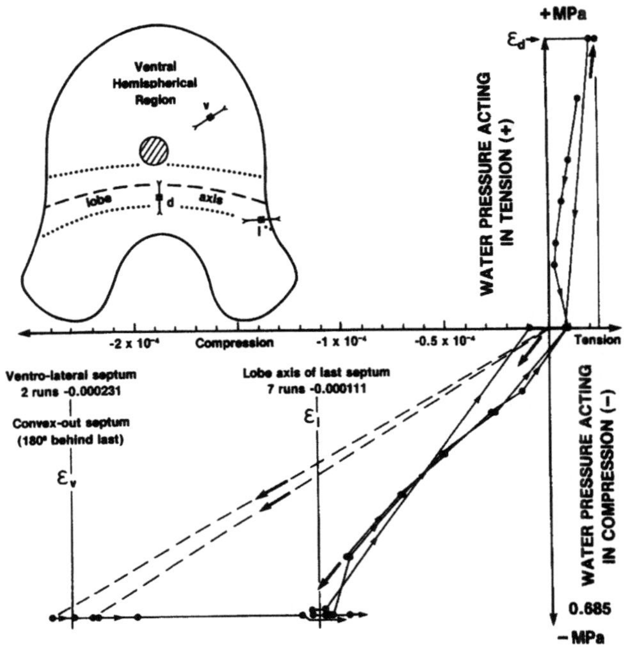

Figure 1. Study of the mature last septum and an immature septum exposed in a convex-out orientation (see Figs. 4 and 7) using three EA-06-030TY-120 gauges (Micro-Measurements Gp., North Carolina) with a smaller gauge length and width of 762 μm. The septa were protected from the surrounding tap water and measured by strain boxes (changes in electrical resistance) affixed externally to a WF 11004 soil-testing cell compressed with a 100 psi air supply. The large negative coefficient of thermal linear expansion defined by the hysteresis loops suggests deformation of the rubber protective coating or resistance changes, rather than events in the organic matrix of the nacre. After correction, the strain from the lobe axis (l) is −0.000118, and the strain from the dorsal median plane (d) is only +0.000011. The value for the convex-out septum represents an approximate strain rate of 4×10^{-5} sec^{-1}, with the loop (l) defined by an apparent adiabatic slope (E = 40 GPa, 2×10^{-5} sec^{-1}) and an apparent isothermal slope (E = 35 GPa, 7×10^{-7} sec^{-1}), shown here with E proportional to the slope. These rates are actually less than those used by Currey (1976) to determine the static E by bending tests.

a rounded notch that caused the shell to fail well away from the shear stress generated by compressive test equipment. Currey (1976; personal communication 1982) obtained a mean strength of 411 MPa in ten tests. We obtained a low mean of 127 MPa in four largely prismatic samples that showed signs of shearing and exfoliation during testing.

The empirical concepts of tensile strength (Currey, 1976) may be studied using Weibull distribution functions that describe the probability of rupture of nautiloid nacre. The Weibull distribution reflects the frequency and distribution of Griffith cracks. It is negatively skewed (Felbeck and Atkins, 1984) and superimposed on Gaussian distributions of random morphological variation.

The tensile strength of approximately 50 mm^3 nacre volume was also tested in the a/b plane of the body chamber. This provided estimates of 63.1 MPa (Currey and Taylor, 1974), 78 ± 12.8 MPa (Currey, 1976, 1977), and 67 MPa inclusive of

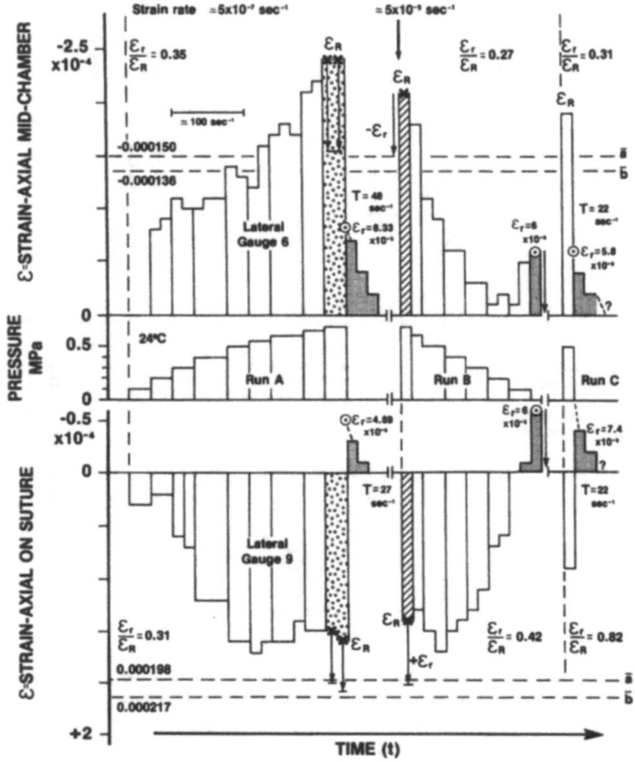

Figure 2. Comparison of the simultaneous change in strain on tensile and compressive parts of the shell situated about 3 mm apart. Series of EA-06-030MF-120 strain gauges with an individual length of 788 μm (width 813 μm) were attached to the exterior of another mature shell loaded as in Fig. 3. Note that the application of any pressure to the cell produces an apparent plastic strain with a decay rate (relaxation time) of about half a minute, defined by a residual strain, ϵ_r, once the cell pressure is entirely removed. The strain was measured by subtracting the residual strain from the maximum strain in each run (average of 16 in run b showing a smaller variance than the apparent mean of 21 in run a).

the outer layer (Currey, personal communication, 1982). This implies a Weibull modulus of 7, according to the formula of Felbeck and Atkins (1984), which is equivalent to the most inhomogeneous of ceramics and an equal volume modulus of rupture of 115 MPa (Felbeck and Atkins, 1984). These predictions are sufficiently inconsistent with the modulus of rupture and shell implosion data to conclude that the tensile testing has introduced numerous flaws (cracks) in these samples.

The maximum tensile strength is derived by plastically deforming the surface layers of the nacre in bending (the modulus of rupture). The mean of 28 three-point bending tests by Currey (1976) was 193 ± 19.8 1σ. The implied Weibull modulus of 12 is average for ceramics and implies a tensile strength of 145 MPa

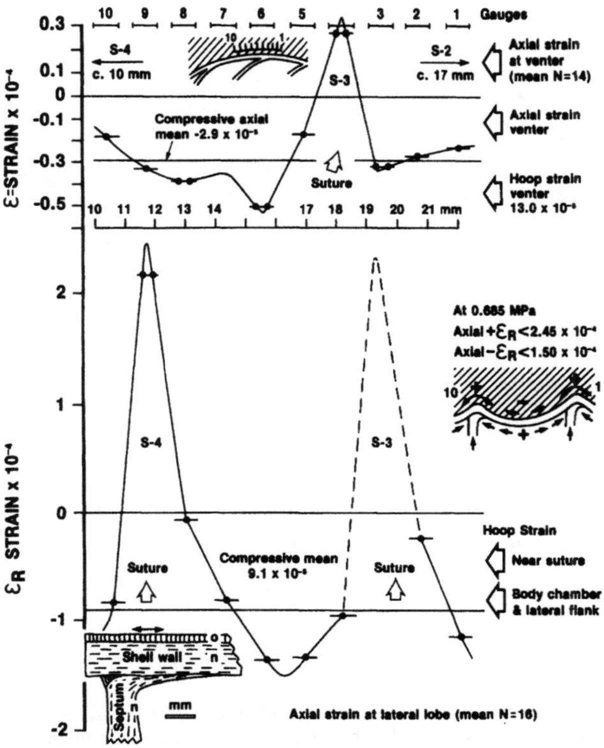

Figure 3. Strain gauge data for the third and fourth sutures (S-3, S-4) behind the mature body chamber, including the series from Fig. 2. Data from larger gauges on the shell wall by Saunders and Wehman (1977) and Kanie and Hattori (1983) are also summarized on the right (all scaled to constant 0.685 MPa pressure). Note the relationship between external bending stresses and the septal sutures. The series from along the principal strain axis of the venter (above) shows a greatly reduced bending moment, reflecting the low angle contact of the central septum and the small compressive membrane stresses generated by R_1 = 65 mm, R_2 = 22 mm, and d_w = 0.8 mm. The series from the principal strain axis along the axial trace of the lateral lobes (left) is dominated by bending stresses over orthogonally aligned and fluted septa (see accurate section below) in an area of potentially destructive compressive membrane stress (R_1 = 230 mm, R_2 = 65 mm, d_h = 1.31 mm).

for about 10 mm^3 in membrane stress. Currey (personal communication, 1982) tested eight samples with the outer shell layer in tension and another set of samples with the inner nacreous surface in tension; a differential ratio of 182.5/109.0 MPa was obtained in favor of the strong nacre.

The curved ventral region of the *Nautilus* septum yields results that are too high or too low, depending on the direction of curvature in the bending test. But the lower strength of the outer shell layer is consistent with data from other molluscan shells composed of prismatic and homogeneous aragonite microstructures (Currey and Taylor, 1974; Currey, 1976). The nacre from the relatively straight, dorsal pillar-flute yielded a modulus of rupture of 184 MPa when loaded in the weak position in four-point bending. Currey (personal communication,

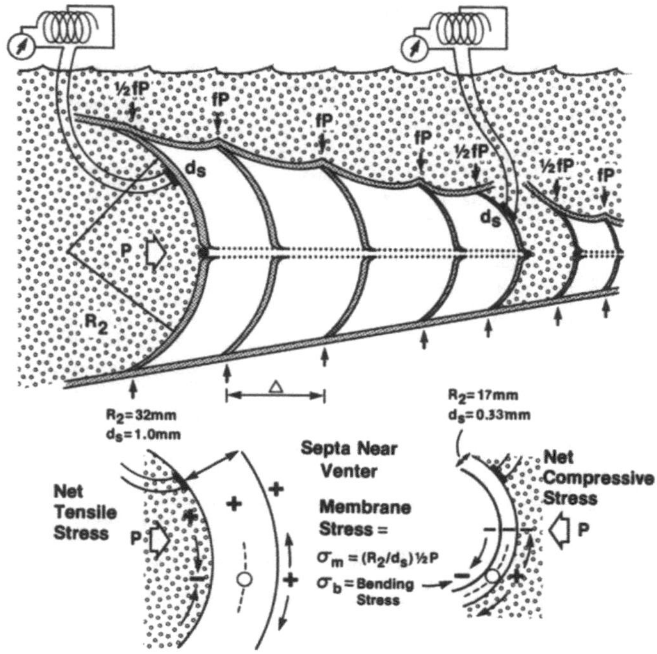

Figure 4. Model orthocone (see Fig. 7 for terminology) illustrating the experiment shown in Fig. 1 and the theoretical relationship between radial bending and membrane stresses in the ventral septum region. Key: (fP) force due to water pressure applied at internal septal sutures; (P) water pressure applied to septum; (R_2) minimum curvature radius.

1982) obtained three-point bending results of 214, 217, and 259 MPa from the septum. This strength is reduced after conversion to membrane tension, as in the case of the body chamber nacre (see above), and is further reduced by the increased incidence of flaws in a large septum volume (Felbeck and Atkins, 1984).

The elastic constants approximate to:

Young's modulus (E) = 45.8 GPa
Shear modulus (G) = 29.1 GPa
Poisson's ratio (ν) = 0.314

The biological significance of the tensile strength is that it indicates the general stress causing the fracture of half a set of identical septa. The elastic strain limit controls the viability of septa exposed to cyclic loading. In the case of a semihemispherical surface, the elastic strain limit is probably equivalent to a stress of about 120 MPa (0.002 strain limit), whereas in a uniaxially strained rod, it would be equivalent to a stress of about 92 MPa. A large Weibull modulus and the implied proportionality between stress and strain within the normal working stress range of *Nautilus* further implies that the stored strain energy and chance of failure within a given shell morphology are predicted by the water depth.

2.2. Finite Element Analysis

One method of computer simulation of stress and strain distributions is termed *finite element analysis*. In the case of nautiloid shells, the loading due to water pressure gives the same stress and strain distributions regardless of their magnitude (e.g., depth), and the analysis can be based on the elastic constants given above. Finite element analysis is used here to deduce the stresses within various idealized and straight nautiloid morphologies at a given constant water pressure on the last septum and shell wall.

The stresses in a generalized shell and in a breviconic shell that resembles the ancestral nautiloids of the Cambrian Period (Fig. 5) show the influence of membrane bending and marginal discontinuity stresses within real nautiloid septa. Note that the upper (internal) and lower (external) surfaces in the middle of the septum show a similar tensile stress, indicative of the dominant tensile membrane stress component assumed by Westermann (1973). Because they are calculated for a water depth of about 100 m and because the tensile strength of the nacre is about 130 MPa (see below), it can be easily calculated that the middle of the *Nautilus* septum would implode at a depth of about 540 m. The more marginal zone of bending stress presumably would break at a depth of about 525 m, assuming a local tensile strength of 193 MPa in bending. But the weakest part of these crude models is the septal sutures. They would break at about 250 m in the *Nautilus* model, about half the value in a real *Nautilus* septum. The difference is due in part to the fact that real sutures have mural ridges and thickness changes not yet added to the model.

2.3. Mechanical Experiments

The macrostructure ("architecture") of the *Nautilus* shell seems to have evolved independently of the primitive and not particularly strong microstructures in the septa and shell wall. It may be that the selection pressure of strength and buoyancy requirements acted mainly on the macrostructure of the shell and that even in this case it is frequently masked by other functional aspects (e.g., hydrodynamic streamlining, camouflage requirements). If attention is confined to the function of resisting water pressure, then it is obvious that the shell can be subdivided into a number of idealized structural regions:

1. The ventral region of the last septum is exposed to pressure via the body chamber and is subject to minimal circumferential bending or compressive stresses. This region is in a state of tensile membrane stress; the load produces only stresses that act within the plane of the septum and forces that result from the radial expansion of the whole septum.
2. The dorsal septal regions of all septa, which to some extent act as struts to support the flanks of the phragmocone wall. The fluted shape of these

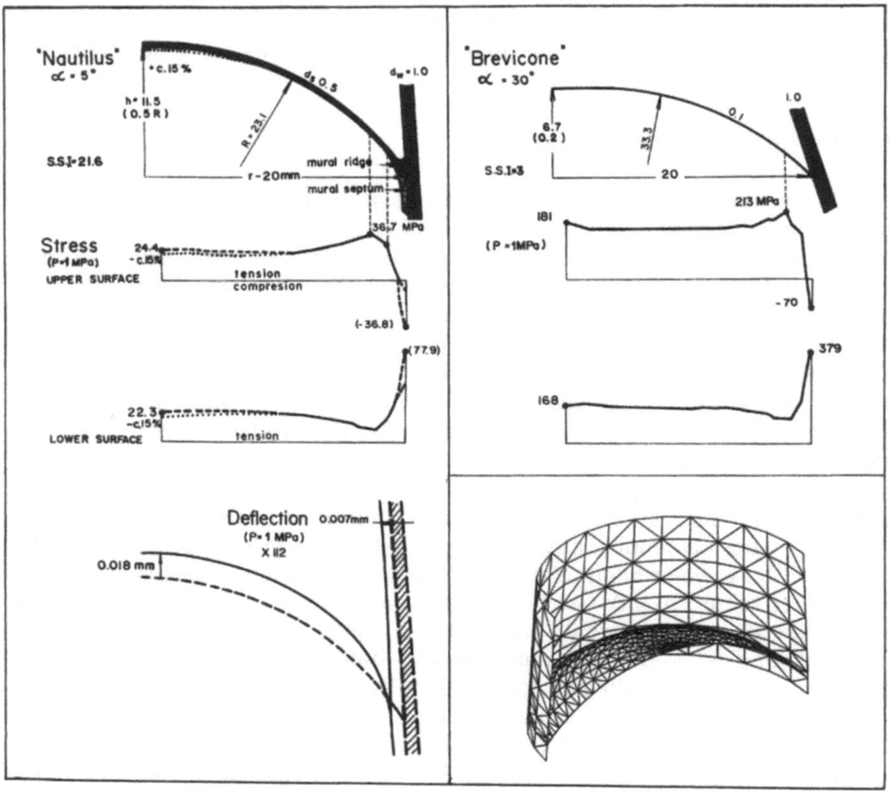

Figure 5. Finite element analysis of two semihemispherical septa indicating tensile and compressive stress and radial deflection at 1 MPa (10 bar ≈ 100 m water depth) hydrostatic pressure. The model on the left represents an uncoiled juvenile *Nautilus* with no thickness variations within the generalized septum; the model on the right resembles a Paleozoic breviconic shell with weaker septa (apical angles = α). Finite element analysis performed by K. Berkkan at McMaster University.

septal regions prevents axial buckling, and the thickness of the septum must be great enough to reduce a resultant tensile stress, due to tensile membrane and axial compressive stresses in the last septum.

3. The outer shell wall of the phragmocone varies from a dominantly compressive membrane stress in the curved ventral region to a flat sheet with bending stresses generated over the supporting septal sutures. The umbilical callus presumably absorbs some of the circumferential strain generated by the hoop stress component of the shell wall.
4. The body chamber wall has an equal load on both sides and is, ideally, free of strain. In reality, it absorbs strain from the adjacent phragmocone and is also subject to point loading by predators and accidents.
5. The coiled siphuncular tube resembles a garden hose loaded internally by water pressure and externally supported against squirming instability by

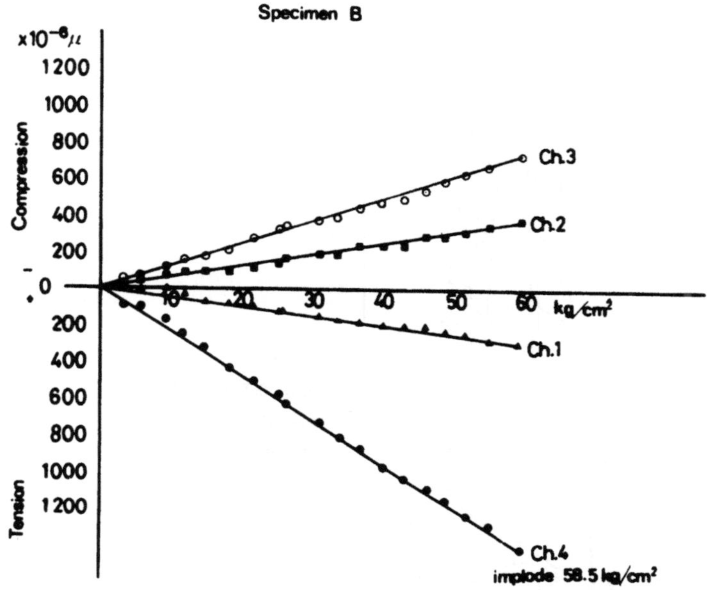

Figure 6. Linear relationship between water pressure (60 kg/cm force = 5.882 MPa pressure) and the tensile strain of +0.0014321 measured during the implosion of the median ventral axis of the last septum within specimen B of Kanie and Hattori (1983). The reduced slope of the other gauges (Ch. 1–3) reflects the averaging of compressive and tensile strains within the outer shell wall, not a variable stiffness E. The linear stress–strain relationship (E) makes it possible to predict the final relative strain energies from low-pressure data. Gauges KFC-5-CL-11 (Kyowa Dengo, Tokyo) and septum strain rate of 3×10^{-6} sec^{-1}. Reprinted with permission from Kanie and Hattori (1983).

the septa. The tube is designed to avoid this and other buckling phenomena and therefore can be considered as a simple cylinder developing major transverse and minor axial tensile membrane stresses.

The complexity of the stresses within the shell is apparent from the strain data in Figs. 1, 2, 3, and 6, but the ventral concave region of the last septum and the exposed convex-out surface of internal septa can be used to estimate the tensile strength at the region of shell failure (Figs. 1, 4, and 7). The membrane stresses and strains are extrapolated from the formulae for synclastic surfaces of revolution developed by Roark (1954) and Den Hartog (1949):

$$S_1 = PR_2/2d_s \quad S_2 = (PR_2/2d_s)(2 - R_2/R_1) \quad S_3 = \text{c axis} = 0$$

$$e_1 = [S_1 - \nu(S_2 + S_3)]/E \quad e_2 = [S_2 - \nu(S_1 + S_3)]/E$$

where: P = water pressure in MPa (force in MNm^{-2})
S = principal orthogonal compressive or tensile stresses in MPa; S_1 is parallel to R_1, not the maximum (S_2); S_3, to the c-axis
R = R_1 maximum local curvature radius measured with rings

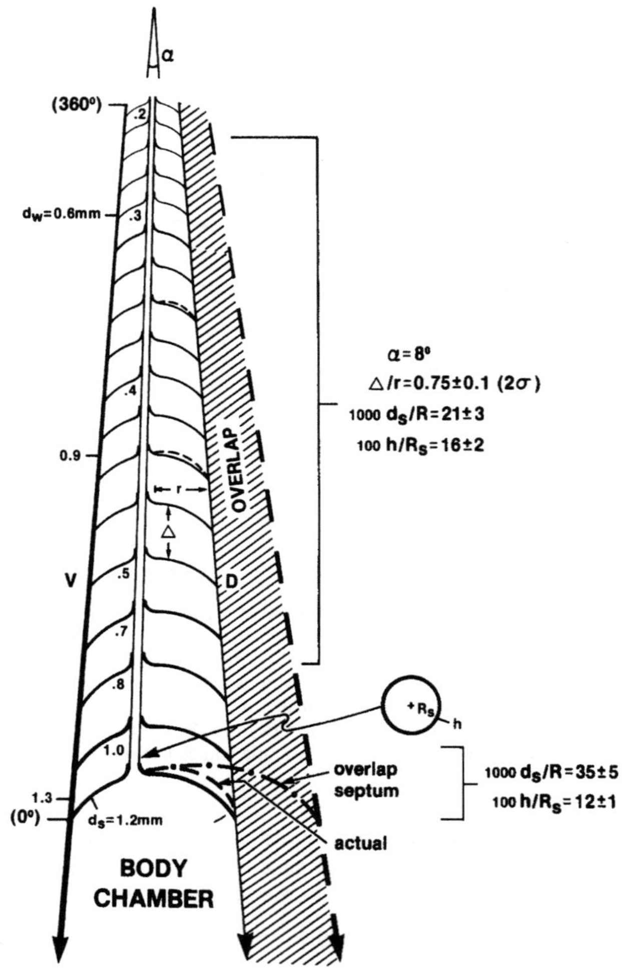

Figure 7. Model of orthocone resulting from the straightening of the *Nautilus* shell and showing morphological features related to the septal strength. Key: (d_s) minimum septal thickness; (d_w) shell wall thickness; (h) horny-tube thickness; (R) curvature radius of septum; (R_s) curvature radius of horny tube; (α) apical angle; (V) ventral; (D) dorsal; (Δ) septal spacing; (r) radius of orthocone.

R_2 = minimum and orthogonal curvature radius (all data in m)
d_s = minimum septal thickness (sometimes local thickness) in m
ν = Poisson's ratio for flow in the a/b plane (0.314)
E = static or isothermal stiffness in the a/b plane (45800 MPa)
e_1 = principal strain along R_1 (linear expansion of 1% = 0.01 strain)
e_2 = principal strain along R_2 and maximum strain when $R_2 < R_1$

Because most of the variance lies in the measurements of R and d_s, it is not possible to test the strength index [S.S.I. = (1000 $d_s/R_1 = R_2$)] with great precision. However, in the case of mature shells, it was found that R_1 and R_2 are well within measurement error of a mean $R_1 + R_2/2$. It is the minimum septum thickness, d_s, that is appropriate for predicting the implosion pressure, although, during testing, strain gauges were situated over the thicker region toward the siphuncle. The curvature radius varies by several percentage points between individual mature shells, and the minimum septal thickness varies from 0 to 1.3 mm due to growth and intraspecific variation. The strain is measured along a 1-mm-long axis to a precision of 1×10^{-7}. After correction and averaging (Fig. 2), the accuracy remains the same, the standard error of E is 6.2% (Currey, 1976), and the standard error of ν is probably even less (e.g., 0.02ν). The total error of these independent variables is the square root of the sum of their variances.

The mature shell A of Kanie and Hattori (1983) imploded at 6.201 MPa pressure, while showing a strain of 0.0010890 about halfway between the siphuncle and the venter [along the median line (Kanie, personal communication, 1985)]. The more centrally positioned gauge of Saunders and Wehman (1977, Fig. 4) apparently imploded at a strain of about 0.0011979 at 6.294 MPa on a mature septum. It is reasonable to extrapolate these strains to the 8.34 MPa failure pressure of fresh shells within the context of the membrane stress model. This procedure yields an ultimate membrane stress of 97.8 MPa for "A" and 106.0 MPa for the Saunders and Wehman specimen. Assuming that the thicker nonmarginal region of the septum corresponds to a strength index of 40.0 (the maximum for mature septa), the predicted stress would be 104.2 (shell A) to 106.5 MPa (Saunders and Wehman specimen). The existence of the thickness variation within the ventral septum is an indication of the existence of bending stresses, which would increase the internal tension while reducing the measured external tension (Figs. 4 and 5). At least in the case of shell A, it can be seen that the gauge would approximate the real minimum septal thickness potentially equivalent to a local strength index of 30.0 and a maximum membrane stress of 139 MPa at 8.34 MPa. Alternatively, it might range up to an index of 40.0, indicative of a membrane stress of 104 MPa at the same pressure. Let us suppose that the septum thickness precisely matches the local bending stress S_b and that specimen A is the worst possible case for the membrane stress hypothesis. The bending stress would be subtracted from the measured radial S_2 direction, whereas the orthogonal stress, being purely of membrane stress origin, would remain equal to $S_2 = S_1 = 139$ MPa:

$$1 = [S_2 - S_b - (S_2 + 0)]/E = 0.0010890\ (8.34/6.201\ \text{MPa})$$

$$\text{Local bending stress } S_b = (139\ \text{MPa} - 44\ \text{MPa} - 67\ \text{MPa}) = 28.27\ \text{MPa}$$

$$\therefore \text{Constant internal tensile strength} = S_2 + S_b = 167\ \text{MPa at } 8.34\ \text{MPa P}$$

$$\therefore \text{Actual implosion strength of dry shell} = 124.4\ \text{MPa at } 6.20\ \text{MPa P}$$

This example tests the hypothesis of Westermann (1973) in a number of ways. First, the worst possible index is assumed for the unusually variable thickness of mature *Nautilus* septa. Second, shell A was not fresh and was imploded with a plugged siphuncle, which may have increased the bending stresses and altered the strain-related discontinuity stresses. Third, the size of the last septum of *Nautilus*, which lies in the upper range of radius measurements taken from fossil orthocones, implies that the scale-dependent bending stresses will be maximized at the expense of the scale-dependent membrane stresses used in the S.S.I. Finally, it has been assumed that the strain data contain no hidden error of the type seen in Fig. 2 and that the septum is not the weakest region of the shell; these assumptions are discussed in connection with new implosion data given below.

Strain data are also available from the 77-mm-diameter phragmocone B of Kanie and Hattori (1983) and from the exposed convex-out-oriented septum at a diameter of 72 mm within a mature shell (Figs. 1 and 6). These immature growth stages have a more predictable and lower strength index, and two gauges both measured the radial strain on the ventral septum. Using the maximum estimates of R_1 and R_2 taken from the area of the gauge and adding the typical septum thickness at that point (Fig. 4), the index implies a stiffness of 52.5 GPa. This value can be reduced to 45.3 GPa merely by accepting the minimum radius estimate of 14.7 mm. Let S.S.I. = 19.5 and implode both shells at 5.735 MPa like phragmocone B. The observed strains were 0.000231 at 0.685 MPa and 0.0014321 in B. A total stress of 182.3 MPa is then implied for B, and 159.3 MPa is extrapolated for "convex-out."

The anomalously low stress of 95.6 MPa implied by the strain data for B (Fig. 6) and the rather high stress implied by assuming the aforementioned S.S.I. of 19.5 can be reconciled by assuming an S.S.I. of 21 near the maximum range at this ontogenetic stage. There is also likely to be some increases in septum thickness away from the margin. The gauge position along the midline (Kanie, personal communication, 1985) is at right angles to the R_2 radius if they are not the same. If R_2/R_1 is 25 mm/27 mm, and assuming an index of 21 in R_1, the calculation is as follows:

$$
\begin{aligned}
S &= (S_2 = 136.5\ \text{MPa}) + (S_1 = 127\ \text{MPA}) - (Ee_1) - (\nu 136.5) \\
&= 136.5\ \text{MPa}\ S_2 + 18.3\ \text{MPa}\ S_b \\
&= 154.8\ \text{MPa implied for phragmocone B at } 5.735\ \text{MPa}\ P
\end{aligned}
$$

It is unreasonable to seek to reduce these stresses much further by explaining them by septum thickness variation within the median line, although some reductions due to that factor are plausible [see Plate 2, Kanie and Hattori (1983), which shows gauge 4 one third of the way from venter to the siphuncle]. Notice also that all four shells with strain data give an apparent tensile strength of about 100 MPa at the realistic failure pressures discussed above (see below for justification by implosion data) and that this is raised to about 120–140 MPa when calculated from the membrane stress formula used in the index S.S.I. The actual internal net stress is about 150–170 MPa, if the calculations given above are realistic. The thickness variation in a septum appears to match the bending stress variation (Fig. 5). This correspondence implies that the index is still proportional to the tensile stresses all over the concave ventral part of the septum, even though the apparent tensile strength measured by the index is too low.

The failure of Saunders and Wehman (1977, Table 1) to find a correlation between the mature septal thickness and the implosion pressure of variably dried shells was explained by Westermann and Ward (1980). However, by taking the eight strongest shells and assuming a 30-mm curvature radius, it is possible to calculate the regression S.S.I. = 10.8 + 3.0 P (in MPa with r = 0.764). In the case of the strongest shell, E, the membrane stress is 108 MPa by extrapolation as above. Ignoring all bending stress and equating the smallest mature index of 30 with the smallest implosion depth of mature fresh shells (7.5 MPa recorded by Saunders, 1984b), one arrives at a similar estimate of about 125 MPa. The implosion of mature dry shells at pressures of about 6 MPa may be indicative of failure due to alteration of the organic matrix.

Most of the implosion data from 35- to 90-mm diameter phragmocones of Saunders and Wehman (1977) are supported by valid implosion data from fresh shells (Saunders, 1984b). The last septum of most of the dry shells was removed before they were tested, and the relatively low pressure required to break the septa would not place too many demands on the shell wall. Calculations made by treating R_1 and R_2 as separate parameters showed that this group maintained a largely spherical septum curvature ($R_1 = R_2$ on the ventral side yields the index in Tables II and III). An ontogenetic variation in response to pressure is implied by the surface of revolution formula at an apparent tensile strength of 131 MPa. There is a rapid decline in predicted implosion pressure from 8 MPa or more in mature septa to only 4.9 MPa at the 75-mm-diameter stage of the phragmocone. The earlier chambers show a rise to nearly 6.0 MPa at a diameter of 45 mm, followed by a variation around an average of 5.3 until a diameter of 25 mm. The smaller and much more depressed septa at the late prehatching stage (≈22 mm) show the geometry $R_1 = 12$ mm, $R_2 = 5$ mm, $d_s = 0.09$ mm, which is indicative of an implosion pressure of only 3.0 MPa (≈300 m depth). At least one of the smallest *Nautilus* specimens tested by Saunders and Wehman (1977) shows a cracked septum and would have been incompressible once it flooded with water. The lower frequency of flaws in a small volume of nacre and the reduction of bending stresses in small septa might permit them to implode at a slightly higher pressure than 3 MPa.

We imploded three intermediate-size fresh shells in an autoclave containing tap water in the McMaster University Chemical Engineering Department (Autoclave Engineers Inc. WP5000 operated by E. Calverley in the laboratory of Dr. R. B. Anderson). Two shells were kept moist in water after being plugged with epoxy resin immediately after the removal of the soft tissues, and the tests were run 3–4 hr after death. They imploded at pressures of 5.69 and 5.82 MPa (Table II). Another shell was imploded unplugged, with the siphuncle in tension, at a higher strain rate (0 to 6.71 MPa in 5 min). A less fresh shell was kept moist and imploded via the shell wall at 6.99 MPa (much of the body chamber having been filled with epoxy resin). As noted by Raup and Takahashi (1966), the septa appear to be considerably weaker than the shell wall at these low pressures and break by a chain reaction of collapse that leaves the shell wall in large fragments (Figs. 8 and 9). Moreover, even the dorsal pillar-flute region (cf. Table III) survived unbroken in one last septum, whereas the ventral, hemispherical part of the septum was generally reduced to fine powder throughout the whole phragmocone. The suggestion that the septal sutures break first (Chamberlain and Moore, 1982) was disproved by these tests and by the illustrated fragments of imploded mature *Nautilus* (Fig. 9). It is even possible to measure the minimum thickness of the septum still attached to the fragmented shell wall and body chamber (Figs. 8 and 9). Because the septal surface had previously been recorded in Plasticine impressions, it was possible to convert this information into estimates of the membrane

Table II. Implosion Data, Size of Phragmocone and Body Chamber, Septal Strength Indices, and Siphuncle Strength of Freshly Killed Juvenile *Nautilus pompilius*[a]

Implosion, size, and last septum strength						Tensile strength					Siphuncle strength		
Run time (min)	P (MPa)	Dia. (mm)	Wb. (mm)	R_1 (mm)	R_2 (mm)	d_s (μm)	d_s^* (μm)	MPa a	MPa b	SSI 1000 d_s/R_2	100 h/R	R (μm)	h (μm)
1. 5	6,712	43.3	22	12	12	300			134	25	Siphuncle all destroyed		
	body chamber	60.3	32.8										
2. 33	6,986	59.8	33								10.9	500	54
		79.8										522	65
body chamber									last ring		12.5		
3. 30	5,685	51.9	25	12.3	11.8	300	413	121	114	25	14.8	587	87
	body chamber	70.1	37.1										
4. 27	5,822	50.2	25	13.6	12.3	283	478	138	133	22	11.1	489	54
	body chamber	67.5	35.6										
5. not tested		49									13.0	510	72
	body chamber	89									16.7	570	95

[a] SSI, septal strength indices. Note that d_s^* is maximum septum thickness at the septal neck and that the tensile strength "a" is the preferred estimate based on R_1 and R_2 orthogonal curvatures. Tensile strength "b" and SSI based on $R_1 + R_2/2$ average in which R_2 is in the median plane of the shell. Whorl breadth (wb), shell diameter (Dia.), and the siphuncle radius (R) data are for the last chamber of each specimen.

Table III. Ontogenetic Variation in the Septal Strength of *Nautilus pompilius* Assuming a 131 MPa Tensile Strength[a]

Ontogenetic index			Ventral hemispherical septa				Dorsal fluted septa			Rupture P in MPa	
Dia. (mm)	Wh_i (mm)	Wb_i (mm)	R_h (mm)	R_b (mm)	d_s (μm)	SSI $(2000\ d_s/R_h+R_b)$	R_1 (mm)	R_2 (mm)	d_s (μm)	Dorsal	Ventral
"Mature"	49	58	?34	?34	1320	?41	∞	14	1540	14.4	?10.2
	39	43	29	25	640	24	"	—	—	—	5.9
80	34	39	25	24	480	19.6	"	15	650	5.7	5.0
70	29	36	21	20	400	19.5	"	14	570	5.3	5.0
62	24	31	17	17.5	380	22	"	12	510	5.6	5.7
48	19	30	13.5	14.5	330	23.5	35	10	380	5.8	6.0
42	16	22	12	12	270	22.5	25	8	350	7.0	5.9
38	15	20	11.2	11.0	230	20.7	18	8	310	6.5	5.4
28	11	15	7.5	8.5	170	21	—	—	—	—	5.3
23	10	13	5.0	7.0	140	23	—	—	—	—	5.7
S + 5	8	15	5.0	12	90	10.5	—	—	—	—	3.0

[a] Measurements converted into implosion pressures, using the apparent tensile strength (131 MPa) obtained later by experiment and the synclastic surface of revolution formula cited on p. 46. Internal whorl height (Wh_i) contains the median radius (R_h) orthogonal to R_b. The dorsal fluted septum is only slightly stronger, in terms of membrane stress in R_2, and has a circumferential load along its axis R_1. Internal whorl breadth (Wb_i) is measured in the dorsal fluted region. Ontogenetic stages are indicated by the diameter and whorl breadth.

Figure 8. Imploded juvenile *N. pompilius* shells, showing the fine debris zone along the imploded lateral flanks and flanges of the marginal last septum along the septal sutures (part of the scalloped edge is due to subsequent micrometer erosion into the 1.0-mm-wide mural flanges).

Figure 9. Shell debris of mature *N. pompilius* imploded live at approximately 8 MPa during the study of Ward and Martin (1980). Note the septal flanges with a width of 6 mm attached to the 60-mm-long fragment shown in the bottom left-hand corner. Photograph provided by P. D. Ward.

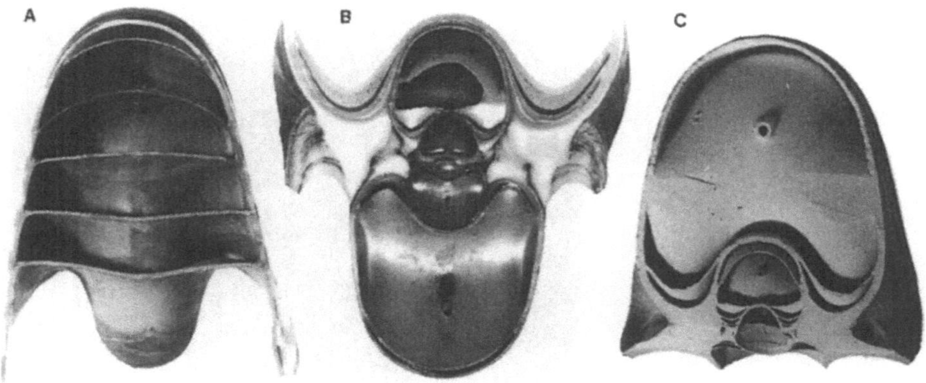

Figure 10. Sectioned specimens of *Nautilus*. (A) *Nautilus belauensis* specimen showing the slightly anticlastic curvature of the mature dorsal-pillar flute. (B, C) Orthogonal views of immature pillar flutes of *N. pompilius* in association with a 20-mm-wide section through the mature umbilical callus (B) used for Table I. (C) The flute is cut out along the axial crest.

stress that broke the less marginal part of the ventral septum up to this point of d_s measurement. The individual results (121, 134, and 138 MPa in Table III) are best considered as a mean minimum tensile strength of 131 MPa. As argued above, it is likely to be raised by 10 or even 20 MPa due to bending moments. It is approximately the same apparent stress found in mature septa when they are imploded at higher pressures of 7.5–8.34 MPa.

The measured compressive strain along the pillar-flute axis of a mature septum (Figs. 1 and 10) was predicted by subtracting the circumferential load on the lateral lobe area of the shell wall (ratio of septum area to wall area being 10.3 : 1 near the lobe axis) from the tensile membrane stress predicted from the semicylindrical geometry of the flute in the area of strain measurement. The result was considerably more successful than the relatively simple case of the ventral septum and supports the precision of elastic constants and strain data used above:

Observed R_2 = 13.4 mm locally; R_1 = infinity; d_s = 1.0 mm
At 0.685 MPa pressure, S_1 = 4.618; MPa S_2 = 9.238 MPa tensile stress
∴ Axial compressive stress $S = (10.3\ P) - 4.618$ MPa = 2.435 MPa
∴ Predicted axial strain = $[(S_1 = 2.435) + (\nu \cdot 9.238)]/E$ = 0.00012 in compression (0.000118 observed)
Tensile S_2 at 8.34 MPa $P = (12.18 \cdot 9.238)$ = 112 MPa

The strain measurements from the shell wall (Fig. 3) do not fit the predictions of the membrane stress formula, although the latter does reduce the stresses and bending moments in the thin ventral shell wall. The lateral lobe region of the wall can be regarded as a sheet, simply supported against ambient water pressure on the more orthogonally aligned sutures of the pillar-flute (Figs. 1, 2, and 3). Rather than speculate on the stresses, it is sufficient to work back from the strain data using the strain formula given previously. Note that there will be an axial bending stress, S_b, aligned normal to the sutures at the crest of the lateral lobe and an orthogonal or hoop bending stress, S_b', parallel to the suture line at that point.

The compressive axial stress found in the underlying septa appears as a finite S_3 term in the strain equation at the sutures, but would not be present midchamber, hence the sharp tensile peaks in Fig. 5. If S_b and S_b' are equal over the suture, then the strain in the axial direction will be reduced as observed midchamber. However, it is more realistic to consider the suture line as being free of circumferential bending moments at this point. Thus, the stress S_b' is taken as 0 over the suture and approximately equal to S_b in the circular area between the sutures of the lateral lobe. The stresses on the outer measured surface are, of course, equal and opposite to those on the interior surface (neglecting complications like mural ridges and septa), so that the apparently safe compressive biaxial bending stress midchamber is equal to the tensile stress on the inside of the nacreous layer. Obviously, tensile stresses are added to compressional stresses in the strain equations, as was illustrated in the lobe calculation. The pressure is 0.685 MPa, and the peak strains of +0.000245 (over suture) and −0.000150 (midchamber) were obtained graphically from Fig. 3 (based on Fig. 2, measured strains being = +0.0002174, −0.0001365):

$$+\ (E \cdot e_1) = S_b - (S_b \nu) + (10.3 \cdot P\nu) = \text{over suture } S_b' = 0?$$
$$-\ (E \cdot e_1) = S_b' - (S_b' \nu) + 0 = \text{midchamber exterior } S_b' = ?$$
$$\therefore\ +11.221 \text{ MPa} - 2.215 \text{ MPa} = 9.005 \text{ MPa estimate of } S_b \text{ (in tension)}$$
$$9.005 \cdot 12.18 = 109 \text{ MPa tensile strength at } 8.34 \text{ MPa} - 6.870 \text{ MPa} = 9.005 \text{ MPa} - (S_b' \cdot \nu)$$
$$\therefore\ S_b = (9.005 \text{ MPa} - 6.870 \text{ MPa})/\nu = -6.80 \text{ MPa at } P = 0.685 \text{ MPa}$$
$$= +83 \text{ MPa at } P = 8.34 \text{ MPa internally}$$
$$\therefore \text{ Critical stress is } S_b = 109 \text{ MPa in both shell layers (bending tension)}$$

It is likely that the critical maximum bending stress of the shell wall over the suture lines is comparable to the three-point bending test yielding the same modulus of rupture of 109 MPa. Although this could be a coincidence, it does illustrate the weakness of the shell wall in these large, mature shells. Umbilicate forms with no callus, such as *N. macromphalus*, probably absorb less compressive strain and appear to have slightly reduced depth ranges (Ward *et al.*, 1980b), with an implosion pressure of less than 7 MPa in the only shell so far tested (Saunders, 1984b). This postulated weakness of the mature shell wall should not be confused with separation along the sutures due to differential radial displacements and tension during the explosion of *Nautilus* shells by Chamberlain and Chamberlain (1985).

Chamberlain and Chamberlain (1985) measured maximum thickness near the siphuncle (d_s^*) and obtained the highly significant regression $R_h = 32.63\ d_s - 0.135$, which actually predicts a constant ontogenetic implosion pressure of 8.06 ± 0.06 MPa at 131 MPa tensile strength. A cyclic variation that produces very weak septa in chambers 13 and 14 and much measurement error is superimposed on the growth trend. Chamberlain and Chamberlain also exploded chambers over a pressure range of 2.1–8.6 MPa and demonstrated that there is no correlation between the S.S.I. and the pressure required to open the septal sutures. They concluded that *Nautilus* septal geometry is too complex to permit implosion depth estimates of simple orthocone septa and, more important, that the shell usually breaks along the sutures of the septa. Both objections are null and void according to Westermann (1985b) and still would not prevent analysis of orthoconic nautiloid depth limits from reasonable estimates of the mechanical properties of *Nautilus* nacre.

We propose that the apparent tensile membrane strength of *Nautilus* nacre is approximately 131 MPa. It is reduced to a maximum working strength of 80–100 MPa, as implied by data of Ward and Martin (1980), Kanie and Hattori (1983), and Saunders (1984b). Chamberlain and Chamberlain (1986) objected that an incomplete thickness of the septum is sometimes exposed to the maximum ambient pressure by the rapid removal of cameral water from the last chamber (Ward *et al.*, 1981). It seems unlikely that immature *N. macromphalus* could remove this liquid at depths in excess of about 250 m (Ward and Westermann, 1985), and they would certainly risk imploding the last septum if they moved down to depths of 400 m while the septum was only 55–60% of its final thickness. Laboratory-grown septa are exposed to pressure when they are only 25–34% of their final thickness (Ward *et al.*, 1981), a possible adaptation for rapid growth near the upper limit of the *Nautilus* depth range (≈100 m). This premature development of a vacuum may also ensure that at least part of the septal nacre is exposed to a similarly large tensile stress regardless of the variable habitat depth of individual shells and populations.

2.4. Siphuncle Strength

Although the nacreous shell is a complex morphology constrained by minimum weight considerations, the small and low-density horny siphuncle tube gives a spurious impression of mechanical and phylogenetic simplicity (Fig. 7). Horny-tube morphology lies at the boundary between short, thick cylinders and longer, thin cylinders, differing in strength criteria (Lam, in Den Hartog, 1949):

$S_1 = P(R + 0.5h)/2h$ = constant axial tensile stress in MPa
$S_t = P(R/h)$ = internal tensile hoop stress used in index
$S_2 = P(R + 0.5h)/h$ = mean tensile hoop stress in thin cylinder
$S_i = P(R + h^2)/(R + h^2 - R^2)$ = critical internal shear stress in MPa
R = inner tube radius in m when pressure (P) is in MPa
h = wall thickness (e.g. outer radius − inner radius)

S_i is the test criterion above an index of about 10 when the index of siphuncular strength (Westermann, 1982) is S.I. = $100\ h/R$. S_2 is strictly the criterion for thin cylinders. The single index is only a crude approximation of the design criteria useful in both cases.

An elegant experiment by Chamberlain and Moore (1982) showed that the rupture pressure of the fresh horny tube of their specimen A varied from the likely implosion pressure of the whole shell in chamber 30 (+8.27 MPa) to only a slightly higher pressure of 8.61 MPa in the thicker rings of the earlier chamber 19. In our opinion, rapid cycles of viscoelastic loading and unloading of the tube provide the only likely explanation for the separation of the posterior horny tube from the "adapertural face of the septum" (Chamberlain and Moore, 1982) and for the reduction of rupture pressures during their experiment (Chamberlain and Moore, 1982). They correctly noted the role of discontinuity stress in setting this nonadaptive limit to the rupture pressure of the siphuncle, but did not make it sufficiently clear that the strains that cause discontinuity stress predict an increase in strength with tube thickness (h). It is clear that a major function of septal necks is to reduce discontinuity stress between the septum and the horny tube and that there is no evolutionary advantage gained by changing the morphology to suit experimental conditions. The difference in axial strain between the stiff neck and the largely unexpandable tube poses little risk (see J. E. Gordon, 1978), but the radial displacement difference (RDD) at the anterior end of the tube is reduced by increased wall thickness:

$$\mathrm{RDD} = [P(R + 0.5h)^2hE_h(1 - \nu_h/2)] - (P(R + 0.5hd_s)^2/d_sE_S)\cdot(1 - \nu_s/2)$$

So, if $E_h/E_s = 0.1$ and $\nu_s/\nu_h = 0.6$, then ideally $h/d'_s = 8.8$ at the interface and the horny tube should be thicker than the septal neck.

We doubt the veracity of the small-tube data measured under a low-power microscope by Chamberlain and Moore (1982). Substituting the data for the strongest index of 18.4 in Westermann (1982) and our Table IV, it can be calculated that chamber 19 broke at $S_t = 46.8$ MPa, $S_1 = 24.6$ MPa, and $S_i = 30.0$ MPa. A better estimate of these parameters is obtained by using the data for chamber 30 presented by Chamberlain and Moore (1982): an index of 10.76 at 8.268 MPa yielding $S_t = 76.8$ MPa, $S_1 = 40.5$ MPa, and $S_i = 44.7$ MPa. The terminal discontinuity stress interpretation of Chamberlain and Moore (1982) implied that these estimates are less than the real tensile and shear strengths of the mature horny tube. In the interpretation of Westermann (1982), the index is an indication of the maximum strength of a variable horny-tube material under natural loading orientations, which, judging from Table IV, varies from 48 to 72 MPa due to ontogenetic factors. Thus, the interpretation of Chamberlain and Moore (1982) might add accuracy to the calibration, provided that the discontinuity stresses in chamber 30 are not much greater than the membrane stresses. The tube would be unable to sustain bending stresses of any size, and there is no axial shear stress or terminal bending moment in thick cylinders (see Den Hartog, 1949). It may be significant that our juvenile shell one (Table II) was loaded via the siphuncle at a nominal S_t of 46 ± 15 MPa and yet yielded none of the long, transverse-cut, horny-tube slices recovered in specimens loaded only via the shell wall and septa.

The siphuncle from a live juvenile animal was fixed in 10% natural buffered formalin, dehydrated through 95% ethanol, embedded in 2-butoxyethanol methacrylate (Polysciences Inc.), and sectioned at 4 μm in a Porter Blum microtome. Strength indices were measured in microtome sections by G. E. G. Westermann and J. E. Mills Westermann (Department of Biology, McMaster University) for comparison with underwater reflected-light measurements made independently by R. A. Hewitt on two specimens kindly provided by J. A. Chamberlain, Jr., and on the imploded debris (Tables II and IV).

Table IV. Ontogenetic Variation in the Outer Radius of the Connecting Ring and the Strength Index of Three Juvenile and One Mature *Nautilus pompilius*[a]

Chamber no.[b]	1 mm	1 100 h/R	2 mm	2 100 h/R	3 mm	3 100 h/R	4[a] mm	4[a] 100 h/R
1	0.27							
2	0.29							
3	0.31							
4	0.30	16.7						
7	0.46							
8	0.48							
9	0.54							
10	0.53	15.0						
11	0.55		0.47	13.2	0.42	12.1		
12	0.60	13.0						
13	0.61				0.44	15.9		
14	0.62	17.4	0.43	14.9				
15	0.62		0.52	14.0	0.50	16.6		
16	0.64	12.2						
17	0.58				0.51	14.1		
18					0.51	13.1	0.69	15.4
20					0.55	14.5	0.61	15.1
21					0.57	16.7		
22							0.68	18.4
24							0.80	15.9
26							0.96	14.3
28							1.04	13.0
30							1.20	11.1
32							1.31	12.9
34							1.36	11.5

[a] Mature *N. pompilius.*
[b] Chamber number counted from protoconch; 1 and 2, moist shells provided by J. A. Chamberlain, Jr.; 3 and 4 are embedded siphuncles fixed by J. E. Mills Westermann.

The microscopic structure of the horny tube was studied by L. Barber (Department of Biology, McMaster University). In light microscopy, unstained sections of siphuncle showed a distinctly dimorphic pattern. The inner layer, thickest in the most recently formed chamber, is composed of very fine, concentric rings. A distinction exists between this inner layer and the outer portion of the siphuncle, in that the unstained concentric rings diffract the light much more noticeably. In stained preparations, the difference is less apparent (Fig. 11). However, histochemical tests on the composition of the siphuncle provide little understanding of its mechanical properties.

There may be a simple solution to the problem of calibrating the siphuncle strength index by a tensile strength of the *Nautilus* horny tube. The scanning-electron-microscopic illustrations of the ruptured tubes of chambers 29 and 30 from shell A (Chamberlain and Moore, 1982, Fig. 4) show large axial splits in addition to terminal transverse splits. Observations by Grégoire (1973, 1985) and Lowenstam *et al.* (1984) indicate that the *α*-chitin fibers of the tube are embedded in a weak aqueous matrix of protein and are aligned within the direction of the hoop stress (S_2 and S_t). This microstructure implies that they have a greater tensile

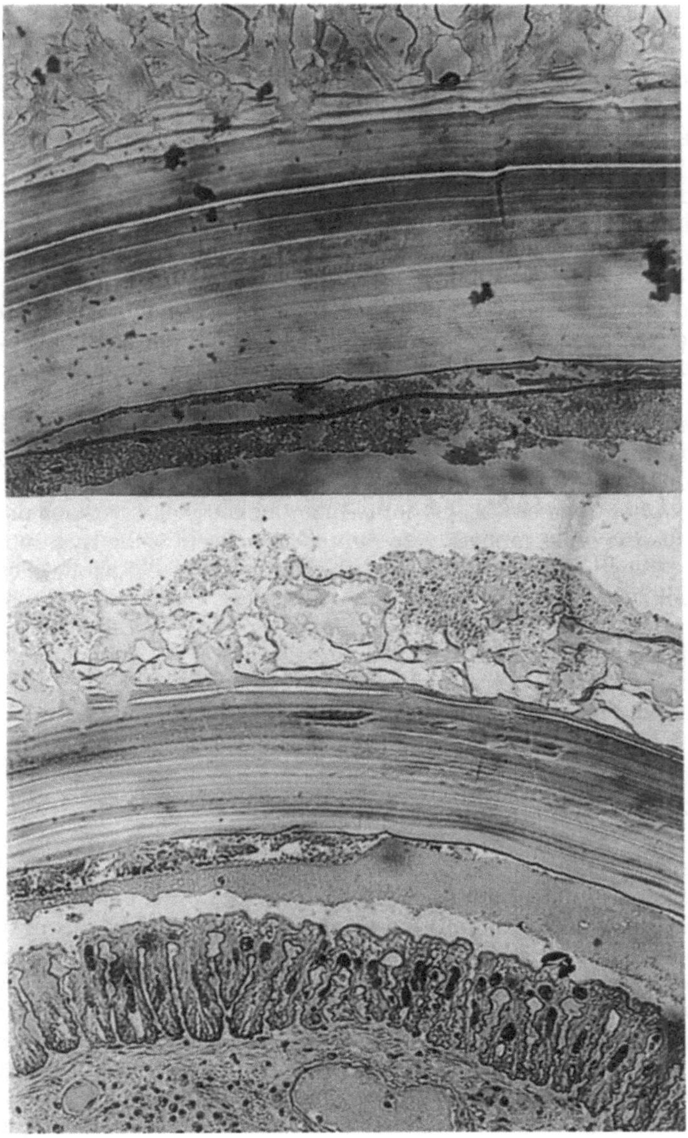

Figure 11. Stained transverse microtome sections (periodic acid–Schiff) of chamber 21, the horny tube in juvenile specimen 5 (Tables III and V). The horny tube (0.01 mm thick) is separated from the coelomic space by a characteristic siphuncular epithelium. Note the outer chalky tube composed of aragonite spicules in a chitinous matrix (*top*) and the development of a prismatic fabric of uncalcified material within the inner half of the horny tube. Sections prepared by L. Barber.

strength in that direction (for the advantages of the design, see Wainwright *et al.*, 1976; J. E. Gordon, 1978). Two tests on unrolled, formalin-fixed horny tubes showed an S_2 tensile strength of 35 ± 1 MPa (corrected from Denton and Gilpin-Brown, 1966), which, despite the deterioration, would be sufficient to bear most of the S_1 membrane stresses deduced above. The membrane strength required by the horny tube is sufficiently great to suggest that the fiber alignment could produce a "sausage," which would burst in all directions at nearly the same internal pressure.

The posterior end of the horny tube forms an advantageous shrink-fit within the septal neck and chalky tube (a material of intermediate stiffness and nearly zero strength), whereas the thicker anterior end is joined to a thin trunk of opaque material lying within the modified aragonitic tissues of the septal neck. Grégoire (1973) and others have described the lateral transition from the nacrelike organic membranes within the horny tube (Grégoire, 1984) into the nacreous layer of the septum itself. The typical necks reduce both the stiffness and the thickness of the septa.

These mechanical adaptations do not invalidate the use of the minimum horny-tube strength index to estimate the habitat depth limit of a fossil nautiloid. Any discontinuity stress would be very localized and could be recognized from marked terminal increase in the horny-tube thickness and perhaps ontogenetic variation in siphuncle strength. The thin posterior part of the ring, in mature or other low-strength chambers, would still yield a reasonable habitat depth limit. This can be calibrated either by the range of apparent working strength in *Nautilus* (S_t = 45 ± 13 MPa, or S_i = 28 ± 7 MPa for work on very thick rings indexed rather differently) or by taking the highest values (S_t = 77 MPa, S_i = 45 MPa) of the *Nautilus* tube for purposes of geological argument. The lower values (S_t ± 45 MPa) seem more consistent with fossil septal strength indices.

3. Conclusions

The septal morphology of nautiloids, including the geometry of the mural ridges and septal sutures, is precisely correlated with water depth. The siphuncle strength index is related to ontogenetic factors and probably sets only a crude limit to habitat depth. The similarity in external shell shape of *Nautilus* to that of Eocene taxa with greatly reduced septal strength indices indicates that external morphology is a conservative character linked to a variety of other functions within the Nautilida. The persistence of the same external morphology in the relatively deep-water *Nautilus* may be cited as further evidence that the thick and dorsolaterally fluted *Nautilus* septa act as internal supports for the lateral shell wall. The resulting risk of bending stress failure of the wall over the septal sutures becomes critical only in the largest specimens and most deep-water habitats of *Nautilus*. Even in mature shells, the weakest part of the shell is determined by the septal strength index. The rupture pressure of the whole siphuncular tube is slightly in excess of the 830-m limit of implosion depth. The following conclusions drawn concern depth limits during *Nautilus* ontogeny:

1. At hatching, the shell would implode at a depth of about 300 m via the

last septum, unless the embryonic chambers remain water-filled until completion of the first juvenile septum. The shell wall or siphuncle is unlikely to break until a higher water pressure is applied to them later in ontogeny.
2. At the stage when the phragmocone is 50 mm in diameter, the septa will implode at a depth of about 580 m, whereas the internally supported lateral shell wall breaks at about 700 m. The siphuncle is presumably adapted for adult loading to a depth of 850 m.
3. In mature animals, the last septum breaks at depths ranging from 700 to 830 m, equivalent to a tensile membrane stress of about 131 MPa, determined from the juvenile septum implosion data. The mature shell wall is evidently stronger than that of juveniles, the dorsal pillar-flute of the septum being adapted for holding the increased load. The siphuncle is known to rupture at a slightly higher pressure than the last septum. The mature shell therefore complies with the principle of synmorphosis; i.e., all parts break at a closely similar hydrostatic loading.

Acknowledgments. We gratefully acknowledge the help of all those cited in the text for assistance and information. Chris Rose attached and soldered the strain gauges during an N.S.E.R.C. summer studentship, and that organization also funded G.E.G.W. Dr. D. H. Collins, Dr. H. Horvath, Peter Koudys, Gerold Westermann, Dr. D. S. Wilkinson, and Dr. J. Weaver and others noted in the text also assisted with the experimental work. Dr. W. B. Saunders kindly lent us previously tested embryonic *Nautilus* shells.

Chapter 31

Ultrastructure of the *Nautilus* Shell

CHARLES GRÉGOIRE

1. Introduction

Studies initiated in the last century, including those of Owen (1832), Valenciennes (1841), Carpenter (1844), Hyatt (1872), and especially Appellöf (1893), produced

CHARLES GRÉGOIRE • Centre de Recherches Métallurgiques, Abbaye du Val Benoît, Liege, Belgium.

fundamental information about the microstructure of the *Nautilus* shell. Since then, it has been studied further with the transmission electron microscope (TEM) and the scanning electron microscope (SEM), in addition to conventional and polarizing microscopes.

Using the TEM, the finest details of surface and internal structure can be observed only in replicas. However, replicas can be prepared without breaking only on very limited surfaces, and the irregularities in relief are greatly leveled. The SEM is especially useful in displaying the relief of large areas of mechanically undisturbed surface. On the other hand, the SEM is inadequate for detection of fine ultrastructure. For example, membranes described as being amorphous on the basis of SEM observations appear in the TEM to be composed of substantial delicate networks or feltworks of fibrils. Therefore, the two instruments must be complementary.

2. Shell Wall

2.1. Structure

Valenciennes (1841), Carpenter (1844), Blake (1882), Appellöf (1893), W. J. Schmidt (1924), Bøggild (1930), and Grégoire (1962) described two aragonitic layers in the shell wall of *Nautilus:* an outer, milky white, porcelaneous layer and an inner, iridescent, nacreous layer. The latter layer is bounded on the inner surface by rows of columnar, prismatic crystals, called "helle Schicht" by W. J. Schmidt (1924). These crystals form the layer called the third layer or inner prismatic layer in the recent literature [Stenzel, 1964; Mutvei, 1964; Erben *et al.*, 1969b (SEM); Wise and Hay, 1968 (SEM); Wise, 1970 (SEM); Blind, 1976].

2.2. Periostracum

Appellöf (1893) did not refer to a periostracum in the *Nautilus* shell. Degens (1967) referred to the periostracum of *Nautilus* in a study on amino acid composition of the organic components of mollusk shells, but gave no information on its aspect and origin. In the literature, the periostracum is commonly illustrated as a black line in ink diagrams. Micrographs of its structure seem to be unavailable.

In a specimen of *N. macromphalus* supplied by Drs. Rancurel and Y. Magnier (Nouméa), the outer surface of the living chamber was uniformly covered with a smooth, orange-yellow deposit. The SEM showed this deposit as shreds or fragments of membrane anchored to the outer surface between and on the growth lines. After this layer was exposed to EDTA, the remaining material, sedimented on formvar-coated screens, appears to be composed of feltworks of fibrils (TEM) (Fig. 1a). This substance might be related to the "drap marin" (Valenciennes, 1841). The uniform coating of the outer surface by this yellow substance and the homogeneous fibrillar composition of the substance strongly suggest that it could possibly be identified as belonging to a periostracal structure and not as an accidental contaminant (e.g., mold). This substance needs further study.

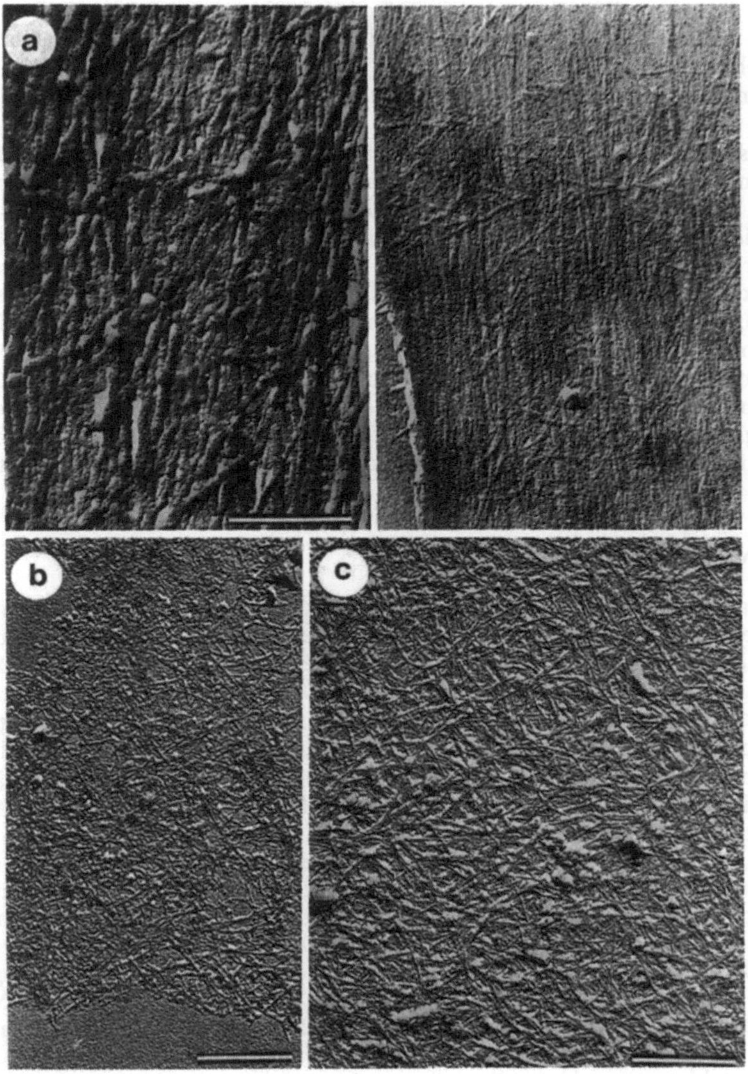

Figure 1. (a) Adult *N. macromphalus* from Nouméa. A bright yellow organic substance covers the outer, adoral surface of the shell (body chamber). Shreds of this substance (periostracum?), which appear amorphous in the SEM, consist of feltworks of fibers and fibrils. Delamination by moderate ultrasonic irradiation yields thinner membranes composed of extremely thin fibrils, unoriented or disposed in parallel arrangements. Platinum-shadowed. TEM, direct print. Scale bar: 0.5 μm. The specimen was a gift of Drs. Y. Magnier and P. Rancurel. (b, c) *Nautilus pompilius*. The EDTA-insoluble organic components of the outer porcelaneous layer of the shell wall (b) and of the umbilical callus (c) consist of feltworks of extremely thin fibrils. Palladium-shadowed. TEM, direct prints. Scale bars: 0.5 μm.

2.3. Outer Layer

A number of names have been applied to this layer, e.g., "äussere Substanz" (Appellöf, 1893), "porcelaneous layer" (W. J. Schmidt, 1924; Grégoire, 1962), "spherulitic–prismatic layer" (Mutvei, 1964), and "outer prismatic layer" (Erben *et al.*, 1969b).

As shown by conventional and polarizing microscopes (Appellöf, 1893; W. J. Schmidt, 1924), this layer consists of an outer sublayer composed of numerous small aragonite grains [Blake, 1882; "Kalkkügelchen" (Appellöf, 1893); W. J. Schmidt, 1924; Bøggild, 1930] and an inner sublayer consisting of elongate crystals or prisms oriented with their long axes at right angles to the shell surface ["Verkalkungsstreifen" (Appellöf, 1893); W. J. Schmidt, 1924; Grégoire, 1962 (TEM); Wise, 1969 (SEM); "prismatische Teilschicht" (Blind, 1976)]. TEM and SEM observations (Grégoire, 1962; Erben *et al.*, 1969b; Wise, 1969) showed that the outer sublayer consists of bulging disks or rounded lenticular corpuscles and the inner sublayer of elongate, sharp-edged tablets, blades, or bars, parallel or diverging in a featherlike arrangement from central elongate stems. The crystals in the innermost part of the inner sublayer are disposed in palisades at right angles to the nacreous stratification (Fig. 2d).

2.4. Nacreous Layer

Nacre (mother-of-pearl), always aragonitic, consists of numerous mineral lamellae, parallel to the inner surface of the shell and superimposed horizontally. A lamella is composed of a single layer of aragonite tablets (001) (010) (110), in which the weakly developed c-axes are oriented at right angles to the inner depositional surface. In tangential view, a mature lamella appears as a flagging of polygonal slabs. Organic sheets (interlamellar "conchiolin" matrices) alternate with mineral lamellae and separate the single crystals in each lamella ("intercrystalline conchiolin"). The columnar stacking pattern ["vertikale Säuligeschichtung" (W. J. Schmidt, 1923)], one of the three modes of disposition of aragonite crystals in successive lamellae of mother-of-pearl, characterizes the mother-of-pearl in *Nautilus* and in physically unaltered fossil nautiloid and ammonoid shell material. The margins of the crystals coincide in successive superimposed lamellae of *Nautilus* (Fig. 2c), but not everywhere in the bulk of the layer. The crystals are piled in columns like stacks of coins [von Nathusius-Königsborn, 1877; W. J. Schmidt, 1923, 1924; Ahrberg, 1935; Grégoire, 1962 (TEM); Mutvei, 1964, 1969 (SEM); Erben *et al.*, 1969b (SEM); Wise, 1969, 1970 (SEM); Erben, 1972a (SEM)]. During growth of the columnar nacre (Fig. 4d), the conical stacks of aragonite tablets expand laterally and come into contact with the crystals of adjacent stacks eventually to form the lamellae (Figs. 4b and 5a–c). At the aperture of the adult shell, the nacreous layer is covered by the prismatic elements of the outer layer, which is progressively thinner in an adapical direction (Blake, 1882; Appellöf, 1893; Erben *et al.*, 1969b, Fig. 1).

2.5. Inner Layer

The inner prismatic layer ["Annulus Substanz" (Appellöf, 1893); "helle Schicht" (W. J. Schmidt, 1924); Grégoire, 1962 (TEM); "annular elevation," "semi-

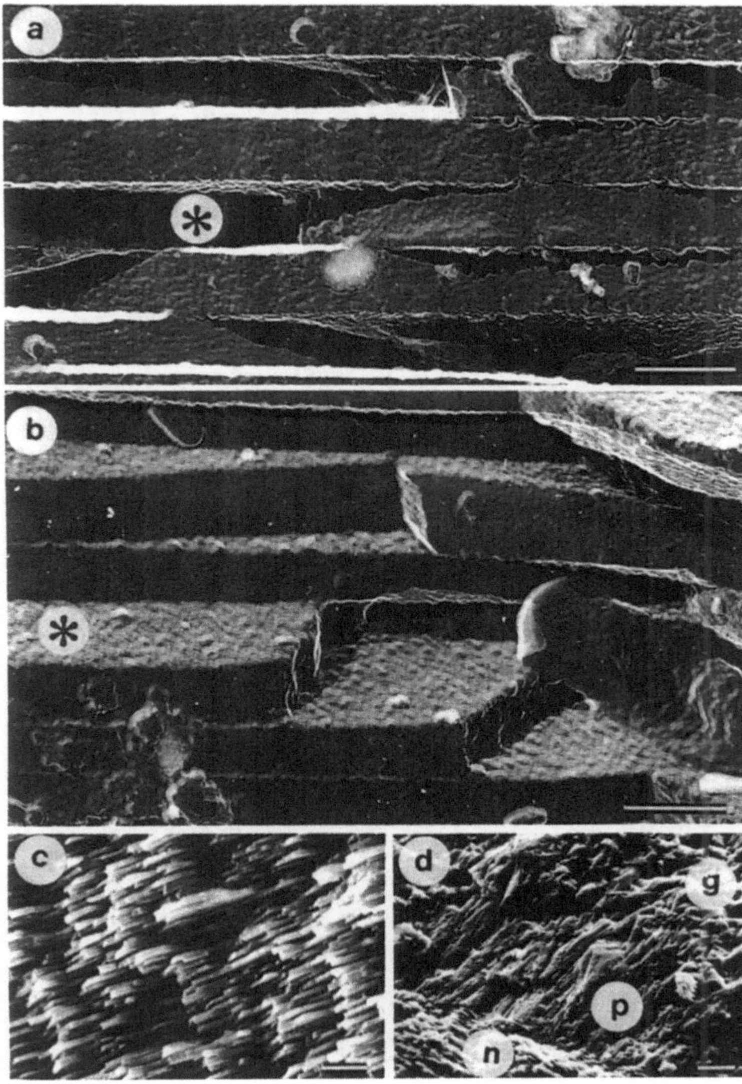

Figure 2. (a, b) *Nautilus macromphalus* from Nouméa. Direct metallic carbon–platinum replicas of the surface of a transverse fracture through the nacreous layer in one of the last adoral septa. Mineral lamellae composed of large aragonite tablets alternate with interlamellar organic sheets (*). The protruding organic latticeworks produce shadows over the lamellae below. TEM, reversed prints. Scale bars: 2 μm. The specimen was a gift of Drs. Y. Magnier and P. Rancurel. (c) *Nautilus pompilius*. Surface of a fracture through the shell wall showing piles of aragonite tablets scattered in the nacreous layer. In *Nautilus*, in contrast with gastropods, the columnar pattern of architecture does not appear everywhere in the nacre. SEM. Scale bar: 10 μm. (d) *Nautilus macromphalus* from Nouméa. Surface of transverse fracture through the shell wall in the living chamber. Key: (n) nacreous layer; (p) spherulitic–prismatic sublayer, with its bundles of prisms; (g) outer sublayer of the outer porcelaneous layer of the shell wall. SEM. Scale bar: 10 μm.

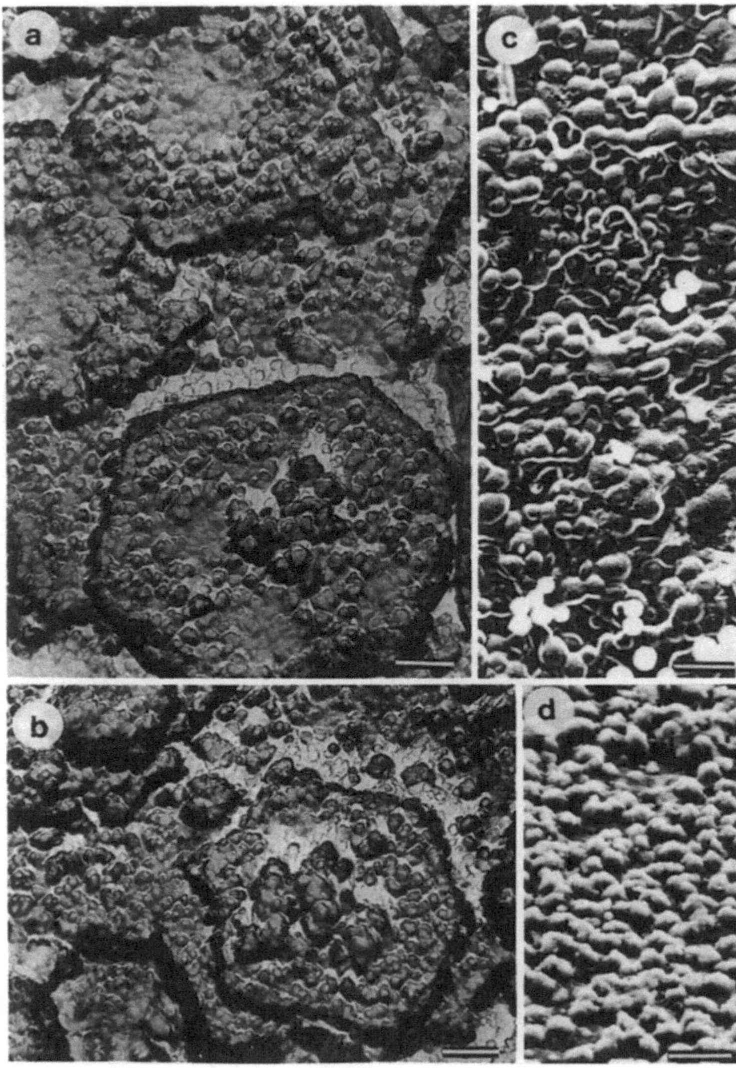

Figure 3. (a, b) *Nautilus macromphalus* from Nouméa. Direct, metallic replicas (carbon–platinum) of the inner surface of the nacreous layer in the living chamber adoral of the anterior ridge of the muscle scars. Scattered euhedral, hexagonal tablets, covered with small crystals, are shown. The orientation of superimposed microcrystals and their elongation parallel to the a-axis of the basal tablets, previously reported in Grégoire (1962, Figs. 3 and 9), are less apparent in these fields. TEM, direct print. Scale bars: 0.5 μm. The specimen was a gift of Dr. R. Catala-Stucki. (c) *Nautilus pompilius*. Double-stage (plastic–palladium) replica of the surface of the crescent-shaped, iridescent, nacreous film covering the umbilical callus. Spheroidal microcrystals appear swarming over the surface. TEM. Scale bar: 0.5 μm. Reprinted with permission from Grégoire (1962, Fig. 7). (d) *Nautilus macromphalus* (Nouméa). Inner surface of the shell wall at the peristome showing intense proliferation of spheroidal microcrystals. TEM. Scale bar: 0.5 μm.

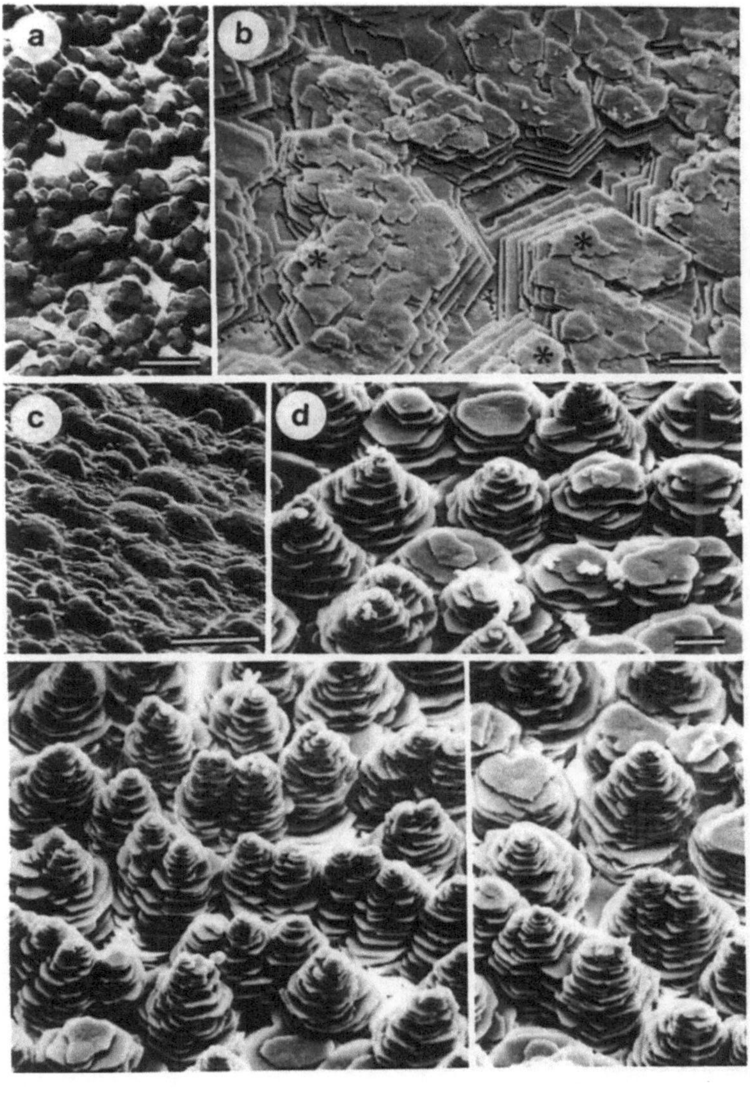

Figure 4. (a) *Nautilus belauensis*. Another aspect of the inner wall surface in the adapical region of muscle insertion, showing proliferation of euhedral or semihedral microcrystals. Direct metallic (carbon–platinum) replica. TEM, direct print. Scale bar: 0.5 μm. (b) *Nautilus macromphalus* from Nouméa. Thin nacreous layer deposited on the black substances covering the outer surface of the shell wall of the next-to-last whorl. Seven superimposed, incomplete lamellae are shown. The disposition in stacks is recognizable. The presence of small tablets instead of microcrystalline seeds on large basal crystals (*) suggests a deceleration or discontinuation in the growth of nacre in this region. SEM. Scale bar: 5 μm. (c) *Nautilus macromphalus* from Nouméa showing another aspect of the inner shell surface at the peristome, namely, the weathered, dome-shaped structures (see Fig. 3d). SEM. Scale bar: 30 μm. (d) Juvenile male *N. pompilius* from Palau (No. 359). The anterior (adoral) concave surface of the last, incompletely formed septum shows an intense proliferation of tightly packed pyramidal crystal stacks, indicative of considerable growth activity. Each of the pyramids is composed of about 20 superimposed crystals. The stacks represent the first stage of deposition of 20 mineral lamellae. SEM. Scale bar: 5 μm. The specimen was a gift of Prof. W. Bruce Saunders.

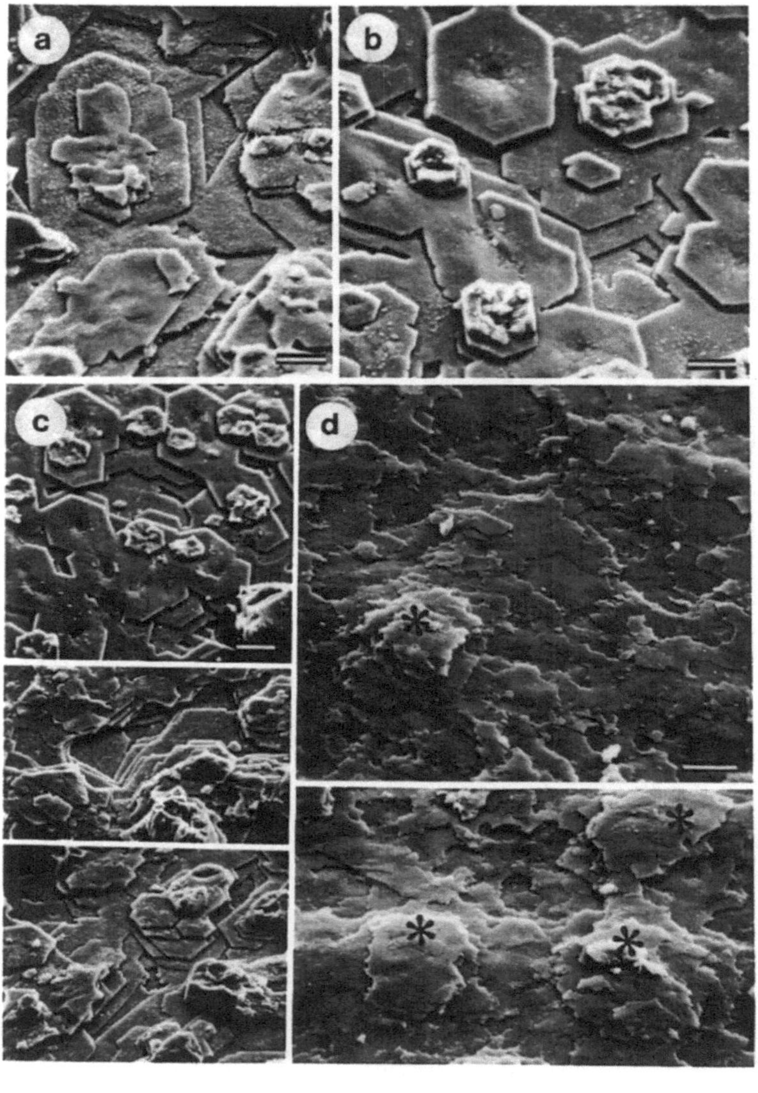

Figure 5. (a–c) *Nautilus macromphalus* from Nouméa. Adoral concave side of the 8th septum in the phragmocone. The incomplete lamellae, low stacks of relatively large tablets, and absence of crystal seeds indicate that the septal growth is slow or has discontinued in this chamber of an adult specimen. SEM. Scale bars: 5 μm. (d) Adult *N. macromphalus* from Nouméa. Adapical, convex side of the 8th septum after the removal of the brown membrane. In contrast with the texture of the adoral (concave) side [see (a–c)], superimposed, complete, smooth lamellae and flat, large crystals (*), piled like pancakes, indicate that the growth on this side has been discontinued for a rather long period. SEM. Scale bar: 5 μm.

prismatic layer" (Mutvei, 1964); "inner prismatic layer" (Erben *et al.*, 1969b)] has long been considered, especially in pelecypods, as a structural variety of the innermost part of the nacreous layer underlying the shell muscles, in which the c-axis of the aragonitic tablets is well-developed and the crystals assume a prismatic structure. According to W. J. Schmidt (1923, 1924), this modification is functional and due to the traction exerted by the shell muscles. The inner prismatic layer, freed by the forward migration of the muscles, is progressively embedded in newly grown nacre. In the texture of the nacreous layer, in transverse sections, the inner prismatic layer appears as a thicker lamella (Grégoire, 1962, Figs. 28 and 31).

2.6. Umbilical Callus

The milky white substance of the umbilical callus of *N. pompilius* is composed of porcelaneous material like that of the outer shell layer. The callus exhibits elongate crystals diverging from a central stem (featherlike disposition), spherulites, sturdy pillars formed by parallel elongate laths, bars, tablets, beams, clustered rods, and spheroidal nodules (Grégoire, 1962, Figs. 16 and 19–26). In some areas, the callus is directly juxtaposed onto the nacreous substance of the septa or of the crescent-shaped layer that overlaps a part of its outer surface; in these areas, wedge-shaped fringes of porcelaneous substance composed of the same elongate crystals form deep indentations in the nacreous substance and disturb the regularity of the lamellar stratification (Grégoire, 1962, Figs. 22 and 23).

3. Shell Surface

3.1. Outer Surface

The following types of structures, generally blurred by weathering, were detected by TEM on the outer surfaces of the living chamber, including areas marked with the flame-colored patterns:

1. Lenticular or spheroidal particles, clustered in considerable numbers, especially along the shell aperture.
2. Dome-shaped mounds, composed of stacked, flat crystals, erected on a base of hexagonal or rounded tablets.
3. Nets of ribbons and granules of unknown nature scattered between the mounds on the surfaces of the *flame-colored streaks*.
4. Granules of unknown nature, 3 mμm in average diameter, adhering to anfractuosities on the surfaces of the black deposits.

In *N. pompilius*, considerable clusters of spheroidal particles [crystal seeds (Fig. 3c)] cover the outer surfaces of the umbilical callus in certain areas.

3.2. Inner Surface

In shells in which growth is still active, formation of new nacre takes place not only along the extending free edges of the lamellae or growth lines ["Wachstumsmaserung" (W. J. Schmidt, 1923)], but also simultaneously within successive, still incomplete, superimposed lamellae (e.g., Figs. 4b and 5a–c). The crystals found on the surfaces are associated in groups disposed in parallel orientation along their a-axes, predominantly at right angles to the growth lines or scattered at random. These crystals are stacked, tabular, euhedral, hexagonal tablets (Grégoire, 1962, Figs. 8 and 9). Weathered microcrystals (crystal seeds) are scattered over the tabular planes of the tablets (Fig. 3a and b). In the whole phragmocone, but not in the living chamber, the mural part of the septa overlaps and conceals the inner surface of the shell wall.

3.3. Region of Insertion of the Shell Muscles

In the zones of attachment of the shell muscles in the living chamber (annulus), rigid, cleavable membranes are interposed between an epithelial layer that lines the shell muscles and the inner shell surface. These semitransparent membranes have been described by various appellations ["horny matter" (Owen, 1832); "pellicule charnue" (Valenciennes, 1841); "dicke Conchiolinplatte" in contrast to the underlying thin conchiolin substance of mother-of-pearl (Waagen, 1867–1870); "dicke unverkalkte Chitinrinde" (Appellöf, 1893); "cementartige Schicht" (Thiele, 1893); "conchiolin" (G. C. Crick, 1898); "Konchinbelag" (Kessler, 1923a,b); W. Lange, 1941; "Konchinschicht der Basalplatte" (M. Schmidt, 1925); "conchin layer, pseudo-tendon" (Mutvei, 1957); "membranous disc" (Grégoire, 1962).

As shown with the TEM (Grégoire, 1962), the texture of the inner surface of the shell wall directly beneath the membrane and bounded by adoral and adapical bulging ridges differs greatly from that of the free inner adoral and adapical shell surfaces. The surface beneath the membrane is covered with swarming lenticular and spheroidal seed crystallites (Fig. 3a and b) (see also Grégoire, 1962, Figs. 10, 12, and 13). The number of these microcrystals per square millimeter of shell surface has been estimated at 2 million on the adoral ridge, 41–126 million on the adapical ridge, and 37–53 million on the shell surface directly underlying the membrane.

This proliferation of microcrystals in the areas of muscle insertion might represent steps of rapid deposition of new nacreous material in the adapical areas of resorption left by the gradual forward (adoral) shifting of the muscles. Along the adoral ridge of the muscle scars, the clusters of seeds might reflect a specially active secretion of new shell substance induced by special environmental conditions in the vicinity of the shifting shell muscles (Grégoire, 1962). As previously reported (Grégoire, 1962), this intense proliferation of spheroidal microcrystals might be related to dense clusters of calcospherules developing into pearls in the muscular tissue and on the surface of the "helle Schicht" along the boundaries of the muscle insertions in pelecypods, as observed in conventional microscopes [*Margaritana vulgaris* (Herdman and Hornell, 1903); Jameson, 1912; *Margaritana*

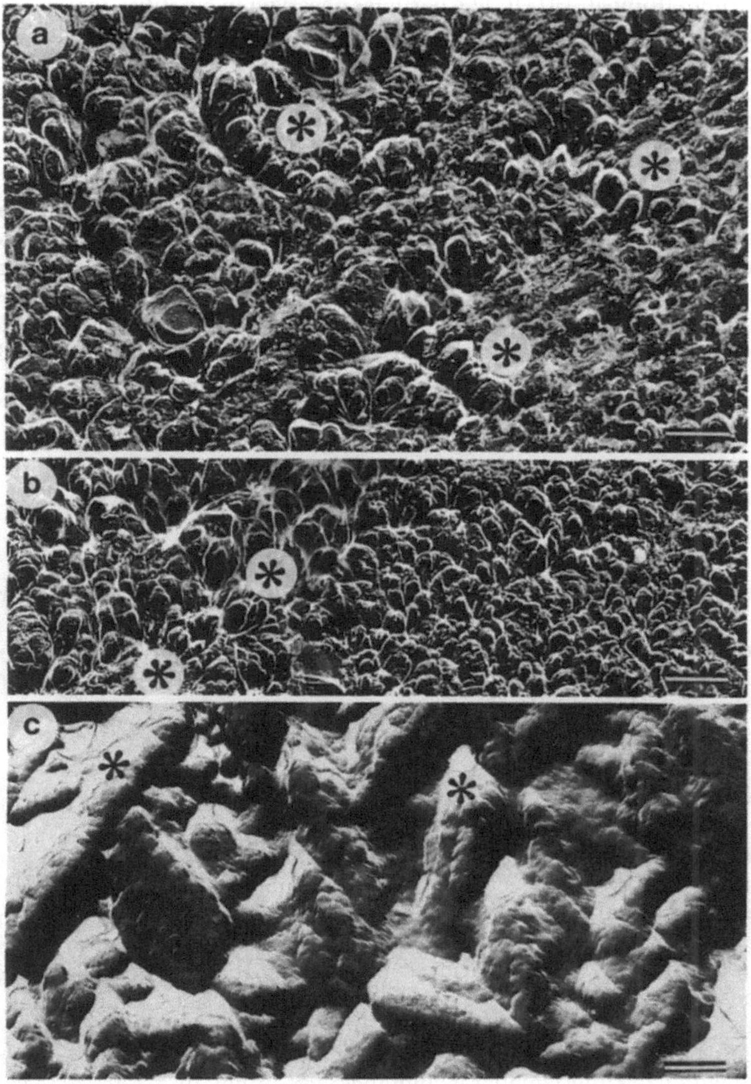

Figure 6. (a, b) *Nautilus pompilius*. Double-stage (plastic–palladium) replica of the inner surface of the nacreous layer near the adapical edge (ridge) of the region of muscle insertion in the living chamber. Among the swarming lenticular or spheroidal microcrystals covering the surface, a number of erect, elongate crystals (*), single or clustered in bundles, illustrate an early stage of formation of the prisms that compose the inner prismatic layer or "helle Schicht." These erect crystals, probably under the influence of the traction exerted by the muscles, are more developed along their c-axis than the tabular crystals in the stratified lamellae of the nacre. Bundles of these elongate crystals resemble fossil conellae (see Grégoire, 1962, Fig. 11). TEM, reversed prints. Scale bars: 0.5 μm. (c) Adult *N. belauensis* from Palau (No. 56). A more advanced stage shows unoriented prismatic structures (*) in another region of the surface near the adapical edge of the muscle scars (adapical ridge). Direct metallic replica (carbon–platinum). TEM, direct print. Scale bar: 0.5 μm. The specimen was a gift of Prof. W. Bruce Saunders.

margaritifera (Rubbel, 1911)]. The particles observed in *Nautilus*, which are about a hundred times smaller than the smallest elements (Hypostrakum muscle pearls) described by Jameson [1912, Plate 39 and Fig. 22 (20 μm)], might be the earliest stages, invisible with a conventional microscope, of development of these calcospherules.

On other surfaces of the shell wall underlying the area of attachment of the shell muscles, the orientation of the elongate crystals of the inner prismatic layer is considerably disturbed. Figure 6c shows pyramidal aggregates of columnar, blunt-ended crystals appearing as warty protuberances pointing in various directions (Grégoire, 1962, Fig. 11). These alterations of the inner prismatic layer resemble the conellae produced by diagenetic transformation of the inner prismatic layer of the shell wall in the floors of hollow spines of ammonites [Hölder and Mosebach, 1950, p. 388; Hölder, 1952; Erben and Reid, 1971 (SEM); Erben, 1972b (SEM)].

4. Organic Components of the Shell Wall

4.1. Outer Prismatic Layer

Decalcification of the outer layer of the shell wall leaves scarce, semirigid shreds that are opaque to electrons and strongly resistant to ultrasonic irradiation. The shreds, observed on their thinnest edges, appear to be composed of microfibrillar feltworks partially embedded in amorphous veils (see Fig. 1b) (Grégoire, 1962, Figs. 33, 34, and 36).

4.2. Nacreous Layer

Decalcification of the nacreous layer of the shell wall yields thousands of soft, iridescent, organic membranes. These membranes collapse because of the dissolution of the mineral lamellae that alternate with them in the nacreous architecture. These membranes are separated by ultrasonic irradiation into single, electron-transparent, strongly iridescent, interlamellar membranes. Examined with the TEM, these organic membranes appear as mosaics of polygonal fields, delimited by straight ridges or cords of intercrystalline conchiolin. These polygonal fields are the outlines of the tabular facets of the dissolved aragonite tablets between which the interlamellar membranes were originally sandwiched ["crystal imprints" (Grégoire, 1959a,b, 1962, 1966); Grégoire and Teichert, 1965; "crystal scars" (Mutvei, 1970)]. Within the polygonal fields, the interlamellar conchiolin

Figure 7. (a, b) *Nautilus pompilius*. EDTA-insoluble fraction of the iridescent, organic, interlamellar matrix of the nacreous layer of the shell wall ("nacre conchiolin"). Robust, irregularly cylindrical trabeculae, studded with hemispheric tuberosities of different sizes and a generally elongate, elliptical fenestration, are the main features of the nautiloid mural pattern. Stained with osmium tetroxide and palladium-shadowed. Reprinted with permission from Grégoire (1962, Fig. 39). TEM, reversed print. Scale bars: 0.5 μm. (b) Same material as in (a) unstained. Palladium-shadowed. TEM, direct print. Scale bar: 0.5 μm. (c) *Nautilus pompilius*. EDTA-insoluble fraction of the organic matrix of the septal nacre.

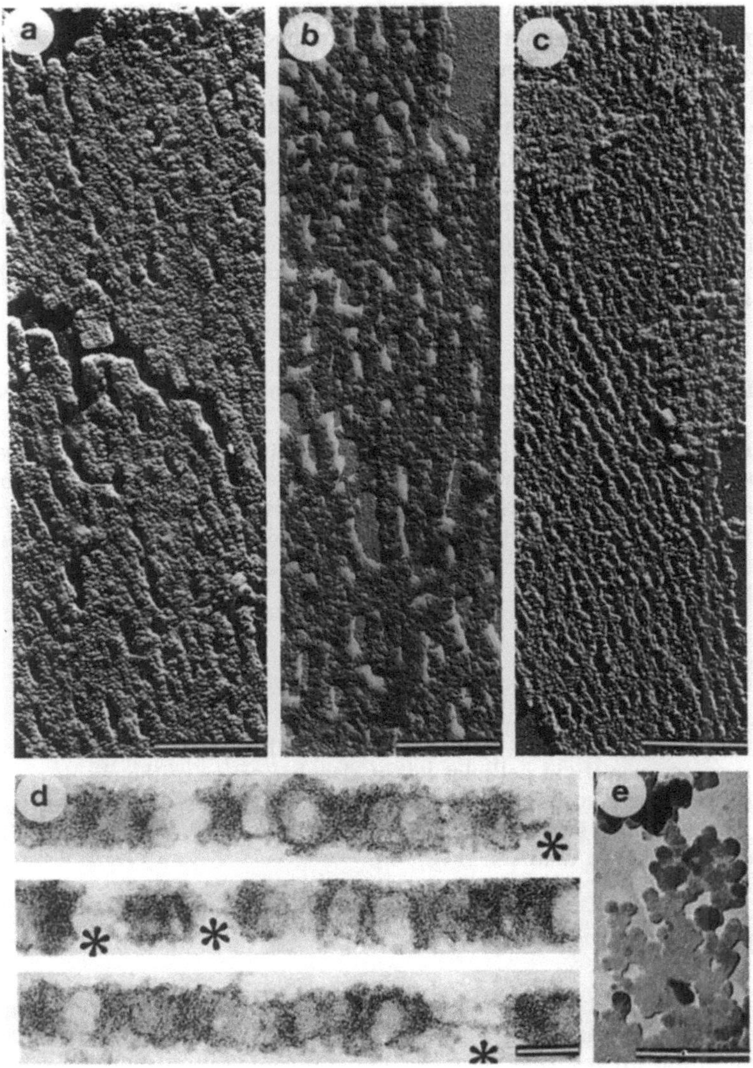

In the nautiloid septal pattern, the texture is tighter, the trabeculae more slender, and the fenestration smaller than in the mural pattern. Numerous tuberosities appear bulging on the trabeculae. Stained with osmium tetroxide and palladium-shadowed. TEM, reversed print. Scale bar: 0.5 μm. Reprinted with permission from Grégoire (1962, Fig. 59). (d) *Nautilus* sp. Ultrathin transverse section of interlamellar membranes of conchiolin [see (a) and (b)] from the nacreous layer of the shell wall (Goffinet *et al.*, 1977, Fig. 6). Material collected after fixation and decalcification of nacre in glutaraldehyde–ascorbic acid. Large, rounded, electron-transparent areas resembling vacuoles are scattered in the membrane. (*) Axial intertrabecular bridges. TEM. Scale bar: 0.1 μm. (e) Decalcified dust from filed orange-tinted portion of the outer porcelaneous layer of the shell wall in the region of the flame-colored streaks. Clustered flat particles, of unknown composition, form the residues of the orange-colored watery suspension. TEM. Scale bar: 0.5 μm.

membranes are composed, as shown in shadow-cast preparations, of networks of sturdy, irregularly cylindrical or varicose knobby cords or trabeculae, studded with hemispheric tuberosities (Fig. 7a,b). The trabeculae are separated by a broad, generally elongate fenestration [elliptic pores (Fig. 7a,b)]. The ellipses seem to be aligned in the direction of the a-axes of the tabular crystals (Grégoire, 1962, Fig. 80). The texture of the organic components of the nacre in *Nautilus* [nautiloid pattern (Grégoire *et al.*, 1955; Grégoire, 1957, 1962; Mutvei, 1969; Iwata, 1975)] differs from the pelecypod and gastropod patterns (see Grégoire, 1972a, b). Extremely thin, brittle, intertrabecular, membranous bridges extend across the fenestration (Mutvei, 1969; Grégoire and Monty, in Grégoire, 1972c; Iwata, 1975) and are especially visible on thin sections of the conchiolin membranes (Fig. 7d) (Goffinet *et al.*, 1977).

4.3. Membranous Disks in the Region of Insertion of the Shell Muscles

The stratified membrane [membranous "disc" (Grégoire, 1962)] that separates the inner layer of the shell wall from the muscles in the region of the muscle scars is extremely resistant to mechanical dissociation, including ultrasonic irradiation. It can be teased into thinner, glassy, rigid shreds composed predominantly of extremely thin microfibrils disposed in dense feltworks (Grégoire, 1962, Fig. 41).

4.4. Umbilical Callus

In contrast to the organic material of the outer porcelaneous layer of the shell wall, the scarce, semirigid pellicles left by decalcification of the umbilical callus are easily disintegrated into fibrils (4–6 mμm in diameter) and fibrillar veils by merely shaking their suspensions. The fibrils seem to be composed of thinner filaments [15–30 Å in diameter (Fig. 1c)].

4.5. Flame-Colored Streaks

The pigment that gives rise to this coloration is confined to the outermost portion of the porcelaneous layer of the shell wall (Valenciennes, 1841). Decalcification of the streaks leaves orange-colored organic shreds. Ultrasonic irradiation dissociates these shreds into rounded particles (Grégoire, 1962, Figs. 35 and 37), scattered or agglutinated into contorted strands (Fig. 7e).

Figure 8. (a) Nautiloid (orthoconic brevicone). Pennsylvanian (about 300 million years old), Buckhorn asphalt, Sulphur, Oklahoma. EDTA-insoluble organic substance from the still aragonitic nacreous layer of the shell wall. Except for slight diagenetic alterations (e.g., shrinkage of fenestration, flattening of the trabeculae), the nautiloid pattern appears to be well preserved. (*) Limit of a polygonal field and the protruding ridges of intercrystalline conchiolin (crystal imprints). Palladium-shadowed. TEM, direct print. Scale bar: 0.5 μm. The specimen was a gift of Prof. C. C. Branson. (b) Nautiloid. Cambrian–Ordovician (about 500 million years old) B and KD 60, boundary section of Dadoushan in Duibian

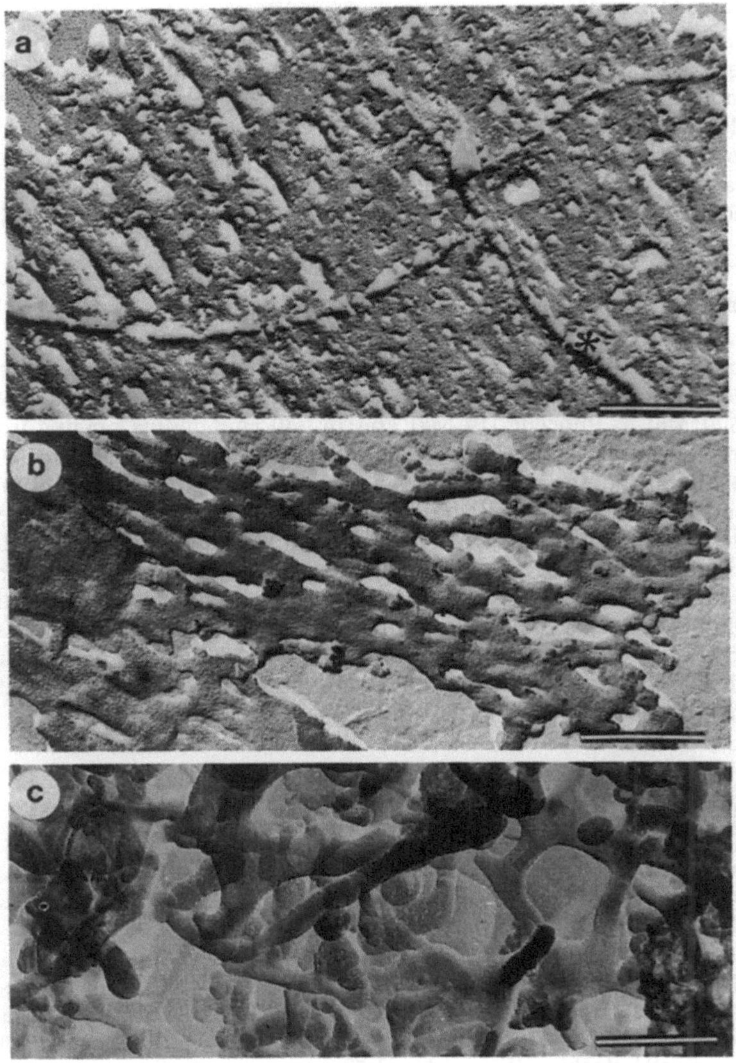

Jiangshan, SW, Zhejiang Province, People's Republic of China. This is the oldest so far recorded EDTA-insoluble organic remnant from the nacreous layer of the shell wall. The flattening of the trabeculae is a diagenetic alteration commonly observed in fossil material and experimentally reproducible in modern material (Grégoire, 1968). The figure demonstrates the great stability throughout geologic times of the nautiloid pattern, which was already established in Late Cambrian. Platinum-shadowed. TEM, direct print. Scale bar: 0.5 μm. (c) *Endolobus clorensis* Collinson (Nautiloidea, Nautilida). Carboniferous (about 300 million years old), Arkansas. The specimen was supplied by Prof. Mackenzie Gordon (U.S.G.S. 15075 PC). EDTA-insoluble remnant of the conchiolin matrix of the thick nacreous layer of the shell wall. The figure shows diagenetic alteration of the nautiloid pattern in the form of loosening and fragmentation of powerful networks of smooth trabeculae. Platinum-shadowed, TEM, direct print. Scale bar: 0.5 μm. Reprinted with permission from Grégoire (1980, Fig. 21).

4.6. Black Deposits

These substances cover the outer shell layer at the aperture (Valenciennes, 1841) and the dorsal shell around the umbilical callus (Stenzel, 1964). They contain melanin (Comfort, 1950). Immersed in decalcifiers, the black substance leaves dark brown particles in aqueous suspensions, in which a pink hue develops during ultrasonic irradiation. This material appears in the TEM in the form of homogeneous particles measuring 3 mμm in diameter (Grégoire, 1962, Fig. 38).

4.7. Organic Components in Fossil Nautiloids

Remnants of organic components of nacre, in the form of membranous shreds, have been detected in the TEM in 150 nautiloids, ranging in age from Late Cambrian (500 Ma) to Pliocene (14 Ma) and in 250 ammonoids ranging from the Cretaceous (115 Ma) back to the Devonian (about 380 Ma) (Grégoire, 1959a,b, 1966, 1980). Most of these shreds have proved biuret-positive. The nautiloid pattern was nearly intact in shells buried in environments like the Pennsylvanian asphaltic sandstones of the Buckhorn Asphalt (300 Ma) (Grégoire, 1959b) (Fig. 8a). As shown in Fig. 8b, the nautiloid pattern already existed in Late Cambrian nautiloids, the oldest material so far investigated for organic components. In samples from all horizons, the texture of the organic membranes was variously degraded by diagenesis (Fig. 8c). Identification was made possible by experimental simulation of diagenesis in modern *Nautilus* (Grégoire, 1964, 1968, 1972b; Grégoire and Lorent, 1972).

5. Structure of the Septa

The septa consist mainly of nacre (Appellöf, 1893). Toward the shell wall and on the adapical side of the septa, infillings from the *sutural substances* cover part of the peripheral septal surface. Toward the siphonal funnel and on the adoral side of the septa, tufts of needle- or spearlike elements ("Kalkpfeilerchen," Appellöf, 1893; Mutvei, 1964; Erben *et al.*, 1969b, Table 12 and Fig. 2; Wise, 1969, Fig. 17) are superimposed on the septal nacre. In transverse sections examined with the TEM, the septa show a brickwork architecture (Fig. 2a and b), with columnar crystal stacking (Figs. 4d and 5a–c), as seen in the nacreous layer of the shell wall. In the last adoral septa, the crystals are distinctly larger than those from adapical septa and the wall (Grégoire, 1962, Figs. 45–47).

As reported by Mutvei (1964) and Blind (1976), each septum in *Nautilus* has fundamentally the same structure as the external shell wall, being made up of four layers: (1) the conchiolin layer (= brown membrane), (2) a spherulitic–prismatic layer, (3) the nacreous layer, and (4) the semiprismatic layer. Identification of the periostracum (first layer) with the brown membrane seems to be supported by the present finding of the fibrillar composition of the periostracum. On the other hand, the existence of the second layer (the spherulitic–prismatic layer in the peripheral part of the adapical side of the septa, which would correspond to the outer layer of shell wall) is explicitly denied by Erben *et al.* (1969b). The second layer might possibly be better identified with the calcareous concretions of the sutural infillings that cannot be considered as belonging to the actual architecture of the septa (Grégoire, 1962). The third layer, composed of nacreous material, makes up the bulk of the septa. The fourth layer, the semiprismatic layer, is limited to a small portion of the septal surface.

6. Septal Surface

6.1. Adapical Convex Surfaces

In the *Nautilus* shell, the surfaces of the chambers are covered with a brown-yellowish organic membrane. This membrane extends over the intracameral siphuncular tube, which it loosely wraps ["brown substance" (Owen, 1832); Edwards, 1849; Woodward, 1880; Barrande, 1857, 1877; Hyatt, 1872; "braune Membran" (Appellöf, 1893)]. In dry shells, this membrane occurs only on the adapical, convex septal surfaces, either in the form of discontinuous, crumpled or curled, brittle, brown, lusterless shreds or as a fused calcified sheet on the underlying mineral background, which then appears as a lusterless, uniformly brown surface (Appellöf, 1893). Between the shreds of the brown membrane, scattered, pyramidal or conical low mounds composed of stacked flat crystals are erected over the septal surface (see Fig. 5d) [Grégoire, 1962, Fig. 42 (TEM)].

After elimination of the brown membrane by ultrasonic irradiation, the adapical septal surfaces appear in the SEM as superimposed flat lamellae (Fig. 5d) in which large, flat, polygonal crystals of aragonite form a continuous flagging. This texture indicates that on the adapical side of the septa, the mineral lamellae are completely developed.

6.2. Adoral Concave Surfaces

The bright, iridescent, concave septal surfaces are composed of a free and a mural part (Foerste and Teichert, 1930; Teichert, 1933). The mural part extends adorally onto the inner surface of the shell wall and progressively thins (Fig. 9). The TEM had previously shown [Grégoire, 1962, Figs. 45 and 46 (confirmed in the SEM by Wise, 1969, 1970)] that the adoral surface of the free part of the septum exhibits low piles of small crystals superimposed on the tabular planes of large, flat, euhedral, hexagonal crystals (001) (010) (110), elongate along their a-axis (see Figs. 5a–c). The same low piles of crystals were seen by Erben [1972a (SEM in adult specimens)]. In immature and juvenile specimens in which the growth of the nacreous layer is still extremely active, the adoral side of each septum is abundantly covered with high pyramidal crystal stacks (see Fig. 4d). This aspect of the adoral side of the septa differs strikingly from that of the adapical side (compare Figs. 4d and 5d). This difference confirms that the growth in septal thickness proceeds in an adoral direction.

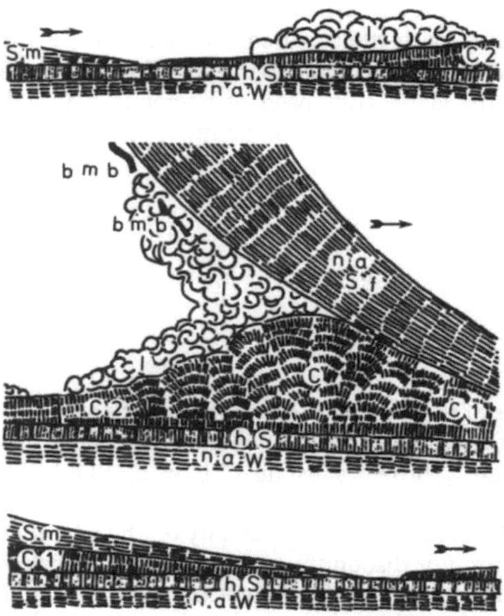

Figure 9. *Nautilus pompilius*. Topography of the sutural substances (see also Grégoire, 1962; text Figs. 4, 5, and 6). Polished and etched longitudinal section of the mural–septal junction of the shell at the last but two adoral septa. The illustrations represent in succession the sutural substances and their continuation in an adoral and adapical direction. Key: (naW) nacreous layer of the shell wall; (naSf) septal nacre; (hS) inner prismatic layer of the shell wall or "helle Schicht"; (C) cements composed of superimposed, festooned layers of palisades of prismatic crystals, interposed between the shell wall and the free part of the septum (Sf). The central part of the sutural plug, triangular in this transverse longitudinal section, extends adapically and adorally in the form of a milky white, linear substance (C1, C2). It can extend over a considerable distance and thereby separate the wedge-shaped adoral end of the mural part of the septum (Sm) from direct contact with the shell wall. The palisades consist of elongate rods, needles, or spindles and alternate with brownish layers of organic substance (in white in the diagram). The cements are covered toward the camerae with yellowish calcareous concretions or infillings (I). These concretions extend over the convex, adapical side of the free part of the septum. The brown membrane (bmb), which coats the convex, adapical side of the septa, intermingles with the calcareous concretions. The disposition of the various layers differs greatly at different junctions. In some adapical junctions, the cements may be reduced to a few palisades. The arrow points in the adoral direction.

7. Organic Components of the Septa

7.1. Nacre

The ultrastructural patterns of the soft, iridescent membranes freed by decalcification of the septa differ from those of the sheets from the mural nacreous layer in possessing more slender trabeculae, tuberosities of smaller size, and a narrower fenestration (see Fig. 7c) (Grégoire, 1962, Figs. 57 and 59). A preferred orientation of the trabeculae along the a-axis of the underlying crystals is recognized in several areas (Grégoire, 1962, Figs. 47 and 80).

7.2. Brown Membrane

The adapical brown membrane is transparent, semirigid, lusterless, and resistant to ultrasonic irradiation. It is composed of microfibrillar feltworks in which microfibrils, single or grouped in fibers, are disposed without definite orientation (Fig. 10a) (Grégoire, 1962, Figs. 60–63; 1973, Fig. 1). The microfibrils seem to be aggregates of microfilaments composed of chains of particles. Irregularly rounded

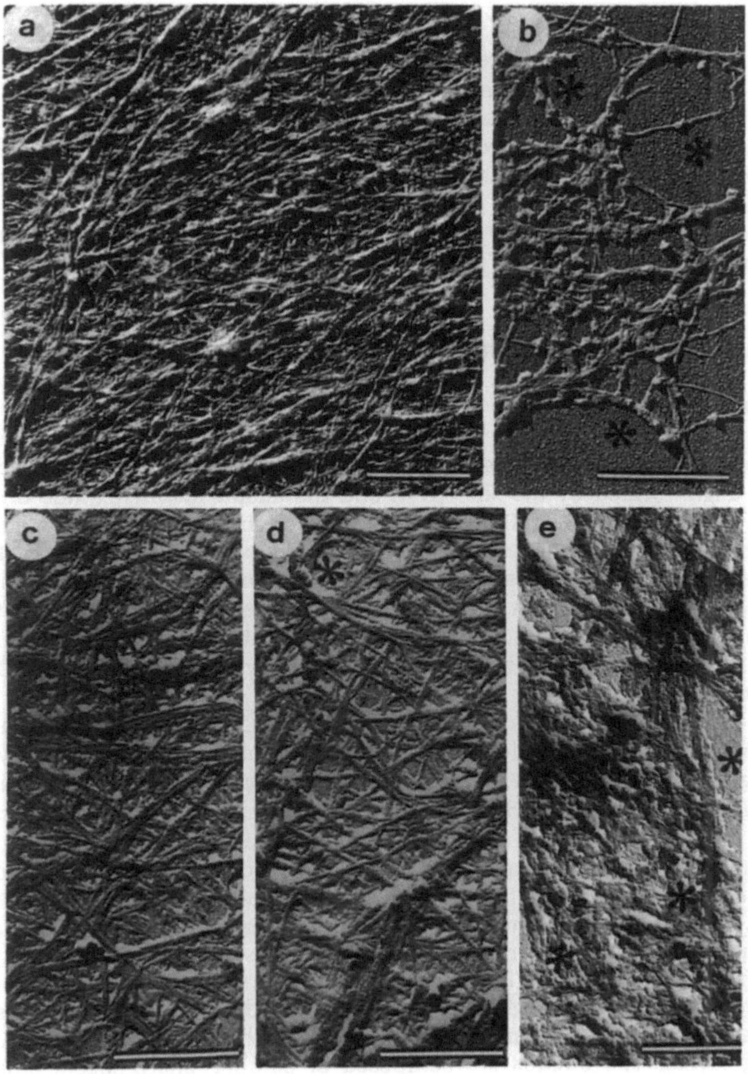

Figure 10. (a) *Nautilus pompilius.* Portion of the brown membrane coating the adapical (convex) side of one of the most adoral septa. The membrane consists essentially of fibrils, scattered or parallel, associated in bundles, forming unoriented feltworks. Irregularly rounded nodules are anchored to the fibrils. Reprinted with permission from Grégoire (1962, Fig. 63). Palladium-shadowed. TEM, reversed print. Scale bar: 0.5 μm. (b) Same material as in (a). The relationships between fibrils and nodules (*) are shown after moderate loosening of the feltworks by ultrasonic irradiation. Stained with osmium tetroxide and palladium-shadowed. TEM, reversed print. Scale bar: 0.5 μm. (c, d) Sexually mature *N. pompilius.* Segment of one of the last connecting rings. Delamination of the concentric sheaths of the horny tube leaves electron-transparent membranes composed of feltworks of unoriented fibers and irregularly parallel, densely packed fibrils, to which nodular material is attached. Stained with osmium tetroxide and platinum-shadowed. TEM, direct print. Scale bar: 0.5 μm. The specimen was a gift of Prof. John M. Arnold. (e) *Nautilus macromphalus* from Nouméa. One of the central concentric sheaths of the horny tube from the last adoral connecting ring (last chamber). Compare with (c) and (d). In this field, the abundant nodular substance (*) conceals the fibers and fibrils to which it is attached. TEM, direct print. Scale bar: 0.5 μm. The specimen was a gift of Dr. R. Catala-Stucki.

nodules [Fig. 10b (*)] are scattered among the fibrils or aligned along them. These nodules disappear after immersion of the brown membrane in hot sodium hydroxide (Grégoire, 1973), whereas the microfibrillar material remains intact. In the protidic saccharidic complex that constitutes the organic substance of mollusk shells, the microfibrillar material seems to represent the chitin and the nodules the protidic fraction.

According to Bayer (1977, p. 327), the fibrillar ultrastructure of the brown membrane, as revealed in the TEM, is ideally resistant to pressure.

8. Sutural Substances: Cements and Infillings

8.1. Cements

A wedge-shaped plug (Figs. 9 and 11a–e), triangular in transverse section, fills the angle formed by the shell wall and the adapical, convex side of each septum [Valenciennes, 1841; Blake, 1882; "dunkle Substanz" on the inner surface of the shell wall (Appellöf, 1893, Plate 11 and Fig. x and x'); Pompeckj, 1912; "erste Kalkausscheidung eines neuen Septums" (M. Schmidt, 1925); "sutural substances" (Grégoire, 1962, Figs. 4–6); "annular supraseptal ridge" (Mutvei, 1964); "Zwickelfülung" (Erben *et al.*, 1969b); "parietaler Stützring und Zwickelfüllung" (Ristedt, 1971); mural ridge, in part "prismatische Zwickelfüllung" (Blind, 1975, 1976, 1980)].

The plug extends in the form of a linear, milky white cement ["X-substance" (Appellöf, 1893, Plate 11 and Fig. 1x; prolongation of the "prismatische Zwickelfüllung" (Blind, 1976)] that is interposed adorally and adapically between the inner prismatic layer ("helle Schicht") of the shell wall and the mural part of the septum. In these regions, the shell wall and septum are not contiguous.

As shown in replicas of polished and etched sections of the phragmocone, examined with TEM and SEM (see Fig. 9) (see also Grégoire, 1962, text Fig. 4 and Figs. 66, 68, and 74), the plug consists of alternating, parallel and undulating cross-striated strands and darker layers, more or less parallel to, or slightly divergent from, the shell wall (Fig. 11a, b, and e). The number of layers in the sutural substances detected in 17 regions of intersection of the shell wall and the septa (Grégoire, 1962, Figs. 4, 5, and 6) varied greatly and sometimes reached 15 alternating layers. The strands are composed of parallel, sharp-edged, elongate needlelike or spindle-shaped imbricate crystals (Fig. 11e) arranged in rows of palisades disposed at right angles to the length of the strands. The darker layers, which alternate with the milky white strands, are composed of an amorphous, probably organic substance. Scattered or piled euhedral or anhedral crystals are embedded at random in this amorphous substance (Grégoire, 1962, text Fig. 6).

The prismatic texture of the plug observed with the TEM ["palisades of elongate crystals" (Grégoire, 1962, Fig. 68)] has also been found by Mutvei [1964 (polarizing microscope)] and Blind [1976, 1980 (thin sections)]. The plug has been identified with a considerable thickening of the inner prismatic layer of the shell wall, on which a new septum is formed. However, the identity and significance of the organic strands that alternate with the prismatic palisades, which are visible only with TEM, are not yet understood.

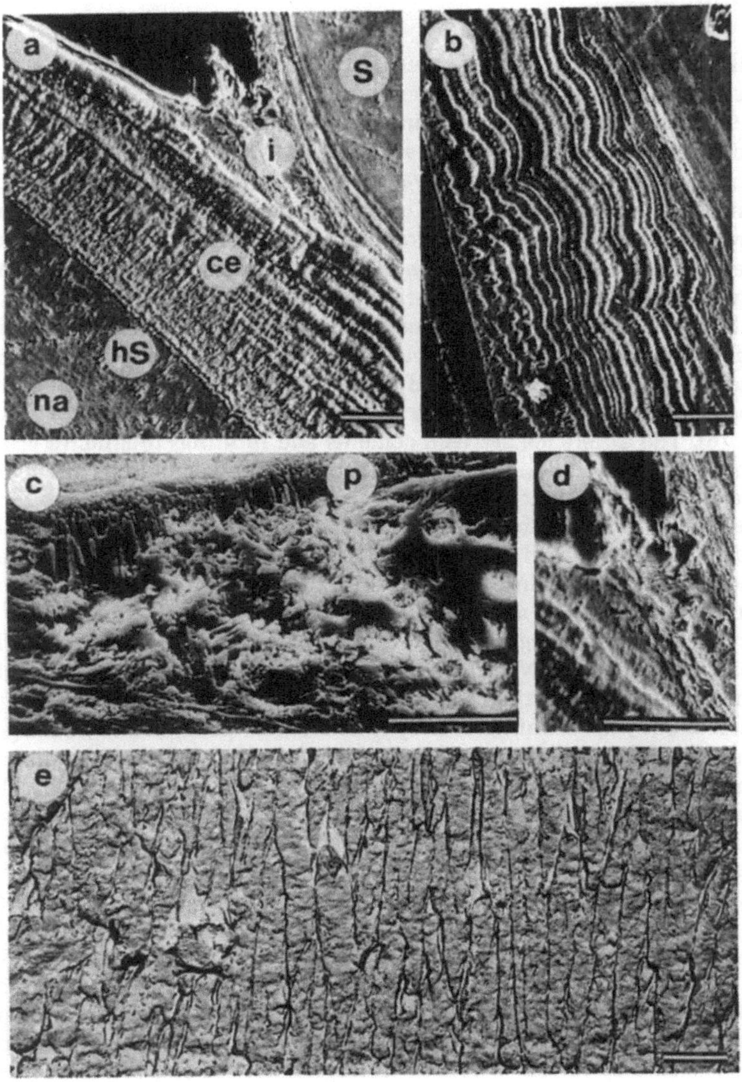

Figure 11. *Nautilus pompilius.* (a) Polished and etched longitudinal section of the phragmocone, showing the sutural substances at one of the last adoral mural–septal junctions (see Fig. 9). Key: (na) nacreous layer of the shell wall; (hS) inner prismatic layer or "helle Schicht"; (ce) cements, composed of a varying number of parallel palisades of prismatic crystals; (i) infillings located between the cements and the convex adapical side of the free septum (S), onto which the infillings extend. SEM. Scale bar: 50 μm. (b) Undulating stratification of the cement palisades is shown in another sutural plug. The organic substance that separates the palisades appears in black in the figure. SEM. Scale bar: 50 μm. (c, d) Calcareous concretions form the infillings (see Grégoire, 1962, TEM, Figs. 66, 67, 69, and 70). Above (p) is a row of prisms of the adjacent cement. SEM. Scale bars: 50 μm. (e) Double–stage (plastic–palladium) replica of a polished and etched transverse section of a cement, showing parallel, elongate crystals forming the palisades. TEM, direct print. Scale bar: 0.5 μm. Reprinted with permission from Grégoire (1962, Fig. 68).

8.2. Infillings

Yellowish, mulberrylike calcareous concretions or infillings [chitinous "Ausfüllungssubstanz des vorderen Kammerwinkels" (Appellöf, 1893); infillings (Grégoire, 1962); "Zwickelfüllung" (Erben *et al.*, 1969b); Ristedt, 1971] overlap the plugs toward the chambers (Figs. 9 and 11c and d). These infillings form protruding ridges or crests that extend over the septal surfaces and over the adjoining mural surfaces, where they overlap the mural parts of the preceding septa. In these areas, they form globular concretions and rounded elevations that locally produce a mulberrylike appearance. In TEM, the ultrastructure of these calcareous infillings is highly heterogeneous. The ultrastructure differs from that of the milky white plugs and linear cements and includes: elongate tablets and blades oriented in bundles; scattered, rounded, pebblelike particles; thinly stratified calcareous substances that form whirlpools; and dense, concentrically festooned stratifications resembling those of chalcedony or agate textures (Grégoire, 1962, Figs. 53, 69, and 76). The difference between cements and infillings (distinctly visible in Appellöf's Plate 11 and Fig. 1) suggests that the infillings should be distinguished from the cements and their extensions.

The configuration of the sutural cements in *Nautilus*, observed in TEM, resembles the texture of fossil cameral deposits. The cameral deposits are calcareous substances developed against the shell wall and septa of the adapical chambers in several groups of fossil nautiloids. As in fossil material [Pennsylvanian *Pseudorthoceras knoxense* (Grégoire, 1962: compare Figs. 6 and 8; Grégoire and Teichert, 1965)], the cement in *Nautilus* is composed of parallel, elongate crystals disposed in palisades alternating with strands of unknown substance. Blake (1882, p. 35) had already noted this similarity in nature and position and had suggested that *Nautilus* infillings and cameral deposits in fossils might have a similar origin. The cements and infillings in *Nautilus* might be the morphological equivalent of the cameral deposits in fossil nautiloids, reduced in *Nautilus* to a wedge-shaped material clogging the angle of intersection of the shell wall and each septum. However, the sutural substances in the *Nautilus* shell fill all the angles of intersection, including those of the ultimate chamber, whereas in fossil phragmocones, several adoral chambers are completely free of these structures (Flower, 1955). Blake's suggestion (see also Grégoire, 1962) was found unrealistic by Ristedt (1971), because the sutural substances are separated from the chambers by the brown membrane. However, as noted by Appellöf (1893), in most of the samples of mural and septal junctions in *Nautilus*, the brown membrane has disappeared from the surfaces by fusion with the infillings, a pattern also found in fossils.

8.3. Organic Components

Appellöf (1893) reported the presence of chitin filaments in the decalcified "Ausfüllungsmasse." In TEM, the organic material of the sutural substances consists of veils, associated with microfibrils and sturdier fibers (Grégoire, 1962, Figs. 78, 79, and 82). These structures might be either fragments of the adjacent brown membrane intermingled with the infillings [see Fig. 9 (bmb)] or possibly remains of soft tissues previously located in the chambers and subsequently trapped and embedded in the concretions.

9. Siphon: Structure and Organic Components

In the *Nautilus* shell, the siphon consists of a series of gray-brown, tubular segments (connecting rings), extending adorally through the successive air chambers from septum to septum, where the septal neck or funnel constitutes the hard part of the siphon (Fig. 12) (Brooks, 1888).

The wall of these tubular segments within a chamber is loosely wrapped in extensions of the brown membrane and is composed of two concentric layers

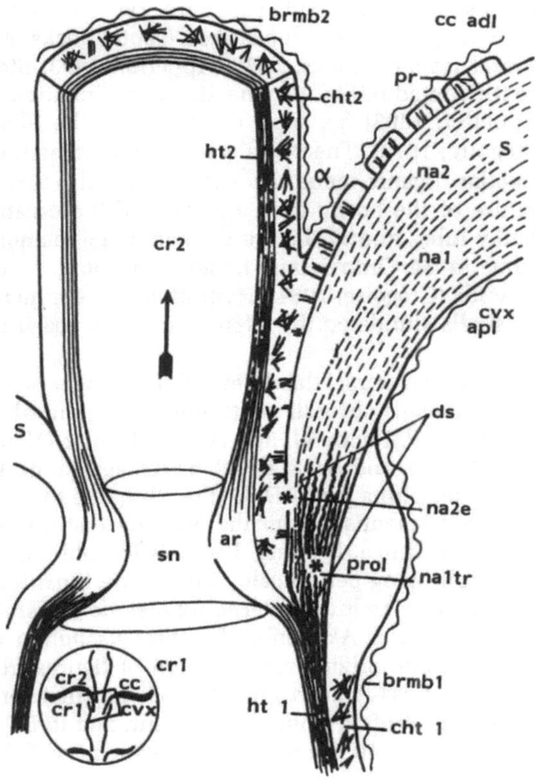

Figure 12. *Nautilus* sp. septal neck and adjacent connecting rings. The diagram is a composite of Brooks (1888, Plate 2), Appellöf (1893, Plate 12 and Fig. 1), Mutvei (1972, Fig. 2), and Grégoire (1973). The adapical end of a connecting ring (cr2), composed of an outer chalky (cht2) and an inner horny (ht2) tube, invaginates (Barrande 1877; Appellöf 1893) into the septal funnel formed by the adapical ends of the nacreous septum (S, na1; na2). (See also the insert.) The chalky tube [cht2 (unoriented bundles of crystallites)] of this connecting ring is fused in the septal neck with the adapical extension of the discontinuous prismatic layer [pr (palisades of crystallites)] that is superimposed on the septal mother-of-pearl (na2) of the adoral (concave) side (cc adl) of the free septum. In the narrowest part of the septal neck (sn), this calcareous layer is covered centrally with a short, ring-shaped calcareous thickening [auxiliary ridge (ar) (Mutvei 1972)]. Through the prolongation of its chalky tube (cht1, prol), the adoral (anterior) end of the preceding adapical connecting ring (cr1, ht1, cht1) (see also the insert) embraces the outer part of the septal funnel. The brown membrane [brmb1, brmb2 (represented schematically in the form of an undulating line)] coats all the septal surfaces, is especially abundant on the convex sides of the septa (cvx apl), and is discontinuous on the concave surfaces (cc adl). Owing to the double curvature of the free part of the septa contiguous to the septal necks, their adapical (convex) surface (cvx apl) and adoral (concave) surface (cc adl) appear on the diagram concave and convex, respectively. Additional key: (ds) dark substance of the septal neck ["dunkle Substanz der Sifonaldüte" (Appellöf, 1893, Plate 12, Fig. 1, ds)]; The arrow points in the adoral direction. Reprinted with permission from the Archives Internationales de Physiologie et de Biochimie.

(Owen, 1832; Valenciennes, 1841; Barrande, 1877; Brooks, 1888; Appellöf, 1893; Mutvei, 1964, 1972; Stenzel, 1964; Denton and Gilpin-Brown, 1966). The first consists of an outer, aragonitic, porous sheath ["chalky-tube" (Denton and Gilpin-Brown, 1966)] composed of unoriented aragonitic, spindle-shaped spicules, or bundles of slender crystallites arranged in stellate figures, and held together by an organic substance. The second layer consists of an inner, mahogany-brown, organic tube ["horny tube" (Buckland, 1837); Denton and Gilpin-Brown, 1966; "couche brune de chitine" (Barrande, 1877); "inner conchiolin tube" (Brooks, 1888); "unverkalkte innere Chitinmembran" (Appellöf, 1893); "inner conchiolin layer" (Mutvei, 1964, 1972)] that is encased in the chalky tube and is composed of a great number of concentric sheaths of organic substance. In the TEM, these sheaths consist of dense microfibrillar feltworks clustered in bundles that form fibers (Grégoire, 1967, 1972c; Mutvei, 1972, Fig. 12c, d, and e) and onto which amorphous and nodular substances adhere. These substances may be dissolved in hot sodium hydroxide (Grégoire, 1967) with no effect on the fibrillar material.

The hard parts of the siphon (septal necks or septal funnels) are made up of several layers (Brooks, 1888; Appellöf, 1893; Stenzel, 1964; Mutvei, 1964) that are illustrated in Fig. 12. The direct continuation of the horny tube into the nacre (Appellöf, 1893) has been confirmed in the SEM (Mutvei, 1972) and the TEM (Grégoire, 1973). There is a gradual transformation of the trabeculae of the interlamellar organic membranes in the septal nacre into the coarse fibrous organic texture of the horny tube. Remains of the organic structures of the horny and chalky tubes and of the brown membrane, namely fibrillar feltworks, have been found in 34 Ordovician-Pliocene nautiloids and 45 Paleozoic-Cretaceous ammonoids (Grégoire, 1984). Hydrothermal treatment of these structures from modern shells simulated the alterations in the fossil material.

ACKNOWLEDGMENTS. I thank several persons for providing fresh material of *Nautilus* and specimens of fossil nautiloids: Prof. John M. Arnold (Hawaii), Dr. R. Catala-Stucki, Dr. P. Rancurel, and Dr. Y. Magnier (Nouméa), Prof. W. Bruce Saunders (Bryn Mawr), and Dr. Mary Wade (Queensland). Dr. Wade was able to bring material from China [bed KD 60 (Fig. 6b)], due to the kindness of Dr. Jun Yuan Chen and other scientists from the Nanjing Institute of Geology and Palaeontology, Academia Sinica, Nanjing, People's Republic of China. I also thank the editors of the following periodicals: *Bulletin de l'Institut Royal des Sciences naturelles de Belgique*, *Archives internationales de Physiologie et de Biochimie*, and *Abhandlungen der Akademie der Wissenschaften und der Literatur*, Mainz, Germany, for authorizing reproduction of figures from former papers. I also thank Dr. M. F. Voss-Foucart and Prof. G. Goffinet for reading the manuscript. I am thankful to Prof. P. Coheur, President, and to Dr. Decker, Director, for their hospitality during 40 years in the Department of Electronmicroscopy of the C.R.M.

IX

Swimming and Buoyancy

Chapter 32

Locomotion of *Nautilus*

JOHN A. CHAMBERLAIN, JR.

1. Introduction

As an agent of rapid transportation, jet propulsion has no significant competitor. Although technologically it is but a latecomer, biologically jet propulsion is of great antiquity, probably having originated in the late Precambrian (≈700 Ma) with the advent of medusoid cnidarians and animals of similar grade.

A wide variety of aquatic organisms propel themselves by jet action (Table I) (for general reviews, see also E. R. Trueman, 1975, 1980), but none is more renowned than the cephalopods. Interest in cephalopod locomotion has centered primarily on squids, undoubtedly because they are common, tolerant of experimentation, and appear to represent the acme of cephalopod locomotory design. Considerable insight has been gained into squid locomotory mechanics and physiology through the admirable efforts of Trueman and his colleagues (E. R. Trueman and Packard, 1968; W. Johnson *et al.*, 1972; Siekmann, 1963; Trueman, 1975), which have focused for the most part on muscle innervation and operation, development of thrust and power, and jet velocity.

Other cephalopods have not received comparable treatment; this is particularly true of the only living ectocochleate cephalopod, *Nautilus*. However, *Nautilus* promises rich rewards, in part because initial reports indicate that *Nautilus* differs markedly from squids in locomotory design and performance (Chamber-

JOHN A. CHAMBERLAIN, JR. • Department of Geology, Brooklyn College of the City University of New York, Brooklyn, New York 11210; Osborn Laboratories of Marine Sciences, New York Aquarium, New York Zoological Society, Brooklyn, New York 11224.

Table I. Representative List of Jet-Propelled Swimmers

Organism	Group	References
Scallops, cardiids	Pelecypods	S. J. Gould (1971), Stanley (1970), Moore and Trueman (1971)
Dolidium	Tunicates	Bone and Trueman (1984), Mackie and Bone (1977)
Sepiids, octopods, squids, *Nautilus*	Cephalopods	E. R. Trueman and Packard (1968), W. Johnson *et al.* (1972), Packard *et al.* (1980)
Aeschna	Dragonflies	G. Hughes and Mill (1966), Mill and Pickard (1975)
Scyphozoans (*Cassiopeia*, *Cyanea*), hydrozoans (*Polyorchis*)	Cnidaria	Gladfelter (1972a,b), Daniel (1983)
Notarchus	Gastropods	R. Martin (1966)
Pike (*Esox*)	Fish	Lagler *et al.* (1977)
Abylopsis, *Chelophyes*	Siphonophores	Bone and Trueman (1982)

lain, 1980a, 1981; Packard *et al.*, 1980). This difference is important because, as S. J. Gould (1971, p. 61) noted in his study of muscle allometry in jet-propelled scallops, "It is not only our intrinsic fascination for peculiarity that motivates the study of unusual adaptations," but that "these adaptations are experiments that test the limits of form." Moreover, *Nautilus* is the only modern analogue of the myriad ammonoids and nautiloids that populated ancient seas; it thus constitutes a prime means of elucidating aspects of cephalopod paleobiology, such as mode of life, that are related to locomotion.

In this chapter, I will examine previous observations pertaining to the locomotion of *Nautilus* and will present new data on the swimming performance of this unique invertebrate. My main goal is to provide a body of information that will serve to strengthen our understanding of the cephalopod locomotory system and, at the same time, provide a basis for interpreting cephalopod paleobiology. The focus of this analysis lies in several areas: drag production, equilibrium control, locomotory mechanism, swimming movements, and propulsive performance.

2. Drag

Knowledge of the drag forces generated by a swimming *Nautilus* and of the role played by such forces in determining locomotory performance derives from work on *Nautilus*-like fossil cephalopods (H. Schmidt, 1930; Kummel and Lloyd, 1955; Chamberlain, 1976, 1980b, 1981; Chamberlain and Weaver, 1978; Chamberlain and Westermann, 1976; J. S. Weaver and Chamberlain, 1976). The drag acting on *Nautilus* can be thought of as consisting of several components: (1) the drag produced by the shell, (2) the effects on drag of the exposed body, and (3) drag resulting from velocity changes and water entrainment during movement. The total drag force acting on an animal at a specified time is determined by a complex and time-variable interplay of these factors.

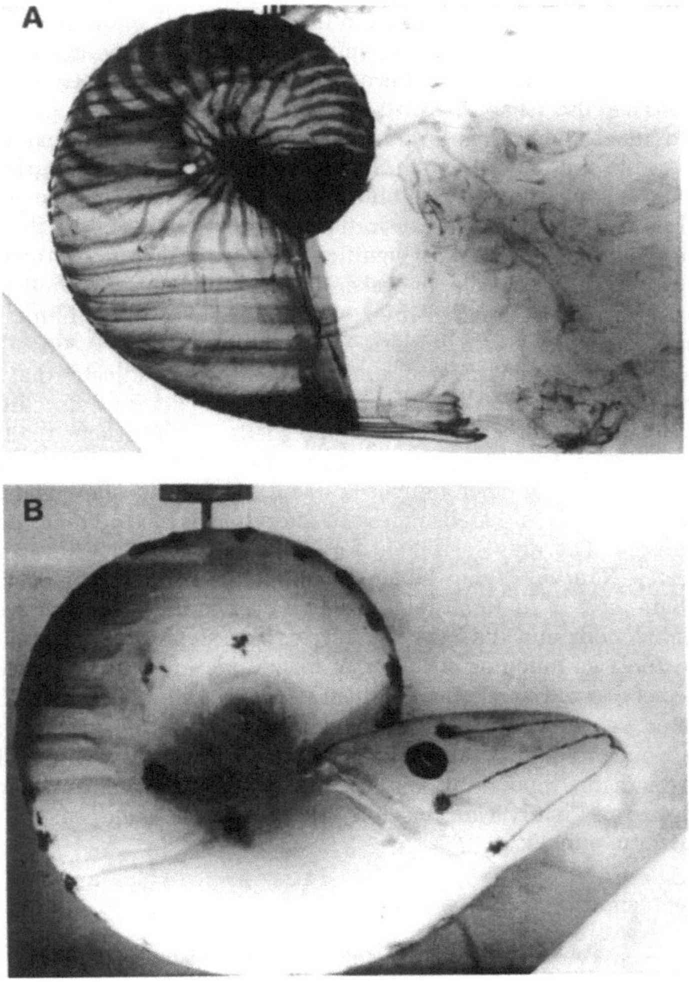

Figure 1. Flow structure for *N. pompilius*. Flow lines are made visible by dye bleeding from crystals of potassium permanganate cemented to the venter. The flow velocity is 10 cm/sec. (A) Shell alone. After Chamberlain (1976). (B) Shell model with body prosthesis. Reprinted with permission from Chamberlain (1980b).

2.1. Shell Drag

Water moves smoothly across the venter and flanks of a moving shell, as shown experimentally by the flow of laminar streams of dye (Fig. 1A). The concentration of dye along the umbilical shoulder indicates that, as the flow reaches this part of the shell, it stalls and separates from the shell surface. Evidently, the shell boundary layer cannot negotiate the sharp inward curve of the shell surface that occurs in the umbilicus. As a result, the umbilicus is filled with essentially stagnant water. This means that variation in umbilical configurations among different species of *Nautilus* (e.g., the open umbilicus of *N. macromphalus* and *N. scrobiculatus* and the umbilical callus of *N. pompilius*) has no real hydrodynamic significance (see also Chapter 33). It also probably explains the propensity of epizoans to attach themselves to these regions of the shell (see Chapter 10). In the region of the shell aperture, the flow remains attached to the surface of the shell until it is forced to separate abruptly at the peristome, where it quickly degenerates into a series of unsteady vortices, visible as swirls of dye downstream behind the shell. Perhaps the most significant feature of this flow structure is the extensive region of turbulence, or wake, developed behind the shell (Fig. 1A), which indicates that in terms of drag, *Nautilus* behaves like a blunt, rounded object, such as a sphere or cylinder. Even though the shell is smooth, gently curved, and to some extent tapered, the prominent wake indicates that *Nautilus* is not as well streamlined as fusiform fish or squids. This flow structure is generally characteristic of all coiled cephalopod shells (Chamberlain, 1976).

Measurement of shell drag force confirms the magnitude of the drag of *Nautilus* compared to that of other animals (Table II). These data show that gyrocone and oxycone ectocochleate shells represent the extremes in shell drag coefficients so far measured. The drag coefficient for *Nautilus* falls in the lower range for ectocochleates. *Nautilus* is thus better streamlined, and hydrodynamically more efficient, than most of its fossil relatives. However, comparing *Nautilus* to soft-bodied cephalopods and fusiform fish reveals that these other swimmers have drag coefficients as much as an order of magnitude lower than *Nautilus*. The significance of this difference can be illustrated with a simple example: To travel at a specified velocity, *Nautilus* will need to expend about ten times as much effort as a trout or squid of equal size; conversely, *Nautilus* will travel at only about one tenth the velocity of squids and fish for equivalent expenditure of propulsive energy. Clearly, *Nautilus* does not belong to the same category of swimmers as fusiform animals.

Table II. Drag Coefficients for Various Swimmers

Organism	C_D[a]	Ref. nos.[b]
Gyroconic cephalopod shell	1.1	1
Nautilus	0.48	1
Compressed, oxyconic cephalopod shell	0.25	1
Fusiform fish (trout)	0.05–0.01	2, 3
Squid		
Dosidicus	0.1–0.05	4
Loligo	0.1–0.05	5

[a] Figures for squid are estimates only. Drag coefficients of ectocochleates are based on (shell volume)$^{2/3}$, whereas drag coefficients for squid and fish are based on plan area. This discrepancy does not induce significant error, however, because plan area and (volume)$^{2/3}$ are roughly equivalent in coiled cephalopod shells.

[b] References: (1) Chamberlain (1976); (2) Hoerner (1965); (3) Hertel (1966); (4) Cole and Gilbert (1970); (5) Trueman and Packard (1968).

2.1.1. Effect of Shell Size

The value of 0.48 given in Table II for the drag coefficient of *Nautilus* varies little (±10%) over the velocity range (30–250 cm/sec) for which I measured drag force. This limited variation results from the occurrence of laminar separated flow (as in Fig. 1A) for these velocities. Since Reynolds numbers for the full size spectrum of *Nautilus* shells can be estimated to have values characteristic of laminar separated flow, it is probable that a drag coefficient of about 0.48 holds throughout ontogeny.

2.1.2. Effect of Shell Orientation

Variation in orientation relative to direction of movement can produce wide fluctuations in drag coefficient. Thus, a flat plate held parallel to flow and perpendicular to flow has radically different drag coefficient values. Configurational change in *Nautilus* is not of the magnitude cited for a plate, but it does vary with the rocking of the shell that accompanies swimming and with swimming backward or forward. I shall deal with the former subject in this section, because the latter involves chiefly body extension, which is treated in Section 2.2.

Bidder (1962), Chamberlain (1976, 1980a, 1981), Packard *et al.* (1980), and many others have described the periodic rotation or rocking of the shell during locomotion. This motion occurs around a horizontal axis, orthogonal to the direction of travel; it is thus actually a pitching movement. The angular displacement of this motion is up to ±15° from the attitude of the shell at rest (Chamberlain, 1980a, 1981).

The effect of shell rotation on drag coefficient for *Nautilus* can be estimated from Fig. 2. The data presented in the figure pertain to a shell similar to *Nautilus* in overall aspect, but having a wider, more circular cross section. This difference in shell geometry produces drag coefficients (≈0.6) somewhat higher than for *Nautilus*. However, both *Nautilus* and this depressed shell have similar whorl expansion rates and relatively large apertures, so that their response to change in orientation will be equivalent.

Varying shell orientation from 0 to 90° has the effect of bringing the aperture in line with the flow direction (left to right in Fig. 2). As a result, the drag contribution of the blunt aperture is minimized, and drag coefficient declines. Because *Nautilus* also has a large aperture, we can expect the drag coefficient changes with orientation to be similar, although they will involve lower absolute C_D values.

In typical rest attitude, the aperture of *Nautilus* is oriented about 30° from vertical. This means that in rotating ±15°, the shell may rotate between 15 and 45° (Fig. 2), the lower limit being approached during the power phase of the propulsive cycle, when water is ejected from the mantle cavity (P in Fig. 2), and the upper limit being approached during the recovery phase, when water is readmitted into the mantle cavity (R in Fig. 2). Such movements will produce an insignificant change in drag coefficient (0.62 ± 0.02 or ±3.5%) in the test shell. A similar situation undoubtedly holds for *Nautilus*: The rocking motion during propulsion is not likely to alter the drag coefficient to any meaningful extent.

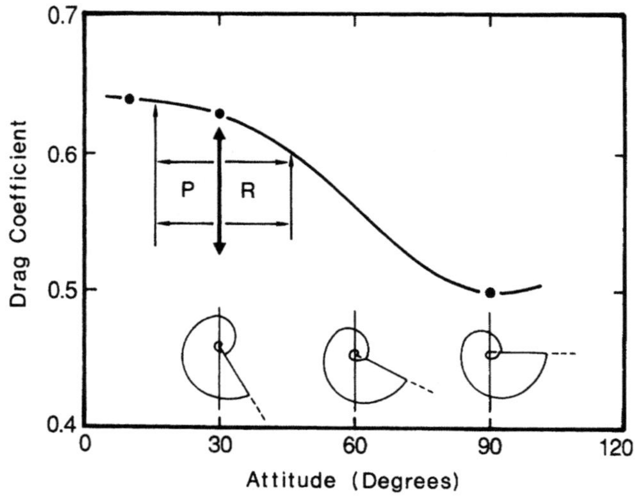

Figure 2. Effect of shell orientation on drag coefficient for a depressed shell geometrically similar to *Nautilus*. Orientation is defined as the angle between peristome and vertical. Each point represents the average drag coefficient of 20–40 individual drag and velocity measurements. The curve was fit by eye. The bold double-headed arrow indicates the approximate attitude of *Nautilus* when at rest; the light single-headed arrows indicate the approximate angular limits of the pitch rotation of the shell that occurs during swimming. Key: (P) forward deflection produced during power phase of propulsive cycle; (R) backward deflection produced during recovery phase of propulsive cycle. After Chamberlain (1976).

2.2. Body Drag

During normal activity, *Nautilus* extends parts of its body outward from the aperture. This extended portion can interact with ambient flow and thus contribute to the overall drag produced by the animal. Two such extensile behaviors are worth considering: (1) exposure of the body behind the aperture and (2) deployment of the digital tentacles in the form of a search web.

2.2.1. Body Exposure

Whether *Nautilus* is at rest or swimming, the cephalic region of the body is always extended to some degree. In the animals I have studied, backward swimming is accompanied by posturing the body so that the eye is completely exposed, its lower rim just clearing the peristome. In this position, the exposed body assumes the shape of an irregular tapering cone with the tentacles retracted into their sheaths. This posture appears to be the customary one when the animal is swimming at peak levels.

Chamberlain (1980b) has approximated the drag effects of this posture by using two shells with attached body prostheses. The main effect of body extension is to reduce vorticity by shifting flow separation to the trailing surfaces of the prosthesis (see Fig. 1B). The effect of this shift on drag coefficient is illustrated in Table III. In the two models examined, drag coefficient was reduced by about

Table III. Effect of Body Exposure on Drag Coefficient for Coiled Cephalopod Shells[a]

Test configuration[b]	C_D	Decrease in C_D (%)
C	0.26	—
C + P	0.23	12
D	0.71	—
D + P	0.64	10

[a] From Chamberlain (1980b). Data are based on drag determinations from two plastic shell models similar to *Nautilus*. The first (Fig. 1B) is more compressed, whereas the second is more depressed.
[b] Test configurations: (C) compressed shell model alone; (C + P) compressed shell model and body prosthesis; (D) depressed shell model alone; (D + P) depressed shell model and body prosthesis.

10%. Because the shell geometry of *Nautilus* is intermediate between those of the two models, and because the size and shape of the artificial bodies are similar to those of *Nautilus*, it seems reasonable to conclude that body extension in *Nautilus* produces a similar reduction in drag coefficient and would generate about a 10% gain in energy economy (C_D = 0.43) compared to the shell alone (C_D = 0.48). Although this gain may be important in the energy balance of a swimming *Nautilus*, it is by no means sufficient to bring *Nautilus* in line with the drag coefficients of fusiform fish and squids.

The arguments detailed in the preceding paragraphs apply to the curious cephalopod habit of swimming backward; in *Nautilus*, this mode of locomotion is accomplished with the exposed body trailing the shell. *Nautilus* also swims forward, i.e., with the body leading the shell. Drag measurements with the shell and body held in this position have not been made, but it is evident from Fig. 3 that only minor differences exist between the two swimming models in the con-

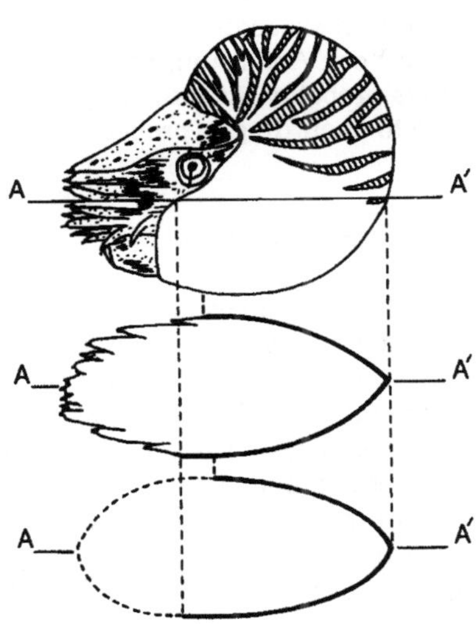

Figure 3. Effect of body extension on configuration presented to flow. *Top:* lateral view of swimming *Nautilus* showing where line of profile (A–A′) in middle is taken. *Middle:* cross section through shell and body along A–A′. *Bottom:* generalized shape of body and shell; configuration is roughly equivalent for both backward swimming (left-to-right) and forward movement (right-to-left).

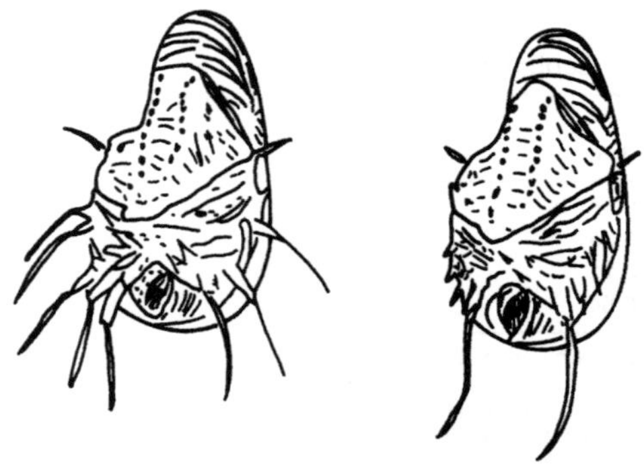

Figure 4. Common modes of tentacle extension exhibited by swimming *Nautilus*. *Left:* search web, in which 10–20 fully extended tentacles radiate into ambient flow. *Right:* two trailing digital tentacles extended downward to maintain tactile contact with substrate.

figuration presented to the flow. As a result, it seems likely that the drag coefficient for forward swimming would be similar to that for the backward orientation, or perhaps slightly higher in view of the somewhat blunter profile of the body and of the irregularities produced by the tentacle sheaths (Fig. 3).

2.2.2. Tentacle Extension

The foregoing analysis presumes retracted tentacles (Fig. 3), but *Nautilus* also swims with its tentacles extended (Fig. 4). When the tentacles are extended, they impinge on the ambient flow, creating their own drag, which increases the total drag.

I have evaluated tentacle drag (Chamberlain, 1980b) by applying the analyses by G. Taylor (1952) and Hoerner (1965) of the drag of cylinders inclined at various angles to the flow. The results show that tentacle drag is considerable and that it is a significant component of total drag (Fig. 5). A search web formed by the extension of 10–20 tentacles produces an increase of about 50% in drag coefficient and hence an equivalent decline in efficiency. In the energetic sense, a search web is an exceedingly costly device. It is restricted, in animals maintained at the New York Aquarium, to slow-moving quests for food. Because tentacle dimensions probably scale directly with shell size, the magnitude of tentacle drag relative to shell drag is probably independent of size.

2.3. Inertial Force

Previous analyses of cephalopod locomotion, and indeed of that of other animals as well, have emphasized the steady-state forces of drag and lift. But movement in cephalopods is by no means steady. Instead, the animals repeatedly accelerate and decelerate, as the pulsing of the propulsive organs periodically supplies thrust. The total hydrodynamic force generated in such situations consists of components that derive from both steady and nonsteady sources.

In the absence of quantitative work on the nonsteady aspects of locomotion in *Nautilus*, some headway can nevertheless be gained by turning to the more general efforts of Wiegel (1964), Batchelor (1967), and Daniel (1984). Following

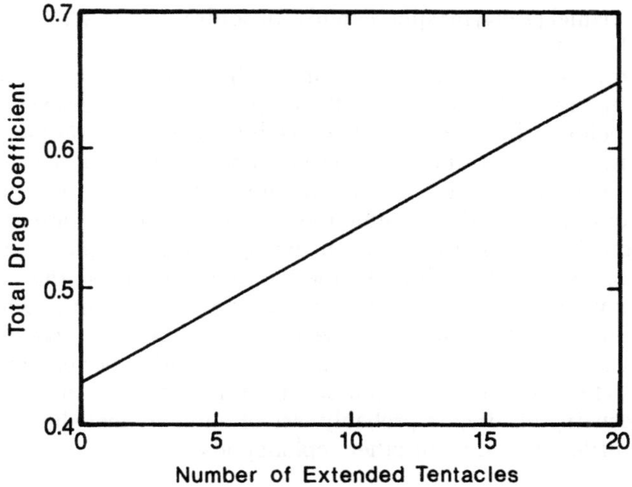

Figure 5. Effect of extended tentacles on drag coefficient for an adult *Nautilus*. With no tentacles extended, the drag coefficient value is that for shell and body. The drag coefficients were calculated following the procedure of Chamberlain (1980b) and based on adult *N. pompilius* in the New York Aquarium, which generally have tentacles approximately 9 cm by 0.3 cm.

arguments given by these authors, the total instantaneous force (F_T) acting on a swimming *Nautilus* can be represented as the sum of the steady-state drag force and a force stemming from change in the animal's inertia resulting from its acceleration:

$$F_T = F_D + F_i$$

where F_D is drag and F_i is the inertial force. When an animal is accelerating, during the active phase of its propulsive cycle, F_i is positive and augments drag. When an animal is decelerating, during the recovery phase of propulsion, F_i is negative and offsets drag. The drag force can be written in the usual way:

$$F_D = \tfrac{1}{2} C_D \rho A v^2$$

where C_D is the drag coefficient, ρ is fluid density, A is an area representative of the animal [I have used $A = (\text{volume})^{2/3}$ for cephalopods], and v is velocity. F_i is defined as

$$F_i = C_m \rho V a$$

where C_m is the added-mass coefficient, V is the volume of the animal, and a is acceleration ($a = dv/dt$). The term $C_m \rho V$ represents the "added mass" of the system, i.e., the mass of the fluid that is accelerated as the animal accelerates. C_m is a dimensionless number that describes the effect of body shape on added mass.

Data given by Daniel (1984) for spheres and ellipsoids indicate that C_m for *Nautilus* is probably about 0.75.

For jet propulsors like *Nautilus*, neglect of inertial forces overestimates performance. Using figures for duration of the propulsive pulse [Packard *et al.* (1980) and reported below], and following the procedure of Daniel (1984) for evaluating the relative importance of drag and inertial forces, I calculate that for an adult *Nautilus* accelerating from rest in a single pulse, the inertial force accounts for the bulk (≈60%) of the hydrodynamic force elicited. Yet it is important to note that for an animal swimming at a constant average velocity (although accelerating and decelerating with each pulse), the inertial forces cancel, so that average performance is governed solely by drag force. However, the extent to which instantaneous velocity departs from average velocity is influenced largely by inertial forces. Because *Nautilus* relies most heavily on extended periods of swimming to find food rather than on the rapid acceleration characteristic of squids, the practice of focusing on drag is probably less of an analytical shortcoming with respect to *Nautilus* than it is for other cephalopods.

3. Equilibrium

The ability of an animal to control its equilibrium during movement is clearly essential to any form of locomotion. Methods employed by aquatic animals for equilibrium maintenance vary considerably; that used by *Nautilus* is unique among modern swimmers, although it is typical of the equilibrium mechanism of fossil ectocochleates.

Buoyancy is the key to equilibrium in *Nautilus*, which, like most other swimmers, maintains approximately neutral buoyancy during locomotion. Neutral buoyancy benefits nektonic animals, because it enables them to maintain their position in the water column without the energetic burden of muscular activity. The means by which *Nautilus* regulates its buoyancy contrasts markedly with those of other swimmers and accounts for the strong disparities in equilibrium control noted above.

The great radiation of Cambrian–Ordovician nautiloids, which established cephalopods as the earliest successful nektonic animals of advance anatomical grade, was brought about by the development of a reliable means of buoyancy control (Teichert, 1967). The innovation that set the stage for the cephalopod expansion was the camerate shell—the structure that is diagnostic of these animals and is still the hallmark of *Nautilus* today. As a buoyancy device, the camerate cephalopod shell performs admirably. The gas-filled phragmocone offsets the deadweight of the animal's body and shell; the epithelial cells of the siphuncle form an osmotic fluid-transport system that provides for sensitive, albeit relatively slow buoyancy adjustments; and the strength and rigidity of the shell preclude the need to equilibrate internal gas pressure with ambient hydrostatic pressure. These properties have not, however, been gained without cost: Ectocochleate equilibrium mechanisms are based on a large, heavy shell.

Because the shell must provide mechanical strength necessary to withstand hydrostatic pressures encountered at normal living depths (up to 80 bars), it is robustly constructed. As a consequence, it is also quite heavy and generally comprises about one third of the total weight of the animal in air. Because the effective density of shell carbonate in seawater is about 1.6, whereas the effective density of the flesh is only about 0.04, the total deadweight (weight in water) that must be compensated by shell buoyancy amounts to about 25% of total air weight. Hence, to accomplish its prime function of generating the uplift necessary for neutral buoyancy, the phragmocone must comprise at least one fourth of an animal's total volume. If any of the camerae contain liquid, as is generally the case in *Nautilus* during ontogeny, the size of the phragmocone must enlarge correspondingly. Ensuring compatibility among body and shell weight, phragmocone volume, and neutral buoyancy has undoubtedly been the primary consideration governing the development of a practicable equilibrium system in shelled cephalopods.

It is worth noting that the shell comprises about 90% of the total deadweight. Thus, the phragmocone serves chiefly to buoy the buoyancy apparatus; relatively little cameral space actually compensates the deadweight of the animal's body. The camerate shell—the structure that impelled ectocochleate cephalopods to their striking initial success—also burdened them with a buoyancy mechanism exceedingly wasteful of space (compared, for example, to the gas bladder of fish). There is perhaps no better example to be found among invertebrates of the constraining nature of bauplan than this.

The force buoying a *Nautilus* acts upward from the center of buoyancy, i.e., from the center of mass of the water displaced by the animal. Weight acts downward through the center of mass. Since weight is concentrated in the body chamber, whereas buoyant force is evenly distributed throughout the displacement volume, the two centers do not coincide. Instead, they are separated by a distance of about 0.6 cm in adult *N. macromphalus* (Denton and Gilpin-Brown, 1966). Because shell geometry of the several species of *Nautilus* is not greatly different, and because gnomonic growth is a strict ontogenetic rule in *Nautilus*, the distance between these two points will, in general, scale directly with some linear dimension of the shell (e.g., diameter or radius). Hence, the distance that separates the two centers in any *Nautilus* shell can be approximated from Denton and Gilpin-Brown's data or directly calculated (see Saunders and Shapiro, 1986) (see also Chapter 33).

Buoyancy and weight, acting vertically through their respective centers, set up a rotational couple when they are not precisely aligned (Fig. 6A and B). The effect of this couple is to restore the animal to a position in which the lines of action of the two forces coincide. This position is the animal's equilibrium position, to which it will always return if displaced in some way. The magnitude of the restoring, or stabilizing, moment is a function of the animal's weight and size (which determine the distance between the centers of buoyancy and mass) and the angle of rotation. The hydrostatic stability of *Nautilus*, and indeed of coiled ectocochleates generally, will be enhanced by separating the two centers as far as possible. In this regard, Raup (1967) and Raup and Chamberlain (1967) have shown that the rapidly expanding shell geometry characteristic of *Nautilus* ranks among the most stable of cephalopod shell types.

The hydrostatic stability of *Nautilus* not only is an interesting mechanical feature, but also forms the operational basis of the equilibrium control system of

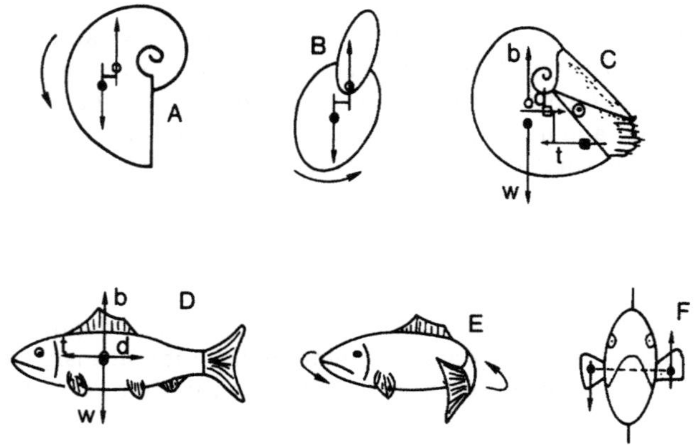

Figure 6. Methods of equilibrium control in *Nautilus* and other shelled cephalopods (A–C) and in fish (D–F). Key: (b) buoyant force; (w) weight; (t) thrust; (d) drag; (●) center of mass; (○) center of buoyancy; (■) point of application of jet thrust (□) center of dynamic pressure (point through which drag force acts). The arrows show the lines of action of forces. (A, B) Stabilizing moment elicited by the moving shell so that centers of buoyancy and mass are not aligned: (A) pitch; (B) roll. (C) Equilibrium position of *Nautilus*. Buoyancy and weight are colinear and have opposite directions. The effect of the jet–drag couple is to rotate the shell to produce the orientation shown in (A). Rotation stops when the stabilizing moment [as in (A)] balances the jet–drag moment. Maximizing the distance between the centers of buoyancy and drag maximizes stability. (D) Equilibrium of fish. The forces are approximately colinear. The hydrostatic stability is low (centers of buoyancy, mass, thrust, and drag are nearly coincident). Attitude control is achieved through movements of the body (E) and the fins (F). After Chamberlain (1981).

Nautilus. During locomotion, thrust developed by the jet, and acting at the position of the hyponome, together with the resulting drag force acting through the center of dynamic pressure, elicits a moment that rotates the animal away from its equilibrium position (Fig. 6C). This moment is resisted by hydrostatic stability. The animal swings away from its equilibrium position until the magnitude of the stabilizing moment increases to the point at which it balances the jet–drag moment. Apart from generating thrust, the animal's soft parts play no role in this stabilization process. The shell encases far too great a portion of the body to allow body flexibility to come into play in suppressing instabilities that arise during movement. Likewise, the tentacles, which trail passively in the wake, exert no control in this process.

With a prominent, heavy shell that lacks flexibility, and without fins, *Nautilus* must rely only on its hydrostatic stability for equilibrium control. High stability, and the uneven distribution of mass that produces it, is an obvious adaptive asset for *Nautilus*. The situation is quite different for fish. Fish lack a heavy skeleton. As a result, their overall deadweight is nearly equal to their buoyancy, rather than 20–25% greater as in *Nautilus*. Only a small flotation device (a gas bladder comprising 2–5% of total volume) is required, or none at all, as is the case, for example, in epipelagic scombroids (e.g., tuna, skipjack, mackerel). Mass distribution is virtually uniform, and hence the centers of buoyancy and mass nearly coincide (Fig. 6D). Stabilizing rotational moments are minimal in this situation, so that instantaneous instabilities that occur during locomotion are suppressed by compensatory movements of the flexible body and fins (Fig. 6E and F). Two factors are especially significant here. First, the axisymmetric fish body ensures that drag and thrust act near the midline of the body, so that their rotational moments are minimized and hence are more readily damped. Second, fins are particularly valuable as stabilizers because they are located at the periphery of the trunk and their moment arms are consequently considerable (Fig. 6F). They can therefore create large stabilizing moments that quickly dampen potentially disequilibrating moments as soon as they appear.

The fin-based control system of fish is unquestionably more sensitive, faster-acting, and therefore more efficient than the hydrostatic stability mechanism of *Nautilus*. This disparity between fish and *Nautilus* (and ectocochleates as a group) accounts in part for the disparity in their swimming speeds [as noted in Chamberlain and Westermann (1976) and below]. *Nautilus* simply lacks an equilibrium system capable of effective functioning at high speeds.

4. Swimming Movements

Observers of *Nautilus* note a seemingly endless array of movements associated with locomotion, yet these distinctive elements can be reduced to a small number of convenient categories similar in origin. First, one can segregate movements that represent responses of relaxed or semiturgid structures to instantaneous fluctuations in the flow. The fluttering of extended tentacles, so often seen as *Nautilus* swims, is an example. Second, some movements are associated with directional change. This is the case for deflection of the funnel, which is apparent to any observer and much discussed in the literature (Bidder, 1962; Packard *et al.*, 1980). A third category consists of the periodic pitching rotations of the shell (Section 2.1.2), inward–outward motions of the body (Section 2.2), and other movements keyed to the production of thrust. While all three types of movements are intrinsically interesting, I focus on this third category in the following discussion.

A gentle rocking, or pitching, motion can be observed in practically all phases of the activity of *Nautilus*, and it is especially strongly developed during active swimming (Bidder, 1962; Denton and Gilpin-Brown, 1966; Packard *et al.*, 1980). Cinematographic analysis of this motion shows that it is tied to an alternating extension and retraction of the body relative to the shell (Chamberlain, 1981). Graphical representations of these movements can be plotted from a frame-by-frame analysis of a film of a swimming animal (Fig. 7). Over an interval of several seconds, the shell first rotates forward (clockwise as seen from the left when swimming from right to left) about 8° from the equilibrium position and then swings backward again about 11°, i.e., 3° past the equilibrium point (Fig. 7). This first rotational couplet is followed by a sequence of eight similar rotations, which do not exceed about 10° in either direction, although the magnitude of the forward component generally exceeds that of its opposite. The duration of each coupled rotation is about 0.8 sec, although there is some variation in this parameter. As shown in Fig. 7B, the extended body mass moves into the body chamber as the shell rotates forward and then outward as the shell swings back. This alternating

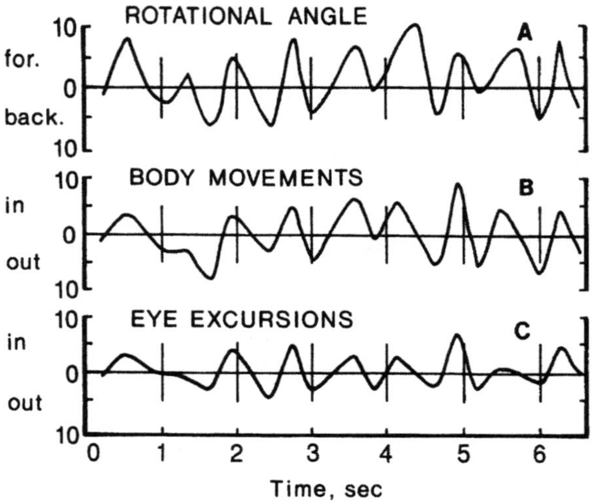

Figure 7. Swimming movements of *Nautilus* transcribed from cine films of swimming specimens. (A) Pitching rotation (rocking) of the shell (degrees, with 0° as the equilibrium position): (for.) forward (clockwise as described in the text) rotational movement; (back.) backward (counterclockwise) rotational movement. (B) Body movements during swimming (mm): (in) inward retraction of body relative to peristome; (out) outward extension of body relative to peristome. (C) Movement of eye relative to peristome (mm): (in) inward movement; (out) outward movement. After Chamberlain (1981).

retraction and extension is also observed in the motion of the eye relative to the shell peristome (Fig. 7C), but within more moderate limits.

Periodic shell rotation is readily attributable to the interaction of the rotational movements outlined above (see Fig. 6A–C). The application of propulsive force at the position of the funnel during the active phase of the propulsive cycle (i.e., when water is actively expelled in the jet) sets up a couple (Fig. 6C) that rotates the shell forward until it is balanced by the stabilizing moment due to buoyancy and weight, which increases with rotational displacement. When jet action terminates, and the mantle cavity expands to receive a new supply of water, this propulsive couple disappears, and the shell rotates back toward its equilibrium position. Angular momentum carries the shell past the equilibrium point, and it acquires the backward orientation noted in Fig. 7A. The generally smaller angles achieved in this backward motion undoubtedly reflect a slowly acting stabilizing couple, which cannot operate quickly enough to reach its full swing before the next jet pushes the shell forward again. Successive applications of propulsive force repeat this process. Variation in rotational angles for each jet pulse probably results from differences in the duration, magnitude, and direction of the force developed during each pulse (Fig. 7). Because the techniques used to measure thrust inhibit free movement of the animal, it is not possible to be more quantitative in correlating these parameters. I have noted, however, that rotational angles during continuous swimming never exceed about 15°.

5. Locomotory Mechanism

The way in which *Nautilus* develops the thrust required for its propulsion is essentially similar to that of soft-bodied cephalopods: Water is taken into the mantle cavity and then expelled under pressure through the funnel. Yet *Nautilus* differs from endocochleates in the structure of the mantle and the hyponomic and cephalic retractor muscles. In fact, one of the more interesting aspects of the locomotion of *Nautilus* has been the question of how *Nautilus* generates thrust at all. It certainly cannot do so in the squid style, because *Nautilus* does not have a heavily muscled mantle.

This question has been resolved by analysis of swimming behavior (Mutvei, 1957; Bidder, 1962; Mutvei and Reyment, 1973; Packard *et al.*, 1980; Chamberlain, 1981). Two observations gained from cinematographic data are relevant. First, the duration of rotation couplets is about 0.8 sec (Section 4). This interval is equivalent to the pulse rate for actively swimming animals reported in Packard *et al.* (1980) and in Section 6.1. Second, as shown in Fig. 7, in most of the individual rotational couplets, the body begins its inward contraction slightly prior to the inception of forward rotation, and maximum inward contraction occurs, in most cases, before maximum rotational displacement is reached; i.e., the body moves before the shell responds.

The correlation in timing among shell rotation, application of thrust, and movement of the body implies that extension and retraction of the body form an integral part of the propulsive system. Because the cephalic and pedal regions of the body overlie the mantle cavity, outward extension of the body, deriving from contraction of the radial fibers of the paired cephalic retractor muscles, causes the mantle cavity to expand and fill with water. Inward contraction of the body, due to activation of the longitudinal fibers of the cephalic retractors, compresses the now-filled mantle cavity and increases its hydrostatic pressure. Release of this pressurized water through the funnel produces thrust.

The cephalic retractors, which insert behind the buccal region of the body and extend to the inner shell surface near the rear of the body chamber, are thus primarily responsible for generating thrust. The hyponomic muscles also function in this process, because they act to control the shape and size of the funnel opening and thus contribute to the elevation of hydrostatic pressure and the velocity of the expelled water. The mantle itself is not highly muscular, as it is in squids, and functions chiefly to seal the mantle margin to the inner shell surface. This seal minimizes water pressure loss during activation of the jet.

Analysis of the retractor and hyponomic muscles further elucidates the propulsive system. The two sets of propulsive muscles are of approximately the same mass in a given animal (Fig. 8). Because muscle strength (i.e., power) is generally proportional to (muscle volume)$^{2/3}$ and hence to (weight)$^{2/3}$, it would seem from this approximate equivalence of mass that the contributions of each muscle pair to the production of thrust may be roughly equivalent, although not necessarily equal. As also shown in Fig. 8, the relative size of the two muscle pairs remains reasonably unchanged during the animal's growth. Together, the propulsive muscles comprise about 9% of the total weight of the animal, and this ratio does not change appreciably during ontogeny (Fig. 9). Comparison of the two species *N. pompilius* and *N. belauensis* intimates that both species are essentially alike with regard to the relative propulsive musculature.

The 9% figure for muscle weight/body weight (mass ratio) of *Nautilus* is quite revealing when compared to values for other swimmers (Table IV); *Nautilus* does not have much propulsive muscle. Except for *Octopus*, all the soft-bodied cephalopods in Table IV have mass ratios two or three times higher, and several of the fish have even higher mass ratios. The high speeds that characterize these

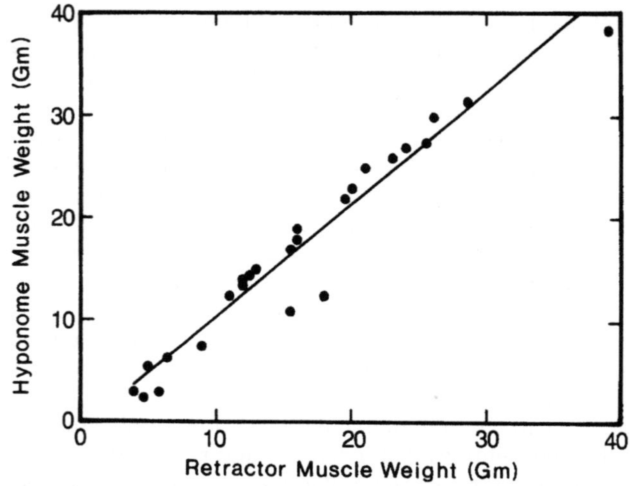

Figure 8. Retractor muscle weight and hyponomic muscle weight, as determined from autopsy for *N. pompilius*. Each point represents paired muscle weights from one animal. Regression parameters: $Y = 1.1\ X - 0.72$; $r = 0.969$.

fish, and cephalopods like *Loligo*, are due in large part to their generous endowment of propulsive muscle. The apparent equivalence in propulsive muscle for *Nautilus* and *Octopus* is misleading, because *Nautilus* is an obligatory swimmer, whereas *Octopus* is, at best, an infrequent swimmer, relying for the most part on its heavily muscled arms to traverse the substrate.

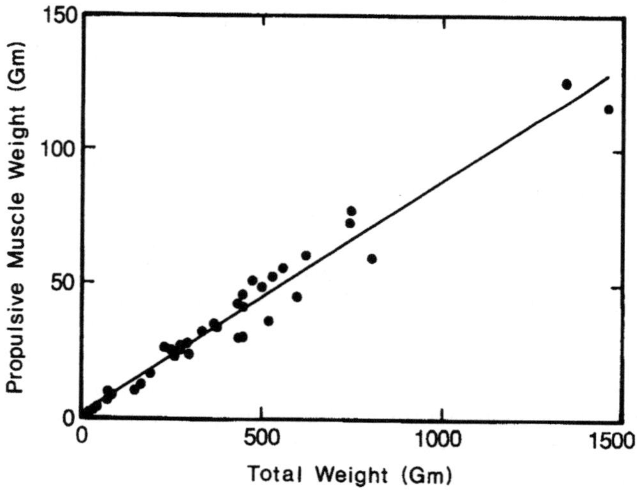

Figure 9. Weight of propulsive muscles (retractors + hyponomic muscles) as a function of total weight (shell + body) as determined from autopsy. Each point represents one animal. All data are for *N. pompilius* except the two points at the upper right corner of the plot, which are for *N. belauensis*. Regression parameters: $Y = 0.086\ X + 1.69$; $r = 0.979$.

Table IV. Propulsive Parameters for *Nautilus* and Other Swimmers

Organism	Mass ratio[a]	Mantle cavity ratio[d]
Ectocochleates[b]		
Nautilus pompilius	0.09	0.15
Nautilus belauensis	0.09	0.15
Endocochleates[c]		
Octopus vulgaris	0.08	0.14
Eledone moschata	0.20	0.50
Loligo vulgaris	0.35	0.57
Sepia officinalis	0.30	0.25
Fish[d]		
Salmo giardneri (rainbow trout)	0.60	
Leuciscus leuciscus (dace)	0.55	
Carassius auratus (goldfish)	0.40	
Perca flavescens (perch)	0.33	

[a] Mass ratio = propulsive muscle weight/total weight. Values for *Nautilus* are based on hyponomic and cephalic retractor musles; for endocochleates, on mantle muscles; for fish, on trunk muscles. Mantle cavity ratio = mantle cavity volume/total weight.
[b] Data are valid for total weights above about 40 g.
[c] Data apply to specific total weights as follows: *Octopus*, 500 g; *Eledone*, 600 g; *Loligo*, 350 g; *Sepia*, 250 g. Data are from Trueman and Packard (1968).
[d] Data refer to adults of average size. Data are from Bainbridge (1960), and Alexander (1977).

The mantle cavity is of considerable importance to locomotion in cephalopods, because it is the reservoir for water contained in the jet. In this respect, perhaps the most significant aspect of the mantle cavity is its volume, which sets an upper limit to the volume of water potentially available as a propellant. I measured the mantle cavity volume in 16 animals by collecting the contents of the mantle cavity in a graduated cylinder. Each animal was allowed to eject water until repeated pulsing of the funnel produced no further accumulation of water in the collector. This technique is similar to that used by Trueman and Packard (1968) to evaluate the mantle cavity capacity of the endocochleates.

The results of this experiment (Fig. 10) show that mantle cavity volume comprises about 15% of the total volume of *Nautilus* (because the animals are neutrally buoyant, volume is approximately equal to weight). As in the case of the propulsive muscles, the ratio of mantle cavity volume to total weight (mantle cavity ratio) remains constant throughout ontogeny. Comparing mantle cavity ratios of *Nautilus* with those of other cephalopods (see Table IV) reveals that *Nautilus* has

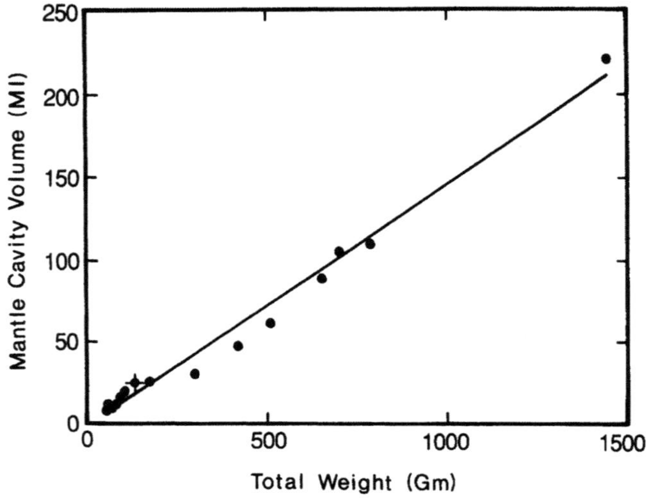

Figure 10. Mantle cavity volume as a function of total weight. Each point represents the average of 5–10 separate trials for a single animal. The dotted cross indicates two coincident data points. All data are for *N. pompilius* except the point in the upper right corner, which is for *N. belauensis*. Regression parameters: $Y = 0.148\ X - 1.8$; $r = 0.992$.

a relatively small mantle cavity. Only in *Octopus* is the size comparable. Species that, like *Nautilus*, are essentially obligatory swimmers have mantle cavities at least twice as large and, in the case of *Loligo*, almost four times as large as *Nautilus*. The output of the propulsive system depends, in part, on the volume of propellant, which is ultimately limited by mantle cavity size; thus, the low mantle cavity ratio observed for *Nautilus* may be construed as a serious constraint on locomotory performance and swimming speed. I return to this point in Section 6.2.1.

6. Performance

Organic power plants are constrained in their operation by the same factors of input, output, and efficiency as are the creations of mechanical engineers, and zoologists, like mechanical engineers, can profit by attending to these aspects of performance. Performance data are especially illuminating in comparing the locomotory capacity of individual specimens that belong to different age, size, or taxonomic groups. With these points in mind, I measured propulsive thrust and mantle cavity pressure and evaluated the time signatures of propulsive pulses for 11 specimens of *Nautilus* ranging in size from 73 to 1616 g total weight.

Force and pressure data were obtained with the apparatus diagrammed in Fig. 11. I measured propulsive force by means of a two-arm strain gauge bridge mounted on a stainless steel flex strip that was secured at one end in the jaws of a vise. An animal was attached to the other end of the strip by a spring clip fastened to the ventral margin of the shell. Propulsive thrust generated by the tethered

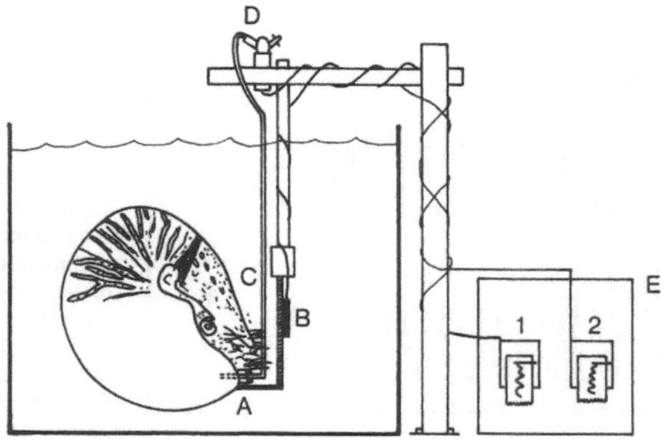

Figure 11. Diagram of apparatus used to measure thrust and mantle cavity pressure of *Nautilus*. Key: (A) specimen tethered to apparatus by spring clip; (B) strain gauge bridge; (C) pressure cannula; (D) pressure transducer; (E) brush recorder with simultaneous thrust (1) and pressure (2) readouts.

animals is monitored as strain by the strain gauge assembly, which in turn sends a signal to a calibrated Gould 13-4615-30 amplifier and Gould 2200 brush recorder. The prime resonant frequency of the flex strip was found to be about 25 Hz. Because this frequency is about 25 times higher than the frequency of the propulsive cycle of *Nautilus*, it seems unlikely that the force signal was significantly augmented by flex strip harmonics. Mantle cavity pressure was measured by inserting a thin cannula into the mantle cavity through the open funnel. The cannula was connected to a Gould-Statham P2310 pressure transducer that carried a signal to the brush recorder via a second amplifier. Most specimens were attached to the apparatus and monitored for periods up to about 10 min. This procedure permitted tracking an animal's behavior from the alarm and hyperactivity associated with its initial restraint in the apparatus to the eventual establishment of essentially respiratory movements as it became conditioned to its situation. Each animal was tested at least three times in this way, with at least one day separating each set of trials. The procedure was patterned after the techniques of Trueman and Packard (1968) and Packard *et al.* (1980).

Figure 12 shows recorder output for a small portion of a typical test run. Each trace records 28 successive pulses of the funnel. The first 22 are propulsive pulses, producing mantle cavity pressures of considerable magnitude that are coupled with correspondingly high thrust peaks; the last 6 pulses are much smaller and are presumably more respiratory in character. The small secondary peaks observed in the thrust record represent rebound of the flex strip and should be disregarded. Within the propulsive segment, it will be noted that the heights of the pressure and corresponding thrust peaks are only loosely correlated. This situation results from the tethered animal twisting its funnel into positions in which the line of action of the jet departs from perpendicularity to the flex strip. When this occurs, the strain gauge measures the thrust component normal to the flex strip rather than the actual thrust. Packard *et al.* (1980) and Trueman and Packard (1968) encountered this same difficulty in controlling their experimental subjects after attaching them to their force-measuring device.

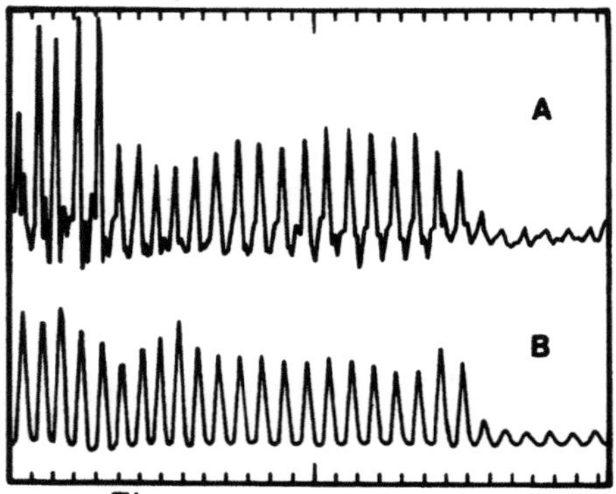

Figure 12. Typical recorder output showing simultaneous recordings of propulsive thrust (A) and mantle cavity pressure (B) for a specimen of *N. pompilius* of 114 g total weight. The divisions on the time axis show 1-sec intervals. Trace (B) gives mantle cavity pressure as a function of time; trace (A) gives the equivalent relationship for thrust. Each trace records 28 successive funnel pulses.

6.1. Pulse Rate

Packard *et al.* (1980) noted that pulse rate declined from about 1 Hz when the animal was actively pumping to about half that value (0.5 Hz) when it was respiring. My data also show a similar relationship to activity (Table V), but not to the extremes reported by Packard *et al.* (1980). Respiratory pulsing is seen to be somewhat slower than the pulsing associated with active jetting. Disparities between this data set and observations of Packard *et al.* (1980) may reflect differences in test intervals. I restrained my test specimens less than half as long as

Table V. Pulse Parameters for Nine Specimens of *Nautilus*[a]

	Respiratory		Jet	
Specimen	Rate (Hz)	Period (msec)	Pulse rate (Hz)	Period (msec)
055	0.90	1117	0.92	1083
322	0.70	1430	0.81	1230
414	1.07	935	1.11	900
415	1.01	989	1.11	900
636	0.83	1210	1.22	820
637	1.17	858	1.46	686
735	0.53	1870	0.81	1230
796	0.72	1390	0.79	1267
888	0.60	1680	0.81	1240

[a] Respiratory rate is the respiratory pulse rate in hertz (Hz); respiratory period is the respiratory period in milliseconds. Jet pulse rate is the pulse rate during active jetting; jet period is the period of jet pulses. Respiration and jetting were determined on the basis of the amplitude of mantle cavity pressure peaks as diagrammed in Fig. 12. Each entry represents an average of 8–15 successive peaks of roughly equal amplitude. In all cases, the standard deviation is ± 10% of the reported value. The data for each specimen were taken from a single test run.

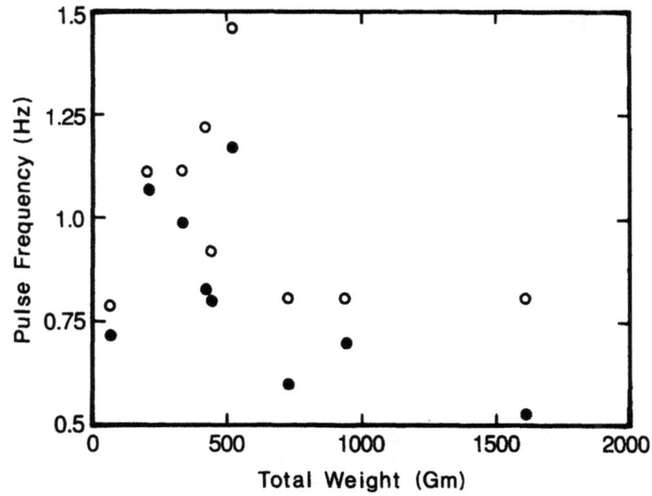

Figure 13. Pulse rate plotted as a function of total weight for specimens in Table V. Key: (●) respiratory frequency; (○) propulsive pulse frequency. Each point represents an average of 8–15 successive funnel pulses. The standard deviation for each average is less than ±10% of plotted value.

they did, so that the data here may not reflect as complete a state of relaxation of the test specimens as did theirs.

It is a standard principle of size scaling in the organic world that large animals operate more slowly than smaller ones. This relationship is particularly true of locomotory mechanisms. In the context of aquatic locomotion, for example, the tail-beat frequency of carangiform fish declines as a function of increasing body length (Bainbridge, 1958; Wardle, 1977), producing lower specific velocity (body lengths traveled per unit time) in larger animals. I examined whether such a scaling rule holds in the case of *Nautilus* by plotting pulse rate data against size (Fig. 13). Respiratory rate [Fig. 13 (●)] appears to vary inversely with size, as expected, because the larger animals have the lowest ventilation rates. But the relationship is a weak one at best, in that its correlation coefficient is only $r = 0.626$, and the slope and *y*-intercept of the least-squares regression fitted to the data cannot be distinguished from 0 at an acceptable level of confidence ($p > 0.05$). Propulsive pulse rate [Fig. 13 (○)] gives even weaker results ($r = 0.127$; $p > 0.05$ for slope and *y*-intercept of least-squares regression). The rate of operation of locomotory systems is commonly compared to linear measures of size, rather than to volumetric ones, as is done in Fig. 13. However, converting weight to a linear measure (weight$^{1/3}$) does not alter this observation, because the correlation coefficients for the new relationship are still unacceptably low ($r = 0.202$ and 0.479 for propulsive and respiratory pulse rate, respectively).

I conclude from this result that the specimens discussed herein do not give strong support to the idea of size scaling in the operation of the *Nautilus* locomotory mechanism. This impression is intensified by data that show a wide range in respiratory rate (0.77–1.19 Hz) and jet pulse rate (0.72–1.62 Hz) for a series of separate test runs on a single animal (Table VI). In fact, it can be seen from Table

Table VI. Pulse Parameters for *Nautilus* Specimen No. 416 Tested on 12 Separate Occasions[a]

Respiratory		Jet	
Rate (Hz)	Period (msec)	Pulse rate (Hz)	Period (msec)
0.84	1189	0.93	1081
0.83	1200	0.79	1267
0.79	1263	0.80	1255
0.83	1200	0.74	1350
0.82	1255	0.95	1050
0.88	1138	0.90	1117
1.19	838	—	—
1.07	936	—	—
0.94	1063	0.97	1030
—	—	1.62	619
0.77	1300	0.72	1390
0.98	1022	1.09	918

[a] See the Table V footnote for explanation of the column headings. Each entry represents an average of 8–15 successive peaks of roughly equal amplitude. The standard deviation is less than ±10% of each reported value. Trials were separated by at least 1 day and together span about 3 months.

VI that one specimen shows as much variation in pulse rate in 12 different test sessions as is seen across the entire size spectrum of the nine animals that comprise Table V.

The data in Tables V and VI are clearly not sufficiently robust to support firm decisions about pulse rate. Yet they may suggest that the pulse rate is reasonably constant for all *Nautilus*, i.e., that *Nautilus* propulsion operates at about the same rate in both large and small animals. If this inference is correct, it certainly departs radically from the typical vertebrate condition. Perhaps we are seeing a product of the unique muscle innervation system and variable contractile behaviors of cephalopod propulsive muscle noted by Packard *et al.* (1980).

The recorder traces are also helpful in evaluating the timing of the thrust and pressure pulses relative to one another. In this regard, I examined two aspects of the phase relationships between thrust and pressure: (1) the onset of the pulses and (2) the pulse peaks. Analysis of the former parameter proved difficult, because the flex strip oscillation dampener often distorted the thrust signal during the recovery phase of propulsion, when thrust was near zero (the small deflections in the troughs of the thrust trace in Fig. 12 derive from this distortion). Although this property of the dampener generally masked the position of thrust initiation, I found the onset of the pressure and thrust pulses to be approximately coincident. This result differs from that of Packard *et al.* (1980), who observed in their experiment that the initiation of the thrust peak was slightly delayed (by 60–200 msec), compared to the pressure peak. They interpreted this delay as deriving from their testing apparatus and considered it artifactual. Although compliance is a very real problem in most physiological testing systems, it is probably less in the experiment presented here, because the flex strip–strain gauge apparatus does not slacken as did the tension line of Packard *et al.* (1980); it can measure

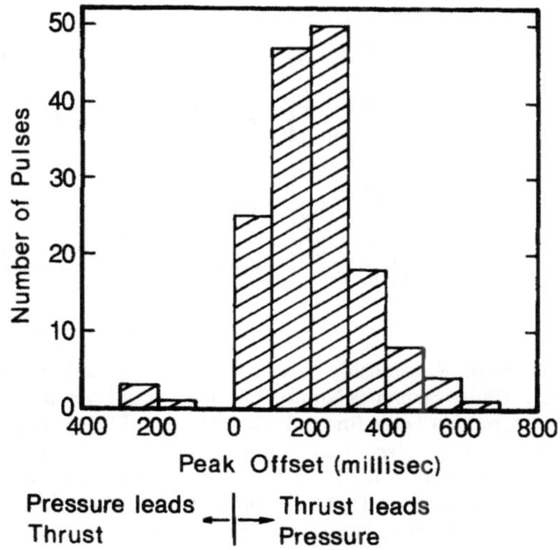

Figure 14. Frequency spectrum of time delays between peaks in thrust–pressure pulses. Key: (Peak Offset) interval (msec) between thrust peak and corresponding pressure peak; (Pressure leads Thrust) pressure peak occurs before thrust peak; (Thrust leads Pressure) thrust peak occurs before pressure peak. The data were recorded during seven test runs reported in Table VI for specimen No. 416.

propulsive force in any position. For this reason, I think that the apparent simultaneity of the onset of the pulses may be reasonably accurate.

Comparison of the time signatures for coeval pressure and thrust peaks shows a frequency spectrum of time delays (peak offset) between a specific thrust peak and the peak of the pressure pulse that produced it (Fig. 14). In the great majority of cases, the two signals are out of phase. Thrust generally attains its peak value before maximum pressure is achieved.

Within individual trial runs, the magnitude of peak offset is commonly correlated with the character of the pressure pulses. Such correlations are seen in Fig. 15, which shows peak offset and pulse length for the sequence of pulses plotted in Fig. 12. Pulse period increases over the duration of the sequence (Fig. 15, top). Because the height of successive thrust peaks declines with time (Fig. 12), the relationship noted in Fig. 15 indicates that active jetting is accomplished with relatively short-lived pulses, whereas during respiration, pulsation occurs at a more leisurely rate. Packard *et al.* (1980) and Kakinuma and Tsukahara (1985) observed similar patterns, although these authors reported a wider range in pulse rate variation. The differences probably reflect the fact that these researchers were dealing with animals that had been at rest for some time, whereas the respiratory peaks reported here (Fig. 12) occurred immediately after a long sequence of locomotory activity and thus were produced by an animal probably not yet fully at rest. Trueman and Packard (1968) reported results for *Sepia* and *Octopus* that are essentially similar to mine in showing relatively little change in pulse rate during short intervals of changing activity; peak offset is greatest during the respiratory portion of the test and is generally lower during active jetting.

The correlations between pulse characteristics and peak offset (suggested in Fig. 15) are weak at best. The reason may lie in several facets of locomotory behavior. First, the character of an animal's response to testing varies considerably from trial to trial, as noted above with regard to pulse rate (Table VI). Combining the results of several tests (as in Fig. 4) could mask individual trial results that

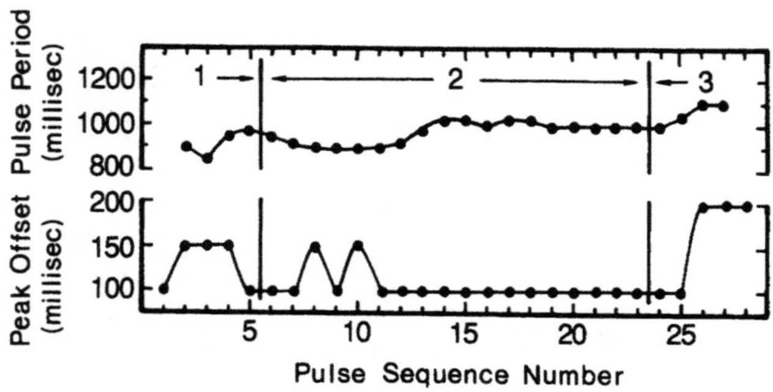

Figure 15. Relationship between pulse length and peak offset for pulse sequence plotted in Fig. 12. Key: (Pulse Sequence Number) position of pulse in pulse sequence in Fig. 12, as counted from left to right; pulse 1 is the first pulse on the left in Fig. 12. *Top:* three-point running average of pulse period for pressure pulses plotted as a function of sequence number. *Bottom:* offset between thrust and pressure peaks plotted as a function of sequence number. Thrust led pressure throughout the entire sequence. Regions: (1) initial series of five highest thrust peaks in Fig. 12; (2) high thrust peaks characteristic of active jetting; (3) low thrust peaks presumably representing respiration.

are actually quite consistent. This may mean that individuals have such a broad response repertoire that generalizations are virtually meaningless. Second, observations regarding funnel movements during jetting indicate that the shape and size of the funnel opening vary considerably during the pulse cycle. This variation typically involves an increase in the size of the funnel aperture during the early phases of a pulse. Because the area of the funnel opening controls the mass flux and velocity of the expelled water, this area will be critical in determining instantaneous values of thrust. Small variations in funnel-opening area from pulse to pulse, and in the rate of change in funnel-opening area during pulsing, could easily cause the peak offsets, as well as the variation in peak offset seen here (Figs. 14 and 15).

It would appear that much work remains before these important aspects of propulsion can be quantified adequately. Nevertheless, present information points to the interesting hypothesis that pulse rate may be largely independent of size and that an animal may fine-tune its propulsive output by controlling the size of its funnel opening during pulse cycles.

6.2. Locomotory Parameters

The principle that governs the *Nautilus* power plant is jet propulsion. In jet engines, propulsive force, or thrust, is generated by expelling pressurized fluid from a nozzle. Fluid enters the engine at low velocity and, due to pressurization, exits at high velocity. The resulting momentum gain of the fluid imparts an equivalent, but oppositely directed, gain to the momentum of the engine—and to whatever is attached to it. The propulsive force thus produced can be found by

$$F = d(m_j v_j)/dt$$

where m_j is the mass of the fluid exhaust, v_j is its velocity, m_jv_j is the propellant's momentum, and t is the time over which momentum varies. In *Nautilus*, thrust production is periodic rather than continuous. The force equation holds during those portions of the pulse cycle when water is expelled from the mantle cavity. It is evident that in defining the propulsive performance of *Nautilus*, several parameters are of some consequence, including: (1) the mass of water contained in the jet, (2) the velocity of water in the jet, and (3) the mantle cavity pressure.

6.2.1. Mass of Jet

The mantle cavity holds a volume of water equal to about 15% of the total weight of the animal (see Fig. 10). This quantity of water represents the upper limit for the mass of water emitted in the jet. To determine how closely this limit is actually approached, I collected water from single locomotory pulses for 16 animals using the same general technique as described in Section 5 for mantle cavity volume. It can be seen that pulse volume comprises about 6% of total weight (Fig. 16). Comparison of the data in Figs. 10 and 16 indicates that during locomotory pulsing, slightly less than half the water in the mantle cavity is ejected in the jet and that this value remains unchanged during ontogeny. Because the animals were quite agitated by being lifted from their tank during the collection procedure, the data undoubtedly reflect maximal levels of activity. Under more quiescent conditions, as during respiration, lower jet mass values should be found.

6.2.2. Mantle Cavity Pressure

The magnitude of the pressure peaks recorded [Tables V and VI (a total of more than 4000 individual pulses)] ranged from about 0.05 kilopascals (kPa) to

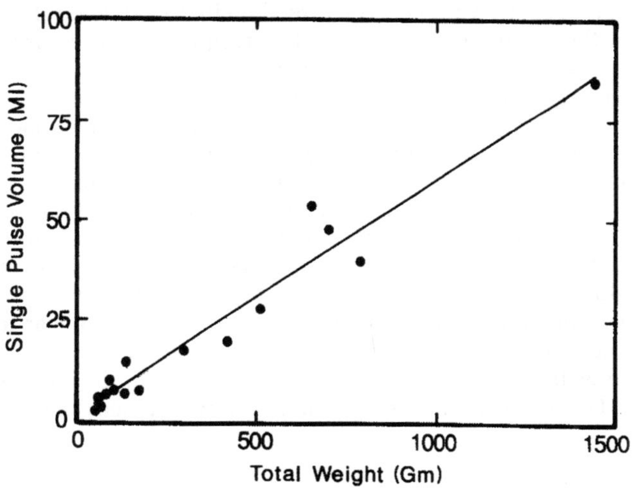

Figure 16. Volume of single locomotory pulse plotted as a function of total weight for *N. pompilius*. Each point represents the average of 6–10 different locomotory pulses for one specimen. The standard deviation for each average is less than ±10% of plotted value. Regression parameters: $Y = 0.059\ X + 1.49$; $r = 0.974$.

Table VII. Propulsion Parameters for Various Cephalopods[a]

Organism	Specimen weight (g)	Maximum pressure (kPa)	Jet velocity (cm/sec)	Thrust ratio	Ref. no.[b]
Sepia	260	9.5	275	0.7	1
Loligo	400	29.6	730	1.3	1
Eledone	600	38.1	940	1.3	1
Octopus	220	16.6	126	0.4	1
Nautilus	687	3.8	150	—	2
	73–1616	11.3	485	0.2	3

[a] Maximum pressure is the peak mantle cavity pressure; jet velocity is the velocity of water expelled in the jet; thrust ratio is peak propulsive thrust/total weight. All values are estimates except those from this chapter (3), which are based on experimental data. Thrust ratio data for *Nautilus* apply to specimens up to 750 g in weight.
[b] References: (1) Trueman and Packard (1968); (2) Packard *et al.* (1980); (3) this chapter.

about 11.3 kPa. These pressures are equivalent to those exerted by columns of water 0.5 to 120 cm in height, respectively. The maximum pressure produced by each of the test specimens, when regressed against specimen weight, produced no significant correlation ($r = 0.390$; t_{slope} not significant at $p \leq 0.05$). It would appear, therefore, that small animals develop about the same mantle cavity pressures as do larger ones. This situation undoubtedly derives from the size-scaling relationships for propulsive muscle and mantle cavity volume (see Figs. 9 and 10). Pressure is force per unit area. The force exerted by muscle tissue scales as the cross-sectional area of the muscle and hence as (muscle weight)$^{2/3}$. The area of application of this force is some reference area of the mantle cavity and will therefore scale as (mantle cavity volume)$^{2/3}$. Because both propulsive muscle weight and mantle cavity volume scale directly with total weight (Figs. 9 and 10), it is clear that the scaling factor for muscle force/(mantle cavity volume)$^{2/3}$, i.e., mantle cavity pressure, is a constant, independent of size. Thus, pressures produced by both large and small animals are approximately equal.

Data on maximum mantle cavity pressure for various cephalopods (Table VII) show some discrepancy, undoubtedly reflecting differences in sample size between data sets. The two short testing sessions used by Packard *et al.* (1980) may not have afforded sufficient opportunity to obtain reliable estimates of extremes in performance; their maximum value may be unrepresentatively low. Extending our analysis to the full complement of species listed in Table VII reveals that soft-bodied cephalopods generally develop much higher mantle cavity pressures than does *Nautilus*. The endocochleate muscular mantle would appear to be a more effective source of motive power than the cephalic retractor system of *Nautilus*. In view of the well-known accelerative capacity of *Sepia*, the value reported for this species (Table VII) seems incongruously low and may not reflect actual capabilities.

6.2.3. Jet Velocity

One can estimate jet velocity on the assumption that it is equivalent to the velocity produced by free fall from a height equal to the pressure load generated by the mantle cavity (≈ 120 cm as discussed in Section 6.2.2). Using this value in the appropriate free-fall equation

$$U_j = (2gh)^{1/2}$$

where g is the acceleration of gravity and h is the pressure head, yields $U_j = 485$ cm/sec. This value is only a rough approximation because it does not take into account the effect of funnel shape on funnel discharge characteristics.

I checked this estimate by measuring jet velocity using a simple technique based on the motion of projectiles. The muzzle velocity of a gun can be obtained from knowledge of the horizontal and vertical distance traversed by a shell fired from it and from the inclination of the barrel to the horizontal. The motion of water expelled in the propulsive pulses of animals held with their funnels just above the water surface adheres to the same principle. Assuming no air resistance, jet velocity can therefore be approximated as

$$U_j = x[(g/2)(y \cos^2\theta - x \sin\theta \cos\theta)^{-1}]^{1/2}$$

where x and y are the horizontal and vertical distances, respectively, and θ is the angle of inclination of the jet as it leaves the funnel.

I measured 20–40 individual pulses for each of 11 animals ranging in weight from 109 to 366 g. The highest jet velocity recorded for each animal is shown in Table VIII. Although my technique yields only rough velocity values, the data nevertheless appear to corroborate the general tenor of the jet velocity estimate made above. Note also that there is no strong correlation in these data between jet velocity and size; a least-squares regression of the two parameters gives r = 0.246. This situation probably results from the fact that mantle cavity pressure is independent of size (see Section 6.2.2). Thus, the thrust required for propelling large animals is produced, not by increasing jet velocity over velocities attained by small animals, but by increasing the mass of the jet (Fig. 16).

Jet velocity data for a variety of cephalopods are given in Table VII. The

Table VIII. Maximum Jet Velocity of 11 *Nautilus*

Total weight (g)	Maximum jet velocity (cm/sec)[a]
109	342
120	288
145	483
165	300
175	304
175	305
182	334
225	341
366	397

[a] The figures for jet velocity represents in each case the maximum value recorded for 20–40 individual locomotory pulses.

evident discrepancy between the jet velocity estimate of Packard *et al.* (1980) and my own experimental data probably stems from disparity in sampling regimens. Their small sample size probably led to underestimating velocity. The experimental data indicate that jet velocity for *Nautilus* falls between that of *Loligo* and *Eledone*, on one hand, and *Sepia* and *Octopus*, on the other. The high mantle cavity pressures developed by *Loligo* and *Eledone* have evidently produced high jet velocities, as one would expect. The jet velocities reported for the other two endocochleates seem too low to account for their observed swimming capacity and may be in need of refinement. In this regard, it should be noted that endocochleate velocities are based on data from only one specimen of each species. Much more work on all these animals is needed.

6.2.4. Funnel Cross-Sectional Area

The characteristics of the funnel are of some interest in the context of this chapter, because of the funnel's importance in controlling the egress of the jet from the mantle cavity. Funnel cross section is particularly interesting, because the mass flux of the jet, and hence the thrust produced, depends in part on cross-sectional area. I measured funnel cross sections from high-speed cine film outtakes of restrained and freely swimming animals and measured their area using a Houston Instruments digitizing pad interfaced with an Apple II+ computer.

Because the funnel flaps flare and contract in concert with the pulsing of the propulsive system, the areal extent of the funnel aperture is by no means constant. Moreover, the animal appears capable of exercising considerable muscular control over aperture shape and size by stiffening or relaxing portions of the funnel crura. The funnel cartilage, however, does not deform during pulsing, and this inelasticity sets an upper limit on the effective opening through which the jet is expelled. I therefore measured funnel cross-sectional area as the area of the funnel aperture, or of the undeformable funnel passage, if aperture flaring was extreme and the funnel opening unusually large.

Plots of funnel cross-sectional area as a function of (specimen volume)$^{2/3}$ (Fig. 17) suggest that effective funnel area remains in a more or less constant proportion to body size during ontogeny.

6.2.5. Thrust

The thrust required to produce a specified velocity can be found easily by noting that in sustained motion, thrust must equal drag. Hence

$$T = \tfrac{1}{2}\rho C_D A v_a^2$$

where T is thrust. Because C_D for *Nautilus* is known (see Section 2) and A is defined as (volume)$^{2/3}$ = (weight/1.026)$^{2/3}$, it becomes a simple matter to find thrust as a function of swimming velocity. Using the estimate of Ward *et al.* (1977) of 20 cm/sec for swimming velocity, for example, yields T = 0.015 newton (N) for the smallest specimen studied here (73 g) and T = 0.12 N for the largest (1616 g) (Table VII). These two animals would therefore need to generate thrust of 0.015 N and 0.12 N, respectively, to swim at this speed.

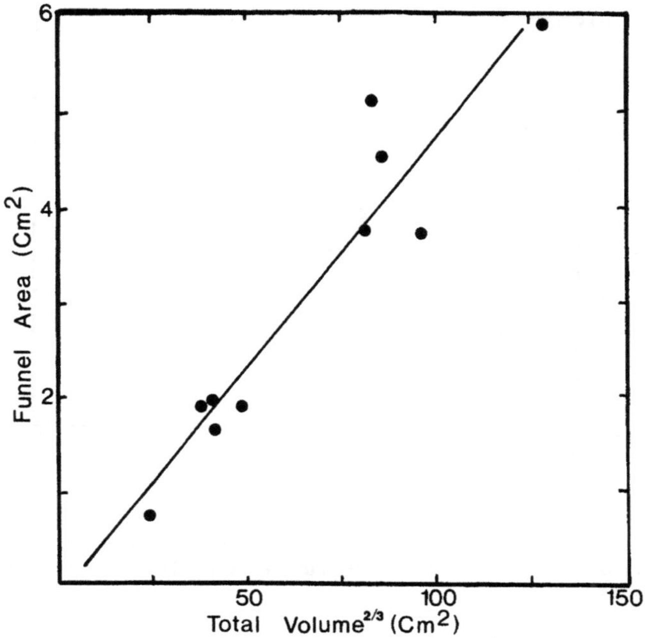

Figure 17. Cross-sectional area of the funnel at peak mantle cavity pressure, plotted as a function of (specimen volume)$^{2/3}$. Volume was determined from the relationship volume = specimen weight (g)/(1.026 g/cm^3). Each point represents one animal. Regression parameters: $Y = 0.049\ X - 0.139$; $r = 0.943$. The latter parameter was chosen for the abscissa to maintain dimensional consistency, but it should be noted that using specimen weight yields a relationship that is equally strong.

The thrust necessary to maintain a specified velocity is not equivalent, of course, to the thrust actually generated by the propulsive musculature. This latter parameter, which, of the two, is intrinsically more interesting and informative, can be obtained by referring to strain gauge data like those depicted in Fig. 12A. I calculated thrust from the strain gauge traces for each of the specimens listed in Tables VI and VII. Figure 18 shows a typical result; maximum thrust is observed to vary *linearly* with peak pressure. This feature typifies the thrust–pressure relationship for all the specimens examined. In fact, the slopes and *y*-intercepts of the other specimens tested cluster around the values calculated for specimen 322. This observation suggests that $T \approx 0.2P$ (when units are N and kPa, respectively). Recalculating to the units used by Trueman and Packard (1968) yields $T \approx 2P$, which compares favorably to thrust–pressure correlations these authors found in endocochleates ($T \approx 0.5P - 3.1P$).

As noted above, the thrust data include measurements for which the funnel was misaligned and the recorded thrust was artifactually low. Incorporating such data into the data set for an animal would have the effect of shifting the thrust–pressure regression downward toward lower pressures. Trueman and Packard (1968) encountered the same difficulty and obtained an estimate of this downward shift by calculating theoretical maximum thrust from the equation

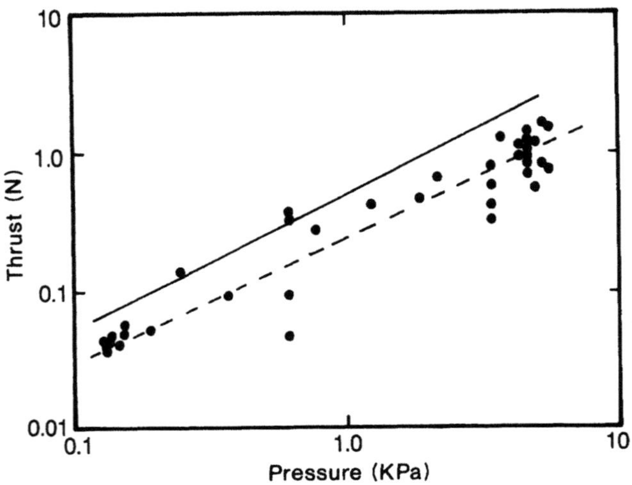

Figure 18. Peak thrust vs. peak mantle cavity pressure for specimen 322. Each point represents one pulse of the propulsive system. Key: (– – –) least-squares regression line for data set (regression parameters: $Y = 0.213\ X + 0.0044$; $r = 0.893$); (——) theoretical maximum thrust calculated from the relationship $T = 2AP$.

$$T = 2AP$$

where A is funnel cross-sectional area and P is mantle cavity pressure. Following this procedure in the case presented herein yields the solid line shown in Fig. 18. The regression line lies below the theoretical maximum, indicating that some downward shift is present in the experimental data. Yet the agreement between the two curves is reasonably good, considering the nature of the assumptions necessary to apply this equation to cephalopods.

Because jet mass is a function of body size (see Figs. 10 and 16) and because jet velocity appears largely independent of size (see Table VIII), one would also expect peak propulsive output to bear a simple relationship to specimen size. I tested this expectation by plotting maximum thrust against a measure of body size for selected animals (Fig. 19). Maximum thrust in Fig. 19 means the highest peak thrust observed among pulses measured. The expected correlation does in fact occur among the specimens tested. It should be noted that maximum recorded thrust must also depend to an extent on the number of pulses measured. Since thrust and sample size are not strongly correlated ($r = 0.205$), I infer that a sufficiently large set of pulses was analyzed to negate this factor.

When the data in Fig. 19 are recalculated to the units used by Trueman and Packard (1968), the result indicates that *Nautilus* generates thrust equivalent to about 20% of its total weight and may thus be said to have a thrust ratio (thrust/weight) of 0.2. Comparison of thrust ratios (see Table VII) for several cephalopods shows that *Nautilus* has the lowest thrust ratio of the species listed. This derives from the fact that, relative to its mass, *Nautilus* expels a smaller proportion of water in its jet than do these other cephalopods (see Table IV) and accelerates

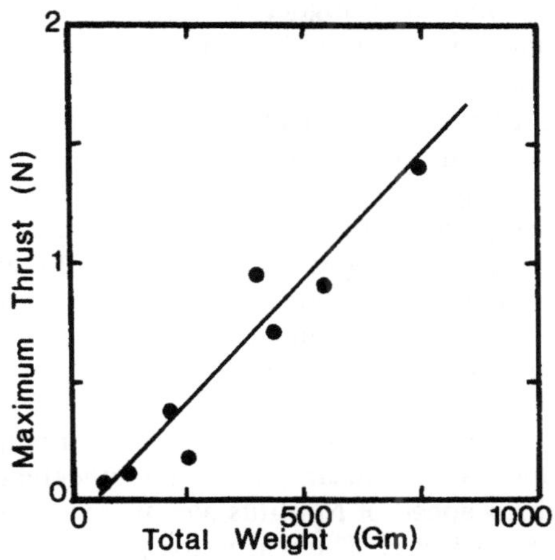

Figure 19. Maximum propulsive thrust plotted as a function of specimen weight. Each point represents the highest recorded thrust pulse for a single animal. Regression parameters: $Y = 0.0021\,X - 0.103$; $r = 0.932$.

this water to comparatively low velocities (Table VII). The propulsive mechanism of *Nautilus* is significantly less powerful than that of endocochleates. Among cephalopods, a fast-acting, heavily muscled mantle is clearly more effective as a propulsive device than are cephalic retractors. The latter situation will result in lower speed and acceleration for *Nautilus*.

6.2.6. Swimming Velocity

Despite its obvious importance in evaluating locomotory performance, quantitative information on the swimming velocity of *Nautilus* is meager. Cinematographic analysis (Chamberlain and Westermann, 1976) indicates a speed on the order of 25 cm/sec for adult specimens. Ward *et al.* (1977) obtained similar speeds in tests on freshly captured animals. These values are thought to represent maximum velocities, because the specimens tested were probably stressed. It is well known that *Nautilus* exhibits a variety of swimming behaviors (Bidder, 1962; Denton and Gilpin-Brown, 1966; Ward *et al.*, 1977). For example, the deployment of a search web occurs when *Nautilus* is swimming at low velocity (5–10 cm/sec), whereas escape movements are associated with the highest observed speeds.

Chamberlain and Westermann (1976) compared the 25 cm/sec maximum-velocity figure of *Nautilus* to the swimming speeds of other aquatic animals. They noted that for animals of equivalent body size (shell diameter for *Nautilus*; body length for other animals), *Nautilus* has a relatively low speed. In fact, its swimming velocity is only 10–25% of that of many fish (trout, dace, goldfish) and squids (*Loligo*) of equal size. This disparity in velocity is readily attributable to the inefficient design of the *Nautilus* propulsion mechanism, particularly with regard to its poor endowment of propulsive muscle (Table IV), small mantle cavity (Table IV), pulse volume, low jet velocity (Table VII), and the consequent low thrust ratio (Table VII). The high drag coefficient of *Nautilus* (Table II) is also a factor.

The velocities attributable to a maximum thrust equal to 20% of weight (Table VII and Fig. 19) can be found from $T = \frac{1}{2}\rho A\, C_D v_a^2$, which yields $v_a \approx 95$ cm/sec, a value much higher than actually observed. This anomalous figure can be explained by noting that it is based on the implicit assumption that the propulsion system in *Nautilus* operates continuously; i.e., thrust is constant—which, of course, it is not. During most of the pulse cycle, instantaneous thrust is much less than peak thrust (see Fig. 12A). This discrepancy can be resolved by determining impulsive force, i.e., by evaluating thrust as a function of time. I numerically integrated representative thrust pulses for several animals. In each case, the time-averaged thrust is about 20–25% of the peak thrust. Thus, if inertial effects can be neglected, actual instantaneous thrust can be approximated by an effective thrust equal to about 23% of the peak value. Inserting this value into the foregoing equation gives a velocity of about 33 cm/sec, which is close to that which has been observed.

In undulatory swimmers, increasing body size leads to higher swimming speeds. The linearity of the thrust/weight relationship (Fig. 19) implies that swimming speed of *Nautilus* also increases with size over the size range of the specimens tested. The reason for this increase may appear to derive from the scaling principles for thrust and drag; thrust scales directly with weight (Fig. 19), whereas from the equation above, drag scales with area, i.e., with $(\text{weight})^{2/3}$. Thus, thrust increases with size faster than drag, which should produce higher velocities in larger animals.

This last result is suspect, because in most animals, propulsive output scales with area rather than weight. The prime reason for this relationship is that muscle strength is limited over time by area-related properties such as the rate of O_2 and CO_2 diffusion across capillary walls. In such cases, higher velocity at larger body size is produced by increased displacement of the power stroke in large animals. One would expect the same principles to apply to *Nautilus*—and in fact several considerations suggest that they do. First, the thrust data in Fig. 19, when plotted against $(\text{weight})^{2/3}$, give a correlation as strong ($r = 0.930$) as that for weight; it would appear that these data cannot actually distinguish the actual size-scaling rule for propulsive thrust in *Nautilus*. Second, one can generate a dimensional argument that supports the notion that swimming velocity scales with size. The argument is based on conservation of momentum, specifically, on the idea that jet force equals drag force. Thus, the impulse of the jet averaged over a single pulse can be written as

$$F_j = m_j\overline{v_j}$$

where F_j is impulsive force and $\overline{v_j}$ is time-averaged jet velocity. Drag force is expressed as

$$F_D = \tfrac{1}{2}\rho_w\overline{v}_a^2 A C_D$$

where ρ_w is the density of seawater and $\overline{v_a}$ is the average swimming velocity over the interval in question. Since $F_j = F_D$, $m_j = 0.06\ m_a$ (Fig. 16), and $A = (m_a/\rho_w)^{2/3}$, we can write

$$\overline{v_a} = [\tfrac{1}{2}\rho_w^{-1/6}\overline{v}_a^{1/2}]m_a^{1/6}$$

Because $\overline{\nu_j}$ is independent of size (Table VIII), largely because mantle cavity pressure is independent of size

$$\overline{\nu_a} = Km_a^{1/6}$$

Thus, under the assumptions stated, swimming velocity in *Nautilus* increases as the $\frac{1}{6}$th power of body mass (or weight). This leads to an approximate twofold variation in swimming velocity over the weight range of the specimens studied.

Although these theoretical considerations are informative, the value of experimental confirmation is apparent. Much more research in the area of jet velocity and swimming velocity is necessary before these ideas can be placed on a firm footing.

6.2.7. Froude Efficiency

Swimming velocity is also of interest because, together with jet velocity, it affords a means of evaluating Froude efficiency. This is a parameter that describes the effectiveness of the propulsion mechanism in driving the animal through the water. It compares the power consumed in accelerating the jet to achieve a requisite thrust to the power consumed in moving the animal forward under the action of the thrust thus produced. Froude efficiency (F.E.) can be defined as

$$\text{F.E.} = \overline{\nu_a}/(\overline{\nu_a} + \tfrac{1}{2}\overline{\nu_j})$$

The reader is referred to Alexander (1977) for further details. Froude efficiency is obviously high, and hence useful power is maximized, when jet velocity is low compared to swimming speed; it is thus more economical to accelerate a large volume of fluid to a low velocity than to accelerate a small volume of a fluid to a high velocity.

Froude efficiency values for a variety of different swimmers are given in Table IX. *Nautilus* has the lowest Froude efficiency, whereas cephalopods as a group have lower values than the trout, a fish fairly representative in performance of carangiform teleosts. The main cause of this distribution can be seen in the figures for exhaust ratio: The ratio of propellant mass ejected per unit time to body mass. Animals with high exhaust ratios accelerate relatively large masses of fluid, so that lower velocities suffice in generating thrust, which leads to higher Froude efficiency.

Among the cephalopods listed in Table IX, *Nautilus* has the smallest exhaust ratio. This low ratio is the result of its small mantle cavity (see Fig. 10) and pulse volume (see Fig. 16), which require comparatively high jet velocities (see Table VII) to produce even low swimming velocities. With the possible exception of *Octopus*, the endocochleates all have much larger mantle cavities than does *Nautilus* (compare Tables IX and IV). This produces high efficiencies, not so much by reducing jet velocity [their jet velocities are not low (Table VII)] as by greatly increasing swimming speed so that jet velocity is relatively small in comparison. The exhaust ratios for trout are significantly higher than are those of cephalopods. This difference is a consequence of the different propulsion systems utilized by these two animals: Cephalopod jet propulsion acts on water held in the mantle

Table IX. Some Propulsive Parameters for *Nautilus* and Other Swimmers

Organism	Froude efficiency[a]	Exhaust ratio[b]
Nautilus	0.09–0.15	0.06
Octopus	0.22	0.22
Sepia	0.16	0.16
Loligo	0.30	0.38
Eledone	0.27	0.33
Salmo	0.61–0.81	0.63–6.91

[a] Froude efficiency is useful power/total power. The figures for *Nautilus* are based on a swimming velocity of 25 cm/sec and maximum jet velocities from Table VIII. The data for endocochleates were calculated from the data of Trueman and Packard (1968) on jet velocity and theoretical maximum swimming velocity during a single pulse, assuming $\frac{2}{3}$rds of mantle water expelled during pulse. The figures for trout are based on fish of mass 220 g swimming at speeds of 10–52 cm/sec; the data are from Alexander (1977).

[b] Exhaust ratio is mass of jet/mass of animal. The data for *Nautilus* are from the volume of a single pulse (Fig. 16). The data for endocochleates are based on single-pulse volume, assuming $\frac{2}{3}$rds of mantle water expelled during pulse. The figures for trout are based on the mass of water accelerated by the caudal fin during an interval equivalent to one pulse period of cephalopods (≈1 sec) and apply to swimming speeds of 10–50 cm/sec. Endocochleates exhaust ratios and fish exhaust ratio ranges were calculated from data in Trueman and Packard (1968) and Alexander (1977), respectively.

cavity, whereas the undulatory propulsion of fish involves ambient water accelerated by the caudal fin. Because the propellant reservoir in cephalopods is internal, its volume, and hence the mass of water accelerated in a pulse, must of necessity be limited to a fraction of total body volume. But because the caudal fin accelerates external water, propellant mass is not limited by body volume. Rather, it is determined by swimming speed and by the span of the caudal fin. Because swimming speed can be substantial, propellant volume may be several times body volume. Carangiform swimmers, like the trout, are much more successful in accelerating large volumes of water than cephalopods, and thus have much higher Froude efficiencies.

7. Evolutionary Implications

To be an effective means of locomotion, swimming has four important requirements: (1) low drag shape, (2) buoyancy control, (3) equilibrium control, and (4) propulsion. The preceding discussion allows evaluation of the locomotory properties of *Nautilus* in light of other swimmers and allows reflection on the role of locomotory design in establishing patterns of evolutionary interaction among ectocochleates, endocochleates, and fish.

These animals differ considerably in bauplan. Ectocochleates and endocochleates are typically jet-propelled swimmers, but the former group requires a large, heavy, external shell. Endocochleates and fish both have soft, flexible bodies, usually strengthened by a lightly constructed internal skeleton. However, they differ markedly with regard to their mode of propulsion. Endocochleates rely on jet propulsion as a locomotory mainstay, with fin propulsion and other forms of locomotion generally of secondary importance. Fishes rely on body and fin undulation and only secondarily (if at all) on jet propulsion. The main evolutionary effect of these differing aspects of design has been to constrain the range of adaptive solutions to problems involving buoyancy, equilibrium, drag, and propulsion, and thus to channel evolutionary development within each group in specific directions.

In ectocochleates like *Nautilus*, this channeling effect is clearly seen to be the consequence of retaining the shell as a major feature of adaptive design. The rigid, external shell of ectocochleates is primarily a buoyancy apparatus. It relieves its owner of the need to expend muscular effort to maintain position in the water column and frees it of the need to equalize internal and external pressure. The gas bladders of fish function effectively in providing uplift, but must always be kept in equilibrium with ambient pressure. This severely restricts the vertical mobility of most fish.

The camerate shell has decided adaptive benefits with regard to buoyancy. Yet the size, weight, and rigidity of the shell have an adverse effect on drag, equilibrium, and propulsion. Because the phragmocone significantly increases the total volume of an animal, drag and drag coefficient are higher than would otherwise be the case. Moreover, the external character of the rigid shell precludes any hydrodynamic benefit that might derive from body flexibility. Thus, such drag-reducing effects as mucous skin secretions and boundary layer vortex dampening (crucial to many fish and probably to some endocochleates as well) lie outside the potentialities of ectocochleates. In addition, the presence of a gas-filled shell forces ectocochleates to rely on hydrostatic stability as a means of equilibrium control. As described in this chapter, this system is a slow-acting and only moderately effective means of providing for fine-tuned attitude and directional control, particularly at high speed. Propulsive capacity is also adversely affected by the shell, because much of the interior of the shell remains devoid of tissue to provide buoyancy. This gas-filled region therefore cannot accommodate propulsive muscle or an enlarged mantle reservoir. As a result, *Nautilus*, and presumably ectocochleates generally, are weakly endowed with propulsive muscle (Table IV) and have a comparatively small mantle cavity (Table IV) and jet mass (Table IX).

Because endocochleates and fish are not saddled with the problem of a heavy, external shell, their buoyancy requirements are in most cases satisfactorily met with small ballast organs that occupy little internal space. As a result, they can pack their bodies with propulsive muscle (Table IV) and, in the case of endocochleates, provide larger mantle cavities (Table IV) and jet mass (Table IX). However, the large mantle cavities of endocochleates, which contribute so importantly to their propulsion, consume the space necessary for accommodating even greater amounts of muscle. Because of their reliance on jet propulsion, cephalopods, and especially endocochleates, are caught in a paradox: Large mantle cavities are needed to provide thrust, but space for propulsive muscle must be sacrificed to reach this goal. Fish have no such problem, because their undulatory propulsion acts on ambient water. A significant portion of their internal volume need not be set aside as a propellant reservoir. Without such constraints, fish achieve the highest mass ratios (Table IV) and propellant volumes (Table IX) of any group of swimmers.

These ideas suggest that, in a general way, cephalopods and fish can be ranked in terms of sophistication in design of their locomotory systems. *Nautilus*, and presumably fossil ectocochleates, would represent the most primitive level of this heirarchy. Their large, external shells prevent the development of a sensitive equilibrium control mechanism and limit propulsive output by severely restricting muscle and mantle cavity volume. Thus, these animals are slow, relatively unmaneuverable swimmers. Endocochleates occupy an intermediate position in this heirarchy. Reduction and loss of the shell have permitted fabrication of a sensitive system of equilibrium control, based on low hydrostatic stability and compensatory body movements. They have also increased muscle volume and mantle cavity capacity, thereby making endocochleate propulsion more powerful and efficient. Fish (at least carangiform and scombroid fish) would seem to occupy the apex of this adaptive ladder. Like endocochleates, they have an effective system of equilibrium control, but they excel endocochleates in possessing a propulsive mechanism that maximizes muscle output and efficient usage of muscle power, while minimizing the effects of fatigue.

These ideas suggest the possibility that locomotive adaptation may have played an important role in the evolution of these three groups (see also Packard, 1972). Invention of the gas-filled, camerate shell seems to have presented the pioneering ectocochleates with both a unique adaptive advantage—an efficient means of buoyancy control—and a unique, insurmountable adaptive dilemma—a weak, inefficient propulsive mechanism. The evolutionary history of ectocochleates may reflect the interweaving of these two great adaptive themes. Their early success in the Paleozoic may have resulted, as Teichert (1967) contended, from the adaptive superiority over contemporary swimmers conferred by buoyancy control. Their demise in the Mesozoic and Cenozoic may have been an inevitable consequence of their inability to match the locomotory accomplishments of fish and endocochleates, which groups, although they appeared later than the ectocochleates, had by the mid-Mesozoic reached their own solutions to buoyancy control—solutions that did not face the severe constraints on propulsion that derive from an external shell. In addition, the expansion of the endocochleates in the late Paleozoic and Mesozoic may well have been spurred by the propulsively beneficial effect that reduction and loss of the shell had on propulsion. The propulsive superiority of fish over endocochleates may have contributed in an important way to the impressive diversity of modern teleosts and to their apparent dominance of neritic and epipelagic realms, where active swimming is a key to ecological success. Viewed in this way, the time of ectocochleates has passed. In a world in which speed is commonly essential, *Nautilus* has been relegated to habits and habitats where it is not.

ACKNOWLEDGMENTS. I thank R. B. Chamberlain for help in carrying forward many aspects of this research. I am grateful to W. B. Saunders and N. H. Landman for providing me the opportunity to prepare this chapter and for their cooperation in allowing me to complete it. I am indebted to W. M. Kier for making the high-speed movies of the animals upon which a significant part of this analysis is based. I am also grateful to the staff of the New York Aquarium, and particularly to G. D. Ruggieri, S. J., Director, and L. Garibaldi, Associate Director, for their long-standing and generous support of my work on *Nautilus*. M. Schomberg typed the manuscript. This research was supported by the National Science Foundation, Grant PCM8019064, and by grants from the PSC-CUNY Award Program of the City University of New York. Acknowledgment is made to the Palaeontological Association and Systematics Association for permitting me to use figures (Figs. 1 and 2 and Figs. 6 and 7, respectively) originally printed by them, for which they retain publication rights.

Chapter 33

Nautilus Shell Hydrostatics

EARL A. SHAPIRO and W. BRUCE SAUNDERS

1. Introduction

In 1897, Arthur Willey remarked that "there is one thing the Nautilus cannot do, namely, turn upside down" (Willey, 1897a, p. 146). The high degree of hydrostatic stability noted by Willey is a direct consequence of the geometry of the *Nautilus* shell, and it raises two questions: (1) Which physical parameters contribute to this stability? (2) How do these factors vary among different species of *Nautilus*?

In paleontology, there are two approaches to the study of functional morphology: the empirical approach and the paradigmatic approach. The empirical approach entails observation of organisms actually performing a function and correlation of success of that function with specific morphological features. This

EARL A. SHAPIRO • Georgia Geologic Survey, Atlanta, Georgia 30334. W. BRUCE SAUNDERS • Department of Geology, Bryn Mawr College, Bryn Mawr, Pennsylvania 19010.

method is useful in dealing with extant organisms; the results can then be projected into the fossil record. The paradigmatic approach, developed by Rudwick (1961), involves several steps: (1) consideration of a structure, (2) proposal of a function (or alternative functions) for that structure, (3) development of a mechanical or mathematical model (a *paradigm*, in Rudwick's terminology) that optimally performs the function, and (4) comparison of the model with the actual structure. Close correspondence between the model and the actual structure is taken as evidence that the proposed function is correct. The paradigmatic approach is essentially an inductive method, and it allows for the direct analysis of extinct organisms.

Numerical analysis provides a deductive variation on the basic paradigmatic approach. This modified methodology proceeds as follows: (1) a structure (in this case, the shell of *Nautilus*) is considered; (2) a function (in this case, flotation) is proposed for the structure; (3) a generalized mathematical model is developed for the structure; (4) the properties of the model relevant to the function are studied through numerical analysis; and (5) comparison is made between the natural variation observed in the structure and the properties of the mathematical model.

The external shell of *Nautilus* serves several functions; it protects the animal's soft parts, provides an attachment site for muscles, and acts as a flotation–buoyancy device that makes the animal weightless in water. The analysis presented in this chapter concerns the latter function: the relationship of shell morphology to the success of the shell as a flotation device.

1.1. Previous Investigations

A mathematical model describing the geometry of coiled shells was developed by the Reverend H. Moseley (1838). Because of its mathematical complexity and the length of time required to make the actual computations, this model was not utilized effectively until the 1960s. Raup (1961, 1962, 1966) and Raup and Michelson (1965) developed a theoretical framework for the mathematical description of coiled shells, using four variables: whorl expansion rate (*W*), distance of the whorl from the axis of coiling (*D*), whorl shape (*S*), and the translation of the whorl along the axis of coiling (*T*). Raup (1967) and Raup and Chamberlain (1967) applied this new framework to planispiral shells of ammonoids, rewriting the equations of Moseley (1838) in terms of the new variables. This new methodology has been applied by Ward (1980) to Jurassic and Cretaceous ammonoids and nautilids, by Chamberlain (1981) to cephalopod hydromechanics, and by Saunders and Swan (1984) to Carboniferous ammonoids. Most recently, Saunders and Shapiro (1986) have used this approach to calculate and simulate the hydrostatic properties of *Nautilus* and ammonoids.

1.2. Objectives

The objectives of this study are: (1) to use the Raup mathematical model of the ectocochleate shell to examine the effects of shell morphology on hydrostatic properties of the shell of *Nautilus* and (2) to examine interspecific variation within

Nautilus in terms of these variables. The analysis is based on four species of *Nautilus*: *N. belauensis* Saunders, 1981, *N. macromphalus* Sowerby, 1849, *N. pompilius* Linnaeus, 1758, and *N. scrobiculatus* [Lightfoot, 1786]. Neither geographic variation nor intraspecific variation is considered, because sample size for the data base is very limited. Therefore, the quantitative results must be considered as preliminary.

1.3. Methods

Measurements of the morphological variables were made from xerographic images of polished axial and sagittal shell cross sections (Figs. 1 and 2). The axial sections were cut approximately one quarter whorl back from the aperture. The measurements were made using a Houston Instruments digitizer interfaced with an Apple II+ microcomputer, using a program for data acquisition written by E. Shapiro. All shells analyzed were mature, as indicated by septal approximation and the presence of a blackened aperture.

On the assumption that the principal function of the shell is to serve as a flotation device, two hydrostatic properties were calculated: (1) the static orientation of the shell in the water and (2) its relative stability (Fig. 1). Calculations involved use of an interactive BASIC microcomputer program written by E.A.S., based on a FORTRAN program written by David Raup and supplied to us by John Chamberlain. The microcomputer version has been extensively modified and rewritten in Microsoft BASIC for use on the Apple II, Apple Macintosh, and IBM Personal Computers (for additional details of procedures, see Saunders and Shapiro, 1986). Copies of the program are available from the authors.

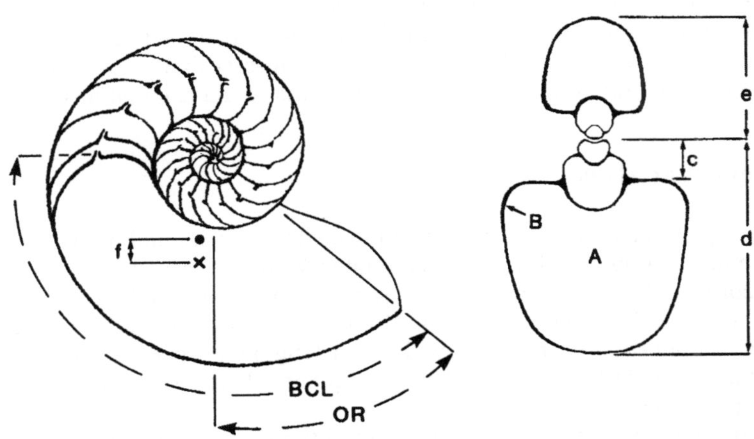

$$W=\left(\frac{d}{e}\right)^2 \quad D=\frac{c}{d} \qquad TH=\left(\frac{B}{A}\right)^{1/2} \quad ST=\left(\frac{f}{d+e}\right)$$

Figure 1. Measurements taken to establish variables: whorl expansion rate (W), distance to the axis of coiling (D), relative thickness (TH), body chamber length (BCL) measured in degrees, orientation (OR) measured in degrees from vertical, and relative stability (ST).

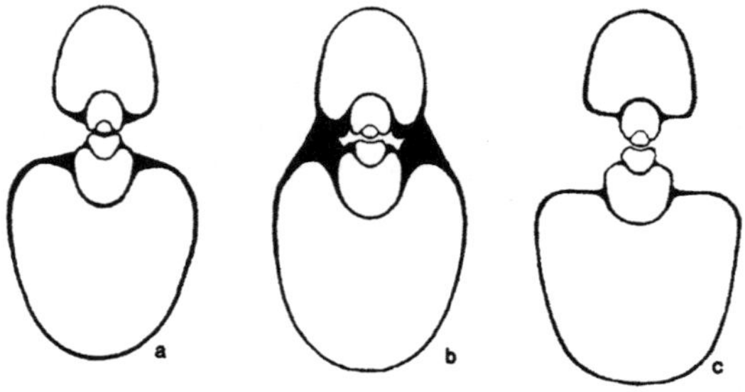

Figure 2. Sagittal cross sections of three species of *Nautilus*: (a) *N. macromphalus* Sowerby, 1849; (b) *N. pompilius* Linnaeus, 1758; (c) *N. scrobiculatus* [Lightfoot, 1786]. As shown here, the major difference among the three species geometrically is the degree of whorl overlap (*D*); this overlap is greatest in *N. pompilius* ($D = 0.08$), which has a heavy umbilical callus (black). The early whorls are partly exposed ($D = 0.118$) in *N. macromphalus* and even more so in *N. scrobiculatus* ($D = 0.20$). Sections represent digital scans of xerographic images of actual shell sections, produced by a digital scanner (Thunderscanner) and an Apple Macintosh microcomputer.

2. Computer Model

2.1. Equations

The equations used in this model (Table I) are those given by Raup and Chamberlain (1967). These equations calculate the center of mass and the center of buoyancy of the animal and make three assumptions that are known to be inappropriate for *Nautilus*: (1) The whorl is circular in axial cross section (i.e., a circular generating curve); (2) the mass of the animal is totally accounted for by the soft tissue within the body chamber (i.e., the weight of the shell is ignored in calculating the center of mass); and (3) the soft tissue of the animal does not extend beyond the margin of the aperture. Despite the violation done to actuality by these assumptions, there is a close fit between calculated and observed hydrostatic properties of *Nautilus*. This close fit suggests that these erroneous assumptions have only a minor effect on the results of the calculations.

2.2. Input Variables

The equations for the calculation of the centers of mass and buoyancy (Table I) utilize four variables: whorl expansion rate (*W*), distance of the whorl from the axis of coiling (*D*), angular length of the shell (Θ), and whorl radius (*R*). Although not specified in the equations, two additional variables are required for the calculation of the center of mass: body chamber length and relative shell thickness (for additional details, see Raup, 1967; Raup and Chamberlain, 1967; Saunders and Shapiro, 1986).

Whorl expansion rate (*W*) is a measure of the increase in shell radius in one half a volution of the shell; in this study, measurements of *W* were taken from polished axial sections.

The distance of the whorl from the axis of coiling (*D*) is defined as the ratio of the radius of the inner periphery of the whorl to the radius of the outer periphery of the whorl (Fig. 1). The radius of the inner periphery is measured from the axis of coiling to the umbilical seam (ignoring the umbilical callus where present).

Table I. Equations for Calculating Hydrostatic Properties of *Nautilus*

$$X = P\left[\frac{\left(\frac{2\ln W}{\pi}\sin\theta - \cos\theta\right)\left(W^{2\theta/\pi}\right)+1}{W^{3\theta/2\pi}-1}\right]$$

$$Y = P\left[\frac{\left(\frac{2\ln W}{\pi}\cos\theta + \sin\theta\right)\left(W^{2\theta/\pi}\right)-\frac{2\ln W}{\pi}}{W^{3\theta/2\pi}-1}\right]$$

where:

$$P = \frac{3}{2}\left(\frac{I}{KR_a}\right)\left(\frac{\pi\ln W}{W^{\theta/2\pi}}\right)\left(\frac{1}{\pi^2+4(\ln W)^2}\right)$$

Evolute forms	Involute forms
$I = \frac{\pi r^4}{4} + \pi r^2 R_a^2$	$I = \left\{\frac{r^4}{4}\left[\frac{\pi}{2} + \sin^{-1}\left(\frac{p}{r}\right)\right] + \frac{r^2}{4W^2}\left[\frac{\pi}{2} - \sin^{-1}\left(\frac{qW}{r}\right)\right]\left[\frac{r^2}{W^2} + 4(p+q)^2\right] + (r^2-p^2)^{1/2}\left[\frac{p(2p^2-r^2)-q\left(2q^2-\frac{r^2}{W^2}\right)}{4} - \frac{4}{3}(p+q)(r^2-p^2) - q(p+q)^2\right] + Kr\left(\frac{1+D}{1-D}\right)\left[2R_a - r\left(\frac{1+D}{1-D}\right)\right]\right\}$
$K = \pi r$	$K = \frac{\pi}{2}r^2\left(1-\frac{1}{W^2}\right) + (p+q)(r^2-p^2)^{1/2} + r^2\sin^{-1}\left(\frac{p}{r}\right) + \frac{r^2}{W^2}\sin^{-1}\left(\frac{qW}{r}\right)$
$R_a = r\left(\frac{1+D}{1-D}\right)$	$R_a = r\left(\frac{1+D}{1-D}\right) + \frac{1}{K}(p+q)\left[\frac{\pi}{2}\left(\frac{r^2}{W^2}\right) - q(r^2-p^2)^{1/2} - \frac{r^2}{W^2}\sin^{-1}\left(\frac{qW}{r}\right)\right]$
	where:
	$p = \frac{r}{2W}\left[\left(\frac{1+D}{1-D}\right)(W-1) + \left(\frac{1-D}{1+D}\right)(W+1)\right]$
	$q = \frac{r}{2W}\left[\left(\frac{1+D}{1-D}\right)(W-1) - \left(\frac{1-D}{1+D}\right)(W+1)\right]$

When considered in conjunction with W, D is an indicator of whorl overlap. If the product of $W \times D$ is greater than 1, a shell is evolute; for involute shells, $W \times D$ is less than 1. Although both W and D can be measured from lateral illustrations of the shell, both the axis of coiling and the umbilical seam can be determined more precisely from polished axial sections.

Body chamber length (*BCL*) is defined as the angular distance from the final septum to the aperture. The actual measurement of *BCL* is taken from a polished sagittal section of the shell. The measurement, in degrees, is made from the posterior most point on the anterior surface of the final septum to the posterior most point of the aperture [the hyponomic sinus (Fig. 1)].

The number of whorls (*NW*) is a measure of the angular length of the shell from the first chamber to the aperture as an integral number of full (360°) volutions. The actual count is taken from a polished sagittal section of the shell. In evolute shells, the number of whorls can be determined from a lateral illustration of the shell.

Whorl radius (*R*) is defined as the radius of the generating curve, assuming a circular whorl cross section (Raup and Chamberlain, 1967). This parameter has no effect on the hydromechanical properties of the shell and can effectively be ignored in studies of shell hydrostatics. In essence, a large *Nautilus* shell will behave in the same fashion as does a small *Nautilus* shell.

Relative shell thickness (*TH*) is defined as the ratio of the absolute thickness of the shell to the radius of the whorl at the place where the thickness is measured. The analyses of Raup (1967) and Raup and Chamberlain (1967) assumed that the whorls are circular and uniform in thickness. However, these two assumptions are not valid for *Nautilus*, and the violation they do presents serious methodological problems. Saunders and Shapiro (1986) utilized a modified thickness index, given as

$$TH \simeq \frac{\text{shell cross-sectional area}}{\text{whorl cross-sectional area}} \times \tfrac{1}{2}$$

This approximation is valid as long as the absolute thickness of the shell is very small compared to the radius of the whorl (Saunders and Shapiro, 1986, Appendix A). It has the advantage of being directly measurable on actual specimens, or measurements can be taken from xerographic images of polished axial sections of the shell. In the case of *Nautilus*, the umbilical callus and the shell material of the preceding overlapped whorl are not included in the measurements.

2.3. Other Variables

If an assumption of neutral buoyancy is made, as was done by Raup (1967), Raup and Chamberlain (1967), and Saunders and Shapiro (1986), four additional variables are required: the density of seawater, the density of shell material, the density of tissue, and a correction factor for the weight of the septa and siphuncle. Because the assumption of neutral buoyancy was relaxed in this analysis, these additional four variables are not considered here.

2.4. Output Variables

The equations shown in Table I are used to calculate the center of buoyancy and the center of mass for a shell with a geometry that is described by the input variables discussed in Section 2.2. The center of buoyancy is the center of gravity of the mass of water displaced by the shell. The center of mass is the center of gravity of the body chamber. Using the centers of buoyancy and mass, it is possible to calculate the orientation of the aperture of the shell in water and the relative stability.

The static orientation of the aperture is determined by alignment of the center of buoyancy directly above the center of mass (Trueman, 1941; Raup, 1967; Raup and Chamberlain, 1967; Saunders and Shapiro, 1986). Orientation is the angle of the aperture from vertical, measured in degrees. If the aperture is vertical and below the axis of coiling, orientation is 0°; if the aperture is vertical but above the axis of coiling, orientation is 180°.

Stability in *Nautilus* is a function of the distance between the centers of buoyancy and mass (Trueman, 1941; Raup, 1967). Relative stability was defined by Raup (1967) as the ratio of the distance between the two centers of gravity to the diameter of the shell.

3. Applying the Model to *Nautilus*

3.1. Observed Input Variables

There is relatively little variation in *W* among the four species of *Nautilus* (Table II); the average value is 3.040. For comparison, most ammonoids have values of *W* between 1.5 and 2.5, with a mode at approximately 2.0 (Raup, 1967; Ward, 1980; Chamberlain, 1981; Saunders and Swan, 1984), whereas most fossil nautilids (Jurassic to Recent) have *W* values between 2.8 and 4.3, with a mode between 3.7 and 3.8 (Ward, 1980).

There is significant variation in *D* among the four species of *Nautilus* (Table II); the mean is 0.1082, indicating a high degree of whorl overlap. By comparison, *D* in most ammonoids ranges from 0.0 to 0.7, with the highest concentration of values between 0.35 and 0.45; most fossil Nautilida have *D* values between 0.0

Table II. Observed Input Variables for Four Species of *Nautilus*[a]

Species	No.	*W*	*D*	*TH*	*BCL*	*NW*
N. belauensis	BMC 19	3.115	0.0643	0.0334	152.8°[b]	3.5
	BMC 368	3.281	0.0769	0.0352	152.8°[b]	—
N. pompilius	BMC 353	2.975	0.0797	0.0448	159.2°[c]	3.5
N. macromphalus	BMC 051	2.813	0.1176	0.0499	139°[d]	3.5
N. scrobiculatus	BMC 7	3.016	0.2025	0.0352	140°[d]	3.5
	Mean:	3.040	0.1082	0.0397	147.75°[e]	
	Standard deviation:	0.17323	0.05635	0.007298	9.8865°	
	Coefficient of variation:	0.05698	0.52077	0.18262	0.06691°	

[a] For explanation of the variables, see Fig. 1, except for *NW*, which is discussed in Section 2.2.
[b] Based on the mean of 10 shells. [c] Based on the mean of 6 shells.
[d] Based on a single specimen, different from the specimen used for *W*, *D*, and *TH*.
[e] Only a single value was used for *N. belauensis*.

Table III. Intraspecific Variation in Body Chamber Length in Mature Specimens of *Nautilus belauensis* and *Nautilus pompilius*

	N. belauensis	*N. pompilius*
	164.5°	167°
	161°	163°
	138°	161°
	140°	154°
	150°	155°
	151.5°	155°
	155°	—
	156°	—
	156°	—
	156°	—
Mean:	152.8°	159.2°
Standard deviation:	8.3805°	5.3071°
Coefficient of variation:	0.05485°	0.03334°

and 0.2, with a modal value between 0.05 and 0.07 (Raup, 1967; Ward, 1980; Chamberlain, 1981; Saunders and Swan, 1984).

The four species of *Nautilus* exhibit relatively low *TH* values, with a mean of 0.0397 (Table II). Although few data are available for ammonoids, values ranging from 0.0198 to 0.0433 have been reported for Middle Carboniferous species (Saunders and Shapiro, 1986).

The *BCL* in *Nautilus* averages 148° (Table II), although a considerable range has been reported [124–167° (Saunders and Shapiro, 1986)]. By comparison, Jurassic ammonoids show *BCL* values ranging from 120 to 480°, with a mode at 270° (A. E. Trueman, 1941; Raup, 1967). Data from a limited number of fossil nautilids show a *BCL* range of 115–163°, with a mean of 140° (Ward, 1980). Although variation in *BCL* among species of *Nautilus* is high in absolute terms (S.D. ≈ 10°), the coefficient of variation is relatively low. Intraspecific variation in *BCL*, as measured in mature specimens of *N. belauensis* and *N. pompilius* (Table III), is comparable to interspecific variation.

All four species of *Nautilus* have approximately 3½ whorls at maturity. This value was rounded to 3 whorls for all calculations because (1) the computer program requires an integral number of whorls and (2) measurements for *W*, *D*, and *TH* were made approximately one quarter whorl back from the aperture.

3.2. Calculated Output Variables

Calculated orientations for the four species are similar (Fig. 3 and Table IV), with the mean orientation 42.3°, range 8°, and standard deviation 3°. These orientations are in close agreement with observed orientations of *N. pompilius* and *N. belauensis* (Saunders and Shapiro, 1986, Fig. 6).

The calculated relative stabilities of the four species of *Nautilus* (Table IV) are also similar, with a mean value of 0.0966. This relatively high stability pro-

Table IV. Calculated Orientation and Relative Stability for Four Species of *Nautilus*

Species	Specimen no.	Orientation	Relative stability
N. belauensis	19	43.4°	0.0892
	368	42.3°	0.0833
N. pompilius	353	46.9°	0.0861
N. macromphalus	051	39.8°	0.1158
N. scrobiculatus	7	38.9°	0.1084
Mean:	42.26°	0.0966	
S.D.:	3.169°	0.01457	
Coefficient of variation:	0.07499°	0.15095	

[a] Orientation is measured in degrees from vertical. Relative stability is a dimensionless variable that relates the distance between the centers of buoyancy and mass to the diameter of the shell. Results are derived from the input variables of Table II.

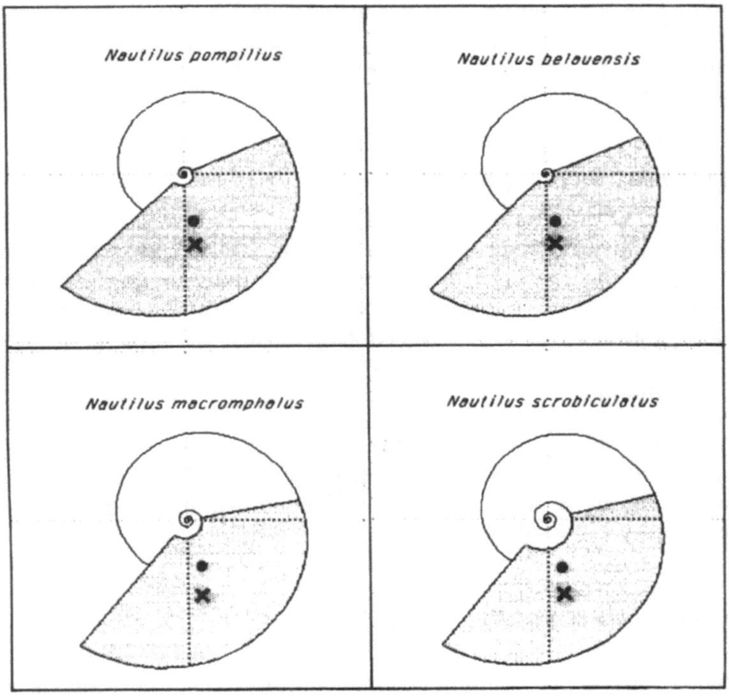

Figure 3. Microcomputer simulation of four species of *Nautilus*, showing aperture orientation, center of buoyancy (●), center of mass (X), and body chamber (stippled). In addition to the geometric parameters listed for each species in Table II, the calculations and simulations also utilized the following physical parameters: tissue density, 1.055 g/cc; seawater density, 1.026 g/cc; shell density, 2.62 g/cc; siphuncle/septal correction factor, 1.3; and whorl radius, 20. Simulations were produced on an Apple Macintosh computer. For additional details, see Saunders and Shapiro (1986).

vides resistance to changes in orientation, but it also has the effect of reducing the agility of the animal in terms of shell orientation (Chamberlain, 1981). It is notable that, although limited data are available, the calculated stabilities for ammonoids are considerably lower, approximately 0.01 in five species of Namurian ammonoids (Saunders and Shapiro, 1986).

The calculated orientations and relative stabilities raise a new question: Given the amount of variability seen in the input variables (Table II), why is there so little variation in the calculated (and observed) orientations and stabilities? Numerical analysis of the equations in Table I provides an answer.

4. Numerical Analysis

Numerical analyses were performed for *W*, *D*, *TH*, *BCL*, and *NW*. Each variable was considered separately, while the other four variables were held constant at their mean values for *Nautilus* (from Table II). For these analyses, the assumption of neutral buoyancy was relaxed, and each variable was considered over a much broader range of values than actually occur in nature.

The results of numerical analyses outside the *Nautilus* range of input values can yield response curves distinctly different from those for *Nautilus*. Therefore, the results of this study should not be applied indiscriminately to other combinations of shell geometry (e.g., ammonoids).

4.1. Whorl Expansion Rate

The relationship between whorl expansion rate (*W*) and orientation is negative and approximately logarithmic (Fig. 4). Within the range of *W* values of 1.1–4.0, orientation decreases from 70.58 to 36.47°, yielding a range of orientations of 34.11°. *Nautilus* is located on the relatively low-slope segment of the response curve. Within the observed *Nautilus* range of *W* values, there is a difference of 3.35° in orientation.

Relative stability also shows a negative, nonlinear response to changes in *W* (Fig. 5). Within the range of *W* values from 1.1 to 4.0, stability decreases from 0.4016 to 0.0687, spanning a range of almost one order of magnitude. The observed range of values for *W* within the four species of *Nautilus* results in a range in relative stability of less than 0.02.

4.2. Distance from the Axis of Coiling

Orientation has a very minor, positive, nonlinear relationship to distance from the axis of coiling (*D*) (Fig. 6). Within the range of *D* values from 0.01 to 0.60, orientation increases less than 0.5°, from 41.89 to 42.3°. The range of observed *D* values for the four species of *Nautilus* results in a range in orientation of 0.13°.

Relative stability has a slightly positive, nonlinear relationship to *D* (Fig. 7). Within the range of *D* values from 0.01 to 0.60, relative stability varies from 0.0964

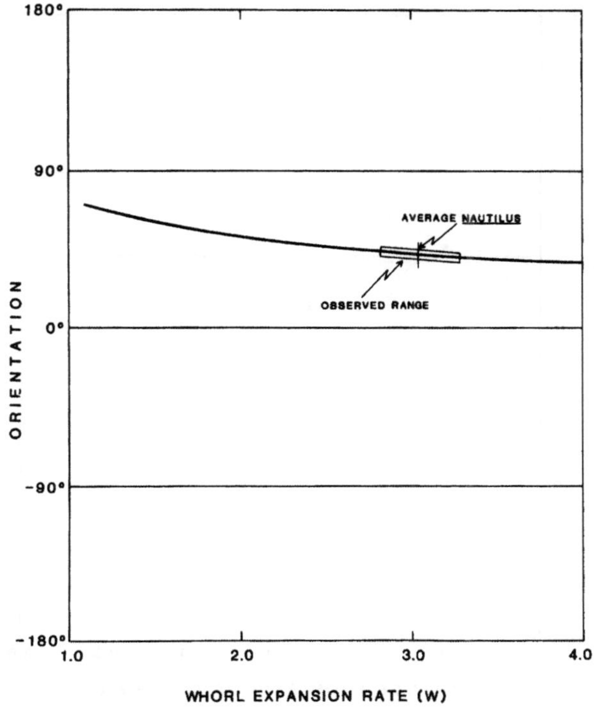

Figure 4. Response curve for orientation vs. whorl expansion rate (W). Other variables are held constant ($D = 0.1082$, $TH = 0.0397$, $BCL = 148°$, $NW = 3$).

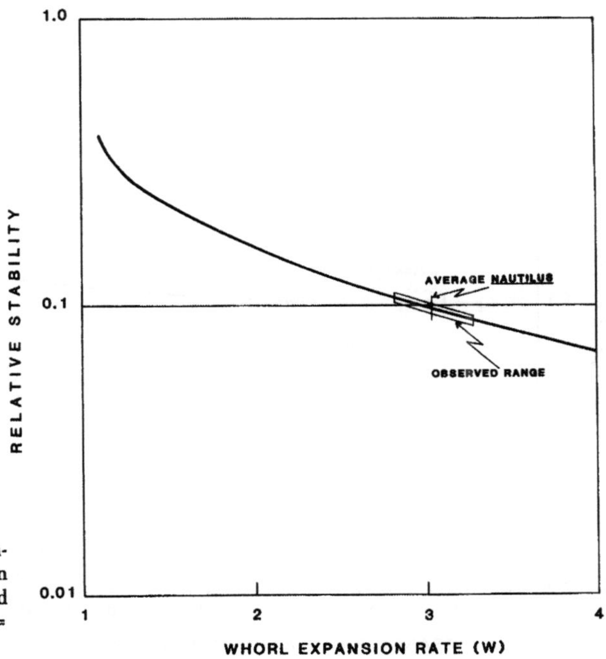

Figure 5. Response curve for relative stability vs. whorl expansion rate (W). Other variables are held constant ($D = 0.1082$, $TH = 0.0397$, $BCL = 148°$, $NW = 3$).

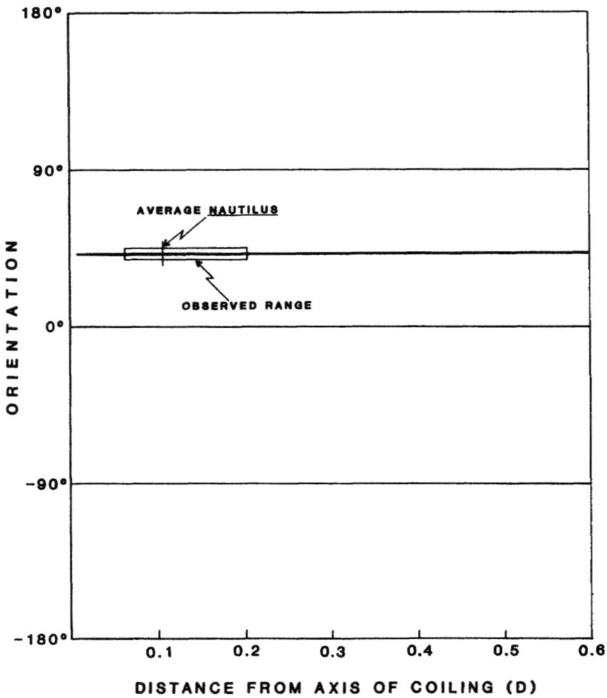

Figure 6. Response curve for orientation vs. distance from the axis of coiling (*D*). Other variables are held constant (*W* = 3.040, *TH* = 0.0397, *BCL* = 148°, *NW* = 3).

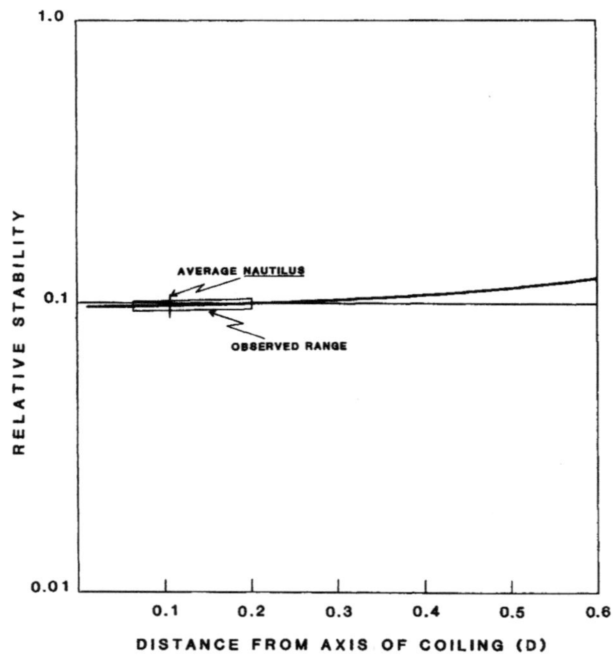

Figure 7. Response curve for relative stability vs. distance from the axis of coiling (*D*). Other variables are held constant (*W* = 3.040, *TH* = 0.0397, *BCL* = 148°, *NW* = 3).

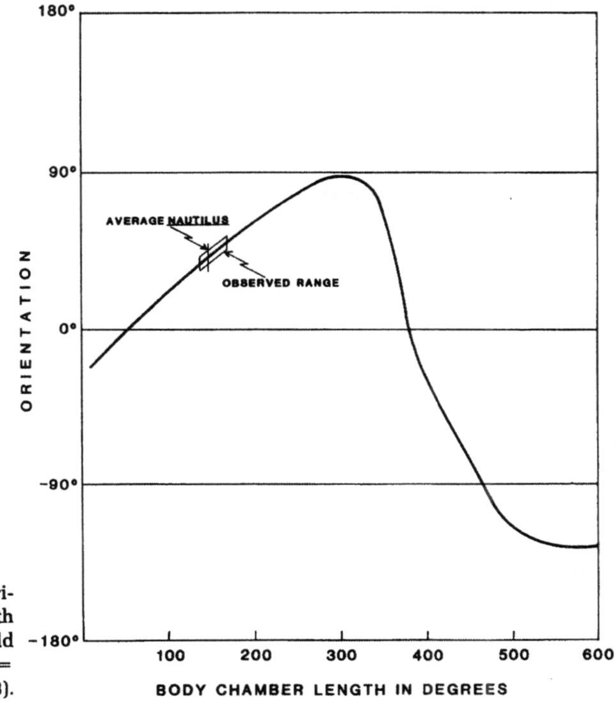

Figure 8. Response curve for orientation vs. body chamber length (*BCL*). Other variables are held constant ($W = 3.040$, $D = 0.1082$, $TH = 0.0397$, $NW = 3$).

to 0.1193, yielding a range of 0.0229. Within the observed range of *D* values for *Nautilus*, the maximum difference in stability is only 0.0012.

4.3. Body Chamber Length

The effect of body chamber length (*BCL*) on orientation is both significant and complex (Fig. 8). Within the range of *BCL* from 10 to 600°, orientation varies by 214°, ranging from 88.62 to −125.38°. Observed *Nautilus BCL* values are located on the moderately sloped, upward limb of the curve; the maximum difference in resultant orientations is 11.43°.

The effect of *BCL* on relative stability is also complex (Fig. 9). Within the range of *BCL* from 10 to 600°, the maximum relative stability value is 0.3348, the minimum value is 0.00078, and the range covers almost three orders of magnitude. Within the range of observed *BCL* for *Nautilus*, relative stability ranges from 0.1071 to 0.0767.

4.4. Relative Thickness

Orientation is a slightly negative function of relative thickness (*TH*) (Fig. 10). Within the range of *TH* from 0.01 to 0.60, maximum orientation ranges from 34.26–

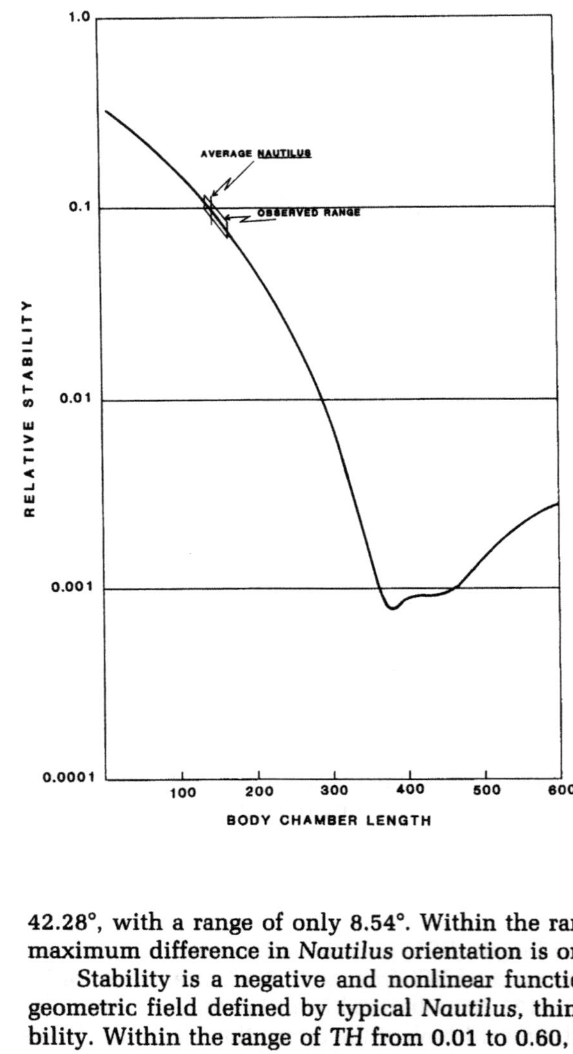

Figure 9. Response curve for relative stability vs. body chamber length (*BCL*). Other variables are held constant ($W = 3.040$, $D = 0.1082$, $TH = 0.0397$, $NW = 3$).

42.28°, with a range of only 8.54°. Within the range of observed *TH* values, the maximum difference in *Nautilus* orientation is only 0.2°.

Stability is a negative and nonlinear function of *TH* (Fig. 11). Within the geometric field defined by typical *Nautilus*, thinner shells produce greater stability. Within the range of *TH* from 0.01 to 0.60, stability decreases from 0.1006 to 0.0511, spanning less than one order of magnitude of values. Within the observed range of *TH*, stability varies by only 0.0022.

4.5. Number of Whorls

The number of whorls (*NW*) has very little effect on orientation (Fig. 12); a range of 1–13 whorls varies orientation by only 0.62°. The number of whorls also has a minimum effect on stability (Fig. 13), which varies only 0.0046 within the range of 1–13 whorls.

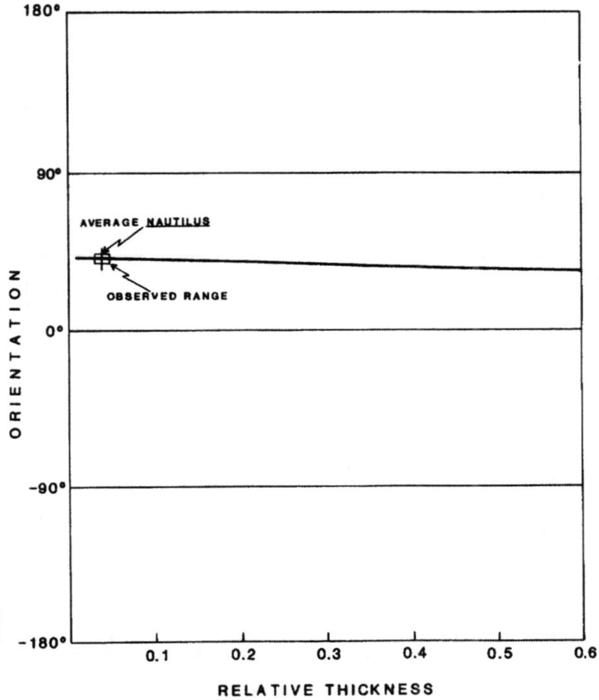

Figure 10. Response curve for orientation vs. relative thickness (*TH*). Other variables are held constant (*W* = 3.040, *D* = 0.1082, *BCL* = 148°, *NW* = 3).

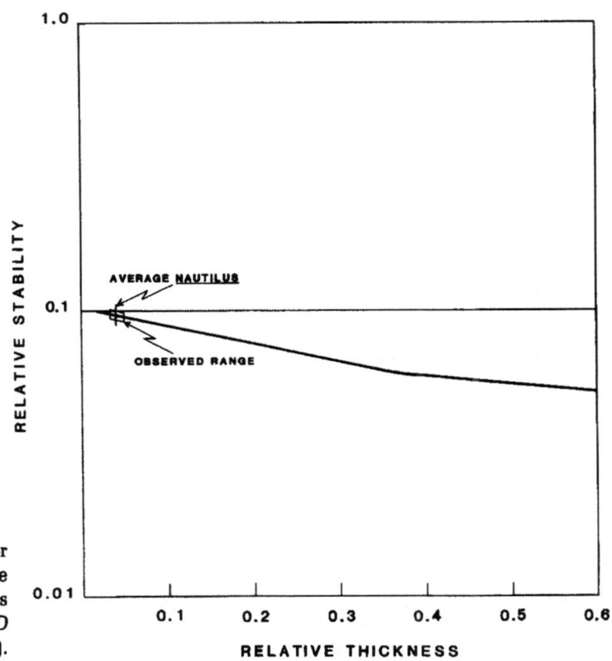

Figure 11. Response curve for relative stability vs. relative thickness (*TH*). Other variables are held constant (*W* = 3.040, *D* = 0.1082, *BCL* = 148°, *NW* = 3).

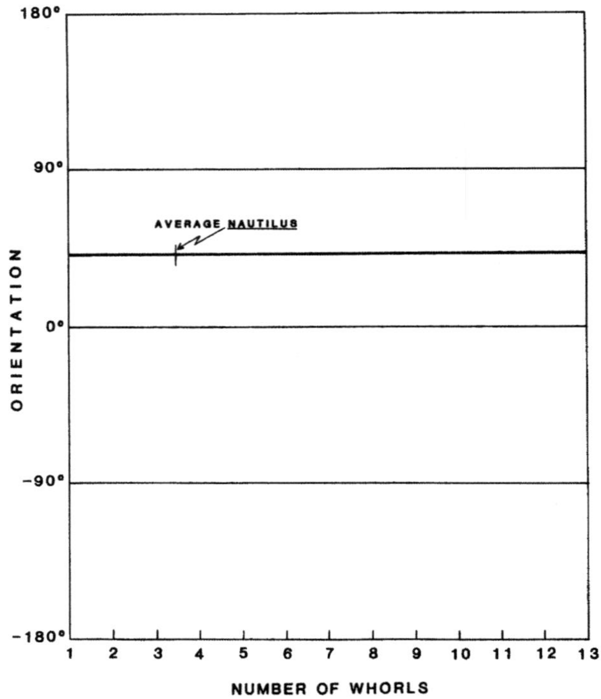

Figure 12. Response curve for orientation vs. number of whorls (*NW*). Other variables are held constant (W = 3.040, D = 0.1082, TH = 0.0397, BCL = 148°).

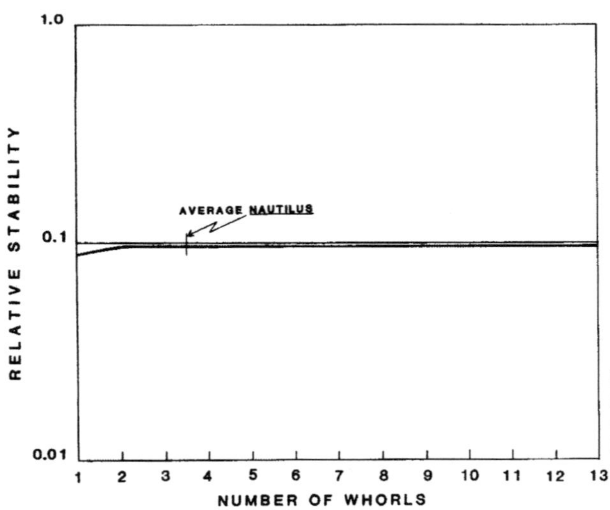

Figure 13. Response curve for relative stability vs. number of whorls (*NW*). Other variables are held constant (W = 3.040, D = 0.1082, TH = 0.0397, BCL = 148°).

5. Conclusions

From the preceding series of response curves, it can be seen that although some variables have almost no effect on *Nautilus* hydrostatics, others have significant and complex influence on such features as orientation and stability. From these response curves, we can answer the following questions:

1. Which variables affect the shell as a flotation device?
2. How do these variables differ among different species of *Nautilus*?
3. Why is there so little variation in the calculated and observed orientation and stability of the four species?

5.1. Variables

The equations used to calculate orientation and relative stability utilize six independent variables: *W*, *D*, *BCL*, *TH*, *NW*, and *R*. The last (*R*) acts as a constant of proportionality and has no effect on orientation or stability. Numerical analysis of the remaining five variables shows their relative effects on orientation and stability (Table V). The results are primarily a function of the shape of the response curves and are relatively independent of the *Nautilus* data base.

The single most important factor that controls relative stability in *Nautilus* is body chamber length (*BCL*). The four species examined showed a range of 28°; this effects a range in orientation values of over 11° and a shift in relative stability of 0.0217. However, it is known that *BCL* also varies as the animals grow. We recorded *BCL* values of 124–167° in immature and mature *N. pompilius* and *N. belauensis* and noted that short *BCL*s tend to occur in immature animals that were in the process of forming new septa (Saunders and Shapiro, 1986, Fig. 6). It is interesting to consider the effects of such variation on orientation and stability. The angular distance between successive septa (≈20° in *Nautilus*) is a measure of the maximum change in *BCL* in an individual specimen during growth. This range of variation (20°) will cause the aperture to shift upward by 8.08° and will reduce relative stability by 0.02082.

Whorl expansion rate (*W*) also produces significant changes in orientation

Table V. Differences in Orientation and Stability That Result from Observed Differences in One Variable When Other Variables Are Held Constant at Their Average Values

Variable	Orientation[a]	Orientation as % of mean	Stability[b]	Stability as % of mean
W	3.35°	7.99%	0.01826	18.90%
D	0.13°	0.31%	0.0012	1.24%
BCL	11.43°	27.25%	0.0217	22.46%
TH	0.20°	0.48%	0.00219	2.27%
NW	0	0	0	0

[a] Mean calculated orientation is 41.94%°.
[b] Mean calculated stability is 0.0966°.

and stability. The range of *W* in the four *Nautilus* species is 0.468, which results in a range of 3.35° in orientation and a range of 0.0183 in relative stability. We have noted elsewhere that in *Nautilus*, *W* typically changes considerably during ontogeny—from 2.4 to 3.4 in the early stages (Saunders and Shapiro, 1986, Fig. 5). These changes, in turn, would affect both orientation and stability during growth. Changes in *D*, *TH*, and *NW* produce no significant changes in orientation and relative stability.

5.2. Variation in Parameters

The descriptive statistics for *Nautilus*, in terms of the four critical variables (see Table II), suggest an inverse relationship between the range of variability and the effectiveness of a variable in controlling orientation and relative stability (Fig. 14). The two variables that have the greatest effect on orientation and stability, *W* and *BCL*, have relatively small coefficients of variation. By contrast, *D* and *TH*, which have negligible effects on orientation and stability, have relatively large coefficients of variation. This suggests that maintaining a viable range of orientation and stability may serve as a functional constraint that limits the range of variability for *W* and *BCL*.

5.3. Hydrostatic Properties

Despite differences in shell geometry for the four species of *Nautilus*, the calculated orientations and relative stabilities show little variation among species. Most of this low variability is explicable in terms of the response curves; only *W* and *BCL* show significant effects on orientation and relative stability. For these two variables, the observed range of values in *Nautilus* is situated on relatively

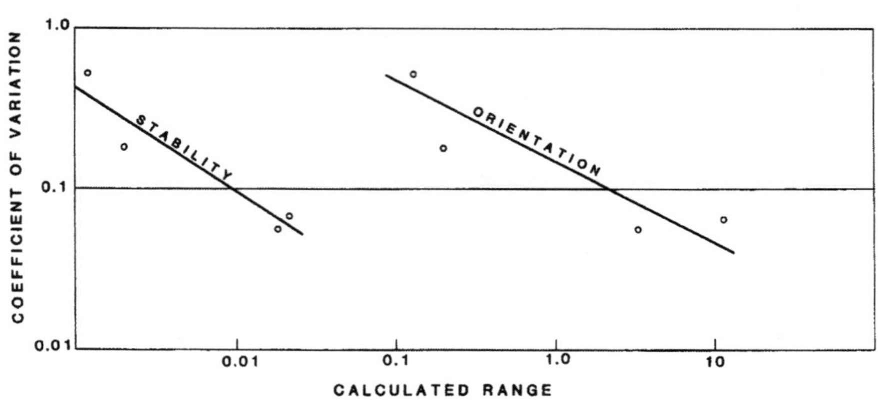

Figure 14. Relationship between the coefficients of variation for observed variables, *W*, *D*, *TH*, and *BCL* (from Table II), and the ranges of orientation and stability that result from the observed maxima and minima of these variables.

low-slope portions of each curve; this means that even large fluctuations in these variables would not change the orientation or stability significantly. The response curves for the remaining three variables are so flat that virtually no amount of increase in the input variables would have a significant effect on orientation and stability.

6. Epilogue

This chapter began with a quote from Arthur Willey regarding the inability of *Nautilus* to turn upside down. However, the numerical analysis presented herein allows us to propose a shell geometry for *Nautilus* that escapes Willey's dictum. If the body chamber length in *Nautilus* were increased approximately 465°, the animal would in fact be oriented upside down. Unfortunately, such an animal would be negatively buoyant. Therefore, if we insist that *Nautilus* be neutrally buoyant, then indeed Willey was correct. There is no combination of variables that will result in a neutrally buoyant, upside-down *Nautilus*.

ACKNOWLEDGMENTS. The authors would like to thank Bruce O'Connor and Paul Huddlestun, of the Georgia Geologic Survey, and John Chamberlain for reviewing this manuscript. The manuscript was approved for publication by the Georgia Geologic Survey and was supported in part by NSF Grant EAR 83-18932 to W. B. Saunders.

Chapter 34

Buoyancy in *Nautilus*

LEWIS GREENWALD and PETER D. WARD

1. Introduction

Since the pioneering efforts of Eric Denton and John Gilpin-Brown (1966), much has been learned about the buoyancy system of the chambered *Nautilus*. It seems timely, therefore, to evaluate critically what is known and not known about the buoyancy system of *Nautilus*, rather than merely to restate what has gone before. In particular, we will discuss what we see as the major outstanding questions on *Nautilus* buoyancy research and what we regard to be the more subtle and poorly understood aspects of buoyancy in *Nautilus*. We will try to make explicit the assumptions that were heretofore implicit in our own and in others' analyses and to present some new data on *Nautilus* buoyancy mechanisms.

2. Cameral Liquid and Cameral Gas

Nautilus that weigh as much as 1 kg or more in air generally weigh only a few grams in seawater (Denton and Gilpin-Brown, 1966; Ward and Martin, 1978). The explanation for this phenomenon is, of course, that much of the *Nautilus* shell consists of a series of chambers from which water has been displaced. That is, because the weight of an object submerged in seawater equals its weight in air minus the weight of the water displaced (which itself equals the volume of seawater displaced times the density of that seawater), the empty chambers displace a sufficient weight of seawater to render the animal nearly neutrally buoyant. The chambers are filled with gas, but it is commonly overlooked that the gas, in a strictly physical sense, decreases buoyancy by a slight though no doubt negligible amount.

LEWIS GREENWALD • Department of Zoology, The Ohio State University, Columbus, Ohio 43210. PETER D. WARD • Department of Geological Sciences, University of Washington, Seattle, Washington 98195.

Although speculation on the buoyancy system of *Nautilus* goes back to at least the writings of Robert Hooke in the 1690s (see Denton, 1974), the modern era of *Nautilus* research began with the careful morphological observations of Willey (1902). Along with his detailed anatomical descriptions, Willey made some intriguing comments relevant to buoyancy in *Nautilus*. First, Willey claimed, but presented no data to show, that *Nautilus* in the Loyalty Islands, near New Caledonia, make nocturnal ascents into "shoal water." In fact, at night, *N. macromphalus* in New Caledonia are occasionally (although rarely) seen in water as shallow as 20 m (Ward and Martin, 1980). Also, the telemetry-based observations of Ward *et al.* (1984) and Carlson *et al.* (1984) show that during daylight hours, *N. belauensis* is typically at depths of 300–500 m, but that during the night, they may ascend to approximately 100 m. The possible contribution of buoyancy changes to such daily vertical migrations is discussed in Section 4.

Willey (1902) also wrote that the pallial veins of *Nautilus* are capable of generating gas, and we presume that he believed that this gas was the source of the gas found in the animal's buoyancy chambers. Willey, who was such a careful observer in anatomical matters, made either an inexplicable error or an observation of critical importance in understanding the buoyancy system of *Nautilus*. We prefer the former choice, because Willey's observation has never been repeated and because a more plausible explanation for gas generation in *Nautilus* exists, as is discussed below. (Perhaps the gas in the pallial veins recorded by Willey resulted from putrefaction or was released from hemocyanin.)

Denton and Gilpin-Brown (1966) described the buoyancy system of *Nautilus* under steady-state conditions. In that study, it was shown that the newest chamber is filled either completely or partially with cameral liquid. When first formed, this liquid closely resembles seawater, but with sufficient differences in divalent ion concentrations relative to seawater to make it likely that cameral liquid is a secreted body fluid and not seawater at all (Denton and Gilpin-Brown, 1966; Greenwald and Ward, 1982; Mangum and Towle, 1982). Also, if there is a pathway by which seawater could be introduced into the space that will become a new chamber, that pathway is not obvious. In hundreds of *Nautilus* examined in the course of our research, we have never noticed any "foreign" material in a chamber. Thus, all evidence suggests that cameral liquid is a secreted body liquid.

Denton and Gilpin-Brown (1966) and Collins *et al.* (1980) pointed out that cameral liquid is a necessary brace against the external hydrostatic pressure pushing on a newly forming septum. Once the septum is sufficiently thickened (i.e., strengthened), emptying of cameral liquid commences. Prior to such emptying, the siphuncle, the vascularized living tissue (encased in horny and in chalky tubes) that runs through each chamber, removes some NaCl from the cameral liquid. After this preemptying salt removal, true emptying begins with both salt and water transport out of the chamber (Denton and Gilpin-Brown, 1966; Ward, 1979). As the cameral liquid level falls to about 50% of the chamber volume (the point at which cameral liquid is no longer in direct contact with the siphuncle—the point of decoupling), the NaCl concentration of the cameral liquid progressively decreases (Fig. 1). After decoupling, although emptying continues, the NaCl concentration of the remaining cameral liquid actually *increases* (Ward, 1979; Ward *et al.*, 1981) to approach, but never exceed, the blood osmotic concentration. This increase suggests that the mechanism of emptying when cameral liquid is

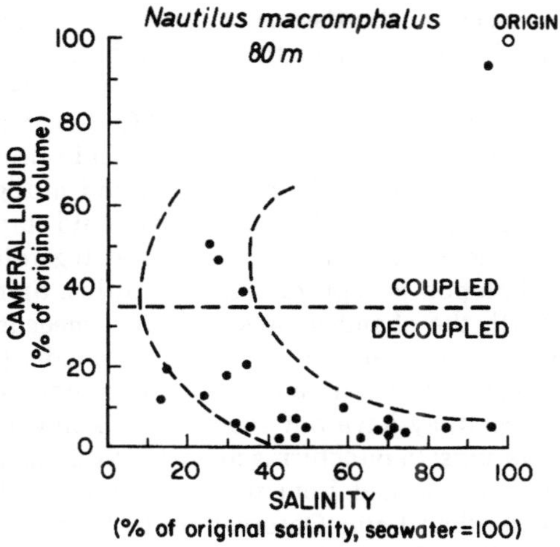

Figure 1. Relationship between the emptying (cameral liquid volume) and the concentration (from chloride concentration) of the cameral liquid left in the shell (data from animals taken from traps at 80 m). Filled chambers contain cameral liquid at a concentration very close to 100% seawater. As cameral liquid is removed from a chamber, the concentration falls until the chamber is about half full. At this stage, the salt concentration of the remaining cameral liquid increases as emptying proceeds. Modified from Ward (1979).

in contact with the siphuncle (coupled emptying) is different from the mechanism of emptying when cameral liquid is not in direct contact with the siphuncle (decoupled emptying). In the latter case, liquid is "wicked" to the siphuncle by the wettable pellicle, which lines each chamber.

As cameral liquid is removed from the chamber, a vacuum is created. Some water vapor equilibrates into this space, but given the concentration of *Nautilus* blood and the temperatures at which *Nautilus* is typically found, this water vapor cannot account for more than about 20 mm Hg. In addition to the water vapor, the various blood gases (nitrogen, argon, carbon dioxide, and oxygen) also diffuse into the partial vacuum of the "empty" chamber.

The chamber gas of older chambers, which have had more time to come to equilibrium, contains nitrogen and argon at partial pressures equal to atmospheric partial pressures (Denton and Gilpin-Brown, 1966). Chamber oxygen is at partial pressures lower than atmospheric partial pressures, and carbon dioxide is at higher pressures. The presence of these gases was explained by Denton and Gilpin-Brown, who recognized that the blood of *Nautilus* is in equilibrium (across the gills) with the nitrogen and argon of well-mixed seawater and that the nitrogen and argon tensions, at any depth, are in equilibrium with surface nitrogen and argon. Nitrogen and argon are metabolically inert; hence, their equilibrium partial pressures in older chambers equal atmospheric partial pressures. Younger chambers are progressively less close to nitrogen and argon equilibrium. This model of nitrogen equilibration with chamber gas was used by Denton and Gilpin-Brown (1966) as the basis of an ingenious attempt to measure the time required by *Nautilus* for chamber formation. This attempt, however, was based on incomplete information about nitrogen permeability of the siphuncle and cameral liquid emptying rates and thus may have yielded a misleading result (see, for example, Ward and Chamberlain, 1983). This approach remains, nevertheless, a creative attack on what then seemed an intractable problem, the rate of chamber formation.

Because the tissues of *Nautilus* consume oxygen, it is reasonable that the chamber PO_2 is less than atmospheric PO_2. Because CO_2 is produced by the tissues of the animal (and is therefore at elevated concentrations in the blood), it is also reasonable that chamber PCO_2 is at higher levels than atmospheric PCO_2.

In summary, the Denton and Gilpin-Brown explanation for the presence and concentrations of chamber argon and nitrogen is based entirely on the passive processes that result from cameral liquid emptying. A critical argument in support of this model is their observation that, regardless of the depth at which a chambered cephalopod is captured [be it *Nautilus*, *Spirula*, or *Sepia* (references in Denton, 1974)], the chamber pressure *when measured within a meter of the surface* is always less than 1 atm. This entire model (and the emptying models that drive it) would be invalidated if chamber pressures greater than 1 atm were ever observed (they have not been, to our knowledge). Also, we can add our own observations: In puncturing many *Nautilus* chambers in seawater, we never observed gas being forced from the chamber—which would indicate chamber pressure greater than 1 atm. Observations by Saunders (1985, p. 676) are in agreement with ours. Saunders perforated the last chamber of two specimens of *N. belauensis* at a depth of 31 m (≈4 atm) immediately after retrieving the animals from a greater depth (≈300 m). Water flowed into the openings, and no gas was observed escaping from the shells as they were slowly taken to the surface. It would still be reassuring to see more pressure measurements made at depth, rather than near the surface, because virtually all our understanding of *Nautilus* emptying physiology hinges on chamber pressure being 1 atm or less.

A consequence of the Denton and Gilpin-Brown gas generation model is that if hyperbaric chambers are to be used in *Nautilus* research, then the results obtained from such studies must be interpreted with care. Most hyperbaric chambers are pressurized with gas, and that gas must dissolve in, and equilibrate with, the water of the hyperbaric chamber, and with the blood of the *Nautilus*. Given sufficient time, the gas would then diffuse into the shell chambers and establish a nonphysiologically high pressure, obviating active cameral liquid transport against a high pressure differential, because the latter would no longer exist. The choice of the gas in the hyperbaric chamber would be a problem. Oxygen and nitrogen are toxic at high pressures. Carbon dioxide could interfere with a variety of physiological processes (respiration, acid–base regulation, calcification). Hyperbaric seawater chambers that are mechanically rather than gas-pressurized could be used, provided gas bubbles were scrupulously avoided.

Ward *et al.* (1981) showed that new chamber formation begins at approximately the point of decoupling (i.e., when the cameral liquid level falls below the siphuncular neck) (Fig. 2). The second-oldest chamber of a *Nautilus* shell thus contains, at most, half a chamber of cameral liquid and may contain only a trace. Collins *et al.* (1980) noted that, although the cameral liquid of the newest chamber initially acts as a brace against external hydrostatic pressure, the cameral liquid of the second chamber serves a buoyancy function. As the septum of the new chamber thickens, and as the aperture of the shell continues to grow, the overall density of the animal increases. This increase is counteracted by compensatory emptying from the second (and possibly older) chambers, the septum of the first chamber being still too weak to permit emptying of that chamber. The third and older chambers contain only traces of cameral liquid, and until the recent ob-

Figure 2. A series of radiographs showing new chamber formation. From left to right and top to bottom, it can be seen that the animal has added a new chamber (*top right*), which is emptied to the point at which the cameral liquid fills about half the chamber (*middle left*). Shortly after this stage, another new chamber is added (*bottom left*), which again begins the emptying process (*bottom right*). Reprinted with permission from Chamberlain (1983).

servations of Ward (1986), these older chambers and their cameral liquid were thought not to play any dynamic role in buoyancy control.

In summary, the observations of Denton and Gilpin-Brown (1966) and Ward and Martin (1978) showed that emptying was a slow and irreversible process. That emptying is slow (less than 1 ml/chamber per day) seemed to be good evidence that *Nautilus* did not use buoyancy changes as an aid to daily vertical migrations. That the volume of cameral liquid in the shell seemed to steadily decrease seemed to suggest that the function of emptying was to compensate for the increase in mass of the animal due to shell growth. At maturity, when shell growth stops, there would be, according to this view, no further need for emptying to occur. As is discussed in Section 4, the work of Ward and Greenwald (1982) on chamber refilling shows that this view must be revised.

3. Mechanism of Emptying

The mechanism by which the siphuncle transports cameral liquid from a chamber can best be understood by reference to the physical and biological factors that impinge on the emptying process (Figs. 1 and 3). First, there is the noncompressible and impermeable chamber containing cameral liquid at a given osmolality denoted as C_c (the thermodynamically correct unit for the osmolality is osmole/kg water). The chamber also contains gas at a pressure (P_{ch} in atm) that is slightly less than 1 atm at equilibrium. Using a value of 1 atm will not introduce appreciable error into the analysis.

The osmolality of *Nautilus* blood is about 1.060 osmole/kg H_2O (Greenwald *et al.*, 1980), and the pressure of the blood in the siphuncle at any depth is equal to the pressure of the blood generated by the heart (less than 0.10 atm) plus the

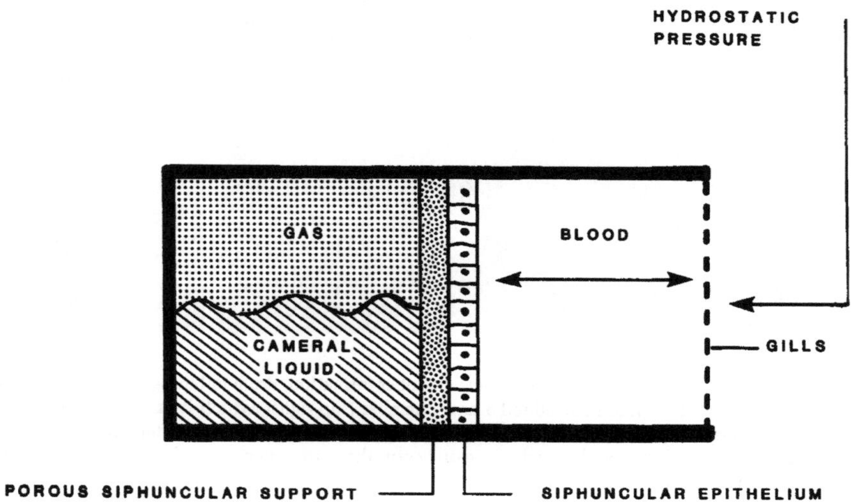

Figure 3. A physical model of the passive pressures that influence the transport of cameral liquid. The open, heavy black box represents the rigid and impermeable chamber wall. The chamber is filled with gas at a pressure of 1 atm or less and cameral liquid. Hydrostatic pressure is transmitted to the siphuncular epithelium via the blood of the animal. The siphuncular epithelium rests on a porous support that is freely permeable to gas and cameral liquid. See the text for an analysis of the various pressures that influence emptying.

ambient hydrostatic pressure (which equals in atm the depth in m/10, plus 1 atm surface pressure). These four parameters determine the pressures that affect emptying across the *Nautilus* siphuncle, which is represented as a single layer of cells separating the blood from the chamber (Fig. 3). The siphuncular cells are supported by the porous horny and chalky tubes. Evidence that the siphuncular tube (or connecting ring) is permeable to gases and liquids has been provided by Denton and Gilpin-Brown (1966), Collins and Minton (1967), and Chamberlain and Moore (1982).

The pressure that leads to emptying of cameral liquid is the sum of the osmotic pressure of the blood (RTC_b) plus the chamber pressure (P_{ch}). The former is a constant of 25 atm at 15°C [$0.082 \times (273 + 15) \times 1.060$]. The pressure that leads to chamber filling is the sum of the osmotic pressure of the cameral liquid (RTC_{ch}) plus the pressure of the blood [$1 + (d/10)$], where d is depth in meters. The blood pressure generated by the heart, being negligible compared to the hydrostatic pressure at any but the shallowest depths, is ignored in this analysis.

The validity of this physical model depends on the siphuncle having low permeability to NaCl, the major osmotic component of *Nautilus* blood and cameral liquid. If the siphuncular epithelial membrane were highly permeable to salt, then no osmotic pressures would be generated. That this membrane is not highly permeable to salt is suggested by our frequent observations that salt solutions different from the blood concentration are maintained for days to weeks with only small changes in NaCl concentration. This qualification is important (and requires further substantiation), because the osmotic pressure of a solution on either side of the siphuncle is not RTC, but σRTC, where σ is the reflection coefficient, a factor that varies from 0 for highly permeant solutes to 1.0 for impermeant solutes.

As emptying proceeds, the osmotic pressure of the cameral liquid falls, until the point of decoupling is reached (Ward, 1979). This drop in osmotic pressure itself can be sufficient to cause emptying. For emptying to occur at any depth, due entirely to passive forces established by the concentrations and pressures impinging on the siphuncle, the emptying pressure must exceed the filling pressure. That is

$$RTC_b + P_{ch} > RTC_{ch} + 1 + d/10$$

If this relationship is written as an equality, then it permits solution for the combination of depth and chamber concentration at which filling and emptying pressures will be equal and for which there will be no change in cameral liquid volume. For example, at a chamber concentration of 0.5 osmole/kg water, the equilibrium depth will be 132 m at 15°C. That is, with a cameral liquid concentration of 0.5 osmole/kg water, an animal at a depth of 132 m will neither gain nor lose cameral liquid from that chamber. Should the animal ascend, then the chamber will empty *automatically* due to the purely passive pressures that exist in the system. No metabolic energy need be expended, because that energy has already been "stored" in the chamber in the form of a low solute concentration. Should the animal descend, then the chamber will fill automatically unless the animal expends energy to counteract the filling.

In the preceding analysis, it is essential to understand that this metabolic expenditure would be not so much in terms of the siphuncle pumping water out of the chamber faster than it enters as it would be in terms of reducing the osmolality of the cameral liquid in the chamber or, perhaps, in the chalky and horny tubes (a small volume of cameral liquid compared to the contents of the entire chamber). As Denton and Gilpin-Brown (1966) pointed out, the latter alternative (changing the osmolality of the unstirred region around the siphuncle) could allow for rapid changes in the equilibrium depth. The rapidity of the change would be due to the small quantities of ions required to be moved in the chalky and horny tubes, compared to the amounts in the total cameral liquid volume.

Thus, the paradox suggested by the results of Chamberlain and Moore (1982) is resolved. They noted that the inorganic siphuncular tube allows water to enter a chamber (under a pressure head) approximately 1000 times more rapidly than the living siphuncle can empty a chamber. It is not necessary for the siphuncle to "bail out" a chamber against flooding allowed by the siphuncular tube under hydrostatic pressure. It is only necessary to establish an "osmotic valve" that stops net filtration by an opposing osmotic pressure. Such a valve is established by reducing the chamber (or chalky tube) osmotic concentration and allowing the osmotic pressure of the blood to "pull" water out of the chamber as fast as it enters.

As noted above, this relationship, when written as an equality, allows the calculation of the maximum depth at which *Nautilus* can empty a chamber by reducing the osmotic pressure of the cameral liquid. Ward *et al.* (1977) found cameral liquid concentrations as low as 10% of seawater, but the limiting minimum value is 0. If a value of 0 is substituted for C_{ch} in the emptying equation, the resulting solution is that depth equals 240 m at 4°C. That is, if an animal with the minimum possible concentration of cameral liquid in a chamber (0) were to submerge deeper than 240 m, then according to this simple osmotic model chamber filling would necessarily occur. This phenomenon was noted by Denton and Gilpin-Brown, who first calculated that 240 m was a critical depth for chambered cephalopods, and this analysis led Greenwald *et al.* (1980) to test the simple osmotic model.

To test the model using animals in an aquarium (depth ≈1 m), Greenwald *et al.* (1980) reversed the osmotic gradient across the siphuncle. By putting 1.5× or 2× seawater (seawater with added NaCl) in a chamber, they increased the filling pressure of the equation to 1.5× and 2× the emptying pressure (d = 0), respectively. An observation of chamber filling would have supported the simple osmotic hypothesis, because the filling pressure was greater than the emptying pressure. An observation of emptying would have demanded rejection of the model. In 12 of the 13 experiments in which 1.5× or 2× seawater was used, emptying did occur (in one case, there was no change in cameral liquid volume). It is notable that the concentrations of these salt-enhanced cameral liquids remaining in the chambers did not decrease by more than 20% in the course of the experiments. These results led Greenwald *et al.* (1980) to reject the simple osmotic model, but it should be noted that only 5 ml of salt-enhanced seawater was added to chambers of 20–25 ml in volume. Hence, the bulk of the cameral liquid was decoupled from the siphuncle. This decoupling would have allowed the bulk of the liquid in the chamber to behave independently of the liquid in the chalky tube.

If decoupled cameral liquid is wicked into the chalky tube, it may be in an unstirred layer, and the siphuncular epithelium may remove salt from this solution, thereby reducing the concentration of the solution in the chalky tube. This desalinization results in a lower osmotic pressure than that of the blood in the siphuncle. This solution would flow (via osmosis) from the chamber into the blood, making it appear as though there were emptying against the osmotic gradient. This would be a continuous process, in which the solution leaving the chalky tube was being constantly replaced by the cameral liquid from the chamber. In summary, although the results in Greenwald *et al.* (1980) suggest emptying against the osmotic gradient (requiring rejection of a simple osmotic mechanism), an alternative interpretation of their results is conceivable.

It is pertinent to note that in attempts we have made to show emptying of salt-enriched seawater from *coupled* chambers (chambers that were filled to about 75% of their volumes), emptying did not occur until *after* there was both filling (consistent with simple osmosis) and dilution of the cameral liquid to about the same concentration as that of the blood. In a coupled chamber, cameral liquid is in direct contact with the chalky tube, perhaps washing out the unstirred layer and preventing the epithelium from diluting the cameral liquid with which it is in contact. Hence, the data relating to the simple osmotic emptying hypothesis are equivocal, and simple osmosis must still be regarded as a viable alternative for explaining emptying.

It is interesting to note that *Nautilus* without flooded chambers are commonly caught well below 240 m (e.g., at 400 or 500 m). However, it is not possible to know how long the animals were at these depths. These animals may be transients, and not long-term, deep-water inhabitants. Moreover, although the shells of these animals were not flooded, no systematic analysis has been made as to whether their shells contained more cameral liquid than the shells of animals caught at shallower depths. The data of Ward *et al.* (1984) on vertical migration of *Nautilus* are relevant to this question. One animal, tracked over a 7-day period, commonly ranged considerably deeper than 240 m, but the time-averaged integrated depth was in fact only 226 m, a depth at which simple osmotic emptying could occur, if the cameral liquid concentrations were quite low and, indeed, lower than has yet been reported.

A final point against the simple osmotic mechanism is the observation that decoupled cameral liquid in wild-caught animals increases in salt concentration during emptying (Ward, 1979; Ward *et al.*, 1981). It is difficult to reconcile this observation with the simple osmotic mechanism, unless one is willing to postulate that these animals spent most of their time before capture near the surface, where the equilibrium emptying concentration would have been high. In retrospect, this observation (the increase in cameral liquid concentration during emptying) is perhaps the best argument that there is more to the emptying mechanism than merely alterations in cameral liquid concentrations, with subsequent passive emptying.

The validity of the simple osmotic hypothesis could be established with holding cage experiments at depths deeper than 240 m. In such an experiment, the filling pressure would be hydrostatic and not osmotic (as in the laboratory simulation), and the unstirred layer phenomena would be irrelevant (even if the chalky tube concentration were zero, passive emptying could not occur). We have performed this experiment on only a few animals, and the results were inconclusive.

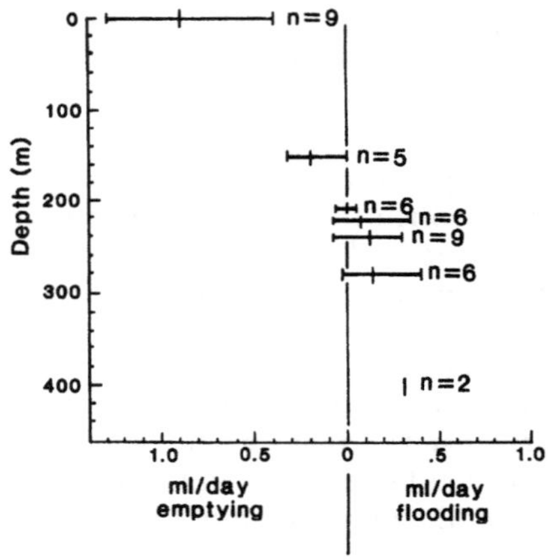

Figure 4. Relationship between spontaneous chamber emptying and flooding in *N. macromphalus* at various depths. At the surface, all animals showed chamber emptying. At depths of about 240 m, both emptying and flooding were observed. At about 400 m, in two cases, only flooding was observed. These results are consistent with the simple osmotic emptying mechanism. Reprinted with permission from Ward and Westermann (1985).

At depths shallower than 240 m, emptying occurred, which could be explained by alterations in cameral liquid concentration and subsequent simple osmosis (passive emptying). At depths close to 240 m, both emptying and filling occurred (Fig. 4). Two animals at 400 m flooded.

In summary, the data that bear on the question of the mechanism of emptying are conflicting. The laboratory simulation experiments using decoupled chambers (Greenwald *et al.*, 1980) did not support the simple osmotic model, but as discussed, these experiments are subject to question. Our similar experiments on coupled chambers yielded results that were consistent with simple osmotic emptying. The observation that animals caught in deep-water traps do not have flooded chambers circumstantially supports a model other than simple osmosis, as does the observation that cameral liquid concentration increases during the later phases of emptying. Deep-water emptying experiments, which in theory could have resolved this question, were inconclusive. In short, insufficient data exist to answer definitively the question of whether emptying occurs by simple osmosis.

A mechanism for emptying in water below depths of 240 m is suggested by the ultrastructure of the cells in the siphuncle. These cells are characterized by numerous infoldings of their basal and lateral membranes, and the cytoplasmic side of each of these infoldings is lined with numerous mitochondria (Fig. 5) (Greenwald *et al.*, 1980, 1982; Mangum and Towle, 1982). This arrangement is typical of nearly all tissues that carry out NaCl solute-coupled water transport, commonly against osmotic (and occasionally against hydrostatic) pressure (Berridge and Oschman, 1972). The transport system is known as the *local osmosis system* (Curran and McIntosh, 1962; Diamond and Bossert, 1967). In this system, the mitochondria supply ATP to ion pumps placed in the basolateral membranes. These ion pumps transport salt into the cul-de-sacs formed by the infoldings. The resulting *local* elevated solute concentration entrains water flow (osmosis—hence, "local osmosis") from the cytoplasm of the cell. This water then leaves the cell through the openings of the basolateral infoldings.

Support for this model in *Nautilus* derives from the ultrastructure of the siphuncular epithelium and from the observation of Mangum and Towle (1982)

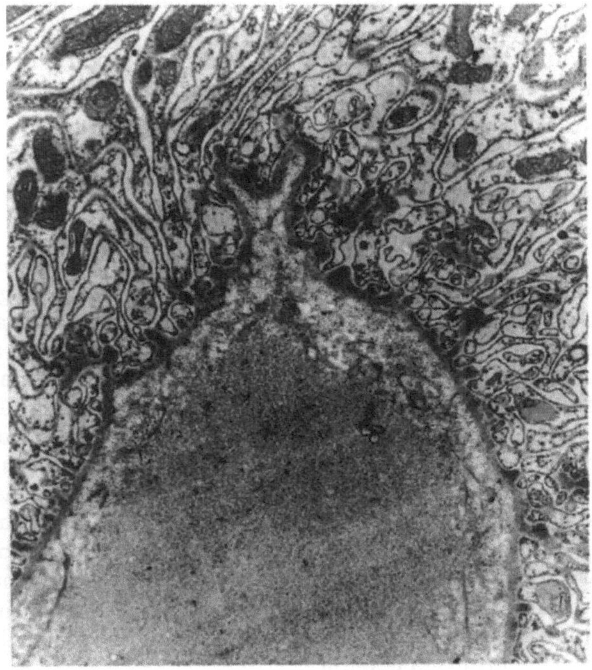

Figure 5. Electron micrograph of the siphuncular epithelium. The large granular area in the center of the picture is hemocyanin-filled extracellular space. Numerous mitochondria-lined infoldings of the cell membrane are apparent. These infolded areas of cell membrane have been shown to contain the sodium transport enzyme Na^+/K^+-ATPase (Greenwald *et al.*, 1984). See the text for a description of how this cytoarchitecture might participate in the emptying process. Reprinted with permission from Greenwald *et al.* (1980).

that the siphuncle is rich in the sodium transport enzyme Na^+/K^+-ATPase (Bonting, 1970). Using an electron histochemical technique, Greenwald *et al.* (1984) were able to show that the Na^+/K^+-ATPase is localized on the membranes of the basolateral infoldings, as would be expected in the local osmosis model. It is also noteworthy that the siphuncular epithelium of chambers with septa too thin to permit emptying do not show these transport specializations (Greenwald *et al.*, 1982). Also, in a shell containing 28 chambers, the 12th most recent chamber contained a siphuncle that showed ultrastructure typical of a recent chamber. The siphuncular tube from the 3rd or 4th oldest chamber of one animal examined by Greenwald and Singley (unpublished observations) contained no living tissue. Perhaps the contribution to buoyancy of such small chambers is so slight that large animals dispense with a living siphuncle in the early chambers.

The local osmosis model, as applied to emptying, would allow for emptying in water at depths greater than 240 m by establishing an elevated salt concentration in the cell cul-de-sacs. For example, a concentration of 2× seawater in the cul-de-sacs would allow emptying at depths of about 480 m, if the osmotic pressure of the cameral liquid was low. A concentration of 2× seawater in the cul-de-sacs would allow emptying at about 240 m, if the osmotic pressure of the cameral liquid was the same as blood, as in the final stages of emptying of some decoupled chambers. The flow of water from the chamber through the cell, and then leaving from the wide openings on the infoldings, depends on the lack of osmotic pressure generation across the wide opening. Such an opening would have a reflection coefficient of 0 (equally permeable to solute and water) and thus would not generate an osmotic pressure leading to water entry. Details of this argument are given in Curran and McIntosh (1962).

4. Control of Buoyancy

Buoyancy adjustments in *Nautilus* are of two kinds. The first are long-term adjustments to compensate for growth. As a *Nautilus* grows, it adds shell material with an average density of 2.7 g/ml and tissue with a density of 1.06 g/ml. The specific gravity of seawater is 1.025 at the surface and at room temperature. Thus, discounting the effect of the emptying process, during the course of growth the animal becomes increasingly heavier in seawater. This long-term increase in weight is compensated for by the removal of liquid from the phragmocone chambers. The rate of growth-related density increase is apparently matched closely by chamber-generated density decreases, because healthy, freshly caught live specimens are always neutrally buoyant. Growth without sufficient chamber formation and emptying would lead to insufficient buoyancy. Chamber emptying without sufficient growth would lead to excessive buoyancy. It is notable that excessive buoyancy is commonly observed in aquarium specimens and may be due to abnormal growth, due perhaps to the constant damage that occurs at the shell aperture in aquaria.

In addition to long-term buoyancy adjustments associated with growth are short-term buoyancy changes, which would be useful to compensate for sudden increases in weight (e.g., from eating) or rapid decreases in weight {e.g., from sudden shell breakage [see Ward and Greenwald (1982) and Chapter 12]}. Rapid buoyancy changes could also be used for vertical movements through the water column, but the observed rates of cameral liquid movement are so slow that such use seems improbable (Ward, 1986). Also, *Nautilus* captured near the surface at night are not positively buoyant. We are aware of no evidence that *Nautilus* uses short-term buoyancy changes as an aid to vertical movements in the water column.

The rates and mechanisms of compensatory buoyancy change have been examined in a number of studies. It has been shown that *Nautilus* can increase its emptying rate in response to artificial flooding of empty chambers with seawater (Greenwald *et al.*, 1980). The emptying rate was depressed when partially filled chambers were artificially emptied and the animal was rendered excessively buoyant. Ward and Greenwald (1982) showed that the response to artificially induced excessive buoyancy was partial refilling of emptied chambers, thereby demonstrating that chamber emptying is not a one-way process. However, these and other studies (e.g., Ward and Martin, 1978) focused on the problem of the volume of liquid in the shell and did not measure the variable of interest, namely, buoyancy (or weight in seawater), per se. This critical variable was specifically dealt with by Ward (1986). Specimens of *N. macromphalus* were suddenly made positively or negatively buoyant. Positive buoyancy was induced by removing cameral liquid from one or two chambers or by removing portions of the shell aperture. Negative buoyancy was induced by adding seawater to empty chambers.

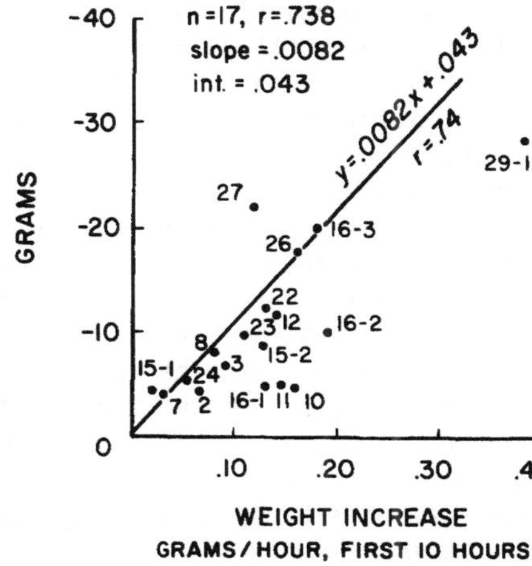

Figure 6. Rates of chamber refilling in animals that have been rendered excessively buoyant by shell removal. The amounts of excessive buoyancy (grams) are plotted against the weight increases (grams/hour) for the first 10 hr of the experiment. After 10 hr, the refilling rate decreased. The total amount of refilling (weight increase) was limited, in this experiment, to a mean of 4 g. Reprinted with permission from Ward (1986).

Within an hour of sudden weight loss, the animals showed a rapid increase in weight in seawater (Fig. 6). The rate at which this increase took place depended on the initial change at the start of the experiment; the greater the initial buoyancy change, the greater the rate of compensatory response. The greatest weight gain occurred during the first 10 hr following initiation of the experiment. Little weight gain occurred after 30 hr, *even if the animal was still highly buoyant*. The total amount of compensation was limited to about 4 g. The mechanism of these weight increases was cameral liquid refilling. More measurements are necessary to determine whether the rates of refilling are more rapid at great depths.

Weight loss in *Nautilus* made artificially heavy was stimulated by the addition of cameral liquid to emptied chambers. Compensatory buoyancy changes (due to emptying) in these heavy animals were noticable within the first hour. Emptying continued until neutral buoyancy was again achieved, even if the emptying took tens of hours. Thus, chamber refilling seems to be limited to about 5 ml (for animals in the range of 500–1000 g mass); chamber emptying appears to be a process that can continue as long as there is liquid to be emptied.

Compensatory buoyancy change in response to a sudden increase in animal weight was most rapid during the first 10 hr, as was the case for the response to induced positive buoyancy.

5. Outstanding Problems

The mechanisms by which *Nautilus* achieves and maintains near-neutral buoyancy in seawater seem, at this point, well understood. Nevertheless, a number of problems remain unresolved, including the following ones:

1. There have been no definitive measurements of cameral liquid emptying at depths greater than 240 m. Deep-water emptying is significant in terms of understanding the mechanism of cameral liquid transport.
2. If, as Ward *et al.* (1981) suggested, buoyancy regulation and new chamber

formation are interrelated, this relationship should be amenable to investigation in surface aquaria using captive animals.

3. It seems clear that *Nautilus* is somehow sensitive to its buoyancy and can make appropriate compensatory adjustments. However, which parameter does the animal monitor as an index of buoyancy, and what is the pathway and signal to the siphuncle to adjust its emptying rate or to initiate refilling? Greenwald *et al.* (1982, Fig. 11) showed what appear to be axons within the infoldings of the siphuncular epithelium. Could these structures be involved in transmitting signals to that epithelium to increase or decrease the emptying rate? Are hormones or neurohormones involved?

4. How often and to what extent does the animal tune its buoyancy in nature? Are there changes in buoyancy that correspond to daily vertical migrations?

5. A puzzling aspect of buoyancy control in *Nautilus* is the refilling process. If cameral liquid is removed from a chamber, causing positive buoyancy, the animal compensates by partially refilling the chamber with cameral liquid of the same osmotic pressure as that which was removed (Greenwald and Ward, 1982). How is such cameral liquid produced and what limits its production to a maximum of about 5 ml? How does this differ from the liquid that initially fills a chamber?

6. What are the details of the cellular mechanism of transport by the siphuncle? Investigation of permeabilities, transport rates, electrochemical potentials, and counter-ions should be quite rewarding (see, for example, Mangum and Towle, 1982). The structure of the tissue (a tube) would make it suitable for perfusion studies such as are done on vertebrate nephrons.

8. How do these various mechanisms vary in young vs. mature *Nautilus* and in the cephalopods with chambered, internal shells, such as cuttlefish and the midocean squid *Spirula?* The latter is of particular interest, because it seems capable of maintaining its buoyancy at depths far in excess of the implosion depth for *Nautilus* (Denton *et al.*, 1967). The recent report of fertile, developing *Nautilus* embryos (Arnold and Carlson, 1986) opens an entire new area of inquiry in this regard.

In summary, with the exception of the question of whether the animal can empty below the thermodynamically critical depth of 240 m, it seems that the major characteristics of the buoyancy system of *Nautilus* are now understood. Information on very young animals and on the neural or neuroendocrine control of buoyancy is entirely lacking, as is information on the relationship among emptying, buoyancy control, and new chamber formation. Thus, considerable progress has been made since the seminal observations of Bidder (1962) and the definitive measurements and analyses of Denton and Gilpin-Brown (1966). One hopes that progress is yet to come.

X

Aquarium Maintenance

Chapter 35

Collection and Aquarium Maintenance of *Nautilus*

BRUCE A. CARLSON

1. Introduction

The first living *Nautilus* to be displayed in a public aquarium was *N. macromphalus*, exhibited at the Nouméa Aquarium in New Caledonia in 1958 (Catala, 1964). In 1961, the Monaco Aquarium obtained one living *N. macromphalus*, which survived 2 months in captivity (Cousteau and Diolé, 1973). Apparently no other successful attempts to keep *Nautilus* in aquaria were made until 1976, when the Yomiuri-Land Aquarium in Tokyo obtained *N. macromphalus* from the Nouméa Aquarium and kept them alive for 8 months (JECOLN, 1980b) (see also Chapter 36). At the same time, specimens of *N. pompilius* were obtained from Fiji by the Waikiki Aquarium in Hawaii, where they were maintained for 8 months (Carlson, 1977, 1979).

Since 1976, many aquaria around the world have maintained *Nautilus*. Considering the great distances between the collecting localities for *Nautilus* and the

BRUCE A. CARLSON • Waikiki Aquarium, University of Hawaii, Honolulu, Hawaii 96815.

laboratories and public aquaria in the United States, Japan, and Europe, it is surprising how well the animals can withstand the rigors of being packed and shipped to these foreign destinations. Perhaps more surprising is the ability of the animals to adapt to completely artificial environments designed essentially for shallow-water animals. Although most aquaria still report some difficulties in the long-term maintenance of *Nautilus* in captivity, great progress has been made since 1976. Several aquaria have kept these animals alive in captivity at least 3 years, and the production of the first viable eggs at the Waikiki Aquarium in 1985 suggests that captive breeding will be possible some day.

The objective of this chapter is to summarize the accumulated experiences of various aquaria that have collected, transported, and maintained living *Nautilus* and to provide guidelines for the construction of aquarium systems for keeping these unusual animals alive and healthy.

2. Collecting and Transporting Live *Nautilus*

Most public aquaria obtain their *Nautilus* from tropical fish importers who in turn obtain them from collectors in the Philippines. However, because Philippine collectors have not, in the past, used chilled seawater for holding and transporting *Nautilus*, stress and subsequent mortality have been high. Some exporters have recently begun using ice packs in their shipping boxes, but mortality still averages 50–80%. As a guide for researchers and aquarists, the collecting and shipping techniques used by the Waikiki Aquarium are described in this section.

2.1. Trapping

Because trapping materials are often unavailable in remote locales, a portable trap was developed and has been used successfully a number of times. It consists of four hoops (1 m in diameter), made from No. 6 fencing wire, spaced and covered with chicken wire so as to make a cylindrical trap 2 m in length. Funnels with tip openings large enough to admit *Nautilus* are mounted in either end of the trap with tips pointing inward, and a 0.5-m screen baffle is mounted midway between the tip openings to deter *Nautilus* from escaping. A steel reinforcing bar fastened along the top of the trap provides rigidity, and weights may be secured along the bottom so that the trap remains upright when set.

Although this type of trap is inexpensive and easy to transport, it can be used only a few times before it becomes so battered it must be discarded. Sturdier traps for long-term use can be constructed on a similar design, but using a welded steel frame instead of wire hoops.

Bait is suspended in the center of the trap by wire. Fresh fish is the usual bait, but almost any meat, including canned fish, will do. The bait is wrapped in chicken wire to prevent *Nautilus* and other marine organisms from devouring it. Additionally, wrapping prevents *Nautilus* from gorging on the bait and later producing copious amounts of waste material, which quickly leads to lethal conditions in a captive environment.

Trapping techniques vary depending on water depth, currents, and slope. In areas where bottom slope is low, a sand-anchor is set at the desired depth, and the trap is then attached to the anchor line by a short length of rope. The location of the trap can be marked by a surface buoy. When the anchor is hauled, the trap is pulled up at the same time.

When working in the vicinity of a steep forereef slope, the trap line can be tied directly to the reef at a shallow depth by divers and stretched to its full extent (generally 400 m) on the surface by a boat moving away from the reef face. When the trap is dropped overboard, it arcs downward on the line to rest suspended against the reef face.

Traps can be left in place for just one night if desired, but three-night sets proved the most productive in Palau. Enough time must be allowed for the animals to locate the trap and find their way inside, but not so much time that all the bait disintegrates or is eaten by smaller organisms such as shrimp. In Palau, catches in 1982 averaged 34 animals per three-night set, and as many as 60 specimens were caught in a single trap (W. B. Saunders, personal communication).

2.2. Transport from Collecting Sites

Raising traps at a rapid pace has no apparent adverse effect on *Nautilus*, even when they are retrieved from depths as great as 400 m. However, mortality can be high during the period between collecting and transporting to shore if the animals are not quickly immersed in chilled seawater on the boat. Newly collected animals maintained at 28°C for 4 hr on board showed poor survival, whereas animals from the same collection maintained at 18°C experienced virtually no mortality. *Nautilus* is susceptible to agitation; animals recovered or transported during rough surface conditions commonly suffer high mortality.

Animals can be transported in water at 18°C, in styrofoam-insulated chests, with about one animal per 4 liters of water. Aeration is usually not provided, since it tends to warm the water. If the transport time exceeds 1 hr, cooler water temperatures should be used. The animals should be observed frequently during transport for signs of stress, which induces production of copious amounts of fecal material that will quickly foul the container of water. Extra chilled seawater should be on hand in case a change of water is needed during transport.

2.3. Temporary Holding Tanks

Most sites where *Nautilus* occurs provide poor aquarium facilities. In Fiji, Palau, Australia, and elsewhere, improvised holding tanks have been developed using existing facilities. At a minimum, a small air-conditioned room may chill water sufficiently to keep a few animals alive in a small aquarium. However, in most instances, this tactic is not adequate. A small portable refrigeration unit is more desirable. Holding tanks of at least 400 liters are preferred for keeping more than 10 animals. In addition, a biofilter, such as a simple airlift device operating through a gravel bed, should be in operation. *Nautilus* should be held for 3–4 days without feeding, prior to shipping, to allow elimination of fecal material. Daily siphoning of waste material and debris from holding tanks to prevent fouling is recommended, as is a water change of at least 25%, especially during the first 4–5 days in captivity.

2.4. Packing and Shipping

Adult *Nautilus* cannot be packed in plastic bags like tropical fish because they will bite their way through the plastic. Instead, insulated, nontoxic, waterproof boxes of at least 20-liter capacity should be used. Each box should be two-thirds filled with chilled seawater (≈15–18°C) and should contain no more than four *Nautilus*. Oxygen should be pumped under the lid to fill the remaining air space, and the lid should then be sealed securely with tape to prevent water and oxygen from escaping during transport. As a precaution against inadvertent leaks, the entire box should be wrapped in a plastic bag and placed inside a cardboard box for shipping. This method works particularly well for large *Nautilus* species, such as *N. belauensis*. Smaller species, such as *N. pompilius*, can be packed in the same manner, or the animals can be placed in individual plastic containers, approximately 20 cm square and 12 cm deep (one-third filled with seawater), which are then placed inside the larger container. Holes should be made in the containers to permit water to circulate through them, and they should then be wrapped in a heavy-gauge plastic bag. Oxygen should be added to the bag in the ratio of two thirds oxygen to one third water by volume, and the bag should be tightly sealed with a rubber band. Several such bags can be placed in styrofoam-lined boxes such as those used in the tropical fish trade. In fact, this method is similar to that employed for triggerfishes and other species that tend to bite and puncture plastic shipping bags.

Shipping *Nautilus* by air cargo, rather than as baggage, is recommended, because baggage usually receives rough treatment and may even arrive upside down and very wet at the baggage carrousel. Cargo shipments are more expensive, but generally prove to be worth the extra cost. When properly packed, *Nautilus* can survive flights of 24 hr or more.

3. Aquarium Systems for *Nautilus*

Most captive *Nautilus* on display were obtained as exhibit animals for public aquaria. After nearly a decade of trial and error, it is possible to begin comparing the success of various aquarium systems designed for *Nautilus*.

Essentially, there are two categories of aquarium systems: closed and open. Home aquarists and inland public aquaria employ closed systems, in which water is continually filtered and recycled and periodic partial water changes are made. Public aquaria located next to the ocean generally operate open systems, with fresh seawater continually pumped from the ocean to the exhibits. In practice, most public aquaria utilize a combination of closed and open systems.

The first aquarium system in the United States to maintain and breed *Nautilus* successfully is at the Waikiki Aquarium in Hawaii. It can serve as a model for comparison with systems elsewhere.

3.1. Waikiki Aquarium *Nautilus* System

In 1976, two *N. pompilius* from Fiji survived 8 months in an open-system exhibit at the Waikiki Aquarium. The 1200-liter tank was fitted with a standard undergravel filter covered with crushed coral gravel and was maintained at a constant 24°C without refrigeration. The tank was illuminated by a 150 W red spotlight. Seawater was obtained from a well and was added to the aquarium at a rate of 600 liters/hr. Despite the simplicity of this system, *Nautilus* survived very well and 2 months after they were put on display began producing eggs.

Today, the *Nautilus* system at the Waikiki Aquarium is more complex due to an increased understanding of the ecology and behavior of these animals and past experiences with them in captivity (Fig. 1). Not surprisingly, the longevity of the animals in the aquarium has also increased. The same 1200-liter display tank is used in an open system as before, but the undergravel filter is now operated in a "reverse-flow" manner. Ordinary undergravel filters operate with airlifts moving water from under the filter to the top of the tank. Unfortunately, *Nautilus* tend to move toward these air tubes and attach themselves to the walls directly in front of the outflow. The bubbles in the stream of water can become trapped under the hood and will frequently enter the eye through the open pupil. Although this occurrence has never been known to cause immediate death, the animals are unable to expel these bubbles and complications can occur later. An aquarist can easily dislodge the bubbles from under the hood and from inside the eyes by tipping the animal from side to side underwater. To avoid this problem, it is simpler to eliminate any stream of air bubbles from the tank environment, hence the use of a reverse-flow undergravel filter.

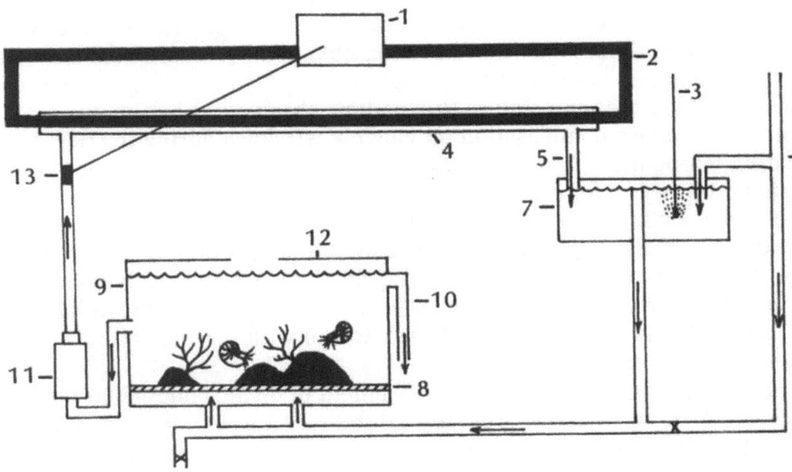

Figure 1. Waikiki Aquarium *Nautilus* system. Key: (1) compressor; (2) titanium tubing with Freon; (3) compressed air input; (4) polyvinylchloride waterjacket–heat exchanger; (5) chilled seawater; (6) ambient (24°C) seawater; (7) aeration sump; (8) undergravel filter; (9) display tank; (10) overflow; (11) filter–pump; (12) tank lid; (13) thermostat.

The exhibit tank is constructed of reinforced concrete 7.5 cm thick, with a glass viewing window on one side. All pipes and the heat exchanger are insulated with neoprene sleeves. A cover on the display tank provides insulation and also darkens the exhibit. Illumination is provided by a 150 W red spotlight shining through a small window in the cover. The aeration sump is constructed of ½-in. marine plywood covered with ½-in. insulating styrofoam that has been covered with fiberglass and resin and fitted with a lid.

Exhibit decoration is sparse; it consists of six basalt rocks and a dead deep-water gorgonian (*Gerardia* sp.), a species that occurs at depths where *Nautilus* lives. The walls of the tank are fitted with ¼-in. sheets of black plexiglass. Fiberglass backdrops were continually chewed by *Nautilus*, as were most plastic items in the tank.

3.1.1. Temperature

Nautilus is unusual in its ability to thrive in a wide range of temperatures. *Nautilus* collected at 400 m, where temperatures are approximately 8°C, can be brought to surface temperatures approaching 30°C in a matter of minutes and show no ill effects. Other invertebrates and fish in the same trap are usually dead, due in part to reduced pressure but also to high temperatures.

Carlson *et al.* (1984) and Ward *et al.* (1984) demonstrated that *N. belauensis* in Palau normally migrate from deep water to shallow water every night and return again at dawn to deep water. They are thus adapted to living under extreme temperature and pressure conditions. Carlson *et al.* (1984) provided data showing that *Nautilus* can move from water as cold as 6°C to water at 24°C in a matter of hours. Interestingly, *N. pompilius* and *N. belauensis* can both survive in captivity for up to 8 months at a constant temperature of 24°C, but A. W. Martin *et al.* (1978) and Carlson (1979) have reported that *Nautilus* spp. maintained at a constant temperature of 27°C for more than 48 hr died. The greatest longevity of *Nautilus* at the Waikiki Aquarium has been obtained by keeping the animals in water that fluctuated between 18°C at night and 21°C in the daytime (a regimen that is opposite to the temperature fluctuations observed in nature but has been maintained to reduce condensation on viewing windows during the daytime). A fluctuating temperature is not essential for maintaining *Nautilus*, but for *N. belauensis* and possibly for other species, it is a natural condition.

The refrigeration unit at the Waikiki Aquarium contains a heat exchanger built around a 6-m length of 2-cm OD titanium tubing containing the refrigerant (Freon 12), which is pumped by a 3 hp Copeland compressor. Although titanium is not a particularly good conductor of heat, it is better than most plastics and nylon, and it is the only metal that is safe to use with marine animals, because it is completely inert.

The titanium pipe runs inside a 4-m length of 8-cm schedule 80 polyvinylchloride (PVC) pipe. Seawater from the exhibit runs through the PVC pipe countercurrent to the refrigerant flow and is cooled as it passes around the titanium pipe. Cold water leaving the heat exchanger flows into a 400-liter sump tank at 14.5 liters/min. In the aeration sump, it mixes with ambient incoming water (24°C) entering at 1.8 liters/min. The sump is strongly aerated to promote mixing and oxygenation, and the cool water then gravity-feeds to the bottom of the display tank through the reverse-flow filter. From the display, the water is pumped by a 2-MD Little Giant pump through an Eheim canister filter packed with nylon and activated carbon and then up 1.5 m to the heat exchanger, where the cycle is repeated. The temperature thermostat is located at the incoming end of the heat exchanger. The flow rate of incoming makeup water produces a flushing time of about 12 hr.

3.1.2. Food and Feeding in Captivity

At the Waikiki Aquarium, each *Nautilus* is fed one whole shrimp (*Heterocarpus laevigatus*, about 18 g each) three times a week. Other food items have met with mixed results. Fresh fish, such as tuna, crab legs, and lobster molts, are all readily accepted. However, some live crabs, such as small Samoan crabs (*Scylla serrata*), and live shrimps, such as *Saron marmoratus*, have never been eaten, even when placed in the grasp of the *Nautilus*. On the other hand, when large, living deep-sea crabs, such as *Thelixiope* sp., are placed in the tank, *Nautilus* will rapidly locate and devour them. Often two or three *Nautilus* will simultaneously devour different parts of the same crab and eat everything including the heavy carapace. Afterward, they are noticeably negatively buoyant.

Curiously, cut squid is rarely accepted by *Nautilus*. In fact, they frequently show a strong negative reaction when offered squid pieces. Live fish have occasionally been displayed with *Nautilus*, but none has been captured and eaten by them.

3.2. Other Aquaria That Maintain *Nautilus*

During the preparation of this chapter, a questionnaire was sent to aquaria around the world requesting information on their *Nautilus* exhibits. A summary of the information received, plus information obtained from a search of the literature, is presented in Table I.

3.2.1. United States

Almost every public aquarium in the United States has displayed *Nautilus* with varying degrees of success. Animals have been obtained from at least four tropical fish importers, from *Nautilus* researchers, from the Waikiki Aquarium, and, in one case, from a local pet shop.

3.2.1a. Steinhart Aquarium (San Francisco). The Steinhart *Nautilus* exhibit tank is constructed of concrete and holds 3400 liters of natural seawater. There is a gravel substrate, but no undergravel filter. The tank is a semiclosed system and receives incoming water at a rate of about 900 liters/hr. The water from the exhibit is returned to the main water system, where it mixes with water from other exhibits, is filtered, mixed with incoming filtered water from San Francisco Bay, and then returned to all the displays, including the *Nautilus* exhibit.

Steinhart Aquarium also operates a research tank for *Nautilus*, consisting of a circular fiber glass tank that holds about 1800 liters of water. This tank is a closed system, with water recirculated through cartridge filters by a $\frac{1}{3}$ hp pump.

Table I. Summary of Public Aquarium Experience with Live *Nautilus* Exhibits[a]

Country	Aquarium	Species	First displayed	Water temperature
New Caledonia	Nouméa Aq.	*macromphalus*	1958	16–19°C
Japan	Yomiuri-Land	*macromphalus*	July 1976	17–20.8°C
		pompilius	Dec. 1978	—
United States	Waikiki Aq.	*pompilius*	Aug. 1976	24°C
		belauensis	July 1977	18–21°C
		macromphalus	Aug. 1984	18–21°C
	Steinhart Aq.	*belauensis*	Aug. 1977	14–23°C
		pompilius	1981	14–23°C
		macromphalus	1983	14–23°C
	New York Aq.	*belauensis*	Aug. 1977	—
		pompilius	?	17°C
	Shedd Aq.	*belauensis*	March 1978	14.4°C
		pompilius	March 1978	14.4°C
	Sea World	*belauensis*	Aug. 1977	14–24°C
		pompilius(?)	June 1978	14–24°C
	Seattle Aq.	*belauensis*	Dec. 1982	16–22°C
	New England Aq.	*belauensis*	Nov. 1979	21–22°C
		pompilius	?	21–22°C
	Baltimore Aq.	*belauensis*	Jan. 1982	16–20°C
		pompilius	Apr. 1982	16–20°C
	Monterey Bay	*macromphalus*	Aug. 1984	16–17°C
Canada	Vancouver Aq.	*belauensis*	Aug. 1982	16–19°C
		pompilius	?	16–19°C
West Germany	Carl Hagenbeck	*pompilius*	Jan. 1981	18–21°C
Monaco	Aq. of Monaco	*macromphalus*	Feb. 1978	—

[a] This summary is based primarily on questionnaire responses.

An airstone is used in the tank for aeration. A 40% water change is made in this tank every week.

Nautilus is fed six days a week on bay shrimp (*Crangon stylirostris*) and occasionally on ghost shrimp (*Callianassa affinis*), turtle crabs (*Cryptolithodes sitchensis*), lobster pieces, slices of various fishes (whitebait, mackerel, rock cod, day smelt, salmon), and even horse liver.

Live bay shrimp have been kept in the tank with the Steinhart *Nautilus*, and *Nautilus* occasionally captured and ate them. Flashlight fish (*Anomalops katoptron*) and pinecone fish (*Monocentrus japonicus*) have also been kept with *Nautilus*, but were removed when they began nipping at the tentacles of *Nautilus*.

3.2.1b. Sea World (San Diego). Sea World operates a 1000-liter fiberglass open-system display tank for *Nautilus* and maintains 12 animals. Filtered, ambient bay water flows continuously into the tank either from the top of the tank or via a reverse-flow undergravel filter. Temperature fluctuates with the temperature of the incoming bay water and ranges from 14.4°C (winter) to 22.2°C [rarely 23.8°C (summer)]; the animals survive well at all temperatures. Illumination is provided by one 75 W floodlight located 0.6 m above the tank. The *Nautilus* are fed every other day, primarily on fish (*Allosmerus* sp.) and shrimp. Clams (*Mercenaria mercenaria*), euphausids, and squid are occasionally used as food, but the *Nautilus* seem to "fear" squid initially.

3.2.1c. New York Aquarium (Brooklyn). The New York Aquarium maintains a 15,100 liter display tank and a 1900 liter research tank. The display tank is steel with a roughened epoxy liner and a glass wall on one side for viewing. The system is a semiclosed system with the water recirculated through an external filter box. A cooling system utilizing plastic coated heat-exchange coils is immersed in the filter box, and a constant temperature of 16.6°C is maintained. Shrimps are fed to the *Nautilus* once a day for the small animals and twice a day for larger ones. A low-wattage red light is used for illumination.

3.2.1d. Shedd Aquarium (Chicago). The Shedd Aquarium operates its *Nautilus* exhibit either as a closed recirculating system or as an "open system." Water for the "open system" is artificial seawater recirculated among many other exhibits. Water in the system is cooled by a 2-cm-PVC-pipe heat exchanger in the tank, and water temperature in the tank is maintained at 14.4°C. Illumination is provided by two fluorescent lights with a blue filter. Krill, smelt, and brown shrimp are used as food, and the *Nautilus* are fed once a day.

3.2.1e. Seattle Aquarium. Seattle maintains a 341-liter open-system *Nautilus* exhibit utilizing water from Puget Sound. Four animals are displayed in this tank. A heater is used to raise the incoming water temperature to 15.5–17.7°C in the daytime and 17.7–22.2°C at night. Incoming water is filtered through a bed of activated carbon; in the tank, it is recirculated through a normal-flow undergravel filter. Small, whole herring, krill (*Euphausia superba*), shrimps, half rock crabs (*Cancer productus*), squid, and rockfish are utilized as food, with feeding two or three times per week. Fluorescent lights, each with a red filter, are utilized for illumination.

3.2.1f. National Aquarium (Baltimore). The National Aquarium uses a closed system for its *Nautilus* display. Artificial seawater (Instant Ocean brand) is mixed with carbon-filtered Baltimore city water. The semicircular exhibit tank has a 1135-liter capacity and is constructed of fiber glass. Water from the exhibit flows into a reservoir tank, where a heat exchanger cools the water. A $2\frac{1}{2}$ hp compressor with an enamel-coated heat exchanger cools the water, but copper (lethal to *Nautilus*) was detected in the system on several occasions. A $\frac{3}{4}$ hp water pump moves the water from the reservoir to a Hayward S-200 high-rate sand filter, to a biofilter filled with plastic biorings, and then returns it to the tank via a reverse-flow undergravel filter. A chemical filter to remove copper and other metals has been installed. Illumination is provided by a 150 W floodlight, masked to allow only a narrow slit of light into the otherwise black tank.

Two live California squat lobsters were put in the tank as food for the *Nautilus*, but were not eaten and have remained on display. Shrimp and krill (*Euphausia superba*), first soaked in an "Aminoplex" solution, are fed to the *Nautilus* one to three times daily. Clams, squid, smelt, and lobster molts are also offered occasionally.

3.2.1g. New England Aquarium (Boston). The New England Aquarium *Nautilus* exhibit is similar in size, shape, and construction to that of the National Aquarium. However, it is operated as an open system with a "trickle" of new water constantly flowing into the tank. A normal undergravel filter is utilized as a biofilter and for circulation. A cooling system is not needed due to the constant inflow of water from Boston Harbor, which maintains the temperature between 20.5 and 21.1°C. Harbor water is settled, filtered, and oxygenated before it enters the *Nautilus* tank. A 100 W red light is used for illumination.

Five *Nautilus* are maintained in this system and are usually fed krill (*Euphausia pacifica*) and occasionally smelt (*Osmerus mordax*) once per day. Two pinecone fish (*Monocentrus* sp.), glasseye snapper, and a short bigeye fish are maintained with the *Nautilus*, but no interactions have been reported.

3.2.1h. Monterey Bay Aquarium. The Monterey Bay Aquarium operates a 1722-liter fiberglass tank with ten *Nautilus*. It is fitted on the inside with natural-looking, fiberglass-reinforced cement rockwork. This is an open system, with new filtered seawater from Monterey Bay entering the tank at 4.5 liters/min, resulting in a flushing rate of once every 6.3 hr. Within the tank, water passes through an undergravel filter covered with crushed coral rock and is airlifted to an external chamber where a 3000 W titanium heater warms the water to 16–17°C. The filter turnover time is 30–45 min. A single 75 W white incandescent bulb provides illumination.

The *Nautilus* are fed daily on whole unshelled prawns, pelagic crabs (*Pleuroncodes* sp.), and krill (*Euphausia superba*). *Tubastrea* corals and several *Oreaster* starfish are exhibited with the *Nautilus*, but no interactions have been observed.

3.2.2. Canada: Vancouver Aquarium

A 1052-liter concrete exhibit tank holding ten *Nautilus* is operated as an open system, with seawater brought in from Burrard Inlet. Incoming water first passes through a sand filter, trickles into the display tank at 10–12.7°C, and is warmed by a 1000 W Glo-Quartz heater to 15.5–18.3°C. An undergravel filter covered with No. 8 brown sand is fitted in the bottom of the exhibit tank. The tank is illuminated with two 75 W lights for 10–12 hr per day.

Nautilus are fed daily on jack herring and whole shrimp. Live coonstripe shrimp (*Pandalus danae*) introduced into the tank as scavengers are occasionally preyed on by the *Nautilus*.

3.2.3. Holland: Artis Aquarium (Amsterdam)

Experiences with two *Nautilus* maintained in a 350 liter aquarium were reported by Graaf (1981). The exhibit tank is connected to other exhibits through a closed recirculating system. An average temperature of 18°C is maintained for the entire system. Once a year, two thirds of the reservoir water is drawn off and replaced with natural seawater. Turnover of water in the exhibit is once per hour, and no aeration is provided within the exhibit itself. Illumination is kept low.

Crabs (*Carcinus moenas*), shrimps (*Crangon crangon*), and pieces of fish (*Merlangus merlangus*) are the main food items. Crabs are preferred over the shrimp, and usually *Nautilus* are fed one crab per day. Graaf noted that live crabs, shrimps, and small fishes are not accepted by the *Nautilus* and that dead crabs would be eaten only if the carapace was first cracked.

3.2.4. New Caledonia: Nouméa Aquarium (Nouméa)

The Nouméa Aquarium was the first aquarium to display living *Nautilus*, beginning in 1958. Its exhibit tank is approximately 800 liter and is an open system. Temperature in the exhibit tank ranges from 16 to 19°C, depending on the time of year. *Nautilus* is fed pieces of fish about 1 cm^3 daily and crab and lobster molts at least once a month. Display animals are rotated with animals in holding tanks every day.

3.2.5. Germany: Aquarium Hagenbeck's Zoo (Hamburg)

The Hagenbeck Aquarium maintains *Nautilus* in a closed-system aquarium. Water is filtered through an exterior filter containing shell and wadding, and one third of the water is changed every 2 weeks. An ultraviolet sterilizer is used in conjunction with the filter. Temperature is maintained between 18 and 21°C by a PVC heat exchanger in the filter box. A 40 W fluorescent light with a blue filter is used for illumination.

Nautilus are fed daily on shrimp (*Crangon vulgaris*) and crabs (*Carcinus maenas*).

3.2.6. Japan: Yomiuri-Land Aquarium (Tokyo)

Yomiuri-Land Aquarium maintains five *Nautilus* in a closed-system aquarium of 2146-liter capacity. Water from the exhibit tank passes into an exterior filter box filled with 80 kg of silicate sand, 5 kg of coral sand, and 2 kg of activated carbon. From there, water is pumped at a rate of 80 liters/min through an ultraviolet sterilizer and back to the top of the tank. A titanium heat exchanger is built into the filter box, and cold seawater at 10°C is used as the coolant with a cryostat controlling the flow of incoming coolant. Water temperature in the exhibit is maintained at 17–21°C. Periodic water changes are made using natural seawater. A 200-W fluorescent light is used for illumination (see also Chapter 36).

Food consists of jack mackerel and shrimp, and the animals are fed 10 g of food per animal each night.

4. Longevity in Captivity

Many factors affect the longevity of animals in aquaria, making it difficult to state with certainty why *Nautilus* survive longer at some institutions than at others. However, it is generally true that aquaria that utilize natural seawater and open systems are more successful in the long-term maintenance of marine animals than institutions that employ artificial seawater and closed systems. This superiority of natural conditions also appears to be true for *Nautilus* (Table II). The use of an open system ensures day-to-day consistency in water quality and minimizes many of the problems associated with the buildup of waste products and other toxins in closed systems. The longevity of *Nautilus* in open-system tanks is greater than in closed systems, although there is considerable variation. The age and condition of animals when received by the aquaria may also be an important factor in determining longevity.

Table II. Longevity of *Nautilus* in Various Aquarium Systems

Institution	Type of system: Open	Closed and semiclosed
Nouméa Aquarium	540 days	—
Waikiki Aquarium	1020 days	—
Steinhart Aquarium	—	390 days
New York Aquarium	≈1095 days	—
Shedd Aquarium	—	289 days
Sea World, San Diego	1140 days	—
New England Aquarium	720 days	—
Baltimore Aquarium	—	240 days
Vancouver Aquarium	120 days	—
Hagenbeck Aquarium	—	962 days
Artis Aquarium	—	883 days
Monaco Aquarium	578 days	—
Yomiuri-Land Aquarium	—	511 days
Average:	745 days	546 days

5. Diseases and Abnormalities

5.1. Shell Abnormalities

There are a number of abnormalities that develop in captive *Nautilus* that have been observed at nearly all aquaria. The most common observation, reported from all aquaria, is abnormal shell growth. All adult and juvenile *Nautilus* brought into captivity quickly develop a black margin on the lip of the shell. In young specimens, new shell material is thinner and more fragile than older shell material, and numerous black lines are interspersed with typical shell material. This condition persists for as long as the animals live in captivity, but although the shell is unattractive in appearance, it does not appear to affect the animal. A. W. Martin *et al.* (1978) have illustrated this abnormal growth. Black deposits also frequently build up on the interior of the body chamber wall. Arnold (1985) has illustrated this condition and discussed possible causes, but no treatment for the condition is known.

Several aquaria have reported an unusual shell growth resulting in a thickened, flared lip on the shell (see Graaf, 1981). No cause for this condition is known; the animals can survive for many months as it develops. This condition seems to occur more often in animals that have survived for several years in captivity.

5.2. Loss of Buoyancy Control

Perhaps the most common ailment among captive *Nautilus* is the apparent loss of buoyancy control. All aquaria report that some or all of their *Nautilus* are unable to maintain neutral buoyancy after living in captivity for more than a month. Most often, the animals become positively buoyant, but a few have become negatively buoyant. A few animals, particularly those obtained as juveniles, seem to have no problem maintaining neutral buoyancy in aquaria.

5.3. Eye Disorders

Because *Nautilus* has an open pupil, the animal is prone to a variety of eye problems. Air can enter the eye and become trapped when the animals are removed from the water, or bubbles from an airstone in the aquarium can collect in the eye or become trapped under the hood. Dirty aquarium water can presumably also enter the eye and cause problems. These factors should be considered when handling the *Nautilus* and designing their aquarium. If air appears to be trapped in the eye or under the hood, the animals can be tilted sideways underwater until the bubbles are gone.

Many public aquaria have reported that eye infections are common. It can be recognized by a white, mucuslike material filling the eye and commonly extruding from the pupil. Gentle pressure on the eye will force out the material. This condition has not been known to be lethal, but it may persist for many months. Frequently, after it clears up, the pupil will be noticeably enlarged and the retina gone. The animal is undoubtedly blind at this point, but this does not seem to affect its behavior or its ability to locate food.

5.4. Other Abnormalities

The Yomiuri-Land Aquarium, the Steinhart Aquarium, and the Waikiki Aquarium have all noticed that occasionally an animal will develop an unusual swelling in the mouth region (JECOLN, 1980b). *Nautilus* with this abnormality always die within a short time, and the cause is unknown.

Loss of hood pigmentation may also occur. In some cases, this decolorization develops soon after capture. If the animal is otherwise healthy and is placed in an adequate aquarium system, the condition will sometimes disappear. At other times, the condition will spread, the entire hood will turn white, and the health of the animal will decline. Again, no specific cause is known.

5.5. Parasites

Newly collected *Nautilus* are often heavily infested with a tiny copepod, *Anchicaligus nautili*. These parasites were first discovered and described by Willey (1897c), and Haven (1972) made a brief mention of their existence in *N. pompilius* in the Philippines. Nothing further was known about them until Ho (1980) published a redescription based on specimens provided by the Waikiki Aquarium from *N. belauensis*. *Anchicaligus* is a monotypic genus belonging to the family Caligidae, which contains about 380 species, nearly all of which are parasites on teleost fish. Caligid copepods are known to kill their fish hosts in captivity. However, it is not clear what effect they may have on *Nautilus*. The copepod is probably not able to complete its life cycle in captivity, as all *Nautilus* appear to be free of these parasites after several months in aquaria.

5.6. Cannibalism

Haven (1972) reported that *Nautilus pompilius* commonly had V-shaped breaks along the anterior margins of the shell, which closely conformed to the contour of the upper beak of *Nautilus*. She surmised that "fighting" within the species is fairly common. In captivity, it is not unusual to find one *Nautilus* in the act of devouring another *Nautilus* and ingesting both soft parts and much of the shell. Juvenile *Nautilus* maintained with adults frequently suffer this fate. Most commonly the animal being eaten is moribund. However, at the Waikiki Aquarium, *N. pompilius* have been observed on several occasions to cannibalize healthy animals. In one case, an animal that had lost a large portion of its hood and tentacles was isolated and later made a complete recovery, although a scar remained on the edge of the hood. The extent to which cannibalism occurs in nature and the effects it has on the ecology and behavior of *Nautilus*, particularly juveniles, are unknown.

6. Reproduction

Nautilus readily copulates and produces eggs in captivity. Of 14 aquaria surveyed, 7 have reported eggs produced by their *Nautilus*. Mikami and Okutani (1977), A. W. Martin *et al.* (1978), Carlson (1979), JECOLN (1980b), and Arnold and Carlson (1986) have reported details of egg production by captive *Nautilus*. *Nautilus* often begin producing eggs within a few weeks after being placed in aquaria. Adult egg-laying has rarely been observed, but photographs of a female depositing an egg on a rock have been published by JECOLN (1980b). Eggs are laid singly and deposited deep within crevices in rocks in the aquarium or occasionally in the corners of the tank.

Haven (1977a) speculated that reproduction in Philippine *Nautilus* may occur during the summer months, and A. W. Martin *et al.* (1978) also suggested that spawning may be seasonal. However, captive *Nautilus* appear to produce eggs almost continuously, but at irregular intervals. Unfortunately, because several females are usually maintained together in aquaria and because egg-laying occurs during the night, most aquaria have been unable to report precisely how many females are producing eggs and their frequency of production.

Carlson (1979) noted that temperature may be important in stimulating egg production and that *Nautilus* kept in constant cold water (12–15°C) cease egg production, but when temperatures are raised to 18–21°C, egg-laying resumes. This observation is anecdotal, but may be relevant to determining whether *Nautilus* lay their eggs in deep or shallow water.

Eggs laid in captivity may be abnormal in appearance and are usually infertile. Frequently, a capsule is produced with no yolk inside, or the yolk is found on the bottom of the tank separate from the capsule. Until recently, all eggs laid in captivity were infertile, until several living embryos were discovered in egg capsules produced by *N. belauensis* at the Waikiki Aquarium (Arnold and Carlson, 1986). Precisely which factors are most important in promoting egg fertility and development cannot be established until more rigorously controlled tests are undertaken, but details of this first successful attempt at obtaining embryos are described below.

Eleven adult *N. belauensis* were obtained in July 1982 from Palau. Four females and two males were kept in one aquarium system, and three females and two males were maintained in a second, separate tank. Both aquaria were open systems, with a flushing rate of 1.8 liters of new water per minute coming from a saltwater well (pH 7.8, O_2 6.8 ppm). The temperature in the two systems fluctuated from 15°C during the daytime to 24°C at night, to simulate the natural temperature variations experienced by *N. belauensis*.

At 5 months after the animals were brought into captivity, they began laying eggs. The number and frequency of eggs laid per individual female were not determined, but egg production was almost continuous throughout 30 months of observations. The eggs were moved to a separate tank and maintained at a temperature of 21°C. By February 1984, 70 eggs had been produced and examined. One of the capsules opened at that time contained an embryonic shell, but the embryo had died and disintegrated.

Between November 1984 and March 1985, 31 more eggs were laid. These eggs were examined after a period of about 4 months in incubation. The exact age of individual eggs was not known, but none was older than 4 months when opened. Four living embryos in various stages of development and a decomposing embryo and shell were discovered among this batch of eggs.

In December 1984, two female and three male *N. macromphalus* from New Caledonia were moved to one of the aquarium systems. Eggs were produced by these animals in subsequent months, and an embryo in an early blastula stage was discovered among the eggs. Embryos from *N. pompilius* have also been obtained from this system.

Until controlled tests can be conducted, it is not possible to state with any certainty why *Nautilus* at the Waikiki Aquarium suddenly began producing fertile eggs. However, temperature may play an important role. Also, the length of time between egg production and examination of the egg (4 months) may have been important. Previously, eggs had been examined within a few weeks of being laid; it may be that development had not proceeded to a detectable stage in such a short time.

7. Juvenile *Nautilus*

Newly hatched *Nautilus* have not yet been produced in captivity, nor, with one reported exception (Davis and Mohorter, 1973), have any living "hatchlings" been collected in the field. However, very young *Nautilus* have been obtained in recent years from fish collectors in the Philippines. No details are known as to where or how these animals have been collected, but they have been obtained by public aquaria, and even some home aquaria, in the United States, Japan, and Europe.

At the Waikiki Aquarium, 11 of these young animals have been reared to maturity, maintained in a 568 liter open-system tank and alternated with other young animals in a 113 liter closed-system display tank. The temperature in the open-system tank was maintained at about 24°C, but the temperature in the closed system frequently reached 27°C (maximum 27.4°C) with no apparent affect on the animals.

The juveniles were fed about 1.5 g of unshelled, fresh shrimp five days a week, and growth was recorded on a weekly basis for the first 4 months in captivity and again a year later for those surviving. Growth was rapid, but the animals appeared stunted, and shell growth was also abnormal. New shell growth had a glossy appearance, was noticeably fragile, and had numerous black lines interspersed among normal growth lines. No obvious periostracum, as frequently found on subadult *Nautilus* collected in the field, was observed on the shells of these animals. Furthermore, at an average diameter of 106 mm (range 96–117 mm), they reached maturity and began laying eggs. At this point, the animals had been in captivity for 15 months and had grown from an average diameter of 66 mm (range 59–78 mm). This is the first time that young *Nautilus* have been raised to maturity, and the exceptionally rapid growth rates are surprising, particularly in comparison with the slower growth rates recorded for *N. belauensis* in nature (Saunders, 1983; Cochran and Landman, 1984).

ACKNOWLEDGMENTS. The information in this chapter represents the combined efforts of numerous aquaria around the world. I especially wish to thank Yves Magnier, Nouméa Aquarium; Christina Slager, Steinhart Aquarium; Carl Jantsch, Sea World, San Diego; John Chamberlain, New York Aquarium and Osborn Laboratory; Roger Klocek, Shedd Aquarium, Chicago; Kitty Nelson, Seattle Aquarium; David Gross, National Aquarium, Baltimore; Paul Sieswerda, New England Aquarium, Boston; David Powell, Monterey Bay Aquarium; Gilbey Hewlett, Vancouver Aquarium; U. Richter, Aquarium Hagenbeck's Zoo; and the Yomiuri-Land Aquarium, Tokyo.

Chapter 36

Experience with Aquarium Rearing of *Nautilus* in Japan

T. HAMADA, S. MIKAMI, and T. OKUTANI

1. Introduction

The aquarium experience with *Nautilus* reported in this chapter was derived over a period of 4 years (1976–1980) at the Yomiuri-Land Marine Aquarium in Tokyo. This aquarium is located on the Tama river, about 110 m above sea level and some 20 km from the sea. Filtered winter seawater from Sagami Bay is stored in a main tank and circulated through the aquarium; the water is recycled for a period of years.

A special tank was installed for *Nautilus*, made of acrylic resin, with a capacity of 2 tons of seawater (Fig. 1). It is a hexagonal column, 160 cm high (filled with 140 cm of water), with faces 70 cm wide. Accessory equipment consists of a filtering unit (a vinylchloride box measuring 150 × 65 × 55 cm), a circulating pump made of plastic, an ultraviolet (UV) sterilizer, a cooling unit, and an aeration system. Except for the titanium heat exchanger, no metal is exposed to the seawater.

The filtering material consists of a mixture of 80 kg of silicate sand (grain diameter 1–2 mm), 5 kg of coral sand of various sizes, and 2 kg of activated carbon. The circulation rate is 50 liters/min. The UV sterilizer is a double 15 W tube light.

The tank is illuminated during the daytime by a 200 W fluorescent mercury light with an intercepting shade to control the intensity. Several dead coral blocks provide a shelter for the animals.

T. HAMADA • Department of Earth Science and Astronomy, College of Arts and Sciences, University of Tokyo, Tokyo 153, Japan. S. MIKAMI • Yomiuri-Land Co. Ltd., Tokyo 206, Japan. T. OKUTANI • Laboratory of Invertebrate Zoology, Department of Fisheries, Tokyo University of Fisheries, Tokyo 108, Japan.

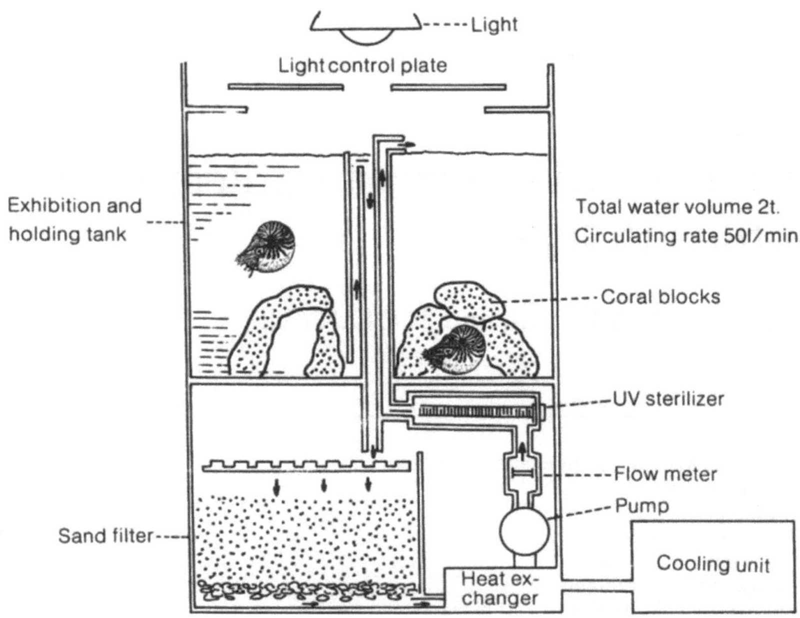

Figure 1. Rearing setup for keeping *Nautilus* at the Yomiuri-Land Marine Aquarium. Modified from Kawamoto *et al.* (1980).

2. Yomiuri-Land Marine Aquarium System for *Nautilus*

2.1. Water Quality

The water was kept oversaturated with oxygen, with chlorinity about 19% and pH ≈ 8.0. The calcium content was 428 ppm (maximum).

Water quality checks were made every 6 hr for temperature and every 3 days for other parameters. Water changes were made when ammonia rose to 0.2 ppm, when pH fell below 8.0, or when nitrate accumulation reached 20 ppm. The interval of water exchange was 3 days at the shortest and 70 days at the longest, with an average of 34 days (Fig. 2).

It is noteworthy that an abrupt rise of ammonia nitrogen (NH_4-N), to 1.17 ppm [due to inadequate biological filtration in the first series of experiments (Fig. 2)], did not affect the animals seriously, although food intake decreased. Water quality returned to normal after the carbonate filtering sand was replaced by silicate sand.

In the same series, the value of nitrate nitrogen (NO_3-N) exceeded the critical value for *Octopus* (i.e., 10 ppm) and reached 21.4 ppm, but no significant change was observed in the respiration of *Nautilus* (Fig. 2). Nitrite nitrogen (NO_2-N) rose to 0.2 ppm, but was generally kept below this level. The phosphate phosphorus (PO_4-P) level rose to 1.48 ppm (Fig. 2), and the CO_2 was measured at 0.61 ppm maximum. These results indicate that *N. macromphalus* is remarkably tolerant to high N content, compared to squids and marine fishes.

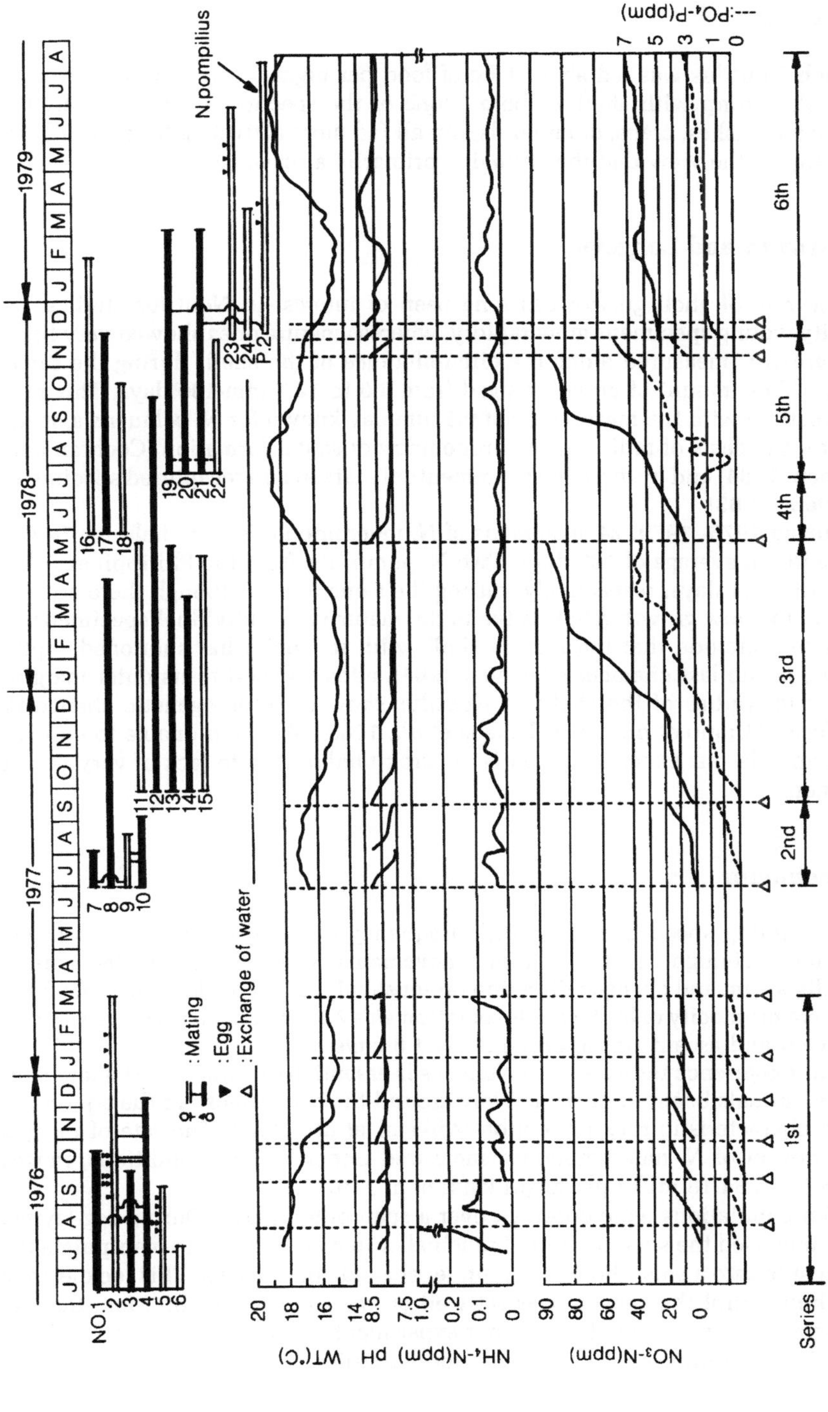

Figure 2. Records of survival and water conditions during rearing experiments 1976–1979, for specimens of *N. macromphalus* (Nos. 1–24) and *N. pompilius* (P.2). Modified from Kawamoto *et al.* (1980).

2.2. Feeding

Each *Nautilus* was fed about 10 g of food per night. Food consisted of jack mackerel, shrimp with shell, or both. None of the specimens reacted to living prey such as fishes, crabs, or annelids, but all fed more actively after dark. These facts support the view that the animal is primarily a scavenger.

2.3. Growth and Longevity

The rate of shell growth has long been of interest in *Nautilus* studies. In captivity, young specimens grew rapidly, whereas mature ones showed very little or no growth (measured along the circumference of the shell) during the same interval. The amount of growth ranged from 8.0 to 12.7 mm/100 days. This rate is compatible with the mean value of 0.1 mm/day known for *N. belauensis*, measured on the basis of radiometric chronometry of septal formation (Cochran and Landman, 1984) and by direct measurements of marked and recaptured specimens (Saunders, 1983).

During 1976–1979, 24 specimens of *N. macromphalus* were held for 7–387 days, with an average of 162 days. Two *N. pompilius* from the Philippines lived for 64 and 224 days, respectively, during the same term. Although there are too few data to show growth rate relative to the maturity of individual specimens, it is safe to conclude that the mature shell, with its somewhat shortened septal distance at the last chamber (Hamada, 1964), adds no shell to its outer margin. Allometric studies of the shells of *Nautilus* show early ontogenetic change in shell form (Hirano *et al.*, 1980; Tanabe *et al.*, 1985). We were unable to observe this change in our experiments on *N. macromphalus*, due to lack of very young specimens.

2.4. Behavior

Swimming speed and respiration frequency were measured under various conditions. Though capable of rapid movements when frightened, the animal normally swam and changed direction rather slowly, using its flexible hyponome. The average velocity of backward locomotion was 2.5 cm/sec. The average velocity of descent and ascent was approximately 1.8 cm/sec.

Our experience indicates that *Nautilus* is essentially nocturnal. At about sunset, the animals begin to move actively around the coral blocks in the aquarium, with a characteristic rocking motion (Mikami *et al.*, 1980). The rate of oxygen consumption of *N. macromphalus*, measured under various conditions, ranges from 6.57 to 34.16 ml/kg per hr (Mikami *et al.*, 1979).

Ten copulations were observed over a span of 9 months. During egg-laying, the female used the smaller, inner branchial crown of tentacles to manipulate the egg capsule into a small hole or gap under shaded coral blocks. This took as long as an hour, until the capsule was set well on the solid basement. More than 40 egg capsules were obtained during our experiments, mainly from underneath the coral blocks. However, all these eggs were infertile.

2.5. Biogeographic Implications

In combining rearing experiments, field data, and isotope analysis, it becomes evident that *Nautilus* lives in cool water, close to the forereef scarp and some 150–450 m deep. We suggest that this habitat is intimately related to the timing and conditions for copulating and spawning.

Hamada and Mikami (1980) proposed a hypothetical diagram to show the seasonal migration of *N. macromphalus* in terms of reproductive activities, with the adults migrating toward the surface during the winter, when water surface temperatures decrease to approximately 20°C. Spawned egg capsules laid in shallow water would remain there until hatched. The unique report of a juvenile *Nautilus* alive in very shallow water near Suva, Fiji (Davis and Mohorter, 1973), suggests the correctness of this speculation.

As Hamada (1980) has pointed out, there should be a great ecological barrier, especially in terms of water temperature, preventing *Nautilus* from inhabiting vast shallow sea areas less than 200 m deep (e.g., the Sahul and Sunda Shelves). Some rare occurrences of *Nautilus* shells in unexpected regions, such as the African coast of the Indian Ocean, could be products of postmortem drift or accidental dispersal (see Chapter 4).

2.6. Application of Our Rearing Technique

The rearing technique for *Nautilus* established by our experiments was efficiently applied to a drifted adult female *N. pompilius* found in Kawajiri, Kagoshima Prefecture, southern tip of Kyushu, Japan, on April 5, 1978. Two days after it was caught, the specimen was brought to the Kagoshima City Fish Market, where it was identified and immediately put into cool water to survive (the water temperature of the Kuroshio Current was estimated to have been about 24°C or less at the surface when the specimen was caught). Although the specimen had been out of water for more than 14 hr, it was subsequently kept alive for 64 days. Kagoshima is located approximately 2000 km from the Philippines, which we regard as the northern limit of *N. pompilius*.

Chapter 37

A Small, Closed Aquarium System for *Nautilus*

CLAUDE SPINOSA

1. Introduction

As described in Chapter 35, aquarium displays of living *Nautilus* are now featured at numerous public aquaria. Although the number of such exhibits has increased in recent years, they are limited primarily to large public institutions, where professional staffing and facilities are possible; these necessities are not available at smaller institutions with limited budgets. The experience at Boise State University has demonstrated that a small, closed-circulation aquarium, customized to specific needs, can be constructed to accommodate long-term maintenance of *Nautilus*, with relatively small expenditure and investment of time. The system described in this chapter was designed and used primarily for educational and display purposes; it has proven worthwhile and easy to construct and maintain.

Aquarium observers, particularly students, are intrigued by the notion that *Nautilus* is a "living fossil," and they appreciate the glimpse, however limited, into the seas of the geologic past. Such displays are especially appropriate as projects in invertebrate paleontology or invertebrate zoology classes. Additionally, greater laboratory access to *Nautilus* for more researchers opens totally new ranges of research possibilities, from physiology to embryology, and from growth to the study of motion (see Muntz and Raj, 1984; Ward and Chamberlain, 1983; Zann, 1984; Arnold and Carlson, 1986).

The basic requirements for the maintenance of a small, closed-circulation

CLAUDE SPINOSA • Department of Geology and Geophysics, Boise State University, Boise, Idaho 83725.

system are: (1) adequate holding and display facilities, (2) a water purification system, and (3) an adequate maintenance protocol.

Following is the description of a closed-system aquarium developed and used successfully for over 5 years at Boise State University to maintain as many as eight *Nautilus* specimens for as long as 14 months.

2. Tank Design and Fabrication

Tank design for *Nautilus* aquaria does not differ from that for other marine species. The design should maximize viewing area and allow sufficient space for specimen movement and observation; location, budget, and materials considerations may impose additional constraints.

A tank approximately 180 cm long, 50 cm wide, and 60–70 cm deep, of 630-liter water capacity, is the largest that can be constructed with inexpensive, generally available stock materials. Larger tanks require thicker glass, which increases prices drastically. Two smaller tanks can be designed into a single display that is as attractive as one larger tank but costs a fraction of the larger tank. In addition, smaller tanks offer greater flexibility in terms of such considerations as mobility and management of leaks, damage, and disease.

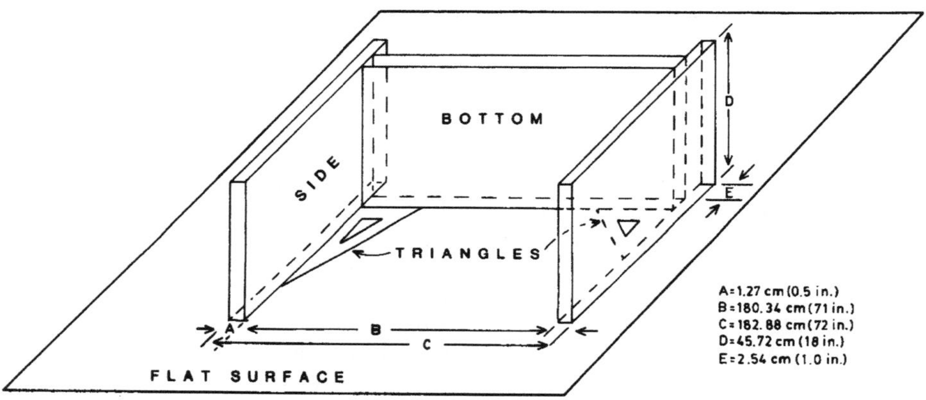

Figure 1. Assembly of an aquarium tank using glass panels. (1) Set the two side panels on edge at right angles to the bottom panel, using triangles or carpenter squares. Apply an even bead of clear silicon sealer along the short edges of the bottom panel. Press the side panels against the beaded edges firmly enough to squeeze out some silicon sealer and assure a tight bond. Note especially dimension "E"; this space assures that the bottom panel will be properly elevated. The cemented panels should be held in position with masking tape for at least 24 hr to cure the silicon sealer. Once cemented, surfaces can be separated only with difficulty; therefore, all surfaces should be checked for correct fit prior to sealing. Minor adjustments should be made before the sealer has cured. (2) After the 24-hr curing time, with the panels still positioned as illustrated, apply a bead of silicon sealer to all top edges and place the back panel gently. Its weight is sufficient to press the silicon and ensure a proper seal. Recheck angles and use masking tape to prevent movement of the panels. (3) After 24 hr, turn the tank over and repeat the procedure with the front panel. (4) After the final curing, scrape off the excess silicon with a razor blade, rinse the tank with dilute acetic acid, and wash it well several times with a baking soda solution. Test the tank for leaks and weakness by filling it completely with water.

A 630-liter tank, as described above, can accommodate as many as four specimens with adequate space for movement. The suggested dimensions provide a large viewing area (≈10, 800 cm^2), and a tank of this size can be constructed with readily available ½-in. stock glazing material. Acceptably strong tanks 45 cm wide can be constructed of ⅜-in. plate glass stock with the following maximum dimensions: (1) 55 cm deep by 180 cm long or (2) 65 cm deep by 120 cm long. With ½-in. plate glass, tanks 45 cm wide can be constructed with the following maximum dimensions: (1) 65 cm deep by 180 cm long or (2) 80 cm deep by 120 cm long. The aforestated dimensions illustrate the approximate largest dimensions for ⅜-in. and ½-in. stock material that will withstand rupturing due to hydrostatic pressure and wall flexing. However, within these limits, actual tank dimensions can be determined by size of display area, material availability, and cost.

Fabrication of the tanks can be accomplished by reasonably skilled or adequately supervised personnel. Local glaziers can generally provide glass cut to required dimensions, or it can be ordered factory-cut, to exact dimensions, with polished edges. Construction of the aquarium is accomplished easily by following the steps outlined in Fig. 1.

3. Filter Design and Construction

In closed, recirculating aquaria, filters remove particulate waste as well as the copious amounts of ammonia produced by *Nautilus*. Particulate matter can be removed mechanically by settling tanks, diatom filters, protein skimmers, or other devices. Ammonia and nitrite are removed by bacterial filtration. The entire filtration apparatus can be separated from the display tank, as illustrated in Fig. 2, or it can be incorporated into the tank as a gravel substrate (Figs. 3 and 4). Tank substrate filters are advantageous for single-unit, smaller aquaria intended primarily for display or classroom purposes, where simplicity of operation and compactness are important considerations. This type of filter consists of a platform elevated approximately 4 cm above the tank bottom and covered with sand to provide the bacterial medium. The filter platform base can be constructed of a variety of materials, such as "egg crate" fluorescent light grid (Fig. 5), which is available from construction supply houses. The platform grid, covered on the upper surface with nylon screening, is placed on the bottom of the tank and covered with calcite or dolomite sand and gravel. A length of stock ¾-in. polyvinylchloride (PVC) pipe connects the underside of the filter to the main part of the tank. In reverse-flow filter systems, the PVC pipe returns water from the underside of the filter to the top of the tank, and water circulation is accomplished by airlifting (Figs. 3A and 4). In normal-flow systems, the subgravel PVC pipe returns water below the filter (Fig. 3B). Where a reliable supply of clean compressed air is available, airlift systems are preferable, because of quiet and dependable operation, requiring little attention; by contrast, pumps require regular maintenance.

The type of system described above (Fig. 2) is recommended for facilities intended to handle large numbers of animals and where mobility, simplicity, or aesthetics are not of primary consideration. Bacterial filter construction for larger

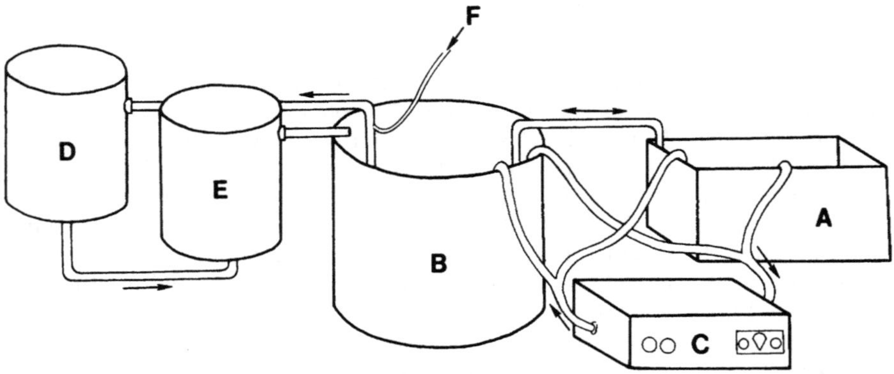

Figure 2. Recirculating aquarium system with separate filtration and holding tanks. Key: (A) specimen display tank connected by siphon to holding tank (B); (C) modular-type water-chilling unit; (D) elevated settling tank for particulate wastes; (E) bacterial filter tank, in which water percolates upward through the filter and spills out into holding tank (B); (F) water airlift; arrows indicate direction of water movement. The system illustrated was developed in cooperation with George Monaco and W. B. Saunders in Palau in 1977 and provided a dependable, working aquarium for many years. Similar models have provided reliable service for over 5 years in Boise, Idaho. This filtration system consists of an elevated 30-gallon settling tank (D), to which water is airlifted from the holding tanks. From the settling tank, water gravity-flows through the bacterial filter (E), from the bottom of the filter to the top, where it spills into the holding tank (B). The bacterial filter consists of several gravel and sand layers. The basal layer is composed of coarse gravel; higher layers are composed of finer coral sand. The separation of filter and holding tanks facilitates changing filtration capacity without altering holding-tank size; it allows utilization of multiple holding or display tanks and facilitates cleaning or replacing faulty filters without disturbing holding tanks. This system, however, is complex and requires more attention than the systems illustrated in Figs. 3 and 4.

units of this type is somewhat more complicated, but accomplishes the same results; i.e., water circulates through the bacterial medium at a rate sufficient to attain purification.

Circulation of water through the filter medium can be accomplished by airlifting or by means of pumps. Both normal and reverse-flow systems (Fig. 3) have worked well, and each has certain advantages.

4. Bacterial Filtration

Nautilus, like most cephalopods, produces copious amounts of ammonia, which, in recirculating aquaria, must be removed by filtration. The gravel-filter method employs *Nitrobacter* and *Nitrosomonas* to purify water through the process of nitrification (Fig. 6). *Nitrosomonas* nitrifies ammonia into nitrite, which is toxic, but which is converted into nitrate by *Nitrobacter*. Nitrate, which is much less toxic than nitrite, is removed by bacteria and algae or is allowed to accumulate to certain levels and is removed by partial water changes. A completely satisfactory method of nitrate removal by organic filtration has not been employed at Boise State University. It was more convenient to make periodic partial water

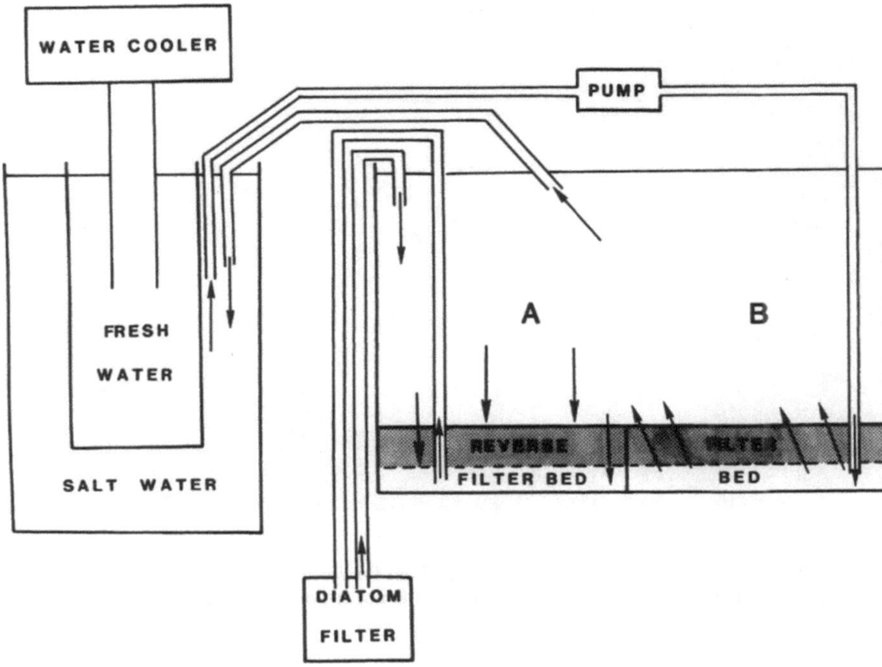

Figure 3. Diagrammatic representation of two types of pump-driven filter systems: (A) a reverse-flow filter bed, driven by the diatom filter pump; (B) a normal-flow, pump-driven system. Only one type of system is used in each tank. The water cooler is immersed in a bath of fresh water for protection against the corrosive action of saltwater.

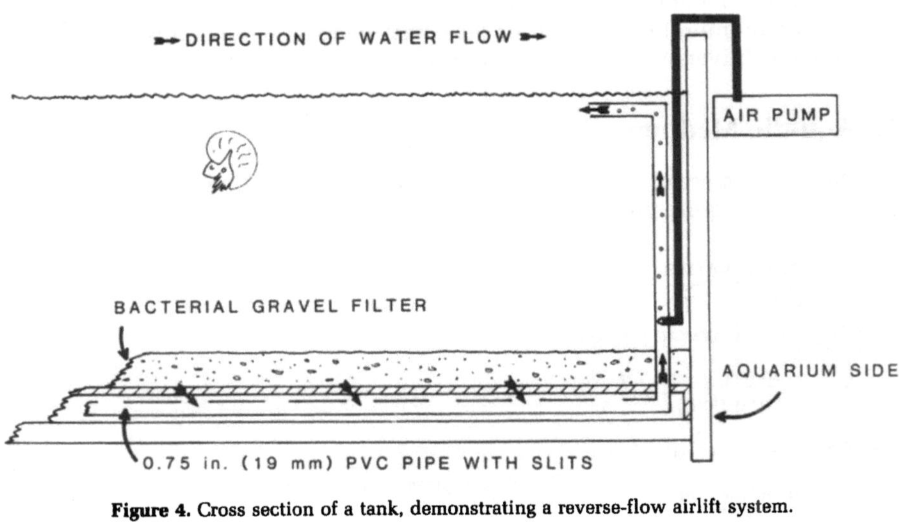

Figure 4. Cross section of a tank, demonstrating a reverse-flow airlift system.

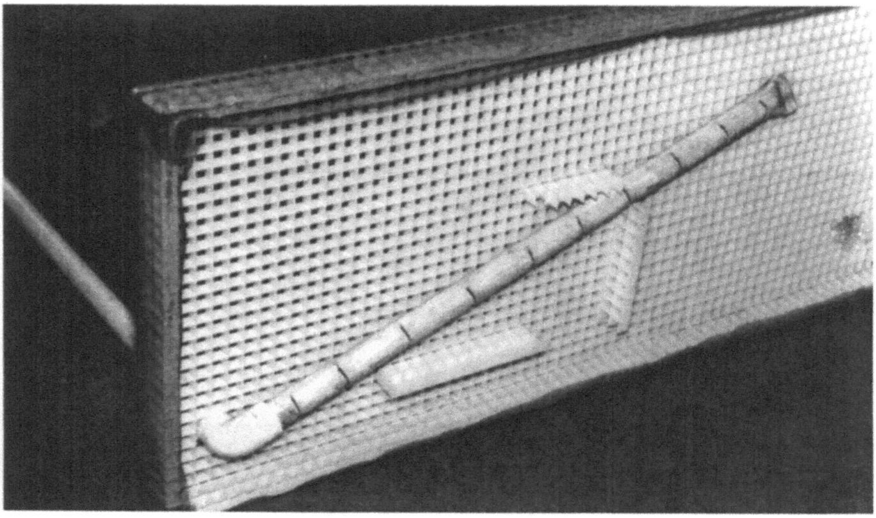

Figure 5. Underside view of a filter base constructed from fluorescent light "egg crate" grating, with slit PVC pipe, to return water from below the filter to the upper portion of the tank.

changes; doing so provided an additional benefit, namely, removing other accumulated organic compounds along with the nitrate.

As shown in Fig. 7, the filter bed matures through time with respect to ammonia, nitrite, and nitrate concentrations. After the introduction of bacteria and ammonia into the system, approximately 30 days are required for the development of a functioning bacterial filtration system. There are alternatives to the *Nitrobacter–Nitrosomonas* ammonia filtration method; one is the use of commercially available clinoptilolite filtration for the direct removal of ammonia, although at Boise State University we have not used this method successfully.

5. Aquatic Medium

Most commercial marine aquaria utilize flow-through of natural seawater; such systems require proximity to an abundant supply of marine water. However,

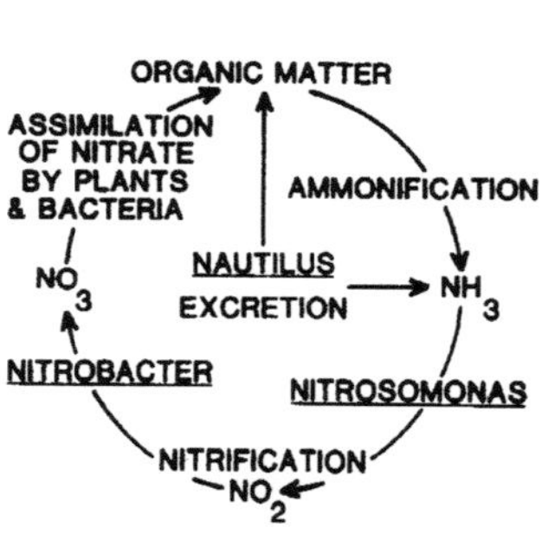

Figure 6. Representation of the nitrification cycle. After Spotte (1979).

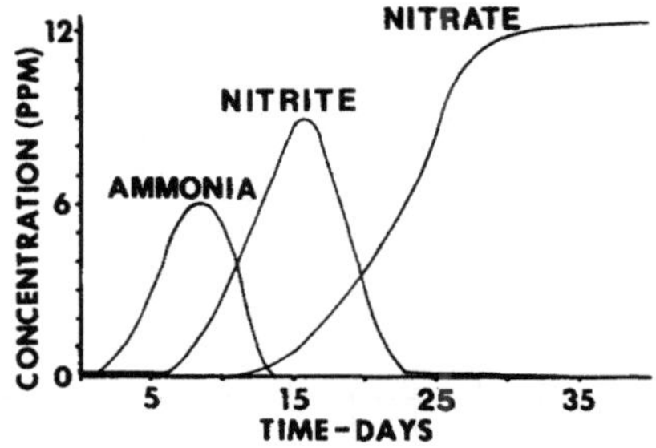

Figure 7. Nitrification cycle with respect to time.

the logistics and expense involved preclude use of flow-through systems by most small laboratory or university aquaria; closed, recirculating systems are logical alternatives. Additionally, utilization of recirculating systems with artificial salts may provide certain advantages, among which is the avoidance of temperature, pollution, parasite, and other problems associated with natural waters. Marine aquaria distant from the ocean must utilize recirculated water, constituted with artificial sea salts. Several brands of salts are commercially available and have proven satisfactory. The brands used at Boise State University are marketed as "Instant Ocean" and "Marine Environment." When mixed with clean water to the manufacturer's specifications, both performed adequately and provided a healthy aquatic medium. Because of price and other considerations, the "Marine Environment" brand was utilized more extensively in Boise, and it proved satisfactory.

6. Maintenance Protocol

Maintenance protocol and regular, consistent monitoring of the aquarium system form the most important link to successful maintenance of *Nautilus* in captivity. One person, or a small team, must assume responsibility and develop a "sixth sense" of aquarium-keeping. This requires an intuitive understanding for the well-being of the animals and for the mechanical operation of the system. Generally, the appropriate person is someone who enjoys the charge and who is willing to be available at odd times. Maintenance is not a difficult task, but it does not lend itself to a nine-to-five, five-day-per-week operation; problems may arise at any time. Carefully selected graduate or undergraduate students are quite often ideally suited.

Maintenance protocol is divided into two distinct phases: (1) the early bacterial filter maturation phase, which usually lasts 30 days, and (2) long-range maintenance. The first phase is critical, because it establishes the bacterial filter colonies. As shown in Fig. 7, the development of a high ammonia level is the necessary precursor of *Nitrosomonas* growth; this step precedes development of high nitrite concentrations. After the filters are mechanically operational, the aquarium is filled with saltwater, and garden soil or commercially available "starter" bacteria sets are introduced into the system. Several marine specimens are then introduced into the tanks to raise the level of ammonia and initiate the

TANK NO. ________

															FEEDING RECORD			
DATE	TIME	INITIALS	TEMP.	AMMONIA	NITRITE	NITRATE	SALINITY	pH	PUMPS	WATER LEVEL	AIR FLOWS	FILTERS	SIPHONS	MISC.	TYPE OF FOOD	AMOUNT	REACTION	REMARKS

ADDITIONAL REMARKS (initial and date)

Figure 8. System check record utilized at Boise State University.

denitrification process (Figs. 6 and 7). These first animals may not survive, and so they should be common, inexpensive marine species. Water chemistry levels for this period of the operation should conform approximately to those indicated in Fig. 7. Approximately 30 days after the bacterial filters are initiated, ammonia and nitrite levels will decrease to near zero and level off. Nitrate levels should increase and then level off. At this point, the bacterial filter should have developed adequate colonies of *Nitrosomonas* and *Nitrobacter* and should be functioning fully. Following the initial phase and barring an accident or major trauma, the filter should continue to operate indefinitely.

Long-term maintenance follows this initial phase; data and observations from this period should be recorded systematically. Observations made regularly as outlined on a checksheet (Fig. 8) foster a systematic maintenance protocol. The checksheet provides a chronological record of water quality, equipment operation, feeding history, and specimen condition, which may be useful in troubleshooting and research analysis. Inexpensive test kits to measure ammonia, nitrite, nitrate, and pH levels are available from scientific supply companies (e.g., Ward's Natural Science Establishment, PO Box 92912, Rochester, NY 14692-9012; Aquarium Pharmaceutical Inc., PO Box 222, Perkasie, PA 18944). Expensive and precise instrumentation to measure water quality parameters is not necessary, because the animals are tolerant to wide fluctuations, especially if they do not occur rapidly or frequently.

7. Cooling

Nautilus does not survive longer than a few days in water above 25°C. The optimal aquarium temperature is 15–17°C. In temperate climates, a small air-conditioned room may be sufficient to maintain correct water temperature on a temporary basis. However, this makeshift arrangement is usually not satisfactory in warm climates or as a long-term approach. The most effective method of maintaining constant water temperature is through the use of a water chiller. Such devices can be constructed from surplus refrigeration parts; one such homemade unit served the system at Boise State University very well for years (see Fig. 3).

Commercial chillers offer greater compactness, as well as simplicity, dependability, and portability. This is the most expensive piece of equipment and may be the only one that needs to be purchased. Excellent chillers are available from Frigid Systems Inc. (3214 Sylvania Avenue, Toledo, OH 43613). One particularly compact and dependable unit, the "Jewell Aquachiller," is manufactured by Jewell Industries (5005 Armitage Avenue, Chicago, IL 60639). These coolers have been used successfully under difficult conditions in Palau, as well as in Boise, and have worked well for years.

8. Specimen Procurement

Of the few extant species of *Nautilus*, *N. pompilius* is the most abundant, the most widely distributed, and the most readily available. Other species, such as *N. scrobiculatus* and *N. belauensis*, should not be considered available under normal circumstances. *Nautilus pompilius* occurs abundantly in the Philippines and, since 1980, has become generally available from commercial marine fish suppliers serving pet markets (at an approximate cost of $100 per specimen). However, this source has been somewhat unpredictable, and in the past, specimens commonly arrived in poor condition because of the supplier's lack of understanding regarding the needs of *Nautilus*. Recently, delivery of specimens in good health has been more regular. This availability is the single most important recent development enabling small laboratory or classroom display of live *Nautilus*.

The alternative to the purchase of specimens is direct procurement of live animals. Although this is possible in the Philippines, New Caledonia, and Palau, it is a financially and logistically demanding operation (see Chapter 35). Another source of live specimens for research purposes may be colleagues involved in *Nautilus* research, including some of the authors of this volume.

9. Space Requirements

The space required for maintenance of *Nautilus* is small, but a few factors are essential. *Nautilus* holding tanks need to be located in dark areas or in subdued light. A small, separate room dedicated exclusively to aquarium functions is best, but a small alcove or corner of a larger room, out of direct light, may suffice. Some additional features are highly recommended: (1) ample electrical outlets, protected by ground-fault circuits; (2) a sink with faucets that can accommodate the use of garden hoses; (3) a concrete or tile floor with floor drain; (4) storage space for salt, other chemicals, test kits, and other equipment and supplies; and (5) space for a desk.

ACKNOWLEDGMENTS. George Monaco, Palo Alto, California, and W. Bruce Saunders were instrumental in developing and maintaining the first recirculating system I used in Palau. Leroy Headley, Boise, maintained the aquaria at Boise State University and developed a "sixth sense" for the well-being of *Nautilus* in his care. Richard A. Davis, Cincinnati Museum of Natural History, devoted much time to making this manuscript readable. I am indebted to these individuals. Boise State University provided space and financial support.

References

Abbott, R. T., and Dance, S. P., 1983, *Compendium of Seashells*, E. P. Dutton, New York, 411 pp.

Ahrberg, P., 1935, Über den feineren Bau der Perlmutter von Schnecken und Cephalopoden, *Arch. Molluskenkd.* **67**:1–20.

Alexander, R. McN., 1977, Swimming, in: *Mechanics and Energetics of Animal Locomotion* (R. McN. Alexander and G. Goldspink, eds.), pp. 222–248, John Wiley, New York.

Amiel, A. J., Friedman, G. M., and Miller, D. S., 1973, Distribution and nature of incorporation of trace elements in modern aragonitic corals, *Sedimentology* **20**:47–64.

Amsellem, J., and Nicaise, G., 1980, Ultrastructural study of muscle cells and their connections in the digestive tract of *Sepia officinalis*, *J. Submicrosc. Cytol.* **12**:219–231.

Appellöf, A., 1893, Die Schalen von Sepia, Spirula und Nautilus: Studien über den Bau und das Wachstum, *K. Sven. Vet. Akad. Handl.* **25**:1–106.

Aprille, J. R., and Brennan, W. A., Jr., 1985, Subcellular distribution of adenine nucleotides as a regulator of metabolic ATP requirements in hypoxic–normoxic transitions, *Mol. Physiol.* **8**:648–649.

Arnold, J. M., 1962, Mating behavior and social structure in *Loligo pealei*, *Biol. Bull. (Woods Hole, Mass.)* **123**:53–57.

Arnold, J. M., 1965, Observations on the mating behavior of *Sepioteuthis sepioidea*, *Bull. Mar. Sci.* **15**:216–222.

Arnold, J. M., 1978, Spermiogenesis in *Nautilus pompilius*. II. Sertoli cell–spermatid junctional complexes, *Anat. Rec.* **191**:261–268.

Arnold, J. M., 1984, Cephalopod reproduction, in: *The Mollusca*, Vol. 7 (P. Biggelar, ed.), pp. 419–454, Academic Press, New York.

Arnold, J. M., 1985, Shell growth, trauma, and repair as an indicator of life history for *Nautilus*, *Veliger* **27**:386–396.

Arnold, J. M., 1987, Some observations on the cicatrix of *Nautilus* embryos (Beobachtungen an der cicatrix von *Nautilus*—Embryonen) in: *Cephalopods—Present and Past*, pp. 181–190, Schweizerbartsche Verlagbuchhandlung, Stuttgart.

Arnold, J. M., and Carlson, B. A., 1986, Living *Nautilus* embryos: Preliminary observations, *Science* **232**:73–76.

Arnold, J. M., and Okerlund-Arnold, K. E., 1969, Some aspects of hole-boring predation by *Octopus vulgaris*, *Am. Zool.* **9**:991–996.

Arnold, J. M., and Williams-Arnold, L. D., 1978, Spermiogenesis of *Nautilus pompilius*. I. General survey, *J. Exp. Zool.* **205**:13–26.

Austin, J., and Aprille, J. R., 1984, Carboxyatractyloside-insensitive influx and efflux of adenine nucleotides in rat liver mitochondria, *J. Biol. Chem.* **259**:154–160.

Avise, J. C., 1974, Systematic value of electrophoretic data, *Syst. Zool.* **23**:465–481.

Avise, J. C., 1983, Protein variation and phylogenetic reconstruction, in: *Protein Polymorphism: Adaptive and Taxonomic Significance* (G. S. Oxford and D. Rollinson, eds.), pp. 103–130, Academic Press, London.

Avise, J. C., and Aquadro, C. F., 1982, A comparative summary of genetic distances in the vertebrates: Patterns and correlations, *Evol. Biol.* **15**:151–185.

Ayala, F. J., 1982, The genetic structure of species, in: *Perspectives on Evolution* (R. Milkman, ed.), pp. 60–82, Sinauer Associates, Sunderland, Massachusetts.

Bachra, B. N., Trautz, O. R., and Simon, S. L., 1965, Precipitation of calcium carbonates and phosphates.

III. The effect of magnesium and fluoride ions on the spontaneous precipitation of calcium carbonate and phosphate, *Arch. Oral Biol.* **10**:731–738.

Bainbridge, R., 1958, The speed of swimming of fish as related to size and to the frequency and amplitude of tail beat, *J. Exp. Biol.* **35**:109–133.

Bainbridge, R., 1960, Speed and stamina in three fish, *J. Exp. Biol.* **37**:129–153.

Baldwin, J., 1982, Correlations between enzyme profiles in cephalopod muscle and swimming behavior, *Pac. Sci.* **36**:349–356.

Bandel, K., and von Boletzky, S., 1979, A comparative study of the structure, development and morphological relationships of chambered cephalopod shells, *Veliger* **21**:313–354.

Bandel, K., and Spaeth, C., 1983, Beobachtungen am rezenten *Nautilus*, *Mitt. Geol.-Palaeontol. Inst. Univ. Hamburg*, **54**:9–26.

Barber, V. C., 1967, The sense organs of *Nautilus*: Preliminary observations on their fine structure, *J. Microsc.* **6**:1067–1072.

Barber, V. C., 1968, The structure of mollusc statocysts, with particular reference to cephalopods, *Symp. Zool. Soc. London* **23**:37–62.

Barber, V. C., and Graziadei, P., 1965, The fine structure of cephalopod blood vessels. I. Some smaller peripheral vessels, *Z. Zellforsch. Microsk. Anat.* **66**:765–781.

Barber, V. C., and Graziadei, P., 1967, The fine structure of cephalopod blood vessels. II. The vessels of the nervous system, *Z. Zellforsch. Mikrosk. Anat.* **77**:147–174.

Barber, V. C., and Wright, D. E., 1969, The fine structure of the sense organs of the cephalopod mollusc *Nautilus*, *Z. Zellforsch. Microsk. Anat.* **102**:293–312.

Barrande, J., 1857, Über die innere Struktur der Nautiliden Schaalen, *Neues Jahrb. Mineral. Geogr. Geol. Petrefak. Jahrg.* **1857**:679–688.

Barrande, J., 1877, Céphalopodes: études générales, *Extraits du système Silurien du Centre de la Bohême*, p. 2, XVI, 253.

Bassot, J.-M., and Gabe, M., 1966, Caractères histologiques généraux des centres nerveux du *Nautile*, *C. R. Acad. Sci.* **263**:1248–1251.

Batchelor, G. K., 1967, *An Introduction to Fluid Dynamics*, Cambridge University Press, London.

Bayer, U., 1977, Cephalopoden-Septen. 1. Konstruktionsmorphologie des Ammonite-Septums, *Neues Jahrb. Geol. Palaeontol. Abh.* **154**:290–366.

Becker, G. L., Chen, C., Greenwalt, J. W., and Lehninger, A. C., 1974, Calcium phosphate granules in the hepatopancreas of the blue crab *Callinectus sapidus*, *J. Cell Biol.* **61**:316–326.

Beis, I., and Newsholme, E. A., 1975, The contents of adenine nucleotides, phosphagens and some glycolytic intermediates in resting muscles from vertebrates and invertebrates, *Biochem. J.* **152**:23–32.

Bennett, G., 1834, *Wanderings in New South Wales*, Vol. 2, pp. 406–411, Richard Bentley, London.

Bennett, G., 1859, Notes on the range of some species of *Nautilus*, on the mode of capture, and on the use made of them as an article of food, *Proc. Zool. Soc. London* **27**:226–229.

Bergstrom, B. H., and Arnold, J. M., 1974, Non-kinetochore association of chromatin and microtubules: A preliminary note, *J. Cell Biol.* **62**:917–920.

Berridge, M. J., and Oschman, J. L., 1972, *Transporting Epithelia*, Academic Press, New York.

Bidder, A. M., 1962, Use of the tentacles, swimming and buoyancy control in the pearly nautilus, *Nature (London)* **196**:451–454.

Birch, F., 1966, Compressibility elastic contants, *Geol. Soc. Am. Mem.* **97**:97–173.

Birchall, J. D., and Thomas N. L., 1983, On the architecture and function of cuttlefish bone, *J. Mater. Sci.* **18**:2081–2086.

Blake, J. F., 1882, *A Monograph of the British Fossil Cephalopoda: Introduction and Silurian Species*, Part I, J. Van Voorst, London, 248 pp.

Blind, W., 1975, Über die Entstehung und Funktion der Lobenlinie bei Ammonoideen, *Palaeontol. Z.* **49**:254–267.

Blind, W., 1976, Die ontogenetische Entwicklung von *Nautilus pompilius* (Linne), *Palaeontogr. Abt. A* **153**:117–160.

Blind, W., 1980, Über Anlage und Ausformung von Cephalopoden-Septen, *Neues Jahrb. Palaeontol. Abh.* **160**:217–240.

Bøggild, O. B., 1930, The shell structure of the mollusks, *Kgl. Dan. Vidensk. Selsk. Skr. Naturvidensk. Math. Afd. 9 Raekke II* **2**:233–326.

Boletzky, S. V., 1974, The "larvae" of Cephalopoda: A review, *Thalassia Jugosl.* **10**:45–76.

Bonaventura, C., Bonaventura, J., Miller, K. I., and van Holde, K. E., 1981, Hemocyanin of the chambered *Nautilus:* Structure–function relationships, *Arch. Biochem. Biophys.* **211**:589–598.

Bone, Q., and Trueman, E. R., 1982, Jet propulsion of the calycophoran siphonophores *Chelophyes* and *Abylopsis*, *J. Mar. Biol. Assoc. U.K.* **62**:263–276.

Bone, Q., and Trueman, E. R., 1984, Jet propulsion in *Dolidium* (Tunicata: Thaliacea), *J. Exp. Mar. Biol. Ecol.* **76**:105–118.

Bone, Q., Pulsford, A., and Chubb, A. D., 1981, Squid mantle muscle, *J. Mar. Biol. Assoc. U.K.* **61**:327–342.

Bonting, S. L., 1970, Sodium–potassium activated adenosine triphosphatase and cation transport. in: *Membranes and Ion Transport* (I. Bittar, ed.), pp. 257–363, John Wiley, New York.

Bottjer, D. J., 1981, Paleoecological implications of antifouling adaptations in shelled molluscs, *Geol. Soc. Am. Abstr. Prog.* **13**:413–414.

Bottjer, D. J., 1982, Morphology and function of projecting periostracal structures in the gastropoda (Mollusca), *Third North Am. Paleontol. Convention Proc.* **1**:51–56.

Bourne, G. B., 1974, Hemodynamics in the pink abalone, *Haliotis corrugata* (Mollusca, Gastropoda), unpublished Ph.D. dissertation, Iowa State University, Ames.

Bourne, G. B., 1982, Blood pressure in the squid, *Loligo pealei*, *Comp. Biochem. Physiol.* **72A**:23–27.

Bourne, G. B., 1983, Chronic examination of the heartbeat in three species of gastropods, *Am. Zool.* **23**:918.

Bourne, G. B., and Redmond, J. R., 1977, Hemodynamics in the pink abalone, *Haliotis corrugata* (Mollusca, Gastropoda). I. Pressure relations and pressure gradients in intact animals, *J. Exp. Zool.* **200**:9–16.

Bourne, G. B., Redmond, J. R., and Johansen, K., 1977, *Nautilus pompilius:* Branchial circulation enhanced by an auxiliary pumping mechanism, *Experientia* **33**:1453.

Bourne, G. B., Redmond, J. R., and Johansen, K., 1978, Some aspects of hemodynamics in *Nautilus pompilius*, *J. Exp. Zool.* **205**:63–70.

Brand, A. R., 1972, The mechanism of blood circulation in *Anodonta anatina* (L), *J. Exp. Biol.* **56**:361–379.

Brassey, A., 1879, *Around the World in the Yacht 'Sunbeam', Our Home on the Ocean for Eleven Months*, H. Holt, New York, 479 pp.

Brennan, W. A., Jr., and Aprille, J. R., 1985, Regulation of hepatic gluconeogenesis in newborn rabbit: Controlling factors in presuckling period, *Am. J. Physiol.* **249**:E498–E505.

Brewer, P. G., 1975, Minor elements in seawater, in: *Chemical Oceanography* (J. P. Riley and G. Skirrow, eds.), pp. 139–147, Academic Press, London and New York.

Britten, R. J., 1986, Rates of DNA sequence evolution differ between taxonomic groups, *Science* **231**:1393–1398.

Brocco, S. L., O'Clair, R. M., and Cloney, R. A., 1974, Cephalopod integument: The ultrastructure of Kolliker's organs and their relationship to setae, *Cell Tissue Res.* **151**:293–308.

Brooks, H., 1888, Preliminary remarks on the structure of the siphon and funnel of *Nautilus pompilius*, *Proc. Soc. Nat. Hist. Boston* **23**:380–382.

Buckland, W., 1837, *Geology and Mineralogy Considered with Reference to Natural History*, Vol. I, William Pickering, London, 618 pp.

Budelmann, B. U., 1975, Gravity receptor function in cephalopods with particular reference to *Sepia officinalis*, *Fortschr. Zool.* **23**:84–96.

Budelmann, B. U., and Young, J. Z., 1984, The statocyst–oculomotor system of *Octopus vulgaris:* Extraocular eye muscles, eye muscle nerves, statocyst nerves and the oculomotor centre in the central nervous system, *Philos. Trans. R. Soc. London Ser. B.* **306**:159–189.

Burton, R. F., 1976, Calcium metabolism and acid–base balance in *Helix pomatia*, in: *Perspectives in Experimental Biology*, Vol. I (P. S. Davies, ed.), pp. 7–16, Pergamon Press, London.

Burton, R. F., 1983, Ionic regulation and water balance, in: *The Mollusca*, Vol. 5, Part 2 (A. S. M. Saleuddin and K. M. Wilbur, eds.), pp. 292–352, Academic Press, New York.

Cain, A. J., 1983, Concluding remarks, in: *Protein Polymorphism: Adaptive and Taxonomic Significance* (G. S. Oxford and D. Rollinson, eds.), pp. 391–397, Academic Press, New York.

Carlson, B. A., 1977, *Nautilus* from Fiji, *Am. Pac. Mag.* 24–26.

Carlson, B. A., 1979, Chambered nautilus: A new challenge for aquarists, *Freshwater Mar. Aquar.* **2**:48–51, 63.

Carlson, B. A., 1985, The first known embryos of the chambered nautilus, *Hawaiian Shell News* **33**:1.

Carlson, B. A., McKibben, J. N., and DeGruy, M. V., 1984, Telemetric investigation of vertical migration of *Nautilus belauensis* in Palau (Western Caroline Islands, Pacific Ocean), *Pac. Sci.* **38**:183–188.

Carpenter, W. B., 1844, On the microscopic structure of shells: Cephalopoda, *Rep. Br. Assoc. Adv. Sci.* **1844**:1–24.

Carson, H. L., and Templeton, A. R., 1984, Genetic revolutions in relation to speciation phenomena: The founding of new populations, *Annu. Rev. Ecol. Syst.* **15**:97–131.

Catala, R. L. A., 1964, *Carnaval sous la Mer*, Les Editions R. Sicard, Paris, 141 pp.

Chamberlain, J. A., Jr., 1976, Flow patterns and drag coefficients of cephalopod shells, *Palaeontology* **19**:539–563.

Chamberlain, J. A., Jr., 1978, Permeability of the siphuncular tube of *Nautilus*: Its ecologic and paleoecologic implications, *Neues Jahrb. Geol. Palaeontol. Monatsh.* **3**:129–142.

Chamberlain, J. A., Jr., 1980a, Motor performance and jet propulsion in *Nautilus*: Implications for cephalopod paleobiology and evolution, *Bull. Am. Malacol. Union Inc.* **1980**:37–42.

Chamberlain, J. A., Jr., 1980b, The role of body extension in cephalopod locomotion, *Palaeontology* **23**:445–461.

Chamberlain, J. A., Jr., 1981, Hydromechanical design of fossil cephalopods, in: *The Ammonoidea* (M. R. House and J. R. Senior, eds.), pp. 289–336, Systematics Association Special Vol. 18, Academic Press, London and New York.

Chamberlain, J. A., Jr., and Chamberlain, R. B., 1985, Septal fracture in *Nautilus*: Implications for cephalopod paleobathymetry, *Lethaia* **18**:261–270.

Chamberlain, J. A., Jr., and Chamberlain, R. B., 1986, Is cephalopod septal strength index an index of cephalopod septal strength? *Alcheringa* **10**:85–97.

Chamberlain, J. A., Jr., and Moore, W. A., Jr., 1982, Rupture strength and flow rate of *Nautilus* siphuncular tube, *Paleobiology* **8**:408–425.

Chamberlain, J. A., Jr., and Weaver, J. S., 1978, Equations of motion for postmortem sinking of cephalopod shells, *Math. Geol.* **10**:675–691.

Chamberlain, J. A., Jr., and Westermann, G. E. G., 1976, Hydrodynamic properties of cephalopod shell ornament, *Paleobiology* **2**:316–331.

Chambers, S. M., 1980, Genetic divergence between populations of *Goniobasis* (Pleuroceridae) occupying different drainage systems, *Malacologia* **20**:63–81.

Chance, B., 1976, Pyridine nucleotide as an indicator of the O_2 requirements for energy-linked functions of mitochondria, *Circ. Res.* **38**:131–138.

Chance, B., and Leigh, J. S., Jr., 1985, The effect of work load upon the operating point for metabolic control in normoxia and hypoxia: "State 4 is best," *News Metab. Res.* **1**:26–33.

Chantler, P. D., 1983, Biochemical and structural aspects of molluscan muscle, in: *The Mollusca*, Vol. 4, *Physiology*, Part 1 (A. S. M. Saleuddin and K. M. Wilbur, eds.), pp. 77–154, Academic Press, New York.

Chapman, G., 1958, The hydrostatic skeleton in the invertebrates, *Biol. Rev.* **33**:338–371.

Chapman, G., 1975, Versatility of hydraulic systems, *J. Exp. Zool.* **194**:249–270.

Chave, K., 1954, Aspects of the biogeochemistry of magnesium. I. Calcareous marine organisms, *J. Geol.* **62**:266–283.

Chen, J. Y., and Teichert, C., 1983, Cambrian cephalopods of China, *Palaeontogr. Abt. A* **181**:1–102.

Clark, R. B., 1964, *Dynamics in Metazoan Evolution: The Origin of the Coelem and Segments*, Clarendon Press, Oxford, 313 pp.

Clark, R. B., 1981, Locomotion and the phylogeny of the Metazoa, *Boll. Zool.* **48**:11–28.

Clark, R. B., and Cowey, J. B., 1958, Factors controlling the change of shape of certain nemertean and turbellarian worms, *J. Exp. Biol.* **35**:731–748.

Clarke, G. L., and Backus, R. H., 1956, Measurements of light penetration in relation to vertical migration and records of luminescence of deep-sea animals, *Deep-Sea Res.* **4**:1–14.

Clarke, G. L., and Denton, E. J., 1962, Light and animal life, in: *The Sea*, Vol. 1 (M. N. Hill, ed.), pp. 456–468, John Wiley, New York.

Cloney, R. A., and Florey, E., 1968, Ultrastructure of cephalopod chromatophore organs, *Z. Zellforsch. Mikrosk. Anat.* **89**:250–280.

Cochran, J. K., and Landman, N. H., 1983, Growth rate of *Nautilus belauensis* in nature, *Geol. Soc. Am. Abstr.* **15**:546.

Cochran, J. K., and Landman, N. H., 1984, Radiometric determination of the growth rate of *Nautilus* in nature, *Nature (London)*, **308**:725–727.

Cochran, J. K., Rye, D. M., and Landman, N. H., 1981, Growth rate and habitat of *Nautilus pompilius* inferred from radioactive and stable isotope studies, *Paleobiology* **7**:469–480.

Cohen, A. I., 1973a, An ultrastructural analysis of the photoreceptors of the squid and their synaptic connections. I. Photoreceptive and non-synaptic regions of the retina, Comp. *Neurol.* **147**:351–378.

Cohen, A. I., 1973b, An ultrastructural analysis of the photoreceptors of the squid and their synaptic connections. II. Intraretinal synapses and plexus, Comp. *Neurol.* **147**:379–398.

Cohnheim, O., 1912, Über den Gaswechsel von Tieren mit glatter und quergestreifter Musculatur, Z. *Physiol. Chem.* **76**:298–313.

Cole, K. S., and Gilbert, D. L., 1970, Jet propulsion of squid, *Biol. Bull. (Woods Hole, Mass.)* **138**:245–246.

Collicutt, J. M., and Hochachka, P. W., 1977, The anaerobic oyster heart: Coupling of glucose and aspartate fermentation, *J. Comp. Physiol.* **115**:147–157.

Collins, D. H., and Minton, P., 1967, Siphuncular tube of *Nautilus*, *Nature (London)* **216**:916–917.

Collins, D. H., and Ward, P. D., 1987, Adolescent growth and maturity in *Nautilus*, *R. Ont. Mus. Life Sci. Contrib.* (in press).

Collins, D. H., Westermann, G. E. G., and Ward, P. D., 1978, The mature *Nautilus*: Its shell and buoyancy, Abstract in Proceedings of the GAC/GSA Joint Annual Meeting, Toronto **10**:382.

Collins, D. H., Ward, P. D., and Westermann, G. E. G., 1980, Function of cameral water in *Nautilus*, *Paleobiology* **6**:168–172.

Comfort, A., 1950, Biochemistry of molluscan shell pigments: Parts 2 and 3, Proc. *Malacol.* Soc. London **28**:79–85.

Conrad, T. A., 1849, Notes on shells, with descriptions of new genera and species, *J. Natl. Acad. Sci. Philadelphia* N. Ser. **1**:210–214.

Cook, A., and Shirbhate, R., 1983, The mucus producing glands and the distribution of the cilia of the pulmonate slug *Limax pseudoflavus*, *J. Zool. (London)* **201**:97–116.

Cotton, B. C., 1957a, Note on *Nautilus* repertus and *Nautilus scrobiculatus*, *Trans. R. Soc. South Aust.* **81**:113.

Cotton, B. C., 1957b, Records of uncommon southern Australian molluscs, *Rec. South Aust. Mus.* **13**:117–130.

Cousteau, J.-Y., and Diolé, P., 1973, *Octopus and Squid: The Soft Intelligence*, Doubleday, Garden City, New York, 304 pp.

Crenshaw, M. A., 1982, Mechanisms of normal biological mineralization of calcium carbonates, in: *Biological Mineralization and Demineralization* (G. H. Nancollas, ed.), pp. 243–257, Springer-Verlag, Berlin.

Crick, G. C., 1898, On the muscular attachment of the animal to its shell in some fossil Cephalopoda (Ammonoidea), *(Trans. Linn. Soc. London 2nd. Ser. 1896–1900) Zoology* **7**:71–113.

Crick, R. E., and Ottensmann, V. M., 1983, Sr, Mg, Ca and Mn chemistry of skeletal components of a Pennsylvanian and Recent nautiloid, *Chem. Geol.* **39**:147–163.

Crick, R. E., Mann, K. O., and Ward, P. D., 1984, Sr, Mg, and Ca chemistry of the skeleton of *Nautilus*, *Geology* **12**:99–102.

Crick, R. E., Burkart, B., Chamberlain, J. A., and Mann, K. O., 1985, Chemistry of calcified portions of *Nautilus pompilius*, *J. Marine Biol. Assoc. U.K.* **65**:415–420.

Crocker, K. C., DeNiro, M. J., and Ward, P. D., 1985, Stable isotopic investigations of early development in extant and fossil chambered cephalopods. I. Oxygen isotopic composition of eggwater and carbon isotopic composition of siphuncle organic matter in *Nautilus*, *Geochim. Cosmochim. Acta* **49**:2527–2532.

Curran, P. F., and McIntosh, J. R., 1962, A model system for biological water transport, *Nature (London)* **193**:347–348.

Currey, J. D., 1975, A comparison of the strength of echinoderm and mollusc shells, *J. Mar. Biol. Assoc. U.K.* **55**:419–424.

Currey, J. D., 1976, Further studies on the mechanical properties of mollusc shell material, *J. Zool. (London)* **180**:445–453.

Currey, J. D., 1977, Mechanical properties of mother of pearl in tension, *Proc. R. Soc. London Ser. B.* **196**:443–463.

Currey, J. D., 1979, The effect of drying on the mechanical properties of molluscan shells, *J. Zool. (London)* **188**:301–308.

Currey, J. D., and Taylor, J. D., 1974, The mechanical behaviour of some molluscan hard tissues, *J. Zool. (London)* **173**:395–406.
Currie, E. D., 1944, Growth stages in some Jurassic ammonites, *Trans. Roy. Soc. Edinburgh* **61**:171–198.
Dall, W. H., 1909, Contributions to the Tertiary paleontology of the Pacific coast. I. The Miocene of Astoria and Coos Bay, Oregon, *U.S. Geol. Surv. Prof. Pap.* **59**:1–278.
Daniel, T. L., 1983, Mechanics and energetics of medusan jet propulsion, *Can. J. Zool.* **61**:1406–1420.
Daniel, T. L., 1984, Unsteady aspects of aquatic locomotion, *Am. Zool.* **24**:121–134.
Dartnall, H. J. A., 1953, The interpretation of spectral sensitivity curves, *Br. Med. Bull.* **9**:24–30.
Dartnall, H. J. A., 1975, Assessing the fitness of visual pigments for their photic environments, in: *Vision in Fishes* (M. A. Ali, ed.), pp. 543–563, Plenum Press, New York.
Dartnall, H. J. A., and Lythgoe, J. N., 1965, The spectral clustering of visual pigments, *Vision Res.* **5**:81–100.
Dauphin, Y., 1979, Coquilles juvéniles de nautiles des îles Loyauté (Pacifique Sud), Cah. *Indo-Pac.* **1**:447–460.
Davis, J. C., 1973, *Statistics and Data Analysis in Geology*, John Wiley and Sons, New York, 500 pp.
Davis, R. A., and Mohorter, W., 1973, Juvenile *Nautilus* from the Fiji Islands, *J. Paleontol.* **47**:925–928.
Dayton, P. K., and Hessler, R. R., 1972, Role of biological disturbance in maintaining diversity in the deep sea, *Deep-Sea Res.* **19**:199–208.
Dean, B., 1901, Notes on living *Nautilus*, *Am. Nat.* **35**:819–837.
Degens, E. T., 1967, Evolutionary trends inferred from the organic tissue variations of mollusc shells, *Medd. Dan. Geol. Foren.* **17**:12–124.
Degens, E. T., Spencer, D. W., and Parker, R. H., 1967, Paleobiochemistry of molluscan shell proteins, *Comp. Biochem. Physiol.* **20**:553–579.
Den Hartog, J. P. (ed.), 1949, *Strength of Materials*, Dover, New York, 323 pp.
Denton, E. J., 1974, On buoyancy and the lives of modern and fossil cephalopods, *Proc. R. Soc. London Ser. B.* **185**:273–299.
Denton, E. J., and Gilpin-Brown, J. B., 1966, On the buoyancy of the pearly *Nautilus*, *J. Mar. Biol. Assoc. U.K.* **46**:723–759.
Denton, E. J., and Gilpin-Brown, J. B., 1973, Flotation mechanisms in modern and fossil cephalopods, *Adv. Mar. Biol.* **11**:197–268.
Denton, E. J., Gilpin-Brown, J. B., and Howarth, J. V., 1967, On the buoyancy of *Spirula spirula*, *J. Mar. Biol. Assoc. U.K.* **47**:181–191.
Diamond, J. M., and Bossert, W. H., 1967, Standing-gradient osmotic flow: A mechanism for coupling of water and solute transport in epithelia, *J. Gen. Physiol.* **50**:2061–2089.
Dilly, P. N., and Nixon, M., 1976, The cells that secrete the beaks in octopods and squids (Mollusca, Cephalopoda), *Cell Tissue Res.* **167**:229–241.
Dilly, P. N., Gray, E. G., and Young, J. Z., 1963, Electron microscopy of optic nerves and optic lobe of *Octopus* and *Eledone*, *Proc. R. Soc. London Ser. B.* **158**:446–456.
Dodd, J. R., 1967, Magnesium and strontium in calcareous skeletons: A review, *J. Paleontol.* **41**:1313–1329.
Doguzhaeva, L., 1982, Rhythms of ammonoid shell secretion, *Lethaia* **15**:385–394.
Donaubauer, H. H., 1979, Adenosine triposphatase localization in the branchial heart of *Sepia officinalis* L. (Cephalopoda), *Histochemistry* **65**:17–30.
Donaubauer, H. H., 1980, Adenosine triphosphatase localization in the branchial heart of *Sepia officinalis* L. (Cephalopoda), *Histochemistry* **69**:27–37.
Drew, G. A., 1911, Sexual activities of the squid *Loligo pealei* (Les.). I. Copulation, egg-laying, and fertilization, *J. Morphol.* **22**:327–360.
Dugdale, H. K. (ed.), 1974–1985, *The Chambered Nautilus Newsletter*, 1 (1974), 2–5 (1975), 6–8 (1976), 9–12 (1977), 13–16 (1978), 17–20 (1979), 21–24 (1980), 25–28 (1981), 29–32 (1982), 33–36 (1983), 37–40 (1984), 41–44 (1985), Delaware Museum of Natural History.
Dugdale, H. K., and Faulkner, D., 1976, The chambered *Nautilus*, exquisite living fossil, *National Geographic Magazine* **149**:38–41.
Dykens, J. A., Mangum, C. P., and Arnold, J. M., 1982, A note on the structural organization of the cardiac myofiber in *Nautilus pompilius*, *Pac. Sci.* **36**:267–271.
Edelstone, D. I., Paulone, M. E., and Holzman, I. R., 1984, Hepatic oxygenation during arterial hypoxemia in neonatal lambs, *Am. J. Obstet. Gynecol.* **150**:513–518.

Edwards, F. E., 1849, *A monograph of the Eocene Mollusca; or descriptions of shells from the older Tertiaries of England, Part 1, Cephalopoda, Palaeontogr. Soc.* **2**:1–56.
Eichler, R., and Ristedt, H., 1966a, Isotopic evidence on the early life history of *Nautilus pompilius* (Linne), *Science* **153**:734–736.
Eichler, R., and Ristedt, H., 1966b, Untersuchungen zur Frühontogenie von *Nautilus pompilius* (Linne), *Palaontol. Z.* **40**:173–191.
Eldredge, N., and Stanley, S. M., 1984, *Living Fossils*, Springer-Verlag, New York, 291 pp.
Ellington, W. R., 1981, Energy metabolism during hypoxia in the isolated, perfused ventricle of the whelk, *Busycon contrarium*, *Conrad. Comp. Physiol.* **142**:457–464.
Emery, D. G., 1975a, Ciliated sensory neurons in the lip of the squid *Lolliguncula brevis* Blainville, *Cell Tissue Res.* **157**:323–329.
Emery, D. G., 1975b, Ciliated sensory cells and associated neurons in the lip of *Octopus joubini* Robson, *Cell Tissue Res.* **157**:331–340.
Emery, D. G., 1975c, The histology and fine structure of the olfactory organ of the squid *Lolliguncula brevis* Blainville, *Tissue Cell* **7**:357–367.
Emery, D. G., 1976, Observations on the olfactory organ of adult and juvenile *Octopus joubini*, *Tissue Cell* **8**:33–46.
Emiliani, C., Kraus, E. B., and Shoemaker, E. M., 1981, Sudden death at the end of the Mesozoic, *Earth Planet. Sci. Lett.* **55**:317–334.
Epstein, S., Buschsbaum, R., Lowenstam, H. A., and Urey, H. C., 1953, Revised carbonate–water isotopic temperature scale, *Geol. Soc. Am. Bull.* **64**:1315–1326.
Erben, H. K., 1972a, Über die Bildung und das Wachstum von Perlmutt, *Biomineralization* **4**:15–46.
Erben, H. K., 1972b, Die Mikro- und Ultrastruktur abgedeckter Hohlelemente und die Conellen des Ammoniten-Gehäuses, *Palaeontol. Z.* **46**:6–19.
Erben, H. K., and Flajs, G., 1975, Über die cicatrix der Nautiloideen, *Mitt. Geol. Palaeont. Inst. Univ. Hamburg* **44**:59–68.
Erben, H. K., and Reid, R. E. H., 1971, Ultrastructure of shell, origin of conellae and siphuncular membranes in an ammonite, *Biomineralisation Forsch.* **3**:22–29.
Erben, H. K., Flajs, G., and Siehl, A., 1969a, Über die Schalenstrucktur von Monoplacophoren, *Akad. Wiss. Lit. Mainz, Math.-Naturwiss. Kl.* **1**:1–24.
Erben, H. K., Flajs, G., and Siehl, A., 1969b, Die frühontogenetische Entwicklung der Schalenstrucktur ectocochleater Cephalopoden, *Palaeontogr. Abt. A* **132**:1–54.
Ernst, S. A., 1972a, Transport adenosine triphosphatase cytochemistry. I. Biochemical characterization of a cytochemical medium for the ultrastructural localization of ouabain-sensitive, potassium-dependent phosphatase activity in the avian salt gland, *J. Histochem. Cytochem.* **20**:13–22.
Ernst, S. A., 1972b, Transport adenosine eriphosphatase cytochemistry. II. Cytochemical localization of ouabain-sensitive, potassium-dependent phosphatase activity in the secretory epithelium of the avian salt gland, *J. Histochem. Cytochem.* **20**:23–28.
Felback, D. K., and Atkins, A. G., 1984, *Strength and Fracture of Engineering Solids*, Prentice-Hall, Englewood Cliffs, New Jersey, 542 pp.
Ferguson, A., 1980, *Biochemical Systematics and Evolution*, John Wiley, New York, 194 pp.
Fernandez, M., 1907, Zur Histologie der Tentakel und des Rhinophors des *Nautilus pompilius*, *Z. Wiss. Zool.* **88**:187–211.
Fields, J. H. A., and Hochachka, P. W., 1982, Glucose and proline metabolism in *Nautilus*, *Pac. Sci.* **36**:337–342.
Finney, D. J., 1962, *Probit Analysis: A Statistical Treatment of the Sigmoid Response Curve*, Cambridge University Press.
Flores, E. E. C., Igarashi, S., and Mikami, T., 1978, Studies on squid behavior in relation to fishing. III. On the optomotor response of squid, *Todarodes pacificus* Steenstrup, to various colors, *Bull. Fac. Fish. Hokkaido Univ.* **29**:131–140.
Florey, E., 1969, Ultrastructure and function of cephalopod chromatophores, *Am. Zool.* **9**:429–442.
Flower, R. H., 1955, Cameral deposits in orthoconic nautiloids, *Geol. Mag.* **92**:89–103.
Foerste, A. F., and Teichert, C., 1930, The Actinoceroids of East-Central North America, *Denison Univ. Bull. J. Sci. Lab.* **25**:201–296.
Foord, A. H., 1888, *Catalogue of the fossil Cephalopoda in the British Museum (Natural History)*, Part I, *Nautiloidea*, Taylor and Francis, London, xxxi and 344 pp.
Foord, A. H., 1891, *Catalogue of the fossil Cephalopoda in the British Museum (Natural History)*, Part II, British Museum (Natural History), London, 407 pp.

Fraenkel, G. S., and Gunn, D. L., 1961, *The Orientation of Animals: Kineses, Taxes and Compass Reactions*, Dover, New York, 376 pp.

Franzen, A., 1955, Investigations into spermiogenesis and sperm morphology among invertebrates, Uppsala, Sweden, 13 pp.

Frazier, J., Margaritoulis, D., Muldoon, K., Potter, C. W., Rosewater, J., Ruckdeschel, C., and Salas, S., 1984, Epizoan communities on marine turtles. I. Bivalve and gastropod mollusks, *Marine Ecology* **6**(2):127–140.

Fukuda, Y., 1980, Observations by SEM, in: *Nautilus macromphalus in Captivity* [T. Hamada, I. Obata, and T. Okutani, (JECOLN) eds.], pp. 23–33, Tokai University Press, Tokyo.

Fukuda, Y., Mikami, S., and Kawamoto, N., 1977a, Electronmicroscopy of tentacles of *Nautilus*, *Rikakyoiku Kenkyu* **16**:2–16 (in Japanese).

Fukuda, Y., Mikami, S., and Kawamoto, N., 1977b, Structure and function of tentacles of *Nautilus macromphalus*, a living fossil (preliminary report), *Bull. Jpn. Fossil Club* **13**:7–15 (in Japanese).

Gasiorowski, S. M., 1973, Les rhyncholites, *Geobios* **6**:127–162.

Gladfelter, W. B., 1972a, Structure and function of the locomotory system of *Polyorchis montereyensis* (Cnidaria, Hydrozoa), *Helgol. Wiss. Meeresunters.* **23**:38–79.

Gladfelter, W. B., 1972b, Structure and function of the locomotory system of the Scyphomedusa *Cyanea capillata*, *Mar. Biol.* **14**:150–160.

Glenner, G. G., Burtner, H. J., and Brown, G. W. 1957, The histochemical demonstration of monoamine oxidase activity by tetrazolium salts, *J. Histochem. Cytochem.* **5**:591–600.

Goffinet, G., Grégoire, C., and Voss-Foucart, M. F., 1977, On ultrastructure of the trabeculae in the interlamellar membranes of nacre conchiolin of the *Nautilus* shell, *Arch. Intern. Physiol. Biochim.* **85**:849–863.

Goldman, B., and Talbot, F. H., 1976, Aspects of the ecology of coral reef fishes, in: *Biology and Geology of Coral Reefs*, Vol. III, *Biology 2* (O. A. Jones and R. Endean, eds.), pp. 125–154, Academic Press, New York.

Gonzalez-Santander, R., and Socastro Garcia-Blanco, E., 1972, Ultrastructure of the obliquely striated or pseudostriated muscle fibres of the cephalopods: *Sepia*, *Octopus* and *Eledone*, *Submicrosc. Cytol.* **4**:233–245.

Gordon, D. P., 1984, The marine fauna of New England: Bryozoa Gymnolaemata from the Kermadec Ridge, *N. Z. Oceanogr. Inst. Mem.* **91**:1–198.

Gordon, J. E., 1978, *Structures*, Penguin, Baltimore, 395 pp.

Gorman, G. C., and Renzi, J., 1979, Genetic distance and heterozygosity estimates in electrophoretic studies: Effects of sample size, *Copeia* **2**:242–249.

Gosline, J. M., and Shadwick, R. E., 1982, The biomechanics of the arteries of *Nautilus*, *Nototodarus*, and *Sepia*, *Pac. Sci.* **36**:283–296.

Gould, A. A., 1857, On the true Nautilus umbilicatus of Lister, *Proc. Zool. Soc. London* **25**:20–21.

Gould, S. J., 1971, Muscular mechanics and the ontogeny of swimming in scallops, *Palaeontology* **14**:61–94.

Gould, S. J., and Lewontin, R. C., 1979, The spandrels of San Marco and the Panglossian paradigm: A critique of the adaptationist programme, *Proc. R. Soc. London Ser. B.* **205**:581–598.

Gould, S. J., and Woodruff, D. S., 1978, Natural history of *Cerion*. VIII. Little Bahama Bank—A revision based on genetics, morphometrics and geographical distribution, *Bull. Mus. Comp. Zool. Harv. Univ.* **148**:371–415.

Gould, S. J., and Woodruff, D. S., 1986, Evolution and systematics of *Cerion* (Mollusca: Pulmonata) on New Providence Island: A radical revision, *Bull. Am. Mus. Nat. Hist.* **182**:389–490.

Graaf, F. de, 1981, *Nautilus pompilius* in captivity, *Vie Mar.* **3**:81–84.

Graham, D. W., Bender, M. L., Williams, D. F., and Keigwin, L. D., Jr., 1982, Strontium–calcium ratios in Cenozoic planktonic foraminifera, *Geochim. Cosmochim. Acta* **46**:1281–1292.

Graham, R. C., Karnovsky, M. J., 1966, The early stages of absorption of injected horseradish peroxidase in the proximal tubules of mouse kidney, *J. Histochem. Cytochem.* **14**:291–302.

Grassle, J. F., and Sanders, H. L., 1973, Life histories and the role of disturbance, *Deep-Sea Res.* **20**:643–659.

Graziadei, P., 1964, Electron microscopy of some primary receptors in the sucker of *Octopus vulgaris*, *Z. Zellforsch. Microsk. Anat.* **64**:510–522.

Graziadei, P., 1965a, Muscle receptors in cephalopods, *Proc. R. Soc. London Ser. B.* **161**:392–402.

Graziadei, P., 1965b, Sensory receptor cells and related neurons in cephalopods, *Cold Spring Harbor Symp. Quant. Biol.* **30**:45–57.

Green, H. D., Rapela, C. E., and Conrad, M. C., 1963, Resistance (conductance) and capacitance phenomena in terminal vascular beds, in: *Handbook of Physiology, Section 2: Circulation*, Vol. II (W. F. Hamilton, ed.), pp. 935–960, American Physiology Society, Washington, D.C.

Greenwald, L., and Ward, P. D., 1982, On the source of cameral liquid in the chambered *Nautilus*, *Veliger* **25**:169–170.

Greenwald, L., Ward, P., and Greenwald, O. E., 1980, Cameral liquid transport and buoyancy control in the chambered nautilus (*Nautilus macromphalus*), *Nature (London)* **286**:55–56.

Greenwald, L., Cook, C. B., and Ward, P. D., 1982, The structure of the chambered nautilus siphuncle: The siphuncular epithelium, *J. Morphol.* **172**:5–22.

Greenwald, L., Verderber, G., and Singley, C., 1984, Localization of Na–K ATPase activity in the nautilus siphuncle, *J. Exp. Zool.* **229**:481–484.

Grégoire, C., 1957, Topography of the organic components in mother-of-pearl, *J. Biophys. Biochem. Cytol.* **3**:797–808.

Grégoire, C., 1959a, A study on the remains of organic components in fossil mother-of-pearl, *Bull. Inst. R. Sci. Nat. Belg. Biol.* **35**:1–14.

Grégoire, C., 1959b, Conchiolin remnants in mother-of-pearl from fossil Cephalopoda, *Nature (London)* **184**:1157–1158.

Grégoire, C., 1962, On submicroscopic structure of the *Nautilus* shell, *Bull. Inst. R. Sci. Nat. Belg. Biol.* **38**:1–71.

Grégoire, C., 1964, Thermal changes in the *Nautilus* shell, *Nature (London)* **203**:868–869.

Grégoire, C., 1966, On organic remains in shells of Paleozoic and Mesozoic cephalopods (nautiloids and ammonoids), *Bull. Inst. R. Sci. Nat. Belg. Biol.* **42**:1–36.

Grégoire, C., 1967, Sur la structure des matrices organiques des coquilles de mollusques, *Biol. Rev.* **42**:653–688.

Grégoire, C., 1968, Experimental alteration of the *Nautilus* shell by factors involved in diagenesis and metamorphism. 1. Thermal changes in conchiolin matrix of mother-of-pearl, *Bull. Inst. R. Sci. Nat. Belg. Biol.* **44**:1–69.

Grégoire, C., 1972a, Structure of molluscan shell, in: *Chemical Zoology*, Vol. VII, *Mollusca* (M. Florkin, and B. T. Scheer, eds.), pp. 45–102, Academic Press, New York.

Grégoire, C., 1972b, Experimental alteration of the *Nautilus* shell by factors involved in diagenesis and in metamorphism. III. Thermal and hydrothermal changes in the organic and mineral components of the mural mother-of-pearl, *Bull. Inst. R. Sci. Nat. Belg. Biol.* **48**:1–83.

Grégoire, C., 1972c, Ultrastructure des composants organiques des coquilles de mollusques, *Haliotis* **2**:51–79.

Grégoire, C., 1973, On the submicroscopic structure of the organic components of the siphon in the *Nautilus* shell, *Arch. Intern. Physiol. Biochem.* **81**:299–316.

Grégoire, C., 1980, The conchiolin matrices in nacreous layers of ammonoids and fossil nautiloids: A survey. I. Shell wall and septa, *Akad. Wiss. Lit. Mainz. Abh. Math.-Naturwiss. Kl.* **2**:1–128.

Grégoire, C., 1984, Remains of organic components in the siphonal tube and in the brown membrane of ammonoids and fossil nautiloids, *Akad. Wiss. Lit. Mainz Abh. Math.-Naturwiss. Kl.* **5**:1–56.

Grégoire, C., and Lorent, R., 1972, Alterations in conchiolin matrices of mother-of-pearl during conversion of aragonite into calcite under experimental conditions of pyrolysis and pressure, *Biomineralisation* **6**:70–83.

Grégoire, C., and Teichert, C., 1965, Conchiolin membranes in shell and cameral deposits of Pennsylvanian cephalopods, Oklahoma, *Okla. Geol. Notes* **25**:175–201.

Grégoire, C., Duchateau, G., and Florkin, M., 1955, La trame protidique des nacres et des perles, *Ann. Inst. Oceanogr. Monaco* **31**:1–36.

Griffin, L. E., 1898a, Notes on the anatomy of Nautilus pompilius, Zool. Bull. **1**:147–161.

Griffin, L. E., 1898b, Notes on the tentacles of Nautilus pompilius, Johns Hopkins Univ. Circ. **18**:11–12.

Griffin, L. E., 1900, The anatomy of *Nautilus pompilius*, *Mem. Natl. Acad. Sci.* **8**:101–230.

Grimpe, G., 1926, Biologische Beobachtungen an *Sepia officinalis*, *Verh. Deutsch. Zool. Ges.* **31**:148–167.

Guary, J. C., and Fowler, S. W., 1982, Experimental studies on the biokinetics of plutonium and americium in the cephalopod *Octopus vulgaris*, *Mar. Ecol. Prog. Ser.* **7**:327–335.

Hallam, A., and Price, N. B., 1968, Further notes on the strontium contents of unaltered fossil cephalopod shells, *Geol. Mag.* **105**:52–55.

Haller, B., 1895, Beiträge zur Kenntnis der Morphologie von *Nautilus pompilius*, in: *Zoologische Forschungsreisen in Australien und dem Malayischen Archipel von Richard Semon*, Vol. 5, *Denkschr. Med. Naturwiss. Gesell. Jena* **8**:184–204.

Hamada, T., 1964, Notes on drifted *Nautilus* in Thailand, *Sci. Pap. Coll. Gen. Educ. Univ. Tokyo* **14**:255–278.

Hamada, T., 1965, Post-mortem drift of *Nautilus*, *Venus* **24**:181–198.

Hamada, T., 1966, The Japanese nautiloids, their past and now, *Chigaku-Kenkyu* **17**:221–238 (in Japanese).

Hamada, T., 1977, Distribution and some ecological barriers on the habitat condition of *Nautilus* and its application to the rearing of *Nautilus macromphalus*, *Sci. Pap. Coll. Gen. Educ. Univ. Tokyo* **27**:89–102.

Hamada, T., 1980, Distribution of modern *Nautilus* species in terms of ecological barriers, in: *Nautilus macromphalus in Captivity* [T. Hamada, I. Obata, and T Okutani (JECOLN), eds.], pp. 53–66, Tokai University Press, Tokyo.

Hamada, T., 1983, Preliminary report on some *Nautilus* drifts and the epifauna on *Nautilus* shells in a living state from the Tañon Strait, the Philippines, in: *Studies on Nautilus pompilius and Its Associated Fauna from Tañon Strait, the Philippines* (S. Hayasaka, ed.), pp. 36–39, *Kagoshima Univ. Res. Center S. Pac. Occas. Pap.*, No. 1.

Hamada, T., 1984, Further notes on *Nautilus* drift, *Geol. Palaeontol. S.E. Asia* **25**:263–265.

Hamada, T., and JECOLN (Japanese Expert Consultation on Living *Nautilus*), 1981, *Nautilus* studies in Japan, *Rec. Adv. Nat. Sci. Jpn.* **6**:95–99.

Hamada, T., and Mikami, S., 1977, A fundamental assumption on the habitat condition of *Nautilus* and its application to the rearing of *N. macromphalus*, *Sci. Pap. Coll. Gen. Educ. Univ. Tokyo* **27**:31–39.

Hamada, T., and Mikami, S., 1980, A fundamental assumption on *Nautilus* ecology and its application to the beginning of the rearing experiments, in: *Nautilus macromphalus in Captivity* [T. Hamada, I. Obata, and T. Okutani (JECOLN), eds.], pp. 44–52, Tokai University Press, Tokyo.

Hamada, K., Deguchi, Y., Fukuda, Y., Habe, T., Hirano, H., Kanie, Y., Kawamoto, N., Mikami, S., Obata, I., Okutani, T., and Tanabe, K., 1978, Recent advancement of the rearing experiments of *Nautilus* in Japan, *Venus Jpn. J. Malacol. (Kairuigaku Zasshi)* **37**:131–136.

Hamada, T., Obata, I., and Okutani, T. (JECOLN) (eds.), 1980a, *Nautilus macromphalus in Captivity*, Tokai University Press, Tokyo, 80 pp.

Hamada, T., Tanabe, K., and Hayasaka, S., 1980b, The first capture of a living chambered nautilus in Japan, *Sci. Pap. Coll. Gen. Educ. Univ. Tokyo* **30**:63–66.

Hamdorf, K., Schwemer, J., and Täuber, U., 1968, Der Sehfarbstaff, die Absorption der Rezeptoren und die spektrale Empfindlichkeit der Retina von *Eledone moschata*, *Z. Vergl. Physiol.* **60**:375–415.

Hammond, L. S., and Poiner, I. R., 1984, Genetic structure of three populations of the "living fossil" brachiopod *Lingula* from Queensland, Australia, *Lethaia* **17**:139–143.

Hara, T., and Hara, R., 1976, Distribution of rhodopsin and retinochrome in the squid retina, *J. Gen. Physiol.* **67**:791–805.

Harmer, S. F., 1957, The polyzoa of the Siboga expedition. 4. Cheilostomata, Ascophora II, *Siboga-Exped.* **28b**:641–1147.

Harris, H., and Hopkinson, D. A., 1976, *Handbook of Enzyme Electrophoresis in Human Genetics*, North-Holland, Oxford.

Harris, R. C., and Pilkey, D. H., 1966, Temperature and salinity control of the concentration of skeletal Na, Mn and Fe in *Dendraster excentricus*, *Pacific Sci.* **20**:235–238.

Harrison, F. M., and Martin, A. W., 1965, Excretion in the cephalopod *Octopus dofleini*, *J. Exp. Biol.* **42**:71–98.

Hartline, P. H., Hurley, A. C., and Lange, G. D., 1979, Eye stabilization by statocyst mediated oculomotor reflex in *Nautilus*, *J. Comp. Physiol.* **132**:117–126.

Haswell, W. A., 1896, Note on certain points in the arrangement and structure of the tentaculiferous lobes in *Nautilus pompilius*, *Proc. Linn. Soc. N.S. Wales Ser. 2* **10**:544–548.

Hattori, M., Tanabe, K., and Ōki, K., 1985, Underwater TV and still camera works in the habitat of *Nautilus* off Suva, *Marine Ecological Studies on the Habitat of Nautilus pompilius in the Environs of Viti Levu*, Fiji (S. Hayasaka, ed.), pp. 31–36, *Kagoshima Univ. Res. Center S. Pac. Occas. Pap.*, No. 4.

Haven, N., 1972, The ecology and behavior of *Nautilus pompilius* in the Philippines, *Veliger* **15**:75–81.

Haven, N., 1977a, The reproductive biology of *Nautilus pompilius* in the Philippines, *Mar. Biol. (Berlin)* **42**:177–184.

Haven, N., 1977b, Cephalopoda, Nautiloidea, *Reproduction of Marine Invertebrates* (C. G. Giese and J. S. Pearse, eds.), pp. 227–241, Vol. 4 Academic Press, New York.

Hayasaka, S. (ed.), 1983, *Studies on Nautilus pompilius and Its Associated Fauna from Tañon Strait, the Philippines, Kagoshima Univ. Res. Center S. Pac. Occas. Pap.*, No. 1.

Hayasaka, S. (ed.), 1985, *Marine Ecological Studies on the Habitat of Nautilus pompilius in the Environs of Viti Levu, Fiji, Kagoshima Univ. Res. Center S. Pac. Occas. Pap.*, No. 4.

Hayasaka, S., Saisho, T., Kakinuma, Y., Shinomiya, A., Ōki, K., Hamada, T., Tanabe, K., Kanie, Y., Hattori, M., Vande Vusse, F., Alcala, L., Cordero, P. A., Jr., Cabrera, J. J., and Garcia, R. G., 1982, Field study on the habitat of *Nautilus* in the environs of Cebu and Negros Islands, the Philippines, *Mem. Kagoshima Univ. Res. Center S. Pac.* **3**:67–137.

Hayasaka, S., Kakinuma, Y., Saisho, T., Tabata, M., and Nagayama, T., 1983, Additional record of observation on *Nautilus pompilius* in the aquarium of the Kamoike marine park, Kagoshima, Japan, in: *Studies on Nautilus pompilius and Its Associated Fauna from Tañon Strait, the Philippines* (S. Hayasaka, ed.), pp. 51–54, *Kagoshima Univ. Res. Center S. Pac. Occas. Pap.*, No. 1.

Hazelhoff, E. H., 1938, Über die Ausnützung des Sauerstoffs bei verschiedenen Wassertieren, *Z. Vergl. Physiol.* **26**:306–327.

Hedrick, P. W., 1983, *Genetics of Populations*, Science Books International, Boston, 629 pp.

Hendrix, J. P., Hulet, W. H., and Greenberg, M. J., 1981, Salinity tolerance and the responses to hypoosmotic stress of the bay squid *Lolliguncula brevis*, a euryhaline cephalopod mollusc, *Comp. Biochem. Physiol.* **69**:641–648.

Hensen, V., 1865, Ueber das Auge einiger Cephalopoden, *Z. Wiss. Zool.* **15**:155–242.

Herdman, W. A., and Hornell, J., 1903, Note on pearl-formation in the Ceylon pearl oyster, *Br. Assoc. Adv. Sci. Rep.*, Southport, 695 pp.

Hertel, H., 1966, *Structure, Form and Movement*, Reinhold, New York, 251 pp.

Hewitt, R. A., 1987, Outline of research on the ecology and evolution of the Eocene nautilid cephalopods from the London Clay, *Tertiary Res.* (in press).

Hewitt, R. A., and Stait, B., 1985, Phosphatic connecting rings and ecology of an Ordovician ellesmerocerid nautiloid, *Alcheringa* **9**:229–243.

Hirano, H., 1977, Biometric characteristics of *Nautilus pompilius*, *Sci. Res. Ser. Biol. Geol. School Educ. Waseda Univ.* **26**:13–23.

Hirano, H., 1981, Growth rates in *Nautilus macromphalus* and ammonoids: Its implications, in: *Inter. Symp. Concpt. Meth. Paleo. Barcelona* (J. Martinell, ed.), pp. 141–146, Universitat de Barcelona, Barcelona.

Hirano, H., and Obata, I., 1979, Shell morphology of *Nautilus pompilius* and *N. macromphalus*, *Bull. Nat. Sci. Mus. Ser. C (Geol.)* **5**:113–130.

Hirano, H., Obata, I., and Tanabe, K., 1980, Biometric characteristics, in: *Nautilus macromphalus in Captivity* [T. Hamada, I. Obata, and T. Okutani (JECOLN), eds.], pp. 34–43, Tokai University Press, Tokyo.

Ho, J. S., 1980, *Anchicaligus nautili* (Willey), a caligid copepod parasitic on *Nautilus* in Palau, with discussion of *Caligulina* Hagaard, 1972, *Publ. Seto. Mar. Biol. Lab.* **25**:157–165.

Hoagland, K. E., 1984, Use of molecular genetics to distinguish species of the gastropod genus *Crepidula* (Prosobranchia: Calyptraeidae), *Malacologia* **25**: 607–628.

Hochachka, P. W., 1985, Assessing metabolic strategies for surviving O_2 lack: Role of metabolic arrest coupled with channel arrest, *Mol. Physiol.* **8**:331–350.

Hochachka, P. W., 1986, Defense strategies against hypoxia and hypothermia, *Science* **231**:234–241.

Hochachka, P. W., and Fields, J. H. A., 1982, Arginine, glutamate, and proline as substrates for oxidation and for glycogenesis in cephalopod tissues, *Pac. Sci.* **36**:325–335.

Hochachka, P. W., and Guppy, M., 1987, *Metabolic Arrest and the Control of Biological Time*, Harvard University Press, Cambridge, Massachusetts, 227 pp.

Hochachka, P. W., and Somero, G. N., 1984, *Biochemical Adaptation*, Princeton University Press, Princeton, New Jersey, 537 pp.

Hochachka, P. W., Moon, T. W., Mustafa, T., and Storey, K. B., 1975, Metabolic sources of power for mantle muscle of a fast-swimming squid, *Comp. Biochem. Physiol.* **52B**:151–158.

Hochachka, P. W., Hartline, P. H., and Fields, J. H. A., 1977. Octopine as an end product of anaerobic glycolysis in the chambered *Nautilus*, *Science* **195**:72–74.

Hochachka, P. W., French, C. J., and Meredith, J., 1978, Metabolic and ultrastructural organization in *Nautilus* muscles, *J. Exp. Zool.* **205**:51–62.

Hochachka, P. W., Fields, J. H. A., and Mommsen, T. P., 1983, Metabolic and enzyme regulation during rest-to-work transition: A mammal versus mollusc comparison, in: *The Mollusca*, Vol. 1 (P. W. Hochachka, ed.), pp. 55–89, Academic Press, New York.

Hoerner, S. F., 1965, *Fluid Dynamic Drag*, Published by the author, Midland Park, New Jersey, 432 pp.

Hoeven, J. van der, 1857, Contributions to the knowledge of the anatomy of *Nautilus pompilius* L. especially with reference to the male animal, *Ann. Nat. Hist.* **19**:58–73.

Hölder, H., 1952, Über den Gehäusebau, insbesondere den Hohlkiel jurassischer Ammoniten, *Palaeontogr. Abt. A* **102**:18–48.

Hölder, H., and Mosebach, R., 1950, Die Conellen auf Ammonitensteinkernen als Schalenrelikte fossiler Cephalopoden, *Neues Jahrb. Geol. Palaeontol. Abh.* **92**:367–414.

Holloszy, J. O., and Booth, F. W., 1976, Biochemical adaptations to endurance exercise in muscle, *Annu. Rev. Physiol.* **38**:273–291.

Holmes, O. W., 1858, *The Autocrat of the Breakfast-Table*, Houghton, Mifflin, Boston and New York, 321 pp.

House, M. R., 1973, An analysis of Devonian goniatite distributions, in: *Organisms and Continents Through Time* (N. F. Hughes, ed.), *Spec. Pap. Palaeontol.* **12**:305–317.

House, M. R., 1975, Faunas and time in the marine Devonian, *Proc. Yorkshire Geol. Soc.* **40**:459–490.

Hughes, G. M., and Mill, P. J., 1966, Patterns of ventilation in dragonfly larvae, *J. Exp. Biol.* **44**:317–333.

Hughes, W. W., 1979, Matters arising: Nautiloid growth rhythms and lunar dynamics, *Nature* **279**:453–454.

Hughes, W. W., 1985, Planetary rotation and invertebrate skeletal patterns: Prospects for extant taxa, *Geophys. Surv.* **7**:169–183.

Humason, G. L., 1979, *Animal Tissue Techniques*, 4th ed., W. H. Freeman, San Francisco, 661 pp.

Hurley, A. C., Lange, G. D., and Hartline, P. H., 1978, The adjustable "pinhole camera" eye of *Nautilus*, *J. Exp. Zool.* **205**:37–44.

Huxley, J. S., 1932, *Problems of Relative Growth*, Dover, New York, 312 pp.

Hyatt, A., 1872, Fossil cephalopods of the Museum of Comparative Zoology: Embryology, *Bull. Mus. Comp. Zool. Harv. Univ.* **3**:59–111.

Hyatt, A., 1884a, Genera of fossil cephalopods, *Proc. Boston Soc. Nat. Hist.* **22**:273–338.

Hyatt, A., 1884b, The protoconch of Cephalopoda, *Am. Nat.* **18**:919–920.

Hyatt, A., 1894, Phylogeny of an acquired characteristic. *Proc. Am. Philos. Soc.* **32**:349–747.

Imbrie, J., 1956, Biometrical methods in the study of invertebrate fossils, *Bull. Am. Mus. Nat. Hist.* **108**:211–252.

Iredale, T., 1944, Australian pearly nautilus, *Aust. Zool.* **10**:294–298.

Ishii, T., 1981, Shells of *Nautilus* drifted ashore after an interval of 11 years, *Malacol. Soc. Jpn.* **12**:37–39 (in Japanese).

Istin, M., and Girard, J. P., 1970, Carbonic anhydrase and mobilisation of calcium reserves in the mantle of lamellibranchs, *Calcif. Tissue Res.* **5**:247–260.

Iwata, K., 1975, Ultrastructure of the conchiolin matrices in molluscan nacreous layer, *J. Fac. Sci. Hokkaido Univ. Ser. IV* **17**:173–229.

Jameson, L. H., 1912, Studies on pearl-oysters and pearls. I. The structure of the shell and pearls of the Ceylon pearl-oyster (*Margaritifera vulgaris* Schumacher): With an examination of the Cestode Theory of Pearl-production, *Proc. Zool. Soc. London* **B22**:260–358.

JECOLN (Japanese Expert Consultation on Living *Nautilus*), 1979, JECOLN annual report for 1977–1978.

JECOLN (Japanese Expert Consultation on Living *Nautilus*), 1980a, JECOLN Annual Report for 1978–1979.

JECOLN (Japanese Expert Consultation on Living *Nautilus*), 1980b, *Nautilus macromphalus in Captivity*, Tokai University Press, Tokyo, 80 pp.

JECOLN (Japanese Expert Consultation on Living *Nautilus*), 1982, JECOLN Annual Report for 1979–1982.

Jerlov, N. G., 1968, *Optical Oceanography*, Elsevier, New York, 199 pp.

Jeuniaux, C., 1963, *Chitine et Chitinolyse*, Masson, Paris, 181 pp.

Johansen, K., 1965, Cardiac output in the large cephalopod *Octopus dofleini*, *J. Exp. Biol.* **42**:475–480.

Johansen, K., and Lenfant, C., 1966, Gas exchange in the cephalopod, Octopus *dofleini*, *Am. J. Physiol.* **210**:910–918.

Johansen, K., and Martin, A. W., 1962, Circulation in the cephalopod, *Octopus dofleini*, *Comp. Biochem. Physiol.* **5**:161–176.

Johansen, K., Redmond, J. R., and Bourne, G. B., 1978, Respiratory exchange and transport of oxygen in *Nautilus pompilius*, *J. Exp. Zool.* **205**:27–35.

Johnson, M. S., Stine, O. C., and Murray, J., 1984, Reproductive compatibility despite large scale genetic divergence in *Cepaea nemoralis*, *Heredity* **53**:655–665.

Johnson, W., Soden, P. D., and Trueman, E. R., 1972, A study of jet propulsion: An analysis of motion of the squid, *Loligo vulgaris*, *J. Exp. Biol.* **56**:155–165.

Jolyet, F., and Regnard, P., 1877, Recherches physiologiques sur la respiration des animaux aquatiques; Deuxième partie, *Arch. Physiol. Ser. 2* **4**:584–633.

Jones, D. P., Kennedy, F. G., Andersson, B. S., Aw, T. Y., and Wilson, E., 1985, When is a mammalian cell hypoxic? Insights from studies of cells versus mitochondria, *Mol. Physiol.* **8**:473–482.

Jones, D. S., and Thompson, I., 1979, Matters arising: Nautiloid growth rhythms and lunar dynamics, *Nature (London)* **279**:454–455.

Jordon, R., and Stahl, W., 1970, Isotopische Palaeotemperatur—Bestimmungen an jurassischen Ammoniten und grundsatzliche Voraussentzungen fuer diese Methode, *Geol. Jb.* **89**:33–62.

Joubin, L., 1890, Recherches sur l'appariel respiratoire des nautiles, *Rev. Biol. Nord. Fr.* **11**:409–429.

Kabamba, M., 1983, Les nautiles du Crétacé supérieur de Madagascar, Université du Dijon, Institut des Sciences de la Terre, Thèse de Doctorat de Troisième Cycle de Géologie, 136 pp.

Kahn, H. R., and Saleuddin, A. S. M., 1979, Effects of osmotic changes and neurosecretory extracts on kidney ultrastructure in the freshwater pulmonate *Helisoma*, *Can. J. Zool.* **57**:1256–1270.

Kahn, P. G. K., and Pompea, S. M., 1978, Nautiloid growth rhythms and dynamical evolution of the earth–moon system, *Nature (London)* **275**:606–611.

Kakinuma, Y., and Tsukahara, J., 1985, A record of observations on *Nautilus pompilius* in laboratory aquariums, in: *Marine Ecological Studies on the Habitat of Nautilus pompilius in the Environs of Viti Levu, Fiji* (S. Hayasaka, ed.), pp. 74–79, *Kagoshima Univ. Res. Center S. Pac. Occas. Pap.*, No. 4.

Kampa, E. M., 1955, Euphausiopsin, a new photosensitive pigment for the eyes of euphausiid Crustaceans, *Nature (London)* **175**:996–998.

Kampa, E. M., and Boden, B. P., 1957, Light generation in a sonic-scattering layer, *Deep-Sea Res.* **4**:73–92.

Kanie, Y., and Hattori, M., 1983, Shell implosion depth of living *Nautilus*, in: Studies on *Nautilus pompilius and Its Associated Fauna from Tañon Strait, the Philippines* (S. Hayasaka, ed.), pp. 30–35, *Kagoshima Univ. Res. Center S. Pac. Occas. Pap.*, No.1.

Kanie, Y., and Tanabe, K., 1979, Preliminary report on the ecology of *Nautilus pompilius* in the Philippine seas, *Ann. Rep. Yokosuka City Mus.* **25**:26–29.

Kanie, Y., Mikami, S., Yamada, T., Hirano, H., and Hamada, T., 1979, Shell growth of *Nautilus macromphalus* in captivity, *Venus Jpn. J. Malacol. (Kairuigaku Zasshi)* **38**:129–134.

Kanie, Y., Fukuda, Y., Nakayama, H., Seki, K., and Hattori, M., 1980, Implosion of living *Nautilus* under increased pressure, *Paleobiology* **6**:44–47.

Kanie, Y., Hattori, M., Nakayama, H., Seki, K., Mizushima, Y., Shidara, F., and Ito, N., 1981, Respiration and shell implosion of living *Nautilus* under increased hydrostatic pressure, *Venus Jpn. J. Malacol. (Kairuigaku Zasshi)* **40**:86–94.

Karnovsky, M. J., and Roots, L. A., 1964, A "direct-coloring" thiocholine method for choline-esterase, *J. Histochem. Cytochem.* **12**:219–221.

Kawaguti, S., 1962, Arrangement of myofilaments in the oblique-striated muscles, in: *Proceedings of the Fifth International Congress for Electron Microscopy*, Vol. 2 (S. S. Breese, ed.), P. M-11, Academic Press, New York.

Kawaguti, S., 1970, Electron microscopy on muscle fibers in blood vessels and capillaries of cephalopods, *Biol. J. Okayama Univ.* **16**:19–28.

Kawaguti, S., and Ikemoto, N., 1957, Electron microscopy of the smooth muscle of a cuttlefish, *Sepia esculenta*, *Biol. J. Okayama Univ.* **3**:196–203.

Kawamoto, N., Deguchi, Y., Mikami, S., Sonoda, S., Yamada, T., Fujiwara, H., Shishida, H., Sakamoto, K., and Hanaue, K., 1980, Long-term rearing of *Nautilus macromphalus*, in: *Nautilus macromphalus in Captivity* [T. Hamada, I. Obata, and T. Okutani (JECOLN), eds.], pp.4–10, Tokai University Press, Tokyo.

Kay, E. A., 1965, The Reverend John Lightfoot, Daniel Solander, and the Portland Catalogue, *Nautilus* **79**:10–19.

Keferstein, W., 1865, Beiträge zur Anatomie des *Nautilus pompilius*, *Nachr. K. Ges. Wiss. Georg-Augusts-Univ.* **1865**:356–376.

Keferstein, W., 1866, Beiträge zur Anatomie des *Nautilus*, *Malakol. Zool. Blatt* **13**:21–38.

Kennedy, W. J., and W. A., Cobban, 1976, Aspects of ammonite biology, biogeography and biostratigraphy, *Spec. Pap. Palaeontol.* **17**:1–94.

Kermack, K. A., and Haldane, J. B. S., 1950, Organic correlation and allometry, *Biometrika* **37**:30–41.

Kerr, J. G., 1895, On some points in the anatomy of *Nautilus pompilius*, *Proc. Zool. Soc. London* **1895**:664–686.

Kessler, P., 1923a, "Konchinbänder," "Haftlinie," "Hohlkiel" und "Streifenbuschel" bei Ammoniten, *Neues Jahrb. Mineral. Centralbl. Jahrg.* **1923**:499–511.

Kessler, P., 1923b, Beiträge zur Kenntnis der Organisation der fossilen Gehäuse-Cephalopoden, *Neues Jahrb. Mineral. Centralbl. Jahrg.* **1923**:689–702.

Kier, W. M., 1982, The functional morphology of the musculature of squid (Lolignidae) arms and tentacles, *J. Morphol.* **172**:179–192.

Kier, W. M., 1985, The musculature of squid arms and tentacles: Ultrastructural evidence for functional differences, *J. Morphol.* **185**:223–240.

Kier, W. M., 1987, The arrangement and function of molluscan muscle, in: *The Mollusca: Molluscan Form and Function*, Vol. 11 (E. R. Trueman and M. R. Clarke, eds.), pp. 211–252, Academic Press, New York.

Kier, W. M., and Smith, K. K., 1985, Tongues, tentacles and trunks: The biomechanics of movement in muscular-hydrostats, *Zool. J. Linn. Soc.* **83**:307–324.

Kinter, D., Fitzpatrick, J. H., Jr., Louie, J. A. and Gilboe, D. D., 1984, Cerebral oxygen and energy metabolism during and after 30 minutes of moderate hypoxia, *Am. J. Physiol.* **247**:E475–482.

Kozur, H., 1980, Die Faunenänderungen nahe der Per/Trias- und Trias/Jura-Grenze und ihre möglichen Ursachen: Teil II, *Freiberg. Forschungsh. C* **357**:111–134.

Kulicki, C., 1974, Remarks on the embryogeny and postembryonal development of ammonites, *Acta Palaeontol. Polonica* **20**:201–224.

Kummel, B., 1953, The ancestry of the family Nautilidae, *Breviora Mus. Comp. Zool. Harv. Univ.* **21**:1–7.

Kummel, B., 1956, Post-Triassic nautiloid genera, *Bull. Mus. Comp. Zool.* **114**:324–494.

Kummel, B., 1959, Triassic–Jurassic cenoceratids from New Zealand, *N. Z. J. Geol. Geophys.* **2**:421–428.

Kummel, B., 1964, Nautiloidea–Nautilidae, in: *Treatise on Invertebrate Paleontology, Part K: Mollusca 3* (R. C. Moore. ed.), pp. K383–K457, Geological Society of America and University of Kansas Press, Lawrence.

Kummel, B., and Lloyd, R. M., 1955, Experiments on relative streamlining of coiled cephalopod shells, *J. Paleontol.* **29**:159–170.

Lagler, K. R., Bardach, J. E., Miller, R. R. and Passino, D. R. M., 1977, *Ichthyology*, 2nd ed., John Wiley, New York, 506 pp.

Lamarck, J. B. P. A. de M., 1822, *Histoire Naturelles des Animaux sans Vertèbres*, No. 7, Paris, 711 pp.

Landman, N. H., 1982a, Ontogeny and evolution of Late Cretaceous (Turonian–Santonian) scaphites, unpublished Ph.D. dissertation, Yale University, New Haven, Connecticut, 341 pp.

Landman, N. H., 1982b, Aristole, Alexander and the pearly nautilus, *Discovery* **16**(1):20–23.

Landman, N. H., 1983a, Barnacle attachment on live *Nautilus*: Implications for *Nautilus* growth rate, *Veliger* **26**:124–127.

Landman, N. H., 1983b, Ammonoid growth rhythms, *Lethaia* **16**:248.

Landman, N. H., 1984, Not to be or to be? *Nat. Hist.* **93**:34–41.

Landman, N. H., 1987, Early ontogeny of Mesozoic ammonites and nautilids, *Neues Jahrb. Geol. Palaeontol. Abh.* (in press).

Landman, N. H., Rye, D. M., and Shelton, K. L., 1983, Early ontogeny of *Eutrephoceras* compared to

Recent *Nautilus* and Mesozoic ammonites: Evidence from shell morphology and light stable isotopes, *Paleobiology* **9**:269–279.

Lange, G. D., Hartline, P. H., and Hurley, A. C., 1979, Retinal responses in *Nautilus*, *J. Comp. Physiol.* **134**:281–285.

Lange, W., 1941, Die Ammonitenfauna der *Psiloceras Stufe* Norddeutschlands, *Palaeontogr. Abt. A* **93**:1–192.

Lankester, E. R., and Bourne, A. G., 1883, On the existence of Spengel's olfactory organ and of paired genital ducts in the pearly *Nautilus*, *Q. J. Microsc. Sci.* **23**:340–348.

Lapan, E. A., 1975, Studies on the chemistry of the octopus renal system and an observation on the symbiotic relationship of the dicyemid Mesozoa, *Comp. Biochem. Physiol.* **52A**:651–657.

Lapan, E. A., and Morowitz, H. J., 1972, The Mesozoa, *Sci. Am.* **227**:94–101.

Lee, S. H., 1970, The possible role of the vesicles in renal ammonia excretion: An implication of concentrated glutamic oxalacetic transaminase, *J. Cell. Biol.* **45**:644–649.

Lenfant, C., and Johansen, K., 1965, Gas transport by hemocyanin-containing blood of the cephalopod *Octopus dofleini*, *Am. J. Physiol.* **209**:991–998.

Levene, H., 1949, On a matching problem arising in genetics, *Ann. Math. Stat.* **20**:91–94.

Lightfoot, J., 1786, A catalogue of the Portland Museum, lately the property of the Duchess Dowager of Portland, Deceased: which will be sold by auction by Mr. Skinner and Co. on Monday 24th of April 1786 and the thirty-seven following days at twelve o'clock, London, 194 pp.

Linnaeus, (Linne), K. (ed.), 1758–1759, *Systema Naturae*, 10. Holm.

Lister, M., 1685, *Historiae conchyliorum Liber Primus*, London.

Lojda, Z., 1965, Remarks on the histochemical detection of dehydrogenases. II. Intracellular localization, *Folia Morph.* (Prague) **13**:84–96.

Lowenstam, H. A., 1963, Biologic problems relating to the composition and diagenesis of sediments, in: *The Earth Sciences: Problems and Progress in Current Research* (T. W. Donnelly, ed.), pp. 137–195, University of Chicago Press.

Lowenstam, H. A., 1964, Sr/Ca ratio of skeletal aragonites from the recent marine biota at Palau and from fossil gastropods, in: *Isotopic and Cosmic Chemistry* (H. Craig, S. L. Miller, and G. J. Wasserburg, eds.), pp. 114–132, North-Holland, Amsterdam.

Lowenstam, H. A., Traub, W., and Weiner, S., 1984, *Nautilus* hard parts: A study of the mineral and organic constituents, *Paleobiology* **10**:268–279.

Luce, J. V., 1969, *The End of Atlantis: New Light on an Old Legend*, Thames and Hudson, London, 224 pp.

Luchtel, D. L., Martin, A. W., and Deyrup-Olsen, I., 1984, The channel cell of the terrestrial slug *Ariolimac columbianus* (Stylomatiphora, Arionidae), *Cell Tissue Res.* **235**:143–152.

Lund, R. D., 1966, Centrifugal fibres to the retina of Octopus *vulgaris*, Exp. *Neurol.* **15**:100–112.

Lykkeboe, G., and Johansen, K., 1982, A cephalopod approach to rethinking about the importance of the Bohr and Haldane effects, *Pac. Sci.* **36**:305–313.

Mackie, G. O., and Bone, Q., 1977, Locomotion and propagated skin impulses in salps (Tunicata: Thaliacea), *Biol. Bull.* **153**:180–197.

Maginniss, L. A., and Wells, M. J., 1969, The oxygen consumption of *Octopus cyanea*, *J. Exp. Biol.* **51**:607–613.

Magnier, Y., and Laboute, P., 1978, *Guides sous-marin de Nouvelle Caledonia*, Les Editions du Pacifique, Paris, 160 pp.

Mallery, C. H., 1983, A carrier enzyme basis for ammonium excretion in teleost gill: Ammonium-stimulated sodium-dependent ATPase activity in *Opsanus beta*, *Comp. Biochem. Physiol.* **74**:889–898.

Mangum, C. P., 1980, Respiratory function of hemocyanin, *Am. Zool.* **20**:19–30.

Mangum, C. P., and Johansen, K., 1975, The colloid osmotic pressures of invertebrate body fluids, *J. Exp. Biol.* **63**:661–671.

Mangum, C. P., and Towle, D., 1977, Physiological adaptation to unstable environments, *Am. Sci.* **65**:67–75.

Mangum, C. P., and Towle, D. W., 1982, The *Nautilus* siphuncle as an ion pump, *Pac. Sci.* **36**:273–282.

Mangum, C. P., and Van Winkle, W., 1973, Responses of aquatic invertebrates to declining oxygen conditions, *Am. Zool.* **13**:529–541.

Mann, K. O., 1983, The Sr, Mg, and Ca chemistry of the mineralized structures of *Nautilus*, M.Sc. thesis, University of Texas at Arlington.

Mapes, R. H., Frest, T. J., and Aronoff, S. M., 1979, Abnormal callus development in *Nautilus pompilius*, *Veliger* **21**:442–443.

Marenko, D. E., 1956, Novi znakhidki Nautilidae iz nizhnotretinnikh vikladiv URSR, *Dopovi Akad. Nauk Ukr. RSR* **2**:162–168 (in Ukrainian with Russian résumé).

Martin, A. W., 1975, Physiology of the excretory organs of cephalopods, *Fortschr. Zool.* **23**:112–123.

Martin, A. W., and Aldrich, F. A., 1970, Comparison of hearts and branchial heart appendages in some cephalopods, *Can. J. Zool.* **48**:751–756.

Martin, A. W., and Meenakshi, V. R., 1974, The conversion of sodium benzoate to hippuric acid by a cephalopod mollusc, *J. Comp. Physiol. A* **94**:287–296.

Martin, A. W., Catala-Stucki, I., and Ward, P. D., 1978, The growth rate and reproductive behavior of *Nautilus macromphalus*, *Neues Jahrb. Geol. Palaeontol. Abh.* **156**:207–225.

Martin, J. H., and Flegal, A. R., 1975, High copper concentrations in squid livers in association with elevated levels of silver, cadmium and zinc, *Mar. Biol.* **30**:51–55.

Martin, K., 1933, Eine neue tertiäre Molluskenfauna aus dem Indischen Archipel, *Leidse Geol. Meded.* **6**:7–32.

Martin, R., 1966, On the swimming behavior and biology of *Notarchus punctatus* Phillipi (Gastropoda, Opisthobranchia), *Pubbl. Stn. Zool. Napoli* **35**:61–75.

Mason, B., and Moore, C. B., 1982, *Principles of Geochemistry*, John Wiley, New York, 344 pp.

Masuda, Y., and Shinomiya, A., 1983, Genetic variation in *Nautilus pompilius*, in: *Studies on Nautilus pompilius and Its Associated Fauna from Tañon Strait, the Philippines* (S. Hayasaka, ed.), pp. 22–25, *Kagoshima Univ. Res. Center S. Pac. Occas. Pap.*, No. 1.

Matsumoto, T., 1983, Cretaceous nautiloids from Hokkaido—I, *Trans. Proc. Palaeontol. Soc. Jpn. N. Ser.* **129**:9–25.

Matsumoto, T., and Muramoto, K., 1983, Cretaceous nautiloids from Hokkaido (Japan). II. Three Nautiloid species from the Santonian and Campanian of Hokkaido, *Trans. Proc. Palaeontol. Soc. Jpn. N. Ser.* **0**:85–95.

Matsumoto, T., Miyauchi, T., Kanie, Y., Miyata, Y., and Ueda, Y., 1984, Cretaceous nautiloids from Hokkaido—V, *Trans. Proc. Palaeontol. Soc. Jpn. N. Ser.* **134**:335–346.

McCance, R. A., and Masters, M., 1937, The chemical composition and acid–base balance of *Archidoris britannica*, *J. Mar. Biol. Assoc. U. K.* **22**:273–279.

McConnell, D., and Ward, P. D., 1978, Nautiloid uroliths composed of phosphatic hydrogel, *Science* **199**:208–209.

McDonald, D. A., 1974, *Blood Flow in Arteries*, 2nd ed., Edward Arnold, London, 512 pp.

Meenakshi, V. R., Martin, A. W., and Wilbur, K. M., 1974, Shell repair in *Nautilus macromphalus*, *Mar. Biol.* **27**:27–35.

Meijer, A. E. F. H., and Bloem, J. H., 1966, Improved histochemical demonstration of carbonate dehydratase, *Acta Histochem.* (Jena) **25**:239–241.

Meischner, D., 1968, Perniciose Epoeke von *Placunopsis* auf *Ceratites*, *Lethaia* **1**:156–174.

Merkt, J., 1966, Ueber Austern und Serpeln als Epoeken auf Ammonitengehaeusen, *Neues Jahrb. Geol. Palaeontol. Abh.* **125**:467–479.

Merton, H., 1905, Über die Retina von *Nautilus* und einigen dibranchiaten Cephalopoden, *Z. Wiss. Zool.* **79**:325–396.

Messenger, J. B., Wilson, A. P., and Hedge, A., 1973, Some evidence for colour-blindness in *Octopus*, *J. Exp. Biol.* **59**:77–94.

Mikami, S., and Okutani, T., 1977, Preliminary observations on maneuvering, feeding, copulating and spawning behaviors of *Nautilus macromphalus* in captivity, *Venus. Jpn. J. Malacol. (Kairuigaku Zasshi)* **36**:29–41.

Mikami, S., and Okutani, T., 1981, A consideration of transfer of spermatophores in *Nautilus macromphalus*, *Venus Jpn. J. Malacol. (Kairuigaku Zasshi)* **40**:57–62.

Mikami, S., Yamada, T., Fujiwara, H., Shishida, H., and Sakamoto, K., 1976, Keeping *Nautilus macromphalus* in the aquarium, *Jpn. Assoc. Zool. Garden Aquar.* **18**:60–66.

Mikami, S., Ichikawa, M., and Deguchi, Y., 1979, Preliminary report on oxygen consumption of *Nautilus macromphalus*, *Bull. Coll. Agric. Vet. Med. Nihon Univ.* **36**:279–284.

Mikami, S., Okutani, T., Hirano, H., Kanie, Y., and Hamada, T., 1980, Behavior in captivity, in: *Nautilus macromphalus in Captivity* [T. Hamada, I. Obata, and T. Okutani (JECOLN), eds.], pp. 11–22, Tokai University Press, Tokyo.

Milkman, R., 1982, Toward a unified selection theory, in: *Perspectives on Evolution* (R. Milkman, ed.), pp. 115–118, Sinauer Associates, Sunderland, Massachusetts.

Mill, P. J., and Pickard, R. S., 1975, Jet propulsion in anisopteran dragonfly larvae, *J. Comp. Physiol.* **97**:329–338.

Miller, A. K., 1947, Tertiary nautiloids of the Americas, *Geol. Soc. Am. Mem.* **23**:1–234.

Miller, A. K., 1949, The last surge of the nautiloid cephalopods, *Evolution* **3**:231–238.

Miller, A. K., 1951, Tertiary Nautiloids of west-coastal Africa, *Ann. Mus. R. Congo Belge Ser. 8° Sci. Géol.* **8**:1–88.

Millman, B. M., 1967, Comparative aspects of muscle: Mechanism of contraction in molluscan muscle, *Am. Zool.* **7**:583–591.

Mommsen, T. P., Ballantyne, J., MacDonald, D., Gosline, J., and Hochachka, P. W., 1981, Analogues of red and white muscle in squid mantle, *Proc. Natl. Acad. Sci. USA* **78**:3274–3278.

Montuori, A., 1913, Les procèsses oxydatifs chez les animaux marins en rapport avec la loi de superficie, *Arch. Ital. Biol.* **59**:213–234.

Moon, T. W., and Hulbert, W. C., 1975, The ultrastructure of the mantle musculature of the squid *Symplectoteuthis oualaniensis*, *Comp. Biochem. Physiol.* **52**:145–149.

Moore, J. D., and Trueman, E. R., 1971, Swimming of the scallop, *Chlamys opercularis (L)*, *J. Exp. Mar. Biol. Ecol.* **6**:179–185.

Moore, R. C., (ed.), 1964, *Treatise on Invertebrate Paleontology, Part K: Mollusca 3. Cephalopoda—General Features—Endoceratoidea—Actinoceratoidea—Nautiloidea—Bactritoidea*, Geological Society of America and University of Kansas, Lawrence, 519 pp.

Morris, G. M., and Baldwin, J., 1984, pH buffering capacity of invertebrate muscle: Correlations with anaerobic muscle work, *Mol. Physiol.* **5**:61–70.

Moseley, H., 1838, On the geometrical forms of turbinated and discoid shells, *Philos. Trans. R. Soc. London* **1838**:351–370.

Moseley, H. N., 1879, *Notes by a Naturalist Made during the Voyage of H.M.S. "Challenger,"* T. Werner Laurie, London, 606 pp.

Moseley, H. N., 1892, *Notes by a Naturalist: An Account of Observations Made during the Voyage of "H.M.S. Challenger" round the World in the Years 1872–1876*, John Murray, London, 540 pp.

Mugglin, F., 1937, Mitteilungen über das optische Leistungsvermögen des *Nautilus*-Auges, *Rev. Suisse Zool.* **44**:401–404.

Müller, A. H., 1974, Über den Kieferapparat fossiler und rezenter Nautilida (Cephalopoda) mit Bemurkungen zur Ökologie, Funktionsweise und Phylogenie, *Freiberger Forsch.* **298**:7–17.

Mulligan, E., and Lahiri, S., 1981, Mitochondrial oxidative metabolism and chemoreception in the carotid body, in: *Arterial Chemoreceptors*, pp. 316–327, Leicester University Press, Leicester, United Kingdom.

Mulvey, M., and Woodruff, D. S., 1985, Genetics of *Biomphalaria glabrata:* Linkage analysis of genes for pigmentation, enzymes, and resistance to *Schistosoma mansoni*, *Biochem. Genet.* **22**:877–889.

Muntz, W. R. A., and Johnson, M. S., 1978, Rhodopsins of oceanic decapods, *Vision Res.* **18**:601–602.

Muntz, W. R. A., and Raj, U., 1984, On the visual system of *Nautilus pompilius*, *J. Exp. Biol.* **109**:253–263.

Mutvei, H., 1957, On the relations of the principal muscles to the shell in *Nautilus* and some fossil nautiloids, *Ark. Mineral. Geol.* **2**:219–254.

Mutvei, H., 1964, On the shells of *Nautilus* and *Spirula* with notes on the shell secretion in non cephalopod molluscs, *Ark. Zool.* **16**:221–278.

Mutvei, H., 1969, On the micro- and ultrastructure of the conchiolin in the nacreous layer of some recent and fossil molluscs, *Stockholm Contr. Geol.* **20**:1–17.

Mutvei, H., 1970, Ultrastructure of the mineral and organic components of molluscan nacreous layers, *Biomineralisation Forsch.* **2**:49–72.

Mutvei, H., 1972, Ultrastructural studies on cephalopod shells. I. The septa and siphonal tube in *Nautilus*, *Bull. Geol. Inst. Univ. Uppsala (N. Ser.)* **3**:237–261.

Mutvei, H., and Reyment, R. A., 1973, Buoyancy control and siphuncle function in ammonoids, *Palaeontology* **16**:623–636.

Muzii, E. O., and de Marca, C., 1970, Localizzazione istochimica e citochimica del rame nell'epatopancreas di *Octopus vulgaris* Lamarck, *Pubbl. Stn. Zool. Napoli* **38**:92–101.

Nachlas, M. M., Monis, B., Rosenblatt, D., and Seligman, A. M., 1960, Improvement in the histochemical localization of leucine aminopeptidase with a new substrate, L-leucyl-4-methoxy-2-naphthylamide, *J. Biophys. Biochem. Cytol.* **7**:261–264.

Naef, A., 1913, Studien zur generellen Morphologie der Mollusken. 2. Das Cölomsystem in seinen topographischen Beziehungen, *Ergebn. Fortsch. Zool.* **3**:329–462.

Naef, A., 1923, Die Cephalopoden: Fauna e Flora del Golfo di Napoli, *Stn. Zool. Napoli Monogr.* **35**:1–821.

Nei, M., 1975, *Molecular Population Genetics and Evolution*, North-Holland, Amsterdam, 288 pp.

Nei, M., 1978, Estimation of average heterozygosity and genetic distance from a small number of individuals, *Genetics* **89**:583–590.

Nei, M., 1981, Genetic distance and molecular taxonomy. in: *Proceedings of the XIVth International Congress on Genetics*, Vol. 2, *Problems in General Genetics* (Y. P. Altukhov, ed.), pp. 7–22, MIA Publishers, Moscow.

Nei, M., and Roychoudhury, A. K., 1974, Sampling variances of heterozygosity and genetic distance, *Genetics* **76**:379–390.

Nei, M., Stephens, C., and Saiton, N., 1985, Standard errors of branching points in an evolutionary tree and their application to molecular data from humans and apes, *Mol. Biol. Evol.* **2**:66–85.

Nevo, E., Bar-El, C., and Bar, Z., 1983, Genetic diversity, climatic selection and speciation of *Sphicterochila* landsnails in Israel, *Biol. J. Linn. Soc.* **19**:339–373.

Nevo, E., Beilles, A., and Ben-Shlomo, R., 1984, The evolutionary significance of genetic diversity: Ecological, demographic and life history correlates, in: *Evolutionary Dynamics of Genetic Diversity: Lecture Notes in Biomath*, 53 (G. S. Mani, ed.), pp. 13–213, Springer-Verlag, New York.

Nicaise, G., and Amsellem, J., 1983, Cytology of muscle and neuromuscular junction, in: *The Mollusca*, Vol. 4, *Physiology: Part 1* (A. S. M. Saleuddin and K. M. Wilbur, eds.), pp. 1–33, Academic Press, New York.

Nixon, M., 1979a, Has *Octopus vulgaris* a second radula?, *J. Zool. (London)* **187**:291–296.

Nixon, M., 1979b, Hole boring in shells by *Octopus vulgaris* Cuvier in the Mediterranean, *Malacalogia* **18**:431–443.

Nixon, M., and Messenger J. B. (eds.), 1977, *The Biology of Cephalopods, Symposium of the Zoological Society, London, 615 pp.*

Noll, W., 1934, *Geochemie des Strontiums: Mit Bemerkungen zur Geochemie des Bariums*, Chem. Erde **8**:508–600.

Oba, T., and Tanabe, K., 1983, Oxygen isotope analysis of the shells of *Nautilus pompilius*, in: *Studies on Nautilus pompilius and Its Associated Fauna from Tañon Strait, the Philippines* (S. Hayasaka, ed.), pp. 26–29, *Kagoshima Univ. Res. Center S. Pac. Occas. Pap.*, No. 1.

O'Dor, R. K., 1982, Respiratory metabolism and swimming performance of the squid, *Loligo opalescens*, *Can. J. Fish. Aquatic Sci.* **39**:580–587.

Odum, H. T., 1957, Biogeochemical deposition of strontium, *Inst. Mar. Sci.* **4**:38–114.

Oechsle, E., 1958, Stratigraphie und Amonitenfauna der Sonninien-Schichten des Filsgebietes unter besonderer Beruecksichtigung der Sowerbyi-Zone (Mittlerer Dogger, Wuerttemberg), *Palaeontogr. Abt. A* **111**:47–129.

Ōki, K., 1983, Preliminary report on foraminifera from the southern part of Tañon Strait, between Cebu and Negros Islands, the Philippines, in: *Studies on Nautilus pompilius and Its Associated Fauna from Tañon Strait, the Philippines* (S. Hayasaka, ed.), pp. 44–50, *Kagoshima Univ. Res. Center S. Pac. Occas. Pap.*, No. 1.

Ōki, K., 1985, Preliminary report on foraminifera from the habitat of *Nautilus* off the southeast coast of Viti Levu, Fiji, in: *Marine Ecological Studies on the Habitat of Nautilus pompilius in the Environs of Viti Levu, Fiji* (S. Hayasaka, ed.), pp. 84–92, *Kagoshima Univ. Res. Center S. Pac. Occas. Pap.*, No. 4.

Okutani, T., and Mikami, S., 1977, Descriptions of beaks on *Nautilus macromphalus* Sowerby, *Venus Jpn. J. Malacol. (Kairuigaku Zasshi)* **36**:115–121.

Okutani, T., and Mikami, S., 1980, Behavior in captivity, in: *Nautilus macromphalus in Captivity*. [T. Hamada, I. Obata, and T. Okutani (JECOLN), eds.], pp. 11–22, Tokai University Press, Tokyo.

Orme, G. R., 1977, The Coral Sea Plateau—A major reef province, in: Biology and Geology of Coral Reefs, Vol. IV, *Geology 2* (O. A. Jones and R. Endean, eds.) pp. 267–306, Academic Press, New York.

Owen, R., 1832, *Memoir on the Pearly Nautilus (Nautilus pompilius*, Linn.) *with Illustrations of Its External Form and Internal Structure*, Richard Taylor, London, 68 pp.

Owen, R., 1843, On the structure and homology of the cephalic tentacles in the pearly nautilus, *Ann. Mag. Nat. Hist.* **12**:305–311.

Oxford, G. S., and Rollinson, D., (eds.), 1983, *Protein Polymorphism: Adaptive and Taxonomic Significance*, Academic Press, New York, 405 pp.

Packard, A., 1972, Cephalopods and fish: The limits of convergence, *Biol. Rev.* **47**:241–307.
Packard, A., Bone, Q., and Hignette, M., 1980, Breathing and swimming movements in a captive *Nautilus*, *J. Mar. Biol. Assoc. U.K.* **60**:313–327.
Palumbo, A., Ito, S., Nardi, G., and Prota, G., 1977, Isolation of a possible biosynthetic precursor of adenochrome from the white bodies of *Octopus vulgaris*, *Comp. Biochem. Physiol.* **58B**:353–356.
Pearse, A. G., 1961, *Histochemistry, Theoretical and Applied*, Vol. 1, Churchill Livingstone, London.
Pearse, A. G., 1985, *Histochemistry Theoretical and Applied*, Vol. 2, 4th ed., Churchill Livingston, New York, 614 pp.
Petit, H., 1977, A scanning electron microscopic (SEM) study of the mantle–shell relationship in the freshwater mussel *Amblema plicata perplicata*, *Conrad Anat. Rec.* **187**:680–681.
Petit, H., Davis, W. L., Jones, R. G., and Hagler, H. K., 1980, Morphological studies on the calcification process in the fresh-water mussel *Amblema*, *Tissue Cell* **12**:13–28.
Pfaff, E., 1911, Über Form und Bau der Ammonitensepten und ihre Beziehungen zur Suturlinie, *Jahrb. Niedersachis. Geol. Ver.* **4**:207–223.
Pia, J., 1914, Untersuchungen über die Liassischen Nautiloidea, *Beitr. Palaeontol. Geol. Oesterr.-Ung.* **27**:19–86.
Pilkey, O. H., and Goodell, H. G., 1964, Comparison of the composition of fossil and recent mollusk shells, *Geol. Soc. Am. Bull.* **75**:217–228.
Pompeckj, J. F., 1912, Cephalopoda: Palaeontologie, in: *Handwörterbuch der Naturwissenschaften*, Vol. 2, pp. 265–296, Verlag G. Fischer, Jena.
Potts, W. T. W., 1965, Ammonia excretion in *Octopus dofleini*, *Comp. Biochem. Physiol.* **14**:339–355.
Potts, W. T. W., and Martin, A. W., 1962, The process of ammonia excretion in an *Octopus*, *Proc. Int. Congr. Zool.* **1**:78.
Price, N. B., and Hallam, A., 1967, Variation of strontium content within shells of Recent *Nautilus* and *Sepia*, *Nature (London)* **215**:1272–1274.
Raup, D. M., 1961, The geometry of coiling in gastropods, *Proc. Natl. Acad. Sci. USA* **47**:602–609.
Raup, D. M., 1962, Computer as aid in describing form in gastropod shells, *Science* **138**:150–152.
Raup, D. M., 1966, Geometric analysis of shell coiling: General problems, *J. Paleontol.* **40**:1178–1190.
Raup, D. M., 1967, Geometric analysis of shell coiling: Coiling in ammonoids, *J. Paleontol.* **41**:43–65.
Raup, D. M., and Chamberlain, J. A., Jr., 1967, Equations for volume and center of gravity in ammonoid shells, *J. Paleontol.* **41**:566–574.
Raup, D. M., and Michelson, A., 1965, Theoretical morphology of the coiled shell, Science **147**:1294–1295.
Raup, D. M., and Takahashi, T., 1966, Experiments on strength of cephalopod shells, *Geol. Soc. Am. Spec. Pap.* **101**:172–173.
Redfield, A. C., and Goodkind, R., 1929, The significance of the Bohr effect in the respiration and asphyxiation of the squid, *Loligo pealei*, *J. Exp. Biol.* **6**:340–349.
Redmond, J. R., 1978, Bohr effect of the pearly nautilus, *Nautilus macromphalus*, *Spec. Sci. Technol.* **1**:325–328.
Redmond, J. R., and Bourne, G. B., 1982, Vascular resistance in the isolated gills of *Octopus macropus* and *Nautilus pompilius*, *Pac. Sci.* **36**: 297–303.
Redmond, J. R., Bourne, G. B., and Johansen, K., 1978, Oxygen uptake by *Nautilus pompilius*, *J. Exp. Zool.* **205**:45–50.
Regneri, H. J., 1981, Untersuchungen zur Struktur eisenhaltiger Systeme im Skelett roter Korallen und in Branchialherzen gemeiner Kraken, Thesis, Saarbrücken.
Rehder, H. A., 1968, Valid zoological names of the Portland Catalogue, *Proc. U.S. Natl. Museum*, **121**:1–51.
Reyment, R. A., 1958, Some factors in the distribution of fossil cephalopods, *Stockholm Contrib. Geol.* **1**:97–184.
Reyment, R. A., 1973, Factors in the distribution of fossil cephalopods. 3. Experiments with exact models of certain shell types, *Bull. Geol. Inst. Univ. Uppsala* (N. Ser.) **4**:7–41.
Richter, I. A., 1980, *The Notebooks of Leonardo da Vinci*, Oxford University Press, Cambridge, 417 pp.
Riddle, A. R., 1920, An adventitious occurrence of *Nautilus pompilius*, Linne, with a short bibliography on ocean currents affecting the Australian coast, *R. Soc. South Aust. Trans. Proc.* **44**:257–262.
Rieber, H., 1963, Ammoniten und Stratigraphie des Braunjura β der Schwaebischen Alb, *Palaeontogr. Abt. A* **122**:1–89.

Ristedt, H., 1971, Zum Bau der orthoceriden Cephalopoden, *Palaeontogr. Abt. A* **137**:155–195.
Roark, R. J., 1954, *Formulas of Stress and Strain*, 3rd ed., McGraw-Hill, New York, 381 pp.
Robertson, W., 1930, Über die Wirkung der Magnesiumsalze, *Biochem. Z.* **221**:33–41.
Robertson, W. G., 1982, The solubility concept, in: *Biological Mineralization and Demineralization* (G. H. Nancollas, ed.), pp. 5–21, Springer-Verlag, New York.
Robertson, W. G., Bach, D. W., Finlayson, B., Francis, M. D., Garside, J., Grynpas, M. D., Hesse, A., Klee, W. E., McCarty, D. J., Moreno, E. C., Smith, L. H., and Vahlensieck, E. W., 1982, Mechanisms of pathological mineralization group report, in: *Biological Mineralization and Demineralization—Dahlem Konferenzen 1982* (G. H. Nancollas, ed.), pp. 367–386, Springer-Verlag, Berlin.
Rogers, J. S., 1972, Measures of genetic similarity and genetic distance, *Univ. Texas. Publ.* **7213**:145–153.
Roinel, N., Morel, F., and Istin, M., 1973, Étude des granules calcifiés du manteau des Lamellibranches à l'aide de la microsonde electronique, *Calcif. Tiss. Res.* **11**:163–170.
Rokop, F. J., 1974, Reproductive patterns in the deep-sea benthos, *Science* **186**:743–745.
Rubbel, A., 1911, Über Perlen und Perlbildung bei *Margaritana margaritifera*, nebst Beiträgen zur Kenntnis ihrer Schalenstruktur, *Zool. Jahrb. Abt. Anat.* **32**:287–366.
Rudwick, M. J. S., 1961, The feeding mechanism of the Permian brachiopod *Prorichthofenia*, *Palaeontology* **3**:450–471.
Rumphius, G. E., 1741, *d'Amboinsche rariteitkamer, Behelzende eene beschryvinge van allerhande zoo weeke als harde Schaalvisschen te weeten raare krabben, kreeften, en diergelyke Zeedieren, als mede allerhande Horntjes en schulpen, die men in d'Amboinsche Zee vindt: Daar benevens zommige mineraalen, gesteenten, en soorten van Aarde, die in d'Amboinsche, en zommige omleggende Eilanden gevonden worden*, pp. 59–62, Jan Roman de Jonge, Amsterdam.
Runcorn, S. K., 1979, Matters arising: Nautiloid growth rhythms and lunar dynamics, *Nature (London)* **279**:452–453.
Saisho, T., and Tanabe, K., 1985, Notes on the esophagus- and stomach-contents of *Nautilus pompilius* in Fiji, in: *Marine Ecological Studies on the Habitat of Nautilus pompilius in the Environs of Viti Levu, Fiji* (S. Hayasaka, ed.), pp. 62–64, *Kagoshima Univ. Res. Center S. Pac. Occas. Pap.*, No. 4.
Sanders, H. L., and Allen, J. A., 1973, Studies on deep-sea Protobranchia (Bivalvia): Prologue and the Pristiglomidae, *Bull. Mus. Comp. Zool. Harv. Univ.* **145**:237–261.
Sarich, V. M., 1977, Rates, sample sizes, and the neutrality hypothesis for electrophoresis in evolutionary studies, *Nature (London)* **265**:24–28.
Satyanarayana Rao, T. S., 1963, On the pattern of surface circulation in the Indian Ocean as deduced from drift bottle recoveries, *Indian J. Meteorol. Geophys.* **14**:1–4.
Saunders, W. B., 1981a, A new species of *Nautilus* from Palau, *Veliger* **24**:1–7.
Saunders, W. B., 1981b, The species of living *Nautilus* and their distribution, *Veliger* **24**:8–17.
Saunders, W. B., 1983, Natural rates of growth and longevity of *Nautilus belauensis*, *Paleobiology* **9**:280–288.
Saunders, W. B., 1984a, *Nautilus belauensis* growth and longevity: Evidence from marked and recaptured animals, *Science* **224**:990–992.
Saunders, W. B., 1984b, The role and status of *Nautilus* in its natural habitat: Evidence from deep-water remote camera photosequences, *Paleobiology* **10**:469–486.
Saunders, W. B., 1985, Studies of living *Nautilus* in Palau, *Natl. Geogr. Soc. Res. Rep.* **18**:669–682.
Saunders, W. B., and Davis, L. E., 1985, A preliminary report on *Nautilus* in Papua New Guinea, *Sci. Papua New Guinea* **11**:60–69.
Saunders, W. B. and Shapiro, E. A., 1986, Calculation and simulation of ammonoid hydrostatics, *Paleobiology* **12**:65–80.
Saunders, W. B., and Spinosa, C., 1978, Sexual dimorphism in *Nautilus* from Palau, *Paleobiology* **4**:349–358.
Saunders, W. B., and Spinosa, C., 1979, *Nautilus* movement and distribution in Palau, Western Caroline Islands, *Science* **204**:1199–1201.
Saunders, W. B., and Swan, A. R. H., 1984, Morphology and morphologic diversity of mid-Carboniferous (Namurian) ammonoids in time and space, *Paleobiology* **10**:195–228.
Saunders, W. B., and Ward, P. D., 1979, Nautiloid growth and lunar dynamics, *Lethaia* **12**:172.
Saunders, W. B., and Ward, P. D., 1987, The occurrence of living *Nautilus* (*N. pompilius* and *N. stenomphalus*) on the Great Barrier Reef, Australia, *The Nautilus* (in press).
Saunders, W. B., and Wehman, D. A., 1977, Shell strength of *Nautilus* as a depth limiting factor, *Paleobiology* **3**:83–89.

Saunders, W. B., Spinosa, C., Teichert, C., and Banks, R. C., 1978, The jaw apparatus of recent *Nautilus* and its palaeontologic implications, *Palaeontology* **21**:129–144.

Saunders, W. B., Davis, L. E., and Knight, R. L., 1987a, Sympatric species of *N. pompilius* and *N. scrobiculatus* in the Admiralty Islands, Papua New Guinea, *The Nautilus* **101**:93–99.

Saunders, W. B., Bond, P. N., Hastie, L. C., and Itano, D., 1987b, On the distribution of *Nautilus pompilius* and associated organisms in the Samoas, Fiji and Tonga, *The Nautilus* (in press).

Schindewolf, O. H., 1934, Über Epöken auf Cephalopoden-Gehäusen, *Palaeontol. Z.* **16**:15–31.

Schipp, R., and Hevert, F., 1978, Distribution of copper and iron in some central organs of *Sepia officinalis* (Cephalopoda): A comparative study of flameless atomic absorption and electron microscopy, *Mar. Biol.* **47**:391–399.

Schipp, R., and Martin, A. W., 1981, Cytology of the renal appendages of *Nautilus* (Cephalopoda, Tetrabranchiata), *Cell Tissue Res.* **219**:585–596.

Schipp, R., and Schafer, A., 1969, Vergleichende elektronenmikroskopische Untersuchungen aus den zentralen Herzorganen von Cephalopoden (*Sepia officinalis*): Feinstruktur und Funktion der Kiemenherzen, *Z. Zellforsch. Mikrosk. Anat.* **101**:367–379.

Schipp, R., and von Boletzky, S., 1975, Morphology and function of the excretory organs in dibranchiate cephalopods, *Fortschr. Zool.* **23**:89–110.

Schipp, R., and von Boletzky, S., 1976, The pancreatic appendages of dibranchiate cephalopods. 1. The fine structures of the "pancreas" in Sepioidea, *Zoomorphologie* **86**:81–98.

Schipp, R., Hohn, P., and Schafer, A., 1971, Elektronenmikroskopische und histochemische Untersuchungen zur Funktion des Kiemenherzanhangs (Pericardialdruse) von *Sepia officinalis*, *Z. Zellforsch. Mikrosk. Anat.* **117**:252–274.

Schipp, R., von Boletzky, S., and Doell, G., 1975, Ultrastructure and cytochemical investigations on the renal appendages and their concrements in dibranchiate cephalopods (Mollusca, Cephalopoda), *Z. Morphol. Tiere* **81**:279–304.

Schipp, R., Martin, A. W., Liebermann, H., and Magnier, Y., 1985, Cytomorphology and function of the pericardial appendages of *Nautilus* (Cephalopoda, Tetrabranchiata), *Zoomorphology* **105**:16–29.

Schmidt, H., 1930, Über die Bewegungsweise der Schalencephalopoden, *Palaeontol. Z.* **12**:194–208.

Schmidt, M., 1925, Ammonitenstudien, *Forschr. Geol. Palaeontol.* **10**:275–363.

Schmidt, W. J., 1923, Bau und Bildung der Perlmuttermasse, *Zool. Jahrb. Abt. Anat.* **45**:1–148.

Schmidt, W. J., 1924, *Die Bausteine des Tierkörpers in polarisiertem Lichte*, Verlag F. Cohen, Bonn, 528 pp.

Schopf, T. J. M., 1982, A critical assessment of punctured equilibria. I. Duration of taxa, *Evolution* **36**:1144–1157.

Schopf, T. J. M., 1984, Rates of evolution and the notion of "living fossils," *Annu. Rev. Earth Planet. Sci.* **12**:245–292.

Schultz, O., 1976, Zur Systematik der Nautilidae, *Anz. Oesterr. Akad. Wiss. Math. Naturwiss. Kl.* **113**:43–51.

Seilacher, A., 1960, Epizoans as a key to ammonoid ecology, *J. Paleontol.* **34**:189–193.

Seilacher, A., 1975, Mechanische Simulation und funktionelle Evolution des Ammoniten-Septums, *Palaeontol. Z.* **49**:268–286.

Seilacher, A., 1982, Ammonite shells as habitats in the Poseidonia Shales of Holzmaden—floats or benthic islands? *Neues Jahrb. Geol. Palaeontol. Abh.* **159**:98–114.

Selander, R. K., and Whittan, T. S., 1983, Protein polymorphism and the genetic structure of populations, *Evolution of Genes and Proteins* (M. Nei and R. K. Koehn, eds.), pp. 89–114, Sinauer Associates, Sunderland, Massachusetts.

Selander, R. K., Yang, S. Y., Lewontin, R. C., and Johnson, W. E., 1970, Genetic variation in the horseshoe crab (*Limulus polyphemus*), a phylogenetic "relic," *Evolution* **24**:402–414.

Selander, R. K., Smith, M. H., Yang, S. Y., Johnson, W. E., and Gentry, J. B., 1971. Biochemical polymorphism and systematics in the genus *Peromyscus*. I. Variation in the old-field mouse (*Peromyscus polionotus*), *Univ. Texas Stud. Genet.* **6**:49–90.

Seligman, A. M., Tsou, K. C., Rutenberg, A. M. and Cohen, R. B., 1954, Histochemical demonstration of β-D-glucuronidase with a synthetic substrate, *J. Histochem. Cytochem.* **2**:209–229.

Selman, K., and Arnold, J. M., 1977, An ultrastructural and cytochemical analysis of oogenesis in the squid *Loligo pealei*, *J. Morphol.* **152**:381–400.

Shaw, C. R. and Prasad, R., 1970, Starch gel electrophoresis of enzymes—A compilation of recipes, *Biochem. Genet.* **4**:297–320.

Shimansky, V. N., 1948, Sovremenniy nautilus i ego znachenie dlya izucheniya iskopaemykh golovonogikh: Moskovskoie gosudarst, *Pedagog. Inst. V. N. Lenin Kaf. Geol.* **52:**78–151.

Shimansky, V. N., 1957, Noviy predstaveli otryada Nautilida v SSSR, *Paleontol. Inst. Materialy Osnovam Paleontol.* **1:**35–44.

Shimansky, V. N., 1962, Otryad Nautilida, in: *Osnovy Paleontologii: Mollyuski-Golovonogie I* (V. E. Ruzhentsev, ed.), pp. 115–154, Akademiya Nauk SSSR, Moscow.

Shimansky, V. N., 1974, Class Cephalopoda, in: *Fundamentals of Paleontology*, Vol. V, *Mollusca—Cephalopoda I* (V. E. Ruzhentsev, ed.), pp. 6–30, English translation, Israel Program for Scientific Translations, Jerusalem.

Shimansky, V. N., 1979, Nautilida (izuchenost, striatigraficheskoe i i geograficheskoe rasprostranenie, etapy razvitiya), *Tr. Paleontol. Inst. Akad. Nauk. SSR* **170:**1–66.

Shimansky, V. N., and Zhuraleva, F. A., 1961, Fundamental questions of the systematics of the nautiloids and of the relationship of this group, *Tr. Paleontol. Inst. Akad. Nauk. SSR Leningrad* **90,** 175 pp.

Shinomiya, A., Raj, U., and Seeto, J., 1985, Studies on the biotic and inorganic factors of environment for *Nautilus*, in: *Marine Ecological Studies on the Habitat of Nautilus pompilius in the Environs of Viti Levu, Fiji* (S. Hayasaka, ed.), pp. 66–73, *Kagoshima Univ. Res. Center S. Pac. Occas. Pap.*, No. 4.

Shnitka, T. K., and Talibi, G. G., 1971, Cytochemical localization by ferricyanide reduction of α-hydroxy acid oxidase activity in peroxisomes of rat kidney, *Histochemie* **27:**137–158.

Sick, T. J., Rosenthal, M., LaManna, J. C., and Lutz, P. L., 1982, Brain potassium ion homeostasis, anoxia, and metabolic inhibition in turtles and rats, *Am. J. Physiol.* **243:**R281–R288.

Siekmann, J., 1963, On a pulsating jet from the end of a tube, with application to the propulsion of certain aquatic animals, *J. Fluid Mechanics* **15:**399–418.

Simkiss, K., 1976, Intracellular and extracellular routes in biomineralization, *Symp. Soc. Exp. Biol.* **30:**423–444.

Simkiss, K., and Mason, A. Z., 1983, Metal ions: Metabolic and toxic effects, in: *The Mollusca*, Vol. 2, *Environmental Biochemistry and Physiology* (P. W. Hochachka, ed.), pp. i–xviii and 1–362, Academic Press, New York.

Simkiss, K., and Wilbur, K. M., 1977, The molluscan epidermis and its secretions, *Symp. Zool. Soc. London* **39:**35–76.

Simmons, G., and Brace, W. F., 1965, Comparison of static and dynamic measurements of compressibility of rocks, *J. Geophys. Res.* **70:**5649–5656.

Simpson, G. G., 1953, *The Major Features of Evolution*, Columbia University Press, New York, 434 pp.

Sminia, T., 1980, Phagocytic cells in molluscs, in: *Aspects of Developmental and Comparative Immunology* (J. B. Solomon, ed.), pp. 124–132, Pergamon Press, Oxford.

Smith, E. A., 1849, Monograph on the genus *Nautilus*, in: *Thesaurus Conchyliorum* (G. B. Sowerby, ed.), pp. 463–465, London.

Smith, E. A., 1887, Notes on the pearly nautilus, *J. Chonchol. London* **5:**226–227.

Smith, S. V., Buddemeier, R. W., Redalje, R. C., and Houck, J. G., 1979, Strontium–calcium thermometry in coral skeletons, *Science* **204:**404–407.

Snyder, G. K., and Mangum, C. P., 1982, The relationship between the capacity for oxygen transport, size, shape and aggregation state of extracellular oxygen carrier, in: *Physiology and Biology of Horseshoe Crabs: Studies on Normal and Environmentally Stressed Animals* (Bonaventura, Bonaventura, and Tesh, eds.), pp. 173–188, Alan R. Liss, New York.

Socastro, M. E., 1969, Observaciones sobre el significado estructural y funcional de la musculatura braquial de los Cefalópodos, *Bol. R. Soc. Esp. Hist. Nat. Secc. Biol.* **67:**181–191.

Solem, A., and Richardson, E. S., 1975, *Paleocadmus*, a nautiloid cephalopod radula from the Pennsylvanian Francis Creek Shale of Illinois, *Veliger* **17:**233–242.

Sowerby, G. B., 1847–1887, *Thesaurus conchyliorum, or monographs of genera of shells*, Vols. 1–5, Sowerby, London.

Sowerby, G. B., 1849, Monograph of the genus *Nautilus*, *Thesaurus Conchyliorum* **2:**463–465.

Spath, L. F., 1927, Revision of the Jurassic cephalopod fauna of Kachh (Cutch), *Palaeontol. Indica N. Ser.* **9:**1–84.

Spotte, S. H., 1979, *Fish and Invertebrate Culture: Water Management in Closed Systems*, 2nd ed., John Wiley, New York 179 pp.

Srinivasan, P. S., 1941, The elastic properties of molluscan shells, *Q. J. Indian Inst.* **4**:189–221.

Stanley, S. M., 1970, Relation of shell form to life habits of the Bivalvia, *Geol. Soc. Am. Mem.* **125**, 296 pp.

Stenzel, H. B., 1957, Nautilus, in: *Treatise on Marine Ecology and Paleoecology*, Vol. I (J. W. Hedgpeth, ed.), *Geol. Soc. Am. Mem.* **67**:1135–1141.

Stenzel, H. B., 1964, Living *Nautilus*, in: *Treatise on Invertebrate Paleontology: Part K (Mollusca 3)* (R. C. Moore, ed.), pp. K59–K93, Geological Society of America and University of Kansas Press, Lawrence.

Storey, K. B., 1985, A re-evaluation of the Pasteur effect: New mechanisms in anaerobic metabolism, *Mol. Physiol.* **8**:439–462.

Storey, K. B., and Storey, J. M., 1983, Carbohydrage metabolism in cephalopod molluscs, in: *The Mollusca*, Vol. 1 (P. W. Hochachka, ed.), pp. 91–136, Academic Press, New York.

Storey, K. B., Fields, J. H. A., and Hochachka, P. W., 1978, Purification and properties of glutamate dehydrogenase from the mantle muscle of the squid *Loligo pealeii*: Role of the enzyme in energy production from amino acids, *J. Exp. Zool* **205**:111–118.

Stumbur, H., 1975, Biometrical characteristics of the shell of a living *Nautilus*, (English title), *Eesti NSV Tead. Akad. Toim. Keem. Geol.* **24**:78–89.

Sugano, T., Oshino, N., and Chance, B., 1974, Mitochondrial functions under hypoxic conditions: The steady states of cytochrome *c* reduction and of energy metabolism, *Biochim. Biophys. Acta* **347**:340–358.

Sundermann, G., 1980, Die Ultrastruktur der vakuolisierten Rundzellen von *Loligo vulgaris Lam.* (Mollusca, Cephalopoda), *Zool. Jahrb. Abt. Anat. Ontog. Tiere* **103**:93–104.

Sutherland, N. S., 1963, Visual acuity and discrimination of stripe widths in Octopus *vulgaris* Lamarck, *Publ. Stn. Zool. Napoli* **33**:92–109.

Swan, A. R. H., and Saunders, W. B., 1987, Function and shape in late Paleozoic (mid-Carboniferous) ammonoids, *Paleobiology* **13**(3):297–311.

Swofford, D. L., and Selander, R. B., 1981, BIOSYS-1: A FORTRAN program for the comprehensive analysis of electrophoretic data in population genetics and systematics, *J. Hered.* **72**:281–283.

Tagawa, K., Nishida, T., Watanabe, F., and Koseki, M., 1985, Mechanism of anoxic damage of mitochondria: Depletion of mitochondrial ATP and concomitant release of free Ca + +, *Mol. Physiol.* **8**.

Talavera, F., and Faustino, L. A., 1931, Industrial shells of the Philippines, *Philipp. J. Sci.* **45**:321–347.

Tanabe, K., 1985, Record of trapping experiment, in: *Marine Ecological Studies on the Habitat of Nautilus pompilius in the Environs of Viti Levu, Fiji* (S. Hayasaka, ed.), pp. 10–18, *Kagoshima Univ. Res. Center S. Pac. Occas. Pap.*, No. 4.

Tanabe, K., and Fukuda, Y., 1983, Buccal mass structure of the Cretaceous ammonite *Gaudryceras*, *Lethaia* **16**:249–256.

Tanabe, K., and Hamada, T., 1978, Keeping record of a drifted living *Nautilus pompilius* specimen caught off Kagoshima Bay and its significance, *Kaiyo Kagaku* **10**:1011–1017 (in Japanese).

Tanabe, K., Fukuda, Y., Kanie, Y., and Lehmann, U., 1980, Rhyncholites and conchorhynchs as calcified jaw elements in some late Cretaceous ammonites, *Lethaia* **13**:157–168.

Tanabe, K., Hayasaka, S., Saisho, T., Shinomiya, A., and Aoki, K., 1983, Morphologic variation of *Nautilus pompilius* from the Philippines and Fiji Islands, in: *Studies on Nautilus pompilius and Its Associated Fauna from Tañon Strait, the Philippines* (S. Hayasaka, ed.), pp. 9–21, *Kagoshima Univ. Res. Center S. Pacific Occas. Pap.*, No. 1.

Tanabe, K., Hayasaka, S., and Tsukahara, J., 1985, Morphologic analysis of *Nautilus pompilius*, in: *Marine Ecological Studies on the Habitat of Nautilus pompilius in the Environs of Viti Levu, Fiji* (S. Hayasaka, ed.), pp. 38–49, *Kagoshima Univ. Res. Center S. Pac. Occas. Pap.*, No. 4.

Taylor, B. E., and Ward, P. D., 1983, Stable isotopic studies of *Nautilus macromphalus* Sowerby (New Caledonia) and *Nautilus pompilius* L. (Fiji), *Palaeogeogr. Palaeoclimatol. Palaeoecol.* **41**:1–16.

Taylor, G., 1952, Analysis of the swimming of long narrow animals, *Proc. R. Soc. London Ser. A* **214**:158–183.

Teichert, C., 1933, Der Bau der actinoceroiden Cephalopoden, *Palaeontogr. Abt. A* **78**:111–234.

Teichert, C., 1964, Glossary of morphological terms used for nautiloids, in: *Treatise on Invertebrate Paleontology*, Part K, Mollusca 3, (R. C. Moore, ed.), pp. K54–59, Geological Society of America and the University of Kansas Press, Lawrence, Kansas.

Teichert, C., 1967, Major features of cephalopod evolution, in: *Essays in Paleontology and Stratigraphy* (C. Teichert and E. Yochelson, eds.), pp. 162–210, Special Publication, Department of Geology, University of Kansas, Lawrence.

Teichert, C., 1970, Drifted nautilus shells in the Bay of Bengal, *J. Paleontol.* **44:**1129–1130.

Teichert, C., Moore, R. C., and Nodine Zeller, D. E., 1964a, Rhyncholites, in: *Treatise on Invertebrate Paleontology, Part K: Mollusca 3* (R. C. Moore, ed.), pp. K59–93, Geological Society of America and University of Kansas Press, Lawrence.

Teichert, C., Kummel, B., Sweet, W. C., Stenzel, H. B., Furnish, W. M., Glenister, B. F., Erben, H. K., Moore, R. C., and Nodine Zeller, D. E., 1964b, *Treatise on Invertebrate Paleontology, Part K: Mollusca 3* (R. C. Moore, ed.), Geological Society of America and University of Kansas Press, Lawrence, 518 pp.

Thiele, J., 1893, Beiträge zur Kenntnis der Mollusken. II. Uber die Molluskenschale, *Z. Wiss. Zool.* **55:**220–251.

Thompson, D., 1917, *On Growth and Form*, Cambridge University Press, London, 1116 pp.

Thompson, T. G., and Chow, T. J., 1955, The strontium–calcium atom ratio in carbonate-secreting marine organisms, *Deep-Sea Res. Suppl.* **3:**20–39.

Thorpe, J. P., 1982, The molecular clock hypothesis: Biochemical evolution, genetic differentiation and systematics, *Annu. Rev. Ecol. Syst.* **13:**139–168.

Thorpe, J. P., 1983, Enzyme variation, genetic distance and evolutionary divergence in relation to levels of taxonomic separation, in: *Protein Polymorphism: Adaptive and Taxonomic Significance* (G. S. Oxford and D. Rollinson, eds.), pp. 131–152, Academic Press, London.

Tintant, H., 1984a, l'Evolution du concept de genre: De la similitude à la parenté, *Bull. Soc. Geol. Fr. Ser. 7* **26:**573–582.

Tintant, H., 1984b, Contribution à la connaissance des nautilacés jurassiques. 1. Le sous-genre *Cenoceras* Hyatt dans le Lias du sud-est de la France, *Geol. Fr.*, No. 1–2, pp. 29–66.

Tintant, H., and Kabamba, M., 1983, Le Nautile, fossile vivant ou forme cryptogène? Essai sur l'évolution et la classification des Nautilacés, *Bull. Soc. Zool. Fr.* **108:**569–579.

Tonosaki, A., 1965, The fine structure of the retinal plexus in *Octopus vulgaris*, *Z. Zellforsch. Microsk. Anat.* **67:**521–532.

Toriyama, R., Sato, T., Hamada, T., and Komalarjun, P., 1965, *Nautilus pompilius* drifts on the west coast of Thailand, *Jpn. J. Geol. Geogr.* **36:**149–161.

Trueman, A. E., 1941, The ammonite body-chamber, with special reference to the buoyancy and mode of life of the living ammonite, *Q. J. Geol. Soc. London* **96:**339–383.

Trueman, E. R., 1975, *The Locomotion of Soft-bodied Animals*, Arnold, London, 208 pp.

Trueman, E. R., 1980, Swimming by jet propulsion, in: *Aspects of Animal Movement* (H. Y. Elder and E. R. Trueman, eds.), pp. 93–105, Cambridge University Press, Cambridge.

Trueman, E. R., and Packard, A., 1968, Motor performance of some cephalopods, *J. Exp. Biol.* **49:**495–508.

Tsukahara, J., 1985, Histological and histochemical studies of gonads of *Nautilus pompilius* from Fiji, in: *Marine Ecological Studies on the Habitat of Nautilus pompilius in the Environs of Viti Levu, Fiji* (S. Hayasaka, ed.), pp. 50–60, *Kagoshima Univ. Res. Center S. Pac. Occas. Pap.*, No. 4.

Tucker, J. K., and Mapes, R. H., 1978, Possible predation on *Nautilus pompilius*, *Veliger* **21:**95–98.

Turton, W., 1806, *A general system of nature through the three grand kingdoms of animals, vegetables, and minerals, systematically divided into their several classes, orders, genera, species, and varieties, with their habitations, manners, economy, structure, and peculiarities by Sir Charles Linne: translated from Gmelin, Fabricius, Willdenow, & c.*, Vol. 4, *Animal Kingdom: Worms*, Lackington, Allen, London, 727 pp.

Tyler, J. E., 1960, Radiance as a function of depth in an underwater environment, *Bull. Scripps Inst. Oceanogr. Univ. Calif.* **7:**363–412.

Tyler, J. E., and Smith, R. C., 1970, *Measurements of Spectral Irradiance under Water*, Gordon and Breach, New York, 116 pp.

Valenciennes, M. A., 1841, Nouvelles recherches sur le Nautile flambé (*Nautilus pompilius* Lam.), *Arch. Mus. Hist. Nat.* **2:**257–314.

Vayssière, A., 1896, Étude sur l'organisation du Nautile (caractères zoologiques, dimorphisme sexuel, tentacules et spadice), *Ann. Sci. Nat. Zool. Paleontol.* **8:**137–186.

Vernon, H. M., 1896, The respiratory exchange of the lower marine invertebrates, *J. Physiol. (London)* **19:**18–70.

Vialleton, L., 1888, Recherches sur les premières phases du développement de la Seiche (*Sepia officinalis*), *Ann. Sci. Nat.* **7**:165–280.

Vigelius, W. J., 1880, Über das Exkretionssystem der Cephalopoden, *Niederl. Arch. Zool.* **5**:115–184.

Vinogradov, A. P., 1945, Geochemistry and biochemistry, *Uspek. Khim.* **7**:639–686.

Von Nathusius-Königsborn, W., 1877, *Untersuchungen über nicht-celluläre Organismen, namentlich Crustacean Panzer Molluskenschalen und Eihüllen*, Berlin, 144 pp.

Waage, K. M., 1968, The Type Fox Hills Formation, Cretaceous (Maestrichtian, South Dakota). 1. Stratigraphy and paleoenvironments, *Peabody Mus. Nat. Hist. Yale Univ. Bull.* **27**:1–175.

Waagen, W., 1867–1870, Ueber die Ansatzstelle der Haftmuskeln beim *Nautilus* und den *Ammoniden*, *Palaeontogr. Abt. A.* **17**:185–210.

Wada, K., and Fujinuki, T., 1976, Biomineralization in bivalve molluscs with emphasis on the chemical composition of the extrapallial fluid, in: *The Mechanisms of Mineralization in the Invertebrates and Plants* (N. Watanabe and K. M. Wilbur, eds.), pp. 175–190, University of South Carolina Press, Columbia.

Wainwright, S. A., 1970, Design in hydraulic organisms, *Naturwissenschaften* **57**:321–326.

Wainwright, S. A., 1982, Structural systems: Hydrostats and frameworks, in: *A Companion to Animal Physiology*, (C. R. Taylor, K. Johansen, and L. Bolis, eds.), pp. 325–338, Cambridge University Press, New York.

Wainwright, S. A., Briggs, W. D., Currey, J. D., and Gosline, J. M., 1976, *Mechanical Design in Organisms*, John Wiley, New York, 423 pp.

Waller, R. A., and Wicklund, R. I., 1968, Observations from a research submersible—Mating and spawning of the squid *Doryteuthis plei*, *Bioscience* **18**:110–111.

Ward, D. V., and Wainwright, S. A., 1972, Locomotory aspects of squid mantle structure, *J. Zool.* **167**:437–449.

Ward, P. D., 1979, Cameral liquid in *Nautilus* and ammonites, *Paleobiology* **5**:40–49.

Ward, P. D., 1980, Restructuring the chambered *Nautilus*, *Paleobiology* **6**:247–249.

Ward, P. D., 1982, The relationship of siphuncle size to emptying rates in chambered cephalopods: Implications for cephalopod paleobiology, *Paleobiology* **8**:426–433.

Ward, P. D., 1983a, *Nautilus macromphalus*, in: *Cephalopod Life Cycles*, Vol. 1 (P. Boyle, ed.), pp. 11–28. Academic Press, London.

Ward, P. D., 1983b, The extinction of the ammonites, *Sci. Am.* **249**:114–124.

Ward, P. D., 1984, Is *Nautilus* a living fossil?, in: *Living Fossils* (N. Eldredge and S. M. Stanley, eds.), pp. 247–256, Springer Verlag, New York.

Ward, P. D., 1985, Periodicity of chamber formation in chambered cephalopods: Evidence from *Nautilus macromphalus* and *Nautilus pompilius*, *Paleobiology* **11**:438–450.

Ward, P. D., 1986, Rates and processes of compensatory buoyancy change in *Nautilus macromphalus*, *Veliger* **28**:356–368.

Ward, P. D., 1987, *Natural History of Nautilus*, Allen and Unwin, London (in press).

Ward, P. D., and Chamberlain, J. A., Jr., 1983, Radiographic observations of chamber formation in *Nautilus pompilius*, *Nature (London)* **304**:57–59.

Ward, P. D., and Greenwald, L., 1982, Chamber refilling in *Nautilus*, *J. Mar. Biol. Assoc. U.K.* **62**:469–475.

Ward, P. D., and Martin, A. W., 1978, On the buoyancy of the pearly *Nautilus*, *J. Exp. Zool.* **205**:5–12.

Ward, P. D., and Martin, A. W., 1980, Depth distribution of *Nautilus pompilius* in Fiji and *Nautilus macromphalus* in New Caledonia, *Veliger* **22**:259–264.

Ward, P. D., and von Boletzky, S., 1984, Shell implosion depth and implosion morphologies in three species of *Sepia* (Cephalopoda) from the Mediterranean Sea, *J. Mar. Biol. Assoc. U.K.* **64**:955–966.

Ward, P. D., and Westermann, G. E. G., 1985, Cephalopod paleoecology, in: *Mollusk Notes for a Short Course* (T. W. Broadhead, ed.), pp. 215–229, University of Tennessee Press, Knoxville.

Ward, P. D., and Wicksten, M. K., 1980, Food sources and feeding behavior of *Nautilus macromphalus*, *Veliger* **23**:119–124.

Ward, P. D., Stone, R., Westermann, G., and Martin, A., 1977, Notes on animal weight, cameral fluids, swimming speed, and color polymorphism of the cephalopod *Nautilus pompilius* in the Fiji Islands, *Paleobiology* **3**:377–388.

Ward, P. D., Greenwald, L., and Greenwald, O. E., 1980a, The buoyancy of the chambered *Nautilus*, *Sci. Am.* **243**:190–205.

Ward, P. D., Greenwald, L., and Rougerie, F., 1980b, Shell implosion depth for living *Nautilus macromphalus* and shell strength of extinct cephalopods, *Lethaia* **13**:182.

Ward, P. D., Greenwald, L., and Magnier, Y., 1981, The chamber formation cycle in *Nautilus macromphalus*, *Paleobiology* **7**:481–493.

Ward, P. D., Carlson, B., Weekley, M., and Brumbaugh, B., 1984, Remote telemetry of daily vertical and horizontal movement by *Nautilus* in Palau, *Nature (London)* **309**:248–250.

Wardle, C. S., 1977, Effects of size on swimming speeds of fish, in: *Scale Effects in Animal Locomotion* (T. J. Pedley, ed.), pp. 299–313, Academic Press, New York.

Watabe, N., 1981, Crystal growth of calcium carbonate in the invertebrates, *Prog. Cryst. Growth Charact.* **4**:99–147.

Watabe, N., Meenakshi, V. R., Blackwelder, P. L., Kurtz, E. M., and Dunkelberger, D. G., 1976, Calcareous spherules in the gastropod, *Pomacea paludosa*, in: *The Mechanics of Mineralization in the Invertebrates and Plants* (N. Watabe and K. M. Wilbur, eds.), pp. 283–308, University of South Carolina Press, Columbia.

Weaver, K. S., 1949, A provisional standard observer for low level photometry, *J. Opt. Soc. Am.* **39**:278.

Weaver, J. S., and Chamberlain, J., 1976, Equations of motion for postmortem sinking of cephalopod shells and the sinking of *Nautilus*, *Paleobiology* **2**:8–18.

Webb, P. W., and Skadsen, J. M., 1979, Reduced skin mass: An adaptation for acceleration in some teleost fish, *Can. J. Zool.* **57**:1570–1575.

Wells, M. J., 1978, *Octopus: Physiology and Behaviour of an Advanced Invertebrate*, Chapman and Hall, London, 424 pp.

Wells, M. J., 1979, The heartbeat of *Octopus vulgaris*, *J. Exp. Biol.* **78**:87–104.

Wells, M. J., 1983, Cephalopods do it differently, *New Sci.* **100**:332–338.

Wells, M. J., and Smith, P. J. S., 1985, The ventilation cycle in *Octopus*, *J. Exp. Biol.* **116**:375–383.

Wells, M. J., and Wells, J., 1982, Ventilatory currents in the mantle of cephalopod, *J. Exp. Biol.* **99**:315–330.

Wells, M. J., and Wells, J., 1984, The effects of reducing gill area on the capacity to regulate oxygen uptake and on metabolic scope in a cephalopod (*Octopus vulgaris*), *J. Exp. Biol.* **108**:393–402.

Wells, M. J., and Wells, J., 1985, Ventilation and oxygen uptake by *Nautilus*, *J. Exp. Biol.* **118**:297–312.

Wells, M. J., O'Dor, R. K., Mangold, K., and Wells, J., 1983a, Diurnal changes in activity and metabolic rate in *Octopus vulgaris*, *Mar. Behav. Physiol.* **9**:275–288.

Wells, M. J., O'Dor, R. K., Mangold, K., and Wells, J., 1983b, Oxygen consumption in movement by *Octopus*, *Mar. Behav. Physiol.* **9**:289–303.

Welsch, U., and Storch, V., 1969, Über das Osphradium der prosobranchen Schnecken *Buccinum undatum* L. und *Neptunea antiqua* (L.), *Z. Zellforsch. Microsk. Anat.* **95**:317–330.

Werner, B., 1966, *Stephanoscyphus* (Scyphozoa, Coronatae) und seine direckte Abstammung von den fossilen Conulata, *Helgol. Wiss. Meeresunters.* **13**:317–347.

Werner, B., 1967, Morphologie, Systematik und Lebensgeschichte von *Stephanoscyphus* (Scyphozoa, Coronatae) sowie seine Bedeutung für die Evolution der Schyphozoa, *Zool. Anz. (Suppl.)* **30**:297–319.

Werner, B., 1974, *Stephanoscyphus eumedusoides* n. spec. (Scyphozoa, Coronatae), ein Höhlenpolyp mit einen neuen Entwicklungsmodus, *Helgol. Wiss. Meeresunters.* **26**:434–463.

Westermann, G. E. G., 1971, Form, structure, and function of shell and siphuncle in coiled Mesozoic ammonoids, *Life Sci. Contr. R. Ont. Mus.* **78**:1–39.

Westermann, G. E. G., 1973, Strength of concave septa and depth limits of fossil cephalopods, *Lethaia* **6**:383–403.

Westermann, G. E. G., 1975, Mode of origin, function and fabrication of fluted cephalopod septa, *Palaeontol. Z.* **49**:235–253.

Westermann, G. E. G., 1977, Form and function of orthoconic cephalopod shells with concave septa, *Paleobiology* **3**:300–321.

Westermann, G. E. G., 1982, The connecting rings of *Nautilus* and Mesozoic ammonoids: Implications for ammonoid bathymetry, *Lethaia* **15**:373–384.

Westermann, G. E. G., 1985a, Post-mortem descent with septal implosion in Silurian nautiloids, *Palaeontol. Z.* **59**:79–98.

Westermann, G. E. G., 1985b, Exploding *Nautilus* camerae does not test septal strength index, *Lethaia* **18**:348.

Westermann, G. E. G., and Ward, P. D., 1980, Septum morphology and bathymetry in cephalopods, *Paleobiology* **6**:48–50.

Whalen, W. J., Buerk, D., and Thuning, C., 1973, Blood flow-limited oxygen consumption in resting cat skeletal muscle, *Am. J. Physiol.* **224**:763–768.

Wiedmann, J., 1960, Zur Systematik jungmesozoischer Nautiliden unter besonderer Berücksichtigung der iberischen Nautilinae d'Orb, *Palaeontogr. Abt. A* **115A**:144–206.

Wiegel, R. L., 1964, *Oceanographical Engineering*, Prentice-Hall, Englewood Cliffs, New Jersey, 532 pp.

Wilbur, K. M., 1972, Shell formation in molluscs, in: *Chemical Zoology*, Vol. VII, *Mollusca* (M. Florkin and B. T. Scheer, eds.), pp. 103–145, Academic Press, New York.

Wilbur, K. M., 1976, Recent studies of invertebrate mineralization, in: *The Mechanisms of Mineralization in the Invertebrates and Plants* (N. Watabe and K. M. Wilbur, eds.), pp. 79–108, University of South Carolina Press, Columbia.

Wilbur, K. M., and Owen, G., 1964, Growth, in: *Physiology of Mollusca*, Vol. 1. (K. M. Wilbur and C. M. Yonge, eds.), pp. 211–242, Academic Press, New York.

Wilbur, K. M., and Saleudin, A. S. M., 1983, Shell formation, in: *The Mollusca* (K. M. Wilbur and A. S. M. Saleudin, eds.), pp. 236–287, Academic Press, New York.

Willey, A., 1895, In the home of the *Nautilus*, *Nat. Sci.* **6**: 405–414.

Willey, A., 1897a, Letters from New Guinea on *Nautilus* and some other organisms, *Q. J. Microsc. Sci. N. Ser.* **39**:145–180.

Willey, A., 1897b, The embryology of the *Nautilus*, *Nature* (London) **55**:402–403.

Willey, A., 1897c, Zoological observations in the South Pacific, *Q. J. Microsc. Sci. N. Ser.* **39**:219–231.

Willey, A., 1897d, The oviposition of *Nautilus macromphalus*, *Proc. R. Soc. London* **60**:467–471.

Willey, A., 1898a, The pre-ocular and post-ocular tentacles and osphradia of *Nautilus*, *Q. J. Microsc. Sci. N. Ser.* **40**:197–201.

Willey, A., 1898b, The adhesive tentacles of *Nautilus* with some notes on its pericardium and spermatophores, *Q. J. Microsc. Sci. N. Ser.* **40**:207–209.

Willey, A., 1899, General account of a zoological expedition to the South Seas during the years 1894–1897, *Proc. Zool. Soc. London* **1899**:7–9.

Willey, A., 1902, *Contribution to the natural history of the pearly Nautilus: Zoological results based on material from New Britain, New Guinea, Loyalty Islands and elsewhere, collected during the years 1895, 1896 and 1897: Part* **6**:691–830, University Press, Cambridge, England.

Wilson, D. F., Owen, C. S., and Erecinska, M., 1979, Quantitative dependence of mitochondrial oxidative phosphorylation on oxygen concentration: A mathematical model, *Arch. Biochem. Biophys.* **195**:494–504.

Wilson, T. R. S., 1975, Salinity and the major elements of sea water, in: *Chemical Oceanography*, Vol. 1 (J. P. Riley and G. Skirrow, eds.), pp. 365–413, Academic Press, London.

Winterstein, H., 1909, Zurkenntnis der Blutgase Wirbellosen Seatiere, *Biochem. Z.* **19**:384–424.

Winterstein, H., 1924, Über die chemische Regulierung der Atmung bei den Cephalopoden, *Z. Vgl. Physiol.* **2**:315–328.

Wise, S. W., 1969, Study of molluscan shell ultrastructures, in: *Scanning Electron Microscopy* (O. Johari, ed.), pp. 205–216, Illinois Institute of Technology Research Institute, Chicago.

Wise, S. W., 1970, Microarchitecture and mode of formation of nacre (mother-of-pearl) in pelecypods, gastropods and cephalopods, *Eclogae Geol. Helv.* **63**:775–797.

Wise, S. W., and Hay, W. W., 1968, Scanning electron microscopy of molluscan shell ultrastructures. II. Observations of growth surfaces, *Trans. Am. Microsc. Soc.* **87**:419–430.

Witmer, A., and Martin, A. W., 1973, The fine structure of the branchial heart appeandage of the cephalopod *Octopus dofleini*, *Z. Zellforsch.* **117**:252–274.

Woodhams, P. L., and Messenger, J. B., 1974, A note on the ultrastructure of the *Octopus* olfactory organ, *Cell Tissue Res.* **152**:253–258.

Woodruff, D. S., 1979, Postmating reproductive isolation in *Pseudophryne* and the evolutionary significance of hybrid zones, *Science* **203**:561–563.

Woodruff, D. S., 1981, Towards a genodynamics of hybrid zones: Studies of Australian frogs and West Indian land snails, in: *Essays on Speciation and Evolution in Honor of M. J. D. White* (W. D. Atchley and D. S. Woodruff, eds.), pp. 171–197, Cambridge University Press, United Kingdom.

Woodruff, D. S., and Carpenter, M. P., 1987, Biochemical genetics of *Nautilus* (in prep.).

Woodruff, D. S., and Gould, S. J., 1980, Geographic differentiation and speciation in *Cerion*: A preliminary discussion of patterns and processes, *Biol. J. Linn. Soc. London* **14**:389–416.

Woodruff, D. S., and Gould, S. J., 1987, Fifty years of interspecific hybridization: Genetics and morphometrics of a controlled experiment involving the land snail *Cerion* in Florida, *Evolution* **41**:1022–1045.

Woodruff, D. S., 1987, Genetic differentiation of molluscan populations and species (in prep.).

Woodruff, D. S., Mulvey, M., Saunders, W. B., and Carpenter, M. P., 1983, Genetic variation in the cephalopod *Nautilus belauensis*, *Proc. Acad. Natl. Sci. Phila.* **135**:147–153.

Woodruff, D. S., Mulvey, M., and Yipp, M. W., 1985, Population genetics of *Biomphalaria straminea* in Hong Kong: A neotropical schistosome-transmitting snail recently introduced into China, *J. Hered.* **76**:355–360.

Woodruff, D. S., Carpenter, M. P., Upatham, E. S., and Viyanant, V., 1986a, Genetic studies of medically important snails in southeast Asia, *Isozyme Bull.* **19**:32.

Woodruff, D. S., McMeekin, L. L., Mulvey, M., and Carpenter, M. P., 1986b, Population genetics of *Crepidula onyx*: Variation in a Californian slipper snail recently established in China, *Veliger* **29**:53–63.

Woodruff, D. S., Staub, K., Upatham, E. S., Viyanant, V., and Yuan, H. C., 1987, Schistosome-transmitting snails (*Oncomelania hupensis* complex) from China and the Philippines are distinct species, *Malacologia* (in press).

Woodward, S. P., 1880, *A Manual of the Mollusca*, 4th ed., Crosby, Lockwood, London, 542 pp.

Wright, S., 1978, *Evolution and the Genetics of Populations*, Vol. IV, *Variability within and among Natural Populations*, University of Chicago Press, Chicago, 580 pp.

Yamamoto, T., Tasaki, K., Sugawara, Y., and Tonosaki, A., 1965, Fine structure of the octopus retina, *J. Cell Biol.* **25**:345–359.

Yancey, T. E., 1975, Floated shells of *Nautilus pompilius* in the south part of the Andaman Sea, *Geol. Soc. Malays. Newsl.* **1**:52–55.

Young, J. Z., 1965a, The central nervous system of *Nautilus*, *Philos. Trans. R. Soc. London Ser. B.* **249**:1–25.

Young, J. Z., 1965b, The buccal nervous system of *Octopus*, *Philos. Trans. R. Soc. London Ser. B.* **249**:27–43.

Young, J. Z., 1965c, The organization of a memory system, *Proc. R. Soc. London* **163**:285–320.

Young, J. Z., 1971, *The Anatomy of the Nervous System of Octopus vulgaris*, Clarendon Press, Oxford, 690 pp.

Young, J. Z., 1983, The distributed tactile memory system of Octopus, *Proc. R. Soc. London* **218**:135–176.

Zammit, V., and Newsholme, E. A., 1976, The maximum activities of hexokinase, phosphorylase, phosphofructokinase, glycerol phosphate dehydrogenases, lactate dehydrogenase, octopine dehydrogenase, phosphoenolpyruvate carboxykinase, nucleoside diphosphatekinase, glutamate-oxaloacetate transminase and arginine kinase in relation to carbohydrate utilization in muscles from marine invertebrates, *J. Biochem.* **160**:447–462.

Zann, L. P., 1984, The rhythmic activity of *Nautilus pompilius* with notes on its ecology and behaviour in Fiji, *Veliger* **27**:19–28.

Zann, L. P., Cuffey, R. J., and Kropach, C., 1975, Fouling organisms and parasites associated with the skin of sea snakes, in: *The Biology of Sea Snakes* (W. A. Dunson, ed.), pp. 251–265, University Park Press, Baltimore.

Zernoff, D., 1869, Über das Geruchsorgan der Cephalopoden, *Bull. Soc. Imp. Natl. Moscow* **42**:71–90.

Zorkendorfer, W., 1930, Uber die Wirkung der Magnesiumsalze, *Biochem. Z.* **221**:33–41.

Index

MIX
Papier aus verantwortungsvollen Quellen
Paper from responsible sources
FSC® C105338

If you have any concerns about our products,
you can contact us on
ProductSafety@springernature.com

In case Publisher is established outside the EU,
the EU authorized representative is:
Springer Nature Customer Service Center GmbH
Europaplatz 3, 69115 Heidelberg, Germany

Printed by Libri Plureos GmbH
in Hamburg, Germany